Drug Design and Discovery in Alzheimer's Disease

Drug Design and Discovery in Alzheimer's Disease

Edited by

Atta-ur-Rahman, FRS
Honorary Life Fellow, Kings College, University of
Cambridge, UK

And

Muhammad Iqbal Choudhary
H.E.J. Research Institute of Chemistry, International
Center for
Chemical and Biological Sciences, University of
Karachi, Pakistan

AMSTERDAM • BOSTON • HEIDELBERG • LONDON • NEW YORK • OXFORD
PARIS • SAN DIEGO • SAN FRANCISCO • SINGAPORE • SYDNEY • TOKYO

ELSEVIER

Elsevier
Radarweg 29, PO Box 211, 1000 AE Amsterdam, Netherlands
The Boulevard, Langford Lane, Kidlington, Oxford OX5 1GB, UK
225 Wyman Street, Waltham, MA 02451, USA

ISBN: 978-0-12-803959-5

British Library Cataloguing in Publication Data
A catalogue record for this book is available from the British Library

Library of Congress Cataloging-in-Publication Data
A catalog record for this book is available from the Library of Congress

For Information on all Elsevier publications
visit our website at http://store.elsevier.com/

Working together
to grow libraries in
developing countries

www.elsevier.com • www.bookaid.org

CONTENTS

Contd….

Contd....

PREFACE

Alzheimer's disease (AD) is one of the most common neurological disorders, affecting a large portion of the human population in both the developed and the developing world. The demographic changes in the next few decades are likely to increase the prevalence of AD to epidemic proportions. Unfortunately, despite major advances in molecular and structural biology, neurochemistry, disease genomics and cell physiology, the etiology of the disease is only marginally understood. AD has been the focus of major research efforts, both in the pharmaceutical industry and in academia. The literature is continuously enriched by exciting discoveries on various aspects of AD and its prevention and treatment. It is nearly impossible for any researcher to remain on top of the most recent developments in this dynamic field without access to comprehensive reviews on various aspects.

This volume of the book series *"Frontiers in Drug Discovery and Development"* is focused on key developments in the understanding of the disease at molecular levels, identification and validation of molecular targets, as well as innovative approaches towards drug discovery, development and delivery. The volume comprises 16 scholarly written review articles by leading researchers in the field, covering a broad range of topics.

The review contributed by Korabecny *et al* sets the stage for the subsequent chapters by detailing the main symptomatic strategies available for the treatment of AD and for the improvement of quality of life of AD patients. This review summarizes various classes of current and future therapies of AD.

Rashid and Ansari focus on the major challenges faced in the discovery and development of anti-AD drugs. They provide examples of several blockbuster AD drugs, obtained through various approaches. Revadigar and his team have contributed a chapter reviewing the literature about the identification of various enzymes involved in the on-set and progression of Alzheimer's diseases. They have included numerous examples of inhibitors of these enzymes which has either already developed as drugs or are in various stages of drug development.

Aziz *et al* focus on the significance of small molecules, both natural and synthetic, as possible drug candidates for the treatment of neurodegenerative diseases, including AD.

The four review articles (chapters 5-8), contributed by Joubert *et al*, Grieg *et al*, Saify and Sultana, and Alcolea-Palafox *et al,* specifically focus on the role of cholinesterases in the on-set and progression of AD and their inhibition, which has potential as an effective treatment. It is important to note that most of the current therapies of AD are based on cholinesterase modulation and the field remains active.

Beta-site APP clearing enzyme-1 (BACE-1) plays an important role in the production of β-amyloid proteins, one of the key reasons of the progression of AD. BACE-1 has been identified as the key target for the development of anti-AD therapies. Klaver and Tesco, Decourt *et al*, and Henary *et al* have contributed three excellent reviews (chapter 9-11) on the modulation of BACE-1 activity, as a strategy towards new drug discovery and development. Along with the small molecular inhibitors of BACE-1, the role of metal chelating agents has also been reviewed.

Along with two key targets, cholinesterase and BACl-1, a number of other targets have also been identified for drug discovery. Sandoval *et al* contributed a comprehensive review on various chemical compounds which serve as agonist of somatostatin subtype-4 receptor as a possible treatment regimen for AD.

Neprilysin, a key target in the etiology of AD, has been the focus of recent research. Neprilysin catalyzes the degradation of amyloid beta peptides (Aβ), which are neurotoxins and contribute in the pathophysiology of AD. Pope and Cascio have reviewed the recent work on the inhibition of neprilysin and the prospects of their development as treatment for AD.

GSK3β/β-catenin signaling has been identified as a possible contributor to the on-set and progressive degeneration of neuron in AD. GSK-3 controls proteosomal degradation of catenin, thus slowdown, the neurodegeneration. Zeidán-Chuliá and

Moreira have critically analyzed its validity as a target for anti-AD drug discovery.

Seneci covers the important aspects of AD treatment and inhibition of phosphorylation of tau-proteins, collectively called taupathy. The focus has been to understand the expression of such proteins at the genetic level and to identify the structural characteristics which makes them aggregatory in nature. Another aim is to use small molecules as effective inhibitors of these processes.

Last but not least, the article contributed by Ahmed *et al* comprehensively reviews the applications of nanotechnology in the development of effective drug delivery vehicles for AD drugs. Nanomedicines can serve as novel carrier of cholinesterase inhibitors effectively crossing the anatomical and biochemical barriers.

We wish to express our profound thanks to all the contributors of reviews in this volume for their hard, and scholarly work and critical commentary. We are confident that their scholarly contributions will make this volume of great interest to many researches. We are also grateful to Mr. Mahmood Alam (Director Publication) and Ms. Sara Yaser (Manager Publications) of Bentham Science Publishers for their excellent management skills and secretarial support.

Atta-ur-Rahman, *FRS*
Honorary Life Fellow, Kings College,
University of Cambridge,
UK

Muhammad Iqbal Choudhary
H.E.J. Research Institute of Chemistry,
International Center for Chemical and Biological Sciences,
University of Karachi,
Pakistan

List of Contributors

A. Michael Crider

United Arab Emirates University, Faculty of Medicine and Health Sciences, Department Pharmacology and Therapeutics, Al Ain, UAE

Abdu Adem

Department of Pharmacology and Therapeutics, Faculty of Medicine and Health Sciences, United Arab Emirates University, Al-Ain, UAE

Adel M. Abuzenadah

King Fahd Medical Research Center, King Abdulaziz University, P. O. Box 80216, Jeddah 21589, Saudi Arabia

Ali Jawad

Department of Applied Physics, Zakir Husain College of Engineering and Technology, Aligarh Muslim University, Aligarh UP, India

Benjamin P. Repsold

Pharmaceutical Chemistry, North-West University, Private Bag X6001, Potchefstroom, 2520, South Africa

Boris Decourt

Banner Sun Health Research Institute, Sun City, AZ, USA

Carmen Martínez-Rincón

Nursing Department, Complutense University of Madrid, Spain

Daniel Jun

University Hospital Hradec Kralove, Sokolska 581, 500 05 Hradec Kralove, Czech Republic

Darrick Pope

Department of Chemistry and Biochemistry, Duquesne University, Pittsburgh, PA 15282, USA

David F. Fischer

BioFocus; Leiden, The Netherlands

David William Klaver

Alzheimer's Disease Research Laboratory, Department of Neuroscience, Tufts University School of Medicine, Boston, MA, USA

Dolores Rodríguez-Martínez

Nursing Department, Complutense University of Madrid, Spain

Eric A. Owens

Department of Chemistry, Center for Diagnostics and Therapeutics, Georgia State University, Atlanta, Georgia, 30303, USA

Eugenie Nepovimova

Department of Toxicology, Faculty of Military Health Sciences, Trebesska 1575, 500 01 Hradec Kralove, Czech Republic

Faizul H. Nasim

Department of Chemistry, BJ Campus, The Islamia University of Bahawalpur, Bahawalpur, Pakistan

Fares Zeidán-Chuliá

Center of Oxidative Stress Research, Department of Biochemistry, Postgraduate Program in Biological Sciences: Biochemistry, Institute of Basic Health Sciences, Federal University of Rio Grande do Sul, Porto Alegre, RS, Brazil

Farhan Jalees Ahmad

Nanomedicine Research Lab, Department of Pharmaceutics, Faculty of Pharmacy, Jamia Hamdard, New Delhi 110062, India

Farzana L. Ansari

Department of Chemistry, Quaid-i-Azam University, Islamabad 45320, Pakistan

Filip Zemek

Department of Toxicology, Faculty of Military Health Sciences, Trebesska 1575, 500 01 Hradec Kralove, Czech Republic

Giuseppina Tesco

Alzheimer's Disease Research Laboratory, Department of Neuroscience, Tufts University School of Medicine, Boston, MA, USA

Gjumrakch Aliev

Department of Health Science and Healthcare Administration, University of Atlanta, Atlanta, GA, USA, and GALLY International Biomedical Research Consulting LLC, San Antonio, TX, USA

Iqbal Ahmad

Nanomedicine Research Lab, Department of Pharmaceutics, Faculty of Pharmacy, Jamia Hamdard, New Delhi 110062, India

Ismael Ortuño- Soriano

Institute for Health Research at the San Carlos Clinical Hospital (IdISSC), Madrid, Spain; Nursing Department, Complutense University of Madrid, Spain

Jacobus P. Petzer

Pharmaceutical Chemistry, North-West University, Private Bag X6001, Potchefstroom, 2520, South Africa

Jacques Joubert

School of Pharmacy, University of the Western Cape, Private Bag X17, Bellville 7535, South Africa

Jan Korabecny

Department of Toxicology, Faculty of Military Health Sciences, Trebesska 1575, 500 01 Hradec Kralove, Czech Republic

Javed Ahmad

Nanomedicine Research Lab, Department of Pharmaceutics, Faculty of Pharmacy, Jamia Hamdard, New Delhi 110062, India

José Cláudio Fonseca Moreira

Center of Oxidative Stress Research, Department of Biochemistry, Postgraduate Program in Biological Sciences: Biochemistry, Institute of Basic Health Sciences, Federal University of Rio Grande do Sul, Porto Alegre, RS, Brazil

José L. Pacheco-del-Cerro

Nursing Department, Complutense University of Madrid, Spain

Julie Frearson

BioFocus; Chesterford Research Park, Saffron Walden, Essex, CB10 1XL, UK

Kamil Kuca

University Hospital Hradec Kralove, Sokolska 581, 500 05 Hradec Kralove, Czech Republic

Kamil Musilek

University Hospital Hradec Kralove, Sokolska 581, 500 05 Hradec Kralove, Czech Republic

Karin E. Sandoval

Department of Pharmaceutical Sciences, Southern Illinois University Edwardsville, Edwardsville, Illinois 62026, USA

Katarina Spilovska

Department of Toxicology, Faculty of Military Health Sciences, Trebesska 1575, 500 01 Hradec Kralove, Czech Republic

Kenneth A. Witt

Department of Pharmaceutical Sciences, Southern Illinois University Edwardsville, Edwardsville, Illinois 62026, USA

Lara Pacheco-Cuevas

Nursing Department, Complutense University of Madrid, Spain

Louis H.A. Prins

Department of Pharmacy, University of Namibia, Windhoek, Namibia

Maged M. Henary

Department of Chemistry, Center for Diagnostics and Therapeutics, Center for Biotechnology and Drug Design, Georgia State University, Atlanta, Georgia, 30303, USA

Maria Kontoyianni

Department of Pharmaceutical Sciences, Southern Illinois University Edwardsville, Edwardsville, Illinois 62026, USA

Marwan Sabbagh

Banner Sun Health Research Institute, Sun City, AZ, USA

Mauricio Alcolea-Palafox

Chemical-Physics Department, Chemistry Faculty, Complutense University of Madrid, Spain

Michael Cascio

Department of Chemistry and Biochemistry, Duquesne University, Pittsburgh, PA 15282, USA

Michael D. Wall

BioFocus; Chesterford Research Park, Saffron Walden, Essex, CB10 1XL, UK

MiMi Macias

Banner Sun Health Research Institute, Sun City, AZ, USA

Mohamed A. Embaby

Department of Chemistry, Faculty of Sciences & Arts Khulais, King Abdulaziz University, Jeddah, KSA

Mohammad A. Kamal

King Fahd Medical Research Center, King Abdulaziz University, P. O. Box 80216, Jeddah 21589, Saudi Arabia

Mohammad Zaki Ahmad

Department of Pharmaceutics, College of Pharmacy, Najran University, Saudi Arabia

Nasimudeen R. Jabir

King Fahd Medical Research Center, King Abdulaziz University, P. O. Box 80216, Jeddah 21589, Saudi Arabia

Nigel H. Grieg

Drug Design & Development Section, Laboratory of Neurosciences, Intramural Research Program, National Institute on Aging, National Institutes of Health, Baltimore, MD 21224, USA

Nighat Sultana

Pharmaceutical Research Center, PCSIR Laboratories Complex, Shahrah-e-Dr. Salimuzzaman Siddiqui, Karachi 75280, Pakistan

Omar Aziz

BioFocus; Chesterford Research Park, Saffron Walden, Essex, CB10 1XL, UK

Ondrej Soukup

University Hospital Hradec Kralove, Sokolska 581, 500 05 Hradec Kralove, Czech Republic

Othman Sulaiman

School of Industrial Technology, Universiti Sains Malaysia, Minden–11800, Pulau Pinang, Malaysia

Paloma Posada-Moreno

Institute for Health Research at the San Carlos Clinical Hospital (IdISSC), Madrid, Spain; Nursing Department, Complutense University of Madrid, Spain

Pierfausto Seneci

Department of Chemistry, University of Milan, Via Golgi 19, I-20133 Milan, Italy

Raza Murad Ghalib

Department of Chemistry, Faculty of Sciences & Arts Khulais, King Abdulaziz University, Jeddah, KSA

Rokiah Hashim

School of Industrial Technology, Universiti Sains Malaysia, Minden–11800, Pulau Pinang, Malaysia

Roland W. Bürli

BioFocus; Chesterford Research Park, Saffron Walden, Essex, CB10 1XL, UK

Sarel F. Malan

School of Pharmacy, University of the Western Cape, Private Bag X17, Bellville 7535, South Africa

Sayed Hasan Mehdi

Department of Chemistry, SPG College, UP, Lucknow, India

Shams Tabrez

King Fahd Medical Research Center, King Abdulaziz University, P. O. Box 80216, Jeddah 21589, Saudi Arabia

Sohail Akhter

Nanomedicine Research Lab, Department of Pharmaceutics, Faculty of Pharmacy, Jamia Hamdard, New Delhi 110062, India

Surendra Reddy Punganuru

Department of Chemistry, Georgia State University, Atlanta, Georgia, 30303, USA

Tyler L. Dost

Department of Chemistry, Georgia State University, Atlanta, Georgia, 30303, USA

Umer Rashid

Department of Chemistry, Hazara University, Mansehra 21120, Pakistan

Vageesh Revadigar

Discipline of Pharmacology, School of Pharmaceutical Sciences, Universiti Sains Malaysia, Minden–11800, Pulau Pinang, Malaysia

Vikneswaran Murugaiyah

Discipline of Pharmacology, School of Pharmaceutical Sciences, Universiti Sains Malaysia, Minden–11800, Pulau Pinang, Malaysia

Zafar Saied Saify

International Center for Chemical Sciences, H. E. J. Research Institute of Chemistry, University of Karachi, Karachi 75270, Pakistan

Ziyaur Rahman

Irma Lerma Rangel College of Pharmacy, Texas A&M Health Science Center, Kingsville, Texas, USA

CHAPTER 1

Pharmacotherapy of Alzheimer's Disease: Current State and Future Perspectives

Jan Korabecny[a,b], **Filip Zemek**[a], **Ondrej Soukup**[b], **Katarina Spilovska**[a,b], **Kamil Musilek**[b,c], **Daniel Jun**[b,d], **Eugenie Nepovimova**[a,b] **and Kamil Kuca**[b,d,*]

[a]*Department of Toxicology, Faculty of Military Health Sciences, Trebesska 1575, 500 01 Hradec Kralove, Czech Republic;* [b]*University Hospital Hradec Kralove, Sokolska 581, 500 05 Hradec Kralove, Czech Republic;* [c]*University of Hradec Kralove, Faculty of Science, Department of Chemistry, Rokitanskeho 62, 50003 Hradec Kralove, Czech Republic;* [d]*Centre of Advanced Studies, Faculty of Military Health Sciences, Trebesska 1575, 500 01 Hradec Kralove, Czech Republic;* [e]*Department of Pharmaceutical Chemistry and Drug Control, Faculty of Pharmacy in Hradec Kralove, Charles University in Prague, Heyrovskeho 1203, 500 05 Hradec Kralove, Czech Republic*

Abstract: Alzheimer's disease (AD) is a multifactorial disorder and apparently involves several different etiopathogenetic mechanisms. Up-to-date, there are no curative treatments or effective disease modifying therapies for AD. A strategy to enhance the cholinergic transmission by using acetylcholinesterase inhibitors (AChEIs) has been proposed more than two decades ago. Food and Drug Administration (FDA) gradually marketed these AChEIs: tacrine (1993), donepezil (1997), rivastigmine (2000) and galantamine (2001); tacrine is no longer used because of its high prevalence of hepatotoxicity. In addition to the AD cholinergic hypothesis , there is great evidence that voltage-gated, uncompetitive, N-methyl-D-aspartate (NMDA) antagonist memantine with moderate affinity can protect neurons from excitotoxicity. It was approved by FDA for treatment of moderate to severe stages of AD in 2003. Beyond symptomatic approaches there are anti-amyloid, neuroprotective and neuron-restorative strategies that hold promise of redefining the course of the disease as it is known. This contribution summarizes the main symptomatic strategies available for treating AD and future perspectives of pharmacotherapy for improving the AD course.

Keywords: Acetylcholinesterase, Alzheimer's disease, butyrylcholinesterase, donepezil, galantamine, GSK-3β, inhibitors, memantine, metal chelators, modulators of secretases, M1 agonists, rivastigmine, statins, tacrine.

*****Address correspondence to Kamil Kuca:** Centre of Advanced Studies, Faculty of Military Health Sciences, Trebesska 1575, 500 01 Hradec Kralove, Czech Republic; Tel: +420 973 253; E-mail: kucakam@pmfhk.cz

Atta-ur-Rahman / Muhammad Iqbal Choudhary (Eds.)

1. ALZHEIMER'S DISEASE - HISTORIC OVERVIEW

In 1906, German psychiatrist Alois Alzheimer firstly diagnosed and defined clinical-pathological syndrome that was later named Alzheimer's disease (AD). When diagnosing the disease on one of his forty years old patient, Alois Alzheimer described observed symptoms as a rare pre-senile dementia occurring before 65 years of age. Nor did he or his colleagues distinguish this new disease from well-known and described senile dementia. Nevertheless, several symptoms which had Alzheimer described are common for majority of patients with diagnosed AD *e.g.* progressive memory loss, impaired cognitive function, behavioral changes, disruption of the integrity of the individual, hallucinations, impaired self-control and loss to the decline of spoken and written speech [1, 2].

The invention of the electron microscope in the 20th century led to the clarification of the histological changes in the brain characteristic for the AD particularly neuritic plaques and neurofibrillary tangles [3]. The existence of the neuritic plaques has been described no sooner than in the seventies of the twentieth century. Furthermore, the adverse effect of neuritic plaques on the neurons producing and releasing acetylcholine (ACh) were also observed. Further research led to the conclusion that these adverse effects are related directly to the enzymes associated with ACh. In particular, decrease of concentration and activity of cholineacetyltransferase (ChAT, EC 2.3.1.6) concentration and acetylcholinesterase (AChE, EC 3.1.1.7) in the limbic system and cerebral cortex were associated with the loss of cholinergic neurons in subcortical areas [4]. These findings opened a novel chapter of pharmacological research in an effort to increase ACh brain levels in the synaptic gaps. First experiments concentrated on the inhibition of AChE responsible for the degradation of ACh. The research was successful and it led to the introduction of several novel compounds into clinical practice: tacrine (Cognex®), donepezil (Aricept®), rivastigmine (Exelon®) and galantamine (Reminyl®) [5, 6].

2. CURRENT STATUS AND PREVALENCE OF AD

AD is one of the most common forms of dementia. The unpredictability and yet unknown etiology makes it increasingly disturbing problem for humanity, not just in terms of health, but also in terms of social and economic parameters. Furthermore, AD is a fatal illness as every person dies after 3 - 10 years from being diagnosed [1, 2].

Globally, AD is the fifth cause of death among people over 65 years. Cases of death caused by AD are increasing dramatically. Between 2000 and 2008 there were 66 % of deaths caused by AD alone. However, cases of death due to heart failure, strokes and prostate cancer were only 13 %, 20 %, 8 % respectively [2, 7].

In United States of America, it is estimated that 5.4 million people have been diagnosed with AD and 200000 of them are under 65 years of age (early AD onset). The total number of patients with dementia in the world is currently estimated to be around 35.6 million people. By 2050, this number is expected to increase up to 115.4 million patients [2]. The most vulnerable population are people over 85 years of age (more than 50 % of AD patients). In the European Union it is estimated that from AD suffers approximately 6 million people and this figure is likely to double by 2050 [8-11]. The main reason is principally the aging population and the absence of an effective therapy. The limited understanding of the basic AD pathophysiology was the major drawback in the research of potential therapeutics. Though the tremendous progress was made in the past 30 years in the biological, biochemical, toxicological and pharmaceutical research, the considered therapeutic procedures failed or came too late in the last stage of the disease [2].

3. RISK FACTORS FOR AD

The trigger mechanism of AD is not yet fully understood, although certain risk factors contributing to the development of AD were determined. Aging population is the most threatened by AD, while it is important to note that AD is not a normal component of the aging process. People over 65 years of age display the greatest increase in the AD incidence, however even people below 65 years of age may develop AD known as early-onset AD. [5].

An important role is played by the genetic factors and heredity. It was scientifically proven that individuals with a close relative (brother, sister, parents), who were AD diagnosed are more likely to develop AD also than patients with only distant relatives suffering from AD [12, 13].

AD is genetically linked to the presence of allele ε4 of apolipoprotein E (ApoE ε4). Allele ε4 can occur in three isoforms, ε2, ε3 and ε4 however only ε4 increases the risk of AD later in life. On the other hand the presence of ε2 allele decreases the risk of developing AD [13-16].

Other risk factors include cardiovascular disease, brain injury, trauma, depression, low education level, smoking, low concentrations of folate and vitamin B12, elevated homocysteine concentration in plasma, high cholesterol, diabetes mellitus type 2, high blood pressure, physical inactivity and obesity. These factors not only promote the formation and development of AD but are also involved in the etiopathogenesis of different dementia subtypes and other diseases [17-29].

Former research has also identified protective factors, which in turn reduce the risk of AD development. These include a higher education level, regular use of anti-inflammatory agents and cholesterol lowering agents (statins), estrogen replacement therapy in post-menopausal period, antihypertensive therapy and a diet rich in fish. The big unknown is the use of compounds with antioxidant properties *e.g.* vitamins C and E [30-33].

4. CHOLINERGIC THEORY

The cholinergic hypothesis is the first and still the only widely accepted theory explaining the nature of AD [34, 35]. Moreover, almost all currently used drugs for relieving symptomatic effects of moderate to severe AD forms are based on this hypothesis [36-38]. The cholinergic theory works with the assumption that the loss of cholinergic activity observed in AD patients confirms a close relationship between the neuromediator ACh and learning plus remembering [39]. The blockade of central cholinergic system with scopolamine in young adults produces similar symptoms as observed in individuals affected by AD. Normal function of the central nervous system can be restored by reversible cholinesterase inhibitors such as physostigmine [40]. Based on these experimental studies, clinical trials were conducted for other type of compounds (reversible cholinesterase inhibitors), which hold the most promise for the treatment of deficient memory function in AD patients (see chapter 6.1 Drugs used in clinical practice).

In a comprehensive examination of the deficient cholinergic system in AD impaired signaling from a population of neurons based in the basal forebrain nuclei and in the *Meynert nucleus* basalis to the cerebral cortex and hippocampus was determined. Furthermore, the concentration and ChAT activity responsible for the synthesis of ACh during AD is significantly reduced, particularly in the areas of the cerebral cortex and hippocampus [41-43]. The release of ACh in the same sections of the brain induced by depolarization and choline uptake into the presynaptic neurons was found to be significantly reduced. If impaired memory function is considered as the primary AD indicator, the above-mentioned role of ACh in cholinergic transmission and its importance for cognitive function supports these findings in AD [44]. All these observations launched the the cholinergic theory, which was published by Bartus *et al.* (1982) [36]. Bartus in his work fully reflects the relationship between the cholinergic hypothesis, age-dependent cholinergic dysfunction and AD-type dementia.

5. CHOLINESTERASES

Cholinesterases (ChE) are a group of essential enzymes that can be divided into two subgroups based on their catalytic properties.

An enzyme that favors decomposition (hydrolysis) of small substrates, such as ACh (Fig. (**1**)) is called acetylcholinesterase (AChE, EC 3.1.1.7).

The enzyme, which is able to adapt to the bulkier substrates such as benzoyl- or butyryl-choline (Fig. (**1**)) and catalyze their decomposition, is termed as butyrylcholinesterase (BChE, EC 3.1.1.8). AChE and BChE can both catalyze the hydrolysis of ACh (Fig. (**2**)) [45, 46].

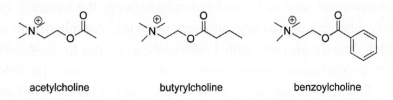

acetylcholine	butyrylcholine	benzoylcholine

Figure 1: Substrates of cholinesterases.

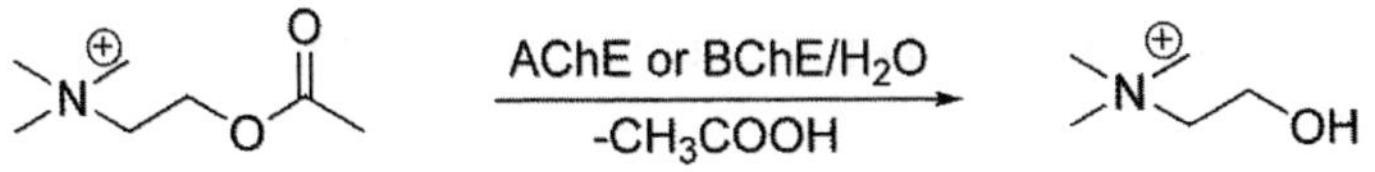

Figure 2: Hydrolysis of ACh catalyzed by AChE or BChE.

In the determination process of the amino acid sequence of both ChE, it was concluded that they belong to a large group of enzymes containing α/β-hydrolase clusters, which are also common for the different lipase, peptidase, haloalkane dehalogenases and even adhesion proteins. AChE and BChE are so similar that it is almost inherent to describe the structure of one without referral to the other. Both have three entrenched disulfide bridges and the amino acid sequence for human BChE (hBChE) and eel AChE isolated from the electrical eel (lat. Electrophorus electricus) is equivalent by 54 %. The main difference between AChE and BChE is in the level of *N*-glycosylation as AChE is less glycosylated than BChE. Glycosylation affects the stability and pharmacokinetic properties, but not the catalytic ability of the enzyme. Differences in the binding affinity of the individual substrates are given by the spatial arrangement of the active site of both enzymes [47].

Active site is located at approximately 20 nm deep straight cavity and is composed of three main clusters - acylation, cation-π and peripheral anionic (aromatic) sites [48].

Hydrolysis of the substrate occurs in the acylation site, which consists of the catalytic triad with amino acids Ser203, Glu202, His447 for human AChE (hAChE) and Ser198, Glu325, His438 for hBChE respectively [48]. Furthermore, entrance to the active site is lined by numerous aromatic residues.

Cation-π site in hAChE is formed by Trp86 and Phe338, which are responsible for weak type interactions such as Columbic forces between the aromatic system and the positively charged quaternary nitrogen moieties. An important role plays Tyr133, which directly interacts with Trp86 and contributes to the stabilization of the cation-π conformation upon binding of the substrate. In hBChE, the stabilization function is represented by Trp82 and Ala328. The absence of phenylalanine in BChE affects affinity to certain inhibitors [49, 50]. Moreover, cation-π site is also capable to interact with ligands without a positive charge [51].

Particularly important are amino acid residues at the peripheral anionic site of hAChE (Tyr72, Tyr124, Trp286), which are all situated at the cavity entrance. However, on hBChE these aromatic residues are missing, what explains the weaker affinity of specific AChE substrates, such as fasciculin (snake venom) or propidium (selective inhibitor of AChE peripheral anionic site). Further studies revealed that the anionic site on BChE consists of Asp70 and Tyr332. (Fig. (**3**)) and (Fig. (**4**)) for the amino acid residues at the peripheral anionic site of hAChE and hBChE, respectively) [49-51].

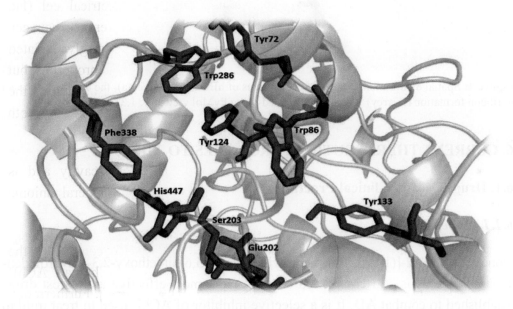

Figure 3: Spatial distribution of amino acid residues of human AChE (in magenta), the enzyme is shown in ribbon formation (in grey). Figure was created using PyMol viewer (v. 1.3).

Since cholinesterases have a catalytic function, they are the main target of various reversible, irreversible or pseudo-irreversible inhibitors. Among reversible inhibitors belong aromatic tertiary amines such as tacrine or donepezil. Pseudo-irreversible binding may interact strongly with the catalytic serine (*e.g.* physostigmine, rivastigmine) and a group of irreversible inhibitors is represented by organophosphate compounds, whose effect on ChE and the entire body is life-threatening (*e.g.* nerve agents sarin, soman, tabun or VX; pesticide paraoxon) [45].

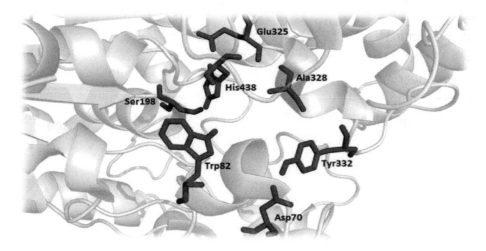

Figure 4: Spatial distribution of amino acid residues of hBChE (in magenta), the enzyme is shown in ribbon formation (in grey). Figure was created using PyMol viewer(v. 1.3).

6. CURRENT THERAPEUTIC APPROACHES TO AD

6.1. Drugs Used in Clinical Practice

6.1.1. Donepezil

Donepezil ((*RS*)-2-[(1-benzyl-4-piperidyl) methyl] -5,6-dimethoxy-2,3-dihydroinden-1 one, Table **1**) is still considered as one of the most effective and safest drugs established to combat AD. It is a selective inhibitor of AChE used to treat mild to moderate AD forms. Nevertheless, donepezil like other AChE inhibitors only slows the progression of the disease to a certain extent, but it cannot completely stop the progression of AD or even cure it [52, 53]. In addition, donepezil improves symptoms of Aβ induced neurotoxicity and positively influences the process of APP cleavage [54]. Donepezil enters cortical neurons and interrupts a mechanism, which leads to the Aβ formation [55]. Moreover, donepezil increases the expression of nicotinic receptors in the cortex and at least partially prevents the AD progression . At the same time, donepezil also reduces the concentration of glutamate, an important neurotransmitter that under AD pathological conditions has neurotoxic effect [56]. Donepezil increases the production of AChE-S isoform, while the synthesis of AChE-R isoform is suppressed, which again is another example of neuroprotective ability [57]. Donepezil has also a significant antioxidant activity, as it is capable to neutralize reactive oxygen species (ROS) and improve blood rheological properties [58-61]. Donepezil is marketed under the commercial brand name Aricept® since 1997. Other features concerning donepezil are summarized in Table **1** and its spatial orientation in AChE from *Torpedo californica* (pdb code: 1e3q) is shown in Fig. (**5**) [62].

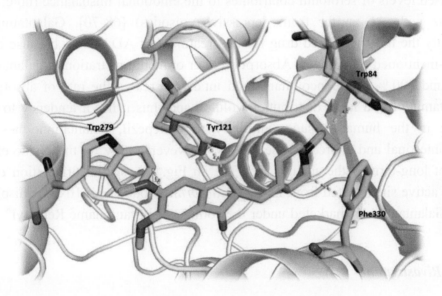

Figure 5: Spatial orientation of donepezil on AChE from *Torpedo californica* and significant interactions with amino acid residues (pdb code: 1e3q). Interactions *via* π-π bonds are predominant. Figure was created using PyMol viewer (v. 1.3) [62].

6.1.2. Galantamine

Galantamine ((4a*S*,6*R*,8a*S*)-5,6,9,10,11,12-hexahydro-3-methoxy-11-methyl-4a*H*-[1]-benzofuro-[3a,3,2-*ef*][2]benzazepin-6-ol, Table **1**) is a selective, reversible inhibitor of AChE. It is a natural alkaloid that was firstly isolated from a plant commonly known as snowdrop (*Galanthus woronowii*) and its presence was confirmed in the bulbs of other various species in the family *Amaryllidaceae* [63]. In addition to its inhibitory activity against AChE galantamine also acts as an allosteric modulator of the N-receptors [64]. Galantamine binds to the opposite side of the receptor than ACh and causes a conformational change of the receptor [65]. The whole process potentiates N-receptors, which results in the enhanced postsynaptic response. Presynaptic N-receptors play an important role in the release of ACh and also regulate concentration of other transmitters such as γ-aminobutyric acid, glutamic acid, serotonin or norepinephrine. All these neuromediators are important for the memory functions and also influence mood and emotions [66, 67]. In the course of AD, pathologically increased levels of glutamate may lead to learning and memory impairment, while pathologically decreased levels of serotonin contributes to the emotional misbalance (note: most patients with AD suffer from depressive episodes) [68-70]. Galantamine is currently the most preferred drug for the treatment of AD mainly because of the above-mentioned advantages. Absorption after oral administration of galantamine is around 100 %, but concomitant food intake decreases the rate of absorption. Galantamine has a large distribution volume and thus it has a tendency to form depots in the human tissues. Similarly to donepezil, galantamine has also gastrointestinal and anorectic side effects. However, rarely do these side effects prevent long-termed therapeutic use [71]. In Fig. (**6**), spatial orientation of the AChE active site isolated from *Torpedo californica* (pdb code: 1dx6) is displayed [72]. Galantamine is marketed under the commercial brand name Reminyl® since 2001.

6.1.3. Rivastigmine

Rivastigmine ((*S*)-3-[1-(dimethylamino)ethyl]phenyl-*N*-ethyl-*N*-methylcarbamate, Table **1**) is a reversible cholinesterase inhibitor intended for symptomatic treatment of moderate to severe AD stages. It is used in the form of a tartrate salt for its better solubility. In the contrary to all other AChE inhibitors, rivastigmine is selective towards both cholinesterases (AChE, BChE), which is particularly useful in later stages of AD. Generally, concentration of BChE increases, while the concentration of AChE decreases in the later stages of AD [73, 74]. Rivastigmine belongs into group of compounds called carbamates, which cause carbamoylation of the active site of the enzyme and impairs its function. The most common side effects are of gastrointestinal origin (*e.g.* vomiting, nausea, diarrhea, anorexia), however frequent dosing schedule can reduce these issues [75, 76].

Fig. (**7**) shows the spatial arrangement of rivastigmine on AChE from *Torpedo californica* (pdb code: 1gqr) [77]. Rivastigmine is marketed under the commercial brand name Exelon® since 2000.

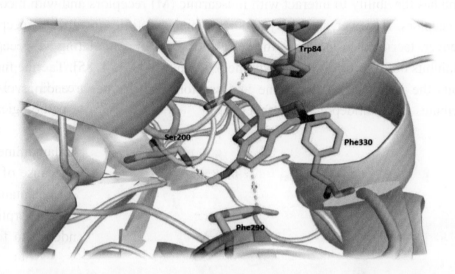

Figure 6: Spatial orientation of galantamine bound to the AChE active site and significant interactions with amino acid residues (pdb code: 1dx6). Besides the π-π interaction (aromatic part of galantamine with Phe290), there is also hydrogen bridge between Ser200 and the hydroxyl group of galantamine and an aliphatic-π interaction with Phe330 and Trp84. Figure was created using PyMol viewer (v. 1.3) [72].

6.1.4. Tacrine

Tacrine (1,2,3,4-tetrahydroakridin-9-amine, Table **1**) is reversible, intermediate-acting cholinesterase inhibitor of AChE and BChE, which is known for more than 60 years [78]. It was originally developed as an antibacterial agent, but when it was thoroughly tested, it demonstrated to be very weak bactericide. However, other surprising characteristics were discovered [79]. The first study described the pharmacological effects of tacrine on analeptic animals sedated by morphine [80]. Nevertheless, the cholinergic properties have not been recognized until 1961 by Heilbronn *et al*. They concluded that tacrine is a stronger inhibitor of BChE than AChE. Previously, it was often used as an antidote for curare and for relieve of intractable pain. Moreover, other previously used or considered indications included Myasthenia gravis, antidepressant effects (similarity with tricyclic antidepressants), tardive dyskinesia or overdosing by anticholinergics [81].

Tacrine was introduced into clinical practice in 1993. It was often used in combination with lecithin [82, 83]. Crystallographic studies of tacrine bounded to AChE demonstrated that it binds to the catalytic AChE site [51]. Importantly, tacrine has the ability to interact with muscarinic (M) receptors and with nicotinic (N) receptors. However, it displays 100-fold higher affinity towards N-receptors. Moreover, tacrine also increases the release of ACh by stimulating M_1-receptors and inhibits both isoforms of monoamine oxidase (MAO) [84, 85]. Tacrine further inhibits the re-uptake of dopamine and serotonin in the nerve endings, which contributes to its antidepressant effect [86].

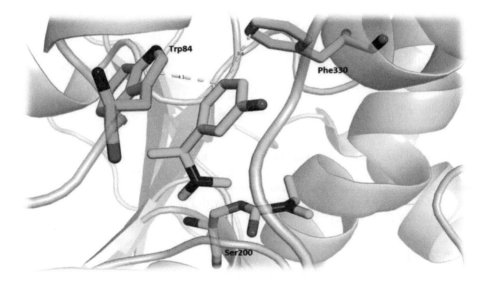

Figure 7: Spatial orientation of rivastigmine on AChE from *Torpedo californica* and significant interactions with amino acid residues (pdb code: 1gqr). Ser200 is carbamoylated and the rest of the molecule inhibitor remains in the cavity stabilized by π-π interactions with Trp84 and Phe330. Figure was created using PyMol viewer (v. 1.3) [77].

The use of tacrine requires more frequent dosing schedule (Table **1**), prolonged titration of blood concentration and monitoring of liver enzymes [87]. Tacrine is no longer used for its narrow therapeutic index, frequent cases of severe hepatotoxicity and gastrointestinal toxicity [88, 89]. In Fig. (**8**) tacrine is superimposed in the active site of AChE from *Torpedo californica* (pdb code: 1acj) [51]. Tacrine was marketed under the commercial brand name Cognex[®] in 1993.

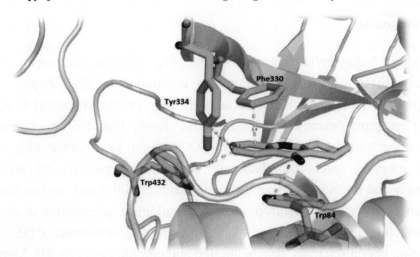

Figure 8: Spatial orientation of tacrine in the active site of AChE from *Torpedo californica* and significant amino acid residues interactions. Tacrine is located in the peripheral anionic site between Trp84 and Phe330. The amino group in position 9 is stabilized by binding to the two water molecules (not shown). Figure was created using PyMol viewer (v. 1.3) (pdb code: 1acj) [51].

Table 1:   AChE Inhibitors in Clinical Practice and their Properties

Name	Donepezil	Galantamine	Rivastigmine	Tacrine
Structure				
Chemical classification	Piperidine derivate	Isochinoline alkaloid	Carbamate	Acridine derivative
Target enzyme	AChE	AChE	AChE/BChE	AChE/BChE
Type of inhibition	Non-competitive, fast reversible	competitive, fast reversible	competitive, slowly reversible	Non-competitive, fast reversible
Metabolism	CYP2D6 a 3A4	CYP2D6 a 3A4	AChE and BChE	CYP1A2
Available formulations	Tablets	Tablets, solutions, control-release tablets	Transdermal patches, solutions and capsules	Capsules
Plasmatic half life	70 h	7 h	3h (patches), 1h (capsules)	6 h

6.1.5. Memantine

Memantine (3,5-dimethyladamantan-1-amine, Fig. (**9**)) has mechanism of action completely different to other compounds mentioned in this chapter. It is a noncompetitive antagonist with medium affinity to the *N*-methyl-D-aspartate (NMDA) receptor. Potency of memantine is closely linked with glutamate. At normal concentration of glutamate, NMDA-receptor is inhibited only weakly, however elevated levels of glutamate increase the potency of memantine significantly and the NMDA-receptor is then strongly inhibited. Its mechanism of action can be described as neuroprotective. In addition, memantine is also being considered for the treatment of Parkinson disease, epilepsy, CNS trauma, amyotrophic lateral sclerosis, drug addiction and chronic pain [90-93]. Memantine is marketed under the commercial brand name Ebixa® since 2002.

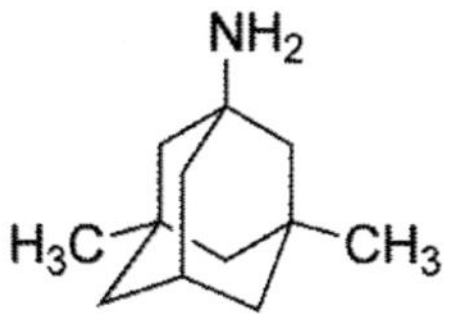

Figure 9: Structure of memantine as a representative of NMDA antagonists in the treatment of AD.

6.2. Potential Therapeutic Approaches to AD

6.2.1. Vaccination and Immunization

Transgenic mice immunized with amyloid-beta (Aβ) bearing human trans-membrane amyloid precursor protein (amyloid precursor protein, APP) exhibited reduced production of amyloid plaques in the brain, which belong to the pathological findings in the brain of AD patients. Furthermore, improved behavioral functions were also detected [94, 95]. This particular finding led to several vaccination clinical trials with AD patients. However, the administration of synthetic Aβ (AN1792) increased the incidence of meningo-encephalitis, which was discovered in six percent of patients and clinical trials were prematurely terminated [96]. Nevertheless, analysis of the clinical trials data has proven that administration of AN1792 produces antibodies against Aβ and produces significant improvements in cognitive functions [97]. Nevertheless, a magnetic resonance after immunization revealed a reduction of the brain volume [98]. Based on these findings, it was concluded that passive immunization should be safer and more effective then tested active form. Currently, several vaccines using selective Aβ monoclonal antibody as passive immunization tool are in the various stages of clinical development [99, 100].

6.2.2. Modulators of Secretases

Aβ is formed from APP by two enzymes known as β-secretase and γ-secretase. Connected research is mainly focused on the possibility to inhibit these secretases. Compound labeled KMI-429 (Fig. (**10**)) is one of the novel compounds that have experimentally proven inhibition ability towards β-secretase. *In vivo* experiments on transgenic mice have proven reduced formation of Aβ after its application [101]. However, one of the encountered problems is the molecular weight of these inhibitors, which does not allow a penetration of the blood-brain barrier. Furthermore, the application of KMI-429 to mice deficient in β -secretase produced problems with learning [102, 103]. For these reasons, the attention is paid to the development of small molecule inhibitors of β-secretase [104-106].

Decreased levels of Aβ in the brain, cerebrospinal fluid (CSF) and plasma were observed in rodents who received inhibitors of γ - secretase labeled Dapto, LY450139 dihydrate and BMS-299897 (Fig. (**10**)) [107-110]. The acquired results lead to the conclusion that Aβ may be associated with cognitive impairment in AD patients and also that the administration of an γ-secretase inhibitor can lead (particularly in the early stage of AD) to reversible changes during the formation of amyloidal plaques. The encouraging results from clinical trial of these inhibitors are on the other hand clouded by detrimental effects in the gastrointestinal compartment, thymus and spleen. This could be suggested by insufficient selectivity of tested compounds [111, 112].

Tarenflurbil (MPC-7869) belongs to a group of γ -secretase modulators, whose mechanism of action is to shift the production of longer $A\beta_{42}$ towards short and less amyloidogenic peptides (*e.g.* $A\beta_{37}$) without affecting any other physiological substrates (Fig. (**10**)) [113]. Results from tarenflurbil second phase clinical trials indicated possible benefit for patients with mild AD. Analysis of the trial data suggested that the best results were achieved in the group receiving the highest dose (800 mg twice a day). The third phase clinical trials reported adverse effects such as eosinophilia, mild anemia, hypertension or rash and tarenflurbil investigation concluded its withdrawal from other clinical trials [114, 115].

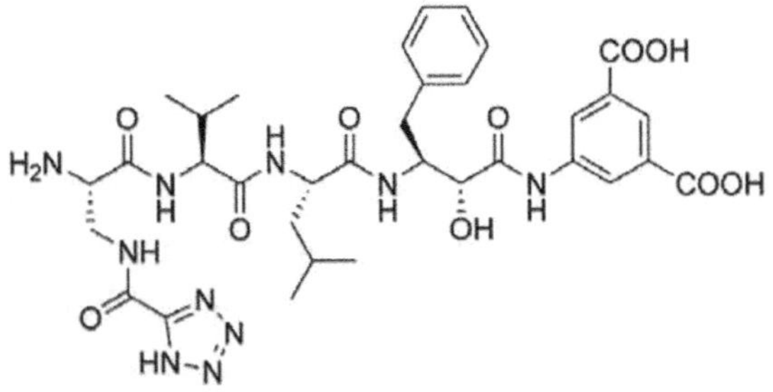

KMI-429

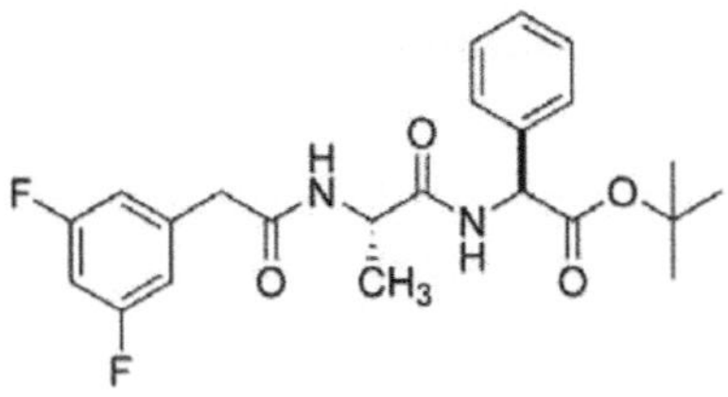

DAPT

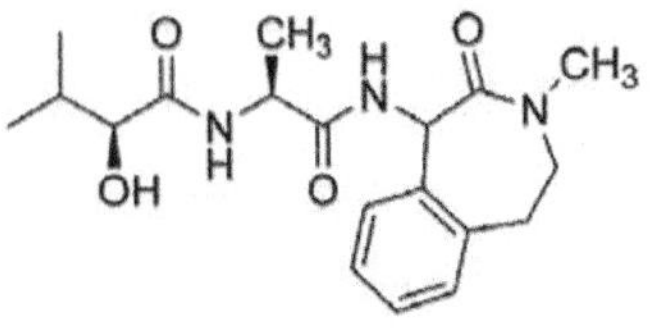

LY450139

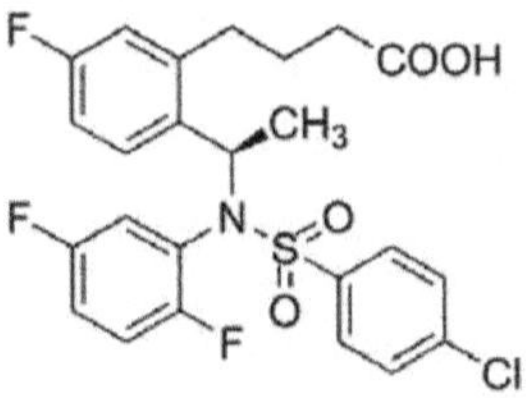

BMS-299897

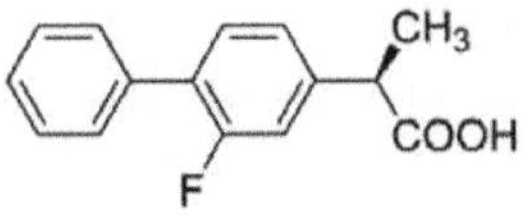

tarenflurbil

Figure 10: Modulators of secretases.

6.2.3. Amyloid-β Antagonists

Aβ forms fibrillar aggregates that cause destruction of neurons. Tramiprosate, which is the main representative of this class of compounds, inhibits the formation of fibrillar aggregates by binding to the soluble Aβ and thereby mimicking the effect of glycosaminoglycans (Fig. (**11**)). A reduction of soluble Aβ levels in the CSF reduces a chance of the possible formation of amyloid plaques in neurons. Open-label clinical study suggested that tramiprosate slows down deterioration of cognitive functions in early stage AD [116]. Tramiprosate is currently undergoing phase three clinical trials in North America and Europe, however its potential benefit in AD treatment remains unknown [117].

The next compound that belongs among amyloid-β antagonists is colostrinin, which is a mixture of proline rich polypeptides derived from colostrum. It showed a reasonable improvement in cognitive function in patients with mild AD (note: for moderate AD test results were inconclusive, when compared with placebo). Nevertheless, colostrinin did not demonstrate up-to-date any long term benefits [118-120].

Particularly interesting drug in this group is scyllo-inositol (AZD-103, Fig. (**11**)). It stabilizes un-clustered, non-toxic Aβ complexes and reduces the harmful impact of Aβ clusters on neurons. Furthermore, in terms of long-term use it is even able to recover some memory functions [121, 122].

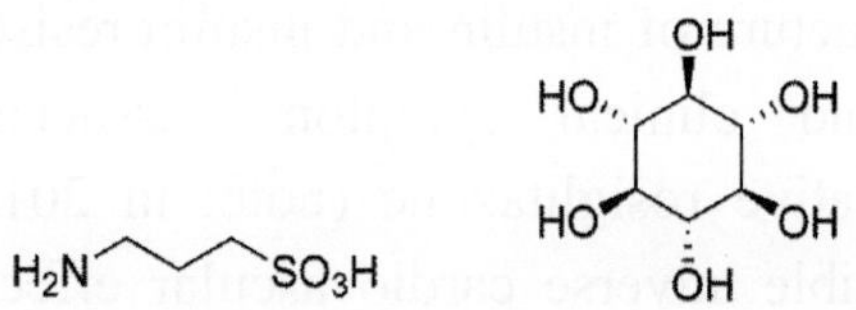

tramiprosate   scyllo-inositol

Figure 11: Amyloid-β antagonists.

6.2.4. Statins in the Treatment of AD

The statin group is primarily used for the treatment of hyperlipidemia acting through the inhibition of 3-hydroxy-3-methyl-glutaryl-CoA reductase (HMG CoA reductase). However, some studies have also showed decrease of Aβ levels after statin therapy *in vivo* [123, 124]. The effect of statins may be particularly seen, when they are regularly used by AD patients before 80 years of age [125]. Moreover, several epidemiological studies suggest statins reduce the AD risk [126-130]. However, large cohort and randomized, placebo-controlled studies on the prevention of coronary heart disease do not confirm that statins have positive effect on cognitive functions in patients with AD [131-133]. Based on these findings, it can be concluded that statins in AD may slow the neurodegenerative processes, but cannot help those with already manifested AD [134]. Two basic structural types of statins are shown in Fig. (**12**).

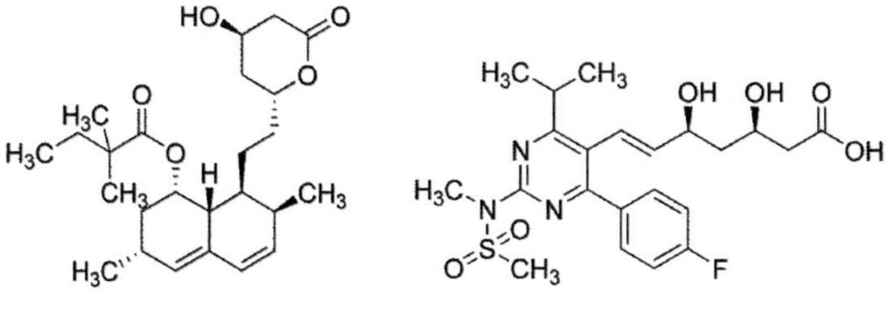

simvastatin   rosuvastatin

Figure 12: Simvastatin (first-generation statin), rosuvastatin (third generation statin).

6.2.5. Peroxisome Proliferator-Activated Receptor Agonists

Abnormalities in the structure of insulin and insulin resistance may contribute to the neuropathology and clinical symptoms associated with AD [135]. Thiazolidinedione derivative rosiglitazone (note: in 2010 withdrawn from the market because of possible adverse cardiovascular effects, Fig. (**13**)) increases peripheral insulin sensitivity by agonistic effect on peroxisome proliferator-activated receptor (PPAR-γ). A study conducted on 30 selected individuals diagnosed with AD suggested a possible benefit in the rosiglitazone group particularly concerning cognitive symptoms, when compared to the placebo group [136]. In addition, individuals without the ApoE ε4 gene had considerably better cognitive functions than patients carrying this gene. Another agonist acting on PPAR-γ tested in clinical trials is pioglitazone (Fig. (**13**)) [137]. Furthermore, it has been observed that the actual administration of intranasal insulin improves memory function [138]. The close correlation between the misbalance of insulin and cerebral glucose metabolism in the brain of AD diagnosed patients indicate recent studies with metformin. Data shows that metformin reduces insulin resistance and could be beneficial in the treatment of AD [139].

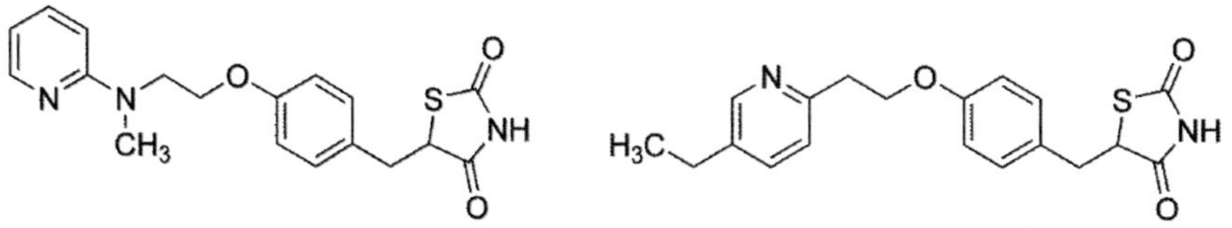

rosiglitazone   pioglitazone

Figure 13: Peroxisome proliferator-activated receptor agonists.

6.2.6. Metal Chelators

Aβ interacts with biogenic elements such as zinc, copper and iron, and together form a cytotoxic aggregates [140-142]. Clioquinol was the first drug in this group, which has been tested against the formation of Aβ aggregates (Fig. (**14**)). Within the 36-week trial, it demonstrated Aβ decrease in the brain by 49 %. However, two serious side effects appeared during this period (syncopal episodes and cardiac arrhythmia) [143, 144]. Other chelators contemplated for use in AD connection contain compounds labeled XH1, DP-109 and (-)-epigallocatechin-3-gallate. All consistently demonstrated significant reduction of Aβ caused brain damage (Fig. (**14**)) [145-148].

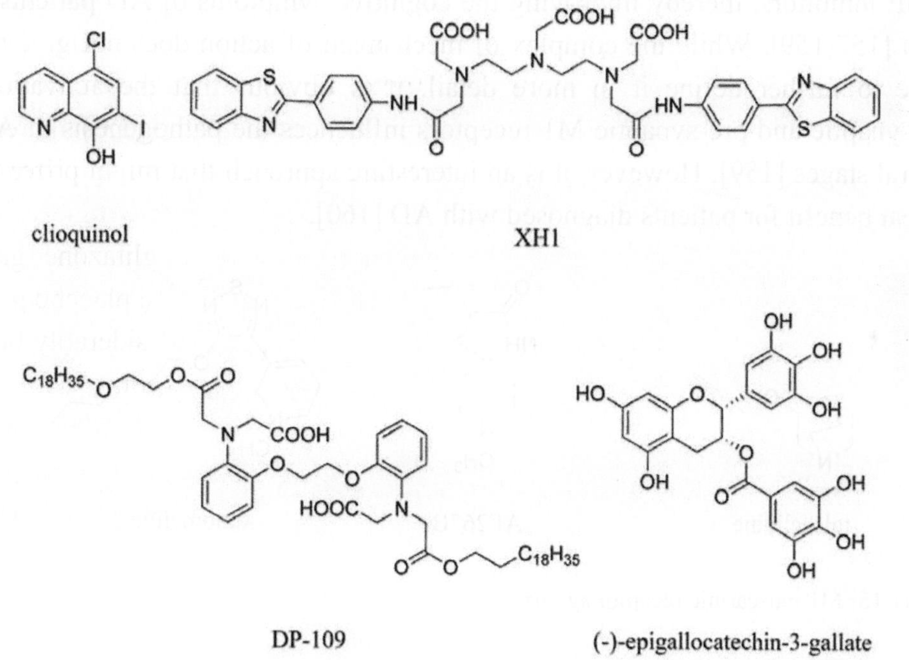

clioquinol   XH1

DP-109   (-)-epigallocatechin-3-gallate

Figure 14: Metal chelators.

6.2.7. M1 Muscarinic Receptor Agonists

Presynaptic M1 muscarinic receptors influence the AD at several levels of its pathogenesis [149]. Based on results from *in vitro* studies, it has been concluded that the muscarinic receptor subtype 1 (M1) plays a vital role in the AD pathogenesis. Furthermore, the M1 muscarinic agonist (AF267B) supported these findings in the *in vitro* and *in vivo* tests, where it was found to mediate cleavage of APP by α-secretase through a non-amyloidogenic pathway. Thus, the soluble APPsα with neuroprotective character is produced instead of the toxic Aβ (Fig. (**15**)). Moreover, the M1 agonist also reduced Aβ levels in the CSF and the incidence of inflammation and cognitive disorders [150-154]. Other studies have demonstrated that AF267B also prevents hyperphosphorylation of the τ-protein and thus retains its normal physiological function [155]. Talsaclidine is another novel compound introduced into clinical testing (Fig. (**15**)). Like AF267B, talsaclidine also reduced levels of CSF Aβ [156]. Similarly, xanomeline is capable to reduce Aβ levels, positively affect τ-protein and also potentiate the effect of AChE inhibitors, thereby improving the cognitive symptoms of AD patients(Fig. (**15**)) [157-159]. While the complex of mechanism of action does not give much space to further define it in more detail, it is obvious that the activation of postsynaptic and pre-synaptic M1 receptors influences the pathogenesis of AD at several stages [159]. However, it is an interesting approach that might prove to be a great benefit for patients diagnosed with AD [160].

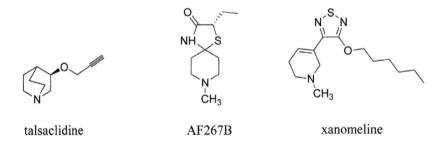

| talsaclidine | AF267B | xanomeline |

Figure 15: M1 muscarinic receptor agonists.

6.2.8. RAGE Receptor Modulation

RAGE (Receptor for Advanced Glycation Endproducts) is a transmembrane receptor that belongs to the immunoglobulin family. It was firstly described in 1992 by Neeperem *et al.* [161]. Its name is derived from the ability to bind glycation end products derived from glycoproteins. RAGE in the presence of Aβ and in contact with endothelial cells of the pallets can induce migration of monocytes through endothelial cells of the brain. The monocyte diapedesis plays an important role in the inflammatory processes associated with AD [162]. The ligands for this receptor are currently being produced, which could suppress the Aβ accumulation and thereby contribute to the improvement of the AD prognosis [163, 164].

6.2.9. Peripherally Acting Scavenger of AB

The reduction of Aβ levels is still considered as one of the therapeutic targets in the AD treatment. While active immunization can directly reduce Aβ levels in the brain, peripherally acting scavengers like gelsolin (actin-binding protein) have a high affinity for peripheral Aβ, which also contribute to the overall AD prognosis [165]. An interesting group of compounds are also dihydropyridine antihypertensives. In AD relation and besides alleviating Aβ production, they also reduce Aβ caused brain damage, while they facilitate the clearance of this pathological protein at same time [166].

6.2.10. Glycogen Synthase Kinase-3 (GSK-3)

Glycogen synthase kinase-3 (GSK-3) is a cellular serine/threonine protein kinase. This enzyme regulates countless number of cellular processes and its dysregulation is crucial to the pathogenesis of diverse diseases such as AD, Parkinson disease, diabetes mellitus type 2, bipolar disorder and cancer [167].

There are two isoforms of this enzyme: GSK-3α and GSK-3β, both are expressed ubiquitously in the brain with an increased incidence in hippocampus, cerebral cortex and in the Purkinje neurons in the cerebellum. These two isoforms have similar catalytic domains and almost the same substrate specificity [168].

Activity of GSK-3 depends on phosphorylation of specific sites. However, deactivation of this enzyme has bigger importance than its activation, because the enzyme is active all the time in varying degrees as it undergoes the process of autophosphorylation. GSK-3 is the enzyme which plays an important role in both accumulation of extracellular deposits, so called senile plaques (the main component is Aβ) and formation of intracellular neurofibrillary tangles (the major element is τ-protein) [169, 170].

GSK-3β is important in physiological and also pathological phosphorylation of τ-protein. Under the physiological conditions, phoshorylation of τ -protein determines its affinity for microtubule binding, whereas during the pathogenesis comes to the hyperphosphorylation, which leads to the dissociation of τ-protein from the microtubules and its subsequent aggregation. *In vivo* tests showed that overexpression of GSK-3β results in τ-neurodegeneration, while inhibition of this kinase decreases τ-toxicity [167].

Furthermore, the relationship between A*β* production and the GSK-3 activity was found out, when the enhanced activity of GSK-3 was induced by increased Aβ production. It was also discovered that APP and presenelin 1 are substrates for GSK-3α and exactly this kinase is considered to be a regulator of the Aβ production by the mechanism of interaction between APP and γ-secretase during the process of APP cleavage [171].

By the above-mentioned reasons, the attention of a huge amount of scientific groups is concentrated exactly on the most potent inhibitors of GSK-3. The inhibitors could be divided into 2 major groups – direct and indirect inhibitors, while the direct ones interact with the enzyme directly and the indirect increase *N*-terminal phosphorylation of GSK-3 (whereby deactivate it) [172].

Direct inhibitors are further subdivided into smaller groups: lithium, small molecule inhibitors and peptide or protein inhibitors. Lithium was the first one and at the same time the only one drug, which was clinically used for the treatment of bipolar disorders. This monovalent cation inhibits both GSK isoforms. The mechanism of action consists in the direct competitive binding to the ATP-dependent magnesium-sensitive catalytic site of the enzyme and also indirectly in the modulation of post-translational modifications of GSK-3. Lithium reduces τ-phosphorylation and Aβ production from APP as well. Lithium established positive influence on the prevention of β-amyloid toxicity, facilitation of neurogenesis and also on rescue of β-amyloid induced cognitive impairments [172].

The small molecule inhibitors could interact by the mechanism of ATP or non-ATP competitive inhibition. Currently, the bigger importance is attached to the non-ATP competitive inhibitors, because of their better ADME properties due to the absence of endogenous ATP competition. They bind outside the ATP pocket, what could show better kinase selectivity and last but not least reason is that they should have lower IC_{50} values. An important group of non-ATP competitive inhibitors are thiadiazolidindions. There are two natural representatives of this group – manzamine, which binds to the allosteric site of the GSK-3β and palinurin, whose mechanism of action has not been yet recognized (Fig. (**16**)) [173, 174]. Among the synthetic analogues, tideglusib is the most promising that is in the IIb phase of AD clinical trials (Fig. (**16**)). The observation of relationship between the structure and the effect discovered that presence of 1,3-dicarbonyl moiety, where the atom of nitrogen is between two carbonyl groups, had the biggest influence on the GSK-3β inhibition [172].

The peptide and protein inhibitors are the last subgroup of direct inhibitors, which were essentially developed for the therapy of early depressive behavior induced by mild traumatic brain injury. The representative of indirect inhibitors of GSK-3 is valproic acid. Nowadays, it is prescribed mostly for epileptic patients, but it also undergoes phase II trials with the AD indication [175].

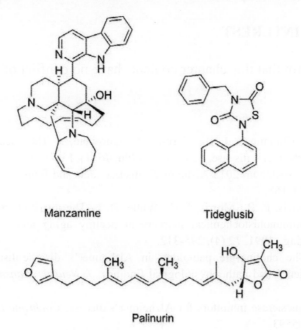

Manzamine   Tideglusib

Palinurin

Figure 16: Structures of GSK-3 inhibitors.

7. CONCLUSIONS

AD therapy is based on the so-called cholinergic theory, which assumes that cholinesterase activity plays an important role in the AD pathogenesis. Up-to-date, there are available almost exclusively only AChE inhibitors and the NMDA receptor antagonist memantine on the market. Here are discussed additional ten different possible therapeutic approaches in the AD treatment. An accurate determination of the AD etiology is still to be determined and thoroughly revised. Despite the worldwide effort, AD is a growing problem of mankind, not only in terms of health, but also in terms of sociological and economic parameters.

ACKNOWLEDGEMENTS

This study was supported by the specific research (SV/FVZ201201, SV/FVZ2011/04), by the Grant Agency of the Czech Republic (No. P303/11/1907 and P303/12/0611), by Post-doctoral project (No. CZ.1.07/2.3.00/30.0044), by Long Term Development plan – 1011, by MH CZ - DRO (University Hospital Hradec Kralove, No. 00179906), by project MSM0021620849 given by Ministry of Education, Youth and Sports.

CONFLICT OF INTEREST

The authors confirm that this chapter contents have no conflict of interest.

REFERENCES

[1]   Mega, M. S.; Cummings, J. L.; Fiorello, T.; Gornbein, J. The spectrum of behavioral changes in Alzheimer's disease. *Neurology,* **1996**, *46* (1), 130-135.

[2]   Alzheimer's Association. 2012 Alzheimer's disease facts and figures. *Alzheimers Dement.,* **2012**, *8* (2), 131-168.

[3]   Price, J. L.; Davis, P. B.; Morris, J. C.; White, D. L. The distribution of tangles, plaques and related immunohistochemical markers in healthy aging and Alzheimer's disease. *Neurobiol.* Aging., **1991**, *12* (4), 295-312.

[4]   Frölich, L. The cholinergic pathology in Alzheimer's disease-discrepancies between clinical experience and pathophysiological findings. *J. Neural. Transm.,* **2002**, *109* (7-8), 1003-1013.

[5]   Birks, J. Cholinesterase inhibitors for Alzheimer's disease. *Cochrane Database Syst. Rev.,* **2006**, *1*, CD005593.

[6]   Corbett, A.; Smith, J.; Ballard, C. New and emerging treatments for Alzheimer's disease. *Expert Rev. Neurother.*, **2012**, *12* (5), 535-543.

[7]   Chopra, K.; Misra, S.; Kuhad, A. Current perspectives on pharmacotherapy of Alzheimer's disease. *Expert Opin. Pharmacother.*, **2011**, *12*(3), 335-350.

[8]   Launer, L. J.; Fratiglioni, L.; Andersen, K.; Breteler, M. M. B.; Copeland, R. J. M.; Dartiques, J. F.; Lobo, A.; Martinez-Lage, J.; Soininen, H.; Hofman, A. Regional differences in the incidence of dementia in Europe: EURODEM collaborative analyses 1999. Collective authors of Alzheimer Europe organization: *Dementia in Europe Yearbook:* **2008**, *20*, 1-178.

[9]   Weiner, M. W.; Aisen, P. S.; Clifford, R. J. Jr.; Jagust, W. J.; Trojanowski, J. Q.; Shaw, L.; Saykin, A. J.; Morris, J. C.; Cairns, N.; Beckett, L. A.; Toga, A.; Green, R.; Walter, S.; Soares, H.; Snyder, P.; Siemers, E.; Potter, W.; Cole, P. E.; Schmidt, M. The Alzheimer's disease neuroimaging initiative: Progress report and future plans. *Alzheimers Dement.*, **2010**, *6* (3), 202-211.

[10]   Green, R. C.; Cupples, L. A.; Go, R.; Benke, K. S.; Edeki, T.; Griffith, P. A.; Williams, M.; Hipps, Y.; Graff-Radford, N.; Bachman, D.; Farrer, L. A.; MIRAGE Study Group. Risk of dementia among white and African-American relatives of patients with Alzheimer's disease. *J. Am. Med. Assoc.*, **2002**, *287* (3), 329-336.

[11]   Drtinova, L.; Pohanka, M. Alzheimerova demence: aspekty současné farmakologické léčby. *Ceska Slov. Farm.*, **2011**, *60*, 219-228.

[12]   Lautenschlager, N. T.; Cupples, L. A.; Rao, V. S.; Auerbach, S. A.; Becker, R.; Burke, J.; Chui, H.; Duara, R.; Foley, E. J.; Glatt, S. L.; Green, R. C.; Jones, R.; Karlinsky, H.; Kukull, W. A.; Kurz, A.; Larson, E. B.; Martelli, K.; Sadovnick, A. D.; Volicer, L.; Waring, S. C.; Growdon, J. H.; Farrer, L. A. Risk of dementia among relatives of Alzheimer's disease patients in the MIRAGE study: what is in store for the oldest-old? *Neurology,* **1996**, *46* (3), 641-650.

[13]   Bertram, L.; Tanzi, R.E. The genetics of Alzheimer's disease. *Prog. Mol. Biol. Transl. Sci.*, **2012**, *107*, 79-100.

[14]   Corder, E. H.; Saunders, A. M.; Strittmatter, W. J.; Schmechel, D. E.; Gaskell, P. C.; Small, G. W.; Roses, A. D.; Haines, J. L.; Pericak-Vance, M. A. Gene dose apolipoproteine E type 4 allele and the risk of Alzheimer's disease in late onset-families. *Science,* **1993**, *261* (5123), 921-923.

[15]   Saunders, A. M.; Strittmatter, W. J.; Schmechel, D.; George-Hyslop, P. H.; Pericak-Vance, M. A.; Joo, S. H.; Rosi, B. L.; Gusella, J. F.; Crapper-MacLachlan, D. R.; Alberts, M. J. Association of apolipoprotein E allele 4 with late onset familial and sporadic Alzheimer's disease. *Neurology,* **1993**, *43* (8), 1467-1472.

[16]   Corder, E. H.; Saunders, A. M.; Risch, N. J.; Strittmatter, W. J.; Schmechel, D. E.; Gaskell, P. C. Jr.; Rimmler, J. B.; Locke, P. A.; Conneally, P. M.; Schmader, K. E.; Protective effect of apolipoproteine E allele 2 for late onset Alzheimer's disease. *Nat. Genet.*, **1994**, *7* (2), 180-184.

[17]   Clarke, R.; Smith, A. D.; Jobst, K. A.; Refsum, H.; Sutton, L.; Ueland, P. M. Folate, vitamin B12, and total homocysteine levels in confirmed Alzheimer's disease. *Arch. Neurol.*, **1998**, *55* (11), 1449-1455.

[18]   Green, R. C.; Cupples, L. A.; Kurz, A.; Auerbach, S.; Go, R.; Sadovnick, D.; Duara, R.; Kukull, W. A.; Chui, H.; Edeki, T.; Griffith, P. A.; Friedland, R. P.; Bachman, D.; Farrer, L. Depression as a risk factor for Alzheimer's disease: the MIRAGE study. *Arch. Neurol.*, **2003**, *60* (5), 753-759.

[19]   Launer, L. J.; Andersen, K.; Dewey, M. E.; Letenneur, L.; Ott, A.; Amaducci, L. A.; Brayne, C.; Copeland, J. R.; Dartigues, J. F.; Kragh-Sorensen, P.; Lobo, A.; Martinez-Lage, J. M.; Stijnen, T.; Hofman, A. Rates and risk factors for dementia and Alzheimer's disease: results from EURODEM pooled analyses. EURODEM Incidence Research Group and Work Groups. European Studies of Dementia. *Neurology,* **1999**, *52* (1), 78-84.

[20]   Kivipelto, M.; Ngandu, T.; Fratiglioni, L.; Viitanen, M.; Kareholt, I.; Winblad, B.; Helkala, E. L.; Tuomilehto, J.; Soininen, H.; Nissinen, A. Obesity and vascular risk factors at midlife and the risk of dementia and Alzheimer's disease. *Arch. Neurol.*, **2005**, *62* (10), 1556-1560.

[21]   Yaffe, K. Metabolic syndrome and cognitive decline. *Curr. Alzheimer Res.*, **2007**, *4* (2), 123-126.

[22]   Whitmer, R. A.; Gustafson, D. R.; Barrett-Connor, E.; Haan, M. N.; Gunderson, E. P.; Yaffe, K. Central obesity and increased risk of dementia more than three decades later. *Neurology,* **2008**, *71* (14), 1057-1064.

[23]   Wu, W.; Brickman, A. M.; Luchsinger, J.; Ferrazzano, P.; Pichiule, P.; Yoshita, M.; Brown, T.; DeCarli, C.; Barnes, C. A.; Mayeux, R.; Vannucci, S. J.; Small, S. A. The brain in the age of old: the hippocampal formation is targeted differentially by diseases of late life. *Ann. Neurol.*, **2008**, *64* (6), 698-706.

[24]   Tsivgoulis, G.; Alexandrov, A. V.; Wadley, V. G.; Unverzagt, F. W.; Go, R. C.; Moy, C. S.; Kissela, B.; Howard, G. Association of higher diastolic blood pressure levels with cognitive impairment. *Neurology,* **2009**, *73* (8), 589-595.

[25]   Solomon, A.; Kivipelto, M.; Wolozin, B.; Zhou, J.; Whitmer, R. A. Midlife serum cholesterol and increased risk of Alzheimer's and vascular dementia three decades later. *Dement. Geriatr. Cogn. Disord.*, **2009**, *28* (1), 75-80.

[26]   Pendlebury, S. T.; Rothwell, P. M. Prevalence, incidence, and factors associated with pre-stroke and post-stroke dementia: a systematic review and meta-analysis. *Lancet Neurol.*, **2009**, *8* (11), 1006-1018.

[27]   Raji, C. A.; Ho, A. J.; Parikshak, N. N.; Becker, J. T.; Lopez, O. L.; Kuller, L. H.; Hua, X.; Leow, A. D.; Toga, A. W.; Thompson, P. M. Brain structure and obesity. *Hum. Brain Mapp.*, **2010**, *31* (3), 353-364.

[28]   Rusanen, M.; Kivipelto, M.; Quesenberry, C. P.; Zhou, J.; Whitmer, R. A. Heavy smoking in midlife and long-term risk of Alzheimer's disease and vascular dementia. *Arch. Intern. Med.*, **2010**, *171* (4), 333-339.

[29]   Anstey, K. J.; von Sanden, C.; Salim, A.; O'Kearney, R. Smoking as a risk factor for dementia and cognitive decline: a meta-analysis of prospective studies. *Am. J. Epidemiol.*, **2007**, *166* (4), 367-738.

[30]   Solomon, A.; Kivipelto, M.; Wolozin, B.; Zhou, J.; Whitmer, R. A. Midlife serum cholesterol and increased risk of Alzheimer's and vascular dementia three decades later. *Dement. Geriatr. Cogn. Disord.*, **2009**, *28* (1), 75-80.

[31]   Veld, B. A.; Ruitenberg, A.; Hofman, A.; Launer, L. J.; van Duijn, C. M.; Stijnen, T.; Breteler, M. M.; Stricker, B. H. Nonsteroidal anti-inflammatory drugs and the risk of Alzheimer's disease. *N. Engl. J. Med.*, **2001**, *345* (21), 1515-1521.

[32]   Cummings, J. L.; Cole, G. Alzheimer disease. *J. Am. Med. Assoc.*, **2002**, *287* (18), 2335-2338.

[33]   Qiu, C.; Winblad, B.; Fastbom, J.; Fratiglioni, L. Combined effects of APOE genotype, blood pressure, and antihypertensive drug use on incident AD. *Neuorology,* **2003**, *61* (5), 655-660.

[34]   Craig, L. A.; Hong, N. S.; McDonald, R. J. Revisiting the cholinergic hypothesis in the development of Alzheimer's disease. *Neurosci. Biobehav. Rev.*, **2011**, *35* (6):1397-1409.

[35]   Dumas, J. A.; Newhouse, P. A. The cholinergic hypothesis of cognitive aging revisited again: cholinergic functional compensation. *Pharmacol. Biochem. Behav.*, **2011**, *99* (2), 254-261.

[36]   Bartus, R. T.; Dean, R. L.; Beer, B.; Lippa, A. S. The cholinergic hypothesis of geriatric memory dysfunction. *Science,* **1982**, *217* (4558), 408-414.

[37]   Bartus, R. T.; Dean, R. L.; Goas, J. A.; Lippa, A. S. Age-related changes in passive avoidance retention: modulation with dietary choline. *Science,* **1980**, *209* (4453), 301-303.

[38]   Balin, B.; Abrams, J. T.; Schrogie, J. Toward a unifying hypothesis in the development of Alzheimer's disease. *CNS Neurosci. Ther.,* **2011**, *17* (6), 587-589.

[39]   Perry, E. K.; Tomlinson, B. E.; Blessed, G.; Perry, R. H.; Cross, A. J.; Crow, T. T. Noradrenergic and cholinergic systems in senile dementia of Alzheimer type. *Lancet,* **1981**, *2* (8238), 149.

[40]   Drachman, D. A. Memory and cognitive function in man: does the cholinergic system have a specific role? *Neurology,* **1977**, *27* (8), 783-790.

[41]   Bowen, D. M.; Smith, C. B.; White, P.; Davison, A. N. Neurotransmitter-related enzymes and indices of hypoxia in senile dementia and other abiotrophies. *Brain.*, **1976**, *99* (3), 459-496.

[42]   Davies, P.; Maloney, A. J. F. Selective loss of central cholinergic neurones in Alzheimer's disease. *Lancet,* **1976**, *2* (8000), 1403.

[43]   Perry, E. K.; Gibson, P. H.; Blessed, G.; Perry, R. H.; Tomlinson, B. E. Neurotransmitter enzyme abnormalities in senile dementia. Choline acetyltransferase and glutamic acid decarboxylase activities in necropsy brain tissue. *J. Neurol. Sci.*, **1977**, *34* (2), 247-265.

[44]   Drachman, D. A.; Leavitt, J. Human memory and the cholinergic system. *Arch. Neurol.*, **1974**, *30* (2), 113-121.

[45]   Giacobini, E. *Cholinesterases and Cholinesterase Inhibitors: Basic Preclinical and Clinical Aspects.* 1st ed. Martin Dunitz: London, GB, **2000**, 270 pp.

[46]   Buphendra, P. D. Butyrylcholinesterase: its use for prophylaxis for organophosphate exposure. In *Butyrylcholinesterase: Its Function and Inhibitors.* 1st ed; Giacobini, E. (Ed.); Martin Dunitz: London, GB, **2003**, 163-178.

[47]   Nachon, F.; Masson, P.; Nicolet, Y.; Lockridge, O.; Fontecilla-Camps, J. C. Comparison of the structures of butyrylcholinesterase and acetylcholinesterase. In *Butyrylcholinesterase: Its Function and Inhibitors.* 1st ed; Giacobini, E. (Ed.); Martin Dunitz: London, GB, **2003**, 39-54.

[48]   Sussman, J. L.; Harel, M.; Frolow, F.; Oefner, C.; Goldman, A.; Toker, L.; Silman, I. Atomic-Structure of Acetylcholinesterase from Torpedo-Californica - A Prototypic Acetylcholine-Binding Protein. *Science,* **1991**, *253* (5022), 872-879.

[49]   Masson, P.; Legrand, P.; Bartels, C. F.; Froment, M. T.; Schopfer, L. M.; Lockridge, O. Role of aspartate 70 and tryptophan 82 in binding of succinyldithiocholine to human butyrylcholinesterase. *Biochemistry,* **1997**, *36* (8), 2266-2277.

[50]   Nachon, F.; Ehret-Sabatier, L.; Loew, D.; Colas, C.; van Dorsselaer, A.; Goeldner, M. Trp82 and Tyr332 are involved in two quaternary ammonium binding domains of human butyrylcholinesterase as revealed by photoaffinity labeling with [3H]DDF. *Biochemistry,* **1998**, *37* (29), 10507-10513.

[51]   Harel, M.; Schalk, I.; Ehret-Sabatier, L.; Bouet, F.; Goeldner, M.; Hirth, C.; Axelsen, P. H.; Silman, I.; Sussman, J. L. Quaternary ligand binding to aromatic residues in the active-site gorge of acetylcholinesterase. *Proc. Natl. Acad. Sci. U.S.A.,* **1993**, *90* (19), 9031-9035.

[52]   Sabbagh, M. N.; Richardson, S.; Relkin, N. Disease-modifying approaches to Alzheimer's disease: challenges and opportunities- lessons from donepezil therapy. *Alzheimers Dement.,* **2008**, *4* (1, Suppl 1), S109-S118.

[53]   Sabbagh, M. N.; Farlow, M. R.; Relkin, N.; Beach, T. G. Do cholinergic therapies have disease-modifying effects in Alzheimer's disease? *Alzheimers Dement.,* **2006**, *2* (2), 118-125.

[54]   Jacobson, S. A.; Sabbagh, M. N.; Donepezil: potential neuroprotective and disease-modifying effects. *Expert Opin. Drug Metab. Toxicol.,* **2008**, *4* (10), 1363-1369.

[55]   Beach, T. G.; Potter, P. E.; Kuo, Y. M.; Emmerling, M. R.; Durham, R. A.; Webster, S. D.; Walker, D. G.; Sue, L. I.; Scott, S.; Layne, K. J.; Roher, A. E. Cholinergic deafferentation of the rabbit cortex: a new animal model of Aβ deposition. *Neurosci. Lett.,* **2000**, *283* (1), 9-12.

[56]   Takada-Takatori, Y.; Kume, T.; Sugimoto, M.; Katsuki, H.; Sugimoto, H.; Akaike, A. Acetylcholinesterase inhibitors used in treatment of Alzheimer's disease prevent glutamate neurotoxicity *via* nicotinic acetylcholine receptors and phosphatidylinositol 3-kinase cascade. *Neuropharmacology,* **2006**, *51* (3), 474-486.

[57]   Nordberg, A. Mechanisms behind the neuroprotective actions of cholinesterase inhibitors in Alzheimer disease. *Alzheimer Dis. Assoc. Disord.,* **2006**, *20* (2, Suppl 1), S12-S18.

[58]   Saxena, G.; Singh, S. P.; Agrawal, R.; Nath, C. Effect of donepezil and tacrine on oxidative stress in intracerebral streptozotocin- induced model of dementia in mice. *Eur. J. Pharmacol.,* **2008**, *581* (3), 283-289.

[59]   Tsukada, H.; Sato, K.; Kakiuchi, T.; Nishiyama, S. Age-related impairment of coupling mechanism between neuronal activation and functional cerebral blood flow response was restored by cholinesterase inhibition: PET study with microdialysis in the awake monkey brain. Brain *Res.,* **2000**, *857* (1-2), 158-164.

[60]   Chen, X.; Magnotta, V. A.; Duff, K.; Boles Ponto, L. L.; Schultz, S. K. Donepezil effects on cerebral blood flow in older adults with mild cognitive deficits. *J. Neuropsychiatry Clin. Neurosci.,* **2006**, *18* (2), 178-185.

[61]   Tayeb, H. O.; Yang, H. D.; Price, B. H.; Tarazi, F. I. Pharmacotherapies for Alzheimer's disease: beyond cholinesterase inhibitors. *Pharmacol Ther.,* **2012**, *134* (1), 8-25.

[62]   Kryger, G.; Silman, I.; Sussman, J. L. Structure of acetylcholinesterase complexed with E2020 (Aricept): implications for the design of new anti-Alzheimer drugs. *Structure,* **1999**, *7* (3), 297-307.

[63]   Irwin, R. L.; Smith, H. J. Cholinesterase inhibition by galanthamine and lycoramine. *Biochem. Pharmacol.,* **1960**, *3*, 147-148.

[64]   Samochocki, M.; Zerlin, M.; Jostock, R.; Groot Kormelink, P. J.; Luyten, W. H.; Albuquerque, E. X.; Maelicke, A. Galantamine is an allosterically potentiating ligand of the human alpha4/beta2 nAChR. *Acta Neurol. Scand.,* **2000**, *176*, 68-73.

[65]   Maelicke, A.; Albuquerque, E. X. Allosteric modulation of nicotinic acetylcholine receptors as a treatment strategy for Alzheimer's disease. *Eur. J. Pharmacol.,* **2000**, *393* (1-3), 165-170.

[66]   Alkondon, M., Pereira, E. F.; Eisenberg, H. M.; Albuquerque, E. X. Nicotinic receptor activation in human cerebral cortical interneurons: A mechanism for inhibition and disinhibition of neuronal networks. *J. Neurosci.,* **2000**, *20* (1), 66-75.

[67]   Levin, E. D.; Simon, B. B. Nicotinic acetylcholine involvement in cognitive function in animals. *Psychopharmacology (Berl.),* 1998, *138* (3-4), 217-230.

[68]   Hu, N. W.; Ondrejcak, T.; Rowan, M. J. Glutamate receptors in preclinical research on Alzheimer's disease: update on recent advances. *Pharmacol. Biochem. Behav.,* **2012**, *100* (4), 855-862.

[69]   Geldenhuys, W. J.; Van der Schyf, C. J. Role of serotonin in Alzheimer's disease: a new therapeutic target? *CNS Drugs.,* **2011**, *25* (9), 765-781.

[70]   Revett, T. J.; Baker, G. B.; Jhamandas, J.; Kar, S. Glutamate system, amyloid ß peptides and tau protein: functional interrelationships and relevance to Alzheimer disease pathology. *J. Psychiatry Neurosci.,* **2012**, *37* (5), doi: 10.1503/jpn.110190.

[71]   Raskind, M. A.; Peskind, E. R.; Wessel, T.; Yuan, W. Galantamine in AD: A 6-month randomized, placebo-controlled trial with a 6-month extension. The Galantamine USA-1 Study Group. *Neurology,* **2000**, *54* (12), 2261-2268.

[72]   Greenblatt, H. M.; Kryger, G.; Lewis, T. T.; Silman, I.; Sussman, J. L. Structure of acetylcholinesterase complexed with (-)-galanthamine at 2.3 A resolution. *FEBS Lett.,* **1999**, *463* (3), 321-326.

[73]   Cutler, N. R.; Polinsky, R. J.; Sramek, J. J.; Enz, A.; Jhee, S. S.; Mancione, L.; Hourani, J.; Zolnouni, P. Dose-dependent CSF acetylcholinesterase inhibition by SDZ ENA 713 in Alzheimer's disease. *Acta Neurol. Scand.,* **1998**, *97* (4), 244-250.

[74]   Ballard, C. G. Advances in the treatment of Alzheimer's disease: Benefits of dual cholinesterase inhibition. *Eur. Neurol.,* **2002**, *47* (1), 64-70.

[75]   Birks, J.; Grimley Evans, J.; Iakovidou, V., Tsolaki, M.; Holt, F. E. Rivastigmine for Alzheimer's disease. Cochrane *Database Syst. Rev.,* **2009**, *2*, CD001191.

[76]   Molinuevo, J. L.; Arranz, F. J. Impact of transdermal drug delivery on treatment adherence in patients with Alzheimer's disease. *Expert Rev. Neurother.,* **2012**, *12* (1), 31-37.

[77]   Bar-on, P.; Millard, C. B.; Harel, M.; Dvir, H.; Enz, A.; Sussman, J. L.; Silman, I.; Kinetic and structural studies on the interaction of cholinesterases with the anti-Alzheimer drug rivastigmine. *Biochemistry,* **2002**, *41* (11), 3555-3564.

[78]   Adem, A.; Mohammed, A.; Nordberg, A.; Winblad, B. Tetrahydroaminoacridine and some of its analogues: effects on the cholinergic system. In: Basic, Clinical, and Therapeutic

Aspects of Alzheimer's and Parkinson's Diseases. Nagatsu, T.; Fisher, A.; Yoshida, M. (eds.). Plenum Press: New York, USA. **1990**, 387-393.

[79]   Albert, A. The chemical and biological properties of acridines. *Sci. Prog.,* **1949**, *37* (147), 418-434.

[80]   Shaw, H.; Bentley, G. Some aspect of the pharmacology of morphine with special reference to its antagonism by 5-aminoacridine and other chemically related compounds. *Med. J. Aust.,* **1949**, *2* (25), 868-875.

[81]   Heilbronn, E. Inhibition of cholinesterase by tetrahydroaminoacric. *Acta Chem. Scand.* **1961**, *15*, 1386-1390.

[82]   Summers, W. K.; Majorski, L. V.; Marsh, H. M.; Tachiki, K.; Kling, A. Oral tetrahydroaminoacridine in long term treatment of senile dementia, Alzheimer type. *New Engl. J. Med.,* **1986**, *315* (20), 1241-1245.

[83]   Eagger, S. A.; Levy, R.; Sahakian, B. J. Tacrine in Alzheimer's disease. *Lancet,* **1991**, *337* (8748), 989-992.

[84]   Perry, E. K.; Smith, C. J.; Court, J. A.; Bonham, J. R.; Rodway, M. Interaction of 9-amino-1,2,3,4-tetrahydroaminoacridine (THA) with human cortical nicotinic and muscarinic receptor binding *in vitro. Neurosci. Lett.,* **1988**, *91* (2), 211-216.

[85]   Adem, A.; Jossan, S. S.; Oreland, L. Tetrahydroaminoacridine inhibits human and rat brain monoamino oxidase. *Neurosci. Lett.,* **1989**, *107* (1-3), 313-317.

[86]   Stenstrom, A.; Oreland, L.; Hardy, J.; Wester, P.; Winblad, B. The uptake of serotonin and dopamine by homogenates of frozen rat and human brain tissue. *Neurochem. Res.,* **1985**, *10* (5), 591-599.

[87]   Knapp, M. J.; Knopman, D. S.; Solomon, P. R.; Pendlebury, W. W.; Davis, C. S.; Gracon, S. I. A 30-week randomized controlled trial of high-dose tacrine in patients with Alzheimer's disease. The Tacrine Study Group. *J. Am. Med. Assoc.,* **1994**, *217* (13), 985-991.

[88]   Gracon, S. I.; Knapp, M. J.; Berghoff, W. G.; Pierce, M.; DeJong, R.; Lobbestael, S. J.; Symons, J.; Dombey, S. L.; Luscombe, F. A.; Kraemer, D. Safety of tacrine: clinical trials, treatment IND, and postmarketing experience. *Alzheimer Dis. Assoc. Disord.,* **1998**, *12* (2), 93-101.

[89]   Patocka, J.; Jun, D.; Kuca, K. Possible role of hydroxylated metabolites of tacrine in drug toxicity and therapy of Alzheimer's disease. *Curr. Drug. Metab.,* **2008**, *9* (4), 332-335.

[90]   Koch, H. J.; Uyanik, G.; Fischer-Barnicol, D. Memantine: a therapeutic approach in treating Alzheimer's and vascular dementia. *Curr. Drug Targets. CNS Neurol. Disord.,* **2005**, *4* (5), 499-506.

[91]   Sonkusare, S. K.; Kaul, C. L.; Ramarao, P. Dementia of Alzheimer's disease and other neurodegenerative disorders: memantine, a new hope. *Pharmacological Research.,* **2005**, *51* (1), 1-17.

[92]   Herrmann, N.; Chau, S. A.; Kircanski, I.; Lanctôt, K. L. Current and emerging drug treatment options for Alzheimer's disease: a systematic review. *Drugs,* **2011**, *71* (15), 2031-2065.

[93]   de la Torre, R.; Dierssen, M. Therapeutic approaches in the improvement of cognitive performance in Down syndrome: past, present, and future. *Prog. Brain Res.,* **2012**, *197*, 1-14.

[94]   Schenk, D.; Barbour, R.; Dunn, W.; Gordon, G.; Grajeda, H.; Guido, T.; Hu, K.; Huang, J.; Johnson-Wood, K.; Khan, K.; Kholodenko, D.; Lee, M.; Liao, Z.; Lieberburg, I.; Motter, R.; Mutter, L.; Soriano, F.; Shopp, G.; Vasquez, N.; Vandevert, C.; Walker, S.; Wogulis, M.; Yednock, T.; Games, D.; Seubert, P. Immunization with amyloid-beta attenuates Alzheimer-disease-like pathology in the PDAPP mouse. *Nature*, **1999**, *400* (6740), 173-177.

[95]   Lavie, V.; Becker, M.; Cohen-Kupiec, R.; Yacoby, I.; Koppel, R.; Wedenig, M.; Hutter-Paier, B.; Solomon, B. EFRH-phage immunization of Alzheimer's disease animal model improves behavioral performance in Morris water maze trials. *J. Mol. Neurosci.* **2004**, *24* (1), 105-113.

[96]   Orgogozo, J. M.; Gilman, S.; Dartigues, J. F.; Laurent, B.; Puel, M.; Kirby, L. C.; Jouanny, P.; Dubois, B.; Eisner, L.; Flitman, S.; Michel, B. F.; Boada, M.; Frank, A.; Hock, C. Subacute meningoencephalitis in a subset of patients with AD after A beta 42 immunization. *Neurology,* **2003**, *61* (1), 46-54.

[97]   Hock, C.; Konietzko, U.; Streffer, J. R.; Tracy, J.; Signorell, A.; Müller-Tillmanns, B.; Lemke, U.; Henke, K.; Moritz, E.; Garcia, E.; Wollmer, M. A.; Umbricht, D.; de Quervain, D. J.; Hofmann, M.; Maddalena, A.; Papassotiropoulos, A.; Nitsch, R. M. Antibodies against beta-amyloid slow cognitive decline in Alzheimer's disease. *Neuron.,* **2003**, *38* (4), 547-554.

[98]   Fox, N. C.; Black, R. S.; Gilman, S.; Rossor, M. N.; Griffith, S. G.; Jenkins, L.; Koller, M. Effects of A beta immunization (AN1792) on MRI measures of cerebral volume in Alzheimer disease. *Neurology,* **2005**, *64* (9), 1563-1572.

[99]   Dodel, R. C.; Du, Y.; Depboylu, C.; Hampel, H.; Frölich, L.; Haag, A.; Hemmeter, U.; Paulsen, S.; Teipel, S. J.; Brettschneider, S.; Spottke, A.; Nölker, C.; Möller, H. J.; Wei, X.; Farlow, M.; Sommer, N.; Oertel, W. H. Intravenous immunoglobulins containing antibodies against beta-amyloid for the treatment of Alzheimer's disease. *J. Neurol. Neurosur. Ps.,* **2004**, *75* (10), 1472-1474.

[100]   Panza, F.; Frisardi, V.; Solfrizzi, V.; Imbimbo, B. P.; Logroscino, G.; Santamato, A.; Greco, A.; Seripa, D.; Pilotto, A. Immunotherapy for Alzheimer's disease: from anti-β-amyloid to tau-based immunization strategies. *Immunotherapy,* **2012**, *4* (2), 213-238.

[101]   Asai, M.; Hattori, C.; Iwata, N.; Saido, T. C.; Sasagawa, N.; Szabó, B.; Hashimoto, Y.; Maruyama, K.; Tanuma, S.; Kiso, Y.; Ishiura, S. The novel beta-secretase inhibitor KMI-429 reduces amyloid beta peptide production in amyloid precursor protein transgenic and wild-type mice. *J. Neurochem.* **2006**, *96* (2), 533-540.

[102]   Wong, P. BACE. *Alzheimers Dement.,* **2005,** *1* (Suppl 1), S3.

[103]   Citron, M. Beta-secretase inhibition for the treatment of Alzheimer's disease-promise and challenge. *Trends Pharmacol. Sci.,* **2004**, *25* (2), 92-97.

[104]   Mancini, F.; de Simone, A.; Andrisano, V. Beta-secretase as a target for Alzheimer's disease drug discovery: an overview of *in vitro* methods for characterization of inhibitors. *Anal. Bioanal. Chem.,* **2011**, *400* (7), 1979-1996.

[105]   Probst, G.; Xu, Y. Z. Small-molecule BACE1 inhibitors: a patent literature review (2006 - 2011). *Expert Opin. Ther. Pat.,* **2012**, *22* (5), 511-540.

[106]   Ghosh, A. K.; Brindisi, M.; Tang, J. Developing β-secretase inhibitors for treatment of Alzheimer's disease. *J. Neurochem.,* **2012**, *120* (Suppl 1), 71-83.

[107] Lanz, T. A.; Himes, C. S.; Pallante, G.; Adams, L.; Yamazaki, S.; Amore, B.; Merchant, K. M. The gamma-secretase inhibitor N-[N-(3,5-difluorophenacetyl)- L-alanyl]-S-phenylglycine t-butyl ester reduces A beta levels *in vivo* in plazma and cerebrospinal fluid in young (plaque-free) and aged (plaque-bearing) Tg2576 mice. *J. Pharmacol. Exp. Ther.*, **2003**, *305* (3), 864-871.

[108] El Mouedden, M.; Vandermeeren, M.; Meert, T.; Mercken, M. Reduction of A beta levels in the Sprague Dawley rat after oral administrativ of the functional gamma-secretase inhibitor, DAPT: a novel non-transgenic model for A beta production inhibitors. *Curr. Pharm. Design.*, **2006**, *12* (6), 671-676.

[109] May, P. C.; Yang, Z.; Li, W.; Hyslop, P. A.; Siemers, E.; Boggs, L. N. Multi-compartmental pharmacodynamic assessment of the functional gamma-secretase inhibitor LY450139 dihydrate in PDAPP transgenic mice and non-transgenic mice. *Neurobiol. Aging.*, **2004**, *25* (Suppl 25), S65.

[110] Barten, D. M.; Guss, V. L.; Cosa, J. A.; Loo, A.; Hansel, S. B.; Zheng, M.; Munoz, B.; Srinivasan, K.; Wang, B.; Robertson, B. J.; Polson, C. T.; Wang, J.; Roberts, S. B.; Hendrick, J. P.; Anderson, J. J.; Loy, J. K.; Denton, R.; Verdoorn, T. A.; Smith, D. W.; Felsenstein, K. M. Dynamics of {beta}-amyloid reductions in brain, cerebrospinal fluid, and plasma of {beta}-amyloid precursor protein transgenic mice treated with a {gamma}-secretase inhibitor. *J. Pharmacol. Exp. Ther.*, **2005**, *312* (2), 635-643.

[111] Imbimbo, B. P. Alzheimer's disease: gamma secretase inhibitors. *Drug Discov. Today: Therapeutic Strategies.* **2008**, *5* (3), 169-175.

[112] Prado-Prado, F.; García, I. Review of theoretical studies for prediction of neurodegenerative inhibitors. *Mini Rev. Med. Chem.* **2012**, *12* (6), 452-466.

[113] Weggen, S.; Eriksen, J. L.; Sagi, S. A.; Pietrzik, C. U.; Golde, T. E.; Koo, E. H. A beta 42-lowering nonsteroidal anti-inflammatory drugs preserve intramembrane cleavage of the amyloid precursor protein (APP) and ErbB-4 receptor and signaling through the APP intracellular domain. *J. Biol. Chem.*, **2003**, *278* (33), 30748-30754.

[114] Becker, R. E.; Greig, N. H. Why so few drugs for Alzheimer's disease? Are methods failing drugs? *Curr. Alzheimer Res.*, **2010**, *7* (7), 642-651.

[115] Mintzer, J. E.; Wilcock, G. K.; Black, S. E.; Zavitz, K. H.; Hendrix, S. B. MPC-7869, a selective Abeta42-lowering agent, delays time to clinically significant psychiatric adverse events in Alzheimer's disease: analysis from a 12-month phase 2 trial. Presented as a poster exhibit at the 10th International Conference on Alzheimer's Disease and Related Disorders; **2006** Jul 15-20; Madrid, Spain.

[116] Aisen, P. S.; Gauthier, S.; Vellas, B.; Briand, R.; Saumier, D.; Laurin, J.; Garceau, D. Alzhemed: A potential treatment for Alzheimer's disease. *Curr. Alzheimer Res.*, **2007**, *4* (4), 473-478.

[117] Santa-Maria, I.; Hernández, F.; Del Rio, J.; Moreno, F. J.; Avila, J. Tramiprosate, a drug of potential interest for the treatment of Alzheimer's disease, promotes an abnormal aggregation of tau. *Mol. Neurodegener.* **2007**, *2*, 17.

[118] Bilikiewicz, A.; Gaus, W. Colostrinin (a naturally occurring, proline-rich, polypeptide mixture) in the treatment of Alzheimer's disease. *J. Alzheimer Dis.*, **2004**, *6* (1), 17-26.

[119] Leszek, J.; Inglot, A. D.; Janusz, M.; Byczkiewicz, F.; Kiejna, A.; Georgiades, J.; Lisowski, J. Colostrinin proline-rich polypeptide complex from ovine colostrum-a long-term study of its efficacy in Alzheimer's disease. *Med. Sci. Monitor.,* **2002**, *8* (10), PI93-PI96.

[120] Janusz, M.; Zabłocka, A. Colostral proline-rich polypeptides--immunoregulatory properties and prospects of therapeutic use in Alzheimer's disease. Curr. Alzheimer Res., **2010**, *7* (4), 323-333.

[121] Townsend, M.; Cleary, J. P.; Mehta, T.; Hofmeister, J.; Lesne, S.; O'Hare, E.; Walsh, D. M.; Selkoe, D. J. Orally available compound prevents deficits in memory caused by the Alzheimer amyloid-beta oligomers. *Ann. Neurol.,* **2006**, *60* (6), 668-676.

[122] Salloway, S.; Sperling, R.; Keren, R.; Porsteinsson, A.P.; van Dyck, C. H.; Tariot, P. N.; Gilman, S.; Arnold, D.; Abushakra, S.; Hernandez, C.; Crans, G.; Liang, E.; Quinn, G.; Bairu, M.;, Pastrak, A.; Cedarbaum, J. M. ELND005-AD201 Investigators. A phase 2 randomized trial of ELND005, scyllo-inositol, in mild to moderate Alzheimer disease. *Neurology,* **2011**, *77* (13), 1253-1262.

[123] Fassbender, K.; Simons, M.; Bergmann, C.; Stroick, M.; Lutjohann, D.; Keller, P.; Runz, H.; Kuhl, S.; Bertsch, T.; von Bergmann, K.; Hennerici, M.; Beyreuther, K.; Hartmann, T. Simvastatin strongly reduces levels of Alzheimer's disease beta-amyloid peptides A beta 42 and A beta 40 *in vitro* and *in vivo*. *Proc. Natl. Acad. Sci. U.S.A.,* **2001**, *98* (10), 5856-5861.

[124] Wolozin, B.; Kellman, W.; Ruosseau, P.; Celesia, G. G.; Siegel, G. Decreased prevalence of Alzheimer disease associated with 3-hydroxy-3-methylglutaryl coenzyme A reductase inhibitors. *Arch. Neurol.,* **2000**, *57* (10), 1439-1443.

[125] Li, G.; Higdon, R.; Kukull, W. A.; Peskind, E.; Moore, K. V.; Tsuang, D.; van Belle, G.; McCormick, W.; Bowen, J. D.; Teri, L.; Schellenberg, G. D.; Larson, E. B. Statin therapy and risk of dementia in the elderly: a community-based prospective cohort study. *Neurology,* **2004**, *63* (9), 1624-1628.

[126] Jick, H.; Zornberg, G. L.; Jick, S. S.; Seshadri, S.; Drachman, D. A. Statins and the risk of dementia. *Lancet,* **2000**, *356* (9242), 1627-1631.

[127] Rockwood, K.; Kirkland, S.; Hogan, D. B.; MacKnight, C.; Merry, H.; Verreault, R.; Wolfson, C.; McDowell, I. Use of lipid-lowering agents, indication bias, and the risk of dementia in community-dwelling elderly people. *Arch. Neurol.,* **2002**, *59* (2), 223-227.

[128] McGuiness, B.; O'Hare, J.; Craig, D.; Bullock, R.; Malouf, R.; Passmore, P. Statins for the treatment of dementia (Review). *Cochrane Database Syst. Rev.,* **2010**, *8*, CD007514.

[129] Yaffe, K.; Barrett-Connor, E.; Lin, F.; Grady, D. Serum lipoprotein levels, statin use, and cognitive function in older women. *Arch. Neurol.,* **2002**, *59* (3), 378-384.

[130] Burgos, J. S.; Benavides, J.; Douillet, P.; Velasco, J.; Valdivieso, F. How statins could be evaluated successfully in clinical trials for Alzheimer's disease? *Am. J. Alzheimers Dis. Other Demen.,* **2012**, *27* (3), 151-153.

[131] Rea, T. D.; Breitner, J. C.; Psaty, B. M.; Fitzpatrick, A. L.; Lopez, O. L.; Newman, A. B.; Hazzard, W. R.; Zandi, P. P.; Burke, G. L.; Lyketsos, C. G.; Bernick, C.; Kuller, L. H. Statin use and the risk of incident dementia: the cardiovascular health study. *Arch. Neurol.,* **2005**, *62* (7), 1047-1051.

[132] Shepherd, J.; Blauw, G. J.; Murphy, M. B.; Bollen, E. L.; Buckley, B. M.; Cobbe, S. M.; Ford, I.; Gaw, A.; Hyland, M.; Jukema, J. W.; Kamper, A. M.; Macfarlane, P. W.; Meinders, A. E.; Norrie, J.; Packard, C. J.; Perry, I. J.; Stott, D. J.; Sweeney, B. J.;

Twomey, C.; Westendorp, R. G., PROSPER study group.: PROspective Study of Pravastatin in the Elderly at Risk. Pravastatin in elderly individuals at risk of vascular disease (PROSPER): a randomised controlled trial. *Lancet*, **2002**, *360* (9346), 1623-1630.

[133]   Zandi, P. P.; Sparks, D. L.; Khachaturian, A. S.; Tschanz, J.; Norton, M.; Steinberg, M.; Welsh-Bohmer, K. A.; Breitner, J. C.; Cache County Study investigators. Do statins reduce risk of incident dementia and Alzheimer disease? The Cache County Study. *Arch. Gen. Psychiatry*, **2005**, *62* (2), 217-224.

[134]   Wang, Q.; Yan, J.; Chen, X.; Li, J.; Yang, Y.; Weng, J.; Deng, C.; Yenari, M. A. Statins: multiple neuroprotective mechanisms in neurodegenerative diseases. *Exp. Neurol.*, **2011**; *230* (1), 27-34.

[135]   Craft, S. Insulin resistance syndrome and Alzheimer's disease: age- and obesity-related effects on memory, amyloid, and inflammation. *Neurobiol. Aging*, **2005**, *26* (Suppl 1), S65-S69.

[136]   Pedersen, W. A.; McMillan, P. J.; Kulstad, J. J.; Leverenz, J. B.; Craft, S.; Haynatzki, G. R. Rosiglitazone attenuates learning and memory deficits in Tg2576 Alzheimer mice. *Exp. Neurol.,* **2006**, *199* (2), 265-273.

[137]   Watson, G. S.; Cholerton, B. A.; Reger, M. A.; Baker, L. D.; Plymate, S. R.; Asthana, S.; Fishel, M. A.; Kulstad, J. J.; Green, P. S.; Cook, D. G.; Kahn, S. E.; Keeling, M. L.; Craft, S. Preserved cognition in patients with early Alzheimer disease and amnestic mild cognitive impairment during treatment with rosiglitazone: a preliminary study. *Am. J. Geriat. Psychiat.,* **2005**, *13* (11), 950-958.

[138]   Geldmacher, D. S.; Frolich, L.; Doody, R. S.; Erkinjuntti, T.; Vellas, B.; Jones, R. W.; Banerjee, S.; Lin, P.; Sano, M. Realistic expectations for treatment success in Alzheimer's disease. *J. Nutr. Health Aging*, **2006**, *10* (5), 417-429.

[139]   Gupta, A.; Bisht, B.; Dey, C. S. Peripheral insulin-sensitizer drug metformin ameliorates neuronal insulin resistance and Alzheimer's-like changes. *Neuropharmacology*, **2011**, *60* (6), 910-920.

[140]   Cuajungco, M. P.; Frederickson, C. J.; Bush, A. I. Amyloid-beta metal interaction and metal chelation. *Subcell. Biochem.*, **2005**, *38*, 235-254.

[141]   Schrag, M.; Mueller, C.; Oyoyo, U.; Smith, M. A.; Kirsch, W. M. Iron, zinc and copper in the Alzheimer's disease brain: a quantitative meta-analysis. Some insight on the influence of citation bias on scientific opinion. *Prog. Neurobiol.*, **2011**, *94* (3), 296-306.

[142]   Duce, J. A.; Bush, A. I. Biological metals and Alzheimer's disease: implications for therapeutics and diagnostics. *Prog. Neurobiol.,* **2010**, *92* (1), 1-18.

[143]   Cherny, R. A.; Atwood, C. S.; Xilinas, M. E.; Gray, D. N.; Jones, W. D.; McLean, C. A.; Barnham, K. J.; Volitakis, I.; Fraser, F. W.; Kim, Y.; Huang, X.; Goldstein, L. E.; Moir, R. D.; Lim, J. T.; Beyreuther, K.; Zheng, H.; Tanzi, R. E.; Masters, C. L.; Bush, A. I. Treatment with a copper-zinc chelator markedly and rapidly inhibits beta-amyloid accumulation in Alzheimer's disease transgenic mice. *Neuron*, **2001**, *30* (3), 665-676.

[144]   Ritchie, C. W.; Bush, A. I.; Mackinnon, A.; Macfarlane, S.; Mastwyk, M.; MacGregor, L.; Kiers, L.; Cherny, R.; Li, Q. X.; Tammer, A.; Carrington, D.; Mavros, C.; Volitakis, I.; Xilinas, M.; Ames, D.; Davis, S.; Beyreuther, K.; Tanzi, R. E.; Masters, C. L. Metal-protein attenuation with iodochlorhydroxyquin (clioquinol) targeting Abeta amyloid deposition and

toxicity in Alzheimer disease: a pilot phase 2 clinical trial. *Arch. Neurol.*, **2003**, *60* (12), 1685-1691.

[145] Dedeoglu, A.; Cormier, K.; Payton, S.; Tseitlin, K. A.; Kremsky, J. N.; Lai, L.; Li, X.; Moir, R. D.; Tanzi, R. E.; Bush, A. I.; Kowall, N. W.; Rogers, J. T.; Huang, X. Preliminary studies of a novel bifunctional metal chelator targeting Alzheimer's amyloidogenesis. *Exp. Gerontol.*, **2004**, *39* (11-12), 1641-1649.

[146] Lee, J. Y.; Friedman, J. E.; Angel, I.; Kozak, A.; Koh, J. Y. The lipophilic metal chelator DP-109 reduces amyloid pathology in brains of human beta-amyloid precursor protein transgenic mice. *Neurobiol. Aging.*, **2004**, *25* (10), 1315-1321.

[147] Reznichenko, L.; Amit, T.; Zheng, H.; Avramovich-Tirosh, Y.; Youdim, M. B.; Weinreb, O.; Mandel, S. Reduction of iron-regulated amyloid precursor protein and beta-amyloid peptide by (-)-epigallocatechin-3-gallate in cell cultures: implications for iron chelation in Alzheimer's disease. *J. Neurochem.*, **2006**, *97* (2), 527-536.

[148] Sommer, A. P.; Bieschke, J.; Friedrich, R. P.; Zhu, D.; Wanker, E. E.; Fecht, H. J.; Mereles, D.; Hunstein, W. 670 nm laser light and EGCG complementarily reduce amyloid-β aggregates in human neuroblastoma cells: basis for treatment of Alzheimer's disease? *Photomed. Laser Surg.*, **2012**, *30* (1), 54-60.

[149] Mulugeta, E.; Karlsson, E.; Islam, A.; Kalaria, R.; Mangat, H.; Winblad, B.; Adem, A. Loss of muscarinic M4 receptors in hippocampus of Alzheimer patients. *Brain Res.*, **2003**, *960* (1-2), 259-262.

[150] Fisher, A.; Pittel, Z.; Haring, R.; Bar-Ner, N.; Kliger-Spatz, M.; Natan, N.; Egozi, I.; Sonego, H.; Marcovitch, I.; Brandeis, R. M1 muscarinic agonists can modulate some of the hallmarks in Alzheimer's disease: implications in future therapy. *J. Mol. Neurosci.*, **2003**, *20* (3), 349-356.

[151] Fisher, A.; Brandeis, R.; Bar-Ner, R. H.; Kliger-Spatz, M.; Natan, N.; Sonego, H.; Marcovitch, I.; Pittel, Z. AF150(S) and AF267B: M1 muscarinic agonists as innovative therapies for Alzheimer's disease. *J. Mol. Neurosci.*, **2002**, *19* (1-2), 145-153.

[152] Caccamo, A.; Fisher, A.; Laferla, F. M. M1 Agonists as a potential disease-modifying therapy for Alzheimers disease. *Curr. Alzheimer Res.*, **2009**, *6* (2), 112-117.

[153] Fisher, A. Cholinergic modulation of amyloid precursor protein processing with emphasis on M1 muscarinic receptor: perspectives and challenges in treatment of Alzheimer's disease. *J. Neurochem.*, **2012**, *120* (Suppl 1), 22-33.

[154] Medeiros, R.; Kitazawa, M.; Caccamo, A.; Baglietto-Vargas, D.; Estrada-Hernandez, T.; Cribbs, D. H.; Fisher, A.; LaFerla, F. M. Loss of muscarinic M1 receptor exacerbates Alzheimer's disease-like pathology and cognitive decline. *Am. J. Pathol.*, **2011**, *179* (2), 980-991.

[155] Caccamo, A.; Oddo, S.; Billings, L. M.; Green, K. N.; Martinez-Coria, H.; Fisher, A.; LaFerla, F. M. M1 receptors play a central role in modulating AD-like pathology in transgenic mice. *Neuron*, **2006**, *49* (5), 671-682.

[156] Hock, C.; Maddalena, A.; Raschig, A.; Müller-Spahn, F.; Eschweiler, G.; Hager, K.; Heuser, I.; Hampel, H.; Müller-Thomsen, T.; Oertel, W.; Wienrich, M.; Signorell, A.; Gonzalez-Agosti, C.; Nitsch, R. M. Treatment with the selective muscarinic m1 agonist talsaclidine decreases cerebrospinal fluid levels of A beta (42) in patients with Alzheimer's disease. *Amyloid.*, **2003**, *10* (1), 1-6.

[157] Bodick, N. C.; Offen, W. W.; Levey, A. I.; Cutler, N. R.; Gauthier, S. G.; Satlin, A.; Shannon, H. E.; Tollefson, G. D.; Rasmussen, K.; Bymaster, F. P.; Hurley, D. J.; Potter, W. Z.; Paul, S. M. Effects of xanomeline, a selective muscarinic receptor agonist, on cognitive function and behavioral symptoms in Alzheimer disease. *Arch. Neurol.*, **1997**, *54* (4), 465-473.

[158] Fang, L.; Jumpertz, S.; Zhang, Y.; Appenroth, D.; Fleck, C.; Mohr, K.; Tränkle, C.; Decker, M. Hybrid molecules from xanomeline and tacrine: enhanced tacrine actions on cholinesterases and muscarinic M1 receptors. *J. Med. Chem.*, **2010**, *53* (5), 2094-2103.

[159] Wang, D.; Yang, L.; Su, J.; Niu, Y.; Lei, X.; Xiong, J.; Cao, X.; Hu, Y.; Mei, B.; Hu, J. F. Attenuation of neurodegenerative phenotypes in Alzheimer-like presenilin 1/presenilin 2 conditional double knockout mice by EUK1001, a promising derivative of xanomeline. *Biochem. Biophys. Res. Commun.*, **2011**, *410* (2), 229-234.

[160] Heinrich, J. N.; Butera, J. A.; Carrick, T.; Kramer, A.; Kowal, D.; Lock, T.; Marquis, K. L.; Pausch, M. H.; Popiolek, M.; Sun, S. C.; Tseng, E.; Uveges, A. J.; Mayer, S. C. Pharmacological comparison of muscarinic ligands: historical versus more recent muscarinic M1-preferring receptor agonists. *Eur. J. Pharmacol.*, **2009**, *605* (1-3), 53-56.

[161] Neeper, M.; Schmidt, A. M.; Brett, J.; Yan, S. D.; Wang, F.; Pan, Y. C.; Elliston, K.; Stern, D.; Shaw, A. Cloning and expression of a cell-surface receptor for advanced glycosylation end-products of proteins. *J. Biol. Chem.*, **1992,** *267* (21), 14998-15004.

[162] Mackic, J. B.; Stins, M.; McComb, J. G.; Calero, M.; Ghiso, J.; Kim, K. S.; Yan, S. D.; Stern, D.; Schmidt, A. M.; Frangione, B.; Zlokovic, B. V. Human blood-brain barrier receptors for Alzheimer's amyloid-beta 1 -40. Asymmetrical binding, endocytosis, and transcytosis at the apical side of brain microvascular endothelial cell monolayer. *J. Clin. Invest.*, **1998**, *102* (4), 734-743.

[163] Deane, R.; Du Yan, S.; Submamaryan, R. K.; LaRue, B.; Jovanovic, S.; Hogg, E.; Welch, D.; Manness, L.; Lin, C.; Yu, J.; Zhu, H.; Ghiso, J.; Frangione, B.; Stern, A.; Schmidt, A. M.; Armstrong, D. L.; Arnold, B.; Liliensiek, B.; Nawroth, P.; Hofman, F.; Kindy, M.; Stern, D.; Zlokovic, B. RAGE mediates amyloid-beta peptide transport across the blood-brain barrier and accumulation in brain. *Nat. Med.*, **2003**, *9* (7), 907-913.

[164] Deane, R. J. Is RAGE still a therapeutic target for Alzheimer's disease? *Future Med. Chem.*, **2012**, *4* (7), 915-925.

[165] Matsuoka, Y.; Saito, M.; LaFrancois, J.; Saito, M.; Gaynor, K.; Olm, V.; Wang, L.; Casey, E.; Lu, Y.; Shiratori, C.; Lemere, C.; Duff, K. Novel therapeutic approach for the treatment of Alzheimer's disease by peripheral administration of agents with an affinity to beta-amyloid. *J. Neurosci.*, **2003**, *23* (1), 29-33.

[166] Bachmeier, C.; Beaulieu-Abdelahad, D.; Mullan, M.; Paris, D. Selective dihydropyiridine compounds facilitate the clearance of β -amyloid across the blood-brain barrier. *Eur. J. Pharmacol.*, **2011**, *659* (2-3), 124-129.

[167] Lei, P.; Ayton, S.; Bush, A. I.; Adlard, P. A. GSK-3 in Neurodegenerative Diseases. *Int. J. Alzheimers Dis.*, **2011**, *2011*, 189246.

[168] Gao, C.; Hölscher, C.; Liu, Y.; Li, L. GSK3: a key target for the development of novel treatments for type 2 diabetes mellitus and Alzheimer disease. *Rev. Neurosci.*, **2011**, *23* (1), 1-11.

[169] Hernandez, F.; Lucas, J. J.; Avila, J. GSK3 and Tau: Two Convergence Points in Alzheimer's Disease. *J. Alzheimers Dis.*, **2012**, ahead of print.

[170] Hernández, F.; Gómez de Barreda, E.; Fuster-Matanzo, A.; Lucas, J. J.; Avila, J. GSK3: a possible link between beta amyloid peptide and tau protein. *Exp. Neurol.*, **2010**, *223* (2), 322-325.

[171] Kramer, T.; Schmidt, B.; Lo Monte, F. Small-Molecule Inhibitors of GSK-3: Structural Insights and Their Application to Alzheimer's Disease Models. *Int. J. Alzheimers Dis.*, **2012**, *2012,* 381029.

[172] Hamann, M.; Alonso, D.; Martín-Aparicio, E.; Fuertes, A.; Pérez-Puerto, M. J.; Castro, A.; Morales, S.; Navarro, M. L.; Del Monte-Millán, M.; Medina. M.; Pennaka, H.; Balaiah, A.; Peng, J.; Cook, J.; Wahyuono, S.; Martínez, A. Glycogen synthase kinase-3 (GSK-3) inhibitory activity and structure-activity relationship (SAR) studies of the manzamine alkaloids. Potential for Alzheimer's disease. *J. Nat. Prod.*, **2007**, *70* (9), 1397-1405.

[173] Ermondi, G.; Caron, G.; Pintos, I. G.; Gerbaldo, M.; Pérez, M.; Pérez, D. I.; Gándara, Z.; Martínez, A.; Gómez, G.; Fall, Y. An application of two MIFs-based tools (Volsurf+ and Pentacle) to binary QSAR: the case of a palinurin-related data set of non-ATP competitive glycogen synthase kinase 3β (GSK-3β) inhibitors. *Eur. J. Med. Chem.*, **2011**, *46* (3), 860-869.

[174] Chen, G., Huang, L. D.; Jiang, Y. M.; Manji, H. K. The mood-stabilizing agent valproate inhibits the activity of glycogen synthase kinase-3. *J. Neurochem.*, **1999**, *72* (3), 1327-1330.

[175] Liu, H.-Ch.; Leu, S.-J.; Chuang, D.-M. Roles of glycogen synthase kinase-3 in Alzheimer's disease: from pathology to treatment target. *J. Exp. Clin. Med.*, **2012**, *4*, 135-139.

CHAPTER 2

Challenges in Designing Therapeutic Agents for Treating Alzheimer's Disease-from Serendipity to Rationality

Umer Rashid[a] and Farzana L. Ansari[b,*]

[a]*Department of Chemistry, Hazara University, Mansehra 21120, Pakistan;*
[b]*Department of Chemistry, Quaid-i-Azam University, Islamabad 45320, Pakistan*

Abstract: The process of rational drug designing together with serendipity has played an important role in the search for new drugs, for example, in neurotherapeutics the dopaminergic dysfunction of Parkinson's disease, the dopaminergic hyperfunction of schizophrenia and the acetylcholine deficit in patients suffering from Alzheimer's disease (AD). AD is the most common form of dementia which is a group of disorders that impairs mental functioning. The aim of this chapter is 1) to analyze some examples of discoveries in AD related disorders based on some serendipitous hypotheses and rational drug design approach such as structure based drug designing, ligand based drug designing and multi-target-directed ligand approach ii) to study the drug targets such as AChE, BACE-1, Tau protein for treating AD iii) to design second-generation agents with improved efficacy and safety iv) to illustrate the complexity of problems that have to be overcome for successful targeting and v) to study drug delivery and intercellular characterization of the blood brain barrier.

Keywords: Alzheimer's disease, blood brain barrier, disease modifying drugs, multi-target-directed ligand, rational drug design, serendipity, symptomatic drugs.

INTRODUCTION

"A nearly 13-year-old, FDA-approved skin cancer drug bexarotene, rapidly alleviates molecular signs of Alzheimer's disease" [1].

Such type of news has been buzzing with stories about old drugs that surprisingly have a second life. These stories throw an insight on how the mechanism of

---

*Address correspondence to Farzana L. Ansari: Department of Chemistry, Quaid-i-Azam University, Islamabad 45320, Pakistan; Tel: +92-51-90642109; Fax: +92-51-90642144; E-mail: fla_qau@yahoo.com

modern drug discovery works. Sometimes these accidently linked discoveries, also termed as serendipity, arise as a result of the side effect of a drug already in trials, surprising yet extremely valuable when it comes to curing ailments and other times by knowingly testing a drug in order to check whether it can have an impact on any other illness. Accidental discoveries always played an important role in science especially in the search for new drugs. "*Ein glücklicher Zufall hat uns ein Präparat in die Hand gespielt*" (a lucky accident played a new drug in our hands) were the first words of a publication which described the fortunate discovery of the fever-reducing activity of acetanilide [2, 3].

Throughout history, the treatment of disease has benefited from serendipity. In prospect, we may envision new medications arising from entirely rational design, but in retrospect, we see that they have emerged from discoveries that were, to a greater or lesser extent, serendipitous. In ancient times, presumably all of them were found as a result of chance encounters with substances, mostly of plant origin, that resulted in pharmacological effects. Those effects in turn prompted the follow up observations and actions that were needed to result in their acceptance as drugs. Thus, drug discovery has historically been empirical, based on trial and error or on specific clinical observation. One feature of serendipitous drug discovery involves the discovery of valuable therapeutic agents as a result of testing a candidate drug for an expected pharmacological effect and finding a quite different effect. In such a case, the test substance is then developed as a practical means of achieving the second effect. An example is seen in the history of piperazine. At the end of the 19th century, piperazine (hexahydropyrazine) was introduced as a treatment for gout because it forms a soluble salt with uric acid. In the early 20th century this treatment was abandoned as unsatisfactory. Later its anthelmintic effects were noticed in many patients. Encouraged by this chance observation, piperazine citrate was developed as a widely used anthelmintic of the 20th century [4].

However, there is another aspect to be considered beyond accidental circumstances in serendipitous discoveries as quoted by Pasteur. "*In the field of observation, chance favors only a prepared mind*" [5]. But there must be proper environments that encourage chance observations, allowing prepared minds to pursue promising leads. This is considered as the rational side of a serendipitous discovery. Let's look at the second issue. For years, scientists, both in and out, of the pharmaceutical industry have been talking about "rational" drug design. This discussion began when crystallographic structures of target proteins became more readily available. The argument was that it would be easier to find molecules that would interact more efficiently in an active protein site once its three-dimensional (3D) structure was known because chemists then would be able to see how to design more efficient drugs. Rational design became the calling card of a number of start-up companies and drug discovery groups.

Drug design or '*tailor-made compound*' was established as a founding concept of chemotherapy by Paul Ehrlich more than 100 years ago. He postulated that man should develop "magic bullets" (drugs) aimed at the "targets or receptors" (microbes) inspired generations of scientists to devise powerful molecular cancer therapeutics [6-8]. Over the last decades, the approaches to drug designing and development have transitioned from serendipity to rationality. These approaches to drug discovery, also termed as a "rational way or rational target-oriented approach" to manage empirical or serendipitous discoveries utilize physical techniques such as molecular modeling, NMR and the use of molecular targets (such as enzymes and receptors).

The use of computational approaches in the process of rational drug design has resulted in great advances in the discovery of new drug candidates. The application of computational screening to identify leads is also called "virtual screening (VS)" or *in silico* screening. VS is a term applied to a diverse combination of computational methods to identify and rank potential drug candidates in a database, in part to reduce the magnitude and complexity of the screening problem, and to focus drug discovery and optimization efforts on the most promising leads. The basic goal of the VS is the reduction of the enormous virtual chemical space of small organic molecules, to synthesize and/or screen against a specific target protein, to a manageable number of the compounds that exhibit the highest chance to become a drug candidate [9-11]. There are two fundamental approaches to VS the data bases for molecules fitting either a known pharmacophore [12, 13] or a 3D structure of a macromolecular target [14-16]. Among all VS methods, protein-structure-based docking has received significant attention. This approach is a direct way to use the rapidly increasing number of protein 3D structures that appear in the lead discovery process. Structure-based VS (SBVS) normally involves fast docking of a large number of chemical compounds against a ligand-binding site on the target molecule. The docked conformations are then scored, usually in the form of a relative rank order by a variety of scoring functions as a way to select a small subset of compounds for further analysis, purchase or testing. Compounds with good scores are supposedly indicative of potentially good binders. This technique has been used in a range of discovery-research projects, all with varying degrees of success [14-16].

During the past few decades the advancement of knowledge about the pathogenesis of diseases and introduction of *in vitro* assays followed by advances in genomics and HTS has shifted drug discovery paradigm towards a reductionist approach focused on single molecular targets. This was a challenging goal for researchers to discover small molecules that were able to modulate the biochemical function of a single protein target among an array of the altered ones. Indeed this approach also called as *one-drug-one-target* had led to develop many successful and highly selective 'magic bullets' that are currently used in many pathologies [17-19].

Regardless of these efforts for the treatment of disorders with complex pathological mechanisms such as neurodegenerative diseases (NDs), diabetes, cardiovascular diseases, and cancer many diseases remain inadequately explored [20]. In the last decade, with the failure of the so-called *"one-drug-one-target"* paradigm alternative approaches were investigated for the treatment of these complex disorders. Different new strategies have emerged to challenge the *"one-drug-one-target"* approach [17-19]. Most of these strategies were aimed to develop drugs that can modulate multiple targets in a parallel fashion (polypharmacology). There are three possible approaches to polypharmacology [21-23]. The first approach named multiple-medication therapy (MMT) involves a cocktail of two or three different drugs that combine different therapeutic mechanisms and is administrated in two or more individual dosage forms (Fig. **1a**). However, this approach might be disadvantageous for patients with compliance problems. In the second approach there has been a move towards the use of a multiple-compound medication (MCM) through which two or more agents are co-formulated in a single dosage form to make dosing regimens simpler thereby improving the patient compliance (Fig. **1b**). Finally, an alternative strategy known as "multi-target-directed ligands" (MTDLs) emerged on the basis of the assumption that a single chemical entity may be able to hit multiple targets simultaneously (Fig. **1c**) [17-19, 21-23].

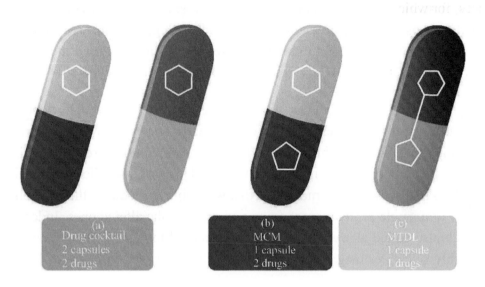

Figure 1: Three possible approaches to polypharmacology.

The success of MTDLs approach is indicated by a large increase in the publications during the last decades [17-24 and references therein]. Conceptually, there are two distinct methods of generating a MTDL: i) a knowledge-based approaches or rational design by a combination of pharmacophore and ii) a screening approach. The former, also termed as framework combination approach, is a frequently reported lead generation strategy. It is based on a series of rational structural modifications by combining frameworks and the underlying pharmacophore of two molecules. These pharmacophores are combined together by a cleavable or non-cleavable metabolically stable linker groups (termed as "conjugates") or overlapped (fused or merged) by taking advantage of structural commonalities (Fig. **2**).

The latter approach, also called serendipitous approach relies on diversity-based screening that involves a high-throughput screening of large and diverse compound collections at one target, and any actives are then triaged on the basis of activity at the second target. The MTDL design strategy is used to develop single chemical entities that are able to simultaneously modulate multiple targets. The development of such compounds might disclose new avenues for the treatment of a variety of pathologies such as cancer, AIDS, neurodegenerative diseases, for which an effective cure is urgently needed.

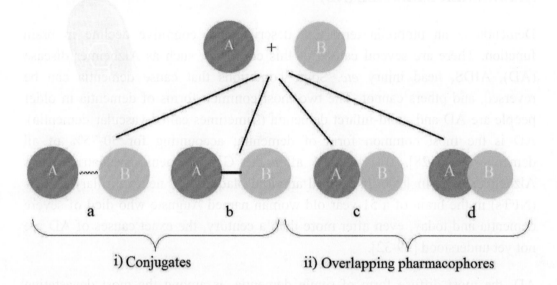

Figure 2: Rational design strategies for framework combination to generate a MTDL. i) frameworks are connected to form conjugates *via* a linker, which in some cases is designed to be cleaved (a) *in vivo* to release two independently-acting drugs. ii) In overlapping pharmacophores the fused MTDLs are obtained when the frameworks are directly attached, or the frameworks are merged together.

Neurodegenerative disorders (NDs) are occasional and congenital. These are characterized by the persistent and progressive loss of neuronal subtypes and include mainly Alzheimer's disease (AD), Parkinson's disease (PD) and Huntington's disease (HD), stroke, Amyotrophic Lateral Sclerosis (ALS), multiple sclerosis and HIV-1 associated neurocognitive disorders. NDs may affect various brain functions such as movement (as in PD and ALS) or memory and cognition (as in AD). Nevertheless the therapeutic options for these NDs are limited to symptomatic treatment only. Most drugs currently available are the result of chance observations rather than a truly comprehensive approach [25, 26].

The discovery efforts directed toward designing disease-modifying drugs (DMD) have increased in recent years. The fundamental challenge to find new and efficacious drugs targeting central nervous system (CNS) is blood brain barrier (BBB) which will be discussed later in a separate section.

ALZHEIMER'S DISEASE (AD)

Dementia is an umbrella term that describes the cognitive decline in brain function. There are several causes of this condition, such as Alzheimer disease (AD), AIDS, head injury *etc*. Some conditions that cause dementia can be reversed, and others cannot. The two most common forms of dementia in older people are AD and multi-infarct dementia (sometimes called vascular dementia). AD is the most common form of dementia; accounting for 50-75% of all dementias [27, 28]. It is named after the German neuropsychiatrist Alois Alzheimer, who in 1906 discovered amyloid plaques and neurofibrillary tangles (NFTs) in the brain of a 51-year old woman named Auguste who died of severe dementia and today, even after more than a century, the exact causes of AD are not yet understood [29-32].

AD, the most diffuse form of senile dementia, is among the most devastating brain disorder an individual can face. It involves progressive and irreversible decline in cognitive functions including memory, judgment, decision-making, orientation to physical surroundings and language. It progressively robs the patients of everything they hold dear such as their memories and relationships, their own personality, their ability to care for themselves, and, ultimately, their life. AD mainly affects elderly individuals and leaves them with end-stage bed-ridden and dependent on custodial care. According to Alzheimer's disease International (ADI), it is estimated that 36 million people in the world are suffering from dementia and this number is expected to increase to 66 million in 2030 and 115 million in 2050. As many as 28 million of the world's 36 million people with dementia have yet to receive a diagnosis and, therefore, do not have access to treatment, information, and care [33].

According to a report of Alzheimer's Association, AD is the sixth-leading cause of death in the USA and the only cause of death among the top 10 in the USA that cannot be prevented, cured or even slowed. There are as many as 5.4 million cases of AD in USA alone. Based on final mortality data from 2000-2008 in USA the death rates have declined for most major diseases *e.g.*, heart disease (-13%), breast cancer (-3%), prostate cancer (-8%), stroke (-20%) and HIV/AIDS (-29%), while deaths from AD have risen to 66% during the same period [34]. It is further estimated that the direct cost of caring of patients suffering from AD or other dementias will rise to 200 billion US $.

According to a report of Alzheimer's Association of Pakistan, there were 377,000 people with dementia in Pakistan in 2010, which will increase to 804,000 in 2030 and by 2050 the number of patients would reach 1,779,000 [35]. AD is placing a considerable and increasing burden on patients, caregivers and society. However, a number of therapeutic options on the market remain severely narrow. No disease-modifying therapy is available yet, despite the intensive efforts to develop innovative medicines. At present, four drugs for AD have been approved by FDA. These drugs, however, are not able to alter or prevent disease progression. They are, instead, palliative *i.e.*, an area of healthcare that focuses on relieving and preventing the sufferings of patients and in alleviating the symptoms of disease.

Before discussing the drug discovery and development process of AD, it is important to understand the mechanism of this disease, which is central to the currently available therapeutic approaches. Since Alois Alzheimer's first report of November 1906, different studies have been undertaken to shed light into the pathophysiological mechanisms of this devastating neurological disorder.

At the most basic level, AD results from neuronal cell death due to many diverse factors such as decreased levels of acetylcholine (ACh), amyloid-β (Aβ) deposits, tau (τ)-protein aggregation and oxidative stress. These factors are thought to play an important role in the pathophysiology of AD. Researchers are also working on understanding the role of other pathological features such as oxidative stress, inflammatory injury, role of free radicals and excitotoxicity. Due to its complexity, AD has been described as a multifactorial disease and its molecular causes of AD are not completely known. The following are the main hypotheses that attempt to explain the onset of AD.

a)   The **cholinergic hypothesis** is the first and the oldest hypothesis which started in mid-1970s. It proposes that AD is associated with cholinergic system impairment and neuronal degeneration due to the loss of cholinergic neurotransmission [36-40]. This cholinergic neurotransmission can be improved by drugs acting at both pre-synaptic and post-synaptic levels which may be choline precursors, ACh releasers, M_2 muscarinic antagonists, M_1 muscarinic agonists, nicotinic agonists *etc.* [41].

b)   In 1991 the **amyloid hypothesis** was postulated that proposed the aggregation of amyloid β-protein (Aβ) into fibriller plaques as the fundamental cause of the disease [42-44]. These Aβ peptides are derived from secretase-mediated cleavages (by β-secretases and γ-secretases) of amyloid precursor protein (APP) into either a 40- or 42-amino acid fragment of Aβ [45-48]. $Aβ_{1-40}$ is the most common form of Aβ, however, $Aβ_{1-42}$ form has the highest tendency to aggregate. Low molecular weight oligomers formed from aggregated Aβ monomers are toxic and exert significant adverse effects on synaptic and cellular functions leading to neurodegeneration and cognitive as well as neuropsychiatric symptoms [49-52].

c)   Another hallmark of AD is the **tau hypothesis.** A study on triple transgenic mice revealed that the deposition of Aβ plaques contributed to AD by stimulating the hyperphosphorylation of tau, thereby, increasing the formation of NFTs. These NFTs are intracellular deposits of the microtubule-associated τ-protein found within dystrophic neurons where it binds tubulin monomers together to form stable polymers that are presumed to be essential in cellular transport and axonal growth. In AD tangles, the τ-protein becomes hyperphosphorylated and this leads to less efficient binding to microtubules. The unbound τ-protein then spontaneously aggregates into insoluble paired-helical filaments (PHFs), which are seen as deposits in the neurons [53-57].

d) Another hypothesis *i.e.*, **oxidative stress hypothesis,** which has gained momentum as a possible cause of neuronal death, is oxidative damage of cell membranes, mitochondria, lipids and proteins [58]. Oxidative stress is defined as the imbalance between biochemical processes leading to the production of reactive oxygen species (ROS) [59-61]. Originating from the free radical theory of aging, this hypothesis has been implicated in the pathogenesis of AD and a wealth of literature exists where the role of oxidative stress in AD has been studied [62-66].

e) **Excitotoxicity** has also been considered to play an important role in the neuronal death. It is triggered by chronic activation of glutamate, a major excitatory neurotransmitter in the brain, in particular of the N-methyl-D-aspartate also known as NMDA [67-69]. As a result of over activation of NMDA, the site disturbance in cellular ion homeostasis (Ca^{2+}, K^+, Na^+) and metabolic activities occur. Dysregulation if Ca^{2+} is considered as the most important as a higher level of Ca^{2+} can enter into the cell which causes a permanent depolarization of the post-synaptic neurons creating reactive species and other substances that cause cell death.

f) Another hypothesis *i.e.*, **neuro-inflammation** has gained considerable evidence with respect to its contributions to AD pathology in the recent decade. The inflammatory reaction is characterized by activation of glial cells, gliosis and the appearance of inflammatory proteins such as complement factors, acute phase proteins, and pro-inflammatory cytokines [70-72].

g) In the last few years, various studies have provided evidence that **metal ions** are critically involved in the pathogenesis of AD. Accordingly, assisting the balance of metal ions back to homeostatic levels has been proposed as a disease-modifying therapeutic strategy for AD as well as other NDs [73, 74].

h) **Among others,** mitochondrial dysfunction is considered as a pivotal component in neurodegeneration. This dysfunction occurs early in AD and promotes synaptic damage and apoptosis. APP and Aβ can be imported into mitochondria, where they can interact with mitochondrial components, impair ATP production, and increase oxidative damage [75, 76].

This understanding, together with further ongoing research about AD pathogenesis, has provided the rationale for therapies directly targeting molecular causes of AD. New drug candidates with disease-modifying potential are now in the pipeline and have reached testing in clinical trials.

CHALLENGES IN DESIGNING ALZHEIMER'S MEDICINES

Regardless of increasing understanding of AD, developing a medicine to prevent, delay, slow, or cure is exceptionally difficult [77]. The design of therapeutics for the treatment of AD has undergone a number of phases in the modern era. As discussed earlier, serendipity is the faculty of making a discovery through a combination of accident and sagacity. In AD, the serendipity played a key role in the discovery of many anti-Alzheimer drugs. Several putative new drugs, some of which are presently under clinical trials, have been discovered *via* a combination of the serendipitous and non-serendipitous approaches. Cholinesterase inhibitors (ChEIs), the most luminous class of drugs used for AD, was traditionally used for medical as well as economical purposes [78]. One of the earliest examples in the first generation ChEIs was the discovery of various organophosphorus compounds, such as diisopropylfluorophosphonate and paraxon *etc.* and the first organophosphorus agent was synthesized in the 1850s [79]. The second group consists of natural alkaloid physostigmine, a carbamate that was the first AChE inhibitor, was introduced in the 1870s for the treatment of glaucoma [80, 81]. The vast majority of compounds of both these series were designed primarily as insecticides. In the first group of carbamate derivatives, almost all of the potent insecticides were monomethyl carbamates lacking a charged nitrogen function. This enabled the molecule to penetrate the insect cuticle and fatty nerve sheath rapidly. The dimethyl derivatives are slightly less potent but they are particularly toxic to houseflies. The monomethyl derivatives tend to be unstable in solution and hydrolyze readily at physiological pH. This greatly limits their biological action in mammals and makes them less suitable as pharmaceutical or therapeutic agents.

The serendipitous discovery of physostigmine led to the development of a second generation carbamate analogue Rivastigmine, an approved drug for AD [82-84]. Recently Latrepirdine, also known as dimebolin or Dimebon, was initially developed as an orally active antihistamine drug which was used clinically in Russia and withdrawn later from the market [85-87]. The interest on Dimebon was renewed when it was found to exhibit neuroprotective effects in preclinical models of AD and PD. The results of a recent study proved that bexarotene (Targretin® by Eisai Inc.), an oncology drug, reduced β-amyloid plaques in the brain of mice suffering from an AD-like condition [1]. Prior to 1990s there was no approved drug for AD. Years of exhaustive research has yielded only six drugs approved by Food and Drug Administration (FDA). These drugs approved for marketing are tacrine, donepezil, rivastigmine and galathamine approved in the years 1993, 1996, 2000 and 2001 respectively. Another drug, memantine (noncompetitive NMDA antagonist), with a different mechanism of action and a limited efficacy was approved in 2002 both by European Medicines Agency (EMA) and FDA in 2000 [88-91], while, Huperzine A (**6**) was approved in China for mild-to-moderate stages of AD. These drugs provide only symptomatic relief in some cases but have not yet resulted in any approved disease-modifying medicine [92].

SYMPTOMATIC DRUGS

The term symptomatic drug refers to an agent that does not alter the progression of the disease, but only decreases the severity of symptoms. The symptomatic effect is usually reversible, such that, if the treatment is interrupted, the treated group might be indistinguishable from the control (placebo) group [93, 94]. On the contrary a disease modifying drug (DMD) is an agent that slows the progression of structural damage, such that its effect is persistent and can be detected even after stopping the treatment, because the cumulative pathological changes would be less severe in the treated group as compared with the control (placebo) group. Current treatment options for AD are limited to drugs that are generally regarded as symptomatic that provide symptomatic relief but poorly affect the progression of the disease.

Cholinergic Drug Therapy

The neuropathology of AD is characterized by an early loss of basal forebrain cholinergic neurons, leading to a deficiency in cholinergic neurotransmission. This deficiency plays an important role in learning and memory impairment of AD patients. During neurotransmission at cholinergic synapsis, acetylcholine (ACh) is released from presynapsis into the synaptic cleft and binds to muscarinic and nicotinic receptors present on postsynapse. This nerve impulse transmission at cholinergic synapsis is terminated by acetylcholinesterase (AChE) which catalyzes the hydrolysis of ACh. The phenomenon of cholinergic neuro-transmission can be enhanced by developing following six classes of drugs acting at both presynaptic and post-synaptic levels [95-100].

a)   AChEIs: which increase the synaptic level of ACh by blocking AChE enzyme.

b)   ACh precursors: such as phosphatidylcholine that increase the bioavailability of choline.

c)   ACh releasers: which should advance the release of ACh from presynapses.

d)   Muscarinic agonists: which activate muscarinic receptors on postsynapse.

e)   Nicotinic agonists: which activate nicotinic receptors on postsynapse.

Among these strategies, the only and the most clinically effective therapeutic strategy for treatment of AD is cholinergic enhancement by inhibition of AChE.

Cholinesterases (ChEs) are a group of chemical compounds which are able to inhibit the hydrolytic activity. There are two types of ChEs *i.e.*, acetyl cholinesterase (AChE) and butyryl-cholinesterase (BuChE). The former is found primarily in the blood and neural synapses and specifically hydrolyses ACh, while the latter is found primarily in the liver and non-specifically hydrolyses any kind of ChE. ChEIs are the first and the most developed group of drugs proposed for AD treatment. Many of the drugs that are available today for the treatment of AD target both AChE and BuChE, but some are more selective than others.

Physostigmine* (1,** eserine),*** the first known AChEI, is a parasympathomimetic major alkaloid found in the seeds of the fabaceous plant *Physostigma venenosum*. It is a tertiary amine that acts as a reversible inhibitor of AChE, however, it has only a small beneficial effect in a subset of AD patients. Despite the advantage of physostigmine for its ability to pass blood brain barrier (BBB), it has exhibited contradictory clinical efficiency because of its short half-life, narrow therapeutic index, and peripheral side effects. Due to these disadvantages as well as its short half-life, it seems to have no advantage over newer lines of AChE drugs and is therefore, no more in clinical use [101-104].

Organophosphates are one of the earliest examples in the first generation ChEIs. By the late 1950s over 50,000 organophosphorous compounds had been synthesized and tested [79]. They were used as agricultural insecticides or as nerve gases in chemical warfare. These compounds react with the active site serine of AChE, forming a very stable covalent phosphoryl-enzyme complex [105]. *Metrifonate* (**2**) is a natural long-acting irreversible ChEI and has a 30-year history to treat schistosomiasis. In late 1980s it was considered as a potential treatment for AD on the basis of its anticholinesterase properties. This drug is unique among ChEIs for AD treatment and it is a prodrug which is nonenzymatically transformed *in vivo* into an active metabolite 2,2-dichlorovinyl dimethylphosphate. Its short term use has low risk of side effects but long-term use causes respiratory paralysis and neuromuscular transmission dysfunction. It was not approved by FDA for AD treatment and clinical trials were discontinued during Phase-III [106].

Tacrine (**3**, THA, Cognex®) was the first drug to receive FDA approval in 1993 specifically for the treatment of mild to moderate AD [107]. It was first synthesized in 1949 and has been used after anaesthesia and in combination with morphine because of its analeptic properties. In Second World War it was used as a safe intravenous antiseptic, while in the 1950s, it was used experimentally to reverse cholinergic coma in animals. Tacrine is a noncompetitive and reversible ChEI, which concentrates in the brain because of its relatively greater lipid solubility. Tacrine's severe side effects such as hepatotoxicity and short biological half-life (1.6 - 3 h), however, limit its clinical use [108].

Donepezil (**4,** E2020, Aricept®) became the second drug to receive FDA approval for the treatment of mild to moderate AD in November 1996. It is a piperidine-based reversible inhibitor that shows high selectivity for AChE relative to BuChE. Donepezil is structurally dissimilar to all other currently available ChEIs [109]. The research on E2020 started in 1983. Following research developments of tacrine, researchers at Eisai Company started to develop tacrine derivatives, however, they failed to develop a non-toxic tacrine derivative. Through random screening N-benzylpiperazine derivatives, originally synthesized in the study of anti-arterial sclerosis, showed a promising AChE inhibitory activity with IC_{50}=12.6 µM in rat brain homogenate. Though it was not a strong inhibitor, however, the compound's novel structure was very intriguing. The researchers decided to use this derivative as the seed compound and synthesized about 700 derivatives. A dramatic increase in anti-AChE activity was noticed when *N*-benzylpiperazine was replaced by *N*-benzylpiperidines and various modifications they obtained Donepezil [110].

Rivastigmine (**5,** SDZ-ENA-713/Exelon®), approved by FDA in April 2000, belongs to a series of miotine derivatives. This carbamyl derivative is structurally related to physostigmine and is a second generation reversible AChEI used in symptomatic treatment of AD [127-129]. Phase-III clinical studies conducted in the USA demonstrated its efficacy and tolerability in patients with AD and under the name of Exelon® has been launched worldwide [111-113].

Galanthamine (**6,** Reminyl®) is a phenanthrene alkaloid that was isolated from snowdrop plant *Galanthus nivalis*. After its first extraction in 1952, it was used by Bulgarian and Russian scientists in postsurgery reversal of tubocurarine-induced muscle relaxation, muscular dystrophy and traumatic brain injury. In 1972, Soviet researchers demonstrated that galanthamine could reverse scopolamine-induced amnesia in mice and this was later on extended to humans. Despite the substantial and long-lasting history of clinical use of this compound in humans, it was not until 1986 when this compound was studied for the treatment of AD [114]. Galanthamine has a dual mode of action and acts as a reversible inhibitor of AChE with competitive action and allosteric modulator of nicotinic acetylcholine receptors in the brain. This modulation improves release of ACh from pre-synapatic terminations. Galanthamine has more than a 10-fold selectivity for AChE relative to BuChE which is in contrast to nonselective agents such as tacrine and physostigmine. This drug was approved by FDA in February 2001 [115].

Huperzine A (7, Hup A), an alkaloid isolated from the herb *Huperzia serrate*, is a potent, selective, reversible inhibitor of AChE. Hup A, the most studied compound of the two, has already been approved as a drug for the symptomatic treatment of AD in China. Hup A has a mechanism of action similar to that of donepezil, rivastigmine, and galanthamine. A pro-drug form of huperzine A (ZT-1) is under development as a treatment for AD [116]. In the USA, huperzine A, a dietary supplement derived from the Chinese club moss *Huperzia serrata*, is sold as a dietary supplement for memory support. This botanical has been used in China for centuries for the treatment of swelling, fever and blood disorders. In the late 1980's clinical trials in China have shown it to be effective in improving cognitive performance in patients with AD and enhancing memory in students [117].

Derivatives of physostigmine, as well as physostigmine itself, inhibit the enzyme by carbamoylating the serine residue of the catalytic triad in a pseudo-irreversible manner. Neostigmine (**8**), an analogue of physostigmine, is another parasympathomimetic agent and has a quaternary amine skeleton that does not enable it to pass BBB after systemic administration. Neostigmine as its bromide salt (Prostigmin®), first synthesized in 1931, was shown to be able to block muscle neuronal ACh receptors by interacting with the open state of ion channels. Besides, it was found to be useful for the treatment of pediatric acute colonic pseudo-obstruction and post-operative analgesia [118-121]. Eptastigmine (**9,** L-693487, MF201), is a more lipophilic carbamate derivative of physostigmine. It was patented by Mediolanum Farmaceutici drug company (Italy) and was advanced in clinical trials [122, 123]. The compound showed less toxicity than physostigmine and a longer period of pseudo-irreversible anticholinesterase effect [124, 125]. In fact, eptastigmine was found to display cognitive enhancement in mice, rats, and monkeys by passive avoidance test as well as eight-arm radial maze performance test in aged rats [126]. Later on, it was proceeded to Phase-II trials, however, after the identification of two out of 96 neutropenia cases, it was withdrawn from the trials. **Quilostigmine** (**10,** NXX066) is a reversible and dose-dependent AChEI whose effects were compared to those of tacrine and donepezil in different *in vitro* and *in vivo* experiments [127]. (-)-Phenserine (**11**), the phenyl carbamate of (-)-eseroline, was prepared in 1916 from natural physostigmine by Polonovsky. This is another compound whose synthesis was based on the chemical structure of physostigmine patented by Axonyx Pharmacuetical Company [128-131]. It is the phenylcarbamate derivative of physostigmine with reversible AChE inhibition and is an anti-amyloid agent, which was shown to improve learning in maze test using aged rats [132, 133]. It has a quick absorption and is less toxic when compared to physostigmine and tacrine [134, 135]. In the years 2003 and 2004, Phase- III clinical trials were initiated , however, it did not achieve significant efficacy in Phase- III trials. Tolserine (**12**) has also been developed as a close derivative of phenserine patented by Axonyx, differing only at 2'-methyl substitution on its phenylcarbamoyl moiety, which shows an approximately 8 hours long duration of action [136]. Preclinical studies with tolserine, as the fourth generation anticholinesterase molecule, were initiated in the year 2000. Tolserine with IC_{50} value of 8.13 nM has 200-fold more selectivity against human AChE than BuChE and easily passes BBB [137].

8　　　**9**

10　　　**11**　R=H
　　　　　12　R= CH_3

Natural Huperzine B (HupB), a Lycopodium alkaloid isolated from the *Huperzia serrate*. It has been demonstrated as an effective and reversible AChEI. HupB is less potent and selective than HupA, but it has higher therapeutic index and other positive benefits [138]. ***Nelumbo nucifera*** is an aquatic plant with numerous medicinal properties. It has recently been demonstrated that the stamens fed to rats performing maze learning tasks improved memory. The compounds show AChE inhibitory effects with the potential to be used for AD treatment [139]. There are no reports of human studies, preclinical and clinical safety and toxicity. ***Himatanthus lancifolius*** is a shrub that contains several indole alkaloids with a number of medicinal properties such as antimicrobial effects, gastroprotection, and the ability to affect the vascular and nonvascular smooth muscle responsiveness. Seidl *et al.* [140] conducted a study to determine if there were any AChE inhibiting properties from the *Himatanthus lancifolius* extract and uliene *in vitro*. The dichloromethane (DCM), and ethyl acetate (EtOAc), fractions showed significant AChE inhibitory effects. **Uliene** was the significant compound present in both fractions [140]. Preclinical and clinical safety and toxicity data are not reported. ***Galangin***, a flavonol isolated from *Alpiniae officinarum*, demonstrated the highest inhibitory effects on AChE activity. Its mechanism is unknown, and its clinical and preclinical toxicities have not yet been established [141].

A variety of AChEIs discussed in the preceding paragraphs have been extensively evaluated in clinical studies. Indeed, the treatment with ChEIs starting from first generation non-specific drugs such as physostigmine to more selective and better tolerated second generation drugs has been limited and resulted in a symptomatic improvement only. AChEIs currently approved by FDA (Table **1**) are limited from mild to moderate AD.

Table 1:   **FDA-approved AChEIs**

Generic Name	Brand Name	Chemical Structure	Mechanism of Action	Disease Stage
Tacrine	Cognex	Acridines	an acetylcholinesterase inhibitors is not commonly used because of a number of side effects	
Donepezil	Aricept®	Piperidine	competitive and non-competitive AChE inhibition	All stages
Rivastigmine	Exelon®	Carbamate	Pseudo-irreverisible AChE inhibition	Mild to moderate
Galanthamine	Razadyne®	Phenantrene alkaloid (Tertiary)	competitive AChE inhibition	Mild to moderate
Huperzine-A	-	Alkaloid	selective, reversible AChE inhibition	Approved in China for mild to moderate stages

NMDA Glutamate Receptor Antagonist

Memantine (**13,** Akatinol®, Axura®, Ebixa/Abixa®, Memox® and Namenda®), a derivative of amantadine, was first synthesized and patented by Eli Lilly in 1968. It was originally developed as an anti-influenza drug and is used presently to treat AD [142]. The serendipitous discovery of amantadine (**14**) and memantine for PD led the scientists to believe that these compounds were dopaminergic or possibly anticholinergic drugs. However, in 1990s, it was observed that memantine has neither dopaminergic nor anticholinergic properties. In early 1990s it was suggested that memantine has NMDA receptor antagonist effects. It protects the synapse and it is anticipated that the drug may keep the nerve cells alive until a better treatment is found and the disease can be managed [143].

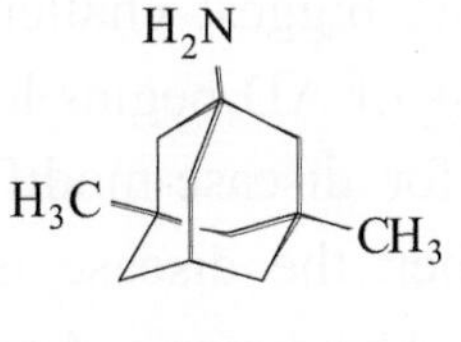

H_2N

H_3C — CH_3

H_2N

13    **14**

Limitations and Future Directions

Currently available drugs for AD are AChEIs or NMDA glutamate receptor antagonists. AChEIs are used to increase synaptic levels of ACh that are reduced as a result of damage to cholinergic neurons, while, memantine is used to prevent/reduce calcium-dependent excitotoxic neuronal cell death. Although, there is some degree of improvement in cognitive functions, but these were confined only to patients with mild to moderate AD during the first year of the treatment. Subsequently, their efficacy declines progressively and disappears entirely after 2 or 3 years. Attempts were made to increase the efficacy of AChEI by combining them with memantine, but it remained to be seen whether these associations are more effective than the single drug alone.

The modest efficacy of the above mentioned symptomatic therapies suggest that there is a dire need of developing new AD drug candidates. Moreover, AD experts also propose that the best strategy for the development of such drugs is to focus on novel targets such as neuronal nicotinic acetylcholine receptors (nAChRs) and muscarinic cholinergic receptors.

DISEASE MODIFYING DRUGS (INVESTIGATIVE APPROACHES)

As discussed earlier, current drugs for AD target cholinergic and glutamatergic neurotransmission, thus improving symptoms but they do not slow or reverse the progression of the disease. Due to enormity of the challenges faced by AD researchers, the drug discovery process has been directed to develop 'disease modifying drugs (DMDs) which are expected to counteract the progression of

AD. Since in a chronic, slow progressing pathological process, an early start of treatment enhances the chances of success, it is crucial to have biomarkers for early detection of AD-related brain dysfunctions, usable before clinical onset. Developing DMDs is one of the biggest challenges for AD researchers because the patho-physiological process of AD begins long before the symptoms develop. Therefore the optimal time for disease-modifying therapy may be in the pre-symptomatic stage of AD when the disease is still hidden. Recently, National Institute on Aging and the Alzheimer's Association workgroup has revised the criteria for the clinical diagnosis of AD. They incorporate biomarkers to identify early stages of AD susceptible to being treated with disease modifying drugs.

In recent years these disease-modifying drugs have advanced to reach late stage of human clinical trials. Based on the mechanism of action these drugs address one of the two controversial hypotheses *i.e.*, β-amyloid or tau, and hence can be divided into the following five major categories with distinct mechanisms of action.

1.   Reduction of Aβ production.

2.   Prevention of Aβ plaque aggregation or disruption of aggregates.

3.   Promotion of Aβ clearance *via* active or passive immunotherapy.

4.   Prevention of τ-protein phosphorylation or aggregation, and

5.   Beyond plaque and tangle-related targets.

The first three categories are driven by the "amyloid hypothesis", which has been the leading focus of research and development for the treatment of AD. Henceforth, the majority of drug candidates, that have advanced into randomized controlled trials (RCTs) for AD, target β-amyloid. In the following section, we will summarize the new strategies for the treatment of AD, focusing mainly on compounds being tested in human beings. Drugs will be discussed according to their main mechanism of action.

Cholinergic Drug Therapy

Despite intense research, there has been also a slight development in the search of cholinergic drugs. Neuroprotective and disease modifying property of AChEIs has also been demonstrated during some investigations [144-146]. Posiphen, the (+)-phenserine enantiomer, having weak activity as AChEI, has now shown its ability to decrease production of APP and τ-protein [147, 148]. Huperzine-A also demonstrated the potential to modify Aβ processing and regulation of NGF expression [149]. A new semi-synthetic drug, ZT-1 (**15**), derived from huperzine-A, has shown an improvement in cognitive function of AD patients in the clinical trials [150]. It also exhibited lower anti-BuChE effect along with less toxicity in mice as compared to huperzine-A.

BuChE may play a role in attention, executive function, emotional memory and behavior, therefore, its inhibition may provide further benefits for the treatment of AD [151]. In comparison to phenserine, its structural analogues proved to be more potent [152, 153]. Moreover, nAChRs which are essential for learning and memory are reduced in AD brains and this reduction has been detected using ligand binding techniques [154]. Abnormalities in nAChR are closely related to other primary pathological changes such as Aβ and NFTs. It is noteworthy that these abnormalities in nAChR may occur in the early stages of AD rendering them an attractive target to develop therapeutics. Ispronicline (AZD-3480), is an orally active and selective nAChR agonist with neuroprotective effects in humans in Phase-II trial [155]. GTS-21 (DMXBA) was also selective nAChR agonist that displayed neuroprotective activity *in vitro* and also showed effective results during Phase-I clinical trials [156]. ABT-089 is selective nAChR modulator which has shown positive effects in rodent and primate cognitive models [157].

Since muscarinic cholinergic irregularities are also detected in AD brain, it is suggested that the role of muscarinic agonists (M1 agonists) could be useful in AD. The M1 agonists are neurotrophic agents that elevate the nonamyloidogenic APP *in vitro*, decrease Aβ levels *in vitro* and *in vivo* and restore cognitive impairments in animal AD models [158, 159]. However, the development of M1 agonists has a very limited success due to their adverse effects. For example, Talsaclidine (**16**), AF-102B, and AF-267B (NGX-267) are M1 muscarinic receptor agonists that can also affect Aβ production, however, these agonists have undesirable cholinergic mediated effects such as increased salivary flow [160, 161].

15

16

Anti-amyloidogenic Therapy

Anti-amyloidogenic therapy primarily involves the reduction of Aβ production, inhibiting secretases, increasing Aβ clearance, or blocking Aβ aggregation (with antibodies, peptides, or small organic molecules that selectively bind and inhibit Aβ aggregate and fibril formation) *via* inhibition of the nucleation-dependent polymerization model [162, 163].

Drugs to Reduce Aβ Production

β-Secretase inhibitors (also known as the β-Site APP-Cleaving Enzyme; BACE-1) has been shown to be a transmembrane aspartic protease which is critically involved in the pathogenesis of AD. This enzyme is responsible for the cleavage of APP and the associated creation of β-amyloid, a neurotoxic peptide believed to be causally linked to AD. The inhibition of β -secretase, is therefore, an appropriate therapeutic strategy for treating this disease.

In the development of BACE-1 inhibitors, the most encouraging news arrived from Oklahoma City based biotech company (CoMentis) in 2008 that presented human data of the company's β -secretase inhibitor CTS-21166 at the 2011 Alzheimer's Association International Conference (AAIC). This transition-state analog inhibitor with cellular IC_{50}=1.2–3.6 nM showed excellent properties in brain penetration, selectivity, metabolic stability, and oral availability; all of these have met the requirements of an ideal oral drug candidate [164-166]. Merck &

Co. Inc. (MRK) announced for the first time the data of BACE-1 inhibitor, MK-8931 which reduced the cerebral spinal fluid (CSF) Aβ peptide by greater than 90% in healthy volunteers without observing dose limiting side effects in this study [167]. Thiazolidinediones such as rosiglitazone (**17**) and pioglitazone (**18**) are oral drugs for type-II diabetes that act as β-secretase inhibitors through the peroxisome proliferator-activated receptor-gamma (PPAR-gamma). Epidemiologic studies have shown a greater prevalence of AD in patients with type-II diabetes [168, 169]. These drugs are under evaluation in AD but current data have shown a potential higher risk of myocardial infarction with these compounds.

17   **18**

Although no Phase-III RCTs in new BACE-1 inhibitors are ongoing, however, several new BACE-1 inhibitors are under investigation.

γ-Secretase inhibitors and modulators: The formation of Aβ is catalyzed by γ-secretase (GS), a protease with numerous substrates. Accordingly, γ-secretase inhibitors are designed for selective inhibition of Aβ synthesis. The development of γ-secretase inhibitors (GSI) presents challenges similar to those for BACE-1 inhibitors. The development of GSI as DMDs presents problems related to potential non-specific effects. This is because GS is not only responsible for Aβ generation but it is also involved in intramembranous cleavage of several proteins, including the Notch receptor, APP and various neuronal substrates [170].

Semagacestat (**19**, LY450139 dihydrate) was the first GSI to undergo extensive clinical testing and has shown to reduce Aβ concentrations in plasma and Aβ production in the CNS [171, 172]. In August 2010, Eli Lilly & Company (LLY) announced that the company would halt development of semagacestat, when preliminary results from two ongoing long-term Phase-III studies showed that it did not slow disease progression and was associated with worsening of clinical measures of cognition and the ability to perform activities of daily living [173-175]. GS enzyme is critical for the processing of Notch, a protein that controls cell differentiation and communication. Notch-sparing GSI, also known as second generation inhibitors, are currently under development. At the 2011 AAIC, Bristol-Myers Squibb Company (BMY) announced the results of a Phase-II study evaluating the safety and tolerability of the investigational oral GSI avagacestat (**20**, BMS-708163) in patients with mild-to-moderate AD [148, 176, 177]. Begacestat (**21**, GSI-953) and ELND006 (**22**) are novel GSIs that were in the clinical trials as a potential treatment for AD. Later the trials for both compounds were halted due to liver toxicity [176-180]. PF-3084014 (**23**) is a novel tetralin imidazole GSI with 10–100 fold selectivity for APP compared with Notch [181].

In comparison to GSIs, γ-secretase modulators (GSM) can selectively block APP proteolysis without any Notch-based adverse effects. The identification of a subset of NSAIDs including ibuprofen (**24**), indomethacin (**25**), and sulindac (**26**), as GSMs in 2001, has led to tarenflurbil (**27,** Flurizan™) as the first NSAID derived GSM to be tested in the clinic [182, 183]. Tarenflurbil, the pure R-enantiomer of flurbiprofen, was the first GSM that was stopped in Phase-III clinical trials. These trials were conducted by Myriad Genetics, Inc. [184, 185].

CHF-5074 (**28**) is a GSM that showed selective $A\beta_{42}$-lowering properties but was devoid of cyclooxygenase (COX) inhibitory activity that rendered it suitable candidate for chronic use in AD patients [186]. E-2012 (**29**), a new imidazole containing GSM discovered by Eisai, has shown some potential to reduce the production of $A\beta_{40/42}$ by modulating the function of GS without interfering with the notch-processing [187]. EVP-0962 is a selective GSM that does not inhibit other GS substrates required for normal function. In animal trials, EVP-0962 appeared to have a better safety profile than other GSIs. In June 2011, EnVivo Pharmaceuticals initiated a Phase-I trial to study multiple doses of EVP-0962 in healthy volunteers to determine its safety profile [188].

Drugs to Prevent Aβ Aggregation

Aggregation of monomeric Aβ species into higher molecular weight oligomers produces the primary neurotoxic species in AD. As AD is characterized by amyloid plaques, scientists have explored drugs that prevent Aβ aggregation as a potential treatment for AD. Dysfunction of cerebral metal ions (Fe^{2+}, Cu^{2+} and Zn^{2+}), and their interactions with Aβ may contribute to AD by playing a role in the precipitation and cytotoxicity of Aβ. Metal ions are required for Aβ protein oligomerisation and recent studies show that metal chelators could produce a significant reversal of Aβ deposition *in vitro* and *in vivo* [189, 190].

Tramiprosate (**30**, Alzhemed) is a glycosaminoglycan that has been designed to block aggregation of Aβ into plaques. Its Phase-III trial was halted in late 2007 due to high data variations among trial sites that invalidated the statistical model for evaluating the drug. The clinical trial was carried out at 67 sites and involved 1,052 individuals with AD. Preclinical data have shown that tramiprosate reduces brain and plasma levels of Aβ and prevents fibril formation [191-193]. PBT2 (**31**, 8-hydroxy quinolone), an experimental drug with good BBB permeability, was developed by Prana Biotechnology Ltd. As zinc and copper are catalysts for Aβ-aggregation and stabilization of Aβ plaques, chelating agents may be effective in dissolving amyloid deposits *in vitro* and *in vivo*. PBT2 was found to remove copper and zinc from CSF, promoted Aβ oligomer clearance and restored cognition in AD mouse models. In a Phase-IIa study, PBT2 lowered $A\beta_{42}$ in CSF and improved cognition, but no correlation was found between Aβ in CSF and cognitive changes [194, 195]. Desferioxamine (**32**, DFO), another chelator, has shown some benefits in AD together with the same severe adverse effects. It is isolated from *Streptomyces pilosus* and this compound was the first to be clinically tested as a metal chelator to treat AD patients. DFO was supposed to chelate aluminum or other metal ions and reduced the neocortical concentration, leading to behavioral improvements in an unknown manner [196].

ELND005 (**33**, Scyllo-inositol, AZD-103), an oral small molecule interfering with the aggregation and fibrillization of Aβ is under evaluation in clinical trial. Developed by Elan Corporation, plc (ELN), ELN005 is stereoisomer of inositol, crosses the BBB using inositol transporters. Scyllo-inositol is thought to bind to Aβ, modulates its misfolding, inhibits its aggregation, and promotes dissociation of aggregates. It is currently being tested in a Phase-II/III RCT in patients with early AD [197-199].

Drugs to Promote Aβ Clearance

Immunotherapy toward Aβ is considered one of the most promising approaches to develop DMDs in AD, because it can potentially affect production, aggregation and deposition of Aβ. Active and passive immunotherapeutic approaches were developed to inhibit generation of toxic Aβ aggregates and to remove soluble and aggregated Aβ [200, 201].

Passive immunotherapy is based on monoclonal antibodies or polyclonal immunoglobulins targeting Aβ to promote its clearance. The results from animal studies have shown that anti-Aβ antibodies can prevent oligomer formation and reduce brain amyloid load with improvement in cognitive functions. Several monoclonal antibodies, generally given intravenously, are being tested in patients with AD [202]. Solanezumab (LY2062430) is a monoclonal antibody used for the treatment of mild to moderate AD. Phase-II studies in AD patients have shown a good safety profile with encouraging indications on cerebrospinal and plasma biomarkers. The drug is currently being investigated in Phase-III trials [203-206]. Bapineuzumab is also a humanized monoclonal antibody that acts on the nervous system and was once believed to have potential therapeutic value for the treatment of mild to moderate AD and possibly glaucoma. However, in 2012 it failed to produce significant improvements in patients in two major trials [207, 208]. GSK-933776, R-1450 (RO-4909832), and crenezumab (MABT-5102A) are monoclonal antibodies that target Aβ and have been tested in patients with AD in Phase-I studies. Another passive immunotherapy in a Phase-III clinical trial is Baxter International Inc.'s (BAX) Gammagard Liquid [Immune Globulin Infusion (Human10%)] [188].

Active immunization uses Aβ peptide or part of the peptide to induce anti-Aβ antibody by human immune system to enhance clearance of Aβ from the brain. At present, at least five active immunotherapies are in development stages to slow or reverse the progression of AD by inhibiting the generation and removal of the soluble and aggregated Aβ. ACC-001, a second-generation Aβ vaccine based on

N-terminal Aβ fragment, intended to induce a highly specific antibody response by the patient's immune system to Aβ. It has been shown to be safe in a Phase-I study and is currently being evaluated in Phase-II clinical studies [209, 210]. Another Aβ vaccine in Phase-II clinical trials is CAD106, which is being evaluated in elderly patients with mild AD [211].

Tau-Related Therapies

NFTs are intracellular aggregates of paired helical filaments whose main constituent is a hyperphosphorylated form of the τ-protein [212]. Tau (τ), a cytoplasmic protein, is another target that is being pursued by researchers to slow or reverse the progression of AD [213]. In AD, τ is abnormally phosphorylated, resulting in the disruption of its normal function in regulating axonal transport and leads to generation of NFTs toxic to neurons. Furthermore, degradation of hyperphosphorylated tau by the proteasome is inhibited by the action of Aβ [214]. There are two main therapeutic approaches to target τ-protein:

i) Modulation of τ phosphorylation with inhibitors of τ - phosphorylating kinases.

ii) Compounds that inhibit τ aggregation and/or promote aggregate disassembly.

Regarding the first approach several kinases are reported to phosphorylate τ *in vitro* including glycogen synthase kinase (GSK-3), cyclin-dependant kinase-5 (Cdk-5), mitogen activated protein kinase family members (MAPK), casein kinase, calcium calmodulin-dependant kinase-II, protein kinase A and others [215]. Reducing abnormal phosphorylation, restoring or stimulating phosphatase activity are promising therapeutic strategies. GSK3β is the main enzyme involved in tau hyperphosphorylation [216]. Lithium and valproate reduce τ-phosphorylation *in vitro*, promotes microtubule assembly through inhibition of GSK-3, and has been shown to reduce tau phosphorylation in APP transgenic mice [217]. Methylene blue (**34**), Rember®, a widely used histology dye, has been shown to interfere with tau aggregation [218]. In 1891, **34** was identified by Ehrlich as a successful drug for treatment of malaria. Due to its reversible side effects it disappeared as an anti-malarial during the Pacific War in the tropics [219]. This first generation tau aggregation inhibitor (TAI), was developed by University of Aberdeen and TauRx Pharmaceuticals Ltd [220]. Its different doses (up to 100 mg) were tested in a Phase-II study in patients with moderate AD. The group given a 60 mg dose had improved cognitive function and, after 1 year, evidence of slower disease progression compared with placebo [221]. The ineffectiveness in the group on the 100 mg dose was attributed to drug formulation defects, limiting its release. A new formulation (leuco-methylthioninium), with a higher bioavailability, was recently announced [222], and Phase-III RCTs are needed to confirm its safety and clinical efficacy. NP-031112 (**35**, NP-12), a thiadiazolidinone-derived GSK3 inhibitor, can reduce concentrations of amyloid deposition and hyperphosphorylated and insoluble tau in brain. This drug has been tested in patients with AD in Phase-II trial and no results have yet been published [223].

34

35

Davunetide (AL-108, NAP), derived from an endogenous brain, is an eight-amino acid peptide fragment (Asn-Ala-Pro-Val-Ser-Ile-Pro-Gln) and is able to cross the BBB thus accumulating in the CNS [224, 225]. Several preclinical studies demonstrated its neuroprotective role due to a significant reduction in both hyperphosphorylated and insoluble tau [226].

Beyond Plaque and Tangle-Related Targets

Neuroprotective Agents

The term neuroprotection refers to mechanisms that protect neurons from degeneration, for example following ischemic injury or as a result of chronic neurodegenerative diseases. AD and other neurodegenerative disorders are associated with oxidative and inflammatory stress and mitochondrial dysfunction, while, trials of antioxidants and anti-inflammatory treatments have provided modest or no beneficial effects, however, efforts to develop effective neuroprotectants continue.

Drugs to Target Mitochondrial Dysfunction

Targeting mitochondria is a new and different approach to AD therapy. Mitochondrial dysfunction occurs early in AD. It can promote synaptic damage and apoptosis and is thought to have a causal role in neurodegeneration. Latrepirdine (**36**) was introduced in Russia as dimebon (or dimebolin). It is a 25-year-old non-selective antihistamine [85-87], is a weak inhibitor of ChEs and a low-affinity NMDA receptor antagonist, which exerts its neuroprotective effects through the stabilization of mitochondria *via* inhibition of mitochondrial permeability transition pores induced by Aβ. However, the ability of **37** to improve cognition in AD is controversial, due to a discrepancy between the positive signal reported in a Phase-II clinical trial and the subsequent null effect observed in a Phase-III trial. Two RCTs are ongoing to assess the clinical efficacy of **37** in combination with donepezil and memantine [226-229].

Antioxidants

Genetic and lifestyle-related evidence has suggested that oxidative damage to neurons plays an important role in the AD pathogenesis. Thus, efforts to reduce oxidative injury may prove beneficial in preventing the onset and progression of AD in patients. Preclinical studies have been conducted with several potential antioxidant drugs that may have therapeutic uses in the treatment of AD. Several antioxidants that have been investigated for their potential to reduce the risk of AD include vitamins A, C and E, coenzyme Q, selenium, polyunsaturated fatty acid and others. Approaches such as omega-3 polyunsaturated fatty acids (*e.g.*, docosahexaenoic acid) or antioxidants (*e.g.*, vitamin E) have been tested in RCTs [230-232]. Some trials have reported beneficial effects of docosahexaenoic acid (DHA) supplementation in elderly people with cognitive decline or AD [233]. Ginkgo biloba (Egb761), an extract from *Ginko biloba*, is a Chinese medicine that has been used for centuries. The medicine is extracted from leaves of the *Ginkgo biloba* tree and is believed to improve brain function. *Ginkgo biloba* special extract is listed in the group of anti-dementia drugs together with cholinesterase inhibitors and memantine. Treated patients showed improvement on the AD Assessment Scale-Cognitive subscale and the Geriatric Evaluation of Relative's Rating Instrument [234-237].

Non-Steroidal Anti-Inflammatory Drugs (NSAIDS)

Destruction of neurons due to inflammation around Aβ plaques is thought to be a major factor in the pathogenesis of AD [238]. NSAIDs inhibit cyclooxygenase-1 and cyclooxygenase-2 (COX-1 and COX-2), which are responsible for the oxidation of arachidonic acid to prostaglandins. Individuals using conventional NSAIDs, on a regular basis, showed a decreased incidence of AD revealing their neuroprotective effect [239]. According to Rogers *et al.*, indomethacin appeared to protect AD patients from cognitive decline but this point of view is not shared by Cochrane reviewers [240, 241]. Another trial with indomethacin failed to show any efficacy in the progression of AD [242]. Ibuprofen, celexocib, rofecoxib and naproxen did not slow the progression of AD [243-245]. Many other anti-inflammatory compounds are now under investigation in AD. Cyclophosphamide is a potent anti-inflammatory and immunomodulatory drug acting primarily by inhibiting proliferation of immune cells [246]. Resveratrol, a component of grapes, berries and other fruits, has been shown to reduce the expression of inflammatory biomarkers and induce antioxidant enzymes [247].

Receptor for Advanced Glycation Endproducts (RAGE)

RAGE is a new pharmacological target proposed for developing neuroprotective drugs in AD. It is a transmembrane protein that belongs to the immunoglobulin superfamily localized in neurons, microglia, astrocytes and BBB [248]. RAGE mediates the effects of Aβ on microglia, the BBB and neurons through different signaling pathways. It enhances generation and accumulation of Aβ in the CNS by modulating BACE-1 besides promoting the transport of Aβ from vascular circulation to the brain [249]. Data from autopsy brain tissues, *in vitro* cell cultures and transgenic mouse models suggest that the Aβ-RAGE interaction exaggerates neuronal stress, impairs learning memory and induces neuroinflammation. A Phase-II trial with PF04494700, a RAGE antagonist, has been recently completed in mild to moderate AD patients [148].

Miscellaneous Therapies

Monoamine oxidase (MAO, EC 1.4.3.4, amine-oxygen oxidoreductase) is a flavoenzyme, localized at the inner side of the outer mitochondrial membrane. Two distinct enzymatic isoforms, namely MAO-A and MAO-B, have been isolated and fully studied. They are characterized by different amino acid sequence, 3D structures [250], organ and tissue distributions [251], sensitivity to inhibitors [252] and substrate specificity. Both MAO-A and MAO-B have been implicated in AD pathogenesis and Rasagiline (**37**), a MAO-B inhibitor which exhibits neuroprotective and anti-apoptotic activity *in vitro* and *in vivo*, is under clinical evaluation [253].

Statins are common drugs used in dyslipidemias which reversibly inhibit the hydroxyl-methyl-glutaryl-CoA (HMG-CoA) reductase thus inhibiting the transformation of HMG-CoA into mevalonate, the first step in the cholesterol biosynthesis. Recent preclinical evidence supported the possible use of atorvastatin (**38**) in AD and the rationale was related to its ability to counteract oxidative stress in specific brain areas, such as parietal cortex, without a significant interaction with the cholesterol biosynthetic pathway [254, 255].

Recent studies have suggested modifications of serotonin cerebral metabolism in mild cognitive impairment and AD. PRX 03140 (**39**), identified by EPIX Pharmaceuticals, is a small molecule with partial agonist activity for the type-IV serotonin (5-HT) receptor [256]. Lecozotan (**40**), a potent antagonist of the 5-HT receptor, is developed for the treatment of AD after having promising results in animal studies [257, 258].

37   38

39   40

Antihypertensive medications are associated with lower incidence of AD and some of them as angiotensin-converting enzyme inhibitors or calcium channel blockers have become a source of interest [259-261]. A Phase-II clinical trial in AD with MEM 1003, the (+)-enantiomer of a dihydropyridine, that has been optimized for CNS activity, is on the way. Neurotrophins are considered as AD therapeutics due to their ability of neuroprotection, restoring neuroplasticity, improving neuronal survival, or even promoting neurogenesis. CERE 110 is a gene therapy that employs an adeno-associated viral vector system to deliver the gene for the nerve growth factor (NGF) to selected brain regions [262, 263]. Preclinical studies in primates demonstrated that NGF gene can remain active for at least 1 year [264]. This gene therapy is not considered as a cure for AD, but it conceivably could protect or restore damaged brain cells and alleviate memory loss. Cerebrolysin, a peptide acting like endogenous neurotrophin was demonstrated to improve global function in clinical trials through intravenous route administration [265].

Limitations

Setbacks have greatly outnumbered successes in Alzheimer's drug development. This is in part related to the complexity and challenges of researching this disease as previously discussed, but also simply reflect the nature of drug discovery and development. In recent years we have seen a steady stream of drug candidates in development accompanied by a long series of unsuccessful outcomes. As research has intensified a higher average of potential medicines pulled out of development each year has been observed. Since 1998, more than 100 medicines in development have become inactive and could not receive regulatory approval (Fig. **3**).

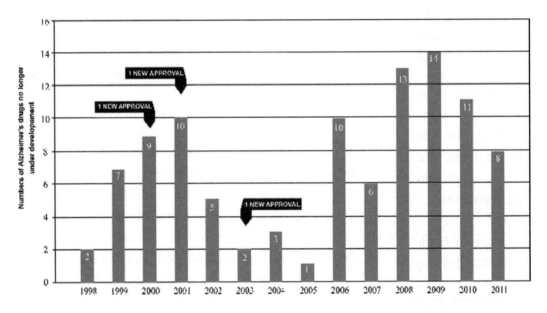

Figure 3: Total number of unsuccessful drugs from 1998-2011 (extracted from [266]).

Since 1998 three new medicines for AD were approved, resulting in a 34 to 1 ratio of "failures" to approval (two other medicines were approved before this timeframe). However, it should not be concluded from these numbers that for every 34 setbacks one new approval will result. This data provides an indication of recent research outcome. Since 2003, when the most recent AD medicine was approved, there have been 66 unsuccessful projects. Recent years have yielded a particularly high number of halted projects, with an average of 10 per year in 2006-2011, compared to 5 per year in 2000-2005. Although this report does not examine trends in the number of medicines in development, it is possible that this increase, at least in part, reflects an increase in the number of research projects in the pipeline.

So far, the research on AD drug development has some success in terms of symptomatic treatments which involves ChEIs and memantine. Although during the last decade there is some success in terms of DMDs developed with the help of preclinical models but none of these drugs has succeeded in Phase-III trial. This failure in AD drug development includes lack of adequate and validated biomarkers and outcome measurements, time course of treatment in relation to the development of disease. Available data from unsuccessful Phase-III studies suggest that mild to moderate AD patients may be too late in the disease process to improve substantively their outcome following drug treatment.

Another problem in development of AD therapeutics is that the design of selective compounds without undesirable and potentially toxic side-effects is difficult, and reaching the stage of clinical testing may take many years. Moreover, the development of experimental models used by the researchers to study the cellular events which characterize AD need to be optimized. Human or rodent cell lines, transgenic mice overexpressing the genes for APP, *ex vivo* studies performed on human tissues are only few examples of the experimental systems used to produce evidence in favor of or against pathogenic hypotheses. Extensive efforts are still needed to develop better animal models.

Furthermore, the drug development process for AD suffers from a peculiar anatomical localization of brain, protected by the BBB. It is estimated that over 400 drugs for AD are being investigated in about 830 clinical trials, while, many more substances are being investigated in animal models of AD. Drugs that must reach deep brain targets (as is the case in AD) must cross the BBB. However, many drug trials fail because of inadequate trial design with one of the chief flaws being a neglect regarding BBB penetration. Successful treatment of AD is likely to be pharmacologically based and, will in almost every case, target the CNS. Therefore, BBB is another formidable obstacle for the failure of a large number of drugs for AD. A scientific innovation is an accumulative result of success and failure. Scientific progress and drug development are deeply connected and setbacks or failures are common in this area. These failures and successes have lessons that guide next steps and new solutions. An analysis of that data from a failure outcome can be applied in the design of new experiments and methodologies until the achievement of a successful therapy [267].

RATIONAL APPROACHES TO DESIGN AD DRUGS

Rational design as applied in the discovery of novel lead drugs, is the rapid development mainly attributed to the tremendous advancements in the computer science, statistics, molecular biology, biophysics, biochemistry, medicinal chemistry, pharmacokinetics and pharmacodynamics experiences in the last few decades. When the 3D structure of a receptor molecule is known, new potential drug molecules may be custom-tailored, by molecular modeling techniques, to precisely fit into the binding pocket of a receptor. This approach has been called "structure-based" or "rational" drug design, as opposed to more random synthesis and screening of a series of substances. Over the last 20 years, 3D structures of many proteins have been determined by X-ray crystallographic methods, which have shed new light on their mode of action. Methods of modern molecular biology have made it possible to produce larger quantities of pure receptor material than did previous conventional methods.

X-ray crystallography has now become a vital tool for medicinal and computational chemists, and the ability to map structure–activity relationships (SAR) and binding interactions directly to protein structure has accelerated the process of lead development. Leads are obtained from various sources, which include HTS, fragment screening, *de novo* design, from inhibitors of enzymes homologous to the target of interest, or from other related activities.

Another tool helpful in drug designing is the prediction of properties of drug-like molecules such as bioavailability and membrane permeability. These properties of molecules have often been connected to simple molecular descriptors such as log P (partition coefficient), molecular mass (MW), and the number of hydrogen bond donors (HBD) and hydrogen bond acceptors (HBA) in a molecule [268]. Lipinski used these molecular properties in formulating his well-known "Rule of Five (Ro5)" which states that most molecules with good membrane permeability are characterized by a log P ≤ 5, a MW ≤ 500, number of HBA ≤ 10, and number of HBD ≤ 5 [269]. In this rule, both oxygen (O) and nitrogen (N) atoms are defined as HBA and N–H or O–H groups are considered as HBD. LogP refers to the octanol/water partition coefficient of a compound and is used as a measure of its lipophilicity which in turn is a measure of a drug's solubility in lipid membranes.

This is usually an important factor in determining how easily a drug passes through cell membranes. Another molecular descriptor which is commonly used in medicinal chemistry for the optimization of cell permeability is the polar surface area (PSA) of a molecule which is defined as the surface sum over all polar atoms, (usually oxygen and nitrogen), including also attached hydrogens. Molecules with a polar surface area of greater than 140 °A are usually believed to be poor at permeating cell membranes. A number of software may be used for predicting drug-like properties of compounds which can be calculated with the help of molecular descriptors. The knowledge of these descriptors helps in improving drug-like properties of the synthesized compounds.

Acetylcholinesterase Inhibitors (AChEIs)

ACh was first synthesized in 1867 and detected in the adrenal gland of human tissue in 1906 as a neurotransmitter [270]. AChE (EC 3.1.1.7) is one of the most widely spread enzymes located in the CNS and in muscles of many living organisms (human beings, vertebrates) and plays very important role in nerve signal transmission. Its inhibition leads to death of an organism, including human beings [271-274]. The knowledge of the 3D structure of AChE is essential for understanding its significant catalytic efficacy for rational drug design and for developing new therapeutic approaches. X-ray crystallographic studies on AChE from a range of species *Drosophila melanogaster*, *Electrophorus electricus*, human, mouse and *Torpedo californica* have now been published [275-292]. Various oligomeric forms of AChE in the electric organ of certain fish, *Electrophorus* and *Torpedo*, are structurally homologous to those in vertebrate muscles and nerves. Highly purified preparations from these abundant sources of AChE have facilitated research on this enzyme.

The crystal structure of *Torpedo californica* (*Tc*AChE) has provided important insights into the catalytic mechanism of this important enzyme. The enzyme monomer, which contains 537 amino acid residues, is α/β protein arranged as a 12-stranded mixed β-sheet surrounded by 14 α-helices. The hydrophobic active site of AChE is subdivided into several subsites. A Ser-His-acid catalytic triad (Ser200, His400 and Glu327) including Trp84, also known as catalytic anionic subsite" (CAS), is located at the bottom of a deep and narrow gorge (around 20 Å long) and is responsible for hydrolyzing ACh. It was suggested that the high catalytic power of the enzyme is due to the formation of a strong hydrogen bond between Glu327 and His400 residues of the catalytic triad. The binding of CAS to the quaternary moiety of choline corresponds to Trp84, which stabilizes the positive charge of choline by forming a cation-π complex [273]. Moreover, the peripheral anionic site (PAS) located at the opening of the gorge, has also been identified and assigned to another tryptophan residue (Trp279) which is involved in the enhanced binding of *bis*-quaternary ligands (Fig. **4**).

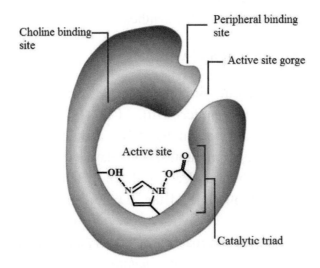

Figure 4: Schematic representation of the binding sites of AChE: esteratic site, anionic binding site as well as the peripheral anionic binding site.

The reaction proceeds by nucleophilic attack of the catalytic Ser200 to the carbonyl carbon of ACh thus acylating the enzyme and liberating choline. This acylated enzyme is then rapidly hydrolysed to yield acetic acid and free enzyme (Fig. **5**).

X-ray crystallographic structures of AChE co-crystallized with various AChEIs provided an understanding of the essential structural features and ideas about its catalytic mechanism and mode of ACh processing. The influence of ligand binding on the conformation of *Tc*AChE was examined to consider the structural changes induced in the 3D structure of un-complexed free enzyme (PDB entry

2ACE) due to the binding of several inhibitor-bound AChE: tacrine (1ACJ), edrophonium (2ACK), decamethonium (1ACL), m-(N,N,N-trimethyl-ammonio) trifluoroacetophenone (1AMN), (-)-huperzine A (1VOT), and donepezil (1EVE). This comparison revealed that the presence of various inhibitors caused very small changes in the overall conformation of the protein. Few remarkable local structural changes were found, such as the peptide flip in residues Gly117 and Gly118 in 1VOT, and the displacement of the loops spanning residues 334-348 and 277-304 in 1ACL [278-292]. A critical inspection of the available AChE complexes with different AChEIs may be valuable to establish SAR and the designing of novel AChEIs. There are a large number of inhibitors (or ligands) of ChEs that have been co-crystallized and deposited at the Protein Data bank (www.pdb.org) in the last couple of years.

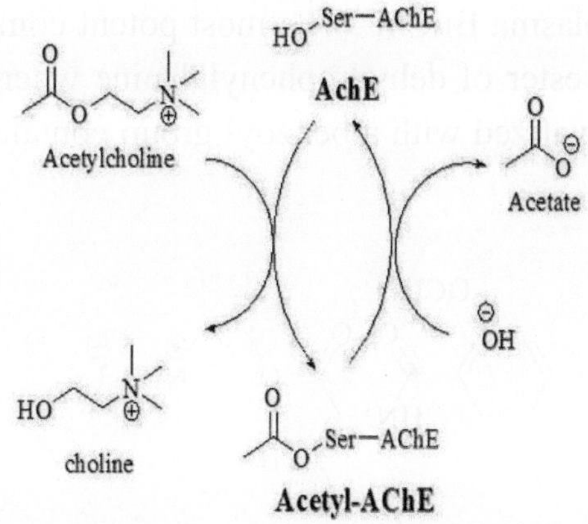

Figure 5: Reactions of AChE at its different sites in the binding pocket.

Structure-activity analysis makes the foundation of understanding the structural features of the inhibitors as well as the target receptors involved in a particular biological process and helps in designing more effective inhibitors. Highly predictive QSAR models CoMFA and CoMSIA have been developed by Zaheer *et al.* using the technique of structure-based alignments of different substrates. These models were used to identify important protein-ligand interactions. The same group also reported an effective docking protocol for virtual screening of AChE using different docking and scoring approaches to determine their ability to reproduce the binding poses in different complexes of AChE [293, 294].

A large number of natural products and their semi-synthetic derivatives been reported as potent AChEIs that exhibited potential for the development of "lead molecule" for the treatment of AD. Danuello *et al.* reported molecular modeling studies on two semi-synthetic acetyl derivatives of the piperidine alkaloids (–)-cassine and (–)-spectaline isolated from *S. spectabilis* [295]. Atta-ur-Rahman *et al.* reported a number of novel natural AChEIs and BuChEIs isolated from various medicinal plants [296-301]. Our research group had reported the synthesis of 3-hydroxy-2,3-dihydro-1,5-benzothiazepines, 2,3-dihydro- and 2,3,4,5-tetrahydrobenzothiazepines as AChEIs and BuChEIs. Moreover, molecular modeling and *in silico* studies were also conducted to study the mode of their interactions in the enzymes active site [302-304]. Grigoryan *et al.* in 2008 reported the synthesis and ChE inhibitory profile of a number of dehydroamino acid choline esters. Their affinity was measured for the inhibition of human red cell AChE and human plasma BuChE. The most potent compound, CBC-171-08-IIIf (**41**), was a choline ester of dehydrophenylalanine where the amine group of the amino acid was derivatized with a benzoyl group containing a methoxy in the 2-position [305].

41

Correa-Basurto *et al.* in 2007, performed and reported molecular docking studies and density functional theory (DFT) of 88 N-aryl derivatives and for some AChE and BuChE. Their docking studies suggest that all the tested compounds bind at the active site of both ChEs [306]. A 3D pharmacophore model has sometimes been used in combination with a structure-based virtual screening approach either as a pre-filtering or a post-processing tool to reduce the massive amount of culling or selection work that needs to be performed in identifying virtual hit-lists. Following a Ligand Based Virtual Screening (LBVS) strategy, a pharmacophore model was developed using HypoGen module [307]. This model consists of five features: one hydrogen bond donor and four hydrophobic features. These pharmacophoric features were used to identify potent AChE inhibitors besides clarifying the quantitative SAR for previously known AChE inhibitors.

β-Secretase (BACE-1) Inhibitors

In the search of new ligands for AD, researchers are faced with the challenge of finding new BACE-1 inhibitors. BACE-1, a member of the aspartyl protease family, is a structurally challenging protein target, featuring multiple sites for effective binding and an induced-fit upon ligand interaction. The great interest in BACE-1 from, both academia and industry, is confirmed by the exponentially growing number of crystal structures deposited in the PDB. Further an ever increasing number of crystal structures of BACE-1 is an evidence to the variety of compounds being tested as AD-modifying drug candidates [308-310]. Since the cloning of β-secretase, in complex with potent peptide inhibitor OM99-2 over a decade ago, its structure and catalytic properties have been thoroughly investigated [311-313]. X-ray crystal structure of BACE-1 is likely to accelerate the discovery of a new inhibitor drug through various rational approaches such as SBDD including VS and *de novo* design methods. Since then, around 185 structures have been deposited in the PDB in which the enzyme was crystallized in complex with different ligands or in its *apo* form.

BACE-1 is a type I transmembrane protein and its catalytic domain is an aspartic protease with a pair of active-site aspartyl residues. These and other structural features that are important for catalysis in the active site of β-secretase are nearly identical to other aspartic proteases of the pepsin family [312]. β-Secretase has an elongated substrate-binding site that can bind up to 11 substrate residues [314, 315]. The amino acid preference in these subsites is somewhat broad [314, 316].

There are various approaches for designing BACE-1 inhibitors ranging from substrate-based inhibitors to fragment based screening. As mentioned earlier, BACE-1 is a member of the aspartyl protease family, hence a common strategy in the designing of BACE-1 inhibitors is to replace the scissile amide bond with a non-cleavable transition-state isostere (TSI) and a large number of BACE-1 inhibitors are their substrates. Standard nomenclature for enzyme-substrate complex as described by Schechter and Berger is shown in Fig. **6** [317].

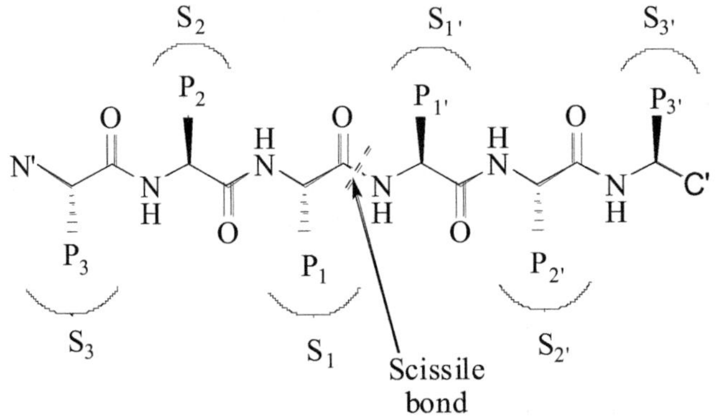

Figure 6: Standard nomenclature for enzyme-substrate complex.

S_1 subsite consists of hydrophobic residues formed by the side chains of Tyr71, Phe108, Trp115, Ile118, and Leu30. Most inhibitors of this class contain aromatic residues at P_1 due to the possibility of aryl ring stacking interactions with aromatic residues of Tyr71 and Phe108 [318, 319]. The hydrophobic subsite S_3 possesses Leu30, Trp115, and Ile110 residues. Moreover, it also contains main chain atoms from Gln12, Gly11, Gly230, Thr231 and Thr232. SAR studies have been focused on simple aliphatic chains in order to exploit the hydrophobic character of S3. However, compared to S_2 and S_4 pockets are more polar and also solvent exposed. In addition to residues Tyr71, Thr72, and Glu73 from the flap, an important feature of the S_2 pocket is the side chain of Arg235, which provides much of the polar character. Therefore, SAR studies in this region have shown a preference for negatively charged groups in order to complement the positive charge of Arg235. Examples of P_2 groups that have been reported in the literature are benzyl carbamates (2HIZ), pyridyl (2HM1), isophthalate (2IQG), sulfones (2VIY), pyrrolidinones (2VIZ), sultams (2VIJ), and aromatics (2EWY). The S4 specificity pocket is highly polar and solvent exposed. In the structure of the OM99-2, the P_4 Glu residue is observed to be interacting primarily with the P_2 Asn residue, along with Arg235 and Arg307 [312].

The prime side of the BACE active site contains the $S_{1'}$ and $S_{2'}$ specificity pockets and these are far less stringent than the non-prime pockets. The rest of the prime side *i.e.*, $S_{3'}$ and $S_{4'}$ pockets do not show any preference for specific interactions. Early work by Hong *et al.* suggested that occupation of these sites formed by Pro70, Tyr71, Arg128, and Tyr198 (S3'), and Gly125, Ile126, Trp197, and Tyr198 (S4') is crucial to achieving high potency inhibitors [314].

The peptidomimetic approach has been effectively applied to the design of the first generation of potent BACE-1 inhibitors and the peptide-based transition state analogues are still the benchmark of BACE-1 inhibition. Peptidomimetics are essentially short peptides in which the scissile peptide bond is replaced with a non-cleavable transition state isostere (TSI). The most common of the BACE-1 protease trasition state isosters (TSI) found are statins (**42**), hydroxyethylene (**43**, HE), hydroxyethylamine (**44**, HEA) and the reduced amide (**45**) (Fig. **7**).

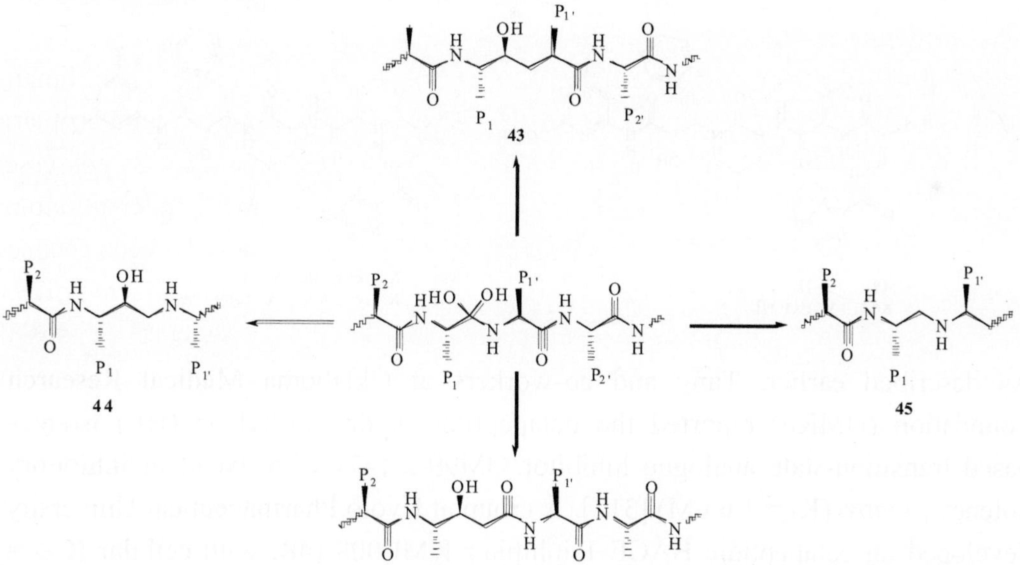

Figure 7: Common BACE-1 TSI scaffolds.

Statin is a TSI derived from natural product pepstatin which is a potent inhibitor of several aspartyl proteases with the exception of BACE-1. Elan Pharmaceuticals developed the first substrate-based BACE-1 peptidic inhibitor, P10–P4' StatVal (**46**). This inhibitor is a P1 (S)-statin substituted substrate analogue with *in vitro* $IC_{50}= \sim 30$ nM. In the design of the P10-P4'StatVal, the scissile bond was replaced with a non-cleavable statin residue and Asp was replaced with Val in the P1' position [320].

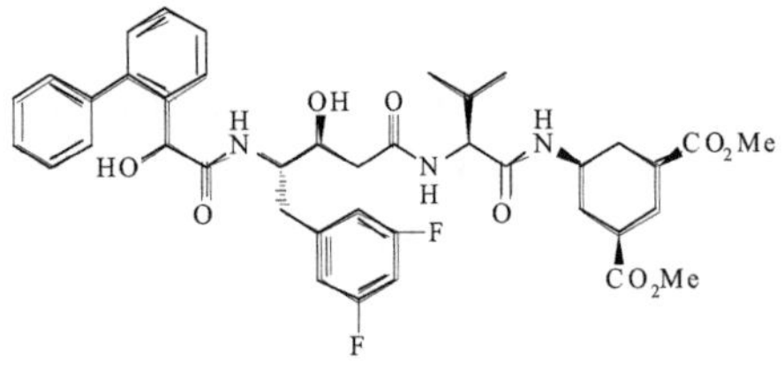

46

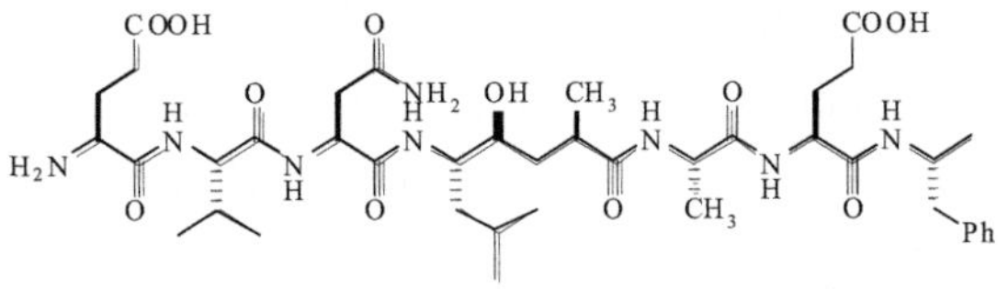

47

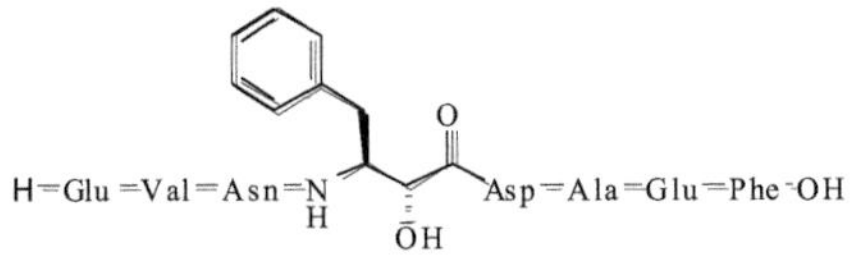

48

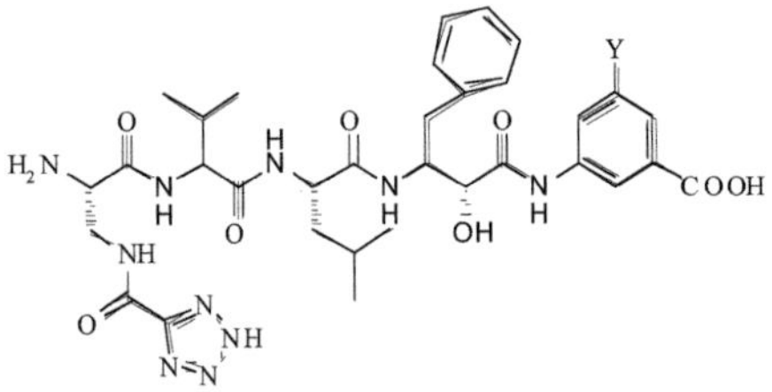

49 Y=H
50 Y=COOH

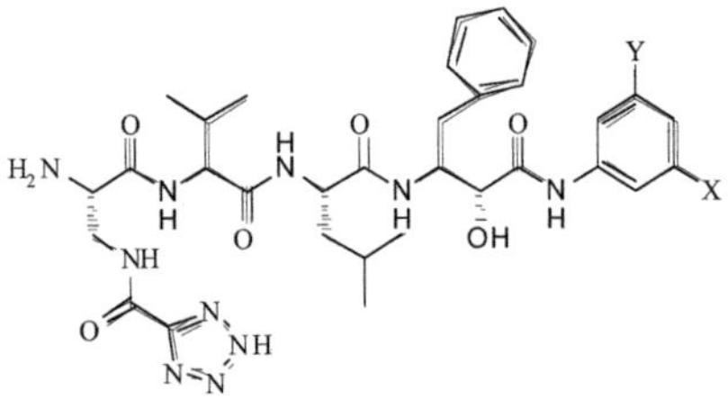

51 X=tetrazole Y=H
52 X=tetrazole Y=tetrazole

As described earlier, Tang and co-workers at Oklahoma Medical Research Foundation (OMRF) reported the octapeptidic, hydroxyethylene (HE) isotere-based transition-state analogue inhibitor OM99-2 (**47**) with excellent inhibitory potency *in vitro* (Ki = 1.6 nM) [312]. A group at Kyoto Pharmaceutical University developed an octapeptidic BACE-1 inhibitor KMI-008 (**48**) with cellular $IC_{50} =$ 413 nM employing a hydroxymethylcarbonyl (HMC) isostere as a transition-state mimic [321]. Further chemical modification of KMI-008, Kimura *et al.* reported more potent pentapeptidic phenyl norstatin based BACE-1 inhibitors KMI-420 (**49**) with *in vitro* $IC_{50}= 8.2$ nM and KMI-429 (**50**) with *in vitro* $IC_{50}=3.9$ nM). KMI-429 appeared to significantly reduce brain Aβ production when directly injected into the hippocampus of both wild-type mice and APP transgenic mice

(Tg2576 mice) [322]. In addition to this, further modifications of **49** and **50** produced tetrazole ring-containing compounds KMI-570 (**51**) (*in vitro* IC_{50}=4.8 nM) and KMI-684 (**52**) (*in vitro* IC_{50}=1.2 nM), Both of them also displayed improved brain permeability [323].

Gosh *et al* developed BACE-1 inhibitors (**53** and **54**) having less-peptidic features *via* chemical modification of **47**. Both these inhibitors exhibited stronger potency with cellular IC_{50} values equal to 39 nM and 1 nM, respectively and impressive *in vivo* efficacy (reduction of plasma Aβ level by 30% and 65%, respectively) when intraperitoneally (*i.p.*) injected into Tg2576 mice [324].

Several other substrate-based peptidomimetic inhibitors were also developed by different pharmaceutical companies and other academic research groups [325-330]. Though these peptidomimetic BACE-1 inhibitors are highly potent *in vitro*, their poor drug-like properties (high molecular weight, poor brain permeability, low oral availability, large number of hydrogen bonding groups and short half-life *in vivo*) have made them unsuitable drug candidates.

These structure-based, first generation inhibitors have laid the foundation for rational drug designing of non-peptidic second generation BACE-1 inhibitors. There are many examples of non-peptidic BACE inhibitors reported in the scholarly literature such as acylguanidine-based inhibitors, amino imidazole and aminohydantoin-based inhibitors and aminoquinazoline-based inhibitors. The acylguanidine–based BACE-1inhibitors were first disclosed at Wyeth [331, 332], *via* a HTS strategy. Optimization of the hit using structure-based design led to compound **55** (IC_{50}= 110 nM). X - ray crystal structures of the complex with analogues of **51** revealed that the N-acylguanidine moiety forms hydrogen bonding interactions with the key catalytic aspartates while the substituents on the acylguanidine nitrogen extended into the S1' pocket, forming hydrogen-bonding interactions with Arg235 and Thr329 *via* bridging water molecules. An aminoimidazole-based inhibitor **56** was synthesized by the addition of either a pyridine or a pyrimidine ring to a previously identified lead at Wyeth [333]. The extension towards the S3 region of the β-secretase binding pocket led to an IC_{50} value for β -secretase of 20 nM [334]. Baxter *et al.* synthesized 2-amino-3,4-dihydroquinazolines as BACE-1 inhibitors **(57-58)** through HTS and SBDD approach [335].

The most detailed account of nonpeptidomimetic BACE-1 inhibitors to date comes from Vertex, *via* a published patent application that disclosed several hundred compounds along with associated Ki ranges of BACE inhibition. These compounds spanned multiple classes of heterocyclic templates, however, among the more potent classes reported (BACE K_i < 3 µM) were the halogen-substituted *bi*-arylnaphthalenes. From these results, Vertex proposed the first 3D pharmacophore map of BACE-1 as a guide for the designing and optimization of inhibitors, wherein HB represents hydrogen-bonding moiety interactions with the active site and other key residues of BACE and HPB represents hydrophobic moiety interactions with BACE subsites [336, 337].

The Vertex pharmacophore is a seven-point pharmacophore that contains four hydrophobic features and three hydrogen-bonding features. Specifically, a hydrogen bonding feature that interacts with one or both catalytic aspartic acids, two hydrogen bond donors with G34 and/or G230, hydrophobic groups in flap/S_1 pocket, S_2, and/or $S_{2'}$ sites, and stacking interactions with Trp71, Phe108, and/or Trp76. In another approach, a set of most active and selective compounds for each class of BACE inhibitors was selected and subjected to an ensemble molecular docking process into five BACE X-ray structures. The superimposition of the calculated bioactive conformations of these inhibitors was then used to generate a common feature pharmacophore. Limongellia *et al.* selected a number of active and structurally diverse inhibitors and subjected them to an ensemble-docking process into five BACE-1 X-ray structures and built a pharmacophore model [338]. In a recent study John *et al.* generated a pharmacophore hypothesis based on key structural features of a training set of 20 compounds with known BACE-1 inhibitory activity. It has provided a rational hypothetical representation of the most important chemical features responsible for activity [339].

Recently, fragment-based drug discovery (FBDD) has emerged as a novel alternative to the traditional HTS method in identifying potent BACE-1 inhibitory drugs. In contrast to HTS, which uses libraries of relatively high molecular weight compounds, the FBDD approach takes advantage of libraries comprising more diverse and smaller-sized compounds (fragments) to identify hits that can be efficiently developed into potent leads with drug-like properties. After the hits identification, chemical modifications may hamper the identification of lead compounds and their further optimization into suitable drug candidates. Together with these methods, a variety of biophysical techniques (NMR, X-ray crystallography, fluorescence resonance energy transfer (FRET) or surface plasmon resonance (SPR), *etc.*), computational tools and biochemical assays can be coupled in fragment screening. In particular, by a fragment screening approach assisted with NMR and followed by X-ray crystallography and FRET assays, AstraZeneca reported a FBDD-based BACE-1 inhibitor **59** with a cellular $IC_{50} \approx$

0.47 µM. The same company in a partnership with Astex discovered several BACE-1 inhibitors by employing a X-ray crystallography based fragment screening [340].

59

The development of an effective and safe β-secretase inhibitor, as a disease-modifying AD treatment, is a critical necessity in interrupting the progress of this devastating disease. Toward this end, the structural evolution of β-secretase inhibitors is an essential path. The availability of 3D structural information for β-secretase in complex with a variety of compounds has enabled great advances in the development of new classes of inhibitors.

Tau Aggregation Inhibitors (TAI)

Tau (τ) protein is a Microtubule-Associated Protein (MAP), that is abundant in both central and peripheral nervous system, being less common elsewhere. A number of studies have revealed that τ is a prototypical "natively unfolded" protein. Owing to its highly flexible structure and variable conformation full-length τ are not amenable to structure analysis by crystallography so far. NMR spectroscopy is the only possible method that allows a description of its conformations and dynamics with high resolution. There are two major approaches for addressing the problem of tau aggregation. First approach is the search for inhibitors of kinases that phosphorylate τ-protein. The other approach addresses the search for direct inhibitors of the τ-aggregation. The former approach is based on the assumption that abnormally phosphorylated τ-protein aggregates more readily. Regarding the first approach one promising strategy to identify new τ-aggregation inhibitors is the screening of compound libraries containing sufficient structural diversity. Several chemical series have been identified that inhibit τ-aggregation, however, with the exception of the phenothiazine methylene blue *i.e.*, Rember® (**34**) no compound has yet been tested *in vivo* for its effect on tau deposition, neurodegeneration and behavioral impairments [341, 342].

Pickhardt *et al.* screened a library of 200,000 compounds for inhibition of τ-aggregation [342]. This screening scheme led to the identification of phenylthiazolylhydrazide (PTH) and rhodanine as lead compounds (Fig. **8**). As a result thereof, 77 compounds were identified as τ-aggregation inhibitors. Rhodanine was identified as one of the active hit classes. An *in silico* screening was also performed by using 21 of the 77 compounds from which the PTH was the first hit identified.

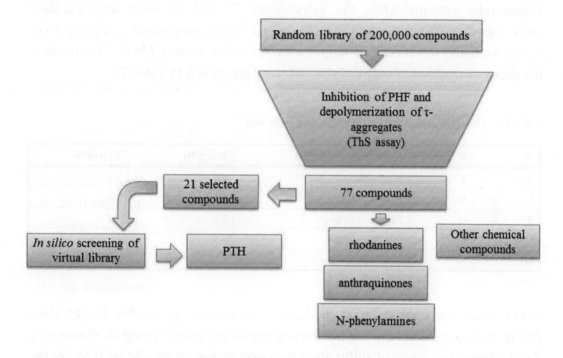

Figure 8: Screening scheme leading to the phenylthiazolylhydrazide and rhodanine lead structures.

The rhodanine scaffold (Fig. **9**) is an appealing hit class that is found in various biologically active compounds such as antimalarial, antituberculotic, antibiotic, anticancer, antitoxin and anti-hypoglycaemic [343].

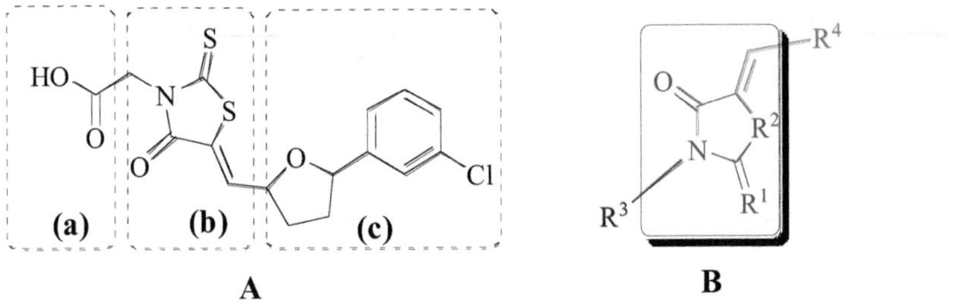

Figure 9: A) Structure of the hit compound, **B)** Variation of the structure of rhodanine inhibitor.

R^1 and R^2 in thiocarbonyl rhodanine core (Fig. **9-B**) were replaced to obtain other heterocyclic systems while the substituents R^3 and R^4 were kept constant (corresponding to parts **a** and **c** in Fig. **9-A**). These heterocyclic systems were synthesized and screened for τ-aggregation inhibition activity [344]. The trends in the depolymerisation of τ-aggregates observed are shown in Table **2**.

Table 2: SAR at R^1 and R^2 in thiocarbonyl rhodanine core

R^1	R^2	Heterocyclic System	IC$_{50}$ (μM)	DC$_{50}$ (μM)
S	S	Rhodanines	0.8	0.1
S	N	Thiohydantoins	6.1	0.4
S	O	Thioxooxazolidines	3.1	2.4
O	O	Oxazolidinediones	3.5	2.2
O	N	Hydantoins	22.6	54.3

It is evident from Table **2** that rhodanine heterocycle appeared to be the most potent compound emphasizing the importance of the thioxo group in rhodanines. The hydrophobic nature of sulfur atom may play an important role, as it leaves the amide part of rhodanine heterocycle free for hydrogen bonding. Besides the central rhodanine core, substitution pattern on R^3 and R^4 of the heterocycle (to parts a and c in Fig. **9**) showed that HBA in the form of a nitro group, carboxylic acids, phenols, sulfonates/sulfonamides may be important for activity. The importance of the carboxylic acid (Part **a**, Fig. **9A**) and the effect of the substitution and variations of the length of the linker connecting the central core to the carboxylic acid have been reported [345]. Furthermore, part **c** consisting of biaryl compounds (Fig. **9A**) was varied to investigate the requirement of aromatic substitution. The heteroaromatic side chain tolerated variations, but modifications on the furan heterocycle showed a trend for potency reduction.

In another *in silico* screening experiment, Larbig *et al.* selected thiazolylhydrazide core (**60**) for lead optimization and SAR investigation [346]. A variety of thiazolylhydrazide derivatives were obtained by synthetic derivatization of R^1, R^2 and R^3. The pharmacophore model possessed two aryl rings at R^1 and R^3, a hydrophobic region on thiazole ring and a HBA on carboxyl amide. A hydrogen binding group at R^3 led to an improved inhibitory potency, while, the hydrophobic or π-stacking interactions on R^1 appeared to contribute to target binding, as confirmed by STD-NMR experiments revealing a strong interaction of the tau construct K18 with R^1 of the aryl ring [347].

Ballatore *et al.* reported the discovery of a novel class of TAIs, known as the aminothienopyridazines (**61**, ATPZ), which exhibited a promising combination of activity in τ-fibrillization assays as well as drug-like physico-chemical properties [348]. The design of the ATPZs employed in these studies took into account the SAR as well as key physico-chemical properties embedded in Ro5 and PSA (Fig. **10**). These compounds were evaluated for efficacy against τ-aggregation *in vitro*, as well as for brain penetration *in vivo*. The results showed that the ATPZs are very promising candidates due to a favorable combination of *in vitro* and physico-chemical properties including excellent brain penetration and oral bioavailability.

Blood Brain Barrier: A Barrier to Cross

The blood brain barrier (BBB) is a separation of circulating blood from the brain extracellular fluid (BECF) in the central nervous system. The relative impermeability of the BBB results from tight junctions between capillary endothelial cells. The human BBB consists of approximately 100 billion capillaries estimated to be 640 km in length with a surface area comprising 20 m and individual neurons are normally no farther than 8–20 μm from a brain capillary. These statistics present a formidable and complex biological barrier for both the large and small molecular exchange in and out of CNS (Fig. **11**).

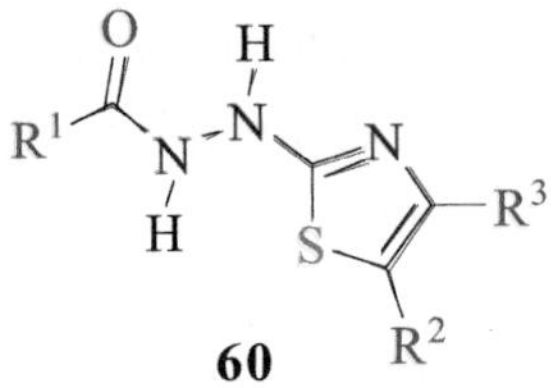

Structure **60**

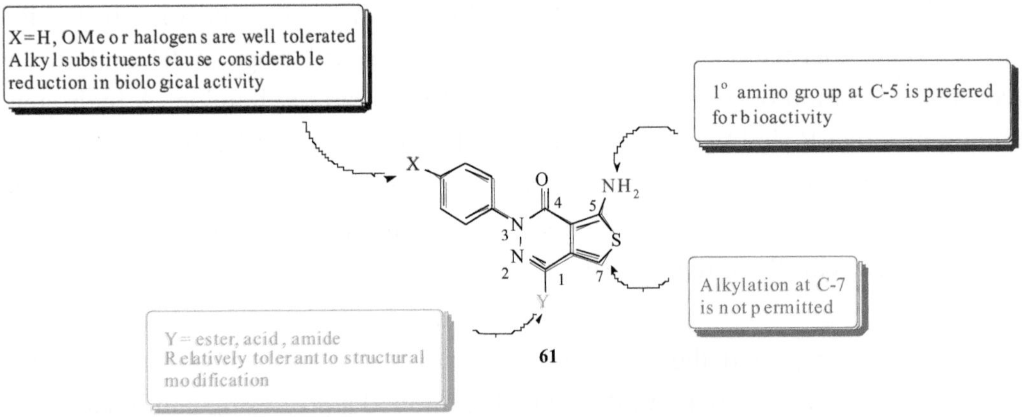

X=H, OMe or halogens are well tolerated
Alkyl substituents cause considerable reduction in biological activity

1° amino group at C-5 is prefered for bioactivity

Y= ester, acid, amide
Relatively tolerant to structural modification

Alkylation at C-7 is not permitted

Structure **61**

Figure 10: Summary of tau anti-fibrillization activity and SAR of ATPZs.

An effective drug design and development program for AD is hampered due to an effective delivery of a drug across BBB. Table **3** illustrates the BBB problem which lists the incompatibilities in the AD drug development process [349].

Owing to the decades of under-development of BBB drug delivery research and to the fact that only so few drugs cross the BBB, there are predictable failures of clinical trials in AD. The following approaches can be used to address the problem of crossing the BBB and to deliver therapeutics to brain [350].

BBB Disruption

Disruption of the BBB can open access of the brain to components in the blood by making the tight junction between the endothelial cells of the brain capillaries leaky. This has attracted many researchers to propose disrupting the BBB for the purposes of drug delivery. The problem in this strategy is that many of the neurotoxic endogenous substances will enter the brain from the blood, hence, disruption of the BBB in the delivery of therapeutics must be carefully controlled.

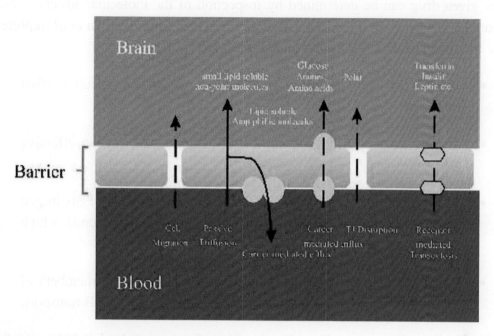

Figure 11: Transport mechanisms at the BBB.

Table 3:   Problems in AD drug development process due to BBB

• Almost 100% of all large molecule drugs, *i.e.*, the products of biotechnology, do not cross the BBB
• No big pharma in the world today has a BBB drug targeting program
• Even if a big pharma wanted to start a BBB drug targeting program, there would be few personnel trained in the BBB to hire, because no academic neuroscience program in the USA emphasizes BBB transport biology, much less BBB drug targeting.
• More than 99% of the drug development effort is devoted to CNS drug discovery
• Less than 1% of drug development is devoted to CNS drug delivery

Pharmacological Approach

The pharmacological approach to crossing the BBB is based on the observation that some molecules freely enter the brain. This ability to passively cross the BBB depends on molecular weight (<400 Daltons (Da), charge and lipophilicity. The lipid solubility of a drug has an inverse relationship to the number of hydrogen bonds that the drug forms with solvent water, which is 55 M. The H-bonding of any given drug can be determined by inspection of the molecular structure, and should be considered in tandem with the MW of the drug. Rishton *et al*. reported the following data on H-bonding of AD drugs [351].

- Current AChEIs form 3-5 H-bonds, which predict effective BBB transport.

- Widely used CNS drugs form 1-5 H-bonds, which predict effective BBB transport.

- γ-Secretase inhibitors for AD have higher degrees of hydrogen bonding and half of the drugs in this class form 8-10 H-bonds, which predicts restricted BBB transport.

- β-Secretase inhibitors for AD are the most polar and all members of this class form 8-12 H-bonds, which predicts restricted BBB transport.

A CNS drug discovery effort for the identification of lead compounds for inhibitors of either γ-secretase or β-secretase for AD will most likely end in termination. Although effective inhibitors will be identified, the polarity of these compounds will preclude effective brain penetration *in vivo*. Decreasing PSA has been another strategy to increase BBB penetration but this approach also requires a careful implementation. In general, PSA discriminates CNS penetrating compounds; increasing logP and minimizing PSA are used to improve brain uptake of small molecules.

Antibodies as Therapeutic Agents

Antibodies as therapeutics have been used in two main ways in AD: 1) as directly targeting pathologic agents and 2) as delivery vehicles. The strategies and the BBB interactions are very different for these two approaches. Since antibody target in AD has usually been Aβ. Animal studies show that active or passive immunization against Aβ can decrease plaque number and improve cognition.

Another proposal is that antibodies may act by binding Aβ in the circulation that would prevent circulating Aβ from contributing to brain levels of Aβ [352].

Transport Systems

The use of endogenous transport systems is an important, untapped strategy in drug delivery to the brain. The vascular BBB and blood-CSF barrier are both richly endowed with known transporters, it is estimated that the majority of BBB transporters have yet to be discovered [350].

Pharmaceutical Technology-Based Strategies

The technological strategies are valuable approaches for enhancing trans-cellular permeability of therapeutic agents and biomacromolecules across BBB. They are based on the use of nanosystems (colloidal carriers), mainly liposomes and polymeric nanoparticles even though other systems such as solid lipid nanoparticles, polymeric micelles and dendrimers are also being investigated.

Nanoparticles (NPs) are solid colloidal particles made of polymeric materials ranging in size from 1-1000 nm. This definition includes both nanocapsules, with a core-shell structure (a reservoir system), and nanospheres (a matrix system). NPs are used as carrier systems in which the drug is dissolved, entrapped, encapsulated, adsorbed or chemically linked to the surface [353]. Brambilla *et al.* suggested some nanocarrier systems to transport AChEIs beyond BBB [354]. Wilson *et al.* reported 3.8 and 4 fold increase in the brain delivery of Rivastigmine and Tacrine by exploiting polysorbated 80-coated polymeric Poly-n-butylcyanoacrylate (PnBCA) nanocapsules [355]. These PnBCA nanocarriers use the LDL receptors to transport through BBB. Yang *et al.* in 2010 introduced a nanocarrier system for delivery of ACh to lysosomes in the brain of mice model of AD through low doses of single-wall carbon nanotubes (SWCN) [356]. Due to their sufficient gastrointestinal absorption these SWCN can be administrated orally. As discussed earlier, metal ions are required for Aβ protein oligomerisation and recent studies show that metal chelators can produce a significant reversal of Aβ deposition both *in vitro* and *in vivo*. Therefore, an efficient metal chelation strategy may significantly decrease both the extracellular oxidative stress and Aβ aggregation, and therefore slow down the progression of AD. However, there are some problems associated with current metal chelators. These problems include difficulty in passing through BBB, risk of non-specific metal chelation from other tissues and pooling of metal ions into amyloid plaques. Nanotechnology approaches are used to study chelator therapeutic systems that overcome these problems [357]. Liu *et al.* reported an efficient chelator nanoparticle system (CNPS) containing conjugated Desferrioxamine, an FDA approved metal chelator, with nanoparticles (through an amide bond between a primary amino group in the chelator and a carboxyl group on the nanoparticle surface) [358]. Cui *et al.* examined the conjugation of D-penicillamine and an FDA approved drug for chelation of copper in Wilson's disease [359].

MULTI-TARGET-DIRECTED LIGAND (MTDL) DESIGN STRATEGY

The MTDL design strategy is used to develop single chemical entities that are able to simultaneously modulate multiple targets. The multifunctional nature of AD provides a logical foundation for the development of an innovative drug design strategy centered on multi-target-directed ligands (MTDLs). In recent years, the MTDL concept has been extensively exploited to design different ligands hitting different biological targets especially for diagnosing AD therapeutics. In preceding sections many approaches and targets have been discovered in the treatment of AD. However, important physiopathological aspects of AD remain unclear. Despite the recognition of many AD relevant biological entities, only four AChEIs and one NMDA receptor antagonist are used in therapy. Unfortunately, such compounds are not DMDs rather only cognition enhancers. Therefore, MTDL strategy is emerging as a powerful drug design paradigm. The starting point for multitarget approach was the discovery of multiple activities of natural products such as curcumin, resveratrol and some flavonoids. These compounds were studied due to their diverse antioxidant and anti-inflammatory properties. Further studies proved their ability to modify Aβ aggregation and metal dyshomeostasis. As described in section **1** there are two complementary approaches of generating multi-targeted lead compounds *i.e.*, knowledge-based (or pharmacophore combination approach) and screening approach. In this section we will summarize the rationality behind the basic design of multiple ligands for next generation of multi-target AD drugs. It is, therefore, appropriate to adopt both strategies in order to enhance the chances of success. The systematic combination of pharmacophoric features extracted from different ligands has many advantages and is currently the predominant technique for the generation of multi-target AD drugs. The degree of pharmacophoric overlap forms a continuum, with high-molecular weight conjugates at one extreme and simpler molecule with highly integrated pharmacophores at the other. A thorough survey of MTDLs designed for AD, revealed that the researchers generally aspire to join the pharmacophores by conjugation or by integrating the two pharmacophores *i.e.*, to maximize the degree of framework overlap (merged) in order to produce smaller and simpler molecules with favorable physicochemical properties.

Dual Binding Site AChEIs

The formation of conjugates of two structurally diverse scaffolds having pharmacophoric features of each target is a classical strategy for designing MTDL for AD. These molecules are structurally joined by a metabolically stable linker group that is not found in either of the selective ligands responsible each for a different biological response. This strategy takes advantage of the chelate effect to design a bifunctional ligand for AChEIs able to simultaneously bind to its dual binding sites *i.e.*, catalytic and peripheral binding. Several classes of dual binding site AChE inhibitors have been developed by joining a suitable linker to the two interacting units, which are generally derived from known AChE inhibitors either commercialized or under development. In past, before the 3D structure of AChE was known, decamethonium, BW284c51 and ambenonium displayed both high affinity and high selectivity for AChE.

Propidium (**62**) was the first reported compound to bind at PAS of AChE [360]. Bourne *et al.* reported the X-ray crystallographic structure of propidium binding at mAChE PAS. The phenanthridinium ring forms a stacking interaction with Trp286, while, the trimethylammoniodecyl moiety fits the shape of the loop Pro290- Phe299.

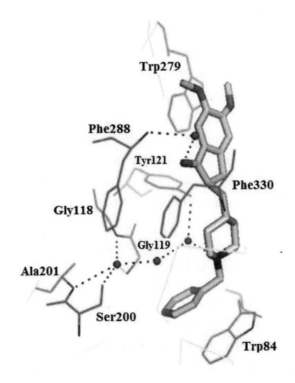

62

Early developments on dual binding site inhibitors were based on the reported structure of donepezil-AChE complex. X-ray structure showed that it interacted with both the anionic subsite at the bottom of the gorge (Trp84), and with the PAS, near its entrance (Trp279), *via* aromatic π -stacking interactions with conserved aromatic residues (Fig. **12**) [361].

Figure 12: Major interactions of donepezil and *Tc*AChE. H-bonds are shown as dashed lines; aromatic stacking and aryl ring H-bonds are shown as black lines.

Consequently attention was directed to design novel N-benzylpiperidine derivatives. In particular, attention has been focused on the replacement of the indanone moiety by a variety of chemical systems, such as dicarboxylic amino acid derivative, thiadiazolidinone, indole and pyrrole [362-364]. Having used rational designing strategies, da Silva *et al.* proposed molecular hybrids by fusion of tacrine with donepezil [365]. They analyzed all the structures by docking, density functional studies and drug-like properties (Fig. **13**).

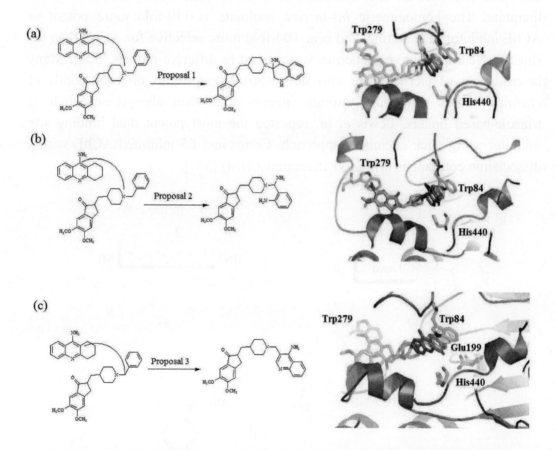

Figure 13: Different proposals of molecular hybridization of tacrine with donepezil. **a)** Superposition of Proposal 1 (carbon atoms in yellow) and the structures of tacrine (carbon atoms in magenta), donepezil (carbon atoms in cyan). **b)** Superposition of Proposal 2 (carbon atoms in light orange) and the structures of tacrine (carbon atoms in magenta), donepezil (carbon atoms in cyan). **c)** Superposition of the Proposal 3 (carbon atoms in orange) and the structures of tacrine (carbon atoms in magenta), donepezil (carbon atoms in cyan).

The structure of tacrine, the first marketed drug for AD, has been widely used as scaffold to provide new MTDLs. The general structure (**63**) of tacrine dimers was used to design many bifunctional analogues of THA. Pang *et al*. designed and synthesized bifunctional analogues of THA using computer modeling of ligand docking with the target protein. The strategy was to connect two THA molecules with an alkylene chain spaced to allow simultaneous binding at the catalytic and peripheral sites. The best results were found with heptylene-linked THA dimer known as *bis*(7)-tacrine (**64**) and was one of the first homodimers reported in the literature. The homodimeric *bis*-tacrine analogue is 149-fold more potent as AChE inhibitor than tacrine, and near 100-fold more selective for AChE than for BuChE. Neuroprotective properties were tested in different models [366]. Many heterodimers of tacrine were also developed linking it to different kinds of chemical motifs sometimes through spacers other than alkyl-chains such as triazole-based linkers. Lewis *et al*. reported the most potent dual binding site inhibitor using click chemistry approach. Compound **65** inhibited AChE with a dissociation constant as low as 77 femtomolar (fM) [367].

Numerous studies have been performed with the aim to exploit the known AChEIs acting at the catalytic site to develop new dual-site inhibitors. These include tacrine-donepezil heterodimers, galanthamine-based *bis*- ligands, Huperzine-A based *bis*-ligands and huprine-tacrine heterodimer [368-375].

Nevertheless, the classification of this dual binding site class of compounds as MTDLs needs some observation. As these inhibitors bind to more than one site along the AChE gorge, they may be termed as "multisite" inhibitors not as MTDLs. However, experiments have proved that multisite inhibitors also have the ability to target Aβ aggregation and other neurotransmitter systems such as MAO, SERT, 5-HT receptor and Ca^{2+} modulators [376].

AChEIs Targeting other Neurotransmitter Systems: MTDLs

Dual Binding Site AChE Targeting Aβ Aggregation

In a study carried out by Inestrosa *et al.* it was pointed out that AChE may also accelerate Aβ deposition, and this noncatalytic action might be facilitated by the interaction of Aβ with residues located in the vicinity of the PAS. All these results support the hypothesis that AChEIs (acting on PAS or on both sites) may have a combined effect of cholinergic neurotransmission enhancement and reduction in Aβ aggregation, thereby, opening new avenues to develop promising therapeutic agents [377, 378].

Caproctamin (**67**), a polyamine based dual inhibitor of AChE designed from benextramine (**66**), represents another example of a successfully designed AChEI. Caproctamine was able to inhibit AChE with IC_{50} = 0.17 μM blocked muscarinic receptor (Kb=0.06 μM), together with PAS inhibition preventing the aggregation of Aβ induced by AChE. The structure of caroctamin was developed through structural manipulations of tetraamine disulfide benextramine (**66,** Fig. **14**), which was initially developed as an irreversible α-adrenoreceptor antagonist and later as a muscarinic M_2 receptor antagonist [379-381].

Piazzi *et al.* developed AP2238 (**68**) an interesting molecule based on the study of crystal structures of different AChE–ligand complexes and molecular modeling studies. This compound was purposely designed to bind at both catalytic and PAS sites of AChE (Fig. **15**). A benzylamino group and a coumarin (2*H*-2-chromene) moiety were selected from previously developed AChEIs and joined with a linker to enable interaction *via* the residues at the gorge of the enzyme. AP2238 is a good AChEI as well as possesses good antiaggregating properties (35% inhibition at 100 mM) [382].

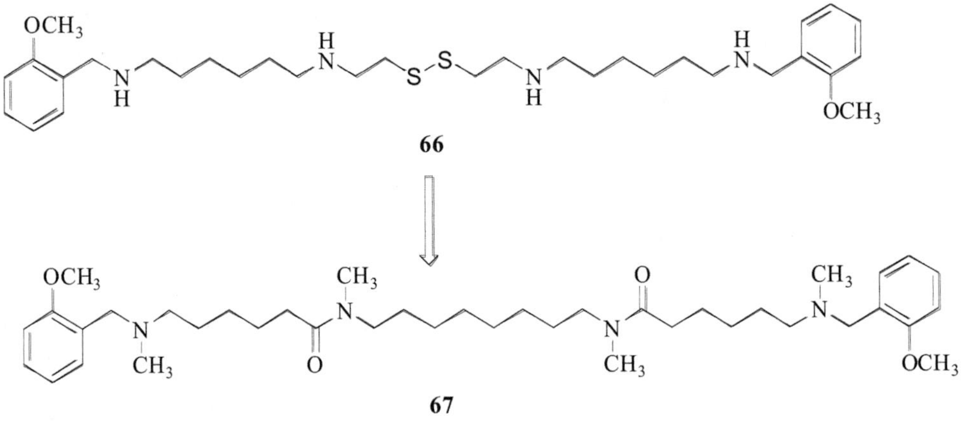

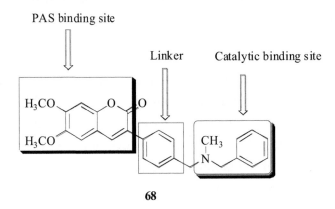

Figure 14: Design strategy leading to caproctamin as MTDL.

Figure 15: Design strategy leading to **68** that effectively inhibited AChE-induced Aβ-aggregation by interacting with both AChE catalytic site and PAS.

Cavalli *et al.* reported the synthesis and pharmacological activity of memoquin **(69)** which is an AChEI with other interesting properties including muscarinic antagonism and Aβ antiaggregating effects [383]. Its design was based on the existing polyamine core [384, 385]. Besides, a 1,4-benzoquinone fragment from coenzyme Q10 (CoQ10) was also introduced in the core [386]. The resulting compound was found to inhibit AChE-induced Aβ aggregation with an $IC_{50}=28.3$ mM and the self-induced $Aβ_{1-42}$ aggregation with an $IC_{50}=5.93$ mM revealing the ability of memoquin to interact with Aβ directly. Moreover, memoquin was also found to inhibit BACE-1 with an $IC_{50}=108$ nM. Finally, its antioxidant properties were investigated and it was found to reduce the formation of free radicals by 44.1% (Fig. **16**) [383].

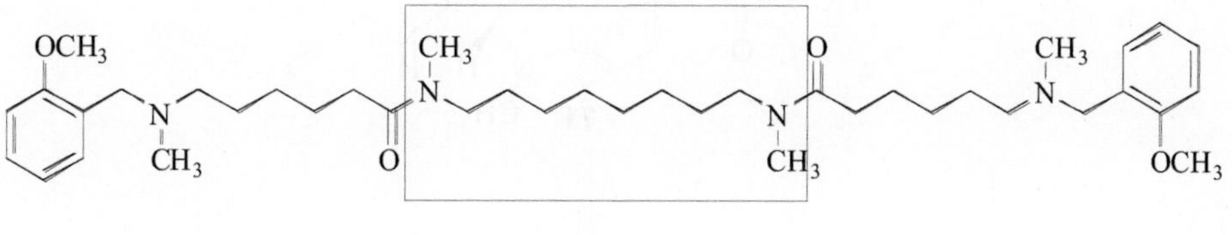

Figure 16: Design strategy leading to the MTDL memoquin.

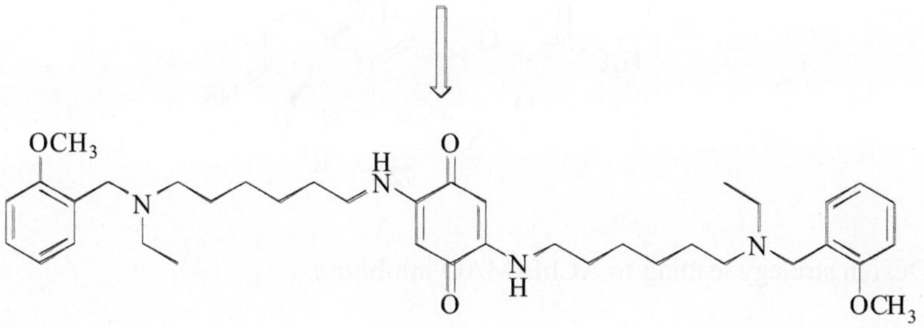

Dual AChE-MAO Inhibitors

The initial design strategy for dual AChE-MAO inhibitors resulted in a series of novel inhibitors (Fig. **17**). This strategy was based on structural modification of physostigmine (**1**) by incorporation of the propargylamine pharmacophore of selegiline (**71**) [387].

Exploiting the pharmacophore overlapping approach a highly integrated Ladostigil (**73**) was designed by combining ChE inhibitory moiety *i.e.*, carbamate found in the rivastigmine molecule with the pharmacophore of rasagiline (**74**, Fig. **18**). Ladostigil is a drug with neuroprotective, multimodal brain-selective inhibitor of both MAO and AChE and has reached Phase-II clinical trials [388, 389].

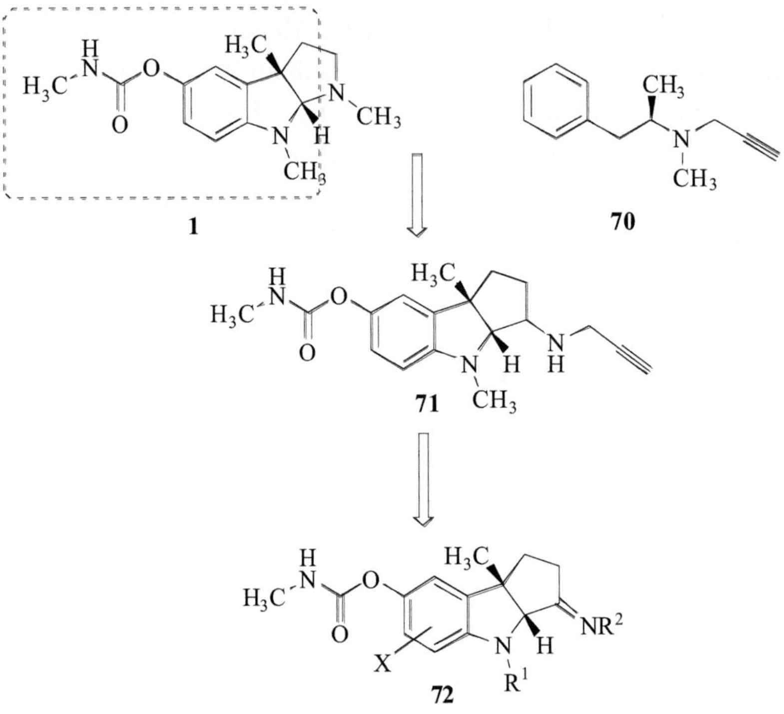

Figure 17: Design strategy leading to AChE-MAO inhibitors.

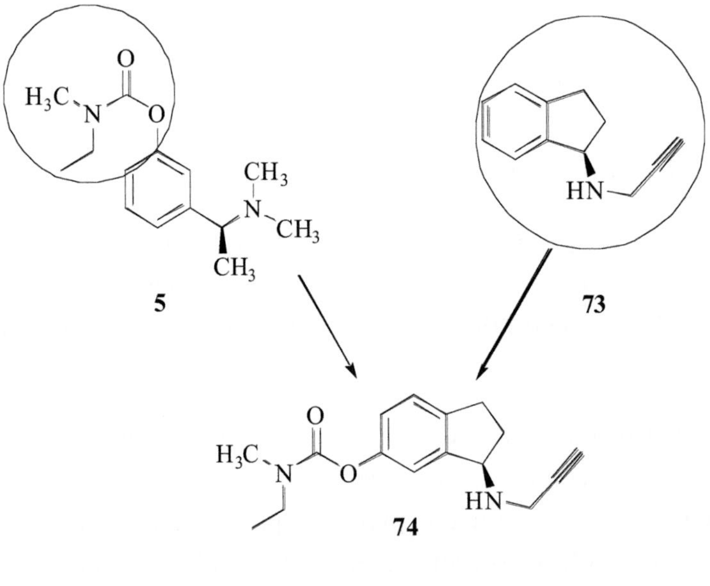

Figure 18: Design strategy leading to multifunctional anti-Alzheimer drug, R-Ladostigil.

AChEIs and Calcium Channel Blockers

The pivotal role of altered calcium homeostasis in the pathogenesis of AD and other age-related dementias is well-supported. Therefore, identification of new compounds that combine both a moderate calcium channel blockade effect and AChE inhibition has been advanced to be useful in the treatment of AD. As a result, tacripyrine (**76**) was developed as MTDL by using tacrine (AChEI) with nimodipine (**75**, Fig. **19**). These hybrids are selective and potent AChE inhibitors with the activity in nanomolar range (IC$_{50}$ = 45 nM) [390].

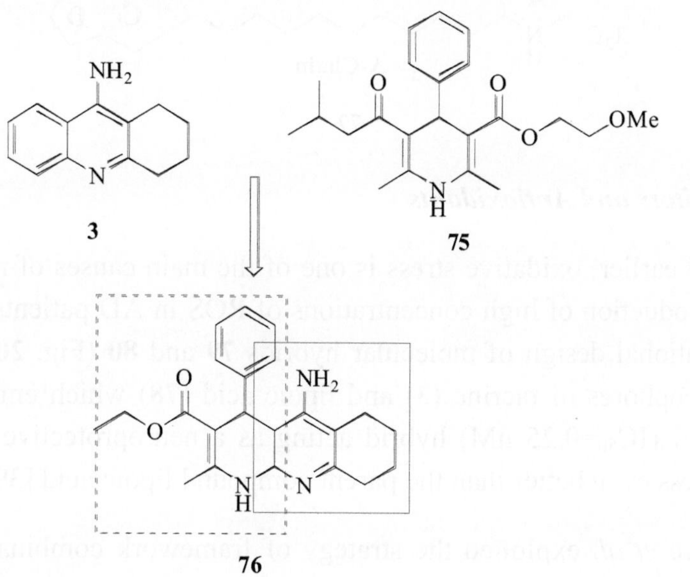

Figure 19: Design strategy of tacripyrines.

Choudhary *et al.* reported juliflorine (**77**), an alkaloid from *Prosopis juliflora* that inhibited both AChE and BuChE in a concentration-dependent fashion with IC$_{50}$ values 0.42 and 0.12 μM, respectively. Moreover, it also showed dose-dependent (30–500 μg/mL) spasmolytic and Ca^{2+}-channel blocking activities in isolated rabbit jejunum preparations [391].

Atta-ur- Rahman *et al.* reported a coumarin and three quinoline alkaloids isolated from the aerial parts of *Skimmia laureola*. One of the compounds isolated showed a dose-dependent (30–500 μg/mL) spasmolytic and Ca^{2+}-channel blocking activities in isolated rabbit jejunum preparations by relaxing the spontaneous (EC_{50}=0.1 mg/mL) and K^+ induced contractions (EC_{50}=0.4 mg/mL), suggesting that the spasmolytic effect of compound is mediated through the blockade of voltage-dependent Ca^{2+} channels [392].

77

AChE Inhibitors and Antioxidants

As discussed earlier, oxidative stress is one of the main causes of neuronal death due to the production of high concentrations of ROS in AD patients. Rosini *et al.* reported a rational design of molecular hybrids **79** and **80** (Fig. **20**) by coupling two pharmacophores of tacrine (**3**) and lipoic acid (**78**) which emerged as most potent AChEI (IC_{50}=0.25 nM) hybrid acting as a neuroprotective agent against oxidative stress even better than the parent compound lipoic acid [393].

In 2009, Arce *et al.* exploited the strategy of framework combination and used dicarboxylic amino acids to investigate a new family of derivatives inspired by compound **81**. They reported L-glutamic acid as a linker for three groups (a) A *ω*-situated N-benzylpiperidine moiety able to bind to the catalytic active site of AChE, based on the structure of donepezil (b) a N-protecting group able to interact with the PAS of AChE proposed to inhibit Aβ-aggregation (c) a lipophilic α-hexyl ester that could favor the crossing of the BBB [394, 395].

Dual AChE and SERT inhibitors

The depression in AD patients has been successfully treated with serotonin transporter (SERT) inhibitors that lack anticholinergic action. However, a combination therapy of SERT and AChE inhibitors produce adverse effects due to the difficulty in managing their pharmacodynamics and pharmacokinetic properties besides causing harmful drug–drug interactions. To overcome these problems, a balanced dual AChE and SERT inhibition is considered to be a promising dual-target approach. Hirai *et al.* reported TAK-147 (**82**) as potent AChE and SERT inhibitor with $IC_{50}=51$ nM and $IC_{50}=1350$ nM respectively [396].

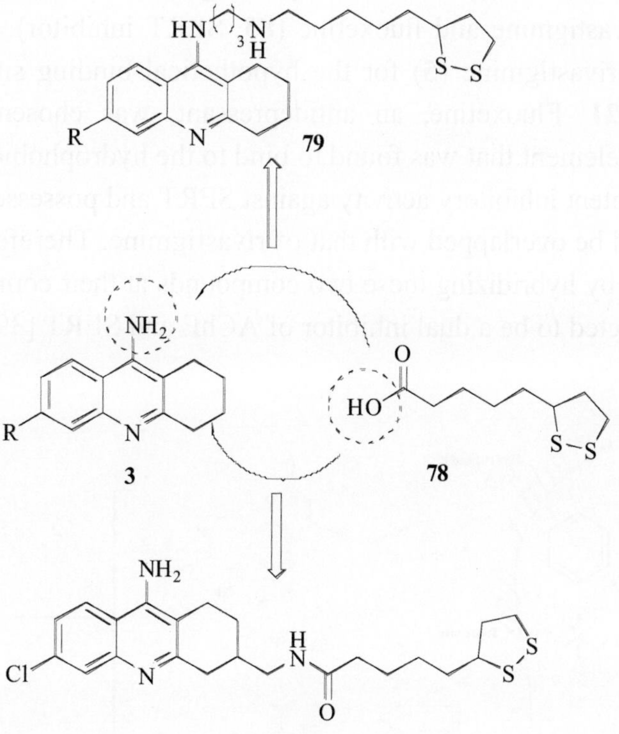

Figure 20: MTDLs **79** and **80** derived from tacrine and lipoic acid.

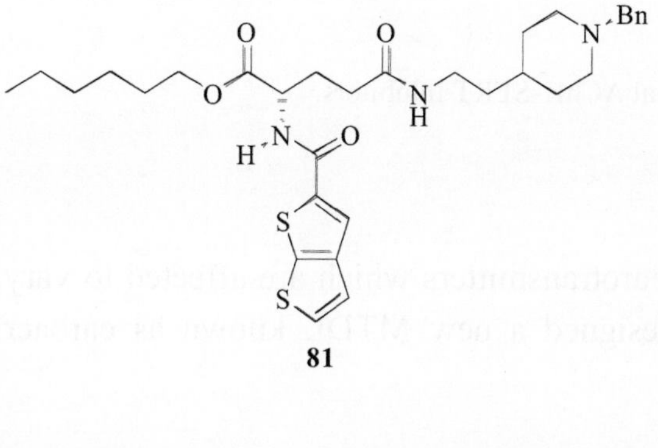

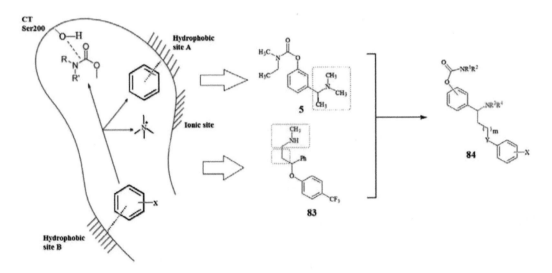

82

A design strategy was proposed for dual AChE-SERT inhibitors (based on a hypothetical model), built by means of crystallographic and molecular modeling studies, using rivastigmine and fluoxetine (**83**, SERT inhibitor). There are three binding sites of rivastigmine (**5**) for the hypothetical binding sites of AChE as shown in Fig. **21**. Fluoxetine, an antidepressant, was chosen as the fourth pharmacophoric element that was found to bind to the hydrophobic binding site B. Fluoxetine has potent inhibitory activity against SERT and possesses an ethylamine moiety that could be overlapped with that of rivastigmine. Therefore, a compound **84** was designed by hybridizing these two compounds at their common ethylamine moiety was expected to be a dual inhibitor of AChE and SERT [397-399].

Figure 21: Design of dual AChE–SERT Inhibitors.

Miscellaneous

There are several neurotransmitters which are affected to varying degrees in AD. Melchiorre *et al*. designed a new MTDL known as carbacrine (**85**) [400]. Its design strategy was based on a vasodilating β-blocker and antioxidant drug carvedilol (**86**) [401, 402]. The tetrahydroacridine moiety from tacrine and the carbazole core from carvedilol were linked. This resulting molecule was expected to inhibit Aβ fibril formation, since carbazols are efficient inhibitors of Aβ aggregation. In addition to the multiple activities, **85** displayed higher potency to inhibit NMDAR than **86** [403]. A bifunctional molecule XH1 (**87**) was designed by using pharmacophore conjugation strategy. This new MTDL has been reported as a metal-chelating agent that specifically targets Aβ. Compound **87** contains in its structure one metal-chelating and two amyloid-binding moieties [404].

MTDL approach has been used to describe the benefits of the use of combination of drugs and is now widely accepted in the treatment of complex diseases such as cancer, where the use of several one-target specific drugs in combination leads to better results. This effect can be explained by the alteration of several interconnected pathological pathways modifying the progression of the disease. This type of therapy has opened a broad line of research for scientists in different fields.

CONCLUDING REMARKS & FUTURE PROSPECTS

Alzheimer's disease is a progressive, neurodegenerative disorder of the brain. It is the most common cause of senile dementia characterized by deterioration in the cognitive functions. Worldwide, nearly thirty six million people are believed to be living with AD or other dementias, estimated to increase to nearly 66 million by 2030 and more than 115 million by 2050 *i.e.*, one in eighty five persons, globally, will be living with this devastating disease. AD is characterized by the presence of insoluble amyloid-β plaques and neurofibrillay tangles which are the key pathological features of the disease. A major therapeutic strategy, adopted to combat AD, is the symptomatic therapy which is based primarily on the cholinergic hypothesis targeting AChE, thereby, degrading acetylcholine in nerve synapse. Many AChE inhibitors (*e.g.*, tacrine, physostigmine, donepzil, rivastigmine, galanthamine *etc.*) and NMDA receptor antagonist, memantine, were the approved drug treatments for AD which were the outcome of the results of serendepitous as well as rational drug designing protocols. These AD therapeutics, however, do not interfere with the neurodegenerative cascade of the progression and counteraction of the disease rather provide only symptomatic short term benefits.

During the last decade, the efforts in the drug discovery paradigm have shifted from symptomatic to disease-modifying drugs (DMDs) which are expected to counteract the slow progressing pathological process of AD. The DMDs developed so far include the drugs that reduce β -amyloid production, prevention of Aβ aggregation, promotion of Aβ clearance and prevention of τ phosphorylation. Unfortunately, none of these drugs has demonstrated efficacy in Phase-III trials. The failure in such drug discovery endeavors may be ascribed either to flaws in fundamental understanding of the mechanism of heterogenic AD pathophysiology, or more importantly, time course of treatment in relation to the development of the disease.

The drugs designed to act against individual molecular targets cannot usually succeed in dealing with complex and mutagenic therapeutic areas such as cancer, diabetes, depression or AD. Hence the drug discovery paradigm shifted from monotherapeutics to combination therapeutics. One such approach, already used to combat AD, is to prescribe several individual drugs *e.g.*, AChEIs in combination with NMDA receptor antagonists and was found to be promising in achieving better symptomatic effects. Alternatively, two or more active ingredients may be delivered in a single pill or capsule. The major problem with both these strategies is that each drug or ingredient needs to be safe and efficient, both individually and in combination. Both these problems, however, could be addressed by designing a single compound with selective polypharmacology *i.e.*, the designing of multi-target-directed ligands (MTDLs) as molecular hybrids. Thereafter, many such synthetic hybrids were designed possessing properties covering various AD-related mechanisms, for example, a simultaneous modulation of AChEIs, antioxidants and APP metabolisms. Many MTDLs are in preclinical trials indicating a shift from a linear *one drug one target* approach to a focus on organelles (*e.g.*, mitochondria) and MTDLs. So far some promising results have been obtained in terms of symptomatic treatment of AD, however, several failures for disease-modifying AD drugs have also been reported. The disappointing results of many RCTs may be associated either with errors in drug choice and development programs or with the designing of selective compounds without undesirable side effects.

Research on pharmacodynamics, biological aspects, or regulatory mechanisms of therapeutic targets is ongoing and it is anticipated to improve drug safety and efficacy. At present, the focus of AD drug development program is redirected from treatment to prevention. The design of future strategy examines the potential neuroprotective activity of DMDs in the presymptomatic stages of AD. Since, an early start of treatment enhances the chance of success, it is more important to have biomarkers that predict disease progression before clinical onset of the disease helping in an early detection of AD-related dysregulation of biological pathways.

ACKNOWLEDGEMENTS

Declared none.

CONFLICT OF INTEREST

The authors confirm that this chapter contents have no conflict of interest.

ABBREVIATIONS

ACh	=	Acetylcholine
Aβ	=	Amyloid-β

APP	=	Amyloid precursor protein
ATPZ	=	Aminothienopyridazines
AD	=	Alzheimer's disease
ALS	=	Amyotrophic lateral sclerosis
ADI	=	Alzheimer's disease international
BBB	=	Blood brain barrier
BuChE	=	Butyrylcholinesterase
BACE-1	=	β-site APP-cleaving enzyme
BMY	=	Bristol Myers Squibb company
BECF	=	Brain extracellular fluid
CADD	=	Computer aided drug design
CNS	=	Central nervous system
ChEIs	=	Cholinesterase inhibitors
COX-1	=	Cyclooxygenase-1
COX-2	=	Cyclooxygenase-2
CAS	=	Catalytic anionic subsite
3D	=	Three-dimensional
DMDs	=	Disease-modifying drugs
DCM	=	Dichloromethane
DFO	=	Desferioxamine

Da	=	Daltons
EMA	=	European medicines agency
FDA	=	Food and drug administration
FBDD	=	Fragment-based drug discovery
GS	=	γ-Secretase
GSI	=	γ-Secretase inhibitor
GSM	=	γ-Secretase modulator
HTS	=	High-throughput screening
HD	=	Huntington's disease
Hup-A	=	Huperzine-A
Hup-B	=	Huperzine-B
HBD	=	Hydrogen bond donor
HBA	=	Hydrogen bond acceptor
HE	=	Hydroxyethylene
HEA	=	Hydroxyethylamine
HMC	=	Hydroxymethylcarbonyl
LBDD	=	Ligand based drug design
MCM	=	Multiple-compound medication
MMT	=	Multiple-medication therapy
MTDLs	=	Multi-target-directed ligands

MW	=	Molecular mass
MPS	=	Multiple parallel synthesis
MAP	=	Microtubule-associated protein
NSAIDs	=	Non-steroidal anti-inflammatory drugs
NGF	=	Nerve growth factor
NDs	=	Neurodegenerative diseases
NFT	=	Neurofibrillary tangle
NMDA	=	N-methyl-D-aspartate
nAChRs	=	Nicotinic acetylcholine receptors
NMDAR	=	NMDA receptor
OMRF	=	Oklahoma medical research foundation
PAS	=	Peripheral anionic site
PPAR-gamma	=	Proliferator-activated receptor-gamma
PD	=	Parkinson's disease
PHFs	=	Paired-helical filaments
PSA	=	Polar surface area
PDB	=	Protein data bank
RAGE	=	Receptor for advanced glycation endproducts
QSAR	=	Quantitative structure–activity relationships
RCTs	=	Randomized controlled trials

Ro5	=	Rule of five
SAR	=	Structure–activity relationships
SBDD	=	Structure based drug design
TcAChE	=	*Torpedo californica*
TSI	=	Transition-state isostere

REFERENCES

[1]   Cramer, P.E.; Cirrito, J.R.; Wesson, D.W.; Lee, C.Y.; Karlo, J.C.; Zinn, A.E.; Casali, B.T.; Restivo, J.L.; Goebel, W.D.; James, M.J.; Brunden, K.R.; Wilson, D.A.; Landreth, G.E. ApoE-directed therapeutics rapidly clear β-amyloid and reverse deficits in AD mouse models. *Science*, **2012**, *335*(6075), 1503-1506.

[2]   Sneader, W. Drug prototypes and their exploitation. *John Wiley & Sons. Chichester*, **1996**.

[3]   Böhm, H.J.; Klebe, G.; Kubinyi, H.; Wirkstoffdesign. *Spektrum Akademischer Verlag, Heidelberg*, **1996**.

[4]   Goodwin, L.G. New drugs for old diseases. *Trans. R. Soc. Trop. Med. Hyg.*, **1980**, *74*, 1-7.

[5]   Kubinyi, H. Chance favors the prepared mind-from serendipity to rational drug design. *J. Recept. Signal Transduct Res.*, **1999**, *19*(1-4), 15-39.

[6]   Riethmiller, S. From Atoxyl to Salvarsan: Searching for the magic bullet. *Chemotherapy*, **2005**, *51*, 234-242.

[7]   Winau, F.; Westphal, O.; Winau, R. Paul Ehrlich: In search of the magic bullet. *Microbes Infect.*, **2004**, *6*, 786-789.

[8]   Schwartz, R.S. Paul Ehrlich's magic bullets. *N. Engl. J. Med.*, **2004**, *350*, 1079-1080.

[9]   Lyne P.D. Structure-based virtual screening: an overview. *Drug Discov. Today*, **2002**, *7*, 1047-1055.

[10]   Joseph-McCarthy, D. Computational approaches to structure based ligand design. *Pharmacol. Ther.*, **1999**, *84*, 179-191.

[11]   Shoichet, B.K. Virtual screening of chemical libraries. *Nature*, **2004**, *432*, 862-865.

[12]   Lengauer, T.; Lemmen, C.; Rarey, M.; Zimmermann, M. Novel technologies for virtual screening. *Drug Discov. Today*, **2004**, *9*, 27-34.

[13]   Oprea, T.I. Virtual screening in lead discovery: A viewpoint. *Molecules*, **2002**, *7*, 51-62.

[14]   Walters, W.P.; Stahl, M.T.; Murcko, M.A. Virtual screening - an overview. *Drug Discov. Today*, **1998**, *3*, 160-178.

[15]   Sotriffer, C. Virtual Screening: Principles, Challenges, and Practical Guidelines. Wiley, **2011**.

[16]   Seifert, M.H.J.; Kraus, J.; Kramer, B. Virtual high-throughput screening of molecular databases. *Curr. Opin. Drug Discov. Dev.*, **2007**, *10*, 298-307.

[17]   Morphy, R.; Rankovic, Z. Designed multiple ligands. An emerging drug discovery paradigm. *J. Med. Chem.*, **2005**, *48*, 6523-6543.

[18]   Morphy, R.; Rankovic, Z. Fragments, network biology and designing multiple ligands. *Drug Discov. Today*, **2007**, *12*, 156-160.

[19]   Morphy, R.; Kay, C.; Rankovic, Z. From magic bullets to designed multiple ligands. *Drug Discov. Today*, **2004**, *9*, 641-651.

[20]   Bishop, T.; Sham, P. Analysis of Multifactorial Diseases. *Academic Press New York*, **2000**, 1-320.

[21]   Korcsmáros, T.; Szalay, M.S.; Böde, C.; Kovács, I.A.; Csermely, P. How to design multi- target drugs, target search options in cellular networks. *Expert Opin. Drug Disc.*, **2007**, *2*, 1-10.

[22]   Cavalli, A.; Bolognesi, M.L.; Minarini, A.; Rosini, M.; Tumiatti, V.; Recanatini, M.; Melchiorre, C. Multi-target-directed ligands to combat neurodegenerative diseases. *J. Med. Chem.*, **2008**, *51*, 347-372.

[23]   Choonara, Y.E.; Pillay, V.; DuToit, L.C.; Modi, G.; Naidoo, D.; Ndesendo, V.M.; Sibambo, S.R. Trends in the molecular pathogenesis and clinical therapeutics of common neurodegenerative disorders. *Int. J. Mol. Sci.*, **2009**, *10*, 2510-2557.

[24]   Leo´, R.; Garcia, A.G.; Marco-Contelles, J. Recent Advances in the Multitarget-Directed Ligands approach for the treatment of Alzheimer's Disease. *Med. Res. Rev.*, **2011**, 1-51.

[25]   Popovic, N.; Brundin, P. Therapeutic potential of controlled drug delivery systems in neurodegenerative diseases. *Int. J. Pharm.*, **2006**, *314*, 120-126.

[26]   Nowacek, A.; Kosloski, L.M.; Gendelman, H.E. Neurodegenerative disorders and nanoformulated drug development: Executive summary. *Nanomedicine*, **2009**, *4*, 541-555.

[27]   Galvin, J.E.; Pollack, J.; Morris, J.C. Clinical phenotype of Parkinson's disease dementia. *Neurology*, **2006**, *67*(9), 1605-1611.

[28]   Welsh-Bohmer K.A.; Warren, L.H. Neurodegenerative dementias. In: Attix DK, Welsh-Bohmer KA, eds. Geriatric neuropsychology: Assessment and Iintervention. New York: Guilford Press; **2006**, 56-88.

[29]   Lage, J.M.M. 100 years of Alzheimer's disease (1906-2006). *J. Alzheimer's Dis.*, **2006**, *9*, 15-26.

[30]   Zilka, N.; Novak. M. The tangled story of Alois Alzheimer. *Bratisl. Lek. Listy.*, **2006**, *107*, 343-345.

[31]   Berchtold, N.C.; Cotman, C.W. Evolution in conceptualization of dementia and Alzheimer's disease: Greco-Roman period to the 1960s, *Neurobiol. Aging*, **1998**, *19*, 173-189.

[32]   Maurer, K.; Volk, S.; Gerbaldo, H. Auguste, D. and Alzheimer's disease. *Lancet,* **1997**, *349*, 1546-49.

[33]   Alzheimer's Association. Alzheimer's disease facts and figures. Alzheimer's and dementia. *J. Alzheimer's Association*, **2012**, *8*, 131-168.

[34]   Alzheimer's Disease Facts and Figures, Alzheimer's Association **2012** (www.alz.org/facts)

[35]   Media Workshop, **2012**. Alzheimer's Pakistan. http://www.alz.org.pk

[36] Giacobini, E. In.; Meyer, E.; Simpkins, J.W.; Yamamoto, J.; Crews, F.T. Eds. Treatment of dementias: A new generation of progress. *Advances in Behavioural Biology. New York, Plenum Press*, **1992**, *40*, 19-34.

[37] Francis, P.T.; Palme, A.M.; Snape, M.; Wilcock, G.K. The cholinergic hypothesis of Alzheimer's disease: A review of progress. *J. Neurol. Neurosurg. Psychiatry*, **1999**, *66*, 137-147.

[38] Bartus, R.; Dean, R.; Beer, B.; Lippa, A. The cholinergic hypothesis of geriatric memory dysfunction. *Science*, **1982**, *217*, 408-414.

[39] Struble, R.; Cork, L.; Whitehouse, P.; Price, D. Cholinergic intervation in neuritic plaques. *Science*, **1982**, *216*, 413-415.

[40] Cummings, J.; Back, C. The cholinergic hypothesis of neuropsychiatric symptoms in Alzheimer's disease. *Am. J. Geriatr. Psychiatry*, **1998**, *6-S*, 64-78.

[41] Cooper, J.R.; Bloom, F.E.; Roth, R.H. The biochemical basis of neuropharmacology. *Oxford: Oxford University Press*. 7th ed. **1996**.

[42] Hardy, J.; Allsop. D. "Amyloid deposition as the central event in the aetiology of Alzheimer's disease". *Trends Pharmacol. Sci.*, **1991**, *12*(10), 383-388.

[43] Carter, M.D.; Simms, G.A.; Weaver, D.F. the Development of New therapeutics for Alzheimer's Disease. *Clin. Pharmacol. Ther.*, **2010**, *88*, 475-486.

[44] Hardy, J. The amyloid hypothesis for Alzheimer's disease: A critical reappraisal. *J. Neurochem.*, **2009**, *110*, 1129-1134.

[45] Haass, C.; Schlossmacher, M. G.; Hung, A.Y.; Vigo-Pelfrey, C.; Mellon, A.; Ostaszewski, B.L.; Lieberburg, I.; Koo, E.H.; Schenk, D.; Teplow, D.B. Amyloid beta-peptide is produced by cultured cells during normal metabolism. *Nature,* **1992**, *359*, 322-325.

[46] Shoji, M.; Golde, T. E.; Ghiso, J.; Cheung, T.T.; Estus, S.; Shaffer, L.M.; Cai, X.D.; McKay, D.M.; Tintner, R.; Frangione, B.; Production of the Alzheimer,s amyloid beta protein by normal proteolytic processing. *Science*, **1992**, *258*, 126-129.

[47] Seubert, P.; Vigo-Pelfrey, C.; Esch, F.; Lee, M.; Dovey, H.; Davis, D.; Sinha, S.; Schlossmacher, M.; Whaley, J.; Swindlehurst, C. Isolation and quantification of soluble Alzheimer's beta-peptide from biological fluids. *Nature*, **1992**, *359*, 325-327.

[48] Querfurth, H.W.; LaFerla, F.M. Alzheimer's disease. *N. Engl. J. Med.* **2010**, *362*, 329-44.

[49] Cappai, R.; Barnham, K.J. Delineating the mechanism of Alzheimer's disease A beta peptide neurotoxicity. *Neurochem Res.*, **2008**, *33*, 526-532.

[50] Giuffrida, M.L.; Caraci, F.; Pignataro, B.; Cataldo, S.; De Bona, P.; Bruno, V.; Molinaro, G.; Pappalardo, G.; Messina, A.; Palmigiano, A.; Garozzo, D.; Nicoletti, F.; Rizzarelli, E.; Copani, A. Beta-amyloid monomers are neuroprotective. *J. Neurosci.*, **2009**, *29*, 10582-10587.

[51] Shankar, G.M.; Mehta, T.H.; Garcia-Munoz, M.; Shepardson, N.E.; Smith, I.; Brett, F.M.; Farrell, M.A.; Rowan, M.J.; Lemerel, C.A.; Regan, C.M.; Walsh, D.M.; Sabatini, B.L.; Selkoe, D.J. Amyloid-beta protein dimers isolated directly from Alzheimer's brains impair synaptic plasticity and memory. *Nat. Med.*, **2008**, *14*, 837-842.

[52] Cerpa, W.; Dinamarca, M.C.; Inestrosa, N.C. Structure-function implications in Alzheimer's disease: Effect of Abeta oligomers at central synapses. *Curr. Alzheimer Res.*, **2008**, *5*, 233-243.

[53]   Weingarten, M.D.; Lockwood, A.H.; Hwo, S.Y.; Kirschner, M.W. A protein factor essential for microtubule assembly. *Proc. Natl. Acad. Sci. USA*, **1975**, *72*(5), 1858-1862.

[54]   Wood, J.G.; Mirra, S.S.; Pollock, N.J.; Binder, L.I. Neurofibrillary tangles of Alzheimer disease share antigenic determinants with the axonal microtubule-associated protein tau (tau). *Proc. Natl. Acad. Sci. USA*, **1986**, *83*(11), 4040-4043.

[55]   Kosik, K.S.; Joachim, C.L.; Selkoe, D.J. Microtubule-associated protein tau (tau) is a major antigenic component of paired helical filaments in Alzheimer's disease. *Proc. Natl. Acad. Sci. USA*, **1986**, *83*(11), 4044-4048.

[56]   Grundke-Iqbal, I.; Iqbal, K.; Tung, Y.C.; Quinlan, M.; Wisniewski, H.M.; Binder, L.I. Abnormal phosphorylation of the microtubule-associated protein tau (tau) in Alzheimer cytoskeletal pathology. *Proc. Natl. Acad. Sci. USA*, **1986**, *83*(13), 4913-4917.

[57]   Iqbal, K.; Grundke-Iqbal, I. Alzheimer neurofibrillary degeneration: Significance, etiopathogenesis, therapeutics and prevention. *J. Cell. Mol. Med.*, **2008**, *12*, 38-55.

[58]   Markesbery, W.R. Oxidative stress hypothesis in Alzheimer's disease. *Free Radic. Biol. Med.* **1997**, *23*,134-147.

[59]   Harman, D. The aging process. *Proc. Natl. Acad. Sci. USA*, **1981**, *78*, 7124-7128.

[60]   Dalle-Donne, I.; Giustarini, D.; Colombo, R.; Rossi, R.; Milzani, A. Protein carbonylation in human diseases. *Trends Mol. Med.,* **2003**, *9*, 169-176.

[61]   Mariani, E.; Polidori, M.C.; Cherubini, A.; Mecocci, P. Oxidative stress in brain aging, neurodegenerative and vascular diseases: An overview. *J. Chromatogr. B Analyt. Technol. Biomed. Life Sci.,* **2005**, *827*, 65-75.

[62]   Praticò, D. Peripheral biomarkers of oxidative damage in Alzheimer's disease: The road ahead. *Neurobiol. Aging*, **2005**, *26*, 581-583.

[63]   Sayre, L.M.; Moreira, P.I.; Smith, M.A.; Perry, G. Metal ions and oxidative protein modification in neurological disease. *Ann. Ist. Super. Sanita.*, **2005**, *41*, 143-164.

[64]   Nunomura, A.; Perry, G.; Pappolla, M.A.; Wade, R.; Hirai, K.; Chiba, S.; Smith, M.A. RNA oxidation is a prominent feature of vulnerable neurons in Alzheimer's disease. *J. Neurosci.,* **1999**, *19*, 1959-1964.

[65]   Nunomura, A.; Perry, G.; Aliev, G.; Hirai, K.; Takeda, A.; Balraj, E.K.; Jones, P.K.; Ghanbari, H.; Wataya, T.; Shimohama, S.; Chiba, S.; Atwood, C.S.; Petersen, R.B.; Smith, M.A. Oxidative damage is the earliest event in Alzheimer disease. *J. Neuropathol. Exp. Neurol.*, **2001**, *60*, 759-767.

[66]   Sayre, L.M.; Perry, G.; Smith, M.A. *In situ* methods for detection and localization of markers of oxidative stress: Application in neurodegenerative disorders. *Methods Enzymol.*, **1999**, *309*, 133-152.

[67]   Danysz, W.; Parsons, A.C. Glycine and N-methyl-D-aspartate receptors: Physiological significance and possible therapeutic applications. *Pharmacol. Rev.,* **1998**, *50*(4), 597-664.

[68]   Petrie, R.X.; Reid, I.C, Stewart, C.A. The N-methyl-D-aspartate receptor, synaptic plasticity, and depressive disorder. A critical review. *Pharmacol. Ther.* **2000**, *87*(1), 11-25.

[69]   Doble, A. The role of excitotoxicity in neurodegenerative disease: Implications for therapy. *Pharmacol. Ther.*, **1999**, *81*(3), 163-221.

[70]   Donna, M.; Wilcock, A. Changing Perspective on the role of neuro inflammation in Alzheimer's disease. *Int. J. Alzheimer's Dis.*, **2012**, 1-7.

[71]   Zotova, E.; Nicoll, J.A.R.; Holmes, C.; Boche, D. Inflammation in Alzheimer's disease: relevance to pathogenesis and therapy. *Alzheimers Res. Ther.*, **2010**, *2*, 1-9.

[72]   Tuppoa, E.E.; Arias, H.R. The role of inflammation in Alzheimer's disease. *Int. J. Biochem. Cell Biol.*, **2005**, *37*, 289-305

[73]   Duce, J.A.; Bush, A.I. Biological metals and Alzheimer's disease: Implications for therapeutics and diagnostics. *Prog. Neurobiol.*, **2010**, *92*, 1-18.

[74]   White, A.R.; Faller, P.; Atwood, C.S.; Zatta, P. Metals and Alzheimer's disease. *Int. J. Alzheimer's Dis.*, **2011**, 1-135.

[75]   Reddy, P.H.; Beal, M.F. Amyloid beta, mitochondrial dysfunction and synaptic damage: Implications for cognitive decline in aging and Alzheimer's disease. *Trends Mol. Med.*, **2008**, *14*, 45-53.

[76]   Petersen, H.C.A.; Alikhani, N.; Behbahani, H.; Wiehager. B.; Pavlov, P.F.; Alafuzoff, I.; Leinonen, V.; Ito, A.; Winblad, B.; Glaser, E.; Ankarcrona, M. The amyloid beta-peptide is imported into mitochondria *via* the TOM import machinery and localized to mitochondrial cristae. *Proc. Natl. Acad. Sci. USA*, **2008**, *105*, 13145-13150.

[77]   Abbott, A. The plaque plan. *Nature*, **2008**, *456*, 161-164.

[78]   Karczmar, A.G. *Intern. Encyclop. Pharmacol. Therap. Pergamon Press: Oxford.* **1970**.

[79]   Andersen, R.A.; Aaraas, I.; Gaare, G.; Fonnum, F. Inhibition of acetylcholinesterase from different species by organophosphorus compounds, carbamates, and methylsulphonyfluoride. *Gen. Pharmacol.*, **1977**, *8*, 331-334.

[80]   Davis, K.L.; Mohs, R.S. Enhancement of memory processes in Alzheimer's disease with multiple-dose intravenous physostigmine. *Am. J. Psychiatry*, **1982**, *139*, 1421-1424.

[81]   Thal, L.; Fuld, P.A.; Masur, D.M.; Sharpless, N.S. Oral physostigmine and lecithin improve memory in Alzheimer disease. *Ann. Neurol.*, **1983**, *13*, 491-496.

[82]   Ogorka, J.; Kalb, O.; Shah, R.; Khanna, S.C. US20036565883. **2003**.

[83]   Goldblum, D. US20046835748, **2004**.

[84]   Desphande, P.B.; Khemani, K.; Boda, B.B.; Shah, T.R.; Acharya, H.H.; Luthra, P.V. WO2008020452, **2008**.

[85]   Galenko-Iaroshevskii, P.A.; Shadurskii, K.S.; Il'iuchenok, T.I.; Anti- allergic properties of dimebon, fencarol and ketotifen in sensitization by common ragweed pollen. *Farmakol Toksikol.*, **1984**, *47*, 75-78.

[86]   Kiseleva, E.E.; Zakharevskii, A.S.; Nikiforova, I.N. The anti-allergic activity and mechanism of action of gamma-carboline derivatives. *Farmakol Toksikol.*, **1990**, *53*, 22-24.

[87]   Bachurin, S.; Bukatina, E.; Lermontova, N.; Tkachenko, S.; Afanasiev, A.; Grigoriev, V.; Grigorieva, I.; Ivanov, Y.; Sablin, S.; Zefirov, N. Antihistamine agent Dimebon as a novel neuroprotector and a cognition enhancer. *Ann. NY Acad. Sci.*, **2001**, *939*, 425-435.

[88]   Musia , A.; Bajda, M.; Malawska, B. Recent developments in cholinsterases inhibitors for Alzheimer's disease treatment. *Curr. Med. Chem.*, **2007**, *14*, 2654-2679.

[89]   Ibach, B.; Haen, E. Acetylcholinesterase inhibition in Alzheimer's disease. *Curr. Pharm. Des.*, **2004**, *10*, 231-251.

[90]   Geerts, H.; Grossberg, G.T. Pharmacology of acetylcholinsterase inhibitors and N-methyl- D-aspartate receptors for combination therapy in the treatment of Alzheimer's disease. *J. Clin. Pharmacol.*, **2006**, *46*, 8S-16S.

[91]   Cosman, K.M.; Boyle, L.L.; Porsteinsson, A.P. Memantine in the treatment of mild-to-moderate Alzheimer's disease. *Expert Opin. Pharmacother.*, **2007**, *8*, 203-214.

[92]   Wang, B.S.; Wang, H.; Wei, Z.H.; Song, Y.Y.; Zhang, L.; Chen, H.Z. Efficacy and safety of natural acetylcholinesterase inhibitor huperzine-A in the treatment of Alzheimer's disease: An updated meta-analysis. *J. Neural. Transm.*, **2009**, *116*, 457-465.

[93]   Salomone, S.; Caraci, F.; Leggio, G.M.; Fedotova J.; Drago. F. New pharmacological strategies for treatment of Alzheimer's disease: Focus on disease modifying drugs. *Br. J. Clin. Pharmacol.*, **2012**, *73*(4), 504-517.

[94]   Galimberti, D. Disease-modifying treatments for Alzheimer's disease. *Ther. Adv. Neurol. Disord.,* **2011**, *4*, 203-216.

[95]   Contestabile, A. The history of the cholinergic hypothesis. *Behav. Brain Res.*, **2011**, *221*, 334-340.

[96]   McGleenon, B.M.; Dynan, K.B.; Passmore, A.P. Acetylcholinesterase inhibitors in Alzheimer's disease. *Br. J. Pharmacol.,* **1990**, *48*, 471-480.

[97]   Dolmella, A.; Bandoli, G.; Nicolini, M. Alzheimer's disease: A pharmacological challenge. *Adv. Drug Res.*, **1994**, *25*, 207-294.

[98]   Wesseling, A.; Agoston S. Effects of 4-aminopyridine in elderly patients with Alzheimer's disease. *N. Engl. J. Med.*, **1984**, *310*, 988-989.

[99]   Greenlee, W.; Clader, J.; Asberom, T.; McCombie, S.; Ford, J.; Guzik, H.; Kozlowski, J.; Li, S.; Liu, C.; Lowe, D.; Vice, S. Zhao, H.; Zhou, G.; Billard, W.; Binch, H.; Crosby, R.; Duffy, R.; Lachowicz, J.; Coffin, V.; Watkins, R.; Ruperto, V.; Strader, C.; Taylor, L.; Cox, K. Muscarinic agonists and antagonists in the treatment of Alzheimer's disease. *Farmaco.*, **2001**, *56*(4), 247-250.

[100]   Maelicke, A.; Albuquerque, E.X. New approach to drug therapy of Alzheimer's dementia. *Drug Discov. Today*, **1996**, *1*, 53-59.

[101]   Davis, K.L.; Mohs, R.S. Enhancement of memory processes in Alzheimer's disease with multiple-dose intravenous physostigmine. *Am. J. Psychiatry*, **1982**, *139*, 1421-1424.

[102]   Proudfoot, A. The early toxicology of physostigmine: A tale of beans, great men and egos. *Toxicol. Rev.*, **2006**, *25*, 99-138.

[103]   Dale, H.H. The action of certain esters and ethers of choline and their relation to muscarine. *J. Pharmacol. Exp. Ther.*, **1914**, *6*, 147-190.

[104]   Mohs, R.C.; Davis, B.M.; Johns, A.A.; Greenwald, B.; Horvath, T.B.; Davis, K.L. Oral physostigmine treatment of patients with Alzheimer's disease. *Am. J. Psychiatry,* **1985**, *142*, 28-33.

[105]   Sussman, J.L.; Harel, M.; Frolow, F.; Oefner, C.; Goldman, A.; Toker, L.; Silman, I. Atomic-structure of acetylcholinesterase from *Torpedo-californica*, a prototypic acetylcholine-binding protein. *Science*, **1991**, *253*, 872-879.

[106]   Opez-Arrieta, J.M.L.; Schneider, L. Metrifonate for Alzheimer's disease, *Cochrane Database of Syst. Rev.*, **2006**, *2*, 1-40.

[107]   Tumiatti, V.; Minarini, A.; Bolognesi, M.L.; Milelli, A.; Rosini, M.; Melchiorre, C.; Tacrine derivatives and Alzheimer's disease. *Curr. Med. Chem.*, **2010**, *17*, 1825-1838.

[108]   Tacrine. Drugs in clinical trials. Alzheimer 's Research Forum, http://www.alzforum.org/drg/drc/detail.asp?id=90.

[109] Rogers S.L.; Doody, R.S.; Mohs, R.C.; Friedho L.T. Donepezil improves cognition and global function in Alzheimer disease: A 15-week, double-blind, placebo-controlled study, *Arch. Intern. Medicine*, **1998**, *158*, 1021-1031.

[110] Sugimoto, H.; Yamanishi, Y.; YouichiIimura.; Kawakami, Y. Donepezil hydrochloride (E2020) and other acetylcholinesterase inhibitors. *Curr. Med. Chem.*, **2000**, *7*, 303-339.

[111] Deleu, D. Rivastigmine in the treatment of Alzheimer's disease. *Eur. Neurol.*, **2001**, *46*, 111-113.

[112] Williams, B.R.; Nazarians, A.; Gill, M.A. A review of rivastigmine: A reversible cholinesterase inhibitor. *Clin. Therap.*, **2003**, *25*, 1634-1653.

[113] Potkin, S.G.; Anand, R.; Fleming, K.; Alva, G.; Keator, D.; Carreon, D.; Messina, J.; Wu, J.C.; Hartman, R.; Fallon, J.H. Brain metabolic and clinical effects of rivastigmine in Alzheimer's disease. *Int. J. Neuropsychopharmacol.*, **2001**, *4*, 223-230.

[114] Heinrich, M.; Lee, H. Galanthamine from snowdrop: The development of a modern drug against Alzheimer's disease from local Caucasian knowledge. *J. Ethnopharmacol.*, **2004**, 147-162.

[115] Tariot, P.N.; Solomon, P.R.; Morris, J.C.; Kershaw, P.; Lilienfeld, S.; Ding, C. A 5-month, randomized, placebo controlled trial of galanthamine in AD, *Neurol.*, **2000**, *54*, 12, 2269-2276.

[116] Wang, B.S.; Wang, H.; Wei, Z.H.; Song, Y.Y.; Zhang, L.; Chen, H.Z. "E cacy and safety of natural acetylcholinesterase inhibitor huperzine A in the treatment of Alzheimer's disease: An updated meta-analysis,". *J. Neural. Transmission*, **2009**, *116*, 4, 457-465.

[117] Li, J. Wu, H.M.; Zhou, R.L.; Liu, G.J.; Dong, B.R. Huperzine A for Alzheimer's disease. *Cochrane Database Syst. Rev.*, **2008**, *2*, CD005592.

[118] Zheng, J.Q.; He, X.D.; Yang, A.Z.; Liu, C.G. Neostigmine competitively inhibited nicotinic acetylcholine receptors in sympathetic neurons. *Life Sci.*, **1998**, *62*, 1171-1178.

[119] Gmora, S.; Poenaru, D.; Tsai, E. Neostigmine for the treatment of pediatric acute colonic pseudo-obstruction. *J. Pediatr. Surg.*, **2002**, *37*, 1-3.

[120] Paran, H.; Silverberg, D.; Mayo, A.; Schwartz, I.; Neufeld, D.; Freund, U. Treatment of acute colonic pseudo-obstruction with neostigmine. *J. Am. Coll. Surg.*, **2000**, *190*, 315-318.

[121] Laureth, G.R.; de Oliviera, R.; Perez, M.W.; Paccola, C.A.J. Postoperative analgesia by intraarticular and epidural neostigmine following knee surgery. *J. Clin. Anest.*, **2000**, *12*, 444-448.

[122] McClellan, K. J.; Benfield, P. Eptastigmine. *CNS Drugs*, **1998**, *9*, 6975.

[123] Terni, P.; Mairani, L.; Mandelli, G.; Pagella, P.G.; Marchesini, D.; Maiorana, S.; Brufani, M. **1997**. US5869484.

[124] Imbimbo, B. P.; Martelli, P.; Troetel, W. M.; Lucchelli, F.; Lucca, U.; Thal, L. J. Efficacy and safety of eptastigmine for the treatment of patients with Alzheimer's disease. *Neurology*, **1999**, *52*, 700-708.

[125] Imbimbo, B.P.; Troetel, W.M.; Martelli, P.; Lucchelli, F.A. 6-month, double-blind, placebo-controlled trial of eptastigmine in Alzheimer's disease. *Dement. Geriatr. Cogn.*, **2000**, *11*, 17-24.

[126] Barida, D.; Ottonello, F.; Sala, M. Eptastigmine improves eight-arm radial maze performance in aged rats. *Pharmacol. Res.*, **2000**, *42*, 299-304.

[127] Sramek, J. J.; Hourani, J.; Jhee, S. S.; Cutler, N. R. NXX-066 in patients with Alzheimer's disease: A bridging study. *Life Sci.,* **1999**, *64*, 1215-1221.

[128] Lahiri, D.K.; Farlow, M.R.; Hintz, N.; Utsuki, T.; Greig, N.H. Cholinesterase inhibitors, beta-amyloid precursor protein and amyloid beta-peptides in Alzheimer's disease. *Acta Neurol. Scand.,* **2000**, *102*, 176, 60-67.

[129] Bruening, J.: US20006495700. **2000**.

[130] Bruening, J.:WO03059909. **2003**.

[131] Bruinsma, G.: CA2553033. **2004**.

[132] Brossi, A.; Pei, X.; F.; Greig, N.; H. Phenserine, a novel anticholinesterase related to physostigmine: Total synthesis and biological properties. *Aust. J. Chem.,* **1996**, *49*, 171-181.

[133] Ingram, D.K.; Spangler, E.L.; Iijima, S.; Ikari, H.; Kuo, H.; Greig, N.H. *London ED.* Rodent models of memory dysfunction in Alzheimer's disease and normal aging: Moving beyond the cholinergic hypothesis. *Life Sci.,* **1994**, *55*, 2037-2049.

[134] Ikari, H.; Spangler, E.L.; Greig, N.H.; Pei, X.F.; Brossi, A.; Speer, D.; Patel, N.; Ingram, D.K. Maze-learning in aged rats is enhanced by phenserine, a novel anticholinesterase. *Neuroreport.,* **1995**, *6*, 481-484.

[135] Greig, N. H.; De Micheli, E.; Holloway, H.W.; Yu, Q.S.; Utsuki, T.; Perry, T.A.; Brossi, A.; Ingram, D.K.; Deutsch, J.; Lahiri, D.K.; Soncrant, T.T. The experimental Alzheimer drug phenserine: Preclinical pharmacokinetics and pharmacodynamics. *Acta Neurol. Scand..* **2000**, *102* (Suppl. 176), 74-84.

[136] Bruinsma, G.B. WO2005123068. **2005**.

[137] Kamal, M.A.; Greig, N.H.; Alhomida, A.S.; Al-Jafari, A.A. Kinetics of human acetylcholinesterase inhibition by the novel experimental Alzheimer therapeutic agent, tolserine. *Biochem Pharm.,* **2000**, *60*, 561-570.

[138] Shi, Y.F.; Zhang, H.Y.; Wang, W.; Yan, F.; Xia, Y.; Tang, X.; Bai, D.; He, X. Novel 16-substituted bifunctional derivatives of huperzine B: Multifunctional cholinesterase inhibitors. *Acta Pharmacologica Sinica.,* **2009**, *30*, 8, 1195-1203.

[139] Jung, H.A.; Jung, Y.J.; Hyun, S.K. Min, B. S.; Kim, D.W.; Jung, J.H.; Choi. J.S. Selective cholinesterase inhibitory activities of a new monoterpenediglycoside and other constituents from *Nelumbonucifera stamens. Biol. Pharm. Bull.,* **2010**, *33*, 2, 267-272.

[140] Seidl, C.; Correia, B. L.; Stinghen, A. E. M.; Santos, C. A. M. Acetylcholinesterase inhibitory activity of uleine from *Himatanthus lancifolius, Z. Naturforsch. C.,* **2010**, *65*, 7-8, 440-444.

[141] De Paula, A.A.N.; Martins, J.B.L.; dos Santos, M.L.; Nascente, L.C.; Romeiro, L.A.; Areas T.F.; Vieira, K.S.; Gambôa, N.F.; Castro, N.G.; Gargano, R. New potential AChE inhibitor candidates. *Eur. J. Med. Chem.,* **2009**, *44* (9), 3754-3759.

[142] Bormann, J. Memantine is a potent blocker of N-methyl-D-aspartate (NMDA) receptor channels. *Eur. J. Pharmacol.,* **1989**, *166*, 591-592.

[143] Lipton, S.A. Failures and Successes of NMDA receptor antagonists: Molecular basis for the use of open-channel blockers like Memantine in the treatment of acute and chronic neurologic insults. *NeuroRx.,* **2004**, *1*, 101-110.

[144] Rees, T.M.; Brimijoin, S. The role of acetylcholinesterase in the pathogenesis of Alzheimer's disease. *Drugs Today (Barc)*, **2003**, *39*, 75-83.

[145] Ballard, C.G.; Greig, N.H.; Guillozet-Bongaarts, A.L.; Enz, A.; Darvesh, S. Cholinesterases: Roles in the brain during health and disease. *Curr. Alzheimer Res.*, **2005**, *2*, 307-318.

[146] Mori, E.; Hashimoto, M.; Krishnan, R. Doraiswamy PM. What constitutes clinical evidence for neuroprotection in Alzheimer's disease: Support for the cholinesterase inhibitors? *Alzheimer Dis. Assoc. Disord.*, **2006**, *20*, S19-26.

[147] Lahiri, D.K.; Chen, D.; Maloney, B.; Holloway, H.W.; Yu, Q.S.; Utsuki, T.; Giordano, T.; Sambamurti, K.; Greig, N.H. The experimental Alzheimer's disease drug posiphen [(+)-phenserine] lowers amyloid-beta peptide levels in cell culture and mice. *J. Pharmacol. Exp. Ther.*, **2007**, *320*, 386-396.

[148] Mangialasche, F.; Solomon, A.; Winblad, B.; Mecocci, P.; Kivipelto, M. Alzheimer's disease: clinical trials and drug development. *Lancet Neurol.*, **2010**, *9*(7), 702-716.

[149] Zhang, H. Y, Tang, X. C. Neuroprotective effects of huperzine A: New therapeutic targets for neurodegenerative disease. *Trends Pharmacol. Sci.*, **2006**, *27*, 619-625.

[150] Wang, R.; Tang, X.C. Neuroprotective effects of huperzine A. A natural cholinesterase inhibitor for the treatment of Alzheimer's disease. *Neurosignals*, **2005**, *14*, 71-82.

[151] Lane, R.M.; Potkin, S.G.; Enz, A. Targeting acetylcholinesterase and butyrylcholinesterase in dementia. *Int. J. Neuropsychopharmacol.* **2006**, *9*, 101-124.

[152] Kamal, M.A.; Al-Jafari, A. A.; Yu, Q.S. Greig, N.H. Kinetic analysis of the inhibition of human butyrylcholinesterase with cymserine. *Biochim. Biophys. Acta.*, **2006**, *1760*, 200-206.

[153] Kamal, M.A.; Klein, P.; Yu Q.S, Greig N.H. Kinetics of human serum butyrylcholinesterase and its inhibition by a novel experimental Alzheimer therapeutic, bisnorcymserine. *J. Alzheimer's Dis.*, **2006**, *10*, 43-51.

[154] Kihara, T.; Shimohama, S.D. Alzheimer's disease and acetylcholine receptors *Acta Neurobiol. Exp.*, **2004**, *64*, 99-105.

[155] Dunbar, G.C.; Inglis, F.; Kuchibhatla, R.; Sharma, T.; Tomlinson, M.; Wamsley, J. Effect of ispronicline, a neuronal nicotinic acetylcholine receptor partial agonist, in subjects with age associated memory impairment (AAMI). *J. Psychopharmacol.*, **2007**, *21*, 171-178.

[156] Kem, W.R. The brain alpha 7 nicotinic receptor may be an important therapeutic target for the treatment of Alzheimer's disease: Studies with DMXBA (GTS-21). *Behav. Brain Res.*, **2000**, *113*, 169-181.

[157] Lin, N.H.; Gunn, D.E.; Ryther, K.B.; Garvey, D.S.; Donnelly-Roberts, D.L.; Decker, M.W.; Brioni, J. D.; Buckley, M.J.; Rodrigues, A. D.; Marsh, K.G.; Anderson, D.J.; Buccafusco, J.J.; Prendergast, M.A.; Sullivan, J.P.; Williams, M.; Arneric, S.P.; Holladay, M.W. Structure-activity studies on 2-methyl-3- (2(S)pyrrolidinylmethoxy) pyridine (ABT-089): An orally bioavailable 3-pyridyl ether nicotinic acetylcholine receptor ligand with cognition-enhancing properties. *J. Med. Chem.*, **199**7, *40*, 385-390.

[158] Fisher, A.; Pittel, Z.; Haring, R.; Bar-Ner, N.; Kliger-Spatz, M.; Natan, N.; Egozi, I.; Sonego, H.; Marcovitch, I.; Brandeis, R. M1 muscarinic agonists can modulate some of the hallmarks in Alzheimer's disease: Implications in future therapy. *J. Mol. Neurosci.*, **2003**, *20*, 349-356.

[159] Caccamo, A.; Oddo, S.; Billings, L. M.;Green, K.N.; Martinez-Coria, H.; Fisher, A.; LaFerla, F.M.; Caccamo, A.; Oddo, S.; Billings, L.M. M1 receptors play a central role in modulating AD-like pathology in transgenic mice. *Neuron.*, **2006**, *49*, 671-682.

[160] Hock, C.; Maddalena, A.; Heuser, I.; Naber, D.; Oertel, W.; von der Kammer, H.; Wienrich, M.; Raschig, A.; Deng, M.; Growdon, J.H.; Nitsch, R.M. Treatment with the selective muscarinic agonist talsaclidine decreases cerebrospinal fluid levels of total amyloid beta-peptide in patients with Alzheimer's disease. *Ann. NY Acad. Sci.*, **2000**, *920*, 285-291.

[161] Fisher, A. M1 muscarinic agonists target major hallmarks of Alzheimer's disease-the pivotal role of brain M1 receptors. *Neurodegener. Dis.,* **2008**, *5*, 237-240.

[162] Haass, C.; Selkoe, D.J. Soluble protein oligomers in neurodegeneration: Lessons from the Alzheimer's amyloid beta-peptide. *Nat. Rev. Mol. Cell Biol.*, **2007**, *8*, 101-112.

[163] Hamaguchi, T.; Ono, K.; Yamada, M. Anti-amyloidogenic therapies: Strategies for prevention and treatment of Alzheimer's disease. *Cell Mol. Life Sci.*, **2006**, *63*, 1538-1552.

[164] Hey, J.A.; Koelsch, G.; Bilcer, G.; Jacobs, A.; Tolar, M.; Tang, J.; Ghosh, A.K.; Hsu, H. H. Single Dose Administration of the β-Secretase Inhibitor CTS21166 (ASP1720) Reduces Plasma Ab40 in Human Subjects. International Conference on Alzheimer's disease (ICAD), Chicago, IL. **2008**.

[165] Koelsch, G. Beta-Secretase inhibitor CTS-21166 reduces plasma abeta-40 in human subjects. Keystone symposium on Alzheimer's Disease. Keystone, Colorado. **2008**.

[166] Hsu, H.H. Clinical trials for disease-modifying drugs such as BACE inhibitors, in BACE: Lead target for orchestrated therapy of Alzheimer's disease (John V., ed.). *John Wiley & Sons, Hoboken.* **2010**, 197-216.

[167] Forman, M.S. The Novel BACE Inhibitor MK-8931 Dramatically Lowers CSF A Peptides in Healthy Subjects: Results from a Rising Single Dose Study. *American Academy of Neurology.* San Diego, **2012**.

[168] Williamson, J.D.; Miller, M.E.; Bryan, R.N.; Lazar, R.M.; Coker, L.H.; Johnson, J.; Cukierman, T.; Horowitz, K.R.; Murray, A.; Launer, L.J. The action to control cardiovascular risk in Diabetes memory in diabetes Study (ACCORD-MIND): Rationale, design and methods. *Am. J. Cardiol.*, **2007**, *99*, 112-122.

[169] Rönnemaa, E.; Zethelius, B.; Sundelöf, J. Sundstrom, J.; Degerman-Gunnarsson, M.; Berne, C.; Lannfelt, L.; Kilander, L. Impaired insulin secretion increases the risk of Alzheimer disease. *Neurology*, **2008**, *71*, 1065-1071.

[170] Okochi, M.; Steiner, H.; Fukumori, A.; Tanii, H.; Tomita, T.; Tanaka, T.; Iwatsubo, T.; Kudo, T.; Takeda, M.; Haass, C. Presenilins mediate a dual intramembranous gamma-secretase cleavage of Notch-1. *Embo. J.,* **2002**, *21*, 5408-5416.

[171] Henley, D.B.; May, P.C.; Dean, R.A.; Siemers, E.R. Development of semagacestat (LY450139), a functional gamma-secretase inhibitor, for the treatment of Alzheimer's disease. *Expert Opin. Pharmacother.,* **2009**, *10*, 1657-1664.

[172] Bateman, R.J.; Siemers, E.R.; Mawuenyega, K.G.; Wen, G.; Browning, K.R.; Sigurdson, W.C.; Yarasheski, K.E.; Friedrich, S.W.; Demattos, R.B.; May, P.C.; Paul S.M.; Holtzman, D.M. A gamma-secretase inhibitor decreases amyloid-beta production in the central nervous system. *Ann. Neurol.*, **2009**, *66*, 48-54.

[173] Yi, P.; Hadden, C.; Kulanthaivel, P.; Calvert, N.; Annes, W.; Brown, T.; Barbuch, R.J.; Chaudhary, A.; Ayan-Oshodi, M.A.; Ring, B.J. Disposition and metabolism of semagacestat, a γ-secretase inhibitor, in humans. *Drug Metab. Dispos.*, **2010**, *38*, 554-565.

[174] Siemers, E.R.; Quinn, J.F.; Kaye, J. Farlow, M.R.; Porsteinsson, A.; Tariot, P.; Zoulnouni, P.; Galvin, J.E.; Holtzman, D.M.; Knopman, D.S.; Satterwhite, J.;

Gonzales, C.; Dean, R.A.; May, P.C. Effects of a gamma-secretase inhibitor in a randomized study of patients with Alzheimer disease. *Neurology.*, **2006**, *66*, 602-604.

[175] Fleisher, A.S.; Raman, R.; Siemers, E. R.; Becerra, L.; Clark, C.M.; Dean, R.A.; Farlow, M.R.; Galvin, J.E.; Peskind, E.R.; Quinn ,J.F.; Sherzai, A.; Sowell, B.B.; Aisen, P.S.; Thal, L.J. Phase-II safety trial targeting amyloid beta production with a gamma-secretase inhibitor in Alzheimer disease. *Arch. Neurol.*, **2008**, *65*, 1031-1038.

[176] Imbimbo, B.P. Alzheimer's disease: γ-secretase inhibitors. *Drug Discov. Today: Ther. Strateg.*, **2008**, *5*, 169-175.

[177] Ereshefsky, L.; Jhee, S.S.; Yen, M.; Moran, S.V. The role of CSF dynabridging studies in developing new therapies for Alzheimer's disease. *Alzheimers Dement.*, **2009**, *5*, 414-415.

[178] Jacobsen, S.; Comery, T.; Kreft, A.; Mayer, S.; Zaleska, M.; Riddell, D.; Bard, J.; Gonzales, C.; Frick, G.; Raje, S.; Forlow, S.; Balliet, C.; Burczynski, M.; Wan, H.; Harrison, B.; Reinhart, P.; Pangalos, M.; Martone R. GSI-953 is a potent APP-selective gamma- secretase inhibitor for the treatment of Alzheimer's disease. *Alzheimer's Dement.*, **2009**, *5*, 139-.

[179] Hopkins, C.R. Molecule Spotlight on ELND006: Another γ-Secretase Inhibitor fails in the Clinic. *ACS Chem. Neurosci.*, **2011**, *2*, 279-280.

[180] LoRusso, P.; Demuth, T.; Heath, E. *et al.* Phase-I Study of the gamma secretase inhibitor MK-0752 in patients with metastatic breast and other advanced solid tumors. 100th Annual Meeting of the American Association for Cancer Research, Denver, CO, April, **2009**, 18-22.

[181] Wood, K.M.; Lanz, T.A.; Coffman, K.J. *et al.* Efficacy of the novel γ-secretase inhibitor, PF-3084014, in reducing Aβ in brain, CSF, and plasma in guinea pigs and Tg2576 mice. *Alzheimer's Dement.*, **2008**, *4*, T482.

[182] Weggen, S.; Eriksen, J.L.; Das, P.; Sagi, S. A.; Wang, R.; Pietrizik, C. U.; Findlay, K. A.; Smith, T.E.; Murphy, M.P.; Bulter, T.; Kang, D.E.; Marquez-sterling, N.; Golde, T.E.; Koo, E.H. A subset of NSAIDs lower amyloidogenic Aβ$_{42}$ independently of cyclooxygenase activity. *Nature,* **2001**, *414*, 6860, 212-216.

[183] Eriksen, J.L.; Sagi, S.A.; Smith, T.E.; Weggen, S.; Das, P.; McLendon, D.C.; Ozols, V. V.; Jessing, K.W.; Zavitz, K.H.; Koo, E.H.; Golde, T.E. NSAIDs and enantiomers of flurbiprofen target γ-secretase and lower A$_{42}$ *in vivo. J. Clin. Invest.*, **2003**, *112*, 3, 440-449.

[184] Green, R.C.; Schneider, L.S.; Amato, D.A. Beelen, A.P.; Wilcock, G.; Swabb, E.A.; Zavitz, K.H. Effect of tarenflurbil oncognitive decline and activities of daily living in patients with mildAlzheimer disease: A randomized controlled trial. *JAMA,* **2009**, *302*, 2557-64.

[185] Imbimbo, B.P. Why did tarenflurbil fail in Alzheimer's disease?. *J. Alzheimer's Dis.*, **2009**, *17*, 757-760.

[186] Imbimbo, B.P.; Hutter-Paier, B.; Villetti, G. Facchinetti, F.; Cenacchi, V.; Volta, R.; Lanzillotta, A.; Pizzi, M.; Windisch, M. CHF5074, a novel gamma-secretase modulator, attenuates brain beta-amyloid pathology and learning deficit in a mouse model of Alzheimer'sdisease. *Br. J. Pharmacol.*, **2009**, *156*, 982-993.

[187] Nagy, C.; Schuck, E.; Ishibashi, A.; Nakatani, Y.; Rege, B.; Logovinsky, V. *AAIC, Honolulu, Hawaii, abstract.* **2010**, 3-415.

[188] Life Science Digest, 2012: A pivitol year for Alzheimer's disease drug development. April **2012**, 1-18. (www.lifesciencedigest.com).

[189] Cuajungco, M.P.; Frederickson, C.J.; Bush, A.I. Amyloid-beta metal interaction and metal chelation. *Subcell Biochem.*, **2005**, *38*, 235-254.

[190] Dedeoglu, A.; Cormier, K.; Payton, S.; Tseitlin, K.A.; Kremsky, J.N.; Lai, L.; Li, X.; Moir, R.D.; Tanzi, R.E.; Bush, A.I.; Kowall, N.W.; Rogers, J.T.; Huang, X. Preliminary studies of a novel bifunctional metal chelator targeting Alzheimer's amyloidogenesis. *Exp. Gerontol.*, **2004**, *39*, 1641-1649.

[191] Gervais, F.; Paquette, J.; Morissette, C.; Krzywkowski, P.; Yu, M.; Azzi, M.; Lacombe, D.; Kong, X.; Aman, A.; Laurin, J.; Szarek, W.A.; Tremblay, P. Targeting soluble A peptide with tramiprosate for the treatment of brain amyloidosis. *Neurobiol Aging.* **2007**, *28*, 537-547.

[192] Gauthier, S.; Aisen, P.S.; Ferris, S.H.; Saumier, D.; Duong, A. Haine, D.; Garceau, D.; Suhy, J.; Oh, J.; Lau, W.; Sampalis, J. Effect of tramiprosate in patients with mild-to-moderate Alzheimer's disease: Exploratory analyses of the MRI sub-group of the Alphase study. *J. Nutr. Health Aging*, **2009**, *13*, 550-557.

[193] Saumier, D.; Duong, A.; Haine, D.; Garceau, D.; Sampalis, J. Domain-specific cognitive effects of tramiprosate in patients with mild to moderate Alzheimer's disease: ADAS-cog subscale results from the Alphase Study. *J. Nutr. Health Aging*, **2009**, *13*, 808-812.

[194] Faux, N.G.; Ritchie, C.W.; Gunn, A.; Rembach, A.; Tsatsanis, A.; Bedo, J.; Harrison, J.; Lannfelt, L.; Blennow, K.; Zetterberg, H.; Ingelsson, M.; Masters, C. L.; Tanzi, R. E.; Cummings, J.L.; Herd, C. M. Bush AI. PBT2 rapidly improves cognition in Alzheimer's disease: Additional Phase-II analyses. *J. Alzheimers Dis.,* **2010**, *20*, 509-516.

[195] Adlard, P.A.; Cherny, R.A.; Finkelstein, D.I.; Gautier, E.; Robb, E.; Cortes, M.; Volitakis, I.; Liu, X.; Smith, J.P.; Perez, K.; Laughton, K.; Li, Q.X.; Charman, S.A.; Nicolazzo, J.A. ; Wilkins, S.; Deleva, K.; Lynch, T.; Kok, G.; Ritchie, C.W.; Tanzi, R. E.; Cappai, R.; Masters, C. L.; Barnham, K.J.; Bush, A.I. Rapid restoration of cognition in Alzheimer's transgenic mice with 8-hydroxy quinolone analogs is associated with decreased interstitial Abeta. *Neuron*, **2008**, *59*, 43-55.

[196] Liu, G.; Garrett, M. R.; Men, P. Zhu, X.; Perry, G.; Smith, M.A. Nanoparticle and other metal chelation therapeutics in Alzheimer's disease. *Biochim. Biophys. Acta.* **2005**, *1741*, 246-252.

[197] Fenili, D.; Brown, M.; Rappaport, R.; McLaurin, J. Properties of scyllo-inositol as a therapeutic treatment of AD-like pathology. *J. Mol. Med. (Berl)*, **2007**, *85*, 603-11.

[198] Amijee, H.; Scopes, D.I. The quest for small molecules as amyloid inhibiting therapies for Alzheimer's disease. *J. Alzheimers Dis.*, **2009**, *17*, 33-47.

[199] Salloway, S.; Sperling, R.; Keren, R.; Porsteinsson, A.P.; van Dyck, C.H.; Tariot, P.N.; Gilman, S.; Arnold, D.; Abushakra, S.; Hernandez, C.; Crans, G.; Liang, E.; Quinn, G.; Bairu, M.; Pastrak, A.; Cedarbaum, J.M. A Phase-II randomized trial of ELND005, scyllo- inositol, in mild to moderate Alzheimer disease. *Neurology*, **2011**, *77*, 1253-1262.

[200] Town, T. Alternative Abeta immunotherapy approaches for Alzheimer's disease. *CNS Neurol. Disord. Drug Targets*, **2009**, *8*, 114-27.

[201] Lemere, C.A. Developing novel immunogens for a safe and effective Alzheimer's disease vaccine. *Prog. Brain Res.*, **2009**, *175*, 83-93.

[202] Wilcock, D. M.; Colton, C. A. Anti-amyloid-beta immunotherapy in Alzheimer's disease: Relevance of transgenic mouse studies to clinical trials. *J. Alzheimers Dis.*, **2008**, *15*, 555-569.

[203] Seubert, P.; Barbour, R.; Khan, K. Motter, R.; Jacobsen, J.S.; Pangalos, M.; Basi, G.; Games, D. Antibody capture of soluble Abeta does not reduce cortical Abeta amyloidosis in the PDAPP mouse. *Neurodegener. Dis.*, **2008**, *5*, 65-71.

[204] Kerchner, G.A.; Boxer, A.L. Bapineuzumab. *Expert Opin. Biol. Ther.*, **2010**, *10*, 1121-1130.

[205] Rinne, J.O.; Brooks, D.J.; Rossor, M.N.; Fox, N.C.; Bullock, R.; Klunk, W.E.; Mathis, C.A.; Blennow, K.; Barakos, J.; Okello, A.A.; Rodriguez, M.L.S.; Liu, E.; Koller, M.; Gregg, K.M.; Schenk, D.; Black, R.; Grundman, M. 11C-PiB PET assessment of change in fibrillar amyloid-beta load in patients with Alzheimer's disease treated with bapineuzumab: A Phase-II, double-blind, placebo-controlled, ascending-dose study. *Lancet Neurol.*, **2010**, *9*, 363-372.

[206] Black, R.S.; Sperling, R.A.; Safirstein, B.; Motter, R.N.; Pallay, A.; Nichols, A.; Grundman, M. A single ascending dose study of bapineuzumab in patients with Alzheimer disease. *Alzheimer Dis. Assoc. Disord.*, **2010**, *24*, 198-203.

[207] Drugs In Clinical Trials http://www.alzforum.org/drg/drc/detail.asp?id=101

[208] News, Johnson & Johnson Announces Discontinuation Of Phase-III Development of Bapineuzumab Intravenous (IV) In mild-to-moderate Alzheimer's Disease. August 6, **2012** (http://www.jnj.com/connect/news/all/johnson-and-johnson-announces-discontinuation-of-Phase-III-development-of-bapineuzumab-intravenous-iv-in-mild-to-moderate-alzheimers- disease)

[209] Muhs, A.; Hickman, D.T.; Pihlgren, M. Chuard, N.; Giriens, V.; Meerschman, C.; van der Auwera, I.; van Leuven, F.; Sugawara, M.; Weingertner, M.C.; Bechinger, B.; Greferath, R.; Kolonko, N.; Nagel-Steger, L.; Riesner, D.; Brady, R.O.; Pfeifer, A.; Nicolau, C. Liposomal vaccines with conformation- specific amyloid peptide antigens define immune response and efficacy in APP transgenic mice. *Proc. Natl. Acad. Sci. USA*, **2007**, *104*, 9810-9815.

[210] Wang, C.Y.; Finstad, C.L.; Walfield A.M, Sia, C,; Sokoll, K.K.; Chang, T.Y.; Fang, X.D.; Hung, C.H.; Hutter-Paier, B.; Windisch, M. Site-specific UBI Thamyloid-beta vaccine for immunotherapy of Alzheimer's disease. *Vaccine,* **2007**, *25*, 3041-3052.

[211] Winblad, B.; Minthon, L.; Floesser A. *et al.* Results of the first-in-manstudy with the active A immunotherapy CAD106 in Alzheimerpatients. *Alzheimers Dement.*, **2009**, *5*, 113-114.

[212] Iqbal, K.; Alonso, A.C.; Gong, C.X.; Khatoon, S.; Pei, J.J.; Wang, J.Z.; Grundke-Iqbal, I. Mechanisms of neurofibrillarydegeneration and the formation of neurofibrillary tangles. *J. Neural. Transm. Suppl.*, **1998**, *53*, 169-180.

[213] Bierer, L.M.; Hof, P.R.; Purohit, D.P.; Carlin, L.; Schmeidler, J.; Davis, K.L.; Perl, D.P. Neocortical neurofibrillary tangles correlate with dementia severity in Alzheimer's disease. *Arch. Neuro.*, **1995**, *52*, 81-88.

[214] Ittner, L.; M, Gotz, J. Amyloid-beta and tau-atoxicpasdedeux in Alzheimer's disease. *Nat. Rev. Neurosci.*, **2011**, *12*, 65-72.

[215] Pei, J.J.; Sjogren, M.; Winblad, B. Neurofibrillary degeneration in Alzheimer's disease: From molecular mechanisms to identification of drug targets. *Curr. Opin. Psychiatry*, **2008**, *21*, 555-561.

[216] Takashima, A.; Honda, T.; Yasutake, K.; Michel, G.; Murayama, O.; Murayama, M.; Ishiguro, K.; Yamaguchi, H.; Activation of tauprotein kinase I/glycogen synthase kinase-3 beta byamyloid beta peptide (25-35) enhances phosphorylation of tau in hippocampal neurons. *Neurosci. Res.*, **1998**, *31*, 317-323.

[217] Tariot, P.N.; Aisen, P.S. Can lithium or valproate untie tangles in Alzheimer's disease? *J. Clin. Psychiatry,* **2009**, *70*, 919-921.

[218] Wischik, C.M.; Edwards, P.C.; Lai, R.Y.; Roth, M.; Harrington, C.R. Selective inhibition of Alzheimer disease-like tau aggregation by phenothiazines. *Proc. Natl. Acad. Sci. USA*, **1996**, *93*, 11213-18.

[219] Schirmer, H.; Coulibaly, B.; Stich, A. Scheiwein, M.; Merkle, H.; Eubel, J.; Becker, K.; Becher, H.; Müller, O.; Zich, T.; Schiek, W.; Kouyaté, B. Methylene blue as an antimalarial agent-past and future. *Redox Rep.*, **2003**, *8* (5), 272-276.

[220] Experimental Alzheimer's drug shows early promise. *Associated Press*, **2008**-07-29.

[221] Gura, T. Hope in Alzheimer's fight emerges from unexpected places. *Nature Medicine*, **2008**, *14*, 9, 894.

[222] Wischik, C.; Staff, R. Challenges in the conduct ofdisease-modifying trials in AD: Practical experience from a Phase-II trial of Tau-aggregation inhibitor therapy. *J. Nutr. Health Aging*, **2009**, *13*, 367-369.

[223] Sereno, L.; Coma, M.; Rodriguez, M. Sánchez-Ferrer, P.; Sánchez, M.B.; Gich, I.;, Agulló, J.M.; Pérez, M.; Avila, J.; Guardia-Laguarta, C.; Clarimón, J.; Lleó, A.; Gómez-Isla, T. A novel GSK-3 inhibitor reduces Alzheimer's pathology and rescuesneuronal loss *in vivo*. *Neurobiol. Dis.*, **2009**, *35*, 3, 359-367.

[224] Gozes, I.; Steingart, R.A.; Spier, A.D. NAP mechanisms of neuroprotection. *J. Mol. Neurosci.,* **2004**, *24*, 67-72.

[225] Gozes, I.; Divinski, I.; Piltzer, I. NAP and D-SAL: Neuroprotection against the beta amyloid peptide (1-42). *BMC Neurosci.,* **2008**, *9*, S3.

[226] Shiryaev, N.; Jouroukhin, Y.; Giladi, E.; Polyzoidou, E.; Grigoriadis, N.C.; Rosenmann, H.; Gozes, I. NAP protects memory, increases soluble tau and reduces tau hyperphosphorylation in a tauopathy model. *Neurobiol. Dis.*, **2009**, *34*, 381-388.

[227] Zhang, S.; Hedskog, L.; Petersen, C.A.; Winblad, B.; Ankarcrona, M. Dimebon (latrepirdine) enhancesmitochondrial function and protects neuronal cells from death. *J. Alzheimers Dis.*, **2010**, *21*, 389-402.

[228] Doody, R. S.; Gavrilova, S. I.; Sano, M.; Thomas, R. G.; Aisen, P. S.; Bachurin, S. O.; Seely, L.; Hung, D. Effect of dimebon on cognition, activities of daily living, behavior, and global function in patients with mild-to-moderate Alzheimer's disease: A randomized, double-blind, placebo-controlled study. *Lancet*, **2008**, *372*, 207-215.

[229] Bezprozvanny, I. The rise and fall of dimebon. *Drug News Perspect.*, **2010**, *23*, 518-523.

[230] Zandi, P.P.; Anthony, J.C.; Khachaturian, A.S. Stone, S.V.; Gustafson, D.; Tschanz, J.T.; Norton, M.C.; Welsh-Bohmer, K.A.; Breitner, J.C. Reduced risk of Alzheimer disease inusers of antioxidant vitamin supplements: The cache county Study. *Arch. Neurol.*, **2004**, *61*, 82-88.

[231] Masaki, K.H.; Losonczy, K.G.; Izmirlian, G.; Foley, D.J.; Ross, G.W.; Petrovitch,H.; Havlik, R.; White, L.R. Association of vitamin E and Csupplement use with cognitive function and dementia in elderly men. *Neurology,* **2000**, *54*, 1265-1272.

[232] Wadsworth, T. L.; Bishop, J. A.; Pappu, A. S.; Woltjer, R.L.; Quinn, J.F. Evaluation of coenzyme Q as anantioxidant strategy for Alzheimer's disease. *J. Alzheimer's Dis.,* **2008**, *14*, 225-234.

[233] Orr, S.K.; Bazinet, R.P. The emerging role of docosahexaenoic acid in neuroinflammation. *Curr. Opin. Investig. Drugs,* **2008**, *9*, 735-743.

[234] Luo, Y.; Smith, J.V.; Paramasivam, V.; Burdick, A.; Curry, K.J.; Buford, J.P.; Khan, I.; Netzer, W.J.; Xu, H.; Butko, P. Inhibition of amyloid-beta aggregation and caspase-3 activation by the Ginkgo biloba extract EGb761. *Proc. Natl. Acad. Sci. USA,* **2002**, *99*, 12197-12202.

[235] Andrieu, S.; Gillette, S.; Amouyal, K.; Nourhashemi, F. Reynish, E.; Ousset, P.J.; Albarede, J.L.; Vellas, B.; Grandjean, H. Association of Alzheimer's disease onset with ginkgo biloba and other symptomatic cognitive treatments in a population of women aged 75 years and older from the EPIDOS study. *J. Gerontol. A Biol. Sci. Med. Sci.,* **2003**, *58*, 372-377.

[236] Schneider, L.S.; DeKosky, S.T.; Farlow, M.R.; Tariot, P.N.; Hoerr, R.; Kieser, M. A randomized, double-blind, placebo controlled trial of two doses of Ginkgo biloba extract in dementia of the Alzheimer's type. *Curr. Alzheimer Res.,* **2005**, *2*, 541-551.

[237] Van, D.M.; van, R.E.; Kessels, A.; Sielhorst, H.; Knipschild, P. Ginkgo for elderly people with dementia and age- associated memory impairment: A randomized clinical trial. *J. Clin. Epidemiol.,* **2003**, *56*, 367-376.

[238] Gupta, A.; Pansari, K. Inflammation and Alzheimer's disease. *Int. J. Clin. Pract.,* **2003**, *57*, 36-39.

[239] Szekely, C.A.; Breitner, J.C.; Fitzpatrick, A.L.; Rea, T.D.; Psaty, B.M.; Kuller, L.H.; Zandi, P.P. NSAID use and dementia risk in the Cardiovascular Health Study: Role of APOE and NSAID type. *Neurology,* **2008**, *70*, 17-24.

[240] Rogers, J.; Kirby, L.C.; Hempelman, S.R.; Berry, D.L.; McGeer, P.L.; Kaszniak, A.W.; Zalinski, J.; Cofield, M.; Mansukhani, L.; Willson, P.l. Clinical trial of indomethacin in Alzheimer's disease. *Neurology,* **1993**, *43*, 1609-1611.

[241] Tabet, N.; Feldman, H. Indomethacin for Alzheimer's disease. *Cochrane Database Syst. Rev.,* **2002**, *2*, CD003673.

[242] de Jong, D.; Jansen, R.; Hoefnagels, W.; Jellesma-Eggenkamp, M.; Verbeek, M.; Borm, G.; Kremer, B. No effect of one-year treatment with indomethacin on Alzheimer's disease progression: a randomized controlled trial. *PLoS ONE,* **2008**, *3*, e1475.

[243] Tabet, N.; Feldmand, H. Ibuprofen for Alzheimer's disease. *Cochrane Database Syst. Rev.,* **2003**, *2*, CD004031.

[244] Reines, S.A.; Block, G.A.; Morris, J.C.; Liu, G.; Nessly, M.L.; Lines, C.R.; Norman, B.A.; Baranak, C.C. Rofecoxib: no effect on Alzheimer's disease in a 1-year, randomized, blinded, controlled study. *Neurology,* **2004**, *62*, 66-71.

[245] Soininen, H.; West, C.; Robbins, J.; Niculescu, L. Long-term efficacy and safety of celecoxib in Alzheimer's disease. *Dement. Geriatr. Cogn. Disord.,* **2007**, *23*, 8-21.

[246] Brode, S.; Cooke, A. Immune-potentiating effects of the chemotherapeutic drug cyclophosphamide. *Crit. Rev. Immunol.,* **2008**, *28*, 109-126.

[247] Harikumar, K.B.; Aggarwal, B.B. Resveratrol: A multitargeted agent for age-associated chronic diseases. *Cell Cycle*, **2008**, *7*, 1020-1035.

[248] Schmidt, A.M.; Sahagan, B.; Nelson, R.B.; Selmer, J.; Rothlein, R.; Bell, J.M. The role of RAGE in amyloid-beta peptide-mediated pathology in Alzheimer's disease. *Curr. Opin. Investig. Drugs,* **2009**, *10*, 672-680.

[249] Cho, H.J.; Son, S.M.; Jin, S.M.; Hong, H.S.; Shin, D.H.; Kim, S.J.; Huh, K.; Mook-Jung, I. RAGE regulates BACE1 and Abeta generation *via* NFAT1 activation in Alzheimer's disease animal model. *FASEB*, **2009**, *23*, 2639-2649.

[250] Edmondson, D.E.; Mattevi, A.; Binda, C.; Li, M.; Hubalek, F. Structure and mechanism of monoamine oxidases. *Curr. Med. Chem.,* **2004**, *11*, 1983-1993.

[251] Grimsby, J.; Ian, N.C.; Neve, R.; Chen, K.; Shih, J.C. Tissue distribution of human monoamine oxidase-A and oxidase-B messenger-RNA. *J. Neurochem.*, **1990**, *55*, 1166-1169.

[252] Fowler, C.J.; Ross, S.B. Selective inhibitors of monoamine oxidase A and B: Biochemical, pharmacological and clinical properties. *Med. Res. Rev.*, **1984**, *4*, 323-358.

[253] Youdim, M.B. The path from anti-Parkinson's drug selegiline and rasagiline to multifunctional neuroprotective anti-Alzheimer's drugs ladostigil and M-30. *Curr. Alzheimer Res.*, **2006**, *3*, 541-550.

[254] Barone, E.; Cenini, G.; Di Domenico, F.; Martin, S.; Sultana, R.; Mancuso, C.; Murphy, M.P.; Head, E.; Butterfield, D.A. Long-term high-dose atorvastatin decreases brain oxidative and nitrosative stress in a preclinical model of Alzheimer disease: A novel mechanism of action. *Pharmacol. Res.,* **2011**, *63*, 172-180

[255] Murphy, M.P.; Morales, J.; Beckett, T.L.; Astarita, G.; Piomelli, D.; Weidner, A.; Studzinski, C.M.; Dowling, A.L.; Wang, X.; Levine, H.; Kryscio, R.J.; Lin, Y.; Barrett, E.; Head, E. Changes in cognition and amyloid-beta processing with long term cholesterol reduction using atorvastatin in aged dogs. *J. Alzheimers Dis.,* **2010**, *22*, 135-150

[256] Mohler, E.G.; Shacham, S.; Noiman, S. *et al*. PRX-03140, A novel 5-HT4 agonist, enhances memory and increases hippocampal acetylcholine efflux. Program No. 653.10. **2005**.

[257] Childers, W.E.; Abou-Gharbia, M.A.; Kelly, M.G.; Andree, T. H.; Harrison, B. L.; Hornby, G.; Potestio, L.; Rosenzweig-Lipson, S. J.; Schmid, J.; Smith, D. L.; Sukoff, S. J.; Zhang, G.; Schecter, L. E. Synthesis and biological evaluation of benzodioxanyl-piperazine derivatives as potent serotonin 5-HT(1A) antagonists: The discovery of Lecozotan. *J. Med. Chem.,* **2005**, *48*, 3467-3470.

[258] Schechter, L.E.; Smith, D.L.; Rosenzweig-Lipson, S.; Rosenzweig-Lipson, S.; Sukoff, S.J.; Dawson, L.A.; Marquis, K.; Jones, D.; Piesla, M.; Andree, T.; Nawoschik, S.; Harder, J.A.; Womack, M.D.; Buccafusco, J.; Terry, A.V.; Hoebel, B.; Rada, P.; Kelly, M.; Abou-Gharbia, M.; Barrett, J.E.; Childers, W. Lecozotan (SRA-333): A selective serotonin 1A receptor antagonist that enhances the stimulated release of glutamate and acetylcholine in the hippocampus and possesses cognitive-enhancing properties. *J. Pharmacol. Exp. Ther.,* **2005**, *314*, 1274-1289.

[259] Khachaturian, A.S.; Zandi, P.P.; Lyketsos, C.G.; Hayden, K.M.; Skoog, I.; Norton, M.C.; Tschanz, J.T.; Mayer, L.S.; Welsh-Bohmer, K.A.; Breitner, J.C. Antihypertensive medication use and incident Alzheimer disease: The Cache County Study. *Arch. Neurol.,* **2006**, *63*, 686-692.

[260] Kehoe, P.G.; Wilcock, G.K. Is inhibition of the renin-angiotensin system a new treatment option for Alzheimer's disease? *Lancet Neurol.,* **2007**, *6*, 373-378.

[261] Yasar, S.; Corrada, M.; Brookmeyer, R.; Kawas, C. Calcium channel blockers and risk of AD: the Baltimore Longitudinal Study of Aging. *Neurobiol. Aging,* **2005**, *26*, 157-163.

[262] Bishop, K.M.; Hofer, E.K.; Mehta, A.; Ramirez, A.; Sun, L.; Tuszynski, M.; Bartus, R.T. Therapeutic potential of CERE-110 (AAV2- NGF): Targeted, stable, and sustained NGF delivery and trophic activity on rodent basal forebrain cholinergic neurons. *Exp. Neurol.,* **2008**, *211*, 574-584.

[263] Mandel, R.J. CERE-110, an adeno-associated virus-based gene delivery vector expressing human nerve growth factor for the treatment of Alzheimer's disease. *Curr. Opin. Mol. Ther.,* **2010**, *12*, 240-247.

[264] Tuszynski, M.H.; Thal. L.; Pay, M.; Salmon, D.P.; U HS, Bakay, R.; Patel, P.; Blesch, A.; Vahlsing, H.L.; Ho, G.; Tong, G.; Potkin, S.G.; Fallon, J.; Hansen, L.; Mufson, E.J.; Kordower, J.H.; Gall, C.; Conner, J. A Phase-1 clinical trial of nerve growth factor gene therapy for Alzheimer disease. *Nat. Med.,* **2005**, *11*, 551-555.

[265] Rockenstein, E,; Mante, M.; Adame, A.; Crews, L.; Moessler, H.; Masliah, E. Effects of Cerebrolysin on neurogenesis in an APP transgenic model of Alzheimer's disease. *Acta Neuropathol.,* **2007**, *113*, 265-275.

[266] Researching Alzheimer's medicines: Setbacks and stepping stones. *PhRMA.* **2012**, 1-20 (www.PhRMA.org).

[267] Lundbert, J.M. Interview with Pharmaceutical Research and Manufacturers of America, October **2011**.

[268] Muegge, I. Selection criteria for drug-like compounds. *Med. Res. Rev.,* **2003**, *23*, 302-321.

[269] Lipinski, C.A.; Lombardo, F.; Dominy, B.W.; Feeney, P.J. Experimental and computational approaches to estimate solubility and permeability in drug discovery and development settings. *Adv. Drug Delivery Rev.,* **1997**, *46*, 3-26.

[270] Hunt, R; Taveau, R.D. On the physiological action of certain cholin derivatives and new methods for detecting cholin. *Br. Med. J.,* **1906**, *2*, 1788.

[271] Massoulie, J.; Pezzementi, L.; Bon, S.; Krejci, E.; Vallette, F.M. Molecular and cellular biology of cholinesterases. *Prog. Neurobiol.,* **1993**, *41*, 31-91.

[272] Won, H.S.; Kenneth S.S.; Yoo, H.S. Therapeutic agents for Alzheimer's Disease. *Curr. Med. Chem.,* **2005**, *5*, 259-269.

[273] Wiesner, J.; Kriz, Z.; Kuca, K.; Jun, D.; Koca, J. Acetylcholinesterases—the structural similarities and differences. *J. Enzyme Inhib. Med. Chem.,* **2007**, *22*, 417-424.

[274] Matthews, G. Neurotransmitter release. *Annu. Rev. Neurosci.,* **1996**, *19*, 219-233.

[275] Harel, M.; Kryger, G.; Rosenberry, T.L.; Mallender, W.D.; Lewis, T.; Fletcher, R.J.; Guss, J.M.; Silman, I.; Sussman J.L. Three-dimensional structures of Drosophila melanogaster acetylcholinesterase and of its complexes with two potent inhibitors. *Protein Sci.,* **2000**, *9* (6), 1063-1072.

[276] Bourne, Y.; Grassi, J.; Bougis, P.E.; Marchot, P. Conformational flexibility of the acetylcholinesterase tetramer suggested by X-ray crystallography. *J. Biol. Chem.,* **1999**, *274* (43), 30370-30376.

[277] Kryger, G.; Harel, M.; Giles, K.; Toker, L.; Velan, B.; Lazar, A.; Kronman, C.; Barak, D.; Ariel, N.; Shafferman, A.; Silman, I.; Sussman, J.L. Structures of recombinant

native and E202Q mutant human acetylcholinesterase complexed with the snake venom toxin fasciculin-II. *Acta. Cryst. D,* **2000**, *56* (11), 1385-1394.

[278] Bourne, Y.; Taylor, P.; Marchot, P. Acetylcholinesterase inhibition by fasciculin: Crystal structure of the complex. *Cell,* **1995**, *83*, 503-512.

[279] Raves, M.L.; Harel, M.; Pang, Y.P.; Silman, I.; Kozikowski, A.P.; Sussman J.L. Structure of acetylcholinesterase complexed with the nootropic alkaloid, (-)-huperzine A. *Nat. Struct. Biol.,* **1997**, *4* (1), 57-63.

[280] Millard, C.B.; Koellner, G.; Ordentlich, A.; Shafferman, A.; Silman, I.; Sussman J.L. Reaction products of acetylcholinesterase and VX reveal a mobile histidine in the catalytic triad. *J. Am. Chem. Soc.,* **1999**, *121*, 9883-9884.

[281] Harel, M.; Kleywegt, G.J.; Ravelli, R.B.G.; Silman, I.; Sussman, J.L. Crystal structure of an acetylcholinesterase-fasciculin complex: Interaction of a three-fingered toxin from snake venom with its target. *Structure,* **1995**, *3*, 1355-1366.

[282] Harel, M.; Quinn, D.M.; Nair, H.K.; Silman, I.; Sussman J.L. The X-ray structure of a transition state analog complex reveals the molecular origins of the catalytic power and substrate specificity of acetylcholinesterase. *J. Am. Chem. Soc.,* **1996**, *118*, 2340-2346.

[283] Sussman, J.L.; Harel, M.; Frolow, F.; Oefner, C.; Goldman, A.; Toker, L.; Silman, I. Atomic Structure of acetylcholinesterase from *Torpedo californica*: A prototypic acetylcholine-binding protein. *Science,* **1991**, *253*, 872-879.

[284] Millard, C.B.; Kryger, G.; Ordentlich, A.; Greenblatt, H.M.; Harel, M.; Raves, M.L.; Segall, Y.; Barak, D.; Shafferman, A.; Silman, I.; Sussman, J.L. Crystal structures of aged phosphonylated acetylcholinesterase: Nerve agent reaction products at the atomic level. *Biochemistry,* **1999**, *38* (22), 7032-7039.

[285] Kryger, G.; Silman, I.; Sussman J.L. Structure of acetylcholinesterase complexed with E2020 (Aricept®): Implications for the design of new anti-Alzheimer 's drugs. *Structure,* **1999**, *7* (3), 297-307.

[286] Greenblatt, H.M.; Kryger, G.; Lewis, T.; Silman, I.; Sussman, J.L. Structure of acetylcholinesterase complexed with (-)-galanthamine at 2.3 angstrom resolution. *FEBS Lett.,* **1999**, *463*, 321-326.

[287] Bartolucci, C.; Perola, E.; Pilger, C.; Fels, G.; Lamba, D. Three-dimensional structure of a complex of galanthamine (Nivalin®) with acetylcholinesterase from *Torpedo californica*: Implications for the design of new anti-Alzheimer drugs. *Proteins,* **2001**, *42* (2), 182-191.

[288] Doucet-Personeni, C.; Bentley, P.D.; Fletcher, R.J.; Kinkaid, A.; Kryger, G.; Pirard, B.; Taylor, A.; Taylor, R.; Taylor, J.; Viner, R.; Silman, I.; Sussman, J.L.; Greenblatt, H.M.; Lewis, T. A Structure-Based Design Approach to the Development of Novel, Reversible AChE Inhibitors. *J. Med. Chem.,* **2001**, *44*, 3203-3215.

[289] Koellner, G.; Steiner, T.; Millard, C.B.; Silman, I.; Sussman J.L. A Neutral Molecule in a Cation-binding Site: Specfic Binding of a PEG-SH to acetylcholinesterase from *Torpedo californica. J. Mol. Biol.,* **2002**, *320* (4), 721-725.

[290] Weik, M.; Ravelli, R.B.G.; Kryger, G.; McSweeney, S.; Raves, M.L.; Harel, M.; Gros, P.; Silman, I.; Kroon, J.; Sussman, J.L. Specific chemical and structural damage to proteins produced by synchrotron radiation. *Proc. Natl. Acad. Sci. USA,* **2000**, *97* (2), 623-628.

[291] Bar-On, P.; Millard, C.B.; Harel, M.; Dvir, H.; Enz, A.; Sussman, J.L.; Silman, I. Kinetic and structural studies on the interaction of cholinesterases with the anti-Alzheimer's drug rivastigmine. *Biochemistry*, **2002**, *41*, 3555-3564.

[292] Dvir, H.; Wong, D.M.; Harel, M.; Barril, X.; Orozco, M.; Luque, F.J.; Munoz-Torrero, D.; Camps, P.; Rosenberry, T.L.; Silman, I.; Sussman, J.L. 3D Structure of *Torpedo californica* acetylcholinesterase complexed with huperine X at 2.1A° resolution: Kinetic and molecular dynamic correlates. *Biochemistry*, **2002**, *41*, 2970-2981.

[293] Ul-Haq, Z.; Halim, S.A.; Uddin, R.; Madura J.D. Benchmarking docking and scoring protocol for the identification of potential acetylcholinesterase inhibitors. *J. Mol. Graphics Modell.*, **2010**, *28*, 870-882.

[294] Ul-Haq, Z.; Uddin, R. Structure Based 3D-QSAR Studies on Cholinesterase Inhibitors. Editor: Prof. Suzanne De La Monte, InTech Publishers Croatia, June 16, **2011**, ISBN: 978-9533076904.

[295] Danuello A.; Romeiro, N.C.; Giesel, G.M.; Pivatto, M.; Viegas, C.; Verli, H.; Barreiro, E. J.; Fraga, C.A.M.; Castro, N.G.; Bolzani. V.S. Molecular docking and molecular dynamic studies of semi-synthetic piperidine alkaloids as acetylcholinesterase inhibitors. *J. Braz. Chem. Soc.*, **2012**, *23* (1), 163-170.

[296] Atta-ur-Rahman, Parveen, S.; Khalid, A.; Farooq, A.; Choudhary, M.I. Acetyl and butyrylcholinesterase-inhibiting triterpenoid alkaloids from *Buxus papillosa*. *Phytochemistry*, **2001**, *58*, 963-968.

[297] Atta-ur-Rahman; Ul-Haq, Z; Feroz, F.; Khalid, A.; Nawaz, S.A.; Khan, M.R.; Choudhary, M.I. New cholinesterase-inhibiting steroidal alkaloids from Sarcococca saligna. *Helv. Chim. Acta,* **2004**, *87*, 439-448.

[298] Atta-ur-Rahman; Ul-Haq, Z; Khalid, A.; Anjum, S.; Khan, M.R.; Choudhary, M.I.; Pregnane-type steroidal alkaloids of Sarcococca saligna: A new class of cholinesterase inhibitors. *Helv. Chim. Acta,* **2002**, *85*, 678-688.

[299] Choudhary, M.I.; Shahnaz, S.; Parveen, S.; Khalid, A.; Ayatollahi, S.A.M.; Atta-ur-Rahman; Parvez, M.; New triterpenoid alkaloid cholinesterase inhibitors from *Buxus hyrcana. J. Nat. Prod.*, **2003**, *66*, 739-742.

[300] Khalid, A.; Choudhary, M.I.; Ul-Haq, Z; Anjum, S.; Khan, M.R.; Atta-ur-Rahman. Kinetics and structure-activity relationship studies on steroidal alkaloids that inhibit cholinesterases, *Bioorg. Med. Chem.*, **2004**, *12*, 1995-2003.

[301] Ul-Haq, Z.; Wellenzohn, B.; Tonmunphean, S.; Khalid, A.; Choudhary, M.I.; Rode, B.M. 3D-QSAR studies on natural acetylcholinesterase inhibitors of Sarcococca saligna by comparative molecular field analysis (CoMFA). *Bioorg. Med. Chem. Lett.*, **2003**, *13*, 4375-4380.

[302] Ul-Haq, Z.; Khan, W.; Kalsoom, S.; Ansari, F.L. *In silico* modeling of the specific inhibitory potential of thiophene-2,3-dihydro-1,5-benzothiazepine against BuChE in the formation of beta-amyloid plaques associated with Alzheimer's disease. *Theor. Biol. Med. Model.*, **2010**, *16*, 7, 22.

[303] Ansari, F.L.; Iftikhar, F.; Ihsan-ul-Haq, Mirza, B.; Baseer, M.; Rashid U. Solid-phase synthesis and biological evaluation of a parallel library of 2,3-dihydro-1,5-benzothiazepines. *Bioorg. Med. Chem.*, **2008**, *16*, 7691-7697.

[304] Ansari, F.L.; Umbreen, S.S.; Hussain, L.; Makhmoor, T.; Nawaz, S.A.; Lodhi, M.A.; Khan, S.N.; Shaheen, F.; Chaudhry M.I.; Atta-ur-Rahman. Syntheses and biological activities of chalcone and 1,5- benzothiazepine derivatives: Promising new free-radical

scavengers, and esterase, urease, and alpha-glucosidase inhibitors. *Chem. Biodivers.* **2005**, *2*(4), 487-96.

[305]   Grigoryan, H.A.; Hambardzumyan, A.A.; Mkrtchyan, M.V.; Topuzyan, V.O.; Halebyan, G.P.; Asatryan, R.S. α,β-Dehydrophenylalanine choline esters, a new class of reversible inhibitors of human acetylcholinesterase and butyrylcholinesterase. *Chem. Biol. Interact.,* **2008**, *171*, 108-116.

[306]   Correa-Basurto, J.; Flores-Sandoval, C.; Marin-Cruz, J.; Rojo-Dominguez, A.; Espinoza-Fonseca, L.M.; Trujillo-Ferrara, J.G. Docking and quantum mechanic studies on cholinesterases and their inhibitors. *Eur. J. Med. Chem.,* **2007**, *42*, 10-19.

[307]   Lu, S.-H.; Wu, J.W.; Liu, H.-L.; Zhao, J.-H.; Liu, K.-T.; Chuang, C.-K.; Lin, H.-Y.; Tsai, W.-B. Ho, Y. The discovery of potential acetylcholinesterase inhibitors: A combination of pharmacophore modeling, virtual screening, and molecular docking studies. *J. Biomed. Sci.,* **2011**, *18*(1), 8-20.

[308]   Citron, M. β-Secretase inhibition for the treatment of Alzheimer's disease-promise and challenge. *Trends Pharmacol. Sci.,* **2004**, *25*, 92-97.

[309]   Vassar, R. β-Secretase (BACE) as a drug target for Alzheimer's disease. *Adv. Drug Deliv. Rev.,* **2002**, *54*, 1589-1602.

[310]   Vassar, R.; Bennett, B. D.; Babu-Khan, S.; Kahn, S.; Mendiaz, E. A.; Denis, P.; Teplow, D. B.; Ross, S.; Amarante, P.; Loeloff, R.; Luo, Y.; Fisher, S.; Fuller, J.; Edenson, S.; Lile, J.; Jarosinski, M. A.; Biere, A. L.; Curran, E.; Burgess, T.; Louis, J.-C.; Collins, F.; Treanor, J.; Rogers, G.; Citron, M. β-Secretase cleavage of Alzheimer's amyloid precursor protein by the transmembrane aspartic protease BACE. *Science*, **1999**, *286*, 735-741.

[311]   Hong, L.; Koelsch, G.; Lin, X.; Wu, S.; Terzyan, S.; Ghosh, A. K.; Zhang, X. C.; Tang, J. Structure of the Protease Domain of Memapsin 2 (beta-Secretase) Complexed with Inhibitor. *Science*, **2000**, *290*, 150-153.

[312]   Patel, S.; Vuillard, L.; Cleasby, A.; Murray, C. W.; Yon, J. Apo and inhibitor complex structures of BACE (β-secretase). *J. Mol. Biol.*, **2004**, *343*, 407-416.

[313]   Hong, L.; Tang, J. Flap position of free Memapsin-2 (β-Secretase), A model for flap opening in aspartic protease catalysis. *Biochemistry,* **2004**, *43*, 4689-4695.

[314]   Turner, R. T.; Koelsch, G.; Hong, L.; Castanheira, P.; Ermolieff, J.; Ghosh, A. K.; Tang, J. Subsite specificity of memapsin-2 (beta-secretase): Implications for inhibitor design. *Biochemistry*, **2001**, *40*, 10001-10006.

[315]   Turner, R.T.; Hong, L.; Koelsch, G.; Ghosh, A.K.; Tang, J. Structural locations and functional roles of new subsites S5, S6, and S7 in memapsin-2 (β-Secretase). *Biochemistry*, **2005**, *44*, 105-112.

[316]   Li, X.; Bo, H.; Zhang, X.C.; Hartsuck, J.A.; Tang, J. Predicting memapsin-2 (beta-secretase) hydrolytic activity. *Protein Sci.*, **2010**, *19*, 2175-2185.

[317]   Schechter, I.; Berger, A. On the size of the active site in proteases. I. Papain. *Biochem. Biophys. Res. Comm.,* **1967**, *27*, 157-162.

[318]   Sauder, J.M.; Arthur, J.W.; Dunbrack, R.L. Modeling of substrate specificity of the Alzheimer's disease amyloid precursor protein beta-secretase. *J. Mol. Biol.,* **2000**, *7*, *300* (2), 241-248.

[319]   Maillard, M.C.; Hom, R.K.; Benson, T.E.; Moon, J.B.; Mamo, S.; Bienkowski, M.; Tomasselli, A.G.; Woods, D.D.; Prince, D.B.; Paddock, D.J.; Emmons, T.L.; Tucker, J.A.; Dappen, M.S.; Brogley, L.; Thorsett, E.D.; Jewett, N.; Sinha, S.; John, V. Design,

synthesis, and crystal structure of hydroxyethyl secondary amine-based peptidomimetic inhibitors of human β-secretase. *J. Med. Chem.*, **2007**, *50* (4), 776-781.

[320]   Tung, J.S.; Davis, D.L.; Anderson, J.P.; Walker, D.E.; Mamo, S.; Jewett, N.; Hom, R K.; Sinha, S.; Thorsett, E.D.; John, V. Design of substrate-based inhibitors of human β-secretase. *J. Med. Chem.*, **2002**, *45*, 259-262.

[321]   Shuto, D.; Kasai, S.; Kimura, T.; Liu, P.; Hidaka, K.; Hamada, T.; Shibakawa, S.; Hayashi, Y.; Hattori, C.; Szabo, B.; KMI-008, a novel beta-Secretase inhibitor containing a hydroxymethylcarbonyl isostere as a transition state mimic: Design and synthesis of substrate-based octapeptides. *Bioorg. Med. Chem. Lett.*, **2003**, *13*, 4273-4276.

[322]   Kimura, T.; Shuto, D.; Hamada, Y.; Igawa, N.; Kasai, S.; Liu, P.; Hidaka, K.; Hamada, T.; Hayashi, Y.; Kiso, Y. Design and synthesis of highly active Alzheimer's [beta]-secretase (BACE1) inhibitors, KMI-420 and KMI429, with enhanced chemical stability. *Bioorg. Med. Chem. Lett.*, **2005**, *15*, 211-215.

[323]   Kimura, T.; Hamada, Y.; Stochaj, M.; Ikari, H.; Nagamine, A.; Abdel-Rahman, H.; Igawa, N.; Hidaka, K.; Nguyen, J.-T.; Saito, K.; Hayashi, Y.; Kiso, Y. Design and synthesis of potent [beta]-secretase (BACE1) inhibitors with carboxylic acid bioisosteres. *Bioorg. Med. Chem. Lett.*, **2006**, *16*, 2380-2386.

[324]   Ghosh, A.K.; Kumaragurubaran, N.; Hong, L.; Kulkarni, S.; Xu, X.; Miller, H.B.; Reddy, S.D.; Weerasena, V.; Turner, R.; Chang, W.; Koelsch, G.; Tang, J. Potent memapsin 2 ([beta]-secretase) inhibitors: Design, synthesis, protein-ligand X-ray structure and *in vivo* evaluation. *Bioorg. Med. Chem. Lett.*, **2008**, *18*, 1031-1036.

[325]   Wångsell, F.; Russo, F.; Savmarker, J.; Rosenquist, Å.; Samuelsson, B.; Larhed, M. Design and synthsis of BACE-1 inhibitors utilizing a tertiary hydroxyl motif as the transition state mimic. *Bioorg. Med. Chem. Lett.*, **2009**, *19*, 4711-4714.

[326]   Wångsell, F.; Gustafsson, K.; Kvarnström, I.; Borkakoti, N.; Edlund, M.; Jansson, K.; Lindberg, J.; Rosenquist, A.; Samuelsson, B. Synthesis of potent BACE-1 inhibitors incorporating a hydroxyethylene isostere as central core. *Eur. J. Med. Chem.*, **2010**, *45*, 870-882.

[327]   Wångsell, F.; Nordeman, P.; Sävmarker, J.; Emanuelsson, R.; Jansson, K.; Lindberg, J.; Rosenquist, Å.; Samuelsson, B.; Larhed. M. Investigation of a phenylnorstatin and α-benzylnorstatin as transition state isostere motifs in the search for new BACE-1 inhibitors. *Bioorg. Med. Chem.*, **2011**, *19*, 145-155.

[328]   Clarke, B.; Demont, E.; Dingwall, C.; Dunsdon, R.; Falle,r A.; Hawkins, J.; Hussain, I.; MacPherson, D.; Maile, G.; Matico, R.; Milner, P.; Mosley, J.; Naylor, A.; O'Brien, A.; Redshaw, S.; Riddell, D.; Rowland, P.; Soleil, V.; Smith, K.J.; Stanway, S.; Stemp, G.; Sweitzer, S.; Theobald, P.; Vesey, D.; Walter, D.S.; Ward, J.; Wayne, G. BACE-1 inhibitors Part-1: Identification of novel hydroxy ethylamines (HEAs). *Bioorg. Med. Chem. Lett.*, **2008**, *18*, 1011-1016.

[329]   Clarke, B.; Demont, E.; Dingwall, C. Dunsdon, R.; Faller, A.; Hawkins, J.; Hussain, I.; MacPherson, D.; Maile, G.; Matico, R.; Milner, P.; Mosley, J.; Naylor, A.; O'Brien, A.; Redshaw, S.; Riddell, D.; Rowland, P.; Soleil, V.; Smith, K.J.; Stanway, S.; Stemp, G.; Sweitzer, S.; Theobald, P.; Vesey, D.; Walter, D.S.; Ward, J.; Wayne, G. BACE-1 inhibitors Part 2: Identification of hydroxyl ethylamines (HEAs) with reduced peptidic character. *Bioorg. Med. Chem. Lett.*, **2008**, *18*, 1017-1021.

[330] Beswick, P.; Charrier, N.; Clarke, B. Demont, E.; Dingwall, C.; Dunsdon, R.; Faller, A.; Gleave, R.; Hawkins, J.; Hussain, I.; Johnson, C.; MacPherson, D.; Maile, G.; Matico, R.; Milner, P.; Mosley, J.; Naylor, A.; O'Brien, A.; Redshaw, S.; Riddell, D.; Rowland, P.; Skidmore, J.; Soleil, V.; Smith, K.; Stanway, S.; Stemp, G.; Stuart, A.; Sweitzer, S.; Theobald, P.; Vesey, D.; Walter, D.S.; Ward, J.; Wayne, G. BACE-1 inhibitors Part 3: Identification of hydroxyethylamines (HEAs) with nanomolar potency in cells. *Bioorg. Med. Chem. Lett.,* **2008**, *18*, 1022-1026.

[331] Cole, D.C.; Stock, J.R.; Chopra, R.; Cowling, R.; Ellingboe, J.W.; Fan, K.Y.; Harrison, B.L.; Hu, Y.; Jacobsen, S.; Jennings, L.D.; Jin, G.; Lohse, P.A.; Malamas, M.S.; Manas, E.S.; Moore, W.J.; O' Donnell, M. M.; Olland, A.M.; Robichaud, A.J.; Svenson, K.; Wu, J.; Wagner, E.; Bard, J. Acylguanidine inhibitors of β -secretase: Optimization of the pyrrole ring substituents extending into the S1 and S3 substrate binding pockets. *Bioorg. Med. Chem. Lett.,* **2008**, *18* (3), 1063-1066.

[332] Fobare, W.F.; Solvibile, W.R.; Robichaud, A.J.; Malamas, M.S.; Manas,E.; Turner, J.; Hu, Y.; Wagner, E.; Chopra, R.; Cowling, R.; Jin, G.; Bard, J. Thiophene substituted acylguanidines as BACE-1 inhibitors. *Bioorg. Med. Chem. Lett.,* **2007**, *17* (19), 5353-5356.

[333] Malamas, M.S.; Erdei, J.; Gunawan, I. Barnes, K.; Johnson, M.; Hui, Y.; Turner, J.; Hu, Y.; Wagner, E.; Fan, K.; Olland, A.; Bard, J.; Robichaud, A. J. Aminoimidazoles as potent and selective human β-secretase (BACE1) inhibitors. *J. Med. Chem.,* **2009**, *52*, 6314-6323.

[334] Malamas, M.S.; Erdei, J.; Gunawan, I. Turner, J.; Hu, Y.; Wagner, E.; Fan, K.; Chopra, R.; Olland, A.; Bard, J.; Jacobsen, S.; Magolda, R.L.; Pangalos M.; Robichaud, A.J. Design and synthesis of 5,5′-disubstituted aminohydantoins as potent and selective human β-secretase (BACE-1) inhibitors. *J. Med. Chem.,* **2010**, *53*, 1146-1158.

[335] Baxter, E.W.; Conway, K.A.; Kennis, L. Bischoff, F.; Mercken, M.H.; Winter, H.L.; Reynolds, C.H.; Tounge, B.A.; Luo, C.; Scott, M.K.; Huang, Y.; Braeken, M.; Pieters, S.M.; Berthelot, D.J.; Masure, S.; Bruinzeel, W.D.; Jordan, A.D.; Parker, M.H.; Boyd, R.E.; Qu, J.; Alexander, R.S.; Brenneman, D.E.; Reitz, A.B. 2-Amino-3,4-dihydroquinazolines as inhibitors of BACE-1 (β-Site APP cleaving enzyme): Use of structure based design to convert a micromolar hit into a nanomolar lead. *J. Med. Chem.,* **2007**, *50*, 4261-4264.

[336] Polgar, T.; Keserue, G.M. Virtual screening for β -secretase (BACE1) inhibitors reveals the importance of protonation states at Asp32 and Asp228. *J. Med. Chem.,* **2005**, *48* (11), 3749-3755.

[337] Bhisetti, G.R.; Saunders, J.O.; Murcko, M.A.; Lepre, C.A.; Britt, S.D.; Come, J.H.; Deninger, D.D.; Wang, T. Preparation of β-carbolines and other inhibitors of BACE-1 aspartic proteinase useful against Alzheimer's and other BACE-mediated diseases. Vertex Pharmaceuticals Incorporated, *WO-2002088101,* **2002**, 208.

[338] Limongelli, V.; Braun, H. A.; Marinelli, L.; Schmidt, B;. Cosconati, S.; Novellino, E. Ensemble-Docking approach on BACE-1: Pharmacophore perception and guidelines for drug design. *ChemMedChem.,* **2007**, *2*, 667-678.

[339] John, S.; Thangapandian, S.; Sakkiah, S.; Lee. K. W. Potent bace-1 inhibitor design using pharmacophore modeling, *in silico* screening and molecular docking studies *BMC Bioinformatics,* **2011**, *12* (Suppl. 1), S28.

[340] Geschwindner, S.; Olsson, L.L.; Albert, J.S.; Deinum, J.; Edwards, P.D.; de Beer, T.; Folmer, R.H.A. Discovery of a novel warhead against β -secretase through fragment-based lead generation. *J. Med. Chem.*, **2007**, *50*, 5903-5911.

[341] Kosik, K.S. Shimura, H. Biochim. Phosphorylated tau and the neurodegenerative foldopathies. *Biochim Biophys Acta*, **2005**, *1739* (2-3), 298-310.

[342] Pickhardt, M.; von Bergen, M.; Gazova, Z.; Hascher, A.; Biernat, J.; Mandelkow, E. M.; Mandelkow, E. Screening for inhibitors of tau polymerization. *Curr. Alzheimer Res.,* **2005**, *2*, 219-226.

[343] Carlson, E.E.; May, J.F.; Kiessling, L.L. Chemical probes of UDP-galactopyranose mutase. *Chem Biol.*, **2006**, *13*, 825-837.

[344] Patani, G.A.; LaVoie, E.J.; Bioisosterism: A rational approach in drug design. *Chem. Rev.,* **1996**, *96*, 3147-3176.

[345] Bulic, B.; Pickhardt, M.; Khlistunova, I.; Biernat, J.; Mandelkow, E.M.; Mandelkow, E.; Waldmann, H. Rhodanine-based tau aggregation inhibitors in cell models of tauopathy. *Angew. Chem. Int. Ed. Engl.*, **2007**, *46*, 9215-9219.

[346] Larbig, G.; Pickhardt, M.; Lloyd, D.G.; Schmidt, B.; Mandelkow, E. Screening for inhibitors of tau protein aggregation into Alzheimer paired helical filaments: A ligand based approach results in successful scaffold hopping. *Curr. Alzheimer Res.,* **2007**, *4*, 315-323.

[347] Pickhardt, M.; Larbig, G.; Khlistunova, I.; Coksezen, A.; Meyer, B.; Mandelkow, E.M.; Schmidt, B.; Mandelkow, E. Phenylthiazolyl-hydrazide and its derivatives are potent inhibitors of tau aggregation and toxicity *in vitro* and in cells. *Biochemistry,* **2007**, *46*, 10016-10023.

[348] Ballatore, C.; Crowe, A.; Piscitelli, F.; James, M.; Rossidivito, G.; Yao, Y.; Trojanowski, J.Q.; Lou, K.; Lee, V.M.Y.; Brunden, K.R.; SmithA.B. Aminothienopyridazine inhibitors of tau aggregation: Evaluation of structure-activity relationship leads to selection of candidates with desirable *in vivo* properties. *Bioorg. Med. Chem.* **2012**, *20*, 4451-4461.

[349] Pardridge, W.M. Alzheimer's disease drug development and the problem of the blood brain barrier. *Alzheimer's Dement.,* **2009**, *5*(5), 427-432.

[350] Banks, W.A. Drug delivery to the brain in Alzheimer's disease: Consideration of the blood-brain barrier. *Adv. Drug Deliv. Rev.,* **2012**, *64*, 629-639.

[351] Rishton, G.M.; LaBonte, K.; Williams, A.J.; Kassam, K.; Kolovanov, E. Computational approaches to the prediction of blood brain barrier permeability: A comparative analysis of central nervous system drugs versus secretase inhibitors for Alzheimer's disease. *Curr. Opin. Drug Discov. Devel.,* **2006**, *9*, 303-13.

[352] Deane, R.; Sagare, A.; Hamm, K.; Parisi, M.; LaRue, B.; Guo, H.; Wu, Z.; Holtzman, D.M.; Zlokovic, B.V. IgG-assisted age-dependent clearance of Alzheimer's amyloid peptide by the blood-brain barrier neonatal Fc receptor. *J. Neurosci.,* **2005**, *25*, 11495-11503.

[353] Kaur, I. P.; Bhandari, R.; Bhandari, S.; Kakkar, V. Potential of solid lipid nanoparticles in brain targeting. *J. Control. Release,* **2008**, *127*, 97-109.

[354] Brambilla, D.; Le Droumaguet, B.; Nicolas, J.; Hashemi, S.H.; Wu, L.-P.; Moghimi, S.M.; Couvreur, P.; Andrieux, K. Nanotechnologies for Alzheimer's disease: Diagnosis, therapy, and safety issues. *Nanomedicine*, **2011**, *7* (5), 521-540.

[355] Wilson, B.; Samanta, M.K.; Santhi, K.; Kumar, K.P.; Paramakrishnan, N.; Suresh, B. Poly(nbutylcyanoacrylate) nanoparticles coated withpolysorbate 80 for the targeted

delivery of rivastigmine into the brain to treat Alzheimer's disease. *Brain Res.,* **2008**, *1200*, 159-168.

[356] Yang, Z,; Zhang, Y.; Yang, Y.; Sun, L.; Han, D.; Li, H.; Wang, C. Pharmacological and toxicological target organelles and safe use of single-walled carbon nanotubes as drug carriers in treating Alzheimer's disease. *Nanomedicine,* **2010**, *6* (3), 427-41.

[357] Curtain, C.; Ali, F.; Volitakis, I.; Cherny, R.; Norton, R.; Beyreuther, K.; Barrow, C.J.; Masters, C.L.; Bush, A.I.; Barnham, K.J. Alzheimer's disease amyloid- binds Cu and Zn to generate an allosterically-ordered membrane- penetrating structure containing SOD-like subunits. *J. Biol. Chem.,* **2001**, *276*, 20466-20473.

[358] Liu, G.; Men, P.; Harris, P.L.; Rolston. R.K.; Perry, G.; Smith, M.A. Nanoparticle iron chelators: A new therapeutic approach in Alzheimer's disease and other neurologic disorders associated with trace metal imbalance. *Neurosci. Lett.,* **2006**, *406* (3), 189-193.

[359] Cui, Z.; Lockman, P.R.; Atwood, C.S.; Hsu, C.H.; Gupte, A.; Allen, D.D.; Mumper, R.J. Novel D- penicillamine carrying nanoparticles for metal chelation therapy in Alzheimer's and other CNS diseases. *Eur. J. Pharm. Biopharm.,* **2005**, *59* (2), 263-272.

[360] Taylor, P.; Lappi, S. Interaction of fluorescence probes with acetylcholinesterase. The site and specificity of propidium binding. *Biochemistry*, **1975**, *14*, 1989-97.

[361] Sussman, J.L.; Harel, M.; Frolow, F.; Oefner, C.; Goldman, A.; Toker, L.; Silman, I. Atomic structure of acetylcholinesterase from Torpedo californica: A prototypic acetylcholine-binding protein. *Science*, **1991**, *253*, 872-879.

[362] Martinez, A.; Lanot, C.; Perez, C.; Castro, A.; López-Serrano, P.; Conde, S. Lipase-catalysed synthesis of new acetylcholinesterase inhibitors: N-benzylpiperidine amino acid derivatives. *Biorg. Med. Chem.*, **2000**, *8*, 731-738.

[363] Martínez, A.; Fernandez, E.; Castro, A.; Conde, S.; Rodríguez-Franco, I.; Baños, J.E.; Badia, A. N-benzylpiperidine derivatives of 1,2,4-thiadiazolidinone as new acetylcholinesterase inhibitors. *Eur. J. Med. Chem.,* **2000**, *35*, 913-922.

[364] Andreani, A.; Cavalli, A.; Granaiola, M.; Guardigli, M.; Leoni, A.; Locatelli, A. Morigi, R.; Rambaldi, M.; Recanatini, M.; Roda, A. Synthesis and screening for antiacetylcholinesterase activity of (1-benzyl-4-oxopiperidin-3- ylidene)methylindoles and pyroles related to donepezil. *J. Med. Chem.,* **2001**, *41*, 4011-4014.

[365] da Silva, C.H.T.P.; Campo, V.L.; Carvalho, I.; Taft, C.A. Molecular modeling, docking and ADMET studies applied to the design of a novel hybrid for treatment of Alzheimer's disease. *J. Mol. Graph. Model.*, **2006**, *25*, 169-175.

[366] Pang, Y.P.; Hong, F.; Quiram, P.; Jelacic, T.; Brimijoin, S. Synthesis of alkylene linked bis-THA and alkylene linked benzyl-THA as highly potent and selective inhibitors and molecular probes of acetylcholinesterase. *J. Chem. Soc. Perkin Trans.,* **1997**, *1*, 171-176.

[367] Lewis, W.G.; Green, L.G.; Grynszpan, F.; Radic, Z.; Carlier, P.R.; Taylor, P.; Finn, M.G.; Sharpless, K.B. Click chemistry in situ: Acetylcholinesterase as a reaction vessel for the selective assembly of a femtomolar inhibitor from an array of building blocks. *Angew Chem. Int. Ed. Engl.,* **2002**, *41*, 1053-1057.

[368] Shao, D.; Zou, C.; Luo, C.; Tang, X.; Li, Y. Synthesis and evaluation of tacrine-E2020 hybrids as acetylcholinesterase inhibitors for the treatment of Alzheimer's disease. *Bioorg. Med. Chem. Lett.,* **2004**, *14*, 4639-4642.

[369] Alonso, D.; Dorronsoro, I.; Rubio, L.; Muñoz, P.; García-Palomero, E.; Del Monte, M.; Bidon-Chanal, A.; Orozco, M.; Luque, F.J.; Castro, A.; Medina, M.; Martínez, A.

Donepezil-tacrine hybrid related derivatives as new dual binding site inhibitors of AChE. *Bioorg. Med. Chem.,* **2005**, *13*, 6588-6597.

[370] Guilliou, C.; Mary, A.; Renko, D.Z.; Gras, E.; Thal, C. Potent acetylcholinesterase inhibitors: Design, synthesis and structure-activity relationships of alkylene linked bis-galanthamine and galanthamine salts. *Bioorg. Med. Chem. Lett.,* **2000**, *10*, 637-639.

[371] Carlier, P.R.; Du, D.M.; Han, Y.; Liu, J.; Pang, Y.P. Potent, easily synthesized huperzine A-tacrine hybrid acetylcholinesterase inhibitors. *Bioorg. Med. Chem. Lett.,* **1999**, *9*, 2335-2338.

[372] Carlier, P.R.; Du, D.M.; Han, Y.; Liu, J.; Perola, E.; Williams, I.D.; Pang, Y.P. Dimerization of an inactive frangment of huperzine A produces a drug with twice the potency of the natural product. *Angew Chem. Int. Ed.,* **2000**, *39*, 1775-1777.

[373] Wong, D.M.; Greenblatt, H.M.; Dvir, H. Carlier, P.R.; Han, Y.F.; Pang, Y.P.; Silman, I.; Sussman, J.L. Acetylcholinesterase complexed with bivalent kigands related to huperzine A: Experimental evidence for species dependent protein-ligand complementarity. *J. Am. Chem. Soc.,* **2003**, *125*, 363-373.

[374] Yan, J.; Sun, L.; Wu, G.; Yi, P.; Yang, F.; Zhou, L.; Zhang, X.; Li, Z.; Yang, X.; Luo, H.; Qiu, M. Rational design and synthesis of highly potent antia- acetylcholinesterase activity huperzine A derivatives. *Bioorg. Med. Chem.,* **2009**, *17*, 6937-6941.

[375] Camps, P.; Formosa, X.; Muños-Torrero, D.; Petrignet, J.; Badia, A.; Clos, M.V. Synthesis and pharmacological evaluation of huprine-tacrine heterodimers: Subnanomolar dual binding site acetylcholinesterase inhibitors. *J. Med. Chem.,* **2005**, *48*, 1701-1704.

[376] Cavalli, A.; Bolognesi, M.L.; Minarini, A.; Rosini, M.; Tumiatti, V.; Recanatini, M.; Melchiorre, C. Multi-target-Directed Ligands to combat neurodegenerative diseases. *J. Med. Chem.,* **2008**, *51*, 347-372.

[377] Inestrosa, N.C.; Alvarez, A.; Calderon, F. Acetylcholinesterase is a senile plaque component that promotes assembly of amyloid beta-peptide into Alzheimer's filaments. *Mol. Psychiatry,* **1996**, *1*, 359-361.

[378] Selkoe, D.J. Translating cell biology into therapeutic advances in Alzheimer's disease. *Nature,* **1999**, *399*, A23-A31.

[379] Melchiorre, C.; Antonello, A.; Banzi, R.; Bolognesi, M.L.; Minarini, A.; Rosini, M.; Tumiatti, V. Polymethylene tetraamine backbone as template for the development of biologically active polyamines. *Med. Res. Rev.,* **2003**, *23*, 200-233.

[380] Melchiorre C, Romualdi P, Bolognesi ML, Donatini A, Ferri S. Binding profile of benextramine at neuropeptide Y receptor subtypes in rat brain areas. *Eur. J. Pharmacol.,* **1994**, *265*, 93-98.

[381] Melchiorre, C.; Andrisano, V.; Bolognesi, M. L.; Budriesi, R.; Cavalli, A.; Cavrini, V.; Rosini, M.; Tumiatti, V.; Recanatini, M. Acetylcholinesterase noncovalent inhibitors based on a polyamine backbone for potential use against Alzheimer's disease. *J. Med. Chem.,* **1998**, *41*, 4186-4189.

[382] Piazzi, L.; Rampa, A.; Bisi, A.; Gobbi, S.; Belluti, F.; Cavalli, A.; Bartolini, M.; Andrisano, V.; Valenti, P.; Recanatini, M. 3-(4[[Benzyl(methyl)amino] methyl] phenyl)-6,7- dimethoxy-2H-2-chromenone (AP2238) inhibits both acetylcholinesterase and acetylcholinesterase-induced beta-amyloid aggregation: A dual function lead for Alzheimer's disease therapy. *J. Med. Chem.,* **2003**, *46*, 2279-2282.

[383] Cavalli, A.; Bolognesi, M. L.; Capsoni, S.; Andrisano, V.; Bartolini, M.; Margotti, E.; Cattaneo, A.; Recanatini, M.; Melchiorre, C. A small molecule targeting the multifactorial nature of Alzheimer's disease. *Angew. Chem., Int. Ed.*, **2007**, *46*, 3689-3692.

[384] Bolognesi, M.L.; Andrisano, V.; Bartolini, M.; Banzi, R.; Melchiorre, C. Propidium-based polyamine ligands as potent inhibitors of acetylcholinesterase and acetylcholinesterase-induced amyloid-beta aggregation. *J. Med. Chem.,* **2005**, *48*, 24-27.

[385] Rosini, M.; Andrisano, V.; Bartolini, M.; Bolognesi, M.L.; Hrelia, P.; Minarini, A.; Tarozzi, A.; Melchiorre, C. Rational approach to discover multipotent anti-Alzheimer drugs. *J. Med. Chem.*, **2005**, *48*, 360-363.

[386] Beal, M.F. Mitochondrial dysfunction and oxidative damage in Alzheimer's and Parkinson's diseases and coenzyme Q10 as a potential treatment. *J. Bioenerg. Biomembr.,* **2004**, *36*, 381-386.

[387] Fink, D.M.; Palermo, M.G.; Bores, G.M.; Huger, F. P.; Kurys, B. E.; Merriman, M. C.; Olsen, G. E.; Petko, W.; O'Malley, G. J. Imino 1,2,3,4-tetrahydrocyclopent[b]indole carbamates as dual inhibitors of acetylcholineesterase and monoamine oxidase. *Bioorg. Med. Chem. Lett.,* **1996**, *6*, 625-630.

[388] Youdim, M. B.; Amit, T.; Bar-Am, O.; Weinreb, O.; Yogev-Falach, M. Implications of co-morbidity for etiology and treatment of neurodegenerative diseases with multifunctional neuroprotective neurorescue drugs; ladostigil. *Neurotoxic Res.,* **2006**, *10*, 181-192.

[389] Bar-Am, O.; Weinreb, O.; Amit, T.; Youdim, M.B. The novel cholinesterase-monoamine oxidase inhibitor and antioxidant, ladostigil, confers neuroprotection in neuroblastoma cells and aged rats. *J. Mol. Neurosci.*, **2009**, *37*, 135-145.

[390] Marco-Contelles, J.; Leon, R.; de Los Rios, C.; Guglietta, A.; Terencio, J.; Lopez, M. G.; Garcia, A. G.; Villarroya, M. Novel multipotent tacrine-dihydropyridine hybrids with improved acetylcholinesterase inhibitory and neuroprotective activities as potential drugs for the treatment of Alzheimer's disease. *J. Med. Chem.*, **2006**, *49*, 7607-7610.

[391] Choudhary, M.I.; Nawaz, S.A.; Ul-Haq, Z.; Azim, M.K.; Ghayur, M.N.; Lodhi, M.A.; Jalil, S.; Khalid, A.; Ahmed, A.; Rode, B.M.; Atta-ur-Rahman; Gilani, A.U.; Ahmad, V.U. Juliflorine: A potent natural peripheral anionic-site-binding inhibitor of acetylcholinesterase with calcium-channel blocking potential, a leading candidate for Alzheimer's disease therapy *Biochem. Biophys. Res. Commun.*, **2005**, *15*, 332(4), 1171-1177.

[392] Atta-ur-Rahman, Khalid, A.; Sultana, N,; Ghayur, M.N.; Mesaik, M.A.; Khan, M.R.; Gilani, A.H.; Choudhary, M.I. New natural cholinesterase inhibition and calcium channel blocking quinoline alkaloids. *J. Enzyme Inhib. Med. Chem.,* **2006**, *21* (6), 703-710.

[393] Bolognesi, M. L.; Minarini, A.; Tumiatti, V.; Melchiorre, C. Lipoic acid, a lead structure for multi-target-directed drugs for neurodegeneration. *Mini-Rev. Med. Chem.*, **2006**, *6*, 1269-1274.

[394] Arce, M.P.; Rodriguez-Franco, M.I.; Gonzalez-Munoz, G.C.; Perez, C.; Lopez, B.; Villarroya, M.; Lopez, M.G.; Garcia, A.G.; Conde, S. Neuroprotective and cholinergic properties of multifunctional glutamic acid derivatives for the treatment of Alzheimer's disease. *J. Med. Chem.*, **2009**, *52*, 7249-7257.

[395] Prokai-Tatrai, K.; Nguyen, V.; Zharikova, A.D.; Braddy, A.C. Stevens SM, Prokai L. Prodrugs to enhance central nervous system effects of the TRH-like peptide pGlu-Glu-Pro-NH$_2$. *Bioorg. Med. Chem. Lett.,* **2003**, *13*, 1011-1014.

[396] Hirai, K.; Kato, K.; Nakayama, T.; Hayako, H.; Ishihara, Y.; Goto, G.; Miyamoto, M. Neurochemical effects of 3-[1-(phenylmethyl)-4-piperidinyl]-1-(2,3,4,5-tetrahydro-1H-1-

benzazepin-8-yl)-1-propanone fumarate (TAK-147), a novel acetylcholinesterase inhibitor, in rats. *J. Pharmacol. Exp. Ther.*, **1997**, *280*, 1261-1269.

[397]   Kogen, H.; Toda, N.; Tago, K.; Marumoto, S.; Takami, K.; Ori, M.; Yamada, N.; Koyama, K.; Naruto, S.; Abe, K.; Yamazaki, R.; Hara, T.; Aoyagi, A.; Abe, Y.; Kaneko, T. Design and synthesis of dual inhibitors of acetylcholinesterase and serotonin transporter targeting potential agents for Alzheimer's disease. *Org. Lett.*, **2002**, *4*, 3359-3362.

[398]   Toda, N.; Tago, K.; Marumoto, S.; Takami, K.; Ori, M.; Yamada, N.; Koyama, K.; Naruto, S.; Abe, K.; Yamazaki, R.; Hara, T.; Aoyagi, A.; Abe, Y.; Kaneko, T.; Kogen, H. Design, synthesis and structure-activity relationships of dual inhibitors of Acetylcholinesterase and serotonin transporter as potential agents for Alzheimer's disease. *Bioorg. Med. Chem.*, **2003**, *11*, 1935-1955.

[399]   Toda, N.; Kaneko, T.; Kogen, H. Development of an efficient therapeutic agent for Alzheimer's disease: Design and synthesis of dual inhibitors of acetylcholinesterase and Serotonin Transporter. *Chem. Pharm. Bull.*, **2010**, *58* (3), 273-287.

[400]   Rosini, M.; Simoni, E.; Bartolini, M.; Cavalli, A.; Ceccarini, L.; Pascu, N.; McClymont, D.W.; Tarozzi, A.; Bolognesi, M.L.; Minarini, A.; Tumiatti, V.; Andrisano, V.; Mellor I.R.; Melchiorre, C. Inhibition of acetylcholinesterase, beta-amyloid aggregation, and NMDA receptors in Alzheimer's disease: A promising direction for the multi-target-directed ligands gold rush. *J. Med. Chem.*, **2008**, *51*, 4381-4384.

[401]   Lysko, P.G.; Lysko, K.A.; Webb, C.L.; Feuerstein, G.; Mason, P.E.; Walter, M.F.; Mason, R.P. Neuroprotective activities of carvedilol and a hydroxylated derivative: Role of membrane biophysical interactions. *Biochem. Pharmacol.*, **1998**, *56*, 1645-1656.

[402]   Howlett, D.R.; George, A.R.; Owen, D.E.; Ward, R.V.; Markwell, R.E. Common structural features determine the effectiveness of carvedilol, daunomycin and rolitetracycline as inhibitors of Alzheimer beta-amyloid fibril formation. *Biochem. J.*, **1999**, *343*, 419-423.

[403]   Bolognesi, M.L.; Rosini, M.; Andrisano, V.; Bartolini, M.; Minarini, A.; Tumiatti, V.; Melchiorre, C. MTDL design strategy in the context of Alzheimer's disease: From lipocrine to memoquin and beyond. *Curr. Pharm. Des.*, **2009**, *15* (6), 601-613.

[404]   Dedeoglu, A.; Cormier, K.; Payton, S.; Tseitlin, K. A.; Kremsky, J. N.; Lai, L.; Li, X.; Moir, R. D.; Tanzi, R. E.; Bush, A. I.; Kowall, N. W.; Rogers, J. T.; Huang, X. Preliminary studies of a novel bifunctional metal chelator targeting Alzheimer's amyloidogenesis. *Exp. Gerontol.*, **2004**, *39*, 1641-1649.

CHAPTER 3

Enzyme Inhibitors Involved in the Treatment of Alzheimer's Disease

Vageesh Revadigar[1], Raza Murad Ghalib[2,*], Vikneswaran Murugaiyah[1], Mohamed A. Embaby[2], Ali Jawad[3], Sayed Hasan Mehdi[4], Rokiah Hashim[5] and Othman Sulaiman[5]

[1]*Discipline of Pharmacology, School of Pharmaceutical Sciences, Universiti Sains Malaysia, Minden–11800, Pulau Pinang, Malaysia;* [2]*Department of Chemistry, Faculty of Sciences & Arts Khulais, King Abdulaziz University, Jeddah, KSA;* [3]*Department of Applied Physics, Zakir Husain College of Engineering and Technology, Aligarh Muslim University, Aligarh UP, India;* [4]*Department of Chemistry, SPG College, UP, Lucknow, India;* [5]*School of Industrial Technology, Universiti Sains Malaysia, Minden–11800, Pulau Pinang, Malaysia*

Abstract: Alzheimer disease (AD), a chronic and progressive neurodegenerative disease is the leading cause of dementia among older people. Several hypotheses exist on the pathogenesis of AD. According to the cholinergic hypothesis, there is an irreversible deficiency in cholinergic functions of brain resulting in substantial reduction of the neurotransmitter, acetylcholine. In contrast, amyloid hypothesis suggests that amyloid beta (Aβ) deposits are fundamental cause of the neurodegeneration. There is also an involvement of another protein, tau protein, which twists into abnormal tangles and lead to the death of brain cells. This chapter aims to provide a comprehensive review on the various enzymes involved in the pathogenic cascade of AD and their potential inhibitors. The chapter starts with a general overview on the enzymes, outlining their morphological and functional features and how they are involved in the pathogenesis of AD. The following section addresses the enzymes inhibitors at various stages of drug development, highlighting their mechanisms of action, advantages, limitations and potential clinical applications.

Keywords: Alzheimer's disease, acetycholinesterase, amyloid precursor protein, α-secretase, aβ-peptide, begacestat (GSI-953), bisnorcymserine, BMS-708163, β-secretase, butyrylcholinesterase, CTS-21166, donepezil, ELND006, galantamine, huperzine A, LY2811376, PF-3084014, (-)-Phenserine, rivastigmine, semagacestat (LY-450139), senile plaques, γ-secretase.

***Address correspondence to Raza Murad Ghalib:** Department of Chemistry, Faculty of Sciences & Arts Khulais, King Abdulaziz University, Jeddah, KSA; Tel: +966-592358844; Fax: +604 6573678; E-mail: raza2005communications@gmail.com

Atta-ur-Rahman / Muhammad Iqbal Choudhary (Eds.)

10.1016/B978-0-12-803959-5.50003-9

INTRODUCTION

Alzheimer's disease (AD) is the leading cause of dementia among older people. It is not a part of normal aging, but is a chronic and progressive neurodegenerative disease involving the progressive loss of brain nerve cells. Since these nerve cells are essential for normal thought, memory and other brain functions, people with AD suffer a decline of mental functions, which eventually interferes with the patient's normal daily activities, and ultimately leads to death [1].

Worldwide, nearly 36 million people are believed to be living with Alzheimer's disease or other types of dementias. By 2030, if breakthroughs are not discovered, the number may increase to nearly 66 million. By 2050, the number could even exceed 115 million [2].

Despite the smaller number of patients affected compared to other diseases such as diabetes mellitus and cardiovascular complications the social and economic impacts of AD are tremendous. If dementia care were portrayed as a country's economy, it would be the world's 18th largest, ranking between Turkey and Indonesia. Total payments for health care, long-term care, and hospice for people with AD and other dementias are projected to increase from $200 billion in 2012 to $1.1 trillion in 2050 [2].

Caring for a person with AD or another dementia is often extremely difficult, and many times their family members and other unpaid caregivers experience high levels of emotional stress and depression as a result. Caring for someone with AD likely has a negative impact on the health, employment, income, and financial security of caregivers. More than 15 million Americans provide unpaid care for someone with AD or another dementia. Unpaid caregivers are usually family members, and they could also be other relatives and friends. In 2011, these people provided an estimated 17.4 billion hours of unpaid care, a contribution to the nation valued at over $210 billion [2].

Finding the solutions for AD in terms of suitable therapy has been a greater challenge and from the past few decades many researchers and pharmaceutical companies are optimistically working towards this direction. The number of therapeutic options for AD remains severely limited. Currently marketed drugs for the treatment of AD do not prevent or reverse this disease and are approved only for the management of the symptoms. The five drugs currently on the market: the cholinesterase inhibitors (ChEI)s donepezil (Aricept; Esai/Pfizer), rivastigmine (Exelon; Novartis), galantamine (Razadyne; Johnson and Johnson) and tacrine (Cognex; first horizon pharmaceuticals), and N-methyl-D-aspartate (NMDA) receptor modulator memantine (Namenda; Forest/Lundbeck) are the currently FDA approved drugs for the symptomatic treatment of AD [3]. Currently, there are no drugs available in the market, which act on the basic pathological mechanism of the disorder thus capable of curing or preventing the progression of the disorder, although many potential molecules have been identified through huge efforts from the last few decades, some have failed in clinical trials and some are still in developmental stages.

This chapter discusses pathophysiology of the disease, enzymes involved in the disease cascade, their potential inhibitors used in current therapy and also an update in the developments concerning enzyme inhibitors as potential AD drugs.

PATHOPHYSIOLOGY OF ALZHEIMER DISEASE

AD is neuropathologically characterized by the extracellular deposition of β - amyloid (Aβ) aggregates (senile plaques) and intraneuronal neurofibrillary tangles (hyperphosphorylated tau protein) and the loss of neurons in the brain. Several hypotheses exist regarding the cause of AD. This chapter will focus on three main hypotheses, namely amyloid hypothesis, tau or tangle hypothesis and cholinergic hypothesis. Amyloid hypothesis suggests that Aβ deposits are a fundamental cause of this disease. It is reported that Aβ induces oxidative stress [4] and that oxidative stress promotes further production of Aβ by potentiating BACE1 gene expression and Aβ generation [5]. Tangle hypothesis describes that the involvement of another protein, tau protein which twists into abnormal tangles and leads to the failure of the nerve cell transport system leading to the death of brain cells. In contrast according to cholinergic hypotheses, there is an irreversible deficiency in the cholinergic functions in brain that leads to memory impairment in AD patients [6].

Amyloid Hypothesis

Amyloid hypothesis (Fig. **1**) argues that the accumulation of the β-amyloid (Aβ) peptide leading to its aggregation and formation of senile plaques (SP)s which triggers pathophysiological changes of the brain and ultimately leads to cognitive dysfunction. This phenomena happens when the protective and compensatory mechanism of the aging brain fails. Aβ is produced by a series of enzymes known as secretases [7]. The Aβ which is insoluble will then be accumulated in extracellular region into senile plaques. The development of the senile plaque is thought to precede and precipitate the formation of neurofibrilliary tangles and oligomeric forms of Aβ which are reported to be the main cause of neuronal death [8, 9].

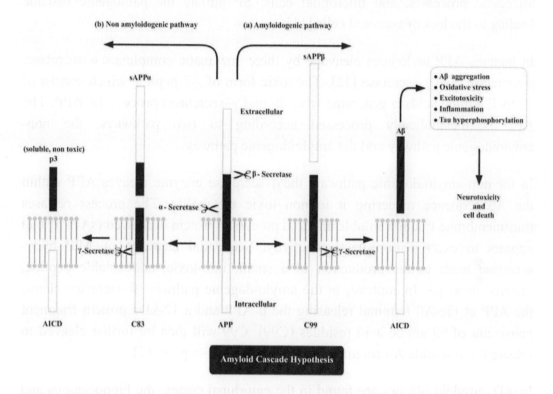

Figure 1: Amyloid cascade hypothesis. Amyloid precursor protein is a transmembrane glyco-protein which undergoes different types of cleavage by different types of enzymes in non amyloid and amyloid pathway. (a) Amyloidogenic pathway –Initially APP will be cleaved by β-secretase to produce sAPPβ (soluble amyloid precursor protein β-cleaved) and Carbon terminal fragment consisting of 99 amino acid residues. Which will further undergo cleavage by γ secretase to produce Aβ peptides and Smaller APP Intra Cellular Domain (AICD) these events further lead to pathological procesess involved in AD. (b) Non amyloidogenic pathway- In this pathway APP will be cleaved by α -secretase enzyme to produce sAPPα (soluble amyloid precursor protein α-cleaved) and Carbon terminal fragment consisting of 83 fragments. Which will be further cleaved by γ-secretase to produce p3 fragment and APP Intra Cellular Domain.

❖ Enzymes involved in both the pathways have been represented as scissors.

The Aβ peptide consists of 38 to 42 amino acids generated by the cleavage of Amyloid precursor protein (APP), a type-1 transmembrane protein. Processing of APP by β- and γ–secretases forms an insoluble and neurotoxic Aβ peptide of 40 to 42 amino acids. The longer form (Aβ42) is more prone to aggregate into fibrils and makes up the major component of senile plaque (SP) [10]. Extracellular Aβ peptides cluster in a β-sheet structure to form SPs. SPs consist of a proteinaceous core composed of 5 to 10 nm amyloid fibrils surrounded by dystrophic neurites, astrocytic processes, and microglial cells. SP initiate the pathogenic cascade leading to the loss of neuronal cells [11].

In humans APP undergoes cleavage by three enzymatic complexes: α-secretase, β-secretase, and γ-secretase [12]. The toxic form of Aβ peptide which consist of 40 to 42 amino acids is generated when β- and γ–secretases process the APP. The APP is metabolically processed according to two pathways, the non-amyloidogenic pathway and the amyloidogenic pathway.

In the non-amyloidogenic pathway, the α-secretase enzyme cleaves APP within the Aβ sequence rendering it to non toxic metabolite. The process releases transmembrane fraction soluble amyloid precursor protein α-cleaved (sAPPα) that appears to exert neuroprotective activity. Sequential cleavage of APP by γ-secretase leads to the production of a small, non-toxic, and soluble peptide, referred to as p3. In contrast, in the amyloidogenic pathway, β-secretase cleave the APP at the Aβ terminal releasing the β APP and a 12-kDa protein fragment consisting of 99 amino acid residues (C99). C99 will then be further cleaved to release the insoluble Ab fragment that deposits into plaques [12].

In AD, amyloid plaques are found in the entorhinal cortex, the hippocampus and neocortical areas [13]. These plaques are also observed in the brains of normal aged individuals and do not correlate with progression of dementia. In fact, many studies showed a weak correlation between Aβ deposits and cognitive status [14]. Some other reports showed that cognitively healthy elderly people could have a substantial amyloid burden [15]. Nevertheless, accumulation of toxic, aggregated forms of Aβ seem to be crucial in the pathogenesis of familial forms of AD [16]. In humans, all individuals with clinical AD have abnormal deposits of Aβ (diffuse or neuritic plaques) [17].

Several arguments support the amyloid-cascade hypothesis as the predominant one in the pathogenesis of AD. Abnormal generation or insufficient clearance of Aβ leads to several secondary events including hyperphosphorylation of the protein *tau* and generation of neurofibrillary tangles, inflammation, oxidation, and excitotoxicity [17-19]. These events lead in turn to the activation of the apoptotic cascade, neuronal cell death and neurotransmitter deficits.

Neuronal loss at the affected regions causes deficit in choline acetyl-transferase leading to a decrease in production of the neurotransmitter acetylcholine, leading to cortical cholinergic dysfunction. The deficit in acetylcholine is primarily responsible for the clinical manifestation of AD. However, deficit in norepinephrine and serotonin was also found in AD patients [20].

Tau or Tangle Hypothesis

Intraneuronal protein clusters composed of paired helical filaments of hyperphosphorylated tau protein is the other neuropathological hallmark of AD. The relationship between the hyperphosphorylated tau and Aβ senile plaques or how they cause neuronal death remains unclear [21]. Neurofibrillary tangles are deposited in a hierarchical and systematic fashion starting from the entorhinal cortex and sequentially spread to the hippocampus, the rest of the temporal lobe, the associated areas of the prefrontal and parietal cortices, and eventually reach all neocortical areas [22]. Deposition of neurofibriallary tangles is suggested as an underlying etiology of the cognitive and memory deficits seen in AD, since it correlates closely with cognitive decline in AD [23, 24].

Cholinergic Hypothesis

The cholinergic system is based on the neurotransmitter acetylcholine (ACh), firstly recognized by Loewi in the 1920s and found widely distributed in both central and peripheral nervous systems [25].

More than two decades of detailed studies of the brains of the people with AD and older age consistently revealed the remarkable damage or abnormalites in the cholinergic pathways of the basal forebrain and rostral forebrain areas including converging projections to thalamus, which were found to have well correlation with the level of cognitive decline in these subjects. The outcome of these studies eventually lead to the so called "Cholinergic hypothesis". This hypothesis mainly emphasised that the cognitive decline associated with AD and older age is the result of that loss of cholinergic function in the central nervous system [26, 27]. According to the cholinergic hypothesis, there is an irreversible deficiency in cholinergic functions in brain that leads to memory impairment in these patients [6]. The most important changes observed in the brain of AD patients are the decrease of the neurotransmitter acetylcholine in hippocampus and cortex levels.

OVERVIEW ON KEY ENZYMES INVOLVED IN ALZHEIMER'S DISEASE

Cholinesterases

Cholinesterases are a family of enzymes that catalyse the hydrolysis of ACh into choline and acetic acid, an essential process allowing for the restoration of the cholinergic neuron. Cholinesterases are divided into two: acetylcholinesterase (AChE; EC 3.1.1.7.) and butyrylcholinesterase (BuChE; EC 3.1.1.8). Cholinesterases are a type α/β hydrolase folded with an α helix bound with β sheet that contains a catalytic domain [28] with catalytic triad Ser – His – Glu [29]. AChE and BuChE are similar, with homology of more than 50% but their significance and localization in the body are very different.

AChE is the key enzyme involved in the metabolic hydrolysis of ACh at cholinergic synapses in the central and peripheral nervous system. This observation led to the introduction of the Acetylcholinesterase inhibitors (AChEI)s to prolong the duration of action of acetylcholine and provide symptomatic treatment in AD [30]. In the mammalian brain most of the AChE is present in the membrane bound form, but its levels decline as the neurons degenerate. Another enzyme, butyrylcholinesterase (BuChE), expressed in selected areas of the central and peripheral nervous systems is also capable of hydrolysing acetylcholine, but its level does not decline, or may even increase in AD [31, 32].

Acetylcholinesterase

AChE is expressed in cholinergic neurons and participates in cholinergic transmission by hydrolyzing ACh (Fig. **2**). It is expressed in nerve and blood cells. The name AChE is derived from the endogenous substrate ACh as opposed to BuChE that has no known endogenous substrate. AChE is also found at high levels in blood cells whereby it breaks down the plasma ACh [33]. Inhibition of AChE causes over-stimulation of ACh receptors [34].

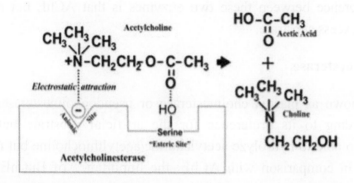

Figure 2: Schematic represention of Acetylcholinesterase with acetylcholine [191].

The AChE structure is evolutionarily conserved. It contains common regions similar to the other serine hydrolases. The structure of AChE has been extensively studied using AChE from the electric eel before the availability of human recombinant AChEs. The electric eel AChE's active site lies on the bottom of a long and narrow cavity (gorge). The active site contains an esteratic site with catalytic domain of amino acids Ser 200, His 440, Glu 327 and an anionic site (also known as α-anionic site) composed of the amino acids Trp 86, Try 337, Phe 338 for murine AChE and Trp 84, Tyr 121 and Phe 330 for the electric eel AChE [35, 36]. The esteratic site hydrolyzes the ester bond of ACh while the anionic site interacts with the ACh quaternary ammonium atom and is responsible for its correct orientation.

There is another site localized on the AChE surface around the cavity entrance known as the peripheral anionic site (also known as β-anionic site). This site was recognized as a target for modulators of AChE. It is also a target site for a number of toxins and promising drugs [37, 38]. This site may be involved in development of AD since amyloid β peptide interacts with the peripheral anionic site resulting in the formation of amyloid plaques and consequently cause damage to cholinergic neurons [39].

AChE has higher affinity for ACh than BuChE. However, AChE is not able to hydrolyze high molecular weight esters unlike BChE. The differences in affinity to substrate are probably caused by changes in the aromatic gorge disposition. Another difference between these two enzymes is that AChE but not BChE is inhibited by excess substrate.

Butyrylcholinesterase

BuChE is known as plasma cholinesterase or pseudocholinesterase. BuChE, is named according to its preference for the artificial substrate butyrylcholine. BuChE is also able to hydrolyze acetylcholine/acetylthiocholine but much slower than AChE. In comparison with AChE, the importance of BuChE is not well understood. Though BuChE activity is prevalent in the human body, its' physiological function is not completely understood. BuChE deficient individuals are generally healthy with no manifest signs of disease [40]. BuChE is not constituted *in situ* but in different organs, mainly in the liver [41]. BuChE is capable of detoxifying a large number of exogenous substances: procaine [42], succinylcholine [43], cocaine [44], heroin, acetylsalicylic acid [45], and it can also protect the body from the impact of organophosphorus AChEIs [46].

There is some agreement that changes in BuChE are associated with the progression of AD. One study has demonstrated that in case of AD; the level of BuChE increases, as that of AChE decreases in affected areas of the brain [47]. The activity of this enzyme in the temporal cortex has also been shown to correlate with the rate of cognitive decline [48].

Secretases

In humans, amyloid precursor protein (APP) found on chromosome 21 undergoes cleavage by three enzymatic complexes: α-secretase, β-secretase, and γ-secretase

[12]. Sequential cleavage of APP by α-secretase and γ-secretase leads to the production of a small, non-toxic, and soluble peptide, referred to as p3. The sequential cleavage of APP by β-secretase and γ-secretase leads to the production of the insoluble Aβ protein [49]. Under appropriate conditions, insoluble Aβ has the ability to aggregate into oligomer and fibrils, culminating in an extracellular SPs formation.

α-Secretases

The α-secretase activity is mediated by a series of membrane bound proteases, which are members of the ADAM (a disintegritin and metalloprotease) family. The α-secretases cleave APP within the Aβ sequence itself, generating a soluble APP α ectodomain and a membrane bound carboxyl terminal fragment, APP-CTF α. The latter fragment is degraded in lysosomes or is further processed by γ-secretases yielding a series of short hydrophobic peptides including Aβ$_{17-40}$ and Aβ$_{17-42}$, which are collectively called p3 fragments [50].

Processing of APP by α-secretase is postulated to be protective in the context of AD because the enzymes cleave within the Aβ sequence, thereby preventing the production of Aβ. Therefore, increasing α-secretase activity would be neuroprotective *via* the increased shedding of growth promoting soluble APPα ectodomains, only if the p3 fragment produced by α-secretase is not pathogenic, and if increasing α-secretase activity actually lowers Aβ production. However, the consequences of chronically upregulating α-secretase mediated cleavage of other substrates remains unknown [7].

β-Secretases

In 1999, five different research groups independently reported the molecular cloning of β-secretase by different isolation methods but identified the same enzyme that possessed all the known characteristics of β-secretase. This enzyme has been given various names including β-site APP cleaving enzyme (BACE), aspartyl protease 2 and membrane associated aspartic protease 2. BACE 1 is a type I transmembrane aspartic protease that is related to the pepsin and retroviral aspartic protease families. BACE 1 is predominantly localized in acidic intracellular compartments (for example, endosomes and the trans-Golgi network). It is highly expressed in the neurons of the brain and has cleavage specificity for APP. On the other hand BACE 2 is expressed at low levels in neurons, it also does not posses the same cleavage activity on APP as β-secretase hence it is considered as a poor β-secretase candidate [51, 52].

γ-Secretases

γ-secretase is an aspartyl protease that cleaves its substrates within the transmembrane region. This enzyme is found to consist of four protein components: presenilin (PS) 1 or 2 (which contains the catalytic domain), nicastrin (which may serve to dock substrates), Aph-1 (anterior pharynx-defective 1), and presenilin enhancer-1 (PS-1) and presenilin enhancer-2 (PS-2) [53].

CLINICALLY USED ENZYMES INHIBITORS FOR ALZHEIMER'S DISEASE

There is no cure for AD. Current management involves the treatment of cognitive, non-cognitive and behavioural symptoms. AChE inhibitors (donepezil, galantamine and rivastigmine) and memantine are the main current pharmacological treatments available specifically for AD [54].

Rationale for Using AChE Inhibitors

The major neurochemical deficit in AD is the reduced availability of ACh. Hence a logical approach is to replenish the lost acetylcholine by replacement or enhancement therapies similar to dopamine replacement therapy for Parkinson's disease patients. However, ACh precursors and cholinergic receptor agonists were used in clinical trials, but abandoned for lack of effectiveness or intolerable side effects [55-57].

Another approach to increase the levels of ACh in the brain and to facilitate cholinergic neurotransmission is *via* inhibition of cholinestrease enzymes which are involved in breaking down ACh after it is released from the vesicles. Cholinesterase inhibitors (ChEI) act by inhibiting the enzymes (acetyl- and butyryl-cholinesterase) responsible for the breakdown of ACh hence increasing its availability at the synaptic cleft. Enhancement or restoration of ACh level in patients with AD or other forms of dementia has provided symptomatic treatment in these patients [58].

On the basis of the cholinergic hypothesis, a number of compounds were studied in preclinical models of cognitive impairment and subsequently tested and clinical trials. Many compounds are still under investigation. As a result of these extensive attempts, currently three cholinesterase inhibitors have been approved by FDA for the symptomatic treatment of AD *viz*, Donepezil, Galantamine, Rivastigmine. Although Tacrine was approved in 1993 as a first drug for AD, it is now rarely been prescribed today because of associated side effects, including possible liver damage [59]. In addition, a newer agent, huperzine A was recently approved as a dietary supplement for memory support.

There seems to be lower interest in BuChE inhibitors probably due its physiological redundancy with AChE in acetylcholine breakdown. Furthermore, its physiological roles are not well characterized, making it more difficult to predict the possible adverse effects of selective BuChE inhibitors. Drugs suppressing the manifestation of Alzheimer's disease through impact on the cholinergic system are predominantly selective inhibitors of AChE [60]. However, selective inhibitors of BuChE have also been investigated as potential drugs for AD [61], but to a lesser degree than AChE. Infact BuChE can temporarily substitute inhibited AChE and is able to slowly hydrolyze accumulated ACh [58].

Three ChEI's currently used in AD will be discussed in detail here: donepezil, rivastigmine and galantamine. These agents are primarily considered for symptomatic treatments as they improve symptoms without modifying the course of the disease. Although all these ChEI's have efficacy for mild to moderate AD, equivalency has not been established in direct comparison. Selection of which agent to be used will be based on adverse effect profile, ease of use, familiarity, and beliefs about the importance of the differences between the agents in their pharmacokinetics and other mechanisms of action [62].

Cholinesterase Enzymes Inhibitors (Table 1)

Tacrine

Tacrine was the first ChEI to be commercialized in 1993. It was quickly taken off the market due to its poor bioavailability (four doses daily) and hepatotoxicity [63].

Table 1:   Summary of the Cholinesterase Inhibitors Currently Used for Treatment of AD

Drug	Donepezil	Rivastigmine	Galantamine
Structure	Fig. (3)	Fig. (4)	Fig. (5)
Serum half life	70 Hours	1-2 Hours	7-8 Hours
Protein binding (%)	96%	40%	18%
Metabolism	Hepatic CYP450-2D6 and 3A4	Non hepatic Urinary	Hepatic CYP450-2D6 and 3A4
Dose	5 mg once daily increased up to 10 mg once daily	1.5 mg twice daily can be increased upto 6 mg twice daily. (maintenance dose 16- 24 mg once depending upon clinical benefit and tolerability)	8 mg once daily increased upto 16 mg once daily

Source – [85, 192].

Donepezil

Donepezil (Fig. **3**) is a piperidine derivative and synthetic drug designed to selectively and reversibly inhibit AChE [64]. It inhibits AChE with a high affinity in a mixed competititive-non competitive way. It is largely a non competitive inhibitor of this enzyme and posses relative selectivity for AChE compared with BuChE [65, 66].

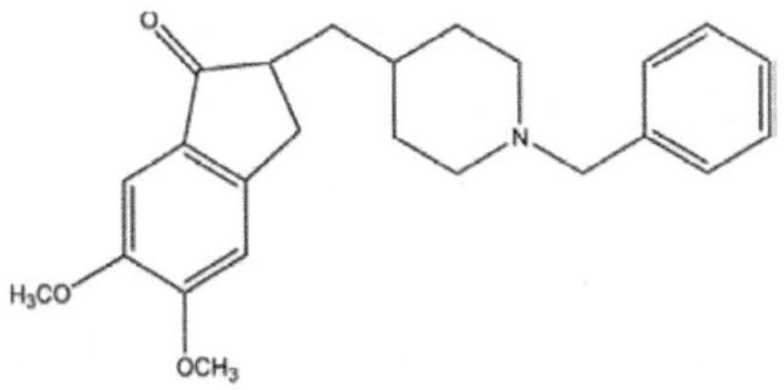

Figure 3: Donepezil.

After oral administration the maximal plasma concentration of the drugs is reached at 3-4 hours. There is no significant food interaction with the absorption of donepezil. Elimination half life of donepezil is approximately 70 hours which allows once a daily dosing, it has 96% of plasma binding. Steady state is reached within 3 weeks. Donepezil is metabolized mainly by the hepatic enzymes namely cytochrome P450 isoenzymes 3A4 and 2D6 and eliminated primarily through the kidneys. Most of the metabolites are mainly found in the urine (*viz* 5-0-desmethyl donepezil, 6-0-desmethyl donepezil and donepezil-cis-N-oxide). About 11-17% of the drug is excreted unchanged in the urine [65-69].

Although both 5 and 10 mg doses are effective, usually the therapy is initiated at 5mg once daily dose and then increased to 10 mg after 4 weeks. The strategy is adopted to avoid or minimize the cholinergic side effects [70]. Donepezil possesses low potential for drug-drug interactions, whether the drug has any potential for the enzyme induction is not known and pharmacokinetic studies have apparently shown no interaction of donepzil with theophylline, cimetidine, warfarin, digoxin and ketoconazole. However the *in vitro* studies revealed that the cytochrome P450 enzyme inducers (*e.g.* phenytoin, carbamazepine, rifampicin) may increase the elimination; whereas inhibitors of CYP450, 3A4 and 2D6 such as ketoconazole and quinidine may inhibit metabolism of donepezil. A synergistic effect may be expected when the drug is given concurrently with neuromuscular blocking agents or cholinergic agonists [71].

Most common adverse reactions reported from clinical studies are nausea, diarrhea, insomnia, vomiting, muscle cramps, fatigue, and anorexia. For the treatment of mild to moderate AD, 5-10 mg as once daily dose is given, and for treatment of moderate to severe AD, 23 mg once daily dose is administered for the patients who have been on a dose of 10 mg once daily dose for at least 3 months [71, 72]. Apart from the AChE inhibition some of the studies have revealed the neuroprotective action of donepezil against glutamate induced toxicity in cortical neurons, as well as neuroprotective action against Aβ induced toxicity. The drug is also found to decrease the Aβ production and increase the expression of nicotinic receptors [73, 74].

Rivastigmine

Rivastigmine (Fig. **4**) is a carbamate derivative and reversible cholinesterase inhibitor [75]. It inhibits the AChE in a pseudo irreversible-non competitive manner. Rivastigmine produces the enzyme inhibition effect that persists much longer than the drug is present in plasma, thus the mode of inhibition is often described as "pseudo irreversible" inhibition. The molecule initially binds to the enzyme in a similar way to that of natural substrate ACh. Then it undergoes cleavage producing an inert phenolic derivative which excretes *via* kidney rapidly. However the esteric site of the enzyme still occupied with carbamyl moiety. The carbamyl moiety of rivastigmine remains bound to its substrate after the acetyl moiety dissociates *via* hydrolysis, resulting in inactivation of the AChE enzyme for more than 24 hours [76]. It inhibits both AChE and BuChE selective for the brain compared with that in peripheral tissue [77]. It is the only ChEI with significant inhibition of BuChE [78, 79]. It is unclear how specific BuChE inhibition relates to rivastigmine's clinical effect.

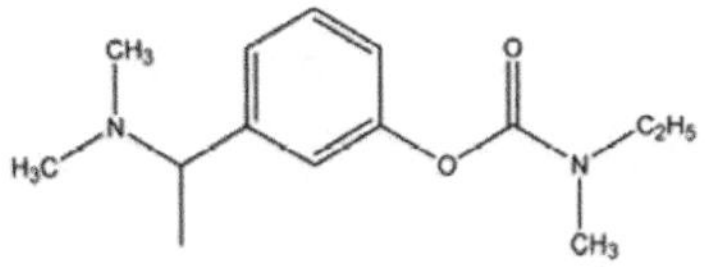

Figure 4: Rivastigmine.

Rivastigmine is absorbed rapidly after oral administration and peak plasma concentration is achieved in one hour. Administration of the drug with food delays the absorption for about 90 minutes, lowers the peak plasma concentration, but increases the bioavailibility for about 30%. Protein binding of the drug is about 40%. The drug is primarily metabolized by choline esterases as a decarbomylated metabolites which further majorly undergoes N-demethylation, sulfate conjugation or both by the hepatic enzymes. Elimination of the drug's metabolites occurs through the kidneys and is virtually complete after 24 hours. The sulfate conjugate is the major metabolite found in the urine [75, 76].

Since the metabolism of the drug does not mainly involve hepatic enzyme to significant degree and the drug is poorly bound to plasma proteins, no clinically relevant drug interactions are anticipated. The fact is apparently also supported by the previous findings [76, 80]. The dose of the drug of 6-12 mg twice daily is known to produce beneficial effects in AD patients. The treatment is started with 1.5 mg twice daily dose and doses are increased with the increment of 1.5 mg up to 6 mg with the interval of 2 weeks considering the tolerance of the patient towards drug [75]. The drug shows adverse effects like nausea, vomiting, loss of appetite and increased frequency of bowel movements [75, 81]. The transdermal patch was first approved for the use in 2007 by FDA and now approved in more than 82 countries for the treatment of mild to moderate dementia of AD [82]. The patch of 5 cm^2 with 4.6 mg/day dose is initiated in the treatment and increased to 10 cm^2 with 9.5 mg/day after 4 weeks of treatment only when the patient well tolerated the previous dose. Peak plasma concentration is attained usually after 8 hours and often found to remain steady upto 16 hours. The patch of 9.5 mg/24 hours demonstrates approximately same exposure as the attained by 6 mg twice a daily dosing by oral route [83].

A main difference with other ChEI's is that the latter agents usually are associated with up-regulation of ChEs. This up-regulation is sometimes invoked to explain the relatively short duration of effect of these medications. However, the relative "lack" of up-regulation of ChE with rivastigmine has not been unequivocally shown to relate to more sustained clinical benefit [84, 85].

Galantamine

Galantamine (Fig. **5**) is a phenanthrene alkaloid originally isolated from bulbs of various flowering plants including daffodils and common snow drop plant *Galanthea woronwii* belonging to family *Amaryllidaceae* [86, 87]. It is a reversible, competitive ChEI.

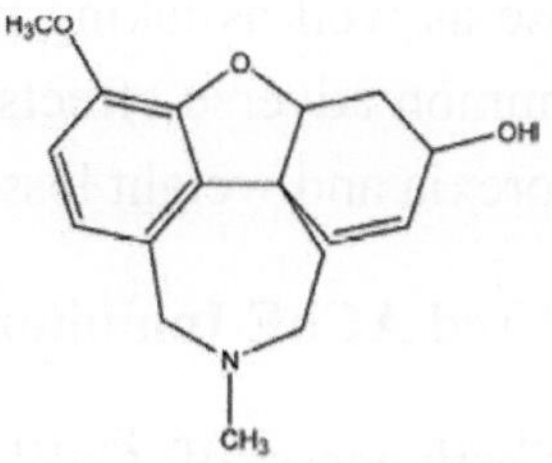

Figure 5: Galantamine.

Among the currently approved three cholinesterase inhibitors galantamine posesses the lowest potency in inhibiting the AChE enzyme, albeit produce the similar therapeutical effects like the other agents [87, 88]. This clinical efficacy of the galantamine is claimed to be because of its additional mechanism, allosteric modulation of the nicotinic cholinergic receptor (nAChR). Galantamine acts as an allosterically potentiating ligand. It binds to an allosteric site on the nAchR and acts synergistically with acetylcholine to facilitate nAchR activity [89]. The drug is approved by FDA to treat mild to moderate AD [59].

Galantamine is rapidly and completely absorbed orally and attains peak plasma concentration in about one hour. It binds poorly with plasma proteins (18%). Although food does not affect the bioavailibility of the drug, it considerably influences the peak plasma concentration (25%) and time to achieve peak plasma concentration will be delayed by 1.5 hour. Galantamine is metabolized by liver cytochrome P450 enzymes mainly by CYP2D6 and CYP3A4 isoenzymes and excreted majorly in glucouronidated and unchanged form. Elimination half-life is between 7 to 8 hours. As liver cytochrome enzymes (CYP450) are involved in the metabolism of galantamine, potent inhibitors of CYP3A4 and CYP2D6 (ketoconazole, paroxetine, erythromycine, amytryptaline, quinidine) remarkably increase the area under the curve (AUC; a parameter to indicate plasma concentration) of galantamine approximately from 10 to 40% [90].

Galantamine immediate release tablets and oral solution are recommended to start at 4 mg twice a day (8 mg/day) up to 4 weeks, further increase to 8 mg and 12 mg twice a day are attempted slowly by keeping the same time period interval between the dose and considering the patient benefit and tolerability. Extended release capsules started at 8 mg/day dose and further increments in doses (16 mg/day and 24 mg/day) will be done after treating patient minimum of 4 weeks with the previous dose as well as taking patients safety and tolerability into an account [91]. Most common adverse effects associated with this drug are nausea, vomiting, diarrhea, anorexia and weight loss [59, 92].

Meta-Analyses of Clinically Used AChE Inhibitors

Six meta-analysis on the effectiveness of ChEIs are available and will be discussed briefly in this section (Birks and Harvey, 2006; Birks *et al.*, 2009;

Hansen *et al.*, 2008; Loy and Schneider, 2006; Raina *et al.*, 2008 Ritchie *et al.*, 2004). However, there appears to be overlapping in the trials included in these meta-analyses.

Many clinical trials evaluated the effectiveness of the currently approved AChEIs in AD. These trials were looking at four main outcomes, namely cognition, global, functional and behavioural domains. Various scales or scoring schemes were employed such as the following:

Donepezil

Ritchie *et al.*, (2004) evaluated 9 trials on cognitive and global outcomes and also completion rates of donepezil treatment (3-10 mg dose) in all kind of severity of AD for duration of 12 to 52 weeks. A dose-dependent improvement was found for ADAS-cog and MMSE, with 10 mg producing a better effect. However the improvements in global outcome measured by CIBIC-Plus, CDR-SB, CGI-C and GDS were dose-independent [93].

A Cochrane database system review on 15 trials concluded that donepezil at 5-10 mg/day was beneficial compared to placebo up to 24 weeks of treatment (Birk and Harvey, 2006). Meta-analysis on cognition tests, ADAS-cog, MMSE and SIB showed improvements in donepezil group with no statistical heterogeneity. Effect on behavior could not be assessed due to insufficient data and wide variability among the studies. The dose of 10 mg/kg showed marginal improvement on the ADA-cog and CIBIC than that of 5 mg/kg [94].

Raina *et al.*, (2008) evaluated 21 trials on effectiveness of donepezil treatment (5-10 mg dose) in all form of AD severity for duration of 12 to 24 weeks on the cognitive (ADAS-cog, MMSE, SIB), global (CDR-SB) and behavior (NPI) outcomes. In general, donepezil produced statistically significant effects compared to placebo group on all those three outcomes. Donepezil-treated patients were more likely to improve in global outcome than the placebo-tretaed group [95].

Another meta-analysis published in the same year by Hansen *et al.*, (2008) analyzed 12 trials that evaluated the effect of donepezil treatment (5-10 mg dose) in all form of AD patients for duration of 3 to 6 months on all four outcomes, cognition (ADAS-cog), global (CIBIC-Plus, CGI-C), functional (DAD, BADLS, ADCS-ADL, IDDD, CMCS, ADFACS, PDS) and behavioral (NPI). Pooled data for donepezil 5 mg and 10 mg showed significant improvement in all the four outcomes [96].

All these analyses demonstrated that Donepezil affords moderately cognitive improvements in AD patients in a dose dependent manner. Although the drug is capable of increasing the likelihood of global improvement, its effect on behavior found to be inconsistent. Hence further research on this matter is necessary.

Rivastigmine

The meta-analysis carried-out on donepezil as mentioned earlier were also investigating the effects of rivastigmine on the four main outcomes.

Ritchie *et al.*, (2004) evaluated 9 trials on cognitive and global outcomes of rivastigmine treatment (1-12 mg dose) in mild to moderate AD patients for the duration of 10 to 26 weeks. Rivastigmine showed a dose-dependent beneficial effect on cognitive outcomes measured by ADAS-cog and MMSE. However, only one study showed similar effects on global outcomes measured by CIBIC-Plus and GDS [93].

Birks *et al.*, (2009) performed a Cochrane database system review on 9 trials of rivastigmine treatment (dose was categorized into low, 1-4 mg and high dose, 6-12 mg of rivastigmine) on the full range of AD severity for duration of 9 to 52 weeks. High dose of rivastigmine resulted in significant improvements in cognition (measured by ADAS-cog) and function (measured by PDS), however behavior outcome measured by NPI, based on only two studies showed no significant benefit compared to those of placebo. Global improvements (measured by GDS, CIBIC-Plus, ADCS-GCIC) were noticed with high dose at 12, 18, 26 and 52 weeks of treatment. As for the low dose of rivastigmine, no significant improvements were found except in cognitive function at weeks 18 and 26 and global improvement at week 26. Generally, there were global improvements associated with both high and low dose rivastigmine treatment on the CIBIC-plus and ADCS-CGIC, but for GDS, only high dose showed improvements. The results of IDEAL (Investigation of transdermal Exelon in Alzheimer's disease) study, comparing the efficacy of once daily rivastigmine given as oral dose (capsule, 6-12 mg) and transdermal patch (small patch, 9.5 mg/day or large patch, 17.4 mg/day) was also highlighted by the author [97].

Raina *et al.*, (2008) evaluated 6 trials on effectiveness of rivastigmine treatments (1-12 mg dose) in all form of AD severity for duration of 14 to 48 weeks on the cognitive (ADAS-cog, MMSE, SIB), global (CIBIC-Plus, GDS) and functional (NOSGER, PDS) outcomes. Inconsistent results were observed for cognitive outcome whereby it produced significant benefit for ADAS-Cog (with heterogeneity) but not for MMSE and SIB measured outcomes. Similar results were seen for functional outcomes as well, whereby it produced significant results for outcomes measured by NOSGER but not for PDS measured outcomes. Nevertheless, treatment with rivastigmine produced benefits in global status [95].

Another meta-analysis by Hansen *et al.*, (2008) analyzed only 3 trials on the effect of rivastigmine treatment (1-12 mg dose) in mild to moderate form of AD for duration of 10 to 26 weeks on cognition (ADAS-cog) global (CIBIC-Plus) and function (NOSGER, PDS) outcomes. Treatment with rivastigmine produced beneficial outcomes in those three domains [96].

Overall outcomes from these studies demonstrated that rivastigmine posesses a dose dependent beneficial effects not only on cognitive but also functional and global domains of the patients (However higher doses were found to be more effective). On contrary, one of the studies reported about inconsistent cognitive and functional effect of the drug across the trials. The drug is also reported to produce insignificant improvement in behavior of the patients.

Galantamine

Ritchie *et al.*, (2004) evaluated 6 trials involving 3,083 subjects receiving donepezil treatment (8-36 mg dose) or placebo in mild to moderate AD for duration of 12 to 24 weeks on all four outcomes, cognition (ADAS-cog), global (CIBIC-Plus, CGI), function (DAD, ADCS-ADL) and behavior (NPI).

Galantamine treatment produced significant effect on cognition and global status, however the improvements were dose-independent [93].

Loy and Schneider (2006) performed a meta-analysis for the Cochrane Collaboration involving 8 trials of galantamine treatment (8-32 mg) in mild to moderate AD for duration of 12 weeks to 2 years on all four main outcomes, cognition (ADAS-cog), global (CIBIC-Plus, ADCS-CGIC, CDR-SB), function (DAD, ADCS-ADL, PDS) and behavior (NPI). Galantamine produced significant improvements in cognition at all dose levels, the effect is more pronounced at 6 months rather than at 3 months of treatment. Treatment with galantamine also caused improvement or no change in global domain in a higher proportion of patients, except those treated with 8 mg daily [98].

Raina *et al.*, (2008) evaluated 8 trials on effectiveness of galantamine treatments (8-36 mg dose) in all form of AD severity for duration of 12 to 48 weeks on the cognitive (ADAS-cog), global (CIBIC-Plus) functional (DAD, ADCS-ADL) and behavior (NPI) outcomes. In general, treatment with galantamine produced beneficial effects in all domains. It produced significant improvement on the ADAS-cog (heterogeneity observed) and produced improvement or stabilization on the CIBIC-Plus [95].

In another meta-analysis, Hansen *et al.*, (2008) analyzed 7 trials on the effect of galantamine treatment (16-24 mg dose) in mild to moderate form of AD for duration of 3 to 6 months on cognition (ADAS-cog) global (CIBIC-Plus, CGI-S), function (ADCS-ADL, DAD) and behavior (NPI) outcomes. Generally, modest improvements were seen on all these main outcome domains. Different doses appear to have no significant differences in the measured outcomes [96].

In general, the drug demonstrated prominent improvements in cognitive, global, functional and behavioural domains. However the effects did not follow in a dose proportionate manner.

Pooled Effects of Cholinesterase Inhibitors

Hansen *et al.*, (2007) carried out pooled effects of ChEIs. In general, standardized effect size for functional measures was small, ($d = 0.1$-0.4), but the effect sizes consistently favoured drug treatment over placebo. For all the drugs, pooled standardized effects sizes were consistent in both short and long trials. The pooled effect size was not significantly affected by parameters such as disease severity, age, gender and drug dose. Adverse events were generally limited to gastrointestinal problems, weight loss and dizziness. The authors conclude that the results support the clinical benefits of these treatment since there is a lack of other effective treatment for AD in current scenario [99].

Birks J. (2006) reviewed the results of 13 randomized double blind, placebo controlled trials and reported that donepezil, galantamine, rivastigmine treated for period of 6 months and one year in the patients with mild to moderate or severe dementia produced on average of -2.7 points on ADAS-Cog scale. Positive improvements on global clinical state were also observed. Treatement also exhibited benefit over activities of daily living and behaviour. No strong evidences were found indicating less effect of these drugs for severe or mild dementia. No evidences found reporting significant differences in terms of efficacy between these drugs [60].

The meta analysis carried out by Krista lanctot *et al.*, (2003) reported that treatment with ChEIs results in modest but significant therapeutic effect which also comprises modestly but significantly higher rates of adverse events and discontinuation of the treatment in comparison with placebo [100].

Robin *et al.*, (2005) carried out meta analysis to compare the efficacy of donepezil and galantamine in reducing cognitive decline in AD patients. After evaluating the mean effect size across a number of studies they reported that neither drug found to be greatly effective. However authors stated that they still have practical value in the treatment. They also reported that galantamine do not possess any advantage over donepezil [101].

Campbelll, *et al.*, (2008) carried out meta analysis to evaluate impact of ChEIs on behavioral and psychological symptoms of AD, they reported that ChEIs produced reduction in behavioral and psychological symptoms in AD patients.

They conclude that drugs produce more beneficial effects in patients with mild AD compared to severe AD [102].

In general it can be concluded that ChEIs offer moderate beneficial improvement on cognitive and global status, as well as function and behavioral domains. More prominently in mild to moderate AD patients than patients with Severe AD.

In contrast, Kaduszkiewicz, *et al.*, (2005) in their systematic review expressed the doubts over scientific basis for recommending ChEIs for the treatment of AD and raised question about the authenticity [103]. However this hypothesis was contested and challenged by few researchers and several suggestions were put forward for systematically analyzing the data. Further more these researchers supported the efficacy of ChEIs in the therapy [104-106]. Similarly, Raschetti, *et al.*, (2007) expressed their views that the use of ChEIs in mild cognitive impairment would not produce any delay in the onset of AD or dementia. furthermore they expressed their concern that risks associated with ChEIs are not negligible and also raised their question against scientific validity of trials [107]. More recently, Russ, *et al.*, (2012) in their meta analysis explicitly stated that ChEIs do not much affect progression of dementia and possess no effect on cognitive test scores. They also expressed that ChEIs associated with increased risk of adverse events, particularly gastrointestinal ChEIs. And concluded that they should not be recommended for mild cognitive impairment [108].

Non-Enzymatic Related Drug

Memantine

Memantine, an uncompetitive NMDA (N-Methyl-D-Aspartate) receptor antagonist is the fifth drug approved by FDA for the treatment of mild to moderate AD [109]. Rather than its utilisation as monotherapy, the combination of this drug with AChEIs is found to be beneficial in mild to moderate AD [110-113]. Detailed discussion about this drug is not possible here because of the scope of this chapter. However we found it worth to atleast mention about this drug to direct the interested readers to refer more on this drug and its utilisation in treatment of mild to moderate AD.

CURRENT ENZYME-TARGETED DRUG DEVELOPMENT FOR ALZHEIMER'S DISEASE

Cholinesterases Inhibitors

Huperzine A

Huperzine A (Fig. **6**) was isolated from the Chinese club moss, *Huperzia serrata* (Thunb) Trevis. in early 1980's [114]. Huperzine A is also found to occur in several families such as Huperziaceae, Lycopodiaceae and Seleginella [115].

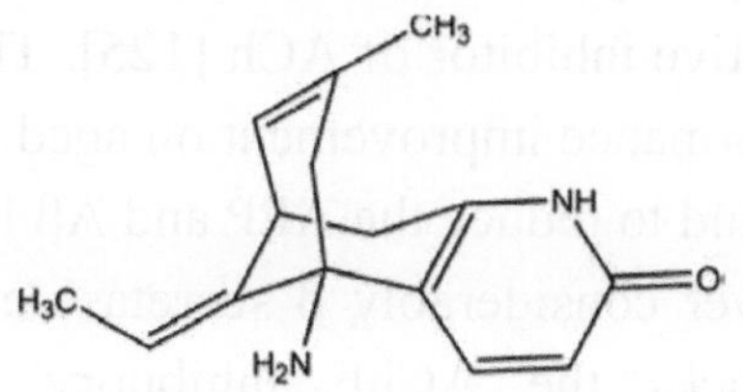

Figure 6: Huperzine A.

Huperzine A is a potent, selective, and well tolerated AChE [116, 117]. It is a linearly competitive and reversible inhibitor of AchE. The evidences suggest that Huperzine A is selective inhibitor of AChE than BuChE and especially more selective on G4 form of acetylcholinesterase compared to G1 form of the enzyme [118, 119].

Huperzine A is also known to enhance the non amyloidogenic pathway through the activation of PKC and MAPK pathways [116]. It has shown protective action in cultured rat cortical neurons from Aβ (25-35) induced toxicity [120]. It also possesses protective action against glutamate induced excitotoxicity in neurons [121]. It has shown protective action in neuroblastoma-spinal motor neuron fusion cell line and rat spinal cord organotypic cultures against various cell death inducers such as staurosporine, thapsigargin, hydrogen peroxide, carbonyl cyanide *m*-chlorophenyl hydrazone and L-(-)-threo-3-hydroxyaspartic acid (THA) [122].

Currently the drug is undergoing phase II clinical trial with an objective to evaluate its efficacy in mild to moderate AD and the study reported that 200 µg twice daily dose did not demonstrate beneficial cognitive effect. However some secondary analysis in this study revealed that a higher 400 µg twice a daily dose may improve cognition [123]. Phase II and III clinical trial of Huperzine A sustained release tablets with 400 µg once a day dose for treatment of mild to moderate AD is still in process [124].

Phenserine

Phenserine (Fig. 7) is a phenyl carbamate derivative of physostigmine. It is a selective and noncompetative inhibitor of ACh [125]. The drug has demonstrated significant cognitive performance improvement on aged rodent and canine models [126-128]. Phenserine found to reduce the APP and Aβ levels in a dose dependent manner as well as to lower considerably β secretase enzyme levels *in vivo*. Its enantiomer, Posiphen lacks the AChE inhibitory action but possess the equipotency on lowering the APP and Aβ production [129].

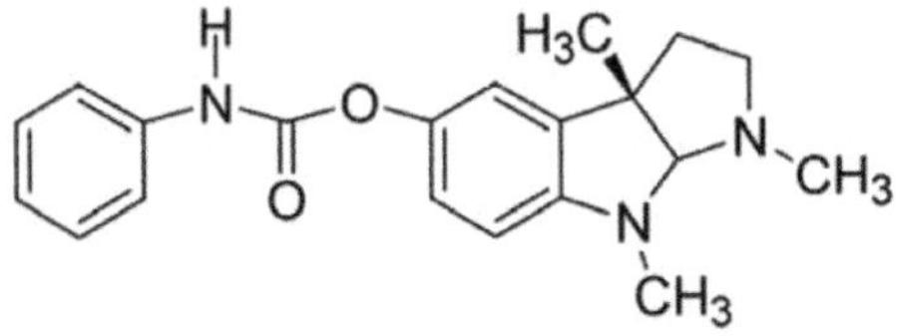

Figure 7: (-)-Phenserine.

Inspite of passing through the phase I study, the drug failed to demonstrate its efficacy both as AChEI and APP inhibitor in several other consecutive series of studies including two phase II and three phase III timed trials. All these developments lead to (-) phenserine being abandoned as a failed drug [130]. However its enantiomer posiphen in a recently conducted a phase 1 randomised, double blind, placebo controlled safety, tolerance and pharmacokinetic study has shown well tolerance in subjects and significantly reduced the CSF levels of sAPPα, sAPPβ, t-τ, p-τ and specific inflammatory markers. In addition to this the molecule demonstrated a trend to lower CSF Aβ. The results of this study demonstrated that pharmacologically relevant drug/metabolite levels reach brain; which support the continued clinical optimisation and evaluation of Posiphen for mild cognitive impairment and AD [131]. Because posiphen devoids of AChEI activity and has a potential to lower Aβ production considerably, it is expected to be one of the promising candidate of drug for AD treatment in future without any cholinergic system related adverse effects.

Bisnorcymserine

Bisnorcymserine (Fig. **8**) is a derivative of Phenylcarbamate. This molecule was developed based on the evidences suggesting an increase in the concentration of BuChE in the AD and BuChE's potentiating role of Aβ peptide toxicity in the progression of AD. It is a potent selective, reversible BuChE inhibitor [132-134]. The compound has shown potent selective BuChE inhibition in rats, improvement in cognitive performance significantly in aged rats. Also found to reduce the APP and Aβ peptide production in cultured human neuroblastoma cells without affecting the cell viability. Reduction in the production of APP and Aβ also in transgenic mice overexpressed human mutant APP considerably [135,136]. The QR Pharma company which is developing this drug has now resumed the phase I clinical trial in collaboration with NIA/NIH [136].

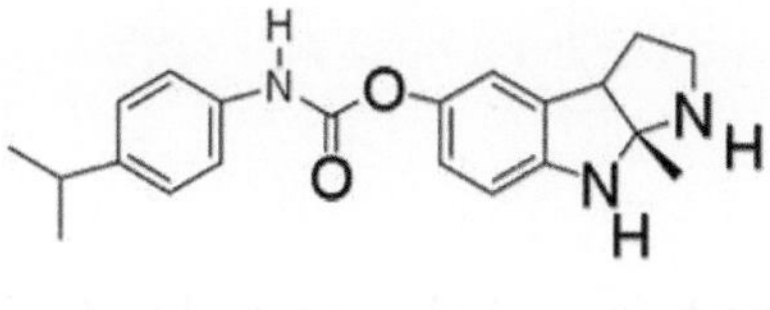

Figure 8: Bisnorcymserine.

Secretases Inhibitors

In the last decade, advances in understanding the neurobiology of AD have translated into an increase in clinical trials assessing various potential AD treatments. Drugs used in the current treatment afford only symptomatic relief for limited period of time and do not affect the main neuropathological hallmarks of the disease, *i.e.* senile plaques and neurofibrillary tangles. Since the accumulation of Aβ peptides is believed to play pivotal role in the development of AD. Huge efforts are being made to identify drugs which are able to interfere with proteases regulating Aβ formation from APP. On one hand compounds that stimulate alpha-secretase, the enzyme responsible for non-amyloidogenic metabolism of APP, are being developed. While on the other hand efforts have also been made to discover the potential of inhibitors of beta-secretase (memapsin-2, beta-amyloid cleaving enzyme-1 [BACE-1]), the enzyme that regulates the first step of amyloidogenic APP metabolism, and also gamma-secretase the pivotal enzyme that generates Aβ [137]. The drugs which have reached upto clinical trials pertaining to these enzymes are only discussed in this chapter.

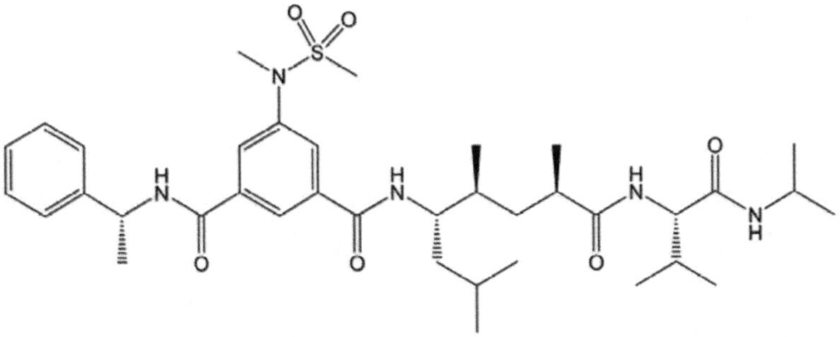

Figure 9: CTS-21166.

β-Secretase Inhibitors

CTS-21166

This is the drug developed in collaboration by Ghosh group at University of Illinois, Chicago and Tang group at Oklahoma medical research foundation Fig. (9). The drug is in clinical development for Alzheimer's disease at Comentis [138]. The drug is reported to have good solution potency (K_i value against β secretase = 2.5 nM), Cellular potency (IC_{50} about 3nM), 1000 times higher selectivity towards Beta secretase than Cathepsin D. A 4 mg/kg dose administration of this drug to 13 month old transgenic AD mice was found to reduce one third of Aβ production in the brain. A phase 1 clinical trial was conducted on healthy young males for the safety and preliminary Aβ responses.

Six different concentrations of drug up to 225 mg were evaluated. The subjects were infused once with either vehicle solvent or vehicle containing the drug at six concentrations up to 225 mg. Drug showed the reduction of plasma Aβ levels in dose dependent manner, a maximum of 80% inhibition of plasma Aβ levels found to be achieved by the highest dose tested at 3 hour. The drug produced significant inhibition of plasma Aβ levels beyond 72 hours. No rebound increase in the Aβ concentrations were found during recovery period. Same kind of result was reported to be obtained from Second phase I trial on subjects receiving oral liquid solution of 200mg of the drug [139].

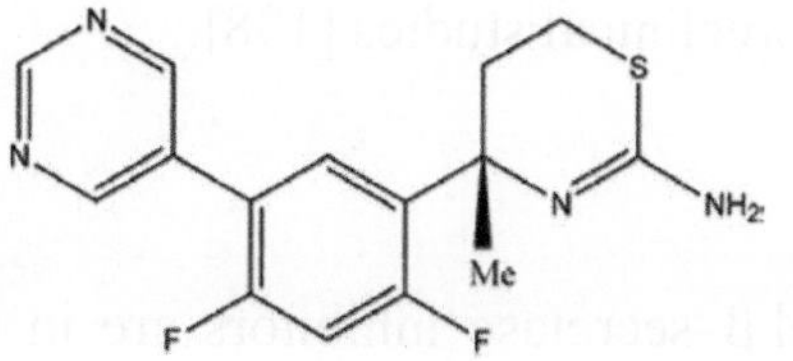

Figure 10: LY2811376.

LY2811376

The drug (Fig. **10**) was developed from fragment based LEAN driven rational drug discovery approaches. The drug is known to possess modest potency in *in vitro* assays. It is also known to possess 10 fold selectivity over BACE2 and 60 fold selectivity over Cathepsin-D. The drug showed dose dependent significant reduction in Aβ, sAPPβ and C99 levels in transgenic APP^{v717F} mouse. A single oral dose administration of 0.5-5 mg/kg of drug in beagle dogs reported to produce dose related reduction in plasma Aβ level which was followed by gradual return to baseline without producing rebound above baseline over 36-48 hours. Upon oral administration of 5mg/kg drug, reduction in the plasma $A\beta_{1-x}$ levels up to 85% was observed from 4-12 hours. Plasma $A\beta_{1-x}$ levels found to be remained below baseline after 24 hours. CSF $A\beta_{1-x}$ levels were found to be reduced 43% by 3 hours and up to 70% by 9 hours. Single Ascending dose study conducted in healthy human subjects with oral administration of drug at doses ranging from 5-90 mg revealed that drug possess dose related reduction in plasma $A\beta_{1-40}$ levels with mean reduction of 64% over the first 24 hours. 90 mg dose administration of drug was known to reduce maximum of 80% reduction in $A\beta_{1-40}$ levels approximately 7 hours after dosing. The reductions of plasma $A\beta_{1-x}$ levels were found to be similar to that of $A\beta_{1-40}$ levels. Single doses also reported to produce reduction in CSF $A\beta_{1-40}$ levels and $A\beta_{1-42}$ levels in a dose dependent manner [140]. In a 14 day multiple dose ascending study, reduction of plasma $A\beta_{1-40}$ levels were found to be achieved steady state with one week of daily dosing of 10 and 25 mg. Similar trend of dose related biomarker changes in CSF of $A\beta$ species as observed in single dose ascending study were found. In spite of these inspiring results the clinical development of the drug has been discontinued since the drug has exhibited toxicity in preclinical studies [138].

Cathepsin B Inhibitors

While the hunt for novel β-secretase inhibitors are in progress; In recent years some of the researchers expressed their opinion that a Cystein protease Cathepsin B (CatB) is also a viable target for developing drug for AD [141]. Although contrast, yet interesting reports are available about the CatB. This enzyme is known to perform similar function as BACE-1. Some of the researchers argue that CatB cut the APP at wild type β-secretase site and thus claim that it is mainly, involved in the progression of sporadic AD, while BACE-1 found to cleave APP at swedish mutation site and thus is claimed to be mainly involved in familial AD [141, 142]. Further more, it is argued that CatB is superior target than BACE-1 since majority of AD patients are found to express wild type APP because it is not linked to genetic mutations [141]. There are reports available in favour of this argument that the CatBs play a vital role in sporadic AD [141, 143, 144]. With these informations as background the CatB inhibitors have been developed, and few among them such as E64 and CA074Me are found to possess promising results in *in vitro* and *in vivo* models [145-148]. In contrast, there are also reports available that CatB possesses neuroprotective role in AD [149-151]. Hence considering inhibition of CatB as therapeutic strategy is remained yet as an unresolved issue and ongoing research in this area in coming years may bring suitable strategy towards effectively utilising this enzyme as a therapeutic target in AD.

Since our concern in this chapter is mainly about enzyme inhibitors employed in current therapy of AD and those reached upto clinical trials and their current status. Here we tried only to give brief introduction about this enzyme considering its significant research developments in recent years claiming its major involvement in the progression of sporadic AD.

γ-Secretase Inhibitor

LY-450139 (Semagacestat)

Semagacestat (Fig. **11**) is one of the prominent γ -secretase inhibitor that has reached clinical trial, it is known to possess 3 fold selectivity in inhibiting the APP than Notch [9]. In a phase I study, semagacestat was administered in single oral doses of 60,100,140 mg to 31 volunteers from 49-53 age groups without neuropsychiatric disease, as an effort to assess its safety, tolerability and effects on plasma and cerebrospinal fluid Aβ. The study observed no significant adverse events or laboratory changes in the subjects. This study also saw a dose proportional increase in drug exposure in plasma and in CSF, as well as dose dependent change in plasma $A\beta_{1-40}$. At the higher dose of 140 mg 72.6% reduction in plasma $A\beta_{1-40}$ was recorded which did not return to baseline for more than 12 hours. Cerebrospinal fluid concentrations of $A\beta_{1-40}$ were found to be unchanged 4 hours after the administration of drug [152].

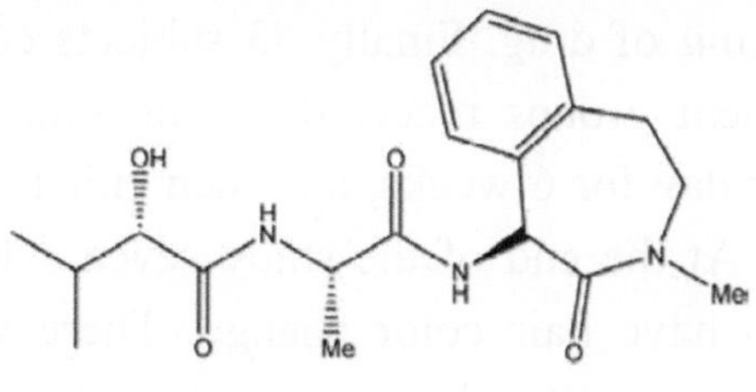

Figure 11: Semagacestat (LY-450139).

Another Phase I study was conducted to determine the safety and tolerability as well as reduction of Aβ in plasma and cerebrospinal fluid after multiple doses. Total of 37 volunteers were analyzed using doses from 5-50 mg/day given for 14 days. Plasma half life of this drug was found to be nearly 2.5 hours. The drug apparently showed linear relationship between dose and plasma concentrations. A dose dependent decrease in the concentrations in plasma Aβ over 6 hours of interval following drug administration with maximum decrease up to 40% relatively to baseline also observed. After returning to baseline, Abeta concentrations were drastically increased. On the other hand CSF Aβ concentrations found to be remain unchanged. Two subjects in the 50 mg dose group possibly developed drug related adverse effects and discontinued the treatment. In the first subject significant increase in the serum amylase and lipase levels were observed which was associated with the moderate abdominal pain. The other subject found to have diarrhoea which was positive for occult blood [153, 161].

In a Phase II randomized, placebo-controlled trial, 70 AD patients received 30 mg of the drug for 6 weeks followed by 40 mg for subsequent 5 weeks. The drug found to decrease $A\beta_{1-40}$ in plasma by 38.2 percent, But $A\beta_{1-40}$ reduction in CSF was not significant. Six subjects recruited in this study complained diarrhoea; an old man had found to developed barret esophagus, eventually developed endocarditis and finally died. Where as in other subjects, significant changes in the levels in circulating CD69, T Lymphocytes, eosinophils, and serum concentrations of potassium and inorganic phosphorus were observed [154, 164].

Semagacestat was evaluated in 51 mild to moderate AD patients in multicenter, randomized, double blind, dose escalation placebo controlled Phase II trial. In this 15 subjects were randomized to receive the placebo 22 to receive 100 mg and 14 subjects to receive 140 mg of drug. Finally 43 subjects completed the treatment phase. The drug treatment groups received 60 mg once daily dosing for two weeks, then 100 mg per day for 6 weeks, and then either 100 or 140 mg per day for 6 additional weeks. At the end of the study seven subjects reported to have skin rashes, 3 found to have hair color change. There were 3 adverse events related discontinuations including 1 transient bowel obstruction. 58.2% and 64.4% reduction in the plasma levels of $A\beta_{1-40}$ was observed with groups treated with 100 mg and 140 mg respectively. In contrast no significant reduction $A\beta_{1-40}$ was seen in cerebrospinal fluid [155].

A study was carried out on 20 healthy volunteers with the objective of evaluating whether this drug could reduce the Aβ production in CNS at single oral doses of

100, 140 or 280 mg (n= 5 per group). The method involved stable isotope labeling ($^{13}C_6$-Leucine) combined with cerebrospinal fluid sampling. At the end of the study significant decrease in the production of CNS Aβ levels were observed in a dose dependent manner. A decrease in the generation of Aβ about 47, 52, and 84% over 12 hour period with doses of 100, 140 and 280 mg observed respectively. The study also recorded rebound effect on CSF Aβ concentrations on later times 24-36 hours in comparison with placebo. Clearance of Aβ was known to be unaffected by the drug [156].

Eli lily initiated two separate Phase III clinical trials named as IDENTITY (100 to 140 mg of once a oral dose of drug) and IDENTITY 2 (60 mg of oral dose titrated upto 140 mg once a daily dose) for evaluating drug efficacy upto 21 months in mild to moderate AD patients in the year 2008 [157]. The trials were estimated to be complete by 2012. But later discontinued its clinical trials in Aug 2010, mainly because the drug not only failed to show any beneficial improvements in cognitive function but also exhibited increased risk of skin cancer in comparison with the placebo group [158, 159].

MK-0752

MK-0752 (structure not disclosed) is a gamma secretase inhibitor developed by Merck. The molecule is nonselective between APP and Notch. A phase I study was conducted with an intention to evaluate the safety, tolerability, pharmacokinetics and pharmacodynamics of single oral doses from 110 to 1000 mg of MK0752 in 27 healthy young men. Tolerability of the drug was generally well. A proportionate increase in the concentrations of drug was observed with respect to dose administered. Drug was peaked at 3-4 hours and declined. The half life of drug was found to be approximately 20 hours. Drug showed good penetration into the CNS. At the doses of 500 mg the drug inhibited 35% $A\beta_{1-40}$ concentrations in CSF upto 12 hours. 1000 mg of drug showed inhibition of $A\beta_{1-40}$ concentrations in CSF effect upto 24 hours. A dose dependent reduction in the plasma $A\beta_{1-40}$ concentrations were also observed but were followed later by rebound over baseline levels. At present the drug has been discontinued for AD. Since the inhibition of gamma secretase is also a viable option in the treatment in the cancer. The clinical trials for the drug's efficacy in the treatment in cancer are on the way but are not discussed here because it is out of scope of this chapter [160,161].

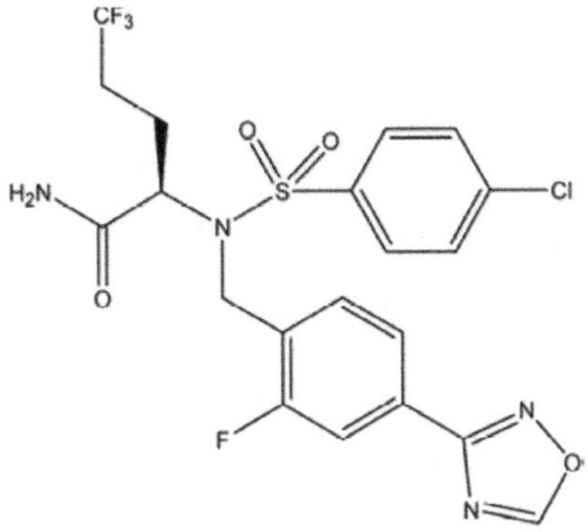

Figure 12: BMS-708163.

BMS-708163 (Avagacestat)

BMS-708163 (Fig. **12**) is a potent inhibitor of γ-secretase developed by Bristol-Myers Squibb. The drug demonstrated 193 fold selectivity against Notch. The drug showed significant reduction in Aβ40 levels for sustained periods in the brain, plasma and CSF of rats and dogs [162]. The drug showed good tolerability up to 400mg after single administration and 150 mg/day after multiple doses for 28 days in healthy young volunteers. The drug showed quick absorption through oral route (T_{max} = 1-2 hours). The elimination half life of the drug was found to be approximately 40 hours. In young subjects, upon single administration of doses 200 and 400 mg of drug a reduction of 37% and 40% CSF Aβ40 levels were seen up to 12 hours [53]. After multiple administrations for 4 weeks of drug from 50 to 150 mg/day, a steady state trough CSF Aβ40 levels were found to be reduced in a dose dependent manner compared to baseline values [163]. A phase II randomized, double blind placebo controlled trial was conducted up to 24 weeks to evaluate the safety, tolerability of different daily dose of the drug from 25 to 125 mg in 209 patients of mild to moderate AD disease. The median age of the patients was 75 years. Discontinuation rates for 25 mg and 50 mg doses of drug were comparable with placebo but were higher in 100 mg and 125 mg groups. 25 and 50 mg daily doses were relatively well tolerated and had low discontinuation rates. The 100 mg and 125 mg doses were poorly tolerated and exhibited the trends for cognitive worsening. Exploratory cerebrospinal fluid biomarker sub studies have provided preliminary support for γ-secretase target engagement, but additional studies are warranted to better characterize pharmacodynamic effects at the 25 and 50 mg doses [164]. A phase II multicentre double blind, placebo controlled safety and tolerability study for the drug planned for 104 weeks in patients in prodromal AD is still recruiting the patients [165]. A study was carried out to evaluate the pharmacokinetics, pharmacodynamics, preliminary tolerability of drug in young and elderly men and women. The subjects received the double blind 50 mg avagacestat capsule and placebo. The capsule of 50 mg was well tolerated and absorbed among young as well old subjects. The T_{max} of the drug was found to be between 1-2 hours. The half life was found to be between 41-71 hours. Aβ40 levels in serum were found to be decreased over a first 4-6 hours followed by a rise above baseline, which was maintained until the end of the assessment period. In general mild adverse events were seen in elderly subjects. Neither dose limiting gastrointestinal effects nor exploratory change in the biomarker inhibition for drug were observed [166].

A multicentre, randomized, double blind, placebo controlled, single ascending dose study for the tolerability, pharmacokinetic and pharmacodynamic markers was carried out in 8 healthy young men per dosing panel from 0.3 mg up to 800 mg and results revealed that the drug concentrations peaked quickly after oral administration and attained biphasic decrease in concentrations with prolonged terminal phase. Exposures were proportional with doses up to 200 mg. The drug was well tolerated at single doses up to 800 mg with biphasic on plasma Aβ40 levels. Mild to moderately severe adverse events were observed with no evidence of dose dependence up to 200 mg. Results suggest that avagacestat is well tolerated at single dose in the range from 0.3 mg up to 800 mg and suitable for further clinical development [167].

PF-3084014

PF-3084014 (Fig. **13**) is a novel γ-secretase inhibitor which, reduces Aβ production with an *in vitro* IC$_{50}$ 1.2nM in whole cell assay to 6.2 in cell free assay. The compound exhibited the inhibition of Notch- related T and B cell maturation in an *in vitro* thymocyte assay with an EC_{50} of 2.1 µM. The drug showed dose related Aβ reduction in brain, cerebrospinal fluid, and plasma in Tg2576 mice. The dose related reduction in Aβ levels in CSF and plasma also observed in guinea pigs at 0.03 to 3 mg/kg/day. Dose dependent reduction in Aβ in brain, CSF and plasma of guinea pigs were also observed when dosed with 0.03 to 10mg/kg. However dose dependent increases in $Aβ_{11-40}$ and $Aβ_{1-43}$ also observed at the dose that potentially inhibited $Aβ_{1-40}$ and $Aβ_{1-42}$ [168]. In a phase I trial single doses were found to be associated with mild and self limiting adverse events. Doses from 1-120 mg were found to be safe and well tolerated by the subjects. Mean pharmacokinetic variables with single 120 mg dose included oral clearance of approximately 16 mL/min/kg, a half life approximately 19 hours and Cmax of approximately 21nM and average steady state concentrations of approximately 4nM. A pharmacokinetic and pharmacodynamic study comprising of multiple oral dosing in humans led to the conclusion to end the development of this drug for the treatment of Alzheimers disease [53]. However the drug is now under Phase 1 Trial to evaluate its safety in patients with Advanced Solid Tumor Malignancy and T-Cell Acute Lymphoblastic Leukemia/Lymphoblastic Lymphoma [169].

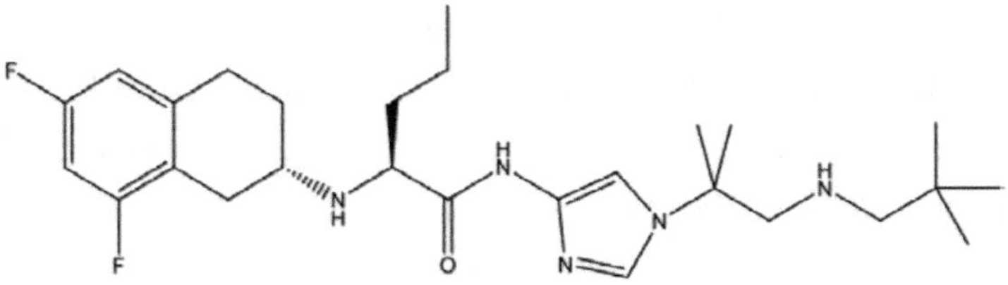

Figure 13: PF-3084014.

ELND-006

ELND-006 (Fig. **14**) is a novel gamma secretase inhibitor developed by Elan Pharmaceuticals. The molecule reported to have prominent selectivity against APP than Notch signaling (IC_{50}-0.34 nM against APP and IC_{50}-5.3nM versus notch signalling). The APP selectivity is found to be higher in cellular assays than Notch. The drug has IC_{50} 1.1 for APP and 82 nM for Notch signaling. The drug also reported to reduce significantly the formation of Aβ in PDAPP mouse model of AD [170, 171]. A study conducted in cynomolgus monkeys revealed that the drug at 0.3 mg/kg was able to reduce Aβ levels upto 25% for approximately 24 hours. After the reduction of Aβ levels subsequent overshoot above the control values were observed at later time points [172]. A phase I study was conducted to evaluate safety, tolerability, pharmacokinetics and pharmacodynamics of multiple oral daily doses of ELND006 in healthy elderly subjects. This was a randomized, double blind, placebo controlled, parallel assignment, ascending multiple dose study in sequential 5 cohorts, 6 subjects received ELND006 and 2 subjects received placebo per cohort. Oral daily doses of 3, 5, 10, 20 and 30 mg were administered for 21 days. Multiple daily doses of ELND006 at 20 mg were considered maximum tolerated dose in this study. The drug showed rapid absorption, large apparent distribution, slow apparent clearance, long apparent half life. ELND at 30 mg showed the most plasma Aβ reduction at 5 hours, Aβ reduction was observed in a dose dependent manner. Three subjects assigned at 30 mg experienced dose limiting toxicity, 2 of which developed symptomatic serious adverse events of abnormal liver laboratory chemistry, and another with grade 2 nausea resulting discontinuation [173]. Finally clinical trial for ELND006 was halted in October 2010 due to liver side effects that were related to unrelated mechanism of action [174].

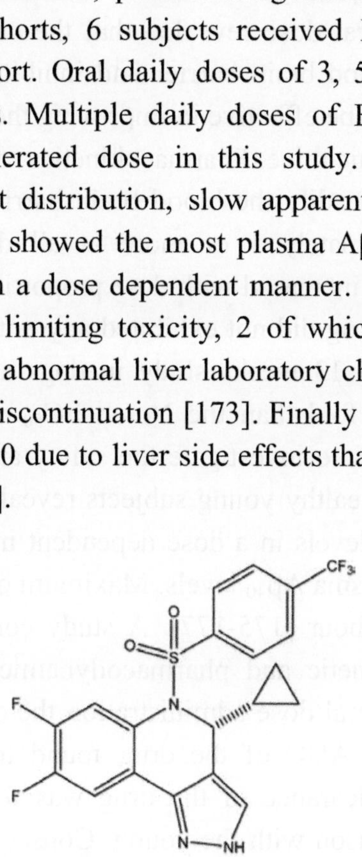

Figure 14: ELND006.

Begacestat

Begacestat (Fig. **15**) is a novel γ-secretase inhibitor developed by Wyeth. The molecule has showed potent dose dependent inhibition against production of Aβ$_{40}$ and Aβ$_{42}$ in cell lines which expresses human recombinant APP. The assays carried in cell free and cellular models confirmed that the molecule readily permeates the cell membranes and inhibits the gamma secretase enzyme. The molecule is found to be 16.8 times more selective to APP than Notch. A single high dose of 100 mg/kg oral administration of drug in transgenic APP (Tg2576) mouse model demonstrated approximately 88% $A\beta_{40}$ reduction in plasma and CSF by 2 hours whereas maximum of 60% reduction of $A\beta_{40}$ levels were observed at 6 hour. Subsequently, a 24 hour time course experiment which was conducted on 30 mg/kg oral dose revealed that the compound undergoes rapid absorption, crosses the blood brain barrier and inhibits the *denovo* synthesis of $A\beta$. The drug also found to be effective in improving the contextual fear condition deficits in Tg2576 transgenic mice. Pharmacokinetics studies in mice showed that the compound penetrates readily the blood brain barrier and achieves brain and plasma ration of approximately 1 or more at all doses evaluated. GSI-953 exposure were found to be increased with dose proportional manner in brain, CSF and plasma of the mice. Drug did not exhibited any notch related toxicity in a 28 days study in rats as well 13 weeks study in dogs at the drug concentrations approximately 56 and 36 fold than the Notch EC_{50} value. An ascending dose study conducted in humans to investigate the safety and tolerability of GSI-953 (3-600 mg oral doses) in healthy young subjects revealed that the drug produced reduction in plasma $A\beta_{40}$ levels in a dose dependent manner, this reduction was followed by increase in plasma $A\beta_{40}$ levels. Maximum of 40% of $A\beta_{40}$ levels were observed at 600 mg at 2 hour [175-177]. A study conducted in human for the evaluation of pharmacokinetic and pharmacodynamic relationships in humans revealed that upon single oral dose administration the drug was rapidly absorbed (t_{max} 1-2 hrs). Cmax and AUC of the drug found to be increased in a dose dependent manner. The clearance of the drug was found to be 40% lower in elderly subjects in comparison with the young. Consequently AUC was found to be 1.6 to 2 fold higher in elderly subjects. Elimination half life was found to be slightly prolonged in elderly than young subjects. In comparison to plasma, the drugs absorption in to CSF, elimination half life also slightly prolonged in AD patients compared to young subjects. AUC ratio of drug between Plasma: CSF were similar between young and AD subjects. The drug was well tolerated and no dose limiting adverse effects were observed. The dose dependent reduction was observed in the plasma $A\beta_{40}$ levels not CSF $A\beta_{40}$ levels in the healthy young, healthy elderly and AD subjects. A mean maximum decrease upto 40% and 14% in plasma Aβ40 and Aβ42 levels were observed at the highest dose tested [178]. A study involving translating the PK/PD biomarker relationship in different biological compartments between rodents and human showed that higher doses of drug are capable of reducing the Aβ40 levels in all the three compartments: brain, CSF and plasma in Tg2576 mice.In contrast, a lower dose of drug was able to reduce significant reduction in Aβ40 levels in brain and plasma not in CSF. In humans only plasma Aβ40 levels were found to be reduced in a dose dependent manner only in plasma in healthy and AD subjects upon single oral dose administration of drug. Plasma: CSF exposure ratio of approximately 10 was observed in both mice and humans. The study concludes that GSI-953 exposure relationship in the biological compartments in mice and humans implies that the concentrations of GSI-953 which lower the Aβ in the mouse model are likely achieved at the doses tested in humans [179].

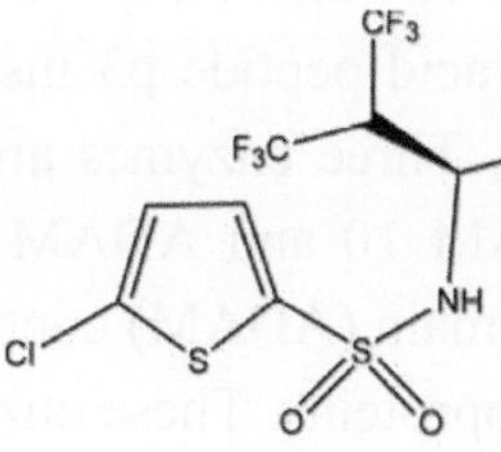

Figure 15: Begacestat (GSI-953).

γ-Secretase Modulators

Recent developments in last decade in developing drugs for AD revealed that some NSAIDS are capable of reducing the Aβ$_{1-42}$ in the cell culture and transgenic animal models devoid of their Cyclooxygenase activity [180, 181]. It was found that NSAIDS are capable of shifting the cleavage of APP to shorter fragments, especially Aβ$_{1-38}$. This is a unique advantage since they do not inhibit the cleavage of other substrates of γ-Secretase, in turn they found to just modulate the activity of γ-Secretase [180, 182, 183]. In the mean time it was also found that some of the NSAIDS and drugs which mainly involved in lipid metabolism are able to increase the Aβ$_{1-42}$ by increasing the catalytic activity of γ-Secretase. These results clearly indicate that these drugs are capable of influencing the γ -Secretase activity. All these observations lead to the development of analogues of the these NSAIDS and Non-NSAIDS as a new class of compounds called γ -Secretase modulators for the treatment of AD. Three of the γ-Secretase modulators namely tarenflurbil, E2012, CHF5074 have entered clinical trials [53, 160]. Search for better γ -Secretase modulators is still on and hopefully in within few years this research will bring fruitful result. Further elaboration about γ -Secretase modulators is not possible here because of the constrain of the scope of this chapter.

α-Secretase Enhancers

In the nonamyloidogenic pathway APP is initially cleaved by α-secretase. The cleavage occurs in the middle of the Aβ and produces soluble N-terminal APPα fragment (SAPPα) and a transmembrane C-terminal fragment (α-CTF). SAPPα has neurotrophic and neuroprotective functions. α -CTF is then cleaved by γ - secretase to generate 23-25 amino acid peptide p3 instead of generating Aβ and APP Intracellular Domain (AICD). Three enzymes are known to possess the α- secretase activity, ADAM 9, ADAM 10 and ADAM 17. They all belong to A Disintigrin and Metallo protease domain (ADAM) containing proteases. These are membrane bound cell surface glycoproteins. These enzymes mainly participate in the cell adhesion, cell ectodomain shedding, matrix protein degradation and cell fusion [184, 185]. The stimulation of α -secretase is a useful approach in the treatment of AD because it not only reduces Aβ formation but also generates the neuroprotective peptides which allow the neuronal survival and cognitive function in brain areas affected by AD [186]. α-Secretase mediated APP cleavage can be activated mainly *via* several G-protein coupled receptors and receptor tyrosine kinases, Protein kinase C, mitogen activated protein kinases [187]. Several α - secretase enhancers drugs are in developmental stage and are not discussed in detail here as it is out of the scope of this chapter.

Amyloid Degrading Enzymes and Their Utilisation in the Treatment of AD

The accumulation of the Aβ peptide in the brain is thought play a central role to the pathogenesis of AD. But it is now believed that accumulation of Aβ reflects its imbalance between production and clearance mechanisms. Recent research strongly suggest that it is the Aβ clearance not the Aβ production is the main contributing factor for the late onset sporadic AD [188]. Unlike metabolism of others biomolecules; Aβ metabolism is also a dynamic process and natural mechanisms exist for amyloid removal. Aβ clearance from the brain is mediated by several multiple processes such as drainage along perivascular basement membranes, mainly to cervical lymph nodes and into the cerebrospinal fluid; low density lipoprotein receptor related protein 1 or P-glycoprotein (PgP/MDR1/ABCB1) efflux pump mediated transport across vessel walls into circulation; the sequestration of Aβ by soluble low density lipoprotein receptor related protein 1 receptor in the circulation to promote the efflux of soluble Aβ out of the CNS; microglial phagocytosis and enzyme mediated degradation of Aβ [189].

Among all these processes Enzyme mediated degradation has received a great deal of attention during past decade. Many Amyloid degrading enzymes (ADEs) have now been identified which are capable of cleaving full length Aβ into smaller, less neuro toxic fragments and more easily cleared. The ADEs, their location and their selectivity towards Aβ is summarised in the form of Table (2) below. Apart from their involvement in Aβ metabolism these proteases also act on multiple substrates and perform physiological functions. At present still it remains unproven that a decline in ADE's activity contributes to the accumulation of Aβ in AD. However over expression of ADEs found to reduce Aβ burden and improve cognitive function in animal models of AD. Based on these results the therapeutic approaches such as upregulating ADE's, Gene therapy towards increasing expression of ADE's in periphery or within the brain, Direct delivery of enzymes into the brain, utilisation of stem cells to express the desired ADE in the brain are ongoing [190]. Some of the researchers also express that it is beneficial to upregulate modestly several of these ADEs than individual modulation, since these enzymes are located in the different subcellular compartments [151]. Research on these ADEs is still in fundamental stage compared to Secretases, hopefully right strategies towards utilising these ADEs will emerge for treatment of AD in coming years.

Table 2:   **Amyloid degrading enzymes, their location and their selectivity towards Aβ**

S.N.	Member	Expression within Human Brain	Aβ Preference
1	NEPRILYSIN (NEP, CD10, CALLA, EC.3.4.24.15)	• Neuronal • Vessel (vascular smooth muscle)	• Soluble Aβ • Oligomeric Aβ
2	NEPRILYSIN-2 (NEP-2, SEP, MMEL1/2)	• Neuronal	• Unspecified
	Endotheline converting enzyme 1 (ECE-1, EC.3.4.24.71)	• Vascular endothelial • Limited neuronal	• Soluble monomeric
	Endothelin converting enzyme 2 (ECE-2, EC.3.4.24.71)	• Pyramidal neurons • Astrocytes/microglia in AD	• Soluble monomeric
3	Angiotensin converting enzyme (ACE) (Dipeptidyl carboxypeptidase, EC.3.4.15.1)	• Pyramidal neurons	• Soluble Aβ
4	Insulin degrading enzyme (Insulysin; IDE, EC.3.4.24.56)	• Neuronal • Astrocyte • Microglia • Vessel (endothelial)	• Soluble monomeric only
5	Matrix metalloproteinase-9 (MMP9, Gelatinase-B, EC.3.4.24.35)	• Neuronal • Astrocyte • Microglia	• Soluble Aβ • Aβ fibrills
5	Matrix metalloproteinase-2 (MMP-2, Gelatinase A, EC.3.4.24.24)	• Blood vessels and glia	• Soluble Aβ • Oligomeric Aβ • Partial digestion of fibrillar Aβ
	Type-1transmembrane MMP (MT1-MMP, EC.3.4.24.80)	• Neuronal • Astrocyte • Microglia • Vessel	• Soluble Aβ • Fibrillar Aβ
6	Myelin basic protein (MBP, EC.2.1.1.126)	• Oligodendrite (White matter) • Neuronal	• Soluble Aβ • Fibrillar Aβ
	Plasmin (Tissue plasminogen Activator for tPA: Urokinase-type plasminogen activator for μPA, EC.3.4.21.7)	• Neuronal	• Aβ monomers and fibers • Oligomers
	Acyl peptidase hydrolase (APEH, EC.3.4.19.1)	• Unknown distribution	• Monomeric and oligomeric Aβ
7	Cathepsin B (CTSB,EC.3.4.22.1)	• Neuronal • Microglia	• Soluble monomeric and nonfibrillar • Fibrillar Aβ

Source: [190].

CONCLUSIONS

AD is a progressive neurodegenerative disorder which involves a complex multiple pathological mechanisms like generation of Aβ peptides, Tau protein hyperphosphorylation, oxidative stress, inflammation, glutamate excitotoxicity, loss of synapses, and death of neurons. Majority of these factors are found to be associated directly or indirectly with functioning of many enzymes. Hence it is postulated that the disorder can be treated effectively by inhibiting or activating/ modulating the key enzymes involved in the pathogenesis of this disorder. Currently approved AChEIs can afford symptomatic relief to the mild and moderately AD patients only for a short span of time but lack their capability in neither slowing down, nor blocking the basic pathological mechanisms. The clinical trial results of secretase inhibitors developed so far is not so encouraging because of their high toxic effects incomparison with their therapeutic advantage. γ - secretase inhibitors developed upto so far although have shown good response in preclinical trials but failed in clinical trials because of their notch related toxicity. The results of β secretase although seems to be appreciated in *in vitro* models and preclinical trials but they also unfortunately exhibited toxicity related to unknown mechanisms in clinical trials. More over among the scientific community it is also argued that the complete pathology involved in the development of AD is not completely understood. All of these factors have made development of effective drug towards treatment of this disease more challenging. But optimistically research in neuroscience is progressing day by day. Many γ and β secretase inhibitors and modulators, α-secretase enhancers are in developmental stages, all these efforts may reveal the clear pathological mechanism of AD in coming years and may bring effective molecules for treatment of this progressive disorder. Since the pathology of this disorder involves complex and multifactorial mechanisms; the molecules which act on multiple targets may be an effective drug for the treatment of AD.

ACKNOWLEDGEMENTS

Vageesh Revadigar expresses gratitude to Institute of Postgraduate Studies, Universiti Sains Malaysia for providing financial support in the form of Graduate Assistanceship for the study.

CONFLICT OF INTEREST

The authors confirm that this chapter contents have no conflict of interest.

ABBREVIATIONS

Cognition

ADAS-cog	=	Alzheimer's Disease Scale-cognitive subscale
MMSE	=	Mini-Mental State Examination
SIB	=	Severe Impairment Battery

Global

ADCS-CGIC	=	Alzheimer's Disease cooperative study –clinical Global Impression of Change
CDR-SB	=	Clinical Dementia Rating- Sum of the Boxes
CGI	=	Clinical global Impression
CGI-C	=	Clinical Global Impression of Change
CIBIC-Plus	=	Clinical Interview Based Impression of Change Plus Care Giver Input
GDS	=	Global Deterioration Scale

Function

ADCS-ADL	=	Alzheimer's Disease Cooperative study-Activities of Daily Living
ADCS-GCIC	=	Alzheimer's Disease Co-operative study-Clinical Global Impression of Change
ADFACS	=	Alzheimer's Disease Functional Assessment and Change Scale

BADLS = Basic Activities of daily Living Scale

CMCS = Care giver rated Modified Crichton Scale

DAD = Disability Assessment for Dementia

IDDD = Interview for Deterioration in Daily Living in Dementia scale

NOSGER = Nurse's Observation Scale for Geriatric Patients

PDS = Progressive Deterioration Scale

Behaviour

NPI = Neuropsychiatric Inventory

REFERENCES

[1]   Alzheimers association. What is alzheimers? http://www.alz.org/alzheimers_disease_what_is_alzheimers.asp (accessed July 23, 2012).

[2]   Alzheimers association. Alzheimers facts and figures. http://www.alz.org/alzheimers_disease facts and figures.asp (accessed July 23, 2012)

[3]   Irena, M. Therapies for Alzheimers disease. *Nature reviewsDrug discovery.*, **2007**, *6*, 341-342.

[4]   Hensley, K.; Carney, J. M.; Mattson, M. P.; Aksenova, M.; Harrus, M.; Wu, J. F.; Floyd, R. A.; Butterfield, D. A. A model for β-amyloid aggregation and neurotoxicity based on free radical generation by the peptide: Relevance to Alzheimer disease. *Proc. Natl. Acad. Sci., USA.*, **1994**, *91*, 3270-3274.

[5]   Tamagno, E.; Guglielmotto, M.; Aragno, M.; Borghi, R.; Autelli, R.; Giliberto, L.; Muraca, G.; Danni, O.; Zhu, X.; Smith, M,A.; Perry, G.; Jo, D,G.; Mattson, M.P.; Tabaton, M. Oxidative stress activates a positive feedback between the γ- and β-secretase cleavages of the β-amyloid precursor protein. *J. Neurochem.*, **2008**, *104*, 683-695.

[6]   Tougu, V. Acetylcholinesterase: mechanism of catalysis and inhibition. *Curr. Med. Chem.-CNS Agents.*, **2001**, *1*, 155-170.

[7]   De Strooper, B.; Vassar, R.; Golde, T. The secretases;enzymes with Therapeutic potential in Alzheimers disease. *Nat. Rev. Drug Discov.*, **2010**, *6*, 99-107.

[8]   Shankar, G. M.; Li, S.; Mehta, H. T.; Munoz, G. A.; Shepardson, E. N.; Smith, I.; Brett, F. M.; Farrell, M. A.; Rowan, M. J.; Lemere, C. A; Regan, M. C.; Walsh, M. D.; Sabatini, B. L.; Selkoe, D. J. Amyloid β-protein dimmers isolated directly from Alzheimers brains impair synaptic plasticity and memory. *Nat. Med.*, **2008**, *14(8)*, 837-842.

[9]   Panza, F.; Frisardi, V.; Imbimbo, B. P.; Capurso, C.; Logroscino, G.; Sancarlo, D.; Seripa, D.; Vendemiale, G.; Pilotto, A.; Solfrizzi, V.γ-secretase inhibitors for the treatment of Alzheimers disease. *CNS Neurosci. ther.*, **2010**, *16*, 272-284.

[10]   Ohyagi, Y.; Asahara, H.; Chui, D. H.; Tsuruta, Y.; Sakae, N.; Miyoshi, K.; Yamada, T.; Kikuchi, H.; Taniwaki, T.; Murai, H.; Ikzoe, K.; Furuya, H.; Kawarabayashi, T.; Shoji, M.; Checler, F.; Iwaki, T.; Makifuchi, T.; Takeda, K.; Kira, J.; Tabira, T. Intracellular Aβ 42 activates p53 promoter; a pathway to neurodegeneration in Alzheimers disease. *The FASEB journal.* [online] **2004**, 1-29. http://www.fasebj.org/content/early/2005/01/27/fj.04-2637fje (accessed July 23,2012).

[11]   Walter, J.; Kaether, C.; Steiner, H.; Haass, C. The cell biology of Alzheimers disease; uncovering the secretes of secretases. *Curr. Opin. Neurobiol.*, **2001**, *11*, 585-590.

[12]   Selkoe, D. J. Alzheimers Disease; genes, proteins and therapy. *Physiol. rev.*, **2001**, *81(2)*, 741-766.

[13]   Wisniewski, T.; Ghiso, J.; Frangione, B. Biology of Aβ amyloid in Alzheimers disease. *Neurobiol Dis.*, **1997**, *4*, 313-328.

[14]   Schmitz, C.; Rutten, B. F.; Pielen, A.; Scha¨fer, S.; Wirths, O.; Tremp, G.; Czech, C.; Blanchard, V.; Multhaup, G.; Rezaie, P.; Korr, H.; Steinbusch, H, M.; Pradier, L.; Bayer, T. A. Hippocampal neuron loss exceeds Amyloid plaque load in a transgenic mouse model of Alzheimers disease. *Am. J. Pathol.*, **2004**, *164(4)*,1495-1502.

[15]   Knopman, D. S.; Parisi, J, E.; Salviati, A.; Floriach, R, M.; Boeve, B.F.; Ivnik, R. J.; Smith, G. E.; Dickson, D. W.; Johnson, K. A.; Petersen, L. E.; McDonald, W. C.; Braak, H.; Petersen, R. C. Neuropathology of Cognitive normal elderly. *J. Neuropathol. Exp.Neurol.*, **2003**, *62 (11)*, 1087-1095.

[16]   Selkoe, D. J. Alzheimers disease results from the cerebral accumulation and cytotoxicity of amyloid β-protein. *J. Alzheimers disease.*, **2001**, 3, 75-80.

[17]   Butterfield, D, A.; Drake, J.; Pocernich, C.; Castegna, A. Evidence of oxidative damage in Alzheimers disease brain; Central role for amyloid β peptide. *Trends. Mol. Med.*, **2001**, *7(12)*, 548-554.

[18]   Harris, M. E.; Hensley, K.; Butterfield, D. A.; Leedle, R. A.; Carney, J. M. Direct evidence of oxidative injury produced by the Alzheimers β amyloid peptide (1-40) in cultured hippocampal neurons. *Exp. Neurolog.*, **1995**, *131*, 193-202.

[19]   Mattson, M. P.; Cheng, B.; Davis, D.; Bryant, K.; Lieberburg, I.; Rydel, R. E. β-amyloid peptide destabilizes calcium homeostasis and render human cortical neurons vulnerable to excitotoxicity. *J. Neurosci.*, **1992**, *12(2)*, 376-389.

[20]   Francisa, P. T.; Palmerb, A. M.; Snapeb, M.; Wilcockc, G. K. The Cholinergic hypothesis of Alzheimers Disease: a review of progress. *J. Neurol. Neurosurg. Psychiatry.*, **1999**, *66*, 137-147.

[21]   Small, S. A.; Duff, K. Linking Aβ and Tau in Late-Onset Alzheimers Disease: A Dual pathway Hypothesis. *Neuron.*, **2008**, *60(26)*, 534-542.

[22]   Delacourte, A.; David, J. P.; Sergeant, N.; Buée, L.; Wattez, A.; Vermersch, P.; Ghozali, F.; Fallet Bianco, C.; Pasquier, F.; Lebert, F.; Petit, H.; Di Menza, C. The biochemical pathway of neurofibrillary degeneration in aging and Alzheimers disease. *Neurology.*, **1999**, *52*,1158-1165.

[23]   Maccioni, R. B.; Farías, G.; Morales, I.; Navarrete, L. The revitalized Tau hypothesis on Alzheimers Disease. *Arc. Med. Res.*, **2010**, *(41)*, 226-231.

[24]   Iqbal, K.; Alonso, A. C.; Gong, C. X.; Khatoon, S.; Pei, J. J.; Wang, J. Z.; Grundke-Iqbal, I. Mechanisms of neurofibrillary degeneration and the formation of neurofibrillary tangles. *J. Neural. Transm. suppl.*, **1998**, *53*,169-80.

[25]　McCormick, D. A. Acetylcholine: distribution, receptors, and actions. *Seminars in Neuroscience.*, **1989**, 91-101.

[26]　Bartus, R. T. On neurodegenerative diseases, models and treatment strategies: Lessons learned and lessons forgotten a generation following the cholinergic hypothesis. *Ex. Neurol.*, **2000**, *163*, 495-529.

[27]　Terry, A.V.; Buccafuso, J.J. The cholinergic hypothesis of Age and Alzheimers Disease-Related cognitive deficits: Recent Challenges and their Implications for novel Drug development. *J. Pharm. Exp. Ther.*, **2003**, *36(3)*, 281-287.

[28]　Ollis, D. L.; Cheah, E.; Cygler, M.; Dijkstra, B.; Frolow, F.; Franken, S.; Harel, M.; Remington, S. J.; Israel, S.; Schragl, J.; Sussman, J. L.; Verschueren, K. G.; Goldmans, A. The α/β hydrolase fold. *Protein Engineering.*, **1992**, *5(3)*, 197-211.

[29]　Brenner, S. The molecular evolution of genes and proteins: a tale of two serines. *Nature*, **1988**, *334*, 528-530.

[30]　Weinstock, M.; Groner, E. Rational design of a drug for Alzheimer's disease with cholinesterase inhibitory and neuroprotective activity. *Chemico-Biol. Interact.*, **2008**, *175*, 216-221.

[31]　Mesulam, M.; Guillozet, A.; Shaw, P.; Quinn, B. Wide spread butyrylcholinesterase can hydrolyse acetylcholine in the normal and Alzheimer brain. *Neurobiol. Dis.*, **2002**, *9*, 88-93.

[32]　Cokugras, A. N. Butyrylcholinesterase: Structure and physiological importance. *Turkish J. Biochem.*, **2003**, *28*, 54-61.

[33]　Fujii, T.; Mori, Y.; Tominaga, T.; Hayasaka, I.; Kawashima, K. Maintenance of constant blood acetylcholine content before and after feeding in young chimpanzees. *Neurosci. Lett.*, **1997**, *227*, 21-24.

[34]　Fukuto, T. R. Mechanism of Action of Organophosphorus and carbamate insecticides. *Environ. Health persp.*, **1990**, *87*, 245-254.

[35]　Bartolucci, C.; Haller, L. A.; Jordis, U.; Fels, G.; Lamba, D. Probing Torpedo Californica Acetylcholinesterase catalytic Gorge with Two novel bis-functional Galanthamine derivatives. *J. Med. Chem.*, **2010**, *53*, 745-751.

[36]　Kreienkamp, H. J.; Weise, C.; Raba, R.; Aaviksaar, A.; Hucho, F. Anionic subsites of the catalytic center of acetylcholinesterase from Torpedo and from Cobra venom. *Proc. Nat. Acad. Sci. USA*, **1991**, *88*, 6117-6121.

[37]　Eichler, J.; Anselment, A.; Sussman, J.L.; Massoulie, J.; Silman, I. Differential effects of peripheral site ligands on Torpedo and Chicken acetylcholinesterase. *Mol. pharmacol.*, **1994**, *45(2)*, 335-340.

[38]　Haviv, D.; Wong, M.; Silman, I.; Sussman, J.L. Bivalent Ligands Derived from Huperizine A as Acetylcholinesterase Inhibitors. *Curr. topics Med. chem.*, **2007**, *7(4)*, 375-387.

[39]　Inestrosa, N.C.; Dinamarca, M. C.; Alvarez, A. Amyloid-cholinesterase interactions implications for alzheimers disease. *FEBS journal.*, **2008**, *275*, 625-632.

[40]　Manoharan, I.; Boopathy, R.; Darvesh, S.; Lockridge, O. A medical health report on individuals with silent butyrylcholinesterase in Vysya community of India. *Clin. Chim. Acta.*, **2007**, *378*, 128-135.

[41]　Iwasaki, T.; Yoneda, M.; Nakajima, A.; Terauchi, Y.Serum butyrylcholinesterase is Strongly associated with Adiposity, the serum lipid profile and Insulin resistance. *Internal Med.*, **2007**, DOI: 10.2169/internalmedicine.46.0049.

[42]   Yuan, J.; Yin, J.; Wang, E. Characterization of Procain metabolism as probe for the butyrylcholinesterase enzyme investigation by simultaneous determination of procaine and its metabolite using capillary electrophoresis with electrochemiluminescence detection. *J. Chromatogr A.*, **2007**, *1154*, 368-372.

[43]   Zelinski, T.; Coghlan, G.; Mauthe, J.; Triggs, R. B. Molecular basis of succinylcholine sensitivity in a prairie Hutterite kindred and genetic characterization of region containing the BchE gene. *Mol. Genet. Metab.*, **2007**, *90*, 210-216.

[44]   Duysen, E. G.; Li, B.; Carlson, M.; Li, Y. F.; Wieseler, S.; Hinrichs, S. H.; Lockridge, O. Increased Hepatotoxicity and Cardiac fibrosis in Cocaine-treated Butyrylcholinesterase knockout mice. *Basic Clin. Pharmacol. Toxicol.*, **2008**, *103*, 514-521.

[45]   Kolarich, D.; Weber, A.; Pabst, M.; Stadlmann, J.; Teschner, W.; Ehrlich, H.; Schwarz, H. P.; Altmann, F. Glycoproteomic characterization of Butyrylcholinesterase from Human plasma. *Proteomics.*, **2008**, *8*, 254-263.

[46]   Saxena, A.; Sun, W.; Luo, C.; Myers, T. M.; Koplovitz, I.; Lenz, D. E.; Doctor, B. P. Bioscavenger for protection from toxicity of organophosphorus compounds. *J. Mol. Neurosci.*, **2006**, *30(1-2)*, 145-8.

[47]   Wright, C. I.; Geuala, C.; Mesulam, M. M. Neurological cholinesterases in the normal brain and in Alzheimers disease: relationship to plaques, tangles, and patterns of selective vulnerability. *Ann. Neurol.*, **1993**, *34(3)*, 373-84.

[48]   Greig, N. H.; Utsuki, T.; Ingram, D. K.; Wang, Y.; Pepeu, G.; Scali, C.; Yu, Q. S.; Mamczarz, J.; Holloway, H. W.; Giordano, T.; Chen, D.; Furukawa, K.; Sambamurti, K.; Brossi, A.; Lahiri, D. K. Selective butyrylcholinesterase inhibition elevates brain acetylcholine, augments learning and lowers Alzheimers β-amyloid peptide in rodent. *Proc. Nat. Acad. Sci.*, **2005**, *102*, 17213-17218.

[49]   Galimberti, E.; Scarpini. E. Disease-modifying treatments for Alzheimers disease. *Ther. Adv. Neurol. Disord.*, **2011**, *4(4)*, 203-216.

[50]   Tomita, T. Secretase inhibitors and modulators for Alzheimers disease treatment. *Expert rev.*, **2009**, *9(5)*, 661-679.

[51]   Vassar, R.; Kovacs, D. M.; Yan, R.; Wong, P. C. The β-Secretase enzyme BACE in health and Alzheimer's disease: Regulation, Cell biology,Function, and Therapeutic potential. *J. Neurosci.*, **2009**, *29(41)*, 12787-12794.

[52]   Citron, M. β -Secretase inhibition for the treatment of Alzheimers disease-promise and challenge. *Trends pharmacol. sci.*, **2004**, *25(2)*, 92-97.

[53]   Imbimbo, B. P.; Giardina, G. A. γ-secretase inhibitors and modulators for the treatment of Alzheimers disease: Disappointments and Hopes. *Curr. Topics Med. Chem.*, **2011**, *11*, 15570-1570.

[54]   Drugs treatment for Alzheimers disease. http://www.alzheimers.org.uk/site/scripts/document_pdf.php?documentID=147 (accessed Aug 26, 2012).

[55]   Bodick, N. C.; Offen, W. W.; Levey, A. I.; Cutler, N. R.; Gauthier, S. G.; Satlin, A.; Shannon, H. E.; Tollefson, G. D.; Rasmussen, K.; Bymaster, F. P.; Hurley, D. J.; Potter, W. Z.; Paul, S. M. Effects of Xanomeline, a selective Muscarinic receptor agonist, on cognitive function and behavioral symptoms in Alzheimers disease. *Arch. Neurol.*, **1997**, *54*, 465-473.

[56]   Thal, L. J.; Rosen,W.; Sharpless, N.S.; Crystal, H. Choline chloride fails to improve cognition of Alzheimers disease. *Neurobiol. Aging.*, **1981**, *2(3)*, 205-208.

[57]   Patel, S. V. Pharmacotherapy of Cognitive impairment in Alzheimers disease: A review. *J. Geriatr. Psychiatry Neurol.*, **1995**, *8*, 81-95.

[58]   Giacobini, E. Cholinesterase inhibitors: new roles and therapeutic alternatives. *Pharmacol. Res.*, **2004**, *50*, 433-440.

[59]   Alzheimers Association. Medications for memory loss. http://www.alz.org/alzheimers_ disease_standard_prescriptions.asp (accessed Aug 26, 2012).

[60]   Birks, J. Cholinesterase inhibitors for Alzheimers disease. *Cochrane Database of Systematic Reviews*.1, [online] **2006**, http://onlinelibrary.wiley.com/doi/10.1002/14651858. CD005593/pdf (accessed July 28, 2012).

[61]   Giacobini, E.; Selective inhibitors of butyrylcholinesterase: a valid alternative for therapy of Alzheimers disease?. *Drugs aging.*, **2001**, *18(12)*, 891-898.

[62]   Hogan, D. B.; Bailey, P.; Carswell, A.; Clarke, B.; Cohen, C.; Forbes, D.; Man-Son-Hing, M.; Lanctôt, K.; Morgan, D.; Thorpe, L. Management of mild to moderate Alzheimers disease and dementia. *Alzheimers dement.*, **2007**, *3*, 355-384.

[63]   Thibodeau, M. P.; Massoud, F. A decade of drug therapies: A review. *Can. Rev. Alzheimers dis. dement.*, **2010**, 4-10.

[64]   Doody, R. S. Update on Alzheimer drugs (Donepezil). *The Neurologist.*, **2003**, *9*, 225-229.

[65]   Tsuno, N. Donepezil in the treatment of patients with Alzheimers disease. *Expert. Rev. Neurther.*, **2009**, *9(5)*, 591-598.

[66]   Shigeta, M.; Homma, A. Donepezil for Alzheimers disease: Pharmacodynamic, Pharmacokinetic, and Clinical profiles. *CNS Drug. Rev.*, **2001**, *7(4)*, 353-368.

[67]   Jackson, S.; Ham, R. J.; Wilkinson, D. The safety and tolerability of donepezil in patients with Alzheimers disease. *Br. J. Clin. Pharmacol.*, **2004**, *58(S1)*,1-8.

[68]   Rogers, S. L.; Friedhoff, L. T. Pharmacokinetic and pharmacodynamic profile of Donepezil Hcl following single oral doses. *Br. J. Clin. Pharmacol.*, **1998**, *46 (suppl-1)*,1-6.

[69]   Tiseo, P. J.; Perdomo, C. A.; Friedhoff, L. T. Metabolism and elimination of ^{14}C-Donepezil in healthy volunteers: a single dose study. *Br. J. Clin. Pharmacol.*, **1998**, *46(Suppl-1)*, 19-24.

[70]   Waldemar, G. Donepezil in the treatment of patients with Alzheimers disease. *Expert Rev. Neurother.*, **2001**, *1(1)*,11-19.

[71]   Aricept prescribing and patient info. http://www.aricept.com/assets/pdf/AriceptCombo FullPIFebruary2012.pdf (accessed Aug 25, 2012).

[72]   Sabbagh, M.; Cummings, J. Progressive cholinergic decline in Alzheimers disease: consideration for treatment with donepezil 23 mg in patients with moderate to severe symptomatology. *BMC Neurology.*, **2011**, 11-21.

[73]   Takada, Y.; Yonezawa, A.; Kume, T.; Katsuki, H.; Kaneko, S.; Sugimoto, H.; Akaike. A. Nicotinic acetylcholine receptor-mediated neuroprotection by Donepezil against glutamate neurotoxicity in rat cortical neurons. *J. Pharmacol. Exp. Ther.*, **2003**, *306(2)*, 772-777.

[74]   Mangialasche, F.; Solomon, A.; Winblad, B.; Mecocci, P.; Kivipelto, M. Alzheimers disease: clinical trials and drug development. *Lancet Neurol.*, **2010**, *9*, 702-16.

[75]   Exelon prescribing information. http://www.pharma.us.novartis.com/product/pi/pdf/exelon. pdf (accessed Aug 28, 2012).

[76]   Polinsky, R. J. Clinical pharmacology of Rivastigmine: A new-generation Acetylcholinesterase inhibitor for the treatment of Alzheimers disease. *Clin. ther.*, **1998**, *20(4)*, 634-647.

[77]   Onor, M. L.; Trevisiol, M.; Aguglia. E. Rivastigmine in the treatment of Alzheimers disease: an update. *Clin. Interv. Aging.*, **2007**, *2(1)*, 17-32.

[78]   Giacobini, E.; Spiegel, R.; Enz, A.; Veroff, A. E.; Cutler, N. R. Inhibition of acetyl and butyryl cholinesterase in the cerebrospinal fluid of patients with Alzheimers disease by Rivastigmine: correlation with cognitive benefit. *J. Neural Transm.*, **2002**, *109*, 1053-1065.

[79]   Standridge, J. B. Pharmacotherapeutic approaches to the treatment of Alzheimers disease. *Clinical therapeutics.*, **2004**, *26(5)*, 615-630.

[80]   Grossberg, G. T.; Stahelin, H. B.; Messina, J. C.; Anand, R.; Veach, J. Lack of adverse pharmacodynamic drug interactions with rivastigmine and twenty-two classes of medications. *Int. J. Geriatr. Psychiatry.*, **2000**, *15*, 242-247.

[81]   Feldman, H. H.; Ferris, S.; Winblad, B.; Sfikas, N.; Mancione, L.; He, Y.; Tekin, S.; Burns, A.; Cummings, J.; del Ser, T.; Inzitari, D.; Orgogozo, J. M.; Sauer, H.; Scheltens, P.; Scarpini, E.; Herrmann, N.; Farlow, M.; Potkin, S.; Charles, H. C.; Fox, N. C.; Lane. R. Effect of rivastigmine on delay to diagnosis of Alzheimers disease from mild cognitive impairment: the InDDEx study. *Lancet neurol.*, **2007**, *6*, 501-12.

[82]   Approval received for Rivastach patch for the treatment of Alzheimers disease. http://www.ono.co.jp/eng/news/pdf/sm_cn110422.pdf (accessed Aug 28, 2012).

[83]   Prescribing information for Excelon patch. http://www.pharma.us.novartis.com/product/pi/pdf/exelonpatch.pdf (accessed Aug 29, 2012)

[84]   Ellis, J. M. Cholinesterase Inhibitors in the treatment of Dementia. *JAOA.*, **2005**, *105(3)*, 145-158.

[85]   Massoud, F.; Gauthier, S. Update on the pharmacological treatment of Alzheimers disease. *Curr. Neuropharmacol.*, **2010**, *8*, 69-80.

[86]   Sarka, S.; Karel, K. Cholinesterases and Cholinesterase Inhibitors. *Curr. Enz. Inhib.*, **2008**, *4*, 160-171.

[87]   Raskind, M. A. Update on Alzheimer drugs (galantamine). *The Neurologist.*, **2003**, *9(5)*, 235-240.

[88]   Thomsen, T.; Kaden, B.; Fischer, J. P.; Bickel, U.; Barz, H.; Gusztony, G.; Cervos-Navarro, J.; Kewitz. H. Inhibition of Acetylcholinesterase activity in Human brain tissue and Erythrocytes by galanthamine, Physostigmine and Tacrine. *Eur. J. Clin. Chem. Cln. Biochem.*, **1991**, *29*, 487-492.

[89]   Samochocki, M.; Höffle, A.; Fehrenbacher, A.; Jostock, R.; Ludwig, J.; Christner, C.; Radina, M.; Zerlin, M.; Ullmer, C.; Pereira, E. F.; Lübbert, H.; Albuquerque, E. X.; Maelicke. A. Galantamine is an allosterically potentiating Ligand of Neuronal nicotinic but not muscarinic Acetylcholine receptors. *J. Pharmacol. Exp. Ther.*, **2003**, *305(3)*, 1024-1036.

[90]   Huang, F.; Fu, Y. A review of clinical pharmacokinetics and Pharmacodynamics of Galantamine, a reversible acetylcholinesterase inhibitor for the treatment of Alzheimers disease, in healthy subjects and patients. *Curr. Clin. Pharmacol.*, **2010**, *5*, 115-124.

[91]   Prescribing information for Razadyne ER. http://www.razadyneer.com/sites/default/files/shared/pi/razadyne_er.pdf#zoom=100 (accessed Aug 29, 2012).

[92]   Shi, J.; Seltzer, B. Galantamine: an update. *Neurodegen. Dis. Manage.*, **2011**, *1(3)*, 227-234.

[93]   Ritchie, C.W.; Ames, D.; Clayton, T.; Lai, R. Meta analysis of randomized trials of the efficacy and safety of donepezil, galantamine, and rivastigmine for the treatment of Alzheimers disease. *Am. J. Geriatr. Psychiatry.*, **2004**, *12(4)*, 358-69.

[94]   Birks, J.; Harvey, J. Donepezil for dementia due to alzheimers disease. Cochrane Database *Syst. Rev.*, [online] **2006**, Issue1. http://onlinelibrary.wiley.com/doi/10.1002/14651858.CD001190.pub2/pdf (accessed July 20, 2012).

[95]   Raina, P.; Santaguida, P.; Ismaila, A.; Patterson, C.; Cowan, D.; Levine, M.; Booker, L.; Oremus. M. Effectiveness of Cholinesterase inhibitors and memantine for treating dementia: Evidence review for a clinical practice guideline. *Ann. Intern. Med.*, **2008**, *148*, 379-397.

[96]   Hansen, R. A.; Gartlehner, G.; Webb, A. P.; Morgan, L. C.; Moore, C. G.; Jonas, D. E. Efficacy and safety of donepezil, galantamine, and rivastigmine for the treatment of alzheimers disease: a systematic review and meta analysis. *Clin. Interv. Aging.*, **2008**, *3(2)*, 211-225.

[97]   Birks, J.; Grimley, E.; Iakovidou, V.; Tsolaki, M.; Holt, F.E. Rivastigmine for Alzheimers disease. *Cochrane Database syst. Rev.*, [online] **2009**, *15(2)*. http://onlinelibrary.wiley.com/ doi/10.1002/14651858.CD001191/pdf (accessed July 21, 2012)

[98]   Loy, C.; Schneider, N. Galantamine for alzheimers disease and mild cognitive impairment review. *Cochrane Database syst. Rev.*, [online] **2009**, Issue 1. http://onlinelibrary.wiley. com/doi/10.1002/14651858.CD001747.pub3/abstract;jsessionid=787B5F781EC9B95AA15 C9E2417A8413A.d04t03 (accessed July 24, 2012)

[99]   Hansen, R. A.; Gartlehner, G.; Lohor, K. N.; Kaufer, D. I. Functional outcomes of drug treatment in Alzheimers disease: A systematic review and meta analysis. *Drugs Aging.*, **2007**, *24(2)*, 155-67.

[100]   Lanctôt, K. L.; Herrmann, N.; Yau, K. K.; Khan, L. R.; Liu, B. A.; LouLou, M. M.; Einarson, T. R. Efficacy and safety of cholinesterase inhibitors in Alzheimers disease: a meta analysis. *CMAJ.*, **2003**, *169(6)*, 557-64.

[101]   Harry, R. D.; Zakzanis, K. K. A comparison of donepezil and galantamine in the treatment of cognitive symptoms of Alzheimers disease: a meta analysis. *Hum. Psychopharmacol. Clin. Exp.*, **2005**, *20*, 183-187,

[102]   Campbell, N.; Ayub, A.; Boustani, M. A.; Fox, C.; Farlow, M.; Maidment, I.; Howards, R. Impact of cholinesterase inhibitors on behavioral and psychological symptoms of Alzheimers disease: A meta-analysis. *Clin. Invest. Aging.*, **2008**, *3(4)*, 719-728.

[103]   Kaduszkiewicz, H.; Zimmermann, T.; Beck-Bornholdt, H.; van den Bussche. H. Cholinesterase inhibitors for patients with Alzheimers disease: systematic review of randomised clinical trials. *BMJ.*, **2005**, 331-321.

[104]   Luckmann, R. Review: Cholinesterase inhibitors may be effective in Alzheimers disease. *Evid. based Med.*, **2006**, 11-23.

[105]   Birks, J. The evidence for the efficacy of cholinesterase inhibitors in the treatment of Alzheimer's disease is convincing. *Int. Psychogeriatr.*, **2008**, *20(2)*, 279-286.

[106]   Lanctôt, K. L.; Rajaram, R. D.; Herrmann, N. Therapy for Alzheimers disease: how effective are current treatments?. *Ther. Adv. Neurol. Disord.*,**2009**, *2(3)*, 163-180.

[107]   Raschetti, R.; Albanese, E.; Vanacore, N.; Maggini, M. Cholinesterase Inhibitors in Mild Cognitive Impairment: A Systematic Review of Randomised Trials. *PLOS MEDICINE.*, **2007**, *4(11)*, 1818-1828.

[108]   Russ, T. C.; Morling, J. R. Cholinesterase inhibitors for mild cognitive impairment. *Cochrane Database Syst Rev.* [Online] **2012**, Sep 12;9. http://onlinelibrary.wiley.com/ doi/10.1002/14651858.CD009132.pub2/pdf (accessed Nov 14, 2012)

[109]   Alzheimer research forum. Drugs in clinical trials. http://www.alzforum.org/dis/tre/drc/ detail.asp?id=51 (accessed Nov 16, 2012)

[110]   Van Marum, R. J. Current and future therapy in Alzheimer's Disease. *Fundam Clin Pharmacol.*, **2008**, *22(3)*, 265-74.

[111] Tariot, P. N.; Farlow, M. R.; Grossberg, G. T.; Graham, S. M.; McDonald, S.; Gergel. I. Memantine treatment in patients with moderate to severe Alzheimer disease already receiving donepezil: a randomized controlled trial. *JAMA.*, **2004**, *21,291(3)*, 317-324.

[112] Grossberg, G. T.; Edwards, K. R.; Zhao, Q. Rationale for combination therapy with galantamine and memantine in Alzheimer's disease. *J. Clin. Pharmacol.*, **2006**, *46(7)*, 17S-26S.

[113] Shah, R. S.; Lee, H. G.; Xiongwei, Z.; Perry, G.; Smith, M. A.; Castellani, R. J. Current approaches in the treatment of Alzheimer's disease. *Biomed Pharmacother.*, **2008**, *62(4)*, 199-207.

[114] Bai, D.L.; Tang, X.C.; He, X.C. Huperzine A, A potential Therapeutic agent for treatment of Alzheimers Disease. *Curr. Med. Chem.*, **2000**, *7*, 355-374.

[115] Bai, D. Development of huperzine A and B for treatment of Alzheimers disease. *Pure Appl.Chem.*, **2007**, *79(4)*, 469-479.

[116] Zhang, H. Y.; Yan, H.; Tang, X. C. Non Cholinergic effects of Huperzine A: Beyond inhibition of Acetylcholinesterase. *Cell. Mol. Neurobiol.*, **2008**, *28*, 173-183.

[117] Sabbagh, M. N. Drug development for Alzheimers disease: Where are we now and where are we headed. *Am. J. Geriatr. Pharmacother.*, **2009**, *7(3)*, 167-185.

[118] Wang, R.; Tang, X. C. Neuroprotective effects of Huperzine A A natural cholinesterase inhibitor for the treatment of Alzheimers disease. *Neurosignals.*, **2005**, *14*, 71-82.

[119] Zhao, Q.; Tang, X. C. Effects of huperzine A on acetylcholinesterase isoforms *in vitro*: comparison with tacrine, donepezil, rivastigmine and physostigmine. *Eur. J. Pharmacol.*, **2002**, *455(2-3)*, 101-107.

[120] Xiao, X. Q.; Zhang, H. Y.; Tang, X. C. Huperzine A attenuates Amyloid β peptide fragment 25-35 induced apoptosis in rat cortical neurons *via* inhibiting reactive oxygen species formation and caspase 3 activation. *J. Neurosci. res.*, **2002**, *67(30)*, 30-36.

[121] Gordon, R. K.; Nigam, S. V.; Weitz, J. A.; Dave, J. R.; Doctor, B. P.; Ved, H. S. The NMDA Receptor ion channer: a site for binding of Huperzine A. *J. Appl. Toxicol.*, **2001**, *21*, S47-S51.

[122] Hemendinger, R. A.; Armstrong, E. J.; Persinski, R.; Todd, J.; Mougeot, J. L.; Volvovitz, F.; Rosenfeld, J. Huperzine A provides neuroprotection against several cell death inducers using *in vitro* model systems of motor neuron cell death. *Neurotox. Res.*, **2008**, *13(1)*, 49-62.

[123] Rafii, M. S.; Walsh, S.; Little, J. T.; Behan, K.; Reynolds, B.; Ward, C.; Jin, S.; Thomas, R.; Aisen, P. S. A Phase II trial of huperzine A in mild to moderate Alzheimer disease. *Neurology.*, **2011**, *76 (19)*, 1389-1394.

[124] Clinical trials.gov. A Multi-Center, Randomized, Double-Blind, Double-Dummy, Placebo- and Active-Controlled, Study to Evaluate the Safety and Efficacy of Huperzine A Sustained-Release Tablets in Patients With Mild to Moderate Alzheimer's Disease. http://clinicaltrials.gov/ct2/show/NCT01282619?term=huperzine+A&rank=5. (accessed Aug 17, 2012).

[125] Greig, N. H.; Ruckle, J.; Comer, P.; Brownell, L.; Holloway, H. W.; Flanagan, DR Jr.; Canfield, C. J.; Burford, R. G. Anticholinesterase and pharmacokinetic profile of Phenserine in healthy elderly human subjects. *Curr. Alzheimer res.*, **2005**, *2*, 483-492.

[126] Ikari, H.; Spangler, E. L.; Greig, N. H.; Pei, X. F.; Brossi, A.; Speer, D.; Patel, N.; Ingram, D. K. Maze learning in aged rats is enhanced by phenserine, a novel anticholinesterase. *Neuroreport,* **1995**, *6(3)*, 481-4.

[127] Janas, A. M.; Cunningham, S. C.; Duffy, K. B.; Devan, B. D.; Greig, N. H.; Holloway, H. W.; Yu, Q. S.; Markowska, A. L.; Ingram, D. K.; Spangler, E. L.The Cholinesterase inhibitor, phenserine, improves morris water maze performance of scopolamine-treated rats. *Life sci.*, **2005**, *76*, 1073-1081.

[128] Arajuo, J.A.; Greig, N.H.; Ingram, D. K. *et al.* Cholinesterase inhibitors improve both memory and complex learning in aged beagle dogs. *J. Alzheimers Dis.*, **2011**, *26(1)*, 143-55.

[129] Lahiri, D. K.; Chen, D.; Maloney, B.; Holloway, H. W.; Yu, Q. S.; Utsuki, T.; Giordano, T.; Sambamurti, K.; Greig, N. H. The experimental Alzheimers disease drug posiphen [(+)-Phenserine] lowers amyloid β peptide levels in cell culture and mice. *J. Pharmacol. Exp. Ther.*,**2007**, *320(1)*, 386-396.

[130] Becker, R. E.; Greig, N. H. Fire in the ashes: Can failed Alzheimers disease drugs succeed with second chances?. *Alzheimers dement.*, **2012**, 1-8.

[131] Maccecchini, M. L.; Chang, M. Y.; Pan, C.; John, V.; Zetterberg, H.; Greig, N, H. Posiphen as a candidate drug to lower CSF amyloid precursor protein, amyloid β peptide and τ levels: target engagement, tolerability and pharmacokinetics in humans. *J. Neurol. Neurosurg. Psychiatry.*, **2012**, 804-902.

[132] Cecilia, B.; Jure, S.; Qian, S; *et al.* Kinetics of Torpedo californica acetylcholinesterase inhibition by bisnorcymserine and crystal structure of the complex with its leaving group. *Biochem. J.*, **2012**, *444*, 269-277.

[133] Kamal, M. A.; Klein, P.; Yu, Q. S.; Tweedie, D.; Li, Y.; Holloway, H. W.; Greig, N. H. Kinetics of human serum butyrylcholinesterase and its inhibition by a novel experimental Alzheimer therapeutic, bisnorcymserine. *J. Alzheimers Dis.*, **2006**, *10(1)*, 43-51.

[134] Greig, N.H.; Yu, Q.; Brossi, A.; Soncrant, T.T.; Hausman, M. Highly selective butyrylcholinesterase inhibitors for the treatment and diagnosis of Alzheimers disease and dementia. US Patent 2002/0094999 A1, July 18 2002.

[135] Greig, N. H.; Utsuki, T.; Ingram, D. K.; Wang, Y.; Pepeu, G.; Scali, C.; Yu, Q.S.; Mamczarz, J.; Holloway, H, W.; Giordano, T.; Chen, D.; Furukawa, K.; Sambamurti, K.; Brossi, A.; Lahiri, D. K. Selective butyrylcholinesterase inhibition elevates brain acetylcholine, augments learning and lowers alzheimers β amyloid peptide in rodent. *Proc. Nat. Acad. Sci.*, **2005**, *102 (47)*, 17213-17218.

[136] QR Pharma.QR pharma announces the start of a phase I clinical study for Bisnorcymserine. http://www.qrpharma.com/pr-BNC-Ph1.html (accessed Aug 22, 2012).

[137] Panza, F.; Solfrizzi, V.; Frisardi, V.; Capurso, C.; D'Introno, A.; Colacicco, A. M.; Vendemiale, G.; Capurso, A.; Imbimbo, B. P. Disease-modifying approach to the treatment of Alzheimers disease: from alpha-secretase activators to gamma-secretase inhibitors and modulators. *Drugs aging.*, **2009**, *26(7)*, 537-55.

[138] Schenk, D.; Basi, G. S.; Pangalos, M. N. Treatment strategies targeting Amyloid β protein. *Cold spring HarbPerspect Med.*, [online] **2012**, http://perspectivesinmedicine.cshlp.org/content/2/9/a006387.full.pdf+html (accessed Aug 8, 2012)

[139] Arun, K.G.; Margherita, B.; Jordan, T. Developing β Secretase inhibitors for treatment of Alzheimers disease. *J. Neurochem.*, **2012**, (suppl, 1), 71-83.

[140] May, P. C.; Dean, R. A.; Lowe, S. L.; Martenyi, F.; Sheehan, S. M.; Boggs, L. N.; Monk, S. A.; Mathes, B. M.; Mergott, D. J.; Watson, B. M.; Stout, S. L.; Timm, D. E.; SmithLabell, E, Gonzales, C. R.; Nakano, M.; Jhee, S. S.; Yen, M.; Ereshefsky, L.; Lindstrom, T. D.; Calligaro, D. O.; Cocke, P. J.; Greg Hall, D.; Friedrich, S.; Citron, M.; Audia, J. E. Robust

central reduction of amyloid β in humans with an orally available, non peptidic β-secretase inhibitor. *J. Neurosci.*, **2011**, *31(46)*, 16506-16516.

[141]   Hook,V.; Hook, G.; Kindy, M. Pharmacogenetic features of cathepsin B inhibitors that improve memory deficit and reduce beta-amyloid related to Alzheimer's disease. *Biol Chem.*, **2010**, *391(8)*, 861-872.

[142]   Hook, V.; Toneff, T.; Bogyo, M.; Greenbaum, D.; Medzihradszky, K. F.; Neveu, J.; Lane, W.; Hook, G.; Reisine, T. Inhibition of cathepsin B reduces beta-amyloid production in regulated secretory vesicles of neuronal chromaffin cells: evidence for cathepsin B as a candidate beta-secretase of Alzheimer's disease. *Biol Chem.*, **2005**, *386(9)*, 931-940.

[143]   Hook, V.; Schechter, I.; Demuth, H. U.; Hook, G. Alternative pathways for production of beta-amyloid peptides of Alzheimer's disease. *Biol Chem.*, **2008**, *389(8)*, 993-1006.

[144]   Hook, V. Y.; Kindy, M.; Reinheckel, T.; Peters, C.; Hook, G. Genetic cathepsin B deficiency reduces beta-amyloid in transgenic mice expressing human wild-type amyloid precursor protein. *Biochem. Biophys. Res. Commun.*, **2009**, 21, *386(2)*, 284-288.

[145]   Hook, V. Y.; Kindy, M.; Hook, G. Inhibitors of cathepsin B improve memory and reduce beta-amyloid in transgenic Alzheimer disease mice expressing the wild-type, but not the Swedish mutant, beta-secretase site of the amyloid precursor protein. *J. Biol. Chem.*, **2008**, 21, *283(12)*, 7745-7753.

[146]   Hook, G.; Hook, V. Y.; Kindy, M. Cysteine protease inhibitors reduce brain beta-amyloid and beta-secretase activity *in vivo* and are potential Alzheimer's disease therapeutics. *Biol. Chem.*, **2007**, *388(9)*, 979-983.

[147]   Hook, V.; Kindy, M.; Hook, G. Cysteine protease inhibitors effectively reduce *in vivo* levels of brain beta-amyloid related to Alzheimer's disease. *Biol. Chem.*, **2007**, *388(2)*, 247-252.

[148]   Hook, G.; Hook, V.; Kindy, M. The cysteine protease inhibitor, E64d, reduces brain amyloid-β and improves memory deficits in Alzheimer's disease animal models by inhibiting cathepsin B, but not BACE1, β -secretase activity. *J. Alzheimers Dis.*, **2011**, *26(2)*, 387-408.

[149]   Mueller-Steiner, S.; Zhou,Y.; Arai, H.; Roberson, E. D.; Sun, B.; Chen, J.; Wang, X.; Yu, G.; Esposito, L.; Mucke, L.; Gan, L. Antiamyloidogenic and neuroprotective functions of cathepsin B: implications for Alzheimer's disease. *Neuron.*, **2006**, *21, 51(6)*, 703-714.

[150]   Strooper, B. Proteases and Proteolysis in Alzheimer Disease: A Multifactorial View on the Disease Process. *Physiol. Rev.*, **2010**, *90*, 2465-2494.

[151]   Nalivaeva, N. N.; Beckett, C.; Belyaev, N. D.; Turner, A. J. Are amyloid-degrading enzymes viable therapeutic targets in Alzheimer's disease?. *J. Neurochem.*, **2012**, *120(1)*, 167-185.

[152]   Siemers, E. R.; Dean, R. A.; Friedrich, S.; Ferguson-Sells, L.; Gonzales,C.; Farlow, M. R.; May, P.C. Safety, tolerability, and effects on plasma and cerebrospinal fluid amyloid-beta after inhibition of gamma-secretase. *Clin. Neuropharmacol.*, **2007**, *30(6)*, 317-25.

[153]   Siemers, E.; Skinner, M.; Dean, R. A.; Gonzales, C.; Satterwhite, J.; Farlow, M.; Ness, D.; May, P. C. Safety, tolerability, and changes in amyloid beta concentrations after administration of a gamma-secretase inhibitor in volunteers. *Clin. Neuropharmacol.*, **2005**, 28(3), 126-132.

[154]   Siemers, E. R.; Quinn, J. F.; Kaye, J.; Farlow, M. R.; Porsteinsson, A.; Tariot, P.; Zoulnouni, P.; Galvin, J. E.; Holtzman, D. M.; Knopman, D. S.; Satterwhite, J.; Gonzales,

C.; Dean, R. A.; May, P. C. Effects of a gamma secretase inhibitor in a randomized study of patients with Alzheimers disease. *Neurology.*, **2006**, *66(4)*, 602-604.

[155] Fleisher, A. S.; Raman, R.; Siemers, E. R.; Becerra, L.; Clark, C. M.; Dean, R. A.; Farlow, M. R.; Galvin, J. E.; Peskind, E. R.; Quinn, J. F.; Sherzai, A.; Sowell, B.B.; Aisen, P. S.; Thal, L.J. Phase 2 safety trial targeting amyloid beta production with a gamma secretase inhibitor in Alzheimer disease. *Arch. Neurol.*, **2008**, *65(8)*, 1031-8.

[156] Bateman, R. J.; Siemers, E. R.; Mawuenyega, K. G.; Wen, G.; Browning, K. R.; Sigurdson, W. C.; Yarasheski, K. E.; Friedrich, S. W.; Demattos, R. B.; May, P. C.; Paul, S. M.; Holtzman, D. M.A gamma secretase inhibitor decreases amyloid-beta production in the central nervous system. *Ann. Neurol.*, **2009**, *66(1)*, 48-54.

[157] Lilly.com. EilLilly Launches Its First Phase III Trial for Treatment of Alzheimer's Disease (NYSE:LLY)   http://newsroom.lilly.com/releasedetail.cfm?releaseid=302104   (accessed Aug 7, 2012)

[158] Extance, A. Alzheimers failure raises question about diseae modifying strategies. *Nat. Rev. Drug Dis.*, **2010**, *9*, 749-751.

[159] Lilly.com. Lilly Halts Development of Semagacestat for Alzheimer's Disease Based on Preliminary Results of Phase III Clinical Trials (NYSE:LLY). http://newsroom.lilly.com/ releasedetail.cfm?releaseid=499794 (accessed Aug 7, 2012)

[160] Panza, F.; Frisardi, V.; Solfrizzi, V.; Imbimbo, B. P.; Logroscino, G.; Santamato, A.; Greco, A.; Seripa, D.; Pilotto, A. Interacting with γ -secretase for treating Alzheimers disease: from inhibition to modulation. *Curr. Med. Chem.*, **2011**, *18*, 5430-5447.

[161] Panza, F.; Frisardi, V.; Imbimbo, B. P.; Seripa, D.; Solfrizzi,V.; Pilotto. A. γ-secretase inhibitors for treating Alzheimers disease: rationale and clinical data. *Clin. Invest.*, **2011**, *1(8)*, 1175-1194.

[162] Kevin W. Gillman, K. W.; Starrett, Jr, J. E.; Parker, M. F.; Xie, K.; Bronson, J. J.; Marcin, L. R.; McElhone, K. E.; Bergstrom, C. P.; Mate, R. A.; Williams, R.; Meredith, Jr, J. E.; Burton, C. R.; Barten, D. M.; Toyn, J. H.; Roberts, S. B.; Lentz, K. A.; Houston, J. G.; Zaczek, R.; Albright, C. F.; Decicco, C. P.; Macor, J. E.; Olson, R. E. Discovery and Evaluation of BMS-708163, a potent selective and orally bioavailable γ-secretase inhibitor. *ACS Med. Chem. Lett.*, **2010**, *1*, 120-124.

[163] Imbimbo, B. P.; Panza, F.; Frisardi,V.; Solfrizzi, V.; D'Onofrio, G.; Logroscino, G.; Seripa, D.; Pilotto, A. Therapeutic intervention for Alzheimers disease with γ-secretase inhibitors: still a viable option. *Expert. opin. Investig. Drugs.*, **2011**, *20(3)*, 325-341.

[164] Coric, V.; van Dyck, C. H.; Salloway, S.; Andreasen, N.; Brody, M.; Richter, R. W.; Soininen, H.; Thein, S.; Shiovitz, T.; Pilcher, G.; Colby, S.; Rollin, L.; Dockens, R.; Pachai, C.; Portelius, E.; Andreasson, U.; Blennow, K.; Soares, H.; Albright, C.; Feldman, H. H.; Berman, R. M. Safety and Tolerability of the γ-secretase inhibitor Avagacestat in a Phase 2 study of mild to moderate Alzheimer Disease. *Arch. Neurol.*, **2012**, *69(11)*, 1430-1440.

[165] Clinical trials.gov. A multicenter, Double Blind, placebo-controlled, Safety and tolerability study of BMS-708163 in patients with prodromal Alzheimersdisease. http://clinicaltrials. gov/ct2/show/NCT00890890?term=NCT00890890&rank=1 (accessed Sep 7, 2012)

[166] Tong, G.; Wang, J. S.; Sverdlov, O.; Huang, S. P.; Slemmon, R.; Croop, R.; Castaneda, L.; Gu, H.; Wong,O.; Li, H.; Berman, R. M.; Smith, C.; Albright, C. F.; Dockens, R. A contrast in safety, pharmacokinetics, and Pharmacodynamics across age groups after a single 50 mg oral dose of the γ -secretase inhibitor avagacestat. *Br. J. Clin. Pharmacol.*, **2012**, *75(1)*,136-145.

[167] Tong, G.; Wang, J. S.; Sverdlov, O.; Huang, S. P.; Slemmon, R.; Croop, R.; Castaneda, L.; Gu, H.; Wong, O.; Li, H.; Berman, R. M.; Smith, C.; Albright, C. F.; Dockens, R. C. Multicenter, Randomized, double-blind, placebo controlled, Single ascending dose study of the oral γ -secretase inhibitor BMS-708163 (Avagacestat): Tolerability profile, Pharmacokinetic parameters, and Pharmacodynamic markers. *Clin. Ther.*, **2012**, *34(3)*, 654-667.

[168] Lanz,T. A.; Wood, K. M.; Richter, K. E.; Nolan, C. E.; Becker, S. L.; Pozdnyakov, N.; Martin, B. A.; Du, P.; Oborski, C. E.; Wood, D. E.; Brown, T. M.; Finley, J. E.; Sokolowski, S. A.; Hicks, C. D.; Coffman, K. J.; Geoghegan, K. F.; Brodney, M. A.; Liston, D.; Tate, B. Pharmacodynamics and Pharmacokinetics of the γ-secretase inhibitor PF-3084014. *J. Pharmacol. Exp. Ther.*, **2010**, *334(1)*, 269-277.

[169] Clinical trials.gov. A Phase 1 trial of PF-03084014 in patients with advanced solid tumor malignancy and T-cell acute lymphoblastic leukemia/Lymphoblastic lymphoma. http://clinicaltrials.gov/ct2/show/NCT00878189 (accessed Sep 9, 2012).

[170] Brigham, E.; Quinn, K.; Kwong, G.; Willits, C.; Goldbach, E.; Motter, R.; Lee, M.; Hu, K.; Wallace, W.; Kholodenko, D.; Tang-Tanaka, P.; Ni, H.; Hemphill, S.; Chen, X.; Eichenbaum, T.; Ruslim, L.; Nguyen, L.; Santiago, P.; Liao, A.; Bova, M.; Probst, G.; Dappen, M.; Jagodzinski, J.; Basi, G.; Ness, D. Pharmacokinetic and pharmacodynamic investigation of ELND006, a novel APP selective gamma secretase inhibitor, on amyloid-β concentrations in the brain, CSF and plasma of multiple nonclinical species following oral administration. *Alzheimers Dement.*, **2010**, *6(4)*, S546.

[171] Schroeter, S.; Brigham, E.; Motter, R.; Nishioka, C.; Guido, T.; Khan, K.; Kholodenko, D.; Tanaka, P.; Soriano,F.; Quinn, K.; Goldbach, E.; Games, D.; Ness,D. APP-selective gamma secretase inhibitor ELND006 effects on brain parenchymal and vascular amyloid beta in the PDAPP mouse model of Alzheimers disease. *Alzheimers dement.*, **2010**, *6(4)*, S546.

[172] Brigham, E.; Quinn,K.; Motter, R.; Hoffman, W.; Goldbach, E.; Kholodenko, D.; Kwong, G.; Willits, C.; Probst, G.; Gunther, J.; Adams, E.; Sauer, J.; Kinney, G.; Ness, D. Effects of single and multiple dose oral administration of ELND006, a novel APP selective gamma secretase inhibitor, on amyloid β concentrations in brain and CSF of cynomolgus monkeys. *Alzheimers dement,* **2010**, *6(4),* S546-S547.

[173] Liang, E.; Liu, W.J.; Lohr, L.; Nguyen, V.; Lin, H.; Munson, M. L.; Crans, G.; Cedarbaum, J. A phase 1, dose escalation study to evaluate the safety, tolerability, pharmacokinetics, and pharmacodynamics of multiple oral daily doses of ELND006 in healthy elderly subjects. *Alzheimers Dement.*, **2011**, *7(4)*, S465.

[174] Hopkins, C. R. Molecule spotlight on ELND006: Another γ-secretase inhibitor fails in the clinic. *ACS Chem. Neurosci.*, **2011**, *2*, 279-280.

[175] Martone, R. L.; Zhou, H.; Atchison, K.; Comery, T.; Xu, J. Z.; Huang, X.; Gong, X.; Jin, M.; Kreft, A.; Harrison, B.; Mayer, S. C.; Aschmies, S.; Gonzales, C.; Zaleska, M. M.; Riddell, D. R.; Wagner, E.; Lu, P.; Sun, S. C.; Sonnenberg-Reines, J.; Oganesian, A.; Adkins, K.; Leach, M. W.; Clarke, D. W.; Huryn, D.; Abou-Gharbia, M.; Magolda, R.; Bard, J.; Frick, G.; Raje, S.; Forlow, S. B.; Balliet, C.; Burczynski, M. E.; Reinhart, P. H.; Wan, H. I.; Pangalos, M. N.; Jacobsen, J. S. Begacestat (GSI-953): A novel, selective thiophene sulfonamide inhibitor of amyloid precursor protein γ-secretase for the treatment of Alzheimers disease. *J. Pharmacol. Exp. Ther.*, **2009**, *331(2)*, 598-608.

[176]   Jacobsen, S.; Comery, T.; Aschmies, S.; Zhou, H.; Jin, M.; Atchison,K.; Xu, J.; Wagner, E.; Sonnenberg-Reines, J.; Kreft, A.; Sun, R.; Liu, P.; Gong, X.; Zaleska, M.; Adkins, K.; Oganesian, A.; Folletti, M.; Wan,H.; Mayer, S.; Hoke, M.; Reinhart, P.; Harrison, B.; Magolda, R.; Pangalos, M.; Martone, R. GSI-953 is a potent APP selective gamma-secretase inhibitor for the treatment of Alzheimers disease. *Alzheimers Dement.*, **2008**, *4(4)*,T461.

[177]   Kreft, A. F.; Abou-Gharbia, M. A.; Aulabaugh, A. E.; Atchison, K. P.; Diamantidis, G.; Harrison, B. L.; Hirst, W. D.; Huang, X.; Kubrak, D. M.; Lipinski, K.; Magolda, R. L.; Martone, R. L.; May, M. K.; Mayer, S. C.; Pangalos, M. N.; Porte, A. M.; Reinhart, P. H.; Resnick, L.; Zhou, H.; Jacobsen, J. S. Characterization of the affinity of GSI-953 for binding to gamma secretase and comparison to benchmark gamma secretase inhibitors. *Alzheimers Dement.*, **2008**, *4(4)*, T471-T472.

[178]   Frick, G.; Raje, S.; Wan, H.; Forlow, S. B.; Balliet, C.; Pastore, A.; Burczynski, M. E.; Jhee, S.; Ereshefsky, L.; Paul, J. GSI-953, a potent and selective gamma-secretase inhibitor:modulation of beta amyloid peptides and plasma and cerebrospinal fluid pharmacokinetic/pharmacodynamic relationships in humans. *Alzheimers Dement.*, **2008**, *4(4)*, T781.

[179]   Wan, H.; Bard, J.; Martone, R,; Raje, S.; Forlow, S.; Kreft, A.; Jacobsen, S.; Silver, P.; Paul, J.; Frick, G. GSI-953, a potent and selective gamma-secretase inhibitor, modulates Abeta peptides in mice and humans:Translating PK/PD biomarker relationships in different biological compartments between rodent and human. *Alzheimers Dement.*, **2008**, *4(4)*, T548.

[180]   Weggen, S.; Eriksen, J. L.; Das, P.; Sagi, S. A.; Wang, R.; Pietrzik, C.U.; Findlay, K. A.; Smith, T. E.; Murphy, M. P.; Bulter, T.; Kang, D. E.; Marquez-Sterling, N.; Golde, T. E.; Koo, E. H. A subset of NSAIDs lower amyloidogenic Abeta42 independently of cyclooxygenase activity. *Nature.*, **2001**, 8, *414(6860), 212-216.

[181]   Sagi, S. A.; Weggen, S.; Eriksen, J.; Golde, T.E.; Koo, H.E. The Non-cyclooxygenase Targets of Non-steroidal Anti-inflammatory Drugs, Lipoxygenases, Peroxisome Proliferator-activated Receptor, Inhibitor of κB Kinase, and NFκB, Do Not Reduce Amyloid β42 Production. *J. Biol. Chem.*, **2003**, 278, *34, (22)*, 31825-31830.

[182]   Eriksen, J. L.; Sagi, S. A.; Smith, T. E.; Weggen, S.; Das, P.; McLendon, D. C.; Ozols, V. V.; Jessing, K. W.; Zavitz, K. H.; Koo, E. H.; Golde, T. E. NSAIDs and enantiomers of flurbiprofen target gamma-secretase and lower Abeta 42 *in vivo. J. Clin. Invest.*, **2003**, *112(3)*, 440-449.

[183]   Takahashi, Y.; Hayashi, I.; Tominari, Y.; Rikimaru, K.; Morohashi, Y.; Kan, T.; Natsugari, H.; Fukuyama, T.; Tomita, T.; Iwatsubo, T. Sulindac sulfide is a noncompetitive gamma-secretase inhibitor that preferentially reduces Abeta 42 generation. *J. Biol. Chem.*, **2003**, 16, *278(20)*, 18664-18770.

[184]   Zhang, C. Therapeutic targeting of the alpha secretase pathway to treat alzheimers disease.http://www.discoverymedicine.com/Can-Zhang/2009/07/29/therapeutic-targeting-of-the-alpha-secretase-pathway-to-treat-alzheimers-disease. (accessed Sep 13, 2012).

[185]   Allinson, T. M.; Parkin, E. T.; Turner, A. J.; Hooper, N. M. ADAMs Family Members As Amyloid Precursur α-secretases. *J. Neurosci. Res.*, **2003**, *74*, 342-352.

[186]   Haas, C. Strategies, Development, and pitfalls of therapeutic options for Alzheimers disease. *J. Alzheimers Dis.*, **2012**, 28, 241-248.

[187]   Postina, R. Activation of α secretase cleavage. *J. Neurochem.* **2012**, *120 (Suppl1)*, 46-54.

[188] Mawuenyega, K. G.; Sigurdson, W.; Ovod, V.; Munsell, L.; Kasten, T.; Morris, J.C.; Yarasheski, K. E.; Bateman, R. J. Decreased Clearance of CNS β-Amyloid in Alzheimer's Disease. [online] **2010**, 1774. http://www.dormivigilia.com/wp-content/uploads/2011/01/Science-2010-Mawuenyega-1774.pdf (accessed Nov 20, 2012).

[189] Miners, J. S.; Baig, S.; Palmer, J.; Palmer, L. E.; Kehoe, P. G.; Love, S.Abeta-degrading enzymes in Alzheimer's disease. *Brain Pathol.*, **2008**, *18(2)*, 240-252.

[190] Miners, J. S.; Barua, N.; Kehoe, P. G.; Gill, S.; Love, S. Aβ-degrading enzymes: potential for treatment of Alzheimer disease. *J. Neuropathol. Exp. Neurol.*, **2011**, *70(11)*, 944-959.

[191] Cholinesterase inhibitors: Including Insecticides and Chemical Warfare Nerve Agents Part 2 What are Cholinesterase inhibitors? http://www.atsdr.cdc.gov/csem/csem.asp?csem=11&po=5 (accessed on Sep 9, 2012).

[192] Cummings, J. C. Alzheimers Disease. *New. Engl. J. Med.*, **2004**, *351*, 56-67.

Drug Design and Discovery in Alzheimer's Disease, 2014, 199-290

Towards Small Molecules as Therapies for Alzheimer's Disease and Other Neurodegenerative Disorders

Omar Aziz[1], Roland W. Bürli[1], David F. Fischer[2], Julie Frearson[1] and Michael D. Wall[1,*]

[1]*BioFocus; Chesterford Research Park, Saffron Walden, Essex, CB10 1XL, UK;*
[2]*BioFocus; Leiden, The Netherlands*

Abstract: Neurodegenerative diseases caused by hereditary or idiosyncratic neuronal dysfunction share some phenotypic commonalities. Intracellular aggregation of proteins, metal dyshomeostasis, generic loss of synaptic connectivity all lead to gradual decline of cognitive or motor neuronal function as patients descend into a clinically symptomatic state. Though significant progress has been made in our understanding of neurological disorders in the past decade, it has yet to translate into therapeutic advancements in disease treatment.

We have chosen to focus this review on Alzheimer's disease (AD) to highlight the main disease modifying mechanisms shared in common with the Huntington's (HD) and Parkinson's disease (PD) phenotypes, specifically, the aggregation of amyloid-β (Aβ) phospho-tau (p-tau), mutant huntingtin (mHtt) and α -synuclein (α-syn) proteins, respectively. We highlight a number of approaches used in pre-clinical drug discovery to identify clinical tools. In addition, we describe a number of less explored alternative hypotheses which have demonstrated good (pre)clinical evidence for a potential therapeutic intervention.

In particular, for AD, we will review the main concepts which have driven drug discovery research in the recent past and for each molecular target, we summarize a rationale and available validation data with commentary on relevant chemical matter and structural biology, then discuss advanced pre-clinical and clinical compounds.

Keywords: Alzheimer's disease, amyloid-β, autophagy, Huntington's disease, lewy bodies, mutant huntingtin, neurodegeneration, Parkinson's disease, physiochemical properties, pre-clinical drug discovery, reactive oxygen species, α-synuclein, tau.

---

*Address correspondence to Michael D. Wall: BioFocus; Chesterford Research Park, Saffron Walden, Essex, CB10 1XL, UK; Tel: 01799 533500; Fax: 01799 531 495; E-mail: michael.wall@glpg.com

Atta-ur-Rahman / Muhammad Iqbal Choudhary (Eds.)
10.1016/B978-0-12-803959-5.50004-0

INTRODUCTION TO NEURODEGENERATIVE DISEASES

Neurodegenerative processes such as reduction of brain mass, loss of neurons, and even formation of low levels of amyloid plaque constitute part of normal ageing. However, in older individuals the probability that an aberrant mechanism adopts a dominant role, is increased, which will likely result in a pathological outcome. Hence, with an increasing average life span in the 21st century combined with the lack of success in developing new therapies, neurodegenerative disorders are likely to become the largest unmet medical need in the developed world. Arguably, neurodegenerative diseases are often very complex in their etiology; for instance, Alzheimer's disease (AD) is characterized by dementia and progressive memory loss. The underlying mechanisms leading to these symptoms may, however, be very diverse, which makes it difficult to develop a 'universal therapy'. In fact, more accurately AD may be considered a multitude of diseases or, perhaps, a syndrome that requires a combination of sophisticated diagnostic tools and biomarkers with specialized therapies.

A hallmark of many neurodegenerative diseases is their late onset, even if the rate of disease progression can be extremely short for instance with amyotrophic lateral sclerosis, or very slow, as is the case for early onset Parkinson's disease (PD). Neuropathological studies have clearly indicated that one characteristic commonality of many neurodegenerative diseases (including AD, PD and Huntington's Disease (HD)), is the accumulation of aberrant protein deposits in the brain. As these deposits gradually accumulate over time, it has been suggested that they could be a cause for the late onset of disease symptoms. Addressing the pathways that are implicated in the generation of aberrant protein deposits is thus a common theme in the development of therapeutics over the past decades. Non-invasive detection of neurodegeneration as measured by brain morphology, can be achieved with MRI technology and has been reviewed recently [1]. In the case of AD, morphological changes to the brain due to atrophy has been correlated to the time when clinical symptoms appear and when clinical diagnosis can be made (Fig. **1**).

In contrast to AD and PD, HD is considered a 'monogenetic disease' as it has a single root cause. HD is caused by an autosomal dominant mutation within the

Huntingtin gene, a polyCAG expansion leading to an extended polyglutamine (polyQ) stretch proximal to the N-terminus of the huntingtin protein (htt). The length of the polyQ expansion inversely correlates with disease onset. However, to date it is not clear how the polyQ extended htt protein translates into the development of HD symptoms though it is clear that brain atrophy follows disease progression (Fig. **2**).

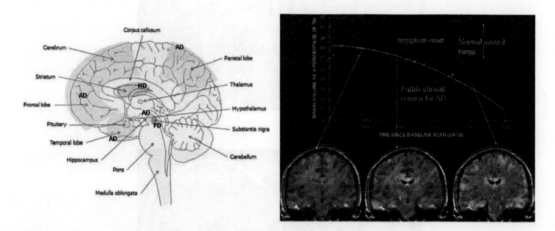

Figure 1: *Left*: Representation of the human brain colored by regions affected by neuro-degeneration caused by AD, PD and HD. *Right*: Serial MRI in pre-symptomatic familial AD patient. Graph shows brain volume changes (from serial T1-weighted brain MRI) over a seven year period spanning symptom onset. Lower images show scans registered to baseline with fluid overlay; green = loss of brain volume; yellow-red = expansion of CSF regions. Note the pre-symptomatic hippocampal atrophy and prominent loss over time. Reproduced with the kind permission of Springer Science + Business Media [1].

Protein aggregation and its relationship to cellular homeostasis is a common theme across AD, PD and HD. One pathway, autophagy, has been implicated in both the formation and also the elimination of aggregates. Autophagy is a key regulator of the maintenance of cytosolic homeostasis in times of cellular stress. Autophagy is defined for the purpose of this review, as a catabolic mechanistic pathway which delivers redundant protein to the lysosome for digestion. As a pre-programmed mechanism for cellular survival during starvation, autophagy provides a means of cellular nutrient provision to meet its energy needs and consequently a mechanism by which non-vital cellular components or excess dysfunctional or aggregated proteins, may be removed from the cytosol. For more details, we refer to a number of reviews on autophagy [2].

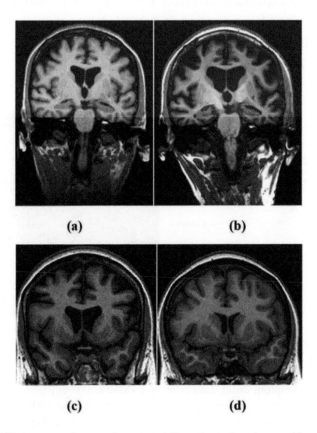

Figure 2: Serial MR images of an early-onset AD patient showing rapid progression over 18 months (the interval between the (**a**) and (**b**) images). There is gross loss of cortical grey matter with marked sulcal widening. *Below*: Coronal MRI of a HD patient. Compare the images of a 43-year-old patient with HD (**c**) (duration 8 years) with the 43-year-old gender-matched control (**d**). Note enlarged ventricles due to volume loss of the putamen and caudate while cortical thinning and sulcal widening is also apparent. Reproduced with the kind permission of Springer Science [1].

Whether neuronal oxidative stress is causative of, or a consequence of neurodegeneration is also open to debate. Herein, oxidative stress is defined as the increased concentration of reactive oxidative species (ROS); highly reactive free radicals in a cytosolic environment leading oxidative damage to lipids, DNA or proteins which then appear to result in cell signaling dysregulation. Either an attenuation of compensatory mechanisms or an increase in concentration of cytosolic ROS result in increased oxidative stress. The role of transition metal ions in promotion of ROS etiology should also be taken into consideration. Thus, antioxidant protection of the brain and metal chelation are considered components for longevity of neuronal function. We refer to several reviews for further reading on this subject [3].

With these three themes in mind, we consider the overarching pathology of neurodegeneration as poly-mechanistic, yet similarities are shared across phenotypes. Whether through aggregated protein accumulation due to genetic pre-disposition, genetically driven loss of function or gain of toxic function, age-related erosion of compensatory mechanisms or oxidative stress, it is clear that an imbalance in neuronal homeostasis, induced either chronically or acutely, results in functional decline.

Herein, we define neurodegeneration as a symphony of aberrant mechanisms leading to catastrophic syndromal dysfunction with resultant detrimental phenotypes. We use AD as a platform to exemplify common pathological themes (Fig. **3**) and review the main concepts which have been the recent focus for drug discovery.

ALZHEIMER'S DISEASE: ETIOLOGY AND THE β-AMYLOID HYPOTHESIS

AD is the most common late-onset neurodegenerative disorder, affecting millions of patients world-wide. The majority of cases are sporadic, only a small proportion of cases show complete penetrance of disease-causing mutations, and those cases typically show an early onset. A large body of work has now revealed a number of genes as risk factors with different levels of penetrance for late-onset AD.

The most studied neuropathological hallmark of AD is the extracellular plaque formation in the brain consisting of β-amyloid aggregates [4]. Aggregated β-amyloid peptide is generated from the amyloid precursor protein (APP) holoprotein by successive processing through concerted action of three secretases

(Fig. **4**). Of the three known secretases, the α-form or the metalloproteinase ADAM10 is anti-amyloidogenic as it cleaves APP within the β-amyloid peptide sequence [5]. Two β-secretases (BACE1 and BACE2) are described, BACE1 is pro-amyloidogenic and is expressed in the CNS [6], whereas BACE2 is expressed outside of the CNS [7]. γ-Secretase is a complex of several subunits, including presenilin 1 (PSEN1) or presenilin 2 (PSEN2) [8]. We refer to a number of reviews on the processing of APP to the β-amyloid peptide [9].

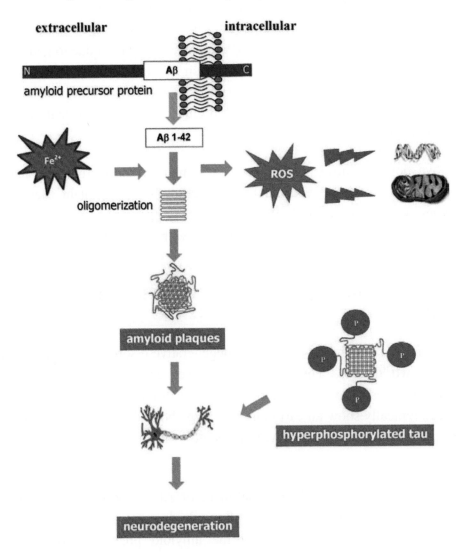

Figure 3: Representation of the symphony of aberrant mechanisms leading to neuronal stress and neurodegeneration in AD.

The amyloid cascade hypothesis summarized in Fig. **4** contends that neural dysfunction is a consequence of chronic aggregation Aβ peptides of length of 40-42 amino acids.

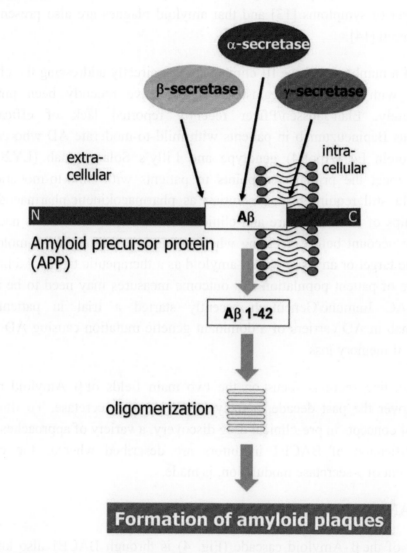

Figure 4: Representation of known amyloid precursor proteins (APP) processing mechanisms; α, β and γ -secretases, leading to the production of β -amyloid peptides and their oligomerization resulting in plague formation.

It is known that mutations in APP [10] and in Presenilin-1/2 [11] genes cause an over-production of Aβ$_{1-42}$ peptide and an early onset of familial AD. Importantly, a recent discovery of a protective mutation at the β-secretase cleavage site in APP genetically proves the relevance of the amyloid cascade to AD [12]. It is known that the formation of amyloid plagues occur very early in the disease course, prior to the onset of symptoms [13] and that amyloid plagues are also present in the non-AD brain [14].

Results of a number of Phase III clinical studies directly addressing the effects of treatment with antibodies against β-amyloid have recently been published. Unfortunately, Elan/Janssen/Pfizer recently reported lack of efficacy for intravenous Bapineuzumab in patients with mild-to-moderate AD who carry the apolipoprotein E4 (ApoE4) genotype and Lilly's Solanezumab (LY2062430) failed to meet the primary endpoints in patients with mild-to-moderate AD. These data still require further scrutiny as pharmacokinetic/pharmacodynamic relationships of CNS exposure and clinical biomarker measurements need to be taken into account before deciding whether this is a failure of the molecule to engage the target or an issue with β-amyloid as a therapeutic target. Additionally, the choice of patient population and outcome measures may need to be revised. Indeed, AC Immune/Genentech recently started a trial in patients with Crenezumab in AD carriers of a dominant genetic mutation causing AD prior to the onset of memory loss.

The proceeding sections focus on the two main fields of β-Amyloid research reported over the past decade, namely BACE1 and γ-secretase. To illustrate a number of concepts in pre-clinical drug discovery, a variety of approaches used in the identification of BACE1 inhibitors are described whereas the case for reassessment of γ-secretase modulation, is made.

BACE1 AND THE β-AMYLOID HYPOTHESIS

Initiation of the β-Amyloid cascade (Fig. **4**) is through BACE1 also known as Membrane anchored protease-2 (Memapsin 2) [15]. It is a membrane-associated aspartic endopeptidase, functionally active on the luminal side of the cell membrane. Its proteolytic activity is the rate determining step in the production of Aβ [16] and is located at the endoplasmic reticulum (ER), Golgi networks, cell surface and endosomes [17]. Following biosynthesis, it is transported to the cell surface *via* the ER and Golgi, though it is unclear whether its proteolytic action occurs at the cell surface or in the cytosol.

Hypothesizing that BACE1 inhibition will prevent production of $A\beta_{42}$, multiple independent research groups reported β -secretase-1 (BACE1) knock-out (KO) mice. Initially, no pathological, biochemical or functionally deleterious effects were observed when compared with wild type [18]. The brain hippocampal regions showed no abnormalities and this was corroborated with histopathological analysis [18b]. Crucially, it was also reported that BACE1 knock-out mice did not display overexpression of compensatory pathways, in particular overexpression of BACE2, and no structural alterations were observed in any central or peripheral organ tissues. Furthermore, BACE1 KO mice, engineered to overexpress APP, did not develop amyloid plagues indicating that BACE1 was a vital molecular requirement in the processing of APP.

However, long-term observations of BACE1 KO mice indicated that apart from being protected against mutant APP-induced neurodegeneration, they could develop deficits often associated with animal models of schizophrenia and were prone to epileptic seizures [18c, 19]. These data indicate that while BACE1 was viewed as an excellent molecular target for therapeutic intervention in the prevention of the formation of Aβ plaques, a full pharmacological knock-out of its activity might not be desirable. Importantly, no adverse phenotype has been described for any of the BACE1 heterozygous lines on the Mouse Genome Informatics website (www.informatics.jax.org). Moreover, recently reported, partial reduction in BACE1 expression by gene silencing or incomplete pharmacological inhibition was sufficient to prevent Aβ accumulation in the brain of pre-clinical AD mouse models and led to an improvement in cognitive function [20].

Structurally, BACE1 is a class I transmembranal protein with an N-terminal pro-domain (21 residues), a connecting strand of 434 residues, a transmembrane region of 22 residues and a short cytosolic domain of 24 residues [21]. Its catalytic domain contains the conserved aspartic acid residues (Asp^{32} and Asp^{228}) associated with aspartate proteases [22]. Recombinant BACE1 (without the transmembranal and intracellular domains) was co-crystallized in complex with an inhibitor OM99-2 (**1**) at 1.9 Å resolution (Fig. **5**) [23]. Since these landmark structures, approximately 157 further structures have been deposited into the protein data bank (PDB).

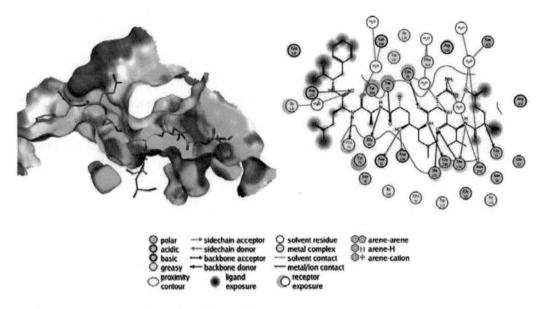

◯ polar	⟶ sidechain acceptor	◯ solvent residue	◎◎ arene-arene	
◯ acidic	⟵ sidechain donor	◯ metal complex	◎H arene-H	
◯ basic	⟶ backbone acceptor	⟋ solvent contact	◎+ arene-cation	
◯ greasy	⟵ backbone donor	⟋ metal/ion contact		
◯ proximity contour	● ligand exposure	◯ receptor exposure		

Figure 5: Image of OM99-2, a peptidomimetic, in the binding site of BACE1. *Left*: Shows the OM99-2 inhibitor bound to the catalytic site. A hydroxyethyl residue interacts with the catalytic aspartates Asp$_{(32, 228)}$. The crystal structure is available from the Protein databank (PDB) as code 2ZHR [23b]. *Right*: Schematic illustration of main residue contacts.

THE EVOLUTION OF BACE1 INHIBITORS

Multiple research groups independently discovered BACE1 and realized its potential as a therapeutic target for AD based on the β-amyloid theory [24]. Early reports of BACE1 inhibitors emerged from *ca.* 1999 when Tang *et al.* identified substrate-based inhibitors, peptidic in nature [25], which contained a hydroxyethyl group. Notably, this moiety was thought to displace a water molecule which is pivotal in the cleavage of the scissile peptide bond, to form a hydrogen bond with the aspartate catalytic diad. This investigation resulted in the identification of the inhibitor OM99-2 (Fig. **6**, **1**) (K$_i$ = 36 nM) for which a crystal structure within BACE1 was obtained (Fig. **5**).

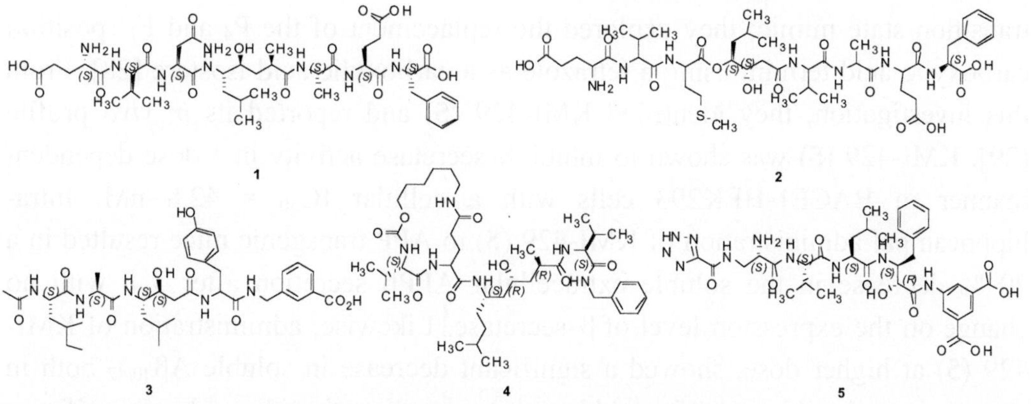

Figure 6: Structures of early reported BACE1 inhibitors: OM99-2 (1); (b) BACE1 inhibitor containing the non-cleavable statine moiety (2) and (3); cycloamide-urethane (4); KMI-429 (5).

Elan Pharmaceuticals described the design and evolution of peptide-based BACE1 inhibitors replacing the P1-Leucine with the non-cleavable statine ((3S,4S)-4-amino-3-hydroxy-6-methyl) moiety (*S*-enantiomer) resulting in the identification of an inhibitor with weak potency [26]. Further fine tuning by N-terminal amino acid modifications led to highly potent inhibitors (**2**) (BACE1 IC_{50}= 0.03 µM) (Fig. **6**). Replacement of the statine moiety with the AHPPA [4(S)-amino 3-hydroxy-5-phenylpentanoic acid] moiety was well tolerated.

The published crystal structure of the BACE1 catalytic site associated with OM99-2 (**1**) allowed the design of non-cleavable statine-based peptide inhibitors through their exploration of the P3, P2 and P2' specificity pockets within the catalytic site. They identified sub-micromolar inhibitors and reported the crystallization of one compound in the catalytic site (**3**, BACE1 IC_{50} = 0.11 µM). Tang *et al.* also progressed inhibitor design by hybridizing peptide-based inhibitors with cycloamide-urethane derived marocyclic structures (**4**) and demonstrated increased ligand affinity which correlated with an increase in the macrocyclic ring size [27].

In an attempt to address the predicted low blood-brain barrier permeability associated with molecules bearing multiple carboxylic acid functions, Kiso *et al.* reported a set of peptidic BACE1 inhibitors which contained the phenyl norstatine transition state mimic; they explored the replacement of the P_4 and P_1' position carboxylic acid termini using a tetrazole as a carboxylic acid isostere [28]. From this investigation, they identified KMI-429 (**5**) and reported its *in vivo* profile [29]. KMI-429 (**5**) was shown to inhibit β-secretase activity in a dose dependent manner in BACE1-HEK293 cells with a cellular IC_{50} = 42.8 nM. Intra-hippocampal administration of KMI-429 (**5**) to APP transgenic mice resulted in a 20 % decrease in the soluble extracellular APPβ secretion after 3 h with no change on the expression level of β-secretase. Likewise, administration of KMI-429 (**5**) at higher dose, showed a significant decrease in soluble $Aβ_{40/42}$ both in transgenic and wild type mice. Although transgenic mice showed no significant reduction of the insoluble $Aβ_{40/42}$ fractions, significant reductions were observed in wild-type animals suggesting that KMI-429 (**5**) was more effective in wild-type mice.

With the increasing availability of peptidomimetic inhibitors and improved protein expression technologies, researchers soon turned to the issue of addressing compound delivery to the brain in preclinical animal models following oral administration. However, this was not readily achieved with peptide-based inhibitors.

By their nature, multi-amino acid peptidic structures are of relatively high molecular weight, low ClogP, high topological polar surface area (tPSA) and bear multiple H-bond donors. These physiochemical parameters are not conducive towards compound delivery across the blood-brain barrier (BBB) by passive diffusion [30]. Concomitantly, the BBB epithelial layer is designed to preclude xenobiotics through lack of para-cellular diffusion, reduced pinocytosis, fenestration and highly effective efflux mechanisms, decreasing the probability of small molecule passage through the epithelial layer. The physiochemical properties of successful CNS penetrant drugs fall into a relatively narrow window of chemical space and have been reviewed in great detail [31]. Most pertinent to CNS penetrant compounds is the review by Mahar-Doan *et al.* who analyzed 48 CNS drugs and 45 non-CNS drugs and concluded their differentiated physiochemical parameters were ClogP, tPSA and the number of rotatable bonds [32]. When compared to peptides therefore, small molecules with right properties offer a higher probability of crossing the blood brain barrier.

BACE1 INHIBITORS: UNCOVERING "CHEMICAL TRUFFLES" IN PRE-CLINICAL DRUG DISCOVERY

Following much effort, medicinal chemists managed to optimize and balance the physiochemical properties and inhibitory activity of molecules required to achieve efficacy in preclinical animal models. Despite its initially worrying intrinsic liabilities, the hydroxyethyl amine (HEA) moiety proved suitable as a transition state mimicking moiety in peptidomimetic inhibitors [33] and molecules possessing this inhibit BACE1.

Scientists at Amgen [34] reported HEA containing BACE1 inhibitors and following a campaign of optimization/stabilization identified compound (**6**), which demonstrated good BACE1 cellular potency and reasonable ADME properties with the exception of efflux; (rER = 43). Oral dosing of the pyrano-pyridine (Fig. **7**, **7**) at 100 mg/kg in rat resulted in a modest decrease of Aβ levels in CSF (35%) and brain (32%).

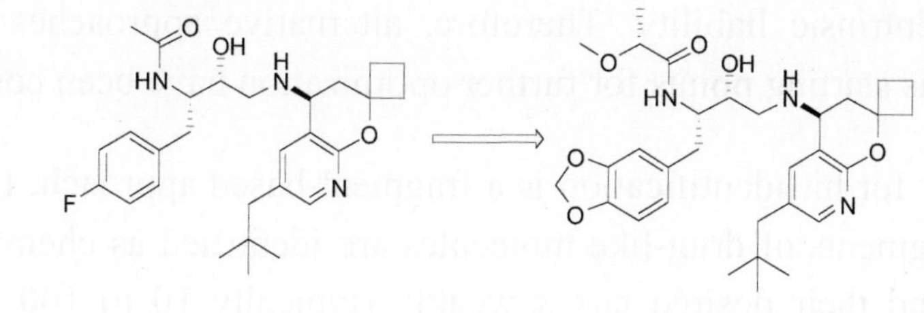

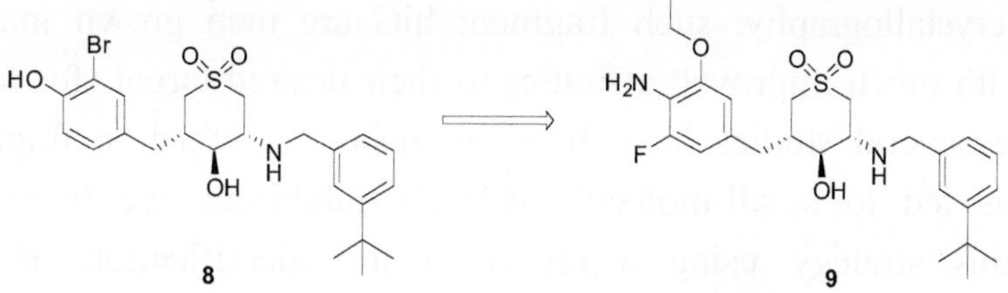

IC_{50} hBACE1 = 0.024 μM,
IC_{50} hCatD = 1.03 μM

Figure 7: Small molecule BACE1 inhibitors containing the hydroxyethyl amine (HEA) moiety reported by investigators at Amgen (**6** and **7**) and Novartis (**8** and **9**).

Similarly, the group at Novartis constrained the flexibility of the HEA functionality in an effort to reduce the number of rotatable bonds and applied temperance to molecular weight [35]. They identified the cyclic sulfone (Fig. **7,** **8**), which demonstrated reasonable cell potency, good permeability and crucially, a low efflux ratio of 2, which translated into a good blood to brain ratio. SAR investigations required a significant synthetic effort and revealed the difficulty in balancing ADME issues such as high efflux ratios and cytochrome P_{450} inhibition. Key to the Novartis' scientists defined success, however, was obtaining selectivity over Cathepsin D which they achieved with a number of analogues and exemplified by the cyclic sulfone (**9**). While the pharmacokinetics for additional analogues were also explored, the cyclic sulfone (**9**) was progressed to an *in vivo* mouse model and showed a 34% reduction in brain $A\beta_{40}$ and 54% reduction in forebrain C99 in after 4h with high doses (180 μM/kg *p.o.*).

Both these examples serve to demonstrate how the liability of carrying the critical pharmacophoric features could be mitigated and represent a new generation of centrally available tool molecules. Although of lower molecular weight than peptides, the properties of both of these chemo-types may still be considered a relatively high intrinsic liability. Therefore, alternative approaches to identify novel scaffolds as starting points for further optimization have been considered.

Another strategy for hit-identification is a fragment-based approach. Here, partial structures or fragments of drug-like molecules are identified as chemical starting points which bind their desired target weakly (typically 10 to 100 μM range). Guided by crystallography, such fragment hits are then grown into drug-like molecules with much improved affinities to their desired target (for reviews, see [36]). While several studies have been published in which a fragment-based approach has led to small-molecule BACE1 inhibitors, we have chosen to exemplify this strategy using a report of the identification of preclinical compounds described by AstraZeneca and Astex Therapeutics [37]. In this case, using a surrogate protein (endothiapepsin) NMR was used to screen fragment libraries. They identified and crystallized the 6-propylisocytosine fragment (Fig. **8, 10**), binding to BACE1, which was subsequently confirmed by surface plasmon resonance (SPR) [38]. A number of close fragment analogues (**11, 12,** **13**) shown in Fig. (**8**) further validated the original fragment hit.

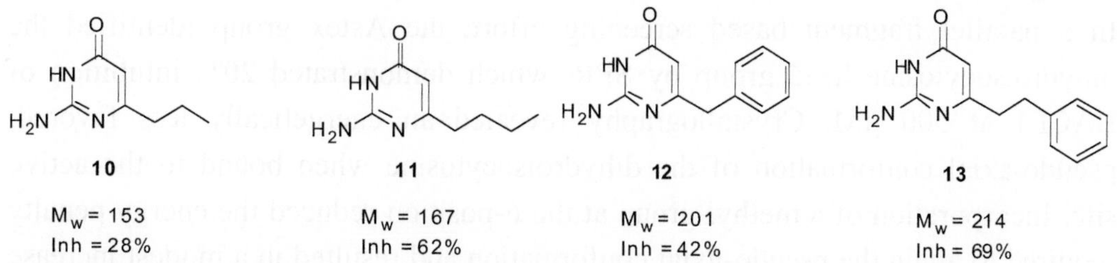

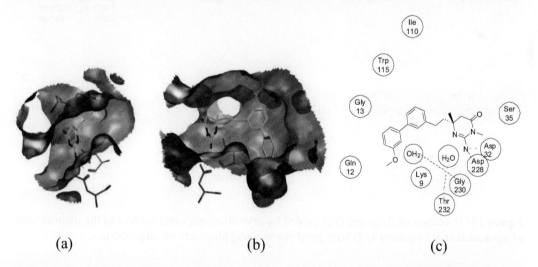

10
M_w = 153
Inh = 28%
11
M_w = 167
Inh = 62%
12
M_w = 201
Inh = 42%
13
M_w = 214
Inh = 69%

Figure 8: Fragments identified by NMR screening with the BACE1 protein and confirmed using SPR at 1 mM.

Similarly, Murray *et al*. described the identification of novel heterocyclic BACE1 inhibitors [39, 40]. The hit fragment (Fig. **10**, **10**) was optimized through maximizing S1/S3 sub-site occupancy and crystal structures demonstrated the hydrogen bonding network of the catalytic aspartates 228 and 32 to the cytosine NH_2 and N (Fig. **9**). Potency increases were achieved through methylation of the 3-position NH (**14**) and introduction of the 3-biphenyl moiety (**15** and **16**). Later, a crystal structure of a more potent analogue (**18**) revealed that the methoxy substituent replaced a water molecule and formed a hydrogen-bond with residue Ser-229.

Ile 110
Trp 115
Gly 13
Ser 35
Gln 12
OH_2
H_2O
Asp 32
Asp 228
Lys 9
Gly 230
Thr 232
(a)
(b)
(c)

Figure 9: (a) Crystal structure of fragment (10) shown in stick representation with green carbon atoms bound to BACE1 (pdb code: 3HVG). Surface is colored by atom type, with carbon in grey, oxygen in red and nitrogen in blue. Catalytic aspartates are shown in stick representation, as is Gln73. (b) Elaborated fragment (18) (in green) crystallized in BACE1 (pdb code: 2VA7). (c) 2D schematic of the elaborated fragment, showing the main protein-ligand interactions. Reference for 2VA7 structure [37].

In a parallel fragment based screening effort, the Astex group identified the dihydroisocytosine head group by SPR, which demonstrated 20% inhibition of BACE1 at 500 μM. Crystallography revealed an energetically less favored, pseudo-axial conformation of the dihydroisocytosine when bound to the active site. Incorporation of a methyl group at the 6-position reduced the energy penalty required to attain the pseudo-axial conformation and resulted in a modest increase in potency.

Using the combined knowledge generated by the two series, the hybrid compound (**17**) was synthesized as the racemate and showed another 30-fold increase in potency while separation of the enantiomers revealed (**18**) which demonstrated a further 2.5 fold potency increase.

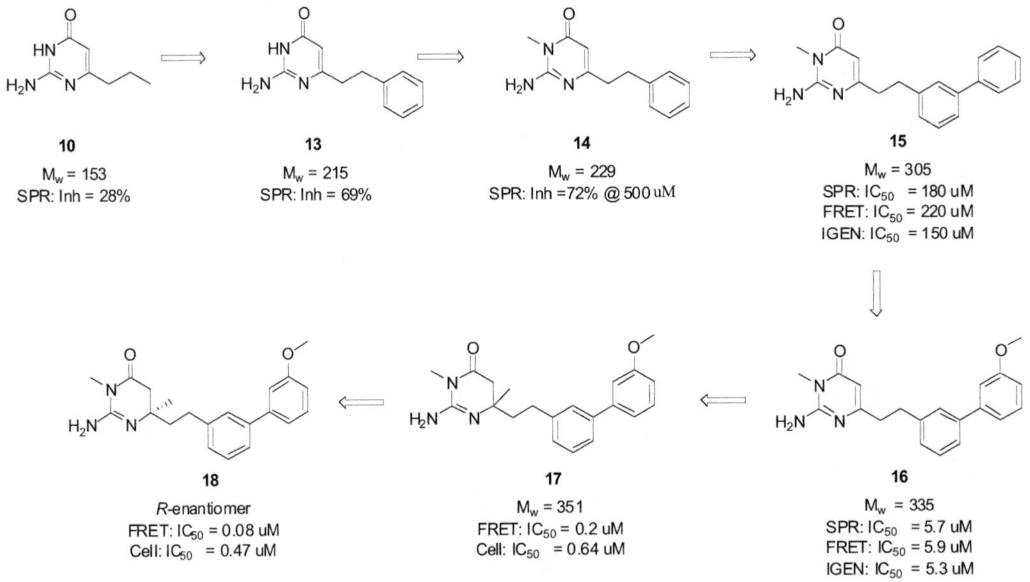

Figure 10: Evolution of fragment (10) guided by crystallography and leading to the identification of a potent BACE1 inhibitor (18) with good potency and physiochemical properties.

This study demonstrates the power of a fragment strategy whereby the intricate understanding gained through crystallographic analysis of co-structures of growing fragments can lead to molecules with improved potency.

Using a different thinking, another example of a fragment-based hit identification strategy has recently been described by Schering-Plough [41]. In this case, *iso*thiourea (Fig. **11**, **19**) was identified by an NMR screen as binding to BACE1 and was confirmed by crystallography. Recognizing the poor developability properties of an *iso*thiourea function due to its propensity towards oxidative degradation, the group identified the gem-disubstituted cyclic acyl guanidines (**20**) as a bio-isosteric replacement.

The basicity of a cyclic guanidine continued to complicate the development of iminohydantoin inhibitors. Substitution at N3 with a benzyl or phenethyl group allowed for the first potent inhibitors from this class. These compounds were labeled type I iminohydantoin inhibitors (**21**). Conversely, un-substituted N3 iminohydantoins demonstrated no measurable inhibition values and these structures were labeled type II iminohydantoin inhibitors (**22**).

The difference in basicity of type I iminohydantoins (pK_a ~7.2) and type II iminohydantoins (pK_a ~4.4) indicated that type I iminohydantoins are predominantly protonated under the conditions of enzyme activity (pH ~5). Further observations from crystallography suggested that the binding modes of the iminohydantoins were dependent on the steric bulk offered by the C5 substituent. Both binding modes offered divergent routes for optimization, however, as exemplified by a crystal structure mode B inhibitors (**23**) required only one hydrogen bond donor in the urea functionality, while mode A inhibitors (**24**) required both.

Guided by molecular modeling and crystal structures, a potent BACE1 acyl guanidine inhibitor (**25**) was obtained by iterative substitutions at the C6 and N3 positions. The X-ray structure of (**25**) showed occupation of the S1 and S2 pockets by the methylcyclohexyl and ethylcyclohexyl moieties, respectively, while the propyl imadazolone function filled the S3 pocket. Broader profiling of compound (**25**) revealed a >30-fold reduction in cellular potency and a modest exposure in rat. However, its physiochemical properties were not optimized for CNS penetration. Therefore, using guidance derived from a crystal structure, truncating (**25**) led to (**26**), a less potent compound but with improved CNS permeability. The smaller acyl guanidine (**26**) showed a modest rat PK oral profile with an AUC = 1 μM·h at 10 mg/kg *p.o.* and equal distribution between blood and brain.

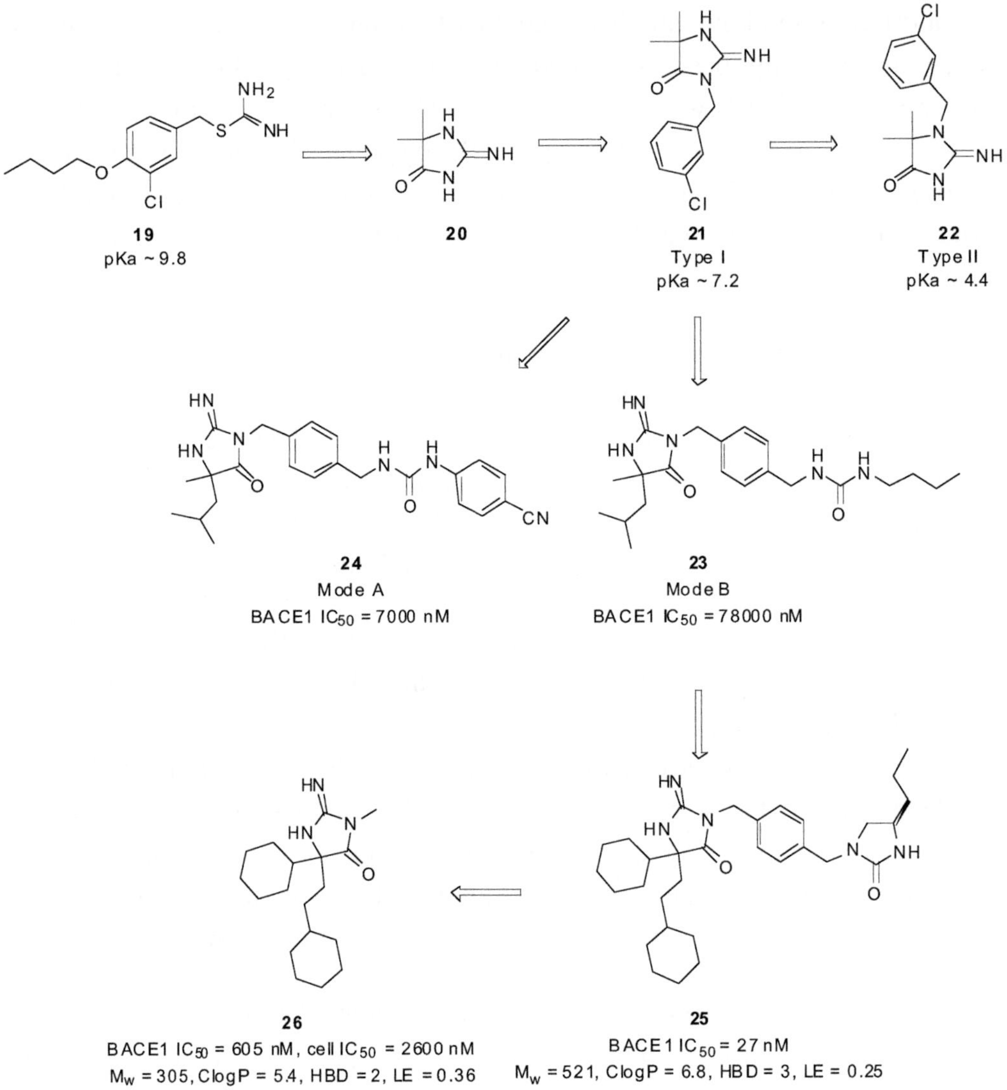

Figure 11: Evolution of iminohydantoin BACE1 inhibitors: original NMR screening hit (19), the C5-gem-substituted acyl guanidine replacement, type I (21) and type II (22) inhibitors and the difference in their basicity, dual binding modes for the type I inhibitors (23 and 24) and the optimized hit (25).

The BACE1, iminohydantoin type inhibitors have been the subject of frenzied activity in terms of patents coverage and publications. Alternative examples have been reviewed and covered extensively in recent literature [33a].

CLINICAL INHIBITORS OF β-SECRETASE

One of the first clinical studies describing the reduction of central Aβ in humans due to BACE1 inhibition has been published by Eli Lilly [42]. The conceptual start points leading to the identification of LY-2811376 (Fig. **12**, **27**) were identified from a fragment screening campaign and following optimization for potency then stabilization of the metabolically labile phenyl moiety, identified LY-2811376.

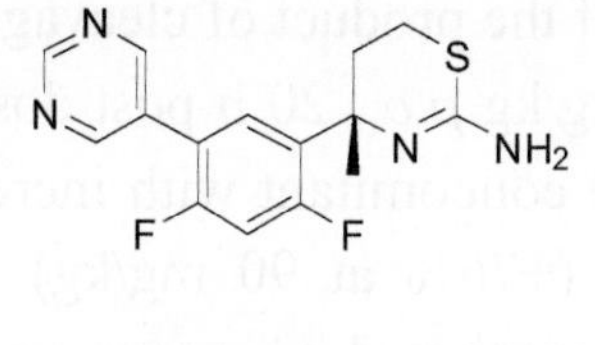

27

Figure 12: Chemical structure of BACE1 inhibitor LY-2811376 (27).

LY-2811376 (**27**) demonstrated moderate inhibitory activity for BACE1 (IC_{50} hBACE1 = 239 nM) and a 10-fold selectivity over the BACE2 form. It demonstrated moderate cell potency in recombinant HEK-APP cell system (EC_{50} HEK293$_{swe}$ = 300 nM) although in primary neuronal cultures from transgenic mice, the compound showed a concentration-dependent decrease in Aβ secretion with an EC_{50} = 100 nM.

LY-2811376 (**27**) demonstrated dose-dependent pharmacokinetics in transgenic mice ensuring significant concentration for target engagement. In the cortex, significant reductions in APP cleavage products were observed when dosed at 30 mg/kg *p.o.*; APP$_{total}$ (64%) soluble Aβ (42%) and C99 (72%). In non-transgenic beagle dogs, plasma and CSF Aβ were assessed; LY-2811376 (**27**) demonstrated significant sustained exposure (C_{max} = 1915 ng/kg; AUC = 12,300 ng/ml/h; $T_{1/2}$ = 6.8 h) when dosed at 5 mg/kg *p.o.* A pharmcodynamic effect was observed in plasma with an 85% reduction of total APP which translated into a 43% reduction in CSF at 3 h. This increased to 70% after 9 h.

In healthy human volunteers, using a single ascending dose (SAD) study, maximal plasma concentrations of 242 ng/ml at 90 mg/kg *p.o.*, were reached 2 h post dosing translating to an AUC of 4560 ng/ml. The AUC relationship between plasma and CSF was claimed to be linear.

Single doses of LY-2811376 (**27**) at concentration ranges from 15-90 mg showed statistically significant reductions in plasma $A\beta_{1-40}$ and total $A\beta$ with a mean reduction of 64% after 24 h at 90 mg/kg *p.o.* Efficacy was also observed in CSF, where the mean reduction of $A\beta_{1-40}$ was 56% as compared to the placebo group. Similarly, the concentration of the product of cleavage of APP by BACE1; sAPPβ was reduced by 42% at 90 mg/kg *p.o.*, 20 h post dosing. These reductions in $A\beta$ fragment concentrations were concomitant with increased product of cleavage of APP by α-secretase; sAPPα (+76% at 90 mg/kg) suggesting that the APP α-cleavage mechanism either continued to occur or compensated to counteract BACE1 inhibition.

Overall, LY-2811376 (**27**) was well tolerated in the CSF study with no serious adverse events reported. However, in three months safety studies in rats, LY-2811376 caused retinal pathology; specifically, cytoplasmic accumulation of finely granular auto-fluorescent material dispersed within the retinal epithelium. These findings were not considered to be mechanism-based but may be related to (partial) inhibition of other aspartyl proteases such as BACE2 or Cathepsin D.

Several small molecule BACE1 inhibitors are being or have been recently evaluated in clinical trials [43] and these are shown in Table **1**.

The most advanced BACE1 inhibitor (Table **1**, MK-8931) is being developed by Merck & Co. for the treatment of AD. The Phase I safety study has been completed and in a two part, randomized, double blind, placebo-controlled single rising dose study, MK-8931 demonstrated a dose-dependent reduction of $A\beta_{40}$ in CSF by 21%, 75% and 92% in CSF following oral dosages of 20, 100 and 550 mg, respectively. Similar reductions were observed for $A\beta_{42}$ and sAPPβ; MK-8931 was generally well tolerated. Phase II/III trials commenced in 2013.

Table 1:   A list of BACE1 inhibitors which recently progressed through clinical trials for AD

Name	Structure	Phase	Sponsor	NCT Number	Status
MK-8931	undisclosed	II/III	Merck	01537757 01496170 01739348 01953601	In Phase II/III
AZD-3839	**28**	I	AstraZeneca	01348737	In Phase 1
AZD-3293	undisclosed	I	AstraZeneca	01739647 01795339	In Phase 1b
E-2609	undisclosed	I	Eisai	01294540 01511783 01600859 01716897	In Phase I
LY-2886721*	**29**	II	Eli Lilly	01367262 01133405 01227252 01561430	Terminated in 2013

*putative structure.

AstraZeneca progressed two BACE1 inhibitors (AZD-3839, **28** and AZD-3293) into clinical development for the treatment of AD [44] and have recently disclosed pre-clinical data for AZD-3839. AZD-3839 was discovered from an iterative scaffold elaboration from the cytosine (**10**), previously described and showed high cell potency (SH-SY5Y/APP, $A\beta_{40}$ readout, $pIC_{50} = 8.3$, mouse primary neurons, $A\beta$ readout; $pIC_{50} = 7.3$, guinea pig primary neurons, $A\beta_{40}$ readout; $pIC_{50} = 7.6$. AZD-3839 showed good CNS exposure with a measured free fraction of 7.9% in brain tissue. In C57BL/6 mice, AZD-3839 demonstrated a dose and time dependent reduction of $A\beta_{40}$ in brain; *ca.* 30% after 1.5 h @ 80 µmol/kg. AZD-3839 reduced brain $A\beta_{40}$ levels for up to 8 h by 20-60% at

200 µmol/kg and reduced Aβ$_{40}$ levels by 50 after 3 h. In cynomologus monkey intravenous fusion of AZD-3839 at 20 µmol/kg resulted in reduced levels of Aβ$_{40,\ 42}$ and sAPPβ in CSF between 3 and 12 h post dose. In addition AZD-3293 (chemical structure undisclosed) is currently being assessed in Phase I trials measuring safety, tolerability and biomarkers in plasma and CSF in a multiple ascending dose escalation study. To date, no pre-clinical data information for AZD-3293 has been published for this molecule.

Eisai is currently developing a BACE1 inhibitor for the treatment of AD. E-2609 completed a Phase I clinical trial probing tolerability, pharmacokinetic and pharmacodynamics. Preclinical evaluation of E-2609 in Sprague-Dawely rats and Hartley guinea pigs demonstrated a dose-dependent decrease in Aβ$_{x-40}$ and Aβ$_{x-42}$ levels in brain and CSF of SD rats at 80% and 57% (6 h post at 10 mg/kg), while reductions of 85% and 74% were observed 4 h post administration, in guinea pigs. In plasma, reductions of 90% were sustained for up to 8 h. A randomized, double-blind, placebo-controlled, SAD Phase I trial was initiated and completed in 2012. E-2609 demonstrated no safety concerns, was well tolerated and showed reductions in Aβ in CSF and plasma following single and repeated dosing, in a dose dependent manner. Plasma Aβ levels were 20-45% below baseline at timepoints of 72 and 144 h post dosing and showed a maximum PD inhibition of 56% and 91% at 5 and 800 mg *p.o.* at 6 and 24 h post dose. A further Phase Ib assessment of E-2609 in patients with mild cognitive impairment is planned.

Eli Lilly terminated the Phase II clinical development of LY-2886721 (putative structure shown in Table **1, 29**) in 2013 due to abnormal liver biochemical tests. This molecule contains a fused 2-amino tetrahydrofurothiazine bicyclic ring system which interacts with the catalytic aspartates of the enzyme [45]. Preclinical data in transgenic mice suggested dose dependent reductions in brain Aβ, C99 and sAPP-β of 20-65%, 3 h following administration of LY-2886721 at 3-30 mg/kg *p.o.* Similarly, in beagle dogs, a sustained reduction of Aβ in plasma was observed following administration at 1 mg/kg [46]. In a multiple ascending dose safety study (5-35 mg *p.o.* for 14 days) in healthy subjects, LY-2886721 was reported to demonstrate significant decreases in sAPPβ and increases in sAPP-α. Decreases in Aβ were also observed in CSF and all doses were well tolerated [47].

In conclusion, it has taken well over a decade of intense research to identify clinical candidates with a suitable balance of BACE1 inhibitory activity in a cellular context, selectivity, metabolic stability and CNS permeability which allow for testing BACE1 inhibition as a potential therapy for AD. Currently, several clinical compounds are at various stages of clinical trials. The availability of pharmacodynamic readouts is critical in the development of these compounds; that is, the proteolytic cleavage products can be analyzed in preclinical species as well as in clinical trials which allows determining target coverage. It still remains to be assessed to what extent a reduction of Aβ will result in improved cognitive abilities in AD patients or delay their disease progression.

γ-SECRETASE AND THE β-AMYLOID HYPOTHESIS

As previously described, APP is fragmented by two sequential proteolytic cleavage steps involving first β - and then γ-secretase. This process ultimately results in multiple Aβ peptide fragments of variable length ranging from 37 to 49 amino acids and it is believed that in particular the hydrophobic Aβ peptide products are prone to form aggregates leading to neurotoxicity. Therapeutic strategies that may lead to a reduced production, aggregation or deposition of such Aβ-based plaques or enhance Aβ clearance are thought to have high potential for slowing down disease progression. Clearly, inhibition of γ -secretase activity would avoid formation of the Aβ peptides, in particular Aβ_{42}, and subsequent aggregates thereof [48]. With this rationale, γ-secretase inhibition has been a heavily investigated approach aiming at a disease modifying therapy of AD.

γ-Secretase is a multimeric protein complex which hydrolyzes the intramembrane portion of its substrates. It is an aspartyl protease comprised of four subunits; presenilin (PS), nicastrin, anterior pharynx-defective-1 (Aph-1), and presenilin enhancer-2 (Pen-2), respectively. The PS unit carries the two aspartate residues that are required for catalytic activity [49]. Importantly, in addition to hydrolysing the BACE cleavage products which are the soluble β-APP fragments, γ-secretase has been shown to recognize and process more than 70 different type-I membrane proteins [50]. In some cases, the intracellular domain of the substrate is released upon cleavage and triggers a signaling cascade whereas other substrates appear to be degraded as part of maintaining the cell membrane state. Understanding the consequence of inhibiting the processing of these substrates is critical to establishing what is required as a therapeutic window for potent γ-secretase inhibitors. For instance, γ-secretase liberates the intracellular portion of the Notch receptors which relocates to the nucleus and regulates gene transcription [51]. Inhibition of Notch signaling has been demonstrated to lead to adverse effects *in vivo*; due to the anti-proliferative effects of inhibitors of the Notch signalling cascade, such molecules are currently being investigated for applications in oncology. Similarly, γ-secretase proteolysis and release of the corresponding intracellular domain has been demonstrated to be essential for the regulatory function of ErbB4 and N-cadherin [52].

Nectin-1α, an immunoglobulin-like cell adhesion molecule and Jagged, a Notch-activating ligand, provide additional examples of γ-secretase substrates with important biological functions [53]. Interestingly, an unbiased proteomic approach to survey the proteome identified new additional γ-secretase substrates [54] raising the possibility of further unwanted side effects from this approach.

Small molecules that interfere with the function and activity of γ-secretase have been categorized into inhibitors and modulators and have been recently reviewed [55]. Inhibitors, often referred to as GSIs, reduce the proteolytic activity of the enzyme, whereas modulators (GSMs) alter the relative production of Aβ fragments of different lengths while the overall rate of APP cleavage is maintained. In particular, the concentration of the $A\beta_{42}$ fragment is reduced, which is thought to be the most neurotoxic degradation product.

The first generation of γ-secretase inhibitors blocked the catalytic site of the protease leading to inhibition of conversion of all substrates. A few of these non-selective inhibitors reached clinical trials and most data is available for Semagacestat (Fig. **13**, **30**), which was developed by Lilly and serves as a prototype for this class of inhibitor.

Semagacestat (**30**) has extensively been studied *in vitro* and in preclinical animal models (mice, guinea pigs, dogs) [56]. Consistent with the mechanism, the compound induced a reduction of Aβ; *e.g.* following a single dose administration (3 and 10 mg/kg *p.o.*), Semagacestat (**30**) caused a significant decrease of $A\beta_{1-40}$

levels in plasma, CSF and hippocampus of PDAPP mice, a transgenic strain over-expressing mutant human amyloid precursor protein V717F. No preclinical data describing an effect of Semagacestat (**30**) on cognition or behaviour has been published [57]. This inhibitor was also studied in several clinical trials; unfortunately its clinical development was halted as a preliminary analysis of the results from two Phase III trials indicated that compared to placebo, Semagacestat (**30**) led to decreased cognitive abilities. In addition, an increased potential for skin cancer was observed [58].

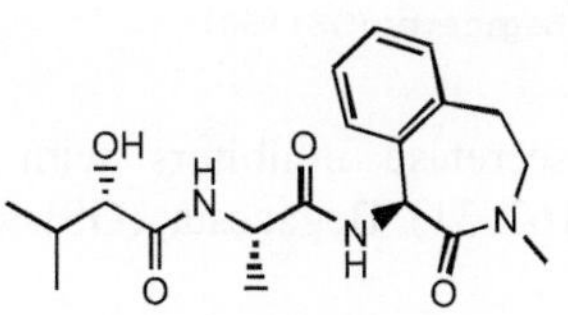

30

Figure 13: The chemical structure of Semagacestat (LY-450139, 30), a prototypic first generation γ-secretase inhibitor which has recently completed clinical evaluation.

It may well be the case that some of these devastating side effects of Semagacestat (**30**) are related to the fact that the compound directly blocks the active site of the protease and therefore non-selectively inhibits processing of all substrates. In fact, it has been reported that deficits in the Notch pathway are associated with mental retardation and dementia, which clearly underpins the necessity for substrate selective GSIs [59]. Based on these and other data, it has clearly been recognized that in particular, good selectivity for APP cleavage over the Notch receptors is required; however, given that γ-secretase processes well over 70 different membrane proteins, it yet remains an open question as to how selectivity is best assessed. Proteomic approaches taking a holistic view of potential substrates may well become critical tools for *in vitro* profiling future generations of γ-secretase inhibitors.

Based on the preclinical and clinical data for the first generation γ -secretase inhibitors, the development of more selective, Notch-sparing compounds became the focus of interest. Indeed, molecules which demonstrate selectivity for APP over Notch processing have been identified. Fig. **14** illustrates three different inhibitors (**31, 32, 33**) which have been or are being investigated in clinical trials.

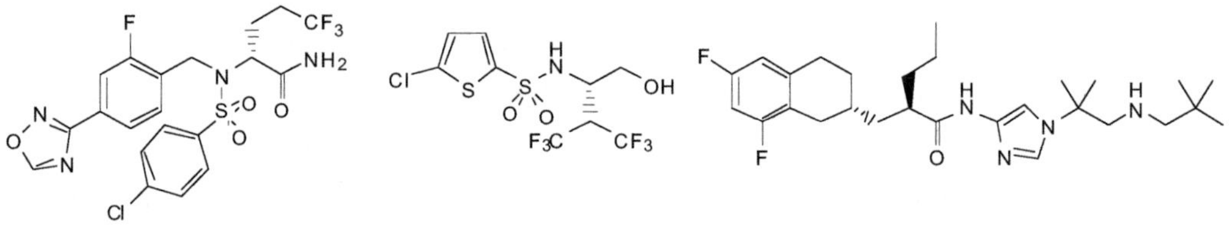

31	32	33
Avagacestat (BMS-708163)	Begacestat (GSI-953)	PF-3084014

Figure 14: Second-generation γ-secretase inhibitors with selectivity for APP over Notch processing; Avagacestat (BMS-708163, 31), Begacestat (GSI-953, 32), PF-3084014 (33).

Avagacestat (**31**) and Begacestat (**32**) are sulfonamides. In 2012, Bristol-Myers Squibb reported that their Phase II profile did not support the advancement of Avagacestat (**31**) into Phase III clinical trials and has abandoned its development programme [60] while favourably, Begacestat (**32**) demonstrated dose dependent Aβ changes in plasma and no notch related safety concerns in rats or dogs [61]. The clinical development of PF-3084014 (**33**) has been discontinued due to an unfavourable pharmacokinetic and pharmacodynamic profile, which may be related to its highly lipophilic nature [62] though analogs of PF-3084014 have also been reported [63].

The observation that long-term usage of non-steroidal anti-inflammatory drugs (NSAIDs) resulted in decreased occurrence of AD constituted a milestone in the γ-secretase field, which led to the discovery of γ-secretase modulators (Fig. **15**). Specifically, NSAIDs like Ibuprofen (**34**), Indomethacin (**35**) and other molecules within these chemotypes have been demonstrated to lower amyloidogenic $A\beta_{42}$ *in vitro* and *in vivo* independently of their cylooxygenase activity [64]. Concomitantly, $A\beta_{38}$ levels were increased and $A\beta_{40}$ concentrations remained constant.

Interestingly, a number of other NSAIDs like celecoxib (**35**) showed the opposite pharmacodynamic effect required for AD phenotype amelioration and raise $A\beta_{42}$

levels [65]. Kinetic studies suggest that these modulators are not competitive to the APP substrate and bind the γ-secretase complex at an allosteric site.

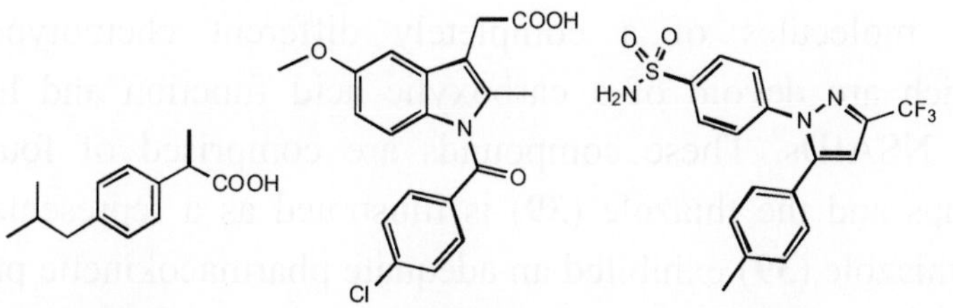

34 Ibuprofen **35** Indomethacin **36** Celecoxib

Figure 15: Structures of selected NSAIDs as γ -secretase modulators. Ibuprofen (34), Indomethacin (35), Celecoxib (36).

Clearly, modulation of γ -secretase is a very active research area since there is potential to reduce the concentration of the most fibrillogenic Aβ fragment ($A\beta_{42}$) with minimal impact on other substrates. GSMs have been classified by their structure into NSAID derived carboxylic acids and non-NSAID derived compounds [66]. In both areas, many modulators with improved potency as compared to the initial NSAIDs, have been reported, and reviewed recently [66]. For illustration, a few representative examples of NSAID- and non-NSAID-derived γ-secretase modulators are shown in Fig. **16**.

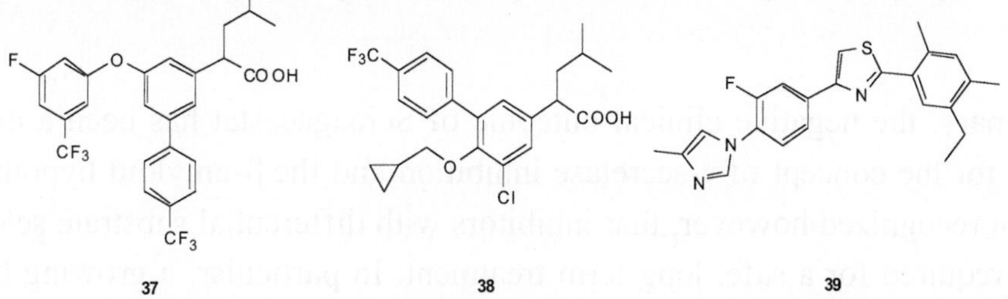

37 IC_{50} = 330 nM **38** IC_{50} = 69 nM **39** IC_{50} = 29 nM

Figure 16: Examples of chemotypes of NSAID and non-NSAID derived γ-secretase modulators.

Promisingly, NSAID derivative (**37**) demonstrated a 43% reduction of $A\beta_{42}$ after 4 hours post oral dosing at 30 mg/kg [67]. However, the high lipophilicity of this molecule may likely hamper its development. As exemplified by the analog (**38**), efforts continue to reduce the lipophilic characteristics of these molecules while maintaining or improving their γ-secretase-modulating properties [68]. Subsequently, molecules of a completely different chemotype have been described which are devoid of a carboxylic acid function and have not been derived from NSAIDs. These compounds are comprised of four consecutive aromatic groups and the thiazole (**39**) is illustrated as a representative example [69]. In fact, thiazole (**39**) exhibited an adequate pharmacokinetic profile to allow for a proof-of-concept study; here, (**39**) has proven to lower $A\beta_{42}$ levels in Tg2576 mice following chronic dosing. This finding correlated well with the histopathological analysis revealing a reduction in plaque in cortex and hippocampus. These promising *in vitro* and *in vivo* data spurred efforts and further search for analogs of modulator (**39**) in various research organizations (for a summary, see [66]).

It should also be noted that GSIs such as LY-450139 (**30**) can cause an increase of plasma Aβ over baseline after a transient reduction, a so-called rebound effect, the mechanism of which is poorly understood to the best of our knowledge. A recent preclinical study in rats indicates that contrary to the inhibitor LY-450139, two chemically distinct modulators did not exert such a rebound effect of Aβ levels in plasma [70]. This may be explained by the lack of accumulation of β-C-terminal fragments. Hence, modulators may be superior to inhibitors in the treatment of AD not only by sparing Notch signaling but also by a more consistent reduction of Aβ.

In summary, the negative clinical outcome of Semagacestat has been a dramatic setback for the concept of γ-secretase inhibition and the β-amyloid hypothesis. It has been recognized however, that inhibitors with differential substrate selectivity will be required for a safe, long-term treatment. In particular, a growing body of evidence suggests that inhibition of conversion of the Notch receptors needs to be avoided. However, given that γ-secretase processes over 70 different substrates, new technologies like a proteomic approach may be required to assess selectivity in a more systematic fashion. On the other hand, the advent of γ-secretase modulators shows a lot of promise, as these compounds appear to down-regulate the levels of $A\beta_{42}$ very selectively and also seem to avoid an Aβ rebound.

Another potential therapeutic rationale may be activation of α-secretase; α-secretase is a metallo-protease which cleaves APP between residues 16 and 17 in the Aβ sequence thus preventing the formation of neurotoxic Aβ$_{42}$. However, from a medicinal chemistry perspective, the feasibility of developing small-molecule activators of a metallo-protease is low.

THE TAU HYPOTHESIS

The second characteristic neuropathological hallmark of Alzheimer's disease, strongly correlating with the progression of clinical symptoms, is the deposition of hyperphosphorylated tau in neurofibrillary tangles in cortical and subcortical areas of the brain [14]. The principal function of tau is to stabilize microtubules, which are required for fast axonal transport of cargo in neurons [71]. Binding of tau to microtubules is disrupted by hyperphosphorylation, upon which, tau can accumulate into paired helical filaments to finally form tangles [72]. It is thus thought that aberrant phosphorylation of tau may be one of the early steps in the pathogenesis of Alzheimer's disease [14, 73].

A number of kinases have been implicated in the hyperphosphorylation of tau: MARK1, CDK5, GSK-3β, CaMKII and MAPK1 [74], but promiscuity between these kinases *in vitro* does not allow validation of a single entity as the most important tau kinase. Furthermore, phosphatases and other regulatory proteins such as glycosyl- and acetyl-transferases complicate the matter as they also modulate tau phosphorylation state [75].

Interestingly, mutations in tau can cause fronto-temporal dementia with parkinsonism (FTDP-17) [76] suggesting an important role for tau in the onset or progression of other complex neurodegenerative processes. Additional neurodegenerative diseases are characterized by tau-positive tangles, including Pick's disease, progressive supranuclear palsy (PSP), and argyrophilic grain disease (AGD) [77].

Direct genetic evidence for tau to be a causal factor in the development of AD is lacking although several studies have suggested that the presence of tau is a requirement for amyloid-induced neurodegeneration *in vitro* [78] and *in vivo* [79].

Taken together, these studies suggest that the β-amyloid cascade [80] is the molecular instigator of AD pathogenesis, and that tau is an essential mediator [81]. To address the pathological function of tau in AD, inhibition of tau-phosphorylating kinases has been explored extensively. Though a number of kinases have been implicated in the hyperphosphorylation of tau, recent literature has focussed on the Glycogen synthase 3 kinase (GSK-3) and Cyclin dependent 5 kinase (CDK5).

ALZHEIMER'S DISEASE: THE TAU HYPOTHESIS AND GSK-3

To put the role of Glycogen Synthase Kinase (GSK-3) into context, it is necessary to understand the cellular function of this kinase. Through its ubiquitous expression and phosphorylating activity, GSK-3 is involved in metabolism, differentiation, proliferation and apoptosis. It has been reported that hyperphosphorylation of tau leads to neurofibrillary tangles [14] and that increased GSK-3 function results from dysfunctional wnt and insulin signaling, suggesting that insulin deficit in the brain may be a cause of AD.

Concomitantly, increased GSK-3 activity is reported to increase Aβ formation via its activity on γ-secretase resulting in the senile plaques observed in AD patients [82]. Further to plague formation, dysregulation of GSK-3 is also implicated in the mechanism of learning/memory and may contribute to cognitive decline [82]. GSK-3 is up-regulated in brains of AD patients [83]. In neurons, GSK-3 is involved in the control of apoptosis, synaptic plasticity, axon formation as well as neurogenesis. Transgenic animal models which overexpress GSK-3 present with brain abnormalities, long-term potentiation impairment leading to phenotypic learning/memory deficits with concomitant hyperphosphorylation of tau and increased production of Aβ. GSK-3 is also implicated in mood disorders with elevated activity being associated with manic and depressive behaviour [84]. It is, however, unclear if inhibition of GSK-3 will alter Aβ deposition and ameliorate the AD clinical phenotype [85].

Due to its complex physiological role, a full pharmacological knock-out of GSK-3 is very likely undesirable. Indeed, it is thought that a range of mechanism-based toxicities through hyperglycemia, result from a >25% inhibition of the kinase [86].

Natural products and small molecules have been described as GSK-3 inhibitors and have been reviewed by Elder-Frinkelman & Martinez as well as Kramer *et al.* [87, 91]. A number of examples for illustration are shown in Fig. **17**. IMX (**40**) is a semi-synthetic, potent, cell-permeable inhibitor of GSK-3 and CDK and inhibited Aβ-induced neurotoxicity in neuroblastoma cells. In APP/PS1 double-cross transgenic mice, IMX (**40**) reduced spatial and memory defects, inhibited Aβ accumulation and demonstrated significant decrease in brain concentrations of both soluble and insoluble $A\beta_{42}$. In addition, IMX (**40**) was shown to enhance Aβ clearance mechanisms and reduce tau hyperphosphorylation with no reduction in tau levels [88].

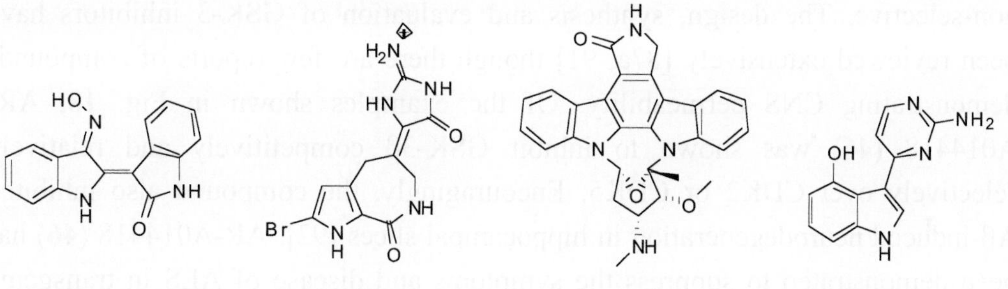

40	**41**	**42**	**43**
IMX	Hymenialdisine	Staurosporine	Meridianine A

Figure 17: A selection of natural products with GSK-3 inhibitory activity. IMX (39), Hymenialdisine (40), Staurosporine (41), Meridianine A (42).

The marine sponge derived Hymenialdisine (**41**) is a non-selective, ATP-competitive GSK-3β, CDK, MEK1, CK1 and Chk1 inhibitor and has been shown to block phosphorylation of tau through downregulation of PAK1 kinase in E18 rat cortical neurons as well as inhibition of GSK-3β through prevention of MAP-1B phosphorylation [89]. The non-specific Staurosporine (**42**) is a potent GSK-3β inhibitor (IC_{50} = 15 nM) while a limited medicinal chemistry campaign failed to identify a selective Meridianine analogue (**43**), a chemotype which instead showed preferential inhibition of the CDK's [90].

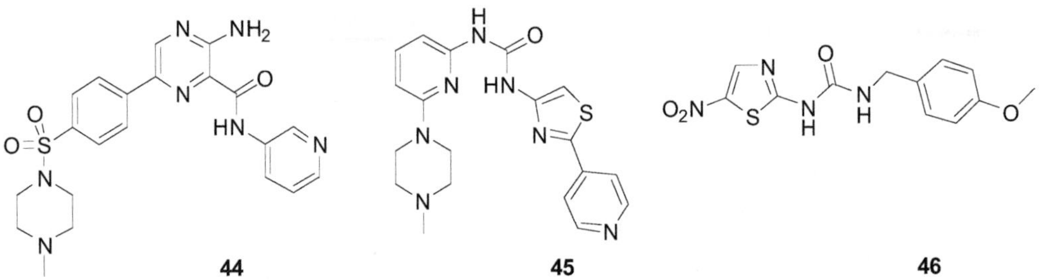

44
$IC_{50} = 76$ nM

45
$IC_{50} = 13$ nM

46
$IC_{50} = 104$ nM

Figure 18: ATP competitive, inhibitors of GSK-3β: 44; BBB = 15 x 10^{-3} cm/min (Bovine Endothelial Cells). 45 exhibited neuroprotective effects in hypoxia-ischemia model of brain injury. 46; (AR-A014418).

Given its ubiquitous nature and the close homology of its catalytic domain compared to the kinome, ATP competitive GSK-3 inhibitors tend to be relatively non-selective. The design, synthesis and evaluation of GSK-3 inhibitors have been reviewed extensively [87a, 91] though there are few reports of compounds demonstrating CNS permeability. Of the examples shown in Fig. **18**, AR-A014418 (**46**) was shown to inhibit GSK-3β competitively and relatively selectively over CDK2 or CDK5. Encouragingly, the compound also inhibited Aβ-induced neurodegeneration in hippocampal slices [92]. AR-A014418 (**46**) has been demonstrated to suppress the symptoms and disease of ALS in transgenic mice [93], provided protection of dopaminergic neurons and restored depletion of striatal dopamine in the MPTP induced parkinsonism model [94].

One GSK-3 inhibitor which has been extensively studied in a variety of paradigms, is the alkali cation lithium (Li^+). Though its selectivity profile has not been clearly defined, lithium salts which facilitate the delivery of the monovalent cation are clinically utilized as effective mood disorder therapeutics. One of the prime targets for lithium therapy is GSK-3 and its apparent selectivity for magnesium containing enzymes has been postulated to be a result of the unique complexation properties of the metal ion [95].

In preclinical AD models, administration of lithium chloride (LiCl) resulted in reduced tauopathy in tau-transgenic mice; specifically, lowering levels of phosphorylation of known AD tau epitopes, diminishing tau levels of aggregation and less axonal degeneration. Postulation that these phenotypic ameliorations were driven mechanistically through GSK-3 has been substantiated by a more selective GSK-3 inhibitor (AR-A014418), which demonstrated a similar profile [96].

The clinical success of GSK-3 inhibitors within AD/p-tau paradigm has yet to be defined. To date, the greatest evidence of efficacy resides in the reports of clinical trials provided by lithium administration, though at best these are contradictory [97]. Lithium has demonstrated beneficial effects in dementia patients, reversed the reduction of BDNF in AD patients, reduced AD in elderly patients with bipolar disorder, and slowed the progression of ALS. Controversially, lithium did not confirm prevention of dementia, improve cognition or reduce tau phosphorylation and Aβ levels in AD patients and, in context of its reported toxicity, was withdrawn from further development. It has been argued that at therapeutic doses, lithium exerts an inhibition of 25% on total GSK-3 without observation of toxicity indicating an available therapeutic window [91e].

Tideglusib (Fig. **19, 47**) from Noscira has recently completed Phase IIb trials for AD. It is reported to be a non-ATP-competitive (GSK-3) inhibitor with potential application for AD, progressive supranuclear palsy and Pick's disease; two additional tauopathies [98]. In the case of AD, Tideglusib (**47**) decreased tau phosphorylation, lowered Aβ plaque, improved learning/memory and prevented neuronal loss in pre-clinical animal models [99].

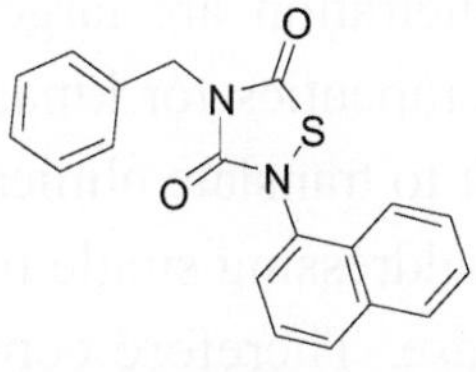

47

Tideglusib

Figure 19: The chemical structure of Tideglusib (47) also known as NYPTA, NP-12 or NP031112.

In a recent Phase IIa trial in patients with mild to moderate AD, Tideglusib (**47**) was reported to show positive effects in four out of the five behavioural measurements and the magnitude of these effects was higher in moderate AD cases, justifying further evaluation [100]. Investigators recently characterized Tideglusib (**47**) as an irreversible inhibitor, which demonstrated lack of enzyme functional recovery after removal of the unbound drug and cited a cysteine residue (Cys-99) as the primary binding site [101]. Unfortunately, Noscira recently announced Tideglusib did not meet the primary or two secondary endpoints of the AD assessment scale-cognitive subscale (ADAS-cog) in a Phase IIb ARGO trial data (n = 306) [102] and further evaluation of the trial data, is required.

FUTURE THERAPEUTIC AVENUES FOR AD

AD is a complicated multi-faceted, poly-dysfunctional disease and therapeutic progress has been compounded by the difficulty of the research over the past decade. Though effective clinical tools are now available to test the β-amyloid hypothesis and clinical trials are underway, it has become apparent that choice of patient population is critical in these trials. Indeed, not addressed previously, pre-symptomatic management of AD is now currently being evaluated in patients known to carry the autosomal dominant mutations in APP, PSEN1 and PSEN2 [103].

While the β-amyloid hypothesis is being addressed, alternative avenues of investigation still lag behind. Selective compounds with the right physiochemical properties to facilitate brain penetration are largely dependent on the target of intervention. Small molecule therapeutics for kinase and protease targets, though tractable *in vitro*, can be difficult to translate clinically to CNS medicines. Also, it has as yet to be established that addressing single mechanisms of intervention will lead to amelioration of the disease. Therefore combination therapies should also be considered, for example, one additional avenue of therapeutic intervention which we have elected to address in the PD section is metal chelation.

Though, our understanding of causative factors in AD etiology continues to increase, it becomes more evident that a multitude of controllable factors including diet, exercise and lifestyle must also be considered. Recently, large scale genome-wide association studies (GWAS) identified SORL1 [104], CLU, CR1, PICALM [105], BIN1 [106], ABCA7, MS4, EPHA1, CD33 and CD2AP [107] as risk factors for late-onset Alzheimer's disease. CLU (also known as ApoJ) has been shown to induce cholesterol efflux in macrophages [108], and APP itself has been proposed to be a receptor for cholesterol [109]. SORL1, also known as LDL receptor 11, has been shown to regulate the secretion of lipoprotein lipase, which hydrolyzes triacylglycerols in circulating lipoproteins [110]. APOE is a crucial regulator of cholesterol metabolism in the brain, ABCA7 is a close homolog of ABCA1, which mediates cholesterol efflux to apolipoproteins, and has been suggested to have a similar functions [111]. These findings, coupled with preclinical and epidemiological data [112] suggest that modulation of brain cholesterol levels could also be a therapeutic intervention pathway in AD.

Though a number of additional genes have recently been identified as risk factors for sporadic Alzheimer's disease, ApoE is the most replicated and significant [113] and this observation has been translated to pre-clinical research. Indeed, investigators recently identified small molecules which "correct" the aberrant 3-dimensional structure of ApoE4, the pro-AD variant, to the 3-dimensional structure of Apo E3, the neutral variant in the population [114]. Close attention to this hypothesis is warranted with the emergence of pre-clinical, transgenic animal data [115] and the recent reporting of secondary retinoic acid activation mechanisms (RXR and RAR) [116].

Both natural and synthetic retinoids, targeting RAR and RXR homo/heterodimers, produce pleotropic responses in many tissues, including anti-proliferative effects and neuronal differentiation. Thus, a number of studies are ongoing using retinoids as candidate drugs for evaluation in Alzheimer's Disease (Fig. **20**). Tamibarotene (**48**) [117], which has selectivity for RARα and RARβ over RARγ, is in Phase II clinical studies for AD, (Clinical trial identifier: NCT01120002). Additionally, *Iso*retinoin (**49**) also known as Accutane, a non-selective retinoid is in Phase I/II clinical studies (Clinical trial identifier: NCT01560585). Bexarotene (**50**), an RXR-selective retinoid, is also in Phase II clinical studies for AD (Clinical trial identifier: NCT01782742).

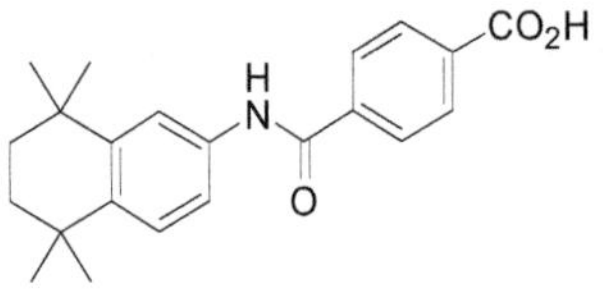

48

Tamibarotene

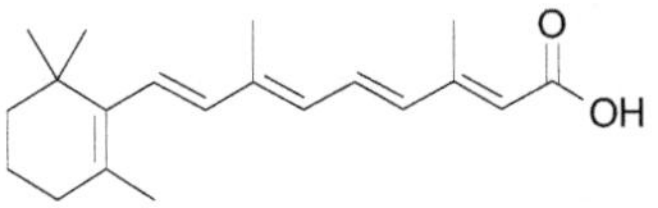

49

Isoretinoin

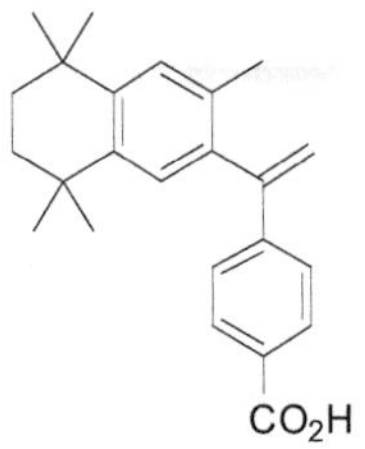

50

Bexarotene

Figure 20: The chemical structures of Tamibarotene (**48**), *Iso*retinoin (**49**) and Bexarotene (**50**); RAR / RXR agonists currently in Phase I and II clinical trials evaluation for AD.

PARKINSON'S DISEASE

Background to Parkinson's Disease

Parkinson's disease (PD) is the most common neurodegenerative movement disorder and is characterized clinically by tremor, bradykinesia, rigidity and postural instability. It is an etiologically and pathologically diverse syndrome which particularly, but not exclusively, affects the neurons in the substantia nigra pars compacta region of the brain. Characterized by three key pathologies, neuronal cell death at the substantia nigra, Lewy Bodies and reactive microgliosis, PD is now understood to be a genetic disorder which can also be induced through environmental stress, or a combination thereof. Although symptoms can be improved using dopamine replacement, therapies that promote neuroprotection and/or disease modifying effects remain an urgent unmet medical need. Many decades of research have implicated the dysfunction of three main pathways in PD pathogenesis: mitochondrial function, oxidative stress and the proteosomal system. The following sections aim to summarize thoughts on targets and mechanisms across these areas that have potential to provide new neuroprotective approaches for this syndrome.

α-SYNUCLEIN, A CENTRAL PLAYER IN PD

On the face of it, α-synuclein, the major component of Lewy bodies, continues to be a protein we understand little about, yet its importance in the pathology of PD is unquestionable. It is expressed at relatively high levels in the brain, at lower levels in other tissues including haematopoetic cells, it associates with lipids and is found specifically binding to synaptic vesicles in neurons. Although its exact function is unknown, it is thought to control vesicular trafficking and release, possibly through transient interactions with lipid vesicular surfaces. Indeed α-synuclein has been implicated in neurotransmitter release, vesicle turnover, channel localization, and dose-dependent dynamic dopamine release [118]. Recently, Chandra *et al.* [119] and Burré *et al.* [120] demonstrated that α-synuclein is required for presynaptic vesicle chaperoning through SNARE-complex assembly by directly binding to VAMP2. These findings further support a neural-specific role for α-synuclein at the level of presynaptic vesicles. The relatively modest effects of α-synuclein knock-out in mice can be explained by compensation of the β- and γ-forms and recent studies on complete synuclein-knock-out animals showed age-dependent neurological impairments with premature death [121].

A key characteristic of this small protein is its tendency to self-associate *in vitro* [122] and α-synuclein is a key component of both Lewy Bodies (LB) and Lewy neuritis, both key markers of PD and Dementia with Lewy Bodies (DLB). Lewy bodies are intracellular deposits of proteins and lipids which are found in many regions of the brain; they contain aggregated forms of α-synuclein, ubiquitin and lipids. The first evidence that α-synuclein was relevant to PD disease progression came with the observation that a mutated version of the SNCA (α-synuclein) gene (A53T) was causal for dominantly inherited PD [123]; up to that point PD was not considered a genetic disease. Since then further SCNA gene disruptions including mutations (E46K, A30P) and multiplications have been linked with various forms of PD and LB diseases [124]. Furthermore, common variants around the SCNA gene, which are likely to increase α-synuclein levels in the brain, have been associated with lifetime risk of sporadic PD.

α-Synuclein can be found as oligomers in brain tissue and in cells [125]. There is mounting evidence that soluble α-synuclein aggregation intermediates (so-called oligomeric species), rather than insoluble aggregated α-synuclein fibrils, are the toxic moieties [126]. There is also an emerging concept of extracellular secreted forms which is supported by evidence of host to graft transmission of LBs [127]. The mechanisms responsible for α -synuclein-mediated toxicity are probably multiple and if there is a common theme between them, it is likely to be its propensity for lipid binding. As such, α-synuclein has been shown to have a detrimental effect on neuronal processes by affecting synaptic vesicle release, ER to Golgi transport, lysosomal function, chaperone-mediated autophagy and mitochondrial function [128].

It is commonly understood that increased levels of α-synuclein lead to increased levels of aggregation and the multiplication mutations in this gene suggest that protein amount is relevant to pathology. It is proposed that although α-synuclein has a fundamental role in neuronal function, a modest reduction of protein levels would be beneficial in a disease context. Numerous mechanisms to manage α-synuclein load can be considered for therapeutic intervention including the regulation of its transcription through GATA transcription factors [129] and signaling events through ERK and PI$_3$K pathways [130]. The SCNA gene has been shown to be hypo-methylated in PD brains [131] hence, epigenetic regulation may also offer therapeutic potential.

The primary degradation pathway of α-synuclein is likely to be dependent upon the pool and species involved, however, it is clear that the majority of wild type α-synuclein in neuronal cells is believed to be cleared by the lysosomal pathways of chaperone mediated autophagy (CMA) and macro-autophagy [128c, 132]. Development of drug-like inducers of autophagy in neurons has been particularly challenging since they appear to be less responsive to classic autophagy inducers than non-neuronal cells. To date, a number of early small molecules, which activate autophagy in cellular models, have been identified (Fig. **21**). Some of these molecules do not target mTOR (molecular target of Rapamycin) and thus should avoid the known side effects associated with chronic administration of Rapamycin [133].

The extracellular homeostasis of α-synuclein is also a consideration for future therapies; indeed the presence of α-synuclein in cerebrospinal fluid, plasma and in interstitial fluid indicates that secretion of this protein is a physiological process. It is therefore plausible that factors disrupting this process may contribute to pathogenesis and perhaps also present therapeutic mechanisms to modulate α-synuclein homeostasis both intra- and extracellularly. Furthermore, understanding the mechanism(s) whereby α-synuclein can apparently propagate between cells could open up avenues for developing novel therapies, not only for Parkinson's disease, but also for other neurodegenerative diseases.

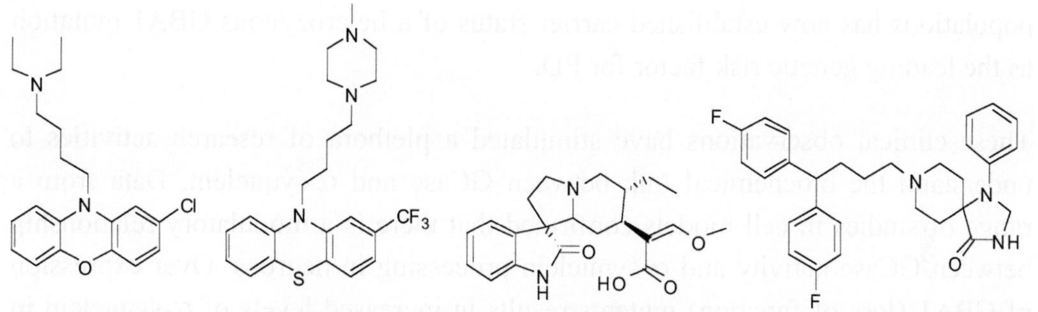

51	52	53	54
10-NCP	Trifluoperazine	Isorhynchophylline	Fluspirilene

Figure 21: The chemical structures of drugs which have been shown to enhance autophagy including 10-NCP **(51)**, Trifluoperazine **(52)**, Isorhynchophylline **(53)**, Fluspirilene **(54)**.

INTERPLAY BETWEEN GLUCOCEREBROSIDASE AND α-SYNUCLEIN PROCESSING - A NEW THERAPEUTIC AVENUE?

Over recent years a clinical link between Gaucher's disease and the development of Parkinsonism has provided evidence that mutations in the Glucocerebrosidase (GCase) gene (GBA1) and alterations in sphingolipid metabolism may contribute to the pathogenesis of synucleinopathies. Gaucher's disease is a rare, autosomal recessive lysosomal storage disorder that results from loss-of-function mutations in GCase, a lysosomal enzyme that cleaves the β-glucosyl linkage of glucosylceramide (GlCer). It can manifest itself both viscerally (type 1) and centrally (types II and III) leading to accumulation of substrate glycosphingolipids in various organs including brain.

The link to Parkinson's disease was originally made when increased Parkinsonism was observed in Gaucher's cohorts, which was further substantiated by the realization of increased frequency of GBA1 mutations in both PD patients and those with dementia with Lewy bodies [134]. It was also noted that relatives of Gaucher's patients with Parkinson's were heterozygous for GBA1 mutations [135]. Notably, neurological evaluation of brains from GD patients with Parkinsonism and PD subjects with GBA1 mutations showed Lewy body pathology in all cases [136]. A series of large genetic analyses in worldwide populations has now established carrier status of a heterozygous GBA1 mutation as the leading genetic risk factor for PD.

These clinical observations have stimulated a plethora of research activities to understand the biochemical link between GCase and α-synuclein. Data from a range of studies in cell models confirmed that there is a modulatory relationship between GCase activity and α-synuclein processing in neurons. Over expression of GBA1 (loss of function) mutants results in increased levels of α-synuclein in neuronal cell lines [137]. In agreement, studies in primary neurons either knocking down GBA1 or over expressing wild type GBA1 led to elevated or reduced levels of α-synuclein, respectively. This link has also been confirmed using the pharmacological chaperone, which stabilizes GBA1 gene product for effective processing and thereby reduces α-synuclein levels [138]. *In vivo* studies have further substantiated this concept: cortical GCase activity is reduced in A53T α–synuclein mice [139], CSF samples from PD and DLB subjects have reduced GCase activity [140] and most recently, it has been reported that total GCase protein levels are reduced by 40% (compared to age matched controls) in the cingulated cortex of PD brains .

A seminal study by Mazzulli *et al.* has confirmed how α-synuclein and GCase might regulate each other's processing; their studies revealed a number of critical observations [141]: (a) loss of function of GCase compromises the lysosomal protein degradation of α-synuclein specifically; (b) neurotoxicity from accumulating α-synuclein develops through aggregation dependent mechanisms; (c) the lipid substrate of GCase (GlcCer) accumulates under condition of repressed GCase activity/levels and acts to stabilize the formation of toxic α-synuclein oligomers; (d) α-synuclein itself inhibits lysosomal activity of wild type GCase by affecting its trafficking from the ER to the lysosome.

Collectively, the inhibitory effect of α-synuclein on GCase trafficking, the stabilization of α-synuclein aggregates as a consequence of accumulation of glycosylceramide, and the reduction in α-synuclein degradation due to compromised GCase activity lead to a self-propagating pathological spiral. Notably, this effect appears specific to α-synuclein since mHtt and Tau proteins were not able to affect GCase function.

Of particular note is the emerging understanding of the relevance of GBA to idiopathic PD. Studies in which wild type GBA1 was over expressed in neurons suggest that elevation of GCase levels in non GBA mutant scenarios could beneficially affect synucleinopathies. In addition, as part of their comprehensive study Mazzulli *et al*. showed that normal subject-to-subject variation in α-synuclein levels appears to modulate lysosomal maturation and activity of GCase *in vivo*. Furthermore, analysis of idiopathic PD brains demonstrated a 40% reduction in wild type GCase protein in the cingulate cortex suggesting that the modulatory link between α-synuclein and GCase is relevant to the most common form of the disease. Therefore, although carrier status of a GBA1 mutation alone is not sufficient to express a PD phenotype, it is clear that correcting GBA function could address some element of α-synuclein's disrupted pathophysiology in PD.

Enhancing proteosomal degradation of α-synuclein through addressing lysosomal GCase activity is an attractive yet presently untested concept for idiopathic PD. Initial proof-of-concept studies in the GBA D409v/D409v mouse, which has a dementia with Lewy bodies-type synucleinopathy, has provided some insight to this approach [142]. These mice display increased brain levels of glycosylsphingosine, α-synuclein aggregates, ubiquitin aggregates and have memory impairment. Administration of adenovirus-associated gene delivered GBA1 into hippocampal neurons resulted in a decrease in glycosphingolipids, reduced α-synuclein and ubiquitin pathologies and critically, corrected memory function. Furthermore, pharmacological chaperones such as derivatives of Isofagomine [143] have been shown to be CNS penetrant and to increase brain GCase activity. It is expected that such molecules will be tested in preclinical models of PD.

LRRK2- A TRACTABLE SMALL MOLECULE TARGET FOR PD?

LRRK2 is a ROCO protein family member characterized by the possession of both a kinase and GTPase domain. Its expression is widespread in the brain, with mRNA levels being highest in areas which are dopamine-receptive. Autosomal dominant mutations in LRRK2 are recognized as the most common genetic cause of PD being responsible for 4.5% of familial and 1-2% of sporadic disease and furthermore, familial PD associated with LRRK2 mutations closely resembles idiopathic PD both clinically and pathologically. Studies in animal and cellular models have shown that certain mutations in the LRRK2 gene can increase its kinase activity, which contributes to neurotoxicity, possibly via the oxidative stress pathway [144]. Although the function of LRRK2 is not fully understood, it has been linked to neurite outgrowth, vesicular trafficking, protein translation, autophagy and neuronal survival and there is evidence that the reduced occurrence of LRRK2 mRNA limits neural progenitor cell differentiation toward dopaminergic neurons and may enhance cell death [145].

It has been consistently demonstrated that dopamine transmission is defective and axonal degeneration evident in LRRK2 mutant mice. The deficits in these mouse models carrying different mutations show a dosage dependence and higher levels of transgene expression are correlated with stronger PD-like phenotypes [146]. Overexpressing either wild type human or mouse LRRK2 produces no phenotypes that are relevant to PD indicating that mutated LRRK2 protein specifically is necessary to drive pathogenic toxicity; however, it is still not clear if enhanced kinase activity is solely responsible for pathogenicity. There are also numerous independently generated LRRK2 null mouse models [147] which indicate that there is no neurodegeneration or neuropathological changes and that the dopaminergic system is normal, but quite possibly as a result of compensation by LRRK1.

The loss of LRRK2 expression does, however, result in a severe kidney phenotype with increased apoptosis, inflammatory responses and increased oxidative damage. In addition, age-dependent accumulation and aggregation of α-synuclein and ubiquitin proteins are found in the kidney; up to six times higher levels than those in brain under normal circumstances. The conclusions drawn from these studies have come together to suggest that the kidney develops a PD-like pathology in these mice and this organ is thought to be particularly affected due to its lack of overlapping LRRK1 and LRRK2 expression in affected areas. The potential for a LRRK2 null mouse to produce a PD-like phenotype, albeit in a peripheral organ, has somewhat confused the straightforward theory that the pathogenic mutations in PD are gain-of-function in their basis and that the increased kinase activity of the GS2019S mutant may be causative for the disease. Data from LRRK2 knock-out, G2019S knock-in and kinase-dead mutant mice now suggest that steady-state abundance of the LRRK2 protein also plays a determining role. A newly emerging theory is that perhaps the clinical mutations in LRRK2 result in a partial loss of function, perhaps via a dominant negative effect through dimerization between mutant and wild type LRRK2 [147]. This theory requires further interrogation and the development of LRRK1/LRRK2 null mice will help address the possibility of the predicted PD-like neurodegenerative phenotype.

These LRRK2 null mouse data create new questions for the popular approach of developing LRRK2 inhibitors for the treatment of PD. Clearly, significant kidney toxicity of a LRRK2 inhibitor would be predicted and the most recent LRRK2 null mouse line showed significant behavioral features including anxiety-related activities and the inability to terminate ongoing behavior [148]. Based upon the knock-out data which suggests that functioning LRRK2 is required to support appropriate protein homeostasis and the possibility that the clinically relevant mutations may in fact be partial loss of function in their nature, it might even be necessary to consider that an activator of the LRRK2 pathway may be required in a disease setting. This assertion is supported to some extent by a study showing that high levels of LRRK2 are well tolerated without induction of neuronal synucleinopathies and even improve motor skills of wild type and α-Synuclein expressing mice [149]. Further studies will be required to determine whether and how LRKK2 protein levels or enzyme activity will impact its function and role in PD and chemical tools can facilitate these investigations.

Despite the fact that we still do not fully understand LRRK2's function and its direct substrates (and thereby important LRRK2 pharmacodynamic biomarkers) have remained elusive, an explosion of LRRK2 inhibitor projects have been performed within the pharmaceutical industry. There has been a number of preclinical LRRK2 inhibitors disclosures (Fig. **22**) demonstrating inhibition of LRRK2$^{(G2019S)}$ (**55**) and wild-type (**56**), [150], for example, the brain penetrant chloropyrimidine (**57**) which demonstrated inhibition of phosphorylation of central LRRK2 at Ser910, Ser935 *in vivo* following administration at 50 mg/kg *i.p.* [151].

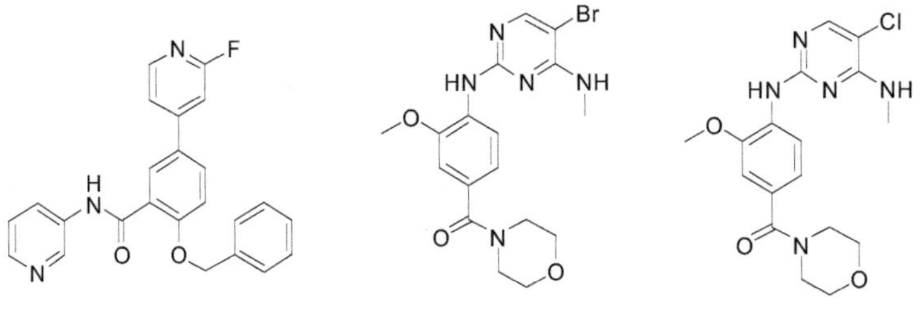

Figure 22: The chemical structures for a selection of recently reported LRRK2 inhibitors.

MITOCHONDRIAL DYSFUNCTION IN PD

Mitochondria are essential for neuronal function, coordinating the balance between life and death of a cell. The limited glycolytic activity and high metabolic requirements of neurons makes them both highly vulnerable to disruption of mitochondrial homeostasis and highly dependent on aerobic oxidative phosphorylation which is a major source of toxic endogenous free radicals. When the levels of these free radicals overwhelm the cell's neutralizing capacity, a propagating cycle of mitochondrial dysfunction and neuronal damage is being established. Reactive species generated by mitochondria have several targets including mitochondrial lipids, protein, and DNA. The lack of histones in mitochondrial DNA and reduced capacity for DNA repair render these organelles particularly vulnerable to cellular oxidative stress. As a result, the mitochondrial network in a neuron can suffer from problems in trafficking, Ca^{2+} homeostasis (fission-fusion balance, biogenesis, shape re-modeling) and quality control through autophagy.

There are now numerous independent lines of evidence that connect oxidative stress and mitochondrial dysfunction with degeneration of dopaminergic neurons in PD. One of the original observations linking PD with mitochondrial function was the fact that the respiratory chain complex I inhibitors (Rotenone, MPTP and paraquat) were shown to reproduce PD-like features in humans and animal models. In addition, complex I activity is clearly reduced in substantia nigra and other PD brain compartments and mitochondria in PD tissue are morphologically abnormal showing fragmentation [152].

MITOCHONDRIAL-TARGETED ANTIOXIDANTS IN PD

Despite the fact that mitochondrial oxidative stress is clearly a cause of cell death in PD, general antioxidant trials have disappointed [153] with only Co-enzyme Q_{10} (Fig. **23**, **58**) showing possible benefit [154]. To enhance the therapeutic potential of this approach, the mitochondria-targeted antioxidant mitoquinone (Fig. **23**, **59**) was developed and comprises the lipophilic triphenylphosphonium (TPP) cation covalently linked to ubiquinone, the active antioxidant moiety of CoQ_{10} [155].

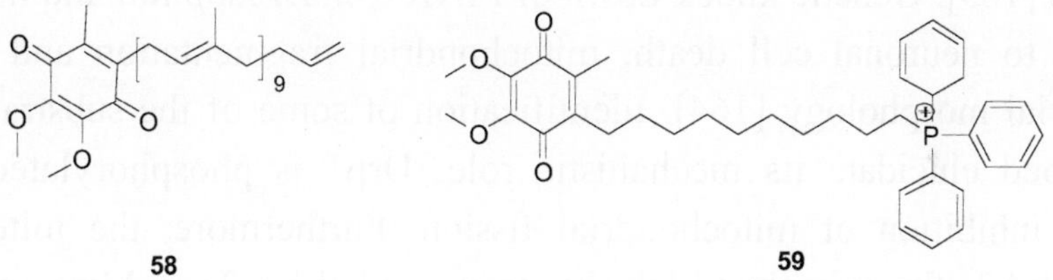

Figure 23: Chemical structures for Co-enzyme Q10 (58) and Mitoquinone (MitoQ, 59).

The TPP cation enables MitoQ (**59**) to cross membranes and to accumulate several-hundred fold within mitochondria due to the large membrane potential [156]. MitoQ (**59**) displayed cardio-protective efficacy in animal models [157]; however, it did not slow down progression of PD as shown by the UPDRS and other measurement scales in patients [158].

MITOCHONDRIAL HOMEOSTASIS TARGETS

The emergence of further genetic-bases to rare forms of PD have re-enforced the central importance of mitochondrial homeostasis and its disruption to both familial and sporadic PD. α-Synuclein itself has been demonstrated to decrease complex I activity, disturb the mitochondrial fusion process, and increase fragmentation in a variety of overexpressing or mutant (A53T) expressing cell and *in vivo* models [159]. α-Synuclein had also been shown to co-locate with mitochondria in neuron models and in the striatum and substantia nigra of PD subjects [160]. Interestingly, the mitochondrial phenotype resultant of over-expressed α-Synuclein can be rescued by the co-expression of a number of putatively neuroprotective genes associated with familial PD, including PINK-1, Parkin and DJ-1 [161]. These gene mutations and polymorphisms associated with early onset PD all have direct or indirect links to mitochondrial biogenesis and clearance (mitophagy).

PINK-1 is a serine threonine kinase and recessive loss-of-function mutations in its gene (Park6) are linked to a specific familial, early onset PD [162]. It contains a mitochondrial targeting sequence and has been shown to fractionate into the mitochondrial fraction of neurons, associating with both inner and outer membranes [163]. Genetic knock-down of PINK-1 in *Drosophila* and mammalian cells leads to neuronal cell death, mitochondrial fragmentation and abnormal mitochondrial morphology [164]. Identification of some of the substrates of this kinase helped elucidate its mechanistic role: Drp1 is phosphorylated at S637 leading to inhibition of mitochondrial fission. Furthermore, the mitochondrial protease HtrA2, the mitochondrial chaperone and the GTase Miro and adapter protein Milton (which have a role in mitochondrial trafficking) have been identified as putative substrates for PINK-1. PINK-1's best understood molecular activities involve its ability to regulate the localization of Parkin specifically to damaged mitochondria. Recently, it has been confirmed that human PINK1 is activated by mitochondrial membrane potential depolarization, enabling it to phosphorylate Parkin at Ser65 [165]. It is proposed that, following recruitment of PINK1 to the mitochondrial membrane via its N-terminal targeting sequence, it is subsequently proteolyzed by the mitochondrial rhomboid protease PARL [166], resulting in a processed form of PINK1which is rapidly degraded by the 20S proteasome [167]. In response to mitochondrial membrane potential depolarization, a marked stabilization of full length PINK1 at the mitochondria has been observed

[168], which may be achieved by trans-localization of PINK1 from the inner to the outer mitochondrial membrane, where it is no longer accessible by PARL.

Parkin, the gene product of Park 2, was first highlighted as a protein of interest when Park 2 was linked to an autosomal recessive version of juvenile PD [169]. It is an E3 ligase whose loss of function in patients results in impaired mitochondrial morphology, decreased complex I activity, and decreased interconnectivity [170]. It is expressed in neurons to high levels but is essentially ubiquitous. While it is predominantly mitochondrial, it has also been observed in the nucleus, ER, Golgi, and synaptic vesicles. Knock-down studies have given rise to an increased sensitivity to oxidative stress, severe structural mitochondrial abnormalities, and dopaminergic neurodegeneration [171]. Parkin's role in tagging and promoting damaged mitochondria for mitophagy in concert with PINK-1 is now well established [168]. Indeed, recent studies demonstrating the phosphorylation of Parkin at Ser65 by PINK-1 also confirmed that this leads to a marked activation of its ubiquitin ligase activity that is prevented by mutation of Ser65 or inactivation of PINK1 [165]. Parkin's role in regulation of mitophagy requires the broad activation of the ubiquitin-proteosome system [172] and is thought be mediated at least in part through ubiquitination and subsequent degradation of the mitofusins, proteins required for fusion [173]. Their loss from the mitochondrial membrane theoretically prevents the damaged mitochondria from fusing with the healthy network. Other substrates have also been tentatively identified through proteomics, proposed as proteins whose removal through the UPS system facilitates entry into the autophagosome via dispersal of mitochondria (which are previously aggregated) or as specific inhibitors of mitophagy and whose removal would enable that process [174]. In addition to its role in mitophagy, data exists to suggest a role for Parkin in mitochondrial biogenesis through regulation of mitochondrial transcription via TFAM [175] and PARIS a transcriptional repressor of PGC-1α, a central controller of proteins involved in mitochondrial biogenesis [176].

DJ-1 is a 189 amino acid, dimeric protein with documented roles in regulation of transcriptional regulation, cellular transformation, redox sensitive chaperone activity and in protecting against oxidative stress [177]. Deletion or point mutations in this gene have been linked to both familial and sporadic PD [178].

Knock-out studies in fly and mouse models have revealed a mitochondrial phenotype comprising reduced levels of mitochondrial DNA, reduced respiration control ratio, and reduced ATP levels [179]. The mitochondrial toxins rotenone and 6-OH DOPA cause an increase in the expression of DJ-1 and its redistribution to mitochondria from the cytosol [180]. Furthermore, loss of DJ-1 expression increases sensitivity to complex I inhibitors [181]. The detailed mechanistic role of DJ-1 is not fully appreciated, but it has been proposed to be a redox-regulated chaperone that prevents misfolding and aggregation of proteins (including α-synuclein) under oxidative stress [182]. A key observation, which links DJ-1 to the targets previously described, is that removal of DJ-1 can be compensated for, or rescued by Parkin and Pink-1 suggesting that they all ultimately co-operate to support continued mitochondrial homeostasis [177b].

The targets discussed above all present theoretical new approaches for intervening in mitochondrial dysfunction. Whilst their relevance to PD has emerged due to their association with relatively rare familial forms of the disease, their roles in mitochondria homeostasis make them valid points of intervention for therapies targeting sporadic PD. Mitochondria damaged either through loss-of-function mutations in mitochondrial homeostasis proteins, α-synuclein toxicity or secondary oxidative damage will liberate free radicals and create an environment of oxidative stress. One clearly logical approach would be to up-regulate clearance of damaged mitochondria to get the mitochondrial homeostasis back towards healthy state and, if possible also improve mitochondrial biogenesis. The PINK-1/Parkin pathway certainly represents new potentially fertile grounds for the discovery of small-molecules activators of mitophagy. However, many challenges have to be overcome to realize this avenue including the need to identify activators of the enzymes involved in this pathway or to identify negative modulators suitable for inhibition.

IRON CHELATION THERAPIES

Increased iron levels and oxidative stress have been heavily implicated in the neurodegenerative process in PD; hence, removal of excess iron by chelation represents a potential approach to neuroprotection. Physiologically, iron plays several critical roles in DNA synthesis, mitochondrial respiration, oxygen transport and, specifically in neurons, in myelination and neurotransmission. It is the most abundant metal ion in the brain and its cellular entry is controlled through active transport by transferrin receptors. Crucially in relation to PD, Fe^{2+} acts as a cofactor for tyrosine hydroxylase (TH), the enzyme at the rate-limiting step in the synthesis of dopamine. PD patients have long been diagnosed with an abnormal accumulation of iron in the substantia nigra [183]. It is still not clear as to whether this is a cause or a consequence of neurodegeneration, but it has been shown that infusion of Fe^{2+} into the substantia nigra can cause progressive Parkinsonism [184]. Further, chronic exposure to high levels of Fe^{2+} over >20 years can increase the risk of developing PD [185].

The mechanisms responsible for accumulation of Fe^{2+} include a compromised blood brain barrier, occupational exposure and age-related disruption in the body's iron storage and transport mechanisms. Excess levels of unbound iron are deleterious to surrounding cells as a result of a shift in the balance between ferric and ferrous ratios and consequential production of free radicals. The Fe^{2+} ion is known to catalyze the production of radicals according to the Fenton reaction increasing the concentrations of reactive oxygen species $(OH^-)^{\cdot}$, $(O_2^-)^{\cdot}$, and $NO^{\cdot}$. Oxidative stress induced by ROS has been associated with the death of dopaminergic pathways through the production of highly reactive and toxic chemical species and as exemplified by the production of 6-hydroxy dopamine (6-OHDA), which is neurotoxic [186, 187]. Iron released from neuromelanin has also been reported to cause mitochondrial dysfunction and to reduce proteosomal function [188]. Overall, increased iron concentrations together with concomitant decreased protective antioxidant enzyme systems in the PD brain may lead to oxidative damage to lipids, proteins and DNA.

Chelator therapies have been used clinically for peripheral indications such as aceruloplasminaemia and thalassaemia major, but their utility for neuro-degeneration has been limited due to their inability to permeate the blood brain barrier. An ideal agent would need to selectively neutralize intracellular reactive ferrous ions and turn it into a safe complex for excretion across the blood brain barrier. The continuing challenge is to incorporate a bioavailable, non-toxic, efficient, metal chelators with CNS permeable properties. Clearly, clinical side effects have to be expected from such a treatment and be managed well [189].

A number of reports document neuroprotective effects of iron chelators in PD models [190]. One such compound class is the substituted 8-hydroxy quinolones, exemplified by the lead compound Clioquinol (Fig. **24, 60**). Clioquinol (**60**) was withdrawn from its use as an intestinal disinfectant in 1970 after it became associated with sub-acute myelo-opticoneuropathy [191]. It was halted in clinical development due to impurities identified from the manufacturing process and with the available data, conflicting reports on the effect of Clioquinol (**60**) in AD patients thwart a clear indication as to whether a clinical benefit was demonstrated [192].

A second analog, believed to be an 8-hydroxyquinoline analogue (PBT-2), is being developed by Prana Biotechnology for AD and HD. In a small cohort of mild to moderate AD patients, investigators reported a significant reduction in CSF Aβ_{42} for those taking PBT-2 when compared to placebo while no effect on AD plasma biomarkers or Zn^{2+}/Cu^{2+} concentrations was observed suggesting a central effect. Cognitive improvements were reported for two executive functions [193] while Phase II trials are underway for HD [194].

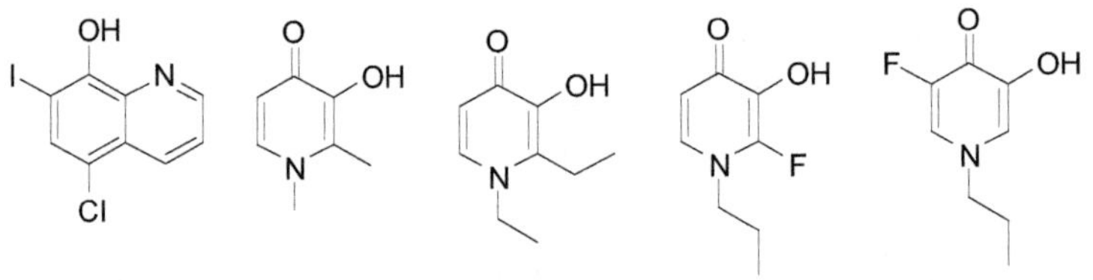

Figure 24: A selection of recently reported CNS penetrant, metal chelating, small molecules. Clioquinol (60), Deferiprone (61), CP94 (62), CP241 (63), CP242 (64).

A second class of metal-chelating compounds has been described: Deferiprone (Fig. **24, 61**) has been clinically approved as a therapy for iron overload and close analogs, designed to overcome its intrinsic metabolically liability, have been reported. Deferiprone (**61**) is currently in Phase II clinical trials for renal injury and PD [195]. It is a bivalent Fe^{2+} chelator with the potential to complex the metal ion in a 2:1 stoichiometry $[Fe(deferiprone)_2]^{2+}$. Investigators have demonstrated that though Deferiprone (**61**) can enter the brain at low concentration [196] and reduce iron content [197], it is prone to rapid metabolism in the liver [198]. In efforts to increase the brain uptake, the alkyl substituents were modified with little success of improving the metabolic clearance. Recently, the effect of pK$_a$ modulation has been reported with two close analogs (Fig. **24**, **63** and **64**) showing increased rate of brain distribution when compared to Deferiprone [199].

The concept of compounds which exhibit dual pharmacology to provide a symptomatic treatment of PD and neuroprotection has also been explored with compounds bearing a metal chelator functionality linked to a pharmacophore known to interact with dopamine receptors [200]. This has been exemplified recently with hydroxyl quinoline (Fig. **25**, **66**), which has been shown to efficiently chelate iron at pH = 7.4 and was an efficient hydroxyl radical scavenger in a deoxyribose antioxidant assay. In the reserpine-induced hypolocomotion rat model, the enantiomer (**66**) demonstrated increased locomotor activity for 6 h post treatment exhibiting a longer duration of action at 10 μmol/kg *i.p.* than the drug Ropinirole (Fig. **25**, **65**). In the 6-OHDA lesioned rat model, (**66**) showed a higher number contralateral rotations when compared to the standard Ropinrole (**65**). Both these data suggest that a proof of concept has been achieved.

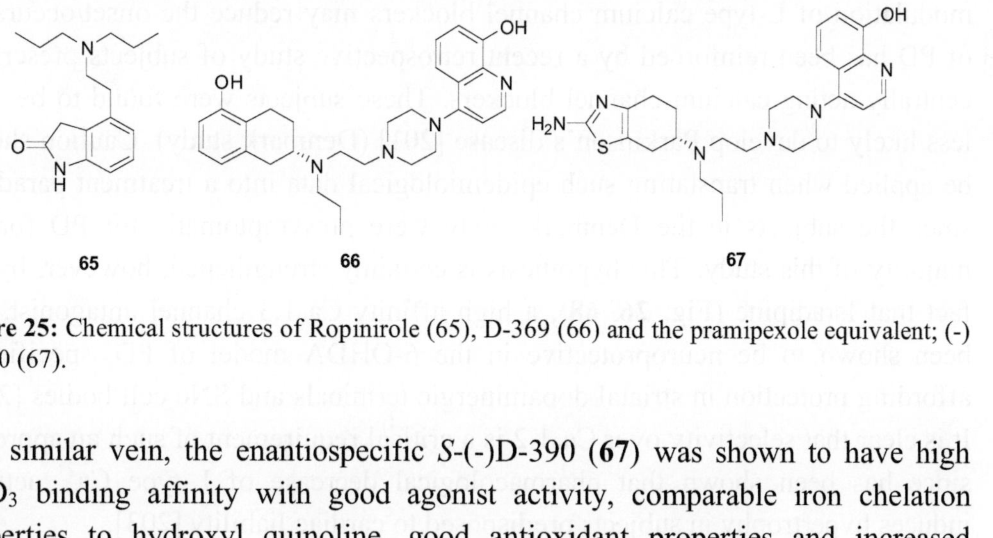

Figure 25: Chemical structures of Ropinirole (65), D-369 (66) and the pramipexole equivalent; (-) D-390 (67).

In a similar vein, the enantiospecific *S*-(-)D-390 (**67**) was shown to have high D$_2$/D$_3$ binding affinity with good agonist activity, comparable iron chelation properties to hydroxyl quinoline, good antioxidant properties and increased contralateral rotation activity in 6-OHDA unilaterally lesioned rats and demonstrated neuroprotection in the MPTP mouse model [200].

In general, the impact of metal chelation in neurodegenerative disorders is still poorly understood and it is not clear how they can be addressed therapeutically. In the case of 8-hydroxy quinolones, the alternatively substituted Clioquinol analog PBT-2 (also in clinical development for HD [194]) may help to provide this guidance in a well-designed study. With the exception of the hydroxyl pyridones, however, there is a lack of centrally available small-molecule chelators ready for clinical trials and research into symptomatic and neurodegenerative dual pharmacophores is only now starting to emerge.

TARGETING CALCIUM HOMEOSTASIS

Unlike most neurons in the brain, adult dopaminergic neurons are L-type calcium (Ca^{2+}) channel-dependent autonomous pacemakers that, in the absence of synaptic input, generate rhythmic action potentials. It has been suggested that reliance on Ca^{2+} channels might result in dopaminergic neurons being especially susceptible to aging. It is thought that perhaps the higher metabolic cost of these cells due to ATP demands for intracellular calcium handling coupled with their dependency on $Ca_v1.3$ channels (rather than the more common $Ca_v1.2$ channels) would create a potential for higher than usual levels of oxidative stress. The theory that chronic modulation of L-type calcium channel blockers may reduce the onset/occurance of PD has been reinforced by a recent retrospective study of subjects prescribed centrally acting calcium channel blockers. These subjects were found to be 27% less likely to develop Parkinson's disease [201] (Denmark study). Caution should be applied when translating such epidemiological data into a treatment paradigm since the subjects in the Denmark study were presymptomatic for PD for the majority of this study. This hypothesis is certainly strengthened, however, by the fact that Isradipine (Fig. **26**, **68**), a high affinity $Ca_v1.3$ channel antagonist, has been shown to be neuroprotective in the 6-OHDA model of PD, specifically affording protection in striatal dopaminergic terminals and SNc cell bodies [202]. It is clear that selectivity over $Ca_v1.2$ is a critical requirement of such an approach since has been shown that pharmacological decrease of L-type Ca^{2+} activity induces hypertrophy in subjects predisposed to cardiac liability [203].

The development of sub-type specific antagonists remains a challenge for this ion channel. Though sequence homology is shared, small binding pockets in $Ca_v1.2$

and 1.3 have been reported [204] suggesting scope for further research. In addition, allosteric binding sites may be explored further [205].

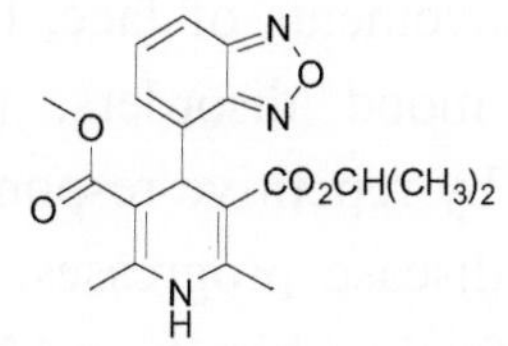

68

Isradipine

Figure 26: The chemical structure of Isradipine (68).

PD: FUTURE PERSPECTIVES

The current clinical development landscape for PD is dominated by agents addressing motor and non-motor symptoms of the disease. Of the potentially neuroprotective mechanisms discussed herein, only iron chelation (Deferiprone), antioxidants (Coenzyme Q10) and calcium modulation (Isradipine) have made it into the clinic for PD. The remaining small-molecule approaches represent early hypotheses which require small-molecule tools as well as considerably improved and predictive disease models for PD to enable robust proof-of-concept study. Not part of this review, but none the less important to mention, are the biological agents with potential for modifying disease progression. These include α-synuclein neutralizing antibodies and growth factor therapies that are currently undergoing preclinical and clinical testing. All of these approaches aiming at a therapy for PD are relevant since regardless of cause, be it genetic, environmental, epigenetic or a combination thereof, the diversity and self-propagating nature of the biological dysfunction in PD means that multiple therapies offering distinct modes of action will ultimately be required. A further significant challenge associated with developing new treatments for PD is that by the time Parkinsonism is clinically evident, ~50% of dopaminergic neurons are already lost and the benefit from neuroprotective agents could be minimal. It is therefore clear that as well as new therapies, new methods of very early diagnosis of PD will also be critical to successful treatment paradigms.

HUNTINGTON'S DISEASE

Huntington's disease (HD) is an autosomal dominant neurodegenerative disorder characterized by involuntary movements of face, hands and body. Other highly alienating symptoms include mood disorders, psychiatric disturbances and cognitive deficits characterized by repetitive response behaviors and an impaired ability to plan ahead. As the disease progresses, motor rigidity and dementia predominate, and the disease is fatal within about 15 to 20 years post onset [206].

Although the entire cortex can show significant signs of degeneration, the most striking neuropathological hallmark of HD is the atrophy of the striatum as observed in post mortem histological evaluation [207] or by non-invasive brain magnetic resonance imaging (MRI) [208] (see Fig. **2**, **C** and **D** for comparison).

The mutation which causes HD is an expansion in the number of repetitions of three nucleotides, cytosine, adenosine, and guanosine (CAG repeat) in exon 1 of the huntingtin gene. The number of CAG repeats in healthy, unaffected individuals is in the range 15-30, whilst a repeat length of over 40 is associated with HD. The CAG codon encodes the expression of the amino acid glutamine (abbreviated to Gln or the letter Q) and as such, leads to the production of a poly-glutamine (or poly-Q) amino acid stretch of protein. There are several forms of poly-glutamine disorders, including spinocerebellar ataxias and dentatorubral-pallidoluysian atrophy (DRPLA).

Broadly speaking there are two forms of HD, one with the more usual mid-life onset, the other, a more aggressive form with juvenile onset. The juvenile form is characterized by an unusually large number of CAG repeats and transmission of the gene from the father. However, both forms of the disease lead to similar neuropathological end-points including huntingtin protein aggregates or inclusion bodies, both found predominantly in the nucleus but also in the cytoplasm of neurons [209]. The role of these aggregates in disease progression is debated, *i.e.* whether they form a coping mechanism, thus neuroprotection [210] or drive the neurodegenerative pathology [211]. The production of these poly-glutamine containing HD proteins and their associated fragments, leads to several pathophysiological consequences for the affected neurons.

HD AND MITOCHONDRIAL DYSFUNCTION

Mitochondria are demonstrably key to cell and neuronal survival through their control of energy metabolism, apoptosis pathways and calcium homeostasis [212]. The role of mitochondria in neurodegenerative diseases originates in the observation of mitochondrial defects and signs of oxidative stress in tissue samples from patients with neurodegenerative conditions. In addition, mitochondria have been shown to play an active role in the events leading to neurodegeneration [213].

Striatal sensitivity to energy metabolism impairment has been well documented in animals and man, where mitochondrial poisoning is often associated with striatal degeneration in both [214]. Genetic mitochondrial defects also demonstrate a similar neurodegenerative profile.

The nuclear co-activator, PGC-1α, plays a major role in mitochondrial biogenesis and is reportedly down-regulated in HD models [215]. In agreement with these findings, mice engineered to express no PGC-1α, display abnormalities clearly linked to energy metabolism as well as striatal lesions [216]. Increasing levels of expressed PGC-1α has demonstrated neuroprotection against mutant huntingtin toxicity. Concomitantly, decreased PGC-1α levels render neurons more susceptible to mutant huntingtin induced toxicity [215b]. Other members of the PGC-1α family are known to regulate aspects of mitochondrial function and biogenesis [217]. Due to this mitochondrial impairment, these neurons also become susceptible to excitotoxicity, that is, toxicity caused by excitatory neurotransmitters, notably glutamate, due to decreased levels of glutamate transporters [218]. Under these conditions, neurons are less able to function under normal conditions and may be damaged further by normal physiological levels of activity.

THE SYNERGY BETWEEN AGGREGATION AND AUTOPHAGY

Autophagy is the process whereby cellular proteins are degraded through delivery to the lysosomal compartment. There are three major forms of this process, macro-autophagy, micro-autophagy and chaperone mediated autophagy. In the context of HD, this review will focus upon macro-autophagy, a process that will simply be referred to as autophagy.

Macro-autophagy is the major pathway for the degradation of long-lived proteins and organelles. This type of autophagy involves the delivery of protein or organelle cargo to the lysosome via a membrane vesicle transport system. In a normal cellular environment, autophagy occurs at a slow and constant rate, enabling normal homeostatic functions. In times of nutrient deprivation, this process is up-regulated to generate intracellular nutrients and energy, however, this process is also up-regulated in response to cell stress, infection, and protein aggregation [219].

One of the major regulators of the autophagic pathway is mTOR kinase, whose role is to enable normalization of the autophagic process in the presence of sufficient nutrients and growth factors. The link between autophagy and neurodegeneration has been clearly demonstrated using animal models genetically designed to have a non-functioning autophagic pathway. One study in particular showed mice with a neurally targeted deficiency in the autophagy regulated protein 5 (Atg5), develop progressive deficits in motor function accompanied by aggregates and inclusion bodies in the affected neurons [220].

In Huntington's disease, the cellular pathways which normally cope with aggregated proteins are overcome by the production of mutant huntingtin, either in its aggregated form or as soluble protein and lead to marked deficits in the autophagosomal and proteasomal systems [221]. This leads to a further build-up of aggregated proteins and a greater likelihood for inclusion of functional proteins within these aggregates. The signature presence of aggregates in several neurodegenerative diseases may indicate a failure of the autophagic pathway to deal with the stresses placed upon it, or an underlying failure of the pathways that recognize these proteins as aberrant and therefore, being in need of break down and clearance from the cell.

The formation of aggregated proteins provides many challenging pathways for disease path elucidation. Once aggregates have been formed, it becomes difficult to decipher which of the constituents are active in disease progression and which have accumulated within the aggregate in a non-specific fashion.

The role of aggregates: In HD patients, the threshold for the onset of aggregation *in vitro* has a similar correlation with the onset of disease in HD patients, leading many to identify the aggregates as the pathophysiological cause of the disease [222]. It is noteworthy that the relationship between aggregate formation and neurodegeneration is not a simple one. Neurons which possess aggregates do not necessarily degenerate. There is a greater density of aggregates in large spiny interneurons which are spared, than in medium spiny interneurons, which are selectively lost [223]. In spite of this finding, the causal link between the number of CAG repeats in HD patients and the density of aggregates is known [223, 224].

As a consequence, many identified HD targets were chosen due their ability to affect aggregation *per se*. The majority of HD drug discovery programs focus on the pathogenic mechanisms, i.e. mutant huntingtin aggregation and proteolysis, autophagic activation, mitochondrial dysfunction, excitotoxicity and transcriptional deregulation and these have been summarized recently by Kaplan and Stockwell [225].

Investigators have described a plethora of approaches to investigate the effects of mutant huntingtin and described mutant hungtingtin driven cytotoxicity or aggregation as endpoints in largely, non-mammalian cell derived phenotypic screens [226]. However, recently, there have been a number of reports which have described the identification of inhibition of mutant huntingtin aggregation in mammalian cell lines and these are summarized in Table **2**. To the best of our knowledge, no output from a mutant huntingtin aggregation screen has yet translated efficaciously into a pre-clinical mammalian model for Huntingtons disease.

However, one small molecule currently in Phase II clinical trials is claimed to reduce mutant huntingtin aggregates (Clinical trial identifier: NCT01357681). (2)-Epigallocatechin-gallate (Fig. **27**, **69**) is a green tea extract which contains a polyphenolic structure and was reported to modulate mis-folding and oligomerization of mutant huntingtin protein *in vitro*. It also reduced polyQ-mediated htt protein aggregation and cytotoxicity in a yeast model of HD [231].

Table 1: **A number of recently reported mutant huntingtin phenotypic screens carried out using mammalian cell lines**

Assay Type & Cell	Compounds Identified	Reference
Protein aggregation & cytotoxicity in PC12 Cells	2033	Titus *et al.* [227]
mHttQ145, FRET in HN10 cells	1 chemotype (HsP90 inhibitor)	Baldo *et al.* [228]
httQ72-Luc 7 Q80-cfp co-aggregation assay in HEK-293 cells	2 chemotypes	Weihl *et al.* [229]
Enhancers of autophagy, no toxicity, confirmed EGFP-HDQ74 aggregation reduction via autophagy in MEF cells	Small molecule enhancers of rapamycin	Schreiber & Rubinsztein *et al.* [230]

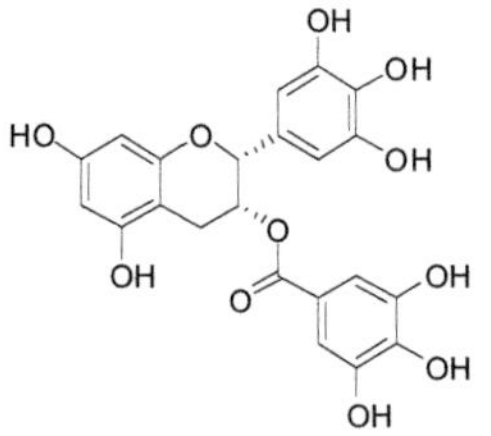

69

Figure 27: The chemical structure of (2)-Epigallocatechin Gallate (69).

One of the unanswered questions concerning aggregation and cellular toxicity is why the striatum represents a far more sensitive brain region than any other? Aggregates themselves are not directly correlated with apoptosis with the greatest density of the aggregate containing neurons found in the least affected areas [223] indicating that soluble mHtt may represent the toxic species involved in HD. Any process that affects the balance of soluble to aggregated mHtt may also affect the extent of neurotoxicity. SUMOylation is a post translational modification involving the attachment of the 11 kDa SUMO group to a protein(s). The role of this SUMO group is multiple; depending upon the protein and its location, it can stabilize proteins, alter their location and/or activity. The SUMOylation of mHtt has been shown to decrease its aggregation and increase or induce neurotoxicity

[232]. In the striatum, increased SUMOylation has been purported to be due to expression of Rhes (Ras homologue enriched in the striatum), with over-expression leading to an increase in mHtt SUMOylation. This was abolished by SUMO-1 knock-down through RNAi methods and was demonstrated to be through its role as an E3 ligase [233] therefore binding a small molecule to Rhes or inhibition of the Rhes-mHtt protein complex, might be an appropriate therapeutic intervention.

Gain of function/loss of function: To address its physiological function, huntingtin knock-out animals have been attempted. Full knock-out is embryonically lethal due to extreme neuronal apoptosis [234] whilst heterozygous knock-out animals have clear cognitive deficits due to decreased neuronal numbers in the sub-thalamic nucleus of the basal ganglia. Crucially, altering these studies such that inactivation of huntingtin was postnatal, led to a progressive degeneration of neurons [235] indicating that the role of huntingtin may be linked to ongoing maintenance and activity of neurons, rather than their genesis. The neuro-protective effect of huntingtin may in part be due to its inhibition of caspase-3 activity [236] as well as the activation of the serine/threonine kinase activated pro-survival pathways [237]. Taken together, this pattern of behavior for huntingtin indicates that pathogenesis is due to both decreased physiological function as well as gain of function in mutant huntingtin. Neuronal survival pathways are clearly important targets for the prevention of neurodegeneration, however, understanding the mechanisms which lead to the selective loss of some neurons and sparing of others may present a more tractable approach.

TRANSCRIPTIONAL REGULATION

Several genes involved with key signaling pathways have been shown to be down-regulated in animal models of HD [238]. These changes in gene expression pattern are comparable to human HD post mortem tissue [239] illustrating that transcriptional deregulation may be an important factor in HD progression [240]. Mutant huntingtin has been shown to induce, either directly or indirectly, changes in protein-protein interactions involving transcription factors or co-factors, including CBP, TAF4, Sp1 and p53, leading to deleterious downstream effects [241]. It is believed that poly-glutamine fragments enter the nucleus and affect transcription by sequestering transcription factors [242] as well as through the disruption of the acetylation status of histones and other proteins [243].

In disease models of HD, there is a demonstrable decrease in brain-derived neurotrophic factor (BDNF) mRNA and protein levels [244], which has been shown to track with disease progression. The role of BDNF is also indicated in studies demonstrating neurodegenerative rescue with BDNF delivery [244, 245]. BDNF and other neuronal proteins may have aberrant transcription, most likely due to altered transcription factor levels. Alternatively, expression profiles may be altered through an epigenetic route, *i.e.* through DNA modification with methyl or acetyl groups which alter the accessibility of distinct DNA stretches for transcription. Indeed, such a process has been implicated in the R6/2 mouse model of HD by the work of the Bates group, who demonstrated a phenotypic rescue in these mice through the knock-down of histone deacetylase-4 (HDAC4) [246].

Ampakines offer one such approach. Indeed CX-929, whose chemical structure is undisclosed was shown to increase BDNF levels in the neocortex and striatium of R6/2 mice as well as phenotypic improvements such as increased body weight, and rotarod control at 5 mg/kg IP, *p.o* over a 16 week period. However, no increased lifespan was observed [247]. RP-103 (Fig. **28**, **70**); a delayed release formulation of Cysteamaine is an FDA approved therapy for nephropathic cystinosis and was shown to increase BDNF brain levels in R6/2, and BDNF / mHtt heterozygous mice at as well as increase levels of BDNF in mouse and primate serum [248] while Uridine triacetate; PN-401 (Fig. **28**, **71**) also showed statistically significant improved rotarod performance and increased lifespan in R6/2 and N171-82Q mice [249]. To our knowledge, PN-401 has not been advanced for clinical evaluation. Phosphodiesterase inhibition leading to increased cAMP and CREB levels is also a recognized exploratory therapeutic intervention strategy. Inhibition of the striatal specific PDE10 in particular, has been shown to reduce the cortico-striatal pathology in R6/2 mice with TP-10 (Fig. **28**, **72**) [250]. Along with increased striatal BDNF and p-CREB levels, TP-10 treated animals showed improved rotarod performance, increased striatal area and volumes and increased lifespan. Though interesting, to the best of our knowledge, no PDE10 inhibitors are currently under clinical development for HD.

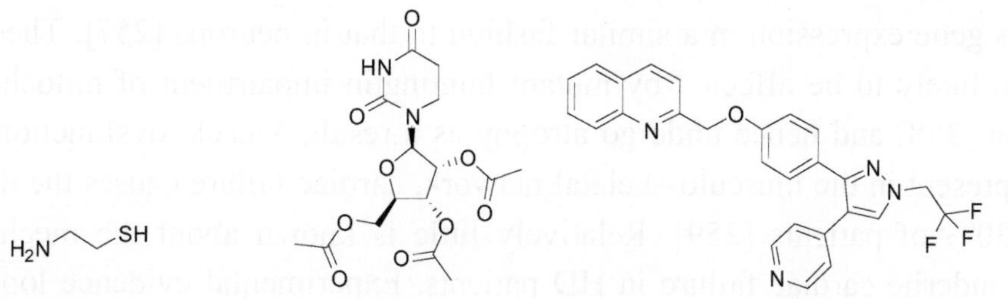

Figure 28: The chemical structure for the active ingredient of RP-103; Cysteamine (70), PN-401; Uridine triacetate (71) and TP-10 (72).

OTHER BIOLOGICAL PATHWAY DEFICITS

The loss of circadian rhythm is a common observation in HD, AD and PD [251]. Although the basal ganglia is the best characterized region for neuronal degeneration in HD, there is also a significant amount of degeneration which occurs in the hypothalamus [208b]. Situated in the anterior portion of the hypothalamus is a cone shaped organ called the suprachiasmatic nucleus (SCN), which contains a circadian pacemaker regulating sleep-wake cycles [252]. In the R6/2 HD model mouse, the expression of two key circadian genes, mPer2 and mBmal1, are reduced compared to wild type animals. This loss of gene expression in the SCN may help to explain the wake-sleep cycle dysfunction observed in HD patients.

In addition to these classic, neurological aspects of the disease, HD is further complicated by other features such as cachexia which is not associated with changes in neurological function [253]. Weight loss is one of the most common abnormalities which affect almost all HD patients. It is progressive, appearing as a minor loss in pre-symptomatic HD and becoming profound in advanced stages [254]. Its underlying cause is unknown, however, it is unlikely to be caused by hyperactivity or anorexia [254a, 255].

Muscle atrophy is another clear hallmark of HD [256] although the mediation mechanisms are unclear. In a similar fashion to that observed within neurons, mutant huntingtin is known to cause aggregates in muscle cells (myocytes) and disrupts gene expression in a similar fashion to that in neurons [257]. These cells are also likely to be affected by mutant huntingtin impairment of mitochondrial function [258] and hence undergo atrophy as a result. Muscle dysfunction is not solely present in the musculo-skeletal network, cardiac failure causes the death of about 30% of patients [259]. Relatively little is known about the mechanisms which underlie cardiac failure in HD patients. Experimental evidence looking at cardiomyocyte specific mutant huntingtin expression (with an 83 CAG repeat expansion) demonstrated aggregate formation and cardiomyocyte necrosis in wild type mice [260]. These data indicate that cardiac failure in HD may be linked to organ specific expression of mutant huntingtin rather than centrally mediated effects of HD.

PERSPECTIVES FOR HD

The hard reality for Huntington's disease is that there is currently no disease specific drug therapy, and although there are several front line treatments for symptoms, these do not affect progression of the disease. In fact, several basic questions remain with regards to disease initiation, its progression and targeting the dysfunctional pathways. The main drivers of dysfunction are still unknown. Loss of autophagic processing of aggregates or mis-folded proteins, altered transcriptional states or reduced mitochondrial function all lead to a loss of protective function. The extent of the functional loss is large but takes several decades to be classically symptomatic, which must lead us to question how effective a treatment will be at a symptomatic stage. Therefore, how early a treatment needs to be administered and whether restoration of any lost function is a viable must be explored. Ultimately, however, prevention of disease progression may be the best end point we can reach.

Therefore, it remains for investigators to continue to evaluate the role of individual degenerative pathways in the hope that through their combined knowledge a link will lead one day to a therapeutic regime for this devastating disease.

CONCLUSIVE REMARKS

Normal ageing involves a natural process of neurodegeneration, the rate of which is governed by genetic pre-disposition as well as environmental stress factors. In the case of PD and AD, the post-translational modifications which occur due to these factors lead to up-regulation of early-phase degenerative processes, most of which do not translate to symptomatic outcomes for several decades. In terms of disease treatment, early diagnosis is highly desirable to establish the most effective therapy. Such a strategy will require screening of the population at risk and identification of early clinically relevant biomarkers. Ideally, the next stage will be to personalize treatment for a given patient according to the biomarker profile exhibited and adapt it during the course of the disease.

Though some symptomatic treatments for AD, PD and HD are available, our understanding of their pathogenesis is incomplete and no disease-modifying therapies are currently available. For instance, of the three main hypotheses which underpin much of AD research, namely β-amyloid, p-tau and ROS driven neurodegeneration, none have delivered clinically viable therapeutics to date. In the case of Alzheimer's disease, having evaluated GSIs, an iterative learning is emerging. It is apparent that full inhibition of γ-secretase actually increases pathology and researchers await the outcome of the next generation of compounds (GSMs) generated, to test the β-amyloid hypothesis. Concomitantly BACE1 inhibitors also offer an alternative angle to this hypothesis and clinical data from on-going trials are eagerly anticipated. Until these data are assessed, the relevance of treating AD patients via this mechanistic hypothesis remains an open question. Disappointingly, the failure of GSK3β inhibitors and metal chelators to show clinical efficacy, has compounded progress towards therapies.

Perhaps targeting a combination of mechanisms may lead to improved treatment strategies. This could either be achieved by administration of several drugs or compounds with well-designed poly-pharmacological properties. Phenotypic screening might be the best approach for identification of such tool molecules.

The advent of efficient DNA sequencing tools coupled with systems analysis has recently enabled the understanding of patient's genetic backgrounds and resulted in new hypotheses and molecular targets as exemplified by Parkin or LRRK2 in PD. Tool molecules are required to study and understand the effect of modulation of such targets as well as the pathways in which they operate. On the other hand, the search for inhibitors of biochemical targets like β - or γ-secretase which combine biochemical activity with suitable CNS exposure has been long though the first compounds are now being clinically evaluated. However, initial results with γ-secretase inhibitors were not encouraging and it remains to be tested whether modulation instead of inhibition will be a safer and more efficacious approach.

In the case of HD, because the disease is caused by a genetic pre-disposition, it reaches a point, when the effect of alleviating the mutant huntingtin polyQ induced stress becomes overbearing in striatal neurons and they rapidly descent into a neurodegenerative state. Indeed there is correlation between length of polyQ and the age of onset. Could it be that the longer polyQ repeat induces enhanced neuronal stress which requires the cell machinery to alleviate this, essentially leading to "corrective burn-out"? It therefore seems reasonable that repression of the mHtt protein transcription should be a primary avenue for exploration.

As our understanding of new branches of biological systems, from epigenetics to transcription factors increases, the shortfall in our knowledge of such progressive diseases becomes increasingly obvious and gives cause to question our understanding of so called "normal" systems. A striking observation concerning neurodegenerative diseases is that all share some resemblance to accelerated ageing, albeit more closely linked to the central nervous system than true ageing, and are for the most part dependent upon ageing as a factor for their prevalence.

Whether studying these diseases will give a greater understanding of the natural, progressive deterioration of our biological systems, remains open for debate. What is clear, however, is that within an ageing population with increasing lifespan, we can clearly see how prevalent these diseases are in our societies. The prevalence of these neurodegenerative diseases may be described more appropriately as a spectrum, rather than a clear division, between affected and unaffected individuals which only becomes apparent as our life spans increase.

ACKNOWLEDGEMENTS

The authors would like to acknowledge and extend a special thanks to Dr. Chris Richardson for providing pictures of the crystal structures, Dr. George McAllister, and Dr. Christopher Newton for their reviews and constructive comments, the Alzheimer's society and Prof. Nicholas Fox from the National Hospital for Neurology and Neurosurgery.

CONFLICT OF INTEREST

The authors confirm that this chapter contents have no conflict of interest..

ABBREVIATIONS

Aβ	=	Amyloid-β or β-amyloid
ADME	=	Absorption, Distribution, Metabolism, Excretion
AD	=	Alzheimer's disease
BACE	=	β-amyloid precursor protein cleaving enzyme
b.i.d.	=	bis in die (twice daily)
CLogP	=	Calculated logP
CSF	=	Cerebrospinal Fluid
HD	=	Huntington's disease
i.v.	=	Intravenous
MAD	=	multiple ascending dose
MRI	=	Magnetic Resonance Imaging
NMR	=	Nuclear Magnetic Resonance
PD	=	Parkinson's disease
ROS	=	reactive oxygen species
p.o.	=	per os (oral)

q.d.   =   quaque die (once daily)

SAD   =   single ascending dose

SAR   =   structure activity relationship

REFERENCES

[1]   Barkhof, F. F., N.C. Bastos-Leite, A.J., Scheltens, P. , *Neuroimaging in dementia.* Springer-Verlag: Berlin Heidelberg, 2011.

[2]   (a) Levine, B.; Kroemer, G., Autophagy in the pathogenesis of disease. *Cell* **2008**, *132* (1), 27-42; (b) Rubinsztein, D. C.; Codogno, P.; Levine, B., Autophagy modulation as a potential therapeutic target for diverse diseases. *Nature reviews. Drug discovery* **2012**, *11* (9), 709-30.

[3]   (a) Gandhi, S.; Abramov, A. Y., Mechanism of oxidative stress in neurodegeneration. *Oxidative medicine and cellular longevity* **2012**, *2012*, 428010; (b) Jimenez-Del-Rio, M.; Velez-Pardo, C., The bad, the good, and the ugly about oxidative stress. *Oxidative medicine and cellular longevity* **2012**, *2012*, 163913; (c) Rivera-Mancia, S.; Perez-Neri, I.; Rios, C.; Tristan-Lopez, L.; Rivera-Espinosa, L.; Montes, S., The transition metals copper and iron in neurodegenerative diseases. *Chemico-biological interactions* **2010**, *186* (2), 184-99; (d) Dexter, D. T.; Carayon, A.; Javoy-Agid, F.; Agid, Y.; Wells, F. R.; Daniel, S. E.; Lees, A. J.; Jenner, P.; Marsden, C. D., Alterations in the levels of iron, ferritin and other trace metals in Parkinson's disease and other neurodegenerative diseases affecting the basal ganglia. *Brain : a journal of neurology* **1991**, *114 (Pt 4)*, 1953-75.

[4]   (a) Alzheimer, A., Über eigenartige Krankheitsfälle des späteren Alters. *Zeitschrift für die gesamte Neurologie und Psychiatrie* **1911**, *4 Originalien*, 356-385; (b) Glenner, G. G.; Wong, C. W., Alzheimer's disease: initial report of the purification and characterization of a novel cerebrovascular amyloid protein. *Biochemical and biophysical research communications* **1984**, *120* (3), 885-90.

[5]   Kuhn, P. H.; Wang, H.; Dislich, B.; Colombo, A.; Zeitschel, U.; Ellwart, J. W.; Kremmer, E.; Rossner, S.; Lichtenthaler, S. F., ADAM10 is the physiologically relevant, constitutive alpha-secretase of the amyloid precursor protein in primary neurons. *EMBO J* **2010**, *29* (17), 3020-32.

[6]   Cai, H.; Wang, Y.; McCarthy, D.; Wen, H.; Borchelt, D. R.; Price, D. L.; Wong, P. C., BACE1 is the major beta-secretase for generation of Abeta peptides by neurons. *Nature neuroscience* **2001**, *4* (3), 233-4.

[7]   Basi, G.; Frigon, N.; Barbour, R.; Doan, T.; Gordon, G.; McConlogue, L.; Sinha, S.; Zeller, M., Antagonistic effects of beta-site amyloid precursor protein-cleaving enzymes 1 and 2 on beta-amyloid peptide production in cells. *J Biol Chem* **2003**, *278* (34), 31512-20.

[8]   De Strooper, B., Aph-1, Pen-2, and Nicastrin with Presenilin generate an active gamma-Secretase complex. *Neuron* **2003**, *38* (1), 9-12.

[9]   (a) Karran, E.; Mercken, M.; Strooper, B. D., The amyloid cascade hypothesis for Alzheimer's disease: an appraisal for the development of therapeutics. *Nat Rev Drug Discov*

2011, *10* (9), 698-712; (b) Citron, M., Alzheimer's disease: strategies for disease modification. *Nat Rev Drug Discov* **2010**, *9* (5), 387-98.

[10]   Suheung T.T, C. X. D., *et al*. An increase in the percentage of long amyloid beta protein secreted by faqmilial amyloid beta protein precursor (beta APP117) mutants. *Science* **1994**, 1336-1340.

[11]   (a) Citron M, O. T., Hass C, *et al*. Mutation of the beta-amyloid precursor protein in familial Alzheimers disease increases beta protein production. *Nature* **1992**, 672-674; (b) Borchelt D. R, T. G., Eckman C.B, *et al*. Familial Alzheimers disease linked Presenlin 1 varients elevate Abeta 1-42/1-40 ratio in-vitro and *in vivo*. *Neuron* **1996**, 1005-1013; (c) Scheuner D, E. C., Jensen M, *et al*. Secreted amyloid beta protein similar to that in the senile plaques in Alzheimers disease is increased *in vivo* by the Presenlin 1 and 2 APP mutations. *Nat. Med.* **1996**, 864-870.

[12]   Jonsson, T.; Atwal, J. K.; Steinberg, S.; Snaedal, J.; Jonsson, P. V.; Bjornsson, S.; Stefansson, H.; Sulem, P.; Gudbjartsson, D.; Maloney, J.; Hoyte, K.; Gustafson, A.; Liu, Y.; Lu, Y.; Bhangale, T.; Graham, R. R.; Huttenlocher, J.; Bjornsdottir, G.; Andreassen, O. A.; Jonsson, E. G.; Palotie, A.; Behrens, T. W.; Magnusson, O. T.; Kong, A.; Thorsteinsdottir, U.; Watts, R. J.; Stefansson, K., A mutation in APP protects against Alzheimer's disease and age-related cognitive decline. *Nature* **2012**, *488* (7409), 96-9.

[13]   Lovell, M. A.; Robertson, J. D.; Buchholz, B. A.; Xie, C.; Markesbery, W. R., Use of bomb pulse carbon-14 to age senile plaques and neurofibrillary tangles in Alzheimer's disease. *Neurobiology of aging* **2002**, *23* (2), 179-86.

[14]   Braak, H.; Braak, E., Neuropathological stageing of Alzheimer-related changes. *Acta Neuropathol (Berl)* **1991**, *82* (4), 239-259.

[15]   Lin, X.; Koelsch, G.; Wu, S.; Downs, D.; Dashti, A.; Tang, J., Human aspartic protease memapsin 2 cleaves the beta-secretase site of beta-amyloid precursor protein. *Proceedings of the National Academy of Sciences of the United States of America* **2000**, *97* (4), 1456-60.

[16]   Sinha, S.; Lieberburg, I., Cellular mechanisms of beta-amyloid production and secretion. *Proceedings of the National Academy of Sciences of the United States of America* **1999**, *96* (20), 11049-53.

[17]   (a) Huse, J. T.; Pijak, D. S.; Leslie, G. J.; Lee, V. M.; Doms, R. W., Maturation and endosomal targeting of beta-site amyloid precursor protein-cleaving enzyme. The Alzheimer's disease beta-secretase. *The Journal of biological chemistry* **2000**, *275* (43), 33729-37; (b) Capell, A.; Steiner, H.; Willem, M.; Kaiser, H.; Meyer, C.; Walter, J.; Lammich, S.; Multhaup, G.; Haass, C., Maturation and pro-peptide cleavage of beta-secretase. *The Journal of biological chemistry* **2000**, *275* (40), 30849-54.

[18]   (a) Roberds S.L, A. J., Basi G, *et al*. BACe knockout mice are healthy despite lacking the primary B-secretase activity in brain: implications for Alzheimers disease therapeutics. *Human Molecular Genetics* **2001**, 1317-1324; (b) Luo Y, B. B., Kahn S, *et al*. Mice deficient in BACE1, the Alzheimers betasecretase, have normal phenotype and abolished beta-amyloid generation *Nat. Neurosci.* **2001**, 231-232; (c) Lou Y, B. B., Damore M.A, *et al*. BACE1 (beta secretase) knockout mice do not aquire compensatory gene expression changes or develop neural lesions over time. *Neurobiol Dis.* **2003**, 81-88.

[19]   (a) Deominguez D, T. J., Hartmann D, *et al*. Phenotypic and biochemical analysis of BACE1- and BACE2- deficient mice. *J. Biol Chem* **2005**, *280*, 30797-30806; (b) Hitt B, J. T., Chetkovich D, *et al*. Bace1(-/-) mice exhibit seizure activity that does not correlate with sodium channel level or axonal localization. *Molo. Neurodegener* **2010**, *5*, 31; (c) Hu X, Z.

X., He W, *et al.* BACE1 deficiency causes altered neuronal activity and neurodegeneration. *J. Neurosci.* **2010**, (30), 8819-8829.

[20] (a) Peng K.A, M. E., Levtivirus-Expressed siRNA Vectors Against Alzheimers Disease. *Methods in Molecular Biology* **2010**, *614*, 215-224; (b) Singer O, M. R., Rockenstein E, *et al.* Targetting BACE1 with siRNAs ameliorates Alzheimer disease neuropathology in a transgenic model. *Nat. Neurosci* **2005**, *8*, 1343-1349; (c) Fukumoto H, T. H., Tarui N, *et al.* A non-competitive BACE1 inhibitor TAK-070 ameliorates AB pathology and behavioural deficits in a mouse model of Alzheimers disease. *J. Neurosci* **2010**, *30*, 11157-11166.

[21] (a) Lin X. Human aspartic protease memapsin 2 cleaves the beta-secretase site of beta-amyloid precursor protein. *Proc. Natl. Acad. Sci. U.S.A.* **2000**, 1456; (b) Venugopal, C.; Demos, C. M.; Rao, K. S.; Pappolla, M. A.; Sambamurti, K., Beta-secretase: structure, function, and evolution. *CNS & neurological disorders drug targets* **2008**, *7* (3), 278-94.

[22] Tang, J.; James, M. N.; Hsu, I. N.; Jenkins, J. A.; Blundell, T. L., Structural evidence for gene duplication in the evolution of the acid proteases. *Nature* **1978**, *271* (5646), 618-21.

[23] (a) Tang J, H. L., Koelsch G, *et al.* Structure of the protease domain of Memapsin 2 (B-secretase) complexed with inhibitor. *Science* **2000**, 150-153; (b) Shimizu, H.; Tosaki, A.; Kaneko, K.; Hisano, T.; Sakurai, T.; Nukina, N., Crystal structure of an active form of BACE1, an enzyme responsible for amyloid beta protein production. *Molecular and cellular biology* **2008**, *28* (11), 3663-71.

[24] (a) Vassar, R.; Bennett, B. D.; Babu-Khan, S.; Kahn, S.; Mendiaz, E. A.; Denis, P.; Teplow, D. B.; Ross, S.; Amarante, P.; Loeloff, R.; Luo, Y.; Fisher, S.; Fuller, J.; Edenson, S.; Lile, J.; Jarosinski, M. A.; Biere, A. L.; Curran, E.; Burgess, T.; Louis, J. C.; Collins, F.; Treanor, J.; Rogers, G.; Citron, M., Beta-secretase cleavage of Alzheimer's amyloid precursor protein by the transmembrane aspartic protease BACE. *Science* **1999**, *286* (5440), 735-41; (b) Yan, R.; Bienkowski, M. J.; Shuck, M. E.; Miao, H.; Tory, M. C.; Pauley, A. M.; Brashier, J. R.; Stratman, N. C.; Mathews, W. R.; Buhl, A. E.; Carter, D. B.; Tomasselli, A. G.; Parodi, L. A.; Heinrikson, R. L.; Gurney, M. E., Membrane-anchored aspartyl protease with Alzheimer's disease beta-secretase activity. *Nature* **1999**, *402* (6761), 533-7; (c) Sinha, S.; Anderson, J. P.; Barbour, R.; Basi, G. S.; Caccavello, R.; Davis, D.; Doan, M.; Dovey, H. F.; Frigon, N.; Hong, J.; Jacobson-Croak, K.; Jewett, N.; Keim, P.; Knops, J.; Lieberburg, I.; Power, M.; Tan, H.; Tatsuno, G.; Tung, J.; Schenk, D.; Seubert, P.; Suomensaari, S. M.; Wang, S.; Walker, D.; Zhao, J.; McConlogue, L.; John, V., Purification and cloning of amyloid precursor protein beta-secretase from human brain. *Nature* **1999**, *402* (6761), 537-40; (d) Hussain, I.; Powell, D.; Howlett, D. R.; Tew, D. G.; Meek, T. D.; Chapman, C.; Gloger, I. S.; Murphy, K. E.; Southan, C. D.; Ryan, D. M.; Smith, T. S.; Simmons, D. L.; Walsh, F. S.; Dingwall, C.; Christie, G., Identification of a novel aspartic protease (Asp 2) as beta-secretase. *Molecular and cellular neurosciences* **1999**, *14* (6), 419-27.

[25] (a) Tang J, G. A. K., Bilcer G, *et al.* Structure Based Design: Potent Inhibitors of Human Brain Memapsin 2 (B-secretase). *J. Med. Chem.* **2001**, *44*, 2865-2868; (b) Tang J, G. A. K., Shin D, *et al.* Design of potent Inhibitors for Human Brain Memapsin 2 (B-secretase). *J. Am. Chem. Soc.* **2000**, *122*, 3522-3523.

[26] Varghese J, A. J. P., Tung J.S *et al.* Design of substrate based inhibitors of Human B-secretase. *J. Med. Chem.* **2002**, *45*, 259-262.

[27] Tang J, G. A. K., Koelsch G, *et al.* Structure based design of cycloamide-urethane derived novel inhibitors of human brain memapsin 2 (B-secretase). *Bioorganic & Medicinal Chemistry Letters* **2005**, *15*, 15-20.

[28] (a) Kiso Y, S. K., Kimura T, *et al.* Design and synthesis of potent B-secretase (BACE1) inhibitors with P1 carboxylic acid bioisoteres. *Biorg. Med. Chem. Lett.* **2006**, *16*, 2380-2386; (b) kiso Y, A. M., Hattori C, *et al.* The novel B-secretase inhibitor KMI-429 reduces amyloid peptide production in amyloid precursor protein transgenic and wild type mice. *Journal of Neurochemistry* **2006**, *96*, 533-540.

[29] Ishuira S, K. Y., Asai M, *et al.* The novel B-secretase inhibitor KMI-429 reduces amyloid B peptide production in amyloid precursor protein transgenic and wild type mice. *Journal of neurochemistry* **2006**, *96*, 533-540.

[30] Pajouhesh H, L. G. R., Medicinal Chemical Properties of Successful Central Nervous System Drugs. *Jnl. Am. Soc. Exp. neuroTher.* **2005**, *2*, 541-553.

[31] (a) Pajoushesh H, L. G. R., Medicinal Chemical Properties of Successful Central Nervous System Drugs. *The Journal of the American Society for Experimental Neurotherapeutics* **2005**, *2*, 541-553; (b) Lipinski C.A, L. F., Dominy B.W, *et al.* Experimental and computational approaches to estimate solubility and permeability in drug discovery and development settings. *ADV. Drug Deliv. Rev.,* **1997**, *23*, 3-25; (c) Wenlock M.C, A. R. P., Leeson P.D, *et al.* A comparison of physiochemical property profiles of development and marketed oral drugs. *J. Med. Chem.* **2003**, *46*, 1250-1256.

[32] Mahar Doan K.M, H. J. E., Webster L.O, *et al.* Passive permeability and P-glycoprotein mediated efflux differentate central nervous system (CNS) and non CNS marketed drugs. *J. Pharmacol. Exp. Ther* **2002**, *303*, 1029-1037.

[33] (a) Evin G, L. G., Wilkins S, BACE Inhibitors as potential Drugs for the treatment of Alzheimers Disease: Focus on Bioactivity. *Recent patents on CNS Drug Discovery* **2011**, *6*, 91-106; (b) Kiso Y, H. Y., Recent progress in the drug discovery of non-peptidic BACE1 inhibitors. *Expert Opin. Drug Discov.* **2009**, 391-416.

[34] (a) Horne D.B, M. H., Bartgerger M.D *et al.* Structure guided P1' modifications of HEA derived B-secretase inhibitors for the treatment of Alzheimers disease. *Bioorg. Medchem. Lett.* **2012**, *22*, 3607-3611; (b) Weiss M. M, W. T., Babu-Khan S *et al.* Design and preparation of a Potent Series of Hydroxyethyloamine Containing B-secretase Inhibitors That Demonstrate Robust Reduction of Central B-amyloid. *J. Med. Chem.* **2012**, *55*(21), 9009-24.

[35] Rueeger H, D. S., Staufenbiel M *et al.* Discovery of cyclic sulfone hydroxyethylamines as potent and selective B-site APP-cleaving Enzyme (BACE1) inhibitors Structure based design and *in vivo* Reduction of Amyloid B-peptides. *J. Med. Chem.* **2012**, *55*, 3364-3386.

[36] (a) Leach, A. R.; Hann, M. M.; Burrows, J. N.; Griffen, E. J., Fragment screening: an introduction. *Molecular bioSystems* **2006**, *2* (9), 430-46; (b) Foloppe, N., The benefits of constructing leads from fragment hits. *Future medicinal chemistry* **2011**, *3* (9), 1111-5; (c) Rees, D. C.; Congreve, M.; Murray, C. W.; Carr, R., Fragment-based lead discovery. *Nature reviews. Drug discovery* **2004**, *3* (8), 660-72; (d) Carr, R. A.; Congreve, M.; Murray, C. W.; Rees, D. C., Fragment-based lead discovery: leads by design. *Drug discovery today* **2005**, *10* (14), 987-92; (e) Erlanson, D. A., Fragment-based lead discovery: a chemical update. *Current opinion in biotechnology* **2006**, *17* (6), 643-52; (f) Murray, C. W.; Verdonk, M. L.; Rees, D. C., Experiences in fragment-based drug discovery. *Trends in pharmacological sciences* **2012**, *33* (5), 224-32; (g) Hann, M. M.; Keseru, G. M., Finding

the sweet spot: the role of nature and nurture in medicinal chemistry. *Nature reviews. Drug discovery* **2012**, *11* (5), 355-65.

[37]　Edwards P.D, Geschwindner S, *et al.* Application of Fragment Based Lead generation to the discovery of Novel, cyclic amidine B-secretase inhibitors with nanomolar potency, cellular activity and high ligand efficiency. *J. Med. Chem.* **2007**, *50*, 5912-5925.

[38]　Folmer R.H.A, Edwards P.D, *et al.* Discovery of a novel warhead against B-secretase through Fragment BAsed Lead Generation. *J. Med. Chem* **2007**, *50*, 5903-5911.

[39]　Murray C.W, Congreve M, *et al.* Application of fragment screening by X-Ray crystallography to B-secretase. *J. Med. Chem.* **2007**, *50*, 1116-1123.

[40]　Hartshorn M.J, Sleasby A, *et al.* Fragment based lead discovery using X-ray crystalography. *J. Med. Chem.* **2005**, *48*, 403-413.

[41]　Zhu Z, Ye Y, *et al.* Discovery of cyclic guanidines as highly potent and selective B-site amyloid cleaving enzyme (BACE) inhibitors: Part I-Inhibitor design and validation. *J. Med. Chem.* **2010**, *53*, 951-965.

[42]　May P.C, Lowe S.L *et al.* Robust Central Reduction of Amyloid-B in humans with an orally available, Non-Peptidic B-secretase Inhibitor. *Jnl. Neurosci.* **2011**, *31* (46), 16507-16516.

[43]　Reuters, T., Source: Thomson Reuters; Search parameter: BACE inhibitors. *Thomsom Reuters Pharma* **2012**.

[44]　Jeppsson, F.; Eketjall, S.; Janson, J.; Karlstrom, S.; Gustavsson, S.; Olsson, L. L.; Radesater, A. C.; Ploeger, B.; Cebers, G.; Kolmodin, K.; Swahn, B. M.; von Berg, S.; Bueters, T.; Falting, J., Discovery of AZD3839, a potent and selective BACE1 clinical candidate for the treatment of Alzheimers Disease. *The Journal of biological chemistry* **2012**.

[45]　Audia, J. E, Mergott, D. J, SHI, C.E, Vaught, G. M, Watson, B. M, Winneroski, L. L, "Bace Inhibitors", WO/2011/005738. 2011.

[46]　Boggs, L. M., P. Brie, R. Calligaro, D. Citron, M. Day, T. Lin, S. Lindstrom, T. Mergott, D. Monk, S. Sanchez-Felix, M. V. Sheehan, S. Vaught, G. Yang, Z. Audia, J. , Preclinical Characterization of LY2886721: A BACE1 inhibitor in clinical development for early Alzheimers disease. In *Alzheimers Association International Conference*, Vancouver, 2012; pp 01-06-03.

[47]　Dean, R. A. L., S. Nakano, M. Monk, S. Willis, B.A. Gonzales, C. Mergott, D. Leslie, D. May, P. James, A. Gevorkyan H. Jhee S. ereshefsky, L. Citron, M. , BACE inhibitor LY2886721 dafety and central and peripheral PK and PD in healthy subjects (HSS). In *Alzheimers Association International Conference*, Vancouver, 2012; pp P3-363.

[48]　Kreft, A. F.; Martone, R.; Porte, A., Recent advances in the identification of gamma-secretase inhibitors to clinically test the Abeta oligomer hypothesis of Alzheimer's disease. *Journal of medicinal chemistry* **2009**, *52* (20), 6169-88.

[49]　Wolfe, M. S., The gamma-secretase complex: membrane-embedded proteolytic ensemble. *Biochemistry* **2006**, *45* (26), 7931-9.

[50]　Lleo, A.; Saura, C. A., gamma-secretase substrates and their implications for drug development in Alzheimer's disease. *Current topics in medicinal chemistry* **2011**, *11* (12), 1513-27.

[51]　De Strooper, B.; Annaert, W.; Cupers, P.; Saftig, P.; Craessaerts, K.; Mumm, J. S.; Schroeter, E. H.; Schrijvers, V.; Wolfe, M. S.; Ray, W. J.; Goate, A.; Kopan, R., A presenilin-1-dependent gamma-secretase-like protease mediates release of Notch intracellular domain. *Nature* **1999**, *398* (6727), 518-22.

[52]   (a) Sardi, S. P.; Murtie, J.; Koirala, S.; Patten, B. A.; Corfas, G., Presenilin-dependent ErbB4 nuclear signaling regulates the timing of astrogenesis in the developing brain. *Cell* **2006**, *127* (1), 185-97; (b) Marambaud, P.; Wen, P. H.; Dutt, A.; Shioi, J.; Takashima, A.; Siman, R.; Robakis, N. K., A CBP binding transcriptional repressor produced by the PS1/epsilon-cleavage of N-cadherin is inhibited by PS1 FAD mutations. *Cell* **2003**, *114* (5), 635-45.

[53]   Lleo, A., Activity of gamma-secretase on substrates other than APP. *Current topics in medicinal chemistry* **2008**, *8* (1), 9-16.

[54]   Hemming, M. L.; Elias, J. E.; Gygi, S. P.; Selkoe, D. J., Proteomic profiling of gamma-secretase substrates and mapping of substrate requirements. *PLoS biology* **2008**, *6* (10), e257.

[55]   (a) Bischoff, F.; Berthelot, D.; De Cleyn, M.; Macdonald, G.; Minne, G.; Oehlrich, D.; Pieters, S.; Surkyn, M.; Trabanco, A. A.; Tresadern, G.; Van Brandt, S.; Velter, I.; Zaja, M.; Borghys, H.; Masungi, C.; Mercken, M.; Gijsen, H. J., Design and Synthesis of a Novel Series of Bicyclic Heterocycles As Potent gamma-Secretase Modulators. *Journal of medicinal chemistry* **2012**; (b) Wolfe, M. S., gamma-Secretase inhibitors and modulators for Alzheimer's disease. *Journal of neurochemistry* **2012**, *120 Suppl 1*, 89-98.

[56]   Lanz, T. A.; Karmilowicz, M. J.; Wood, K. M.; Pozdnyakov, N.; Du, P.; Piotrowski, M. A.; Brown, T. M.; Nolan, C. E.; Richter, K. E.; Finley, J. E.; Fei, Q.; Ebbinghaus, C. F.; Chen, Y. L.; Spracklin, D. K.; Tate, B.; Geoghegan, K. F.; Lau, L. F.; Auperin, D. D.; Schachter, J. B., Concentration-dependent modulation of amyloid-beta *in vivo* and *in vitro* using the gamma-secretase inhibitor, LY-450139. *The Journal of pharmacology and experimental therapeutics* **2006**, *319* (2), 924-33.

[57]   Imbimbo, B. P.; Peretto, I., Semagacestat, a gamma-secretase inhibitor for the potential treatment of Alzheimer's disease. *Current opinion in investigational drugs* **2009**, *10* (7), 721-30.

[58]   Panelos, J.; Massi, D., Emerging role of Notch signaling in epidermal differentiation and skin cancer. *Cancer biology & therapy* **2009**, *8* (21), 1986-93.

[59]   Costa, R. M.; Honjo, T.; Silva, A. J., Learning and memory deficits in Notch mutant mice. *Current biology : CB* **2003**, *13* (15), 1348-54.

[60]   Squibb, B.-M. Statement on Avagacestat Development Status. http://www.bms.com/news/features/2012/Pages/AvagacestatDevelopmentStatus.aspx.

[61]   Martone, R. L.; Zhou, H.; Atchison, K.; Comery, T.; Xu, J. Z.; Huang, X.; Gong, X.; Jin, M.; Kreft, A.; Harrison, B.; Mayer, S. C.; Aschmies, S.; Gonzales, C.; Zaleska, M. M.; Riddell, D. R.; Wagner, E.; Lu, P.; Sun, S. C.; Sonnenberg-Reines, J.; Oganesian, A.; Adkins, K.; Leach, M. W.; Clarke, D. W.; Huryn, D.; Abou-Gharbia, M.; Magolda, R.; Bard, J.; Frick, G.; Raje, S.; Forlow, S. B.; Balliet, C.; Burczynski, M. E.; Reinhart, P. H.; Wan, H. I.; Pangalos, M. N.; Jacobsen, J. S., Begacestat (GSI-953): a novel, selective thiophene sulfonamide inhibitor of amyloid precursor protein gamma-secretase for the treatment of Alzheimer's disease. *The Journal of pharmacology and experimental therapeutics* **2009**, *331* (2), 598-608.

[62]   D'Onofrio, G.; Panza, F.; Frisardi, V.; Solfrizzi, V.; Imbimbo, B. P.; Paroni, G.; Cascavilla, L.; Seripa, D.; Pilotto, A., Advances in the identification of gamma-secretase inhibitors for the treatment of Alzheimer's disease. *Expert opinion on drug discovery* **2012**, *7* (1), 19-37.

[63]   Brodney, M. A.; Auperin, D. D.; Becker, S. L.; Bronk, B. S.; Brown, T. M.; Coffman, K. J.; Finley, J. E.; Hicks, C. D.; Karmilowicz, M. J.; Lanz, T. A.; Liston, D.; Liu, X.; Martin, B.

A.; Nelson, R. B.; Nolan, C. E.; Oborski, C. E.; Parker, C. P.; Richter, K. E.; Pozdnyakov, N.; Sahagan, B. G.; Schachter, J. B.; Sokolowski, S. A.; Tate, B.; Wood, D. E.; Wood, K. M.; Van Deusen, J. W.; Zhang, L., Design, synthesis, and *in vivo* characterization of a novel series of tetralin amino imidazoles as gamma-secretase inhibitors: discovery of PF-3084014. *Bioorganic & medicinal chemistry letters* **2011**, *21* (9), 2637-40.

[64] (a) Weggen, S.; Eriksen, J. L.; Das, P.; Sagi, S. A.; Wang, R.; Pietrzik, C. U.; Findlay, K. A.; Smith, T. E.; Murphy, M. P.; Bulter, T.; Kang, D. E.; Marquez-Sterling, N.; Golde, T. E.; Koo, E. H., A subset of NSAIDs lower amyloidogenic Abeta42 independently of cyclooxygenase activity. *Nature* **2001**, *414* (6860), 212-6; (b) Eriksen, J. L.; Sagi, S. A.; Smith, T. E.; Weggen, S.; Das, P.; McLendon, D. C.; Ozols, V. V.; Jessing, K. W.; Zavitz, K. H.; Koo, E. H.; Golde, T. E., NSAIDs and enantiomers of flurbiprofen target gamma-secretase and lower Abeta 42 *in vivo*. *The Journal of clinical investigation* **2003**, *112* (3), 440-9.

[65] Kukar, T.; Murphy, M. P.; Eriksen, J. L.; Sagi, S. A.; Weggen, S.; Smith, T. E.; Ladd, T.; Khan, M. A.; Kache, R.; Beard, J.; Dodson, M.; Merit, S.; Ozols, V. V.; Anastasiadis, P. Z.; Das, P.; Fauq, A.; Koo, E. H.; Golde, T. E., Diverse compounds mimic Alzheimer disease-causing mutations by augmenting Abeta42 production. *Nature medicine* **2005**, *11* (5), 545-50.

[66] Oehlrich, D.; Berthelot, D. J.; Gijsen, H. J., gamma-Secretase Modulators as Potential Disease Modifying Anti-Alzheimer's Drugs. *Journal of medicinal chemistry* **2011**, *54* (3), 669-98.

[67] Ho, C. Y, "GSM Inhibitors", WO/2009/052341. 2009.

[68] Shapiro, G.; Chesworth, R.; "Tetrasubstituted Benzenes", WO/2009/086277. 2009.

[69] Kounnas, M. Z.; Danks, A. M.; Cheng, S.; Tyree, C.; Ackerman, E.; Zhang, X.; Ahn, K.; Nguyen, P.; Comer, D.; Mao, L.; Yu, C.; Pleynet, D.; Digregorio, P. J.; Velicelebi, G.; Stauderman, K. A.; Comer, W. T.; Mobley, W. C.; Li, Y. M.; Sisodia, S. S.; Tanzi, R. E.; Wagner, S. L., Modulation of gamma-secretase reduces beta-amyloid deposition in a transgenic mouse model of Alzheimer's disease. *Neuron* **2010**, *67* (5), 769-80.

[70] Li, T.; Huang, Y.; Jin, S.; Ye, L.; Rong, N.; Yang, X.; Ding, Y.; Cheng, Z.; Zhang, J.; Wan, Z.; Harrison, D. C.; Hussain, I.; Hall, A.; Lee, D. H.; Lau, L. F.; Matsuoka, Y., Gamma-secretase modulators do not induce Abeta-rebound and accumulation of beta-C-terminal fragment. *Journal of neurochemistry* **2012**, *121* (2), 277-86.

[71] De Vos, K. J.; Grierson, A. J.; Ackerley, S.; Miller, C. C., Role of axonal transport in neurodegenerative diseases. *Annu Rev Neurosci* **2008**, *31*, 151-73.

[72] Alonso, A. C.; Grundke-Iqbal, I.; Iqbal, K., Alzheimer's disease hyperphosphorylated tau sequesters normal tau into tangles of filaments and disassembles microtubules. *Nat Med* **1996**, *2* (7), 783-7.

[73] (a) Hanger, D. P.; Anderton, B. H.; Noble, W., Tau phosphorylation: the therapeutic challenge for neurodegenerative disease. *Trends Mol Med* **2009**, *15* (3), 112-9; (b) Brunden, K. R.; Trojanowski, J. Q.; Lee, V. M., Advances in tau-focused drug discovery for Alzheimer's disease and related tauopathies. *Nat Rev Drug Discov* **2009**, *8* (10), 783-93.

[74] (a) Matenia, D.; Mandelkow, E. M., The tau of MARK: a polarized view of the cytoskeleton. *Trends Biochem Sci* **2009**, *34* (7), 332-42; (b) Drewes, G.; Ebneth, A.; Preuss, U.; Mandelkow, E. M.; Mandelkow, E., MARK, a novel family of protein kinases that phosphorylate microtubule- associated proteins and trigger microtubule disruption. *Cell* **1997**, *89* (2), 297-308; (c) Wang, J. Z.; Grundke-Iqbal, I.; Iqbal, K., Kinases and

phosphatases and tau sites involved in Alzheimer neurofibrillary degeneration. *Eur J Neurosci* **2007**, *25* (1), 59-68; (d) Noble, W.; Olm, V.; Takata, K.; Casey, E.; Mary, O.; Meyerson, J.; Gaynor, K.; LaFrancois, J.; Wang, L.; Kondo, T.; Davies, P.; Burns, M.; Veeranna; Nixon, R.; Dickson, D.; Matsuoka, Y.; Ahlijanian, M.; Lau, L. F.; Duff, K., Cdk5 is a key factor in tau aggregation and tangle formation *in vivo*. *Neuron* **2003**, *38* (4), 555-65.

[75]   (a) Liu, F.; Iqbal, K.; Grundke-Iqbal, I.; Hart, G. W.; Gong, C. X., O-GlcNAcylation regulates phosphorylation of tau: a mechanism involved in Alzheimer's disease. *Proc Natl Acad Sci U S A* **2004**, *101* (29), 10804-9; (b) Mazanetz, M. P.; Fischer, P. M., Untangling tau hyperphosphorylation in drug design for neurodegenerative diseases. *Nat Rev Drug Discov* **2007**, *6* (6), 464-79.

[76]   Hutton, M.; Lendon, C. L.; Rizzu, P.; Baker, M.; Froelich, S.; Houlden, H.; Pickering-Brown, S.; Chakraverty, S.; Isaacs, A.; Grover, A.; Hackett, J.; Adamson, J.; Lincoln, S.; Dickson, D.; Davies, P.; Petersen, R. C.; Stevens, M.; de Graaff, E.; Wauters, E.; van Baren, J.; Hillebrand, M.; Joosse, M.; Kwon, J. M.; Nowotny, P.; Heutink, P.; *et al.* Association of missense and 5'-splice-site mutations in tau with the inherited dementia FTDP-17. *Nature* **1998**, *393* (6686), 702-705.

[77]   Spillantini, M. G.; Goedert, M., Tau protein pathology in neurodegenerative diseases. *Trends Neurosci* **1998**, *21*, 428-433.

[78]   (a) Nussbaum, J. M.; Schilling, S.; Cynis, H.; Silva, A.; Swanson, E.; Wangsanut, T.; Tayler, K.; Wiltgen, B.; Hatami, A.; Ronicke, R.; Reymann, K.; Hutter-Paier, B.; Alexandru, A.; Jagla, W.; Graubner, S.; Glabe, C. G.; Demuth, H. U.; Bloom, G. S., Prion-like behaviour and tau-dependent cytotoxicity of pyroglutamylated amyloid-beta. *Nature* **2012**, *485* (7400), 651-5; (b) Rapoport, M.; Dawson, H. N.; Binder, L. I.; Vitek, M. P.; Ferreira, A., Tau is essential to beta -amyloid-induced neurotoxicity. *Proc Natl Acad Sci U S A* **2002**, *99* (9), 6364-9.

[79]   (a) Roberson, E. D.; Scearce-Levie, K.; Palop, J. J.; Yan, F.; Cheng, I. H.; Wu, T.; Gerstein, H.; Yu, G. Q.; Mucke, L., Reducing endogenous tau ameliorates amyloid beta-induced deficits in an Alzheimer's disease mouse model. *Science* **2007**, *316* (5825), 750-4; (b) Ittner, L. M.; Ke, Y. D.; Delerue, F.; Bi, M.; Gladbach, A.; van Eersel, J.; Wolfing, H.; Chieng, B. C.; Christie, M. J.; Napier, I. A.; Eckert, A.; Staufenbiel, M.; Hardeman, E.; Gotz, J., Dendritic Function of Tau Mediates Amyloid-beta Toxicity in Alzheimer's Disease Mouse Models. *Cell* **2010**, *142*, 387-97.

[80]   (a) Tanzi, R. E.; Bertram, L., Twenty years of the Alzheimer's disease amyloid hypothesis: a genetic perspective. *Cell* **2005**, *120* (4), 545-55; (b) Selkoe, D. J., The molecular pathology of Alzheimer's disease. *Neuron* **1991**, *6* (4), 487-98.

[81]   Ittner, L. M.; Gotz, J., Amyloid-beta and tau - a toxic pas de deux in Alzheimer's disease. *Nat Rev Neurosci* **2011**, *12* (2), 65-72.

[82]   Takishima A, M. M., Murayama O *et al.* Presenilin 1 associated with glycogen synthase Kinase 3 beta and its substrate tau. *PNAS (USA)* **1998**, *95* (16), 9637-9641.

[83]   Leroy K, Y. Z., Brion J.P,, increased level of GSK3beta in Alzheimers disease and accumulation of argyrophic grains and in neurones at different stages of neurofibrillary degeneration. *Neuropathology & applied Neurobiology* **2007**, *33* (1), 43-55.

[84]   (a) Polter, A.; Beurel, E.; Yang, S.; Garner, R.; Song, L.; Miller, C. A.; Sweatt, J. D.; McMahon, L.; Bartolucci, A. A.; Li, X.; Jope, R. S., Deficiency in the inhibitory serine-phosphorylation of glycogen synthase kinase-3 increases sensitivity to mood disturbances.

Neuropsychopharmacology : official publication of the American College of Neuropsychopharmacology **2010**, *35* (8), 1761-74; (b) Prickaerts, J.; Moechars, D.; Cryns, K.; Lenaerts, I.; van Craenendonck, H.; Goris, I.; Daneels, G.; Bouwknecht, J. A.; Steckler, T., Transgenic mice overexpressing glycogen synthase kinase 3beta: a putative model of hyperactivity and mania. *The Journal of neuroscience : the official journal of the Society for Neuroscience* **2006**, *26* (35), 9022-9; (c) O'Brien, W. T.; Harper, A. D.; Jove, F.; Woodgett, J. R.; Maretto, S.; Piccolo, S.; Klein, P. S., Glycogen synthase kinase-3beta haploinsufficiency mimics the behavioral and molecular effects of lithium. *The Journal of neuroscience : the official journal of the Society for Neuroscience* **2004**, *24* (30), 6791-8; (d) Meijer, L.; Skaltsounis, A. L.; Magiatis, P.; Polychronopoulos, P.; Knockaert, M.; Leost, M.; Ryan, X. P.; Vonica, C. A.; Brivanlou, A.; Dajani, R.; Crovace, C.; Tarricone, C.; Musacchio, A.; Roe, S. M.; Pearl, L.; Greengard, P., GSK-3-selective inhibitors derived from Tyrian purple indirubins. *Chemistry & biology* **2003**, *10* (12), 1255-66; (e) Gould, T. D.; Einat, H.; Bhat, R.; Manji, H. K., AR-A014418, a selective GSK-3 inhibitor, produces antidepressant-like effects in the forced swim test. *The international journal of neuropsychopharmacology / official scientific journal of the Collegium Internationale Neuropsychopharmacologicum* **2004**, *7* (4), 387-90.

[85]   Sudduth TL, W. J., Everhart A, Colton CA, Wilcock DM., Lithium treatment of APPSwDI/NOS2-/- mice leads to reduced hyperphosphorylated tau, increased amyloid deposition and altered inflammatory phenotype. *PLoS One.* **2012**, *7* (2), e31993.

[86]   Martinez, A.; Gil, C.; Perez, D. I., Glycogen synthase kinase 3 inhibitors in the next horizon for Alzheimer's disease treatment. *International journal of Alzheimer's disease* **2011**, *2011*, 280502.

[87]   (a) Eldar-Finkelman, H.; Martinez, A., GSK-3 Inhibitors: Preclinical and Clinical Focus on CNS. *Frontiers in molecular neuroscience* **2011**, *4*, 32; (b) Kramer, T.; Schmidt, B.; Lo Monte, F., Small-Molecule Inhibitors of GSK-3: Structural Insights and Their Application to Alzheimer's Disease Models. *International journal of Alzheimer's disease* **2012**, *2012*, 381029.

[88]   Ding Y, Q. A., Guo-Huang Fan, Indirubin-3'-monoxime rescues spatial memory deficits and attenuates β -amyloid-associated neuropathology in a mouse model of Alzheimer's disease. *Neurobiology of disease* **2010**, *39*, 156-168.

[89]   Andreani, A.; Cavalli, A.; Granaiola, M.; Leoni, A.; Locatelli, A.; Morigi, R.; Rambaldi, M.; Recanatini, M.; Garnier, M.; Meijer, L., Imidazo[2,1-b]thiazolylmethylene- and indolylmethylene-2-indolinones: a new class of cyclin-dependent kinase inhibitors. Design, synthesis, and CDK1/cyclin B inhibition. *Anti-cancer drug design* **2000**, *15* (6), 447-52.

[90]   Akue-Gedu, R.; Debiton, E.; Ferandin, Y.; Meijer, L.; Prudhomme, M.; Anizon, F.; Moreau, P., Synthesis and biological activities of aminopyrimidyl-indoles structurally related to meridianins. *Bioorganic & medicinal chemistry* **2009**, *17* (13), 4420-4.

[91]   (a) Kramer, T.; Schmidt, B.; Lo Monte, F., Small-molecule inhibitors of GSK-3: structural insights and their application to Alzheimer's disease models. *International Journal of Alzheimer's Disease* **2012**; (b) Lee, S. C.; Kim, H. T.; Park, C. H.; Lee do. Y; Chang, H. J.; Park, S.; Cho. J. M.; Ro, S.; Suh, Y. G. Design, synthesis and biological evaluation of novel imidazopyridines as potential antidiabetic GSK3b inhibitors. *Bioorg. Med. Chem. Lett* **2012**, 4221-4224; (c) Polychronopoulos, P.; Magiatis, P., Structural Basis for the Synthesis of Indirubins as Potent and Selective Inhibitors of Glycogen Synthase Kinase-3 and Cyclin-Dependent Kinases. *J. Med. Chem.* **2004**, *47* (4), 935-946; (d) Medina, M.; Avila, J.,

Glycogen synthase kinase-3 (GSK-3) inhibitors for the treatment of Alzheimer's disease. *Current Pharmaceutical Design* **2010**, *16* (25), 2790-2798; (e) Martinez, A.; Perez, D. I., GSK-3 Inhibitors: A Ray of Hope for the Treatment of Alzheimer's Disease? *Journal of Alzheimer's Disease* **2008**, *15* (2), 181-191; (f) artinez, A., Preclinical efficacy on GSK-3 inhibitors: towards a future generation of powerful drugs. *Medicinal Research Reviews* **2008**, *28* (5), 773-796.

[92]   Bhat, R.; Xue, Y.; Berg, S.; Hellberg, S.; Ormo, M.; Nilsson, Y.; Radesater, A. C.; Jerning, E.; Markgren, P. O.; Borgegard, T.; Nylof, M.; Gimenez-Cassina, A.; Hernandez, F.; Lucas, J. J.; Diaz-Nido, J.; Avila, J., Structural insights and biological effects of glycogen synthase kinase 3-specific inhibitor AR-A014418. *The Journal of biological chemistry* **2003**, *278* (46), 45937-45.

[93]   Koh, S. H.; Kim, Y.; Kim, H. Y.; Hwang, S.; Lee, C. H.; Kim, S. H., Inhibition of glycogen synthase kinase-3 suppresses the onset of symptoms and disease progression of G93A-SOD1 mouse model of ALS. *Experimental neurology* **2007**, *205* (2), 336-46.

[94]   Wang, W.; Yang, Y.; Ying, C.; Li, W.; Ruan, H.; Zhu, X.; You, Y.; Han, Y.; Chen, R.; Wang, Y.; Li, M., Inhibition of glycogen synthase kinase-3beta protects dopaminergic neurons from MPTP toxicity. *Neuropharmacology* **2007**, *52* (8), 1678-84.

[95]   Dudev T, L. C., Competition between Li^+ and $Mg2^+$ in metalloproteins. Implications for lithium therapy. *J Am Chem Soc.* **2012**, *133* (44), 9505-9515.

[96]   Noble, W.; Planel, E.; Zehr, C.; Olm, V.; Meyerson, J.; Suleman, F.; Gaynor, K.; Wang, L.; LaFrancois, J.; Feinstein, B.; Burns, M.; Krishnamurthy, P.; Wen, Y.; Bhat, R.; Lewis, J.; Dickson, D.; Duff, K., Inhibition of glycogen synthase kinase-3 by lithium correlates with reduced tauopathy and degeneration *in vivo. Proc Natl Acad Sci U S A* **2005**, *102* (19), 6990-5.

[97]   Eldar-Finkelman H, M. A., GSK-3inhibitors: preclinical and clinical focus on CNS. *Fromtiers in Molecular Neuroscience* **2011**, *4* (32), 1-18.

[98]   Lee, V. M.-Y.; Trojanowski, J. Q., Neurodegenerative Tauopathies: Human Disease and Transgenic Mouse Models. *Neuron* **1999**, *24* (3), 507–510.

[99]   (a) Gomez-Isla, T. In *The efficacy results of NP-12 in a double hAPPxTau transgenic mouse model*, International Conference on Alzheimer's Disease, Chigago (USA), Chigago (USA), 2008; (b) Martinez, A. In *Non ATP-competitive inhibitors of GSK3, for the treatment of Alzheimer's Disease*, International Conference on Alzheimer's Disease, Chigago (USA), Chigago (USA), 2008.

[100]   del serono, T., Phase IIa clinical trial on Alzheimers disease with NP12, a GSK3 inhibitor. *Alzheimers amnd dementia* **2010**, *6*, S147.

[101]   Juan Manuel Domínguez, A. F., Leyre Orozco, María del Monte-Milla, Elena Delgado, and Miguel Medina, Evidence for Irreversible Inhibition of Glycogen Synthase. *J. Biol. Chem.* **2012**, 893-904.

[102]   S.A., N. *Noscira announces results from ARGO Phase IIb trial of tideglusib for the treatment of Alzheimers disease*; Zeltia S.A.: www.zeltia.com and www.noscira.com, 2012.

[103]   DIAN, Drug Candidates selected for pioneering Alzheimer's prevention trial. *Nature reviews Drug Discovery* **2012**, *11*, 821.

[104]   Rogaeva, E.; Meng, Y.; Lee, J. H.; Gu, Y.; Kawarai, T.; Zou, F.; Katayama, T.; Baldwin, C. T.; Cheng, R.; Hasegawa, H.; Chen, F.; Shibata, N.; Lunetta, K. L.; Pardossi-Piquard, R.; Bohm, C.; Wakutani, Y.; Cupples, L. A.; Cuenco, K. T.; Green, R. C.; Pinessi, L.; Rainero, I.; Sorbi, S.; Bruni, A.; Duara, R.; Friedland, R. P.; Inzelberg, R.; Hampe, W.; Bujo, H.;

Song, Y. Q.; Andersen, O. M.; Willnow, T. E.; Graff-Radford, N.; Petersen, R. C.; Dickson, D.; Der, S. D.; Fraser, P. E.; Schmitt-Ulms, G.; Younkin, S.; Mayeux, R.; Farrer, L. A.; St George-Hyslop, P., The neuronal sortilin-related receptor SORL1 is genetically associated with Alzheimer disease. *Nat Genet* **2007**, *39* (2), 168-177.

[105]   (a) Lambert, J. C.; Heath, S.; Even, G.; Campion, D.; Sleegers, K.; Hiltunen, M.; Combarros, O.; Zelenika, D.; Bullido, M. J.; Tavernier, B.; Letenneur, L.; Bettens, K.; Berr, C.; Pasquier, F.; Fievet, N.; Barberger-Gateau, P.; Engelborghs, S.; De Deyn, P.; Mateo, I.; Franck, A.; Helisalmi, S.; Porcellini, E.; Hanon, O.; de Pancorbo, M. M.; Lendon, C.; Dufouil, C.; Jaillard, C.; Leveillard, T.; Alvarez, V.; Bosco, P.; Mancuso, M.; Panza, F.; Nacmias, B.; Bossu, P.; Piccardi, P.; Annoni, G.; Seripa, D.; Galimberti, D.; Hannequin, D.; Licastro, F.; Soininen, H.; Ritchie, K.; Blanche, H.; Dartigues, J. F.; Tzourio, C.; Gut, I.; Van Broeckhoven, C.; Alperovitch, A.; Lathrop, M.; Amouyel, P., Genome-wide association study identifies variants at CLU and CR1 associated with Alzheimer's disease. *Nat Genet* **2009**, *41* (10), 1094-9; (b) Harold, D.; Abraham, R.; Hollingworth, P.; Sims, R.; Gerrish, A.; Hamshere, M. L.; Pahwa, J. S.; Moskvina, V.; Dowzell, K.; Williams, A.; Jones, N.; Thomas, C.; Stretton, A.; Morgan, A. R.; Lovestone, S.; Powell, J.; Proitsi, P.; Lupton, M. K.; Brayne, C.; Rubinsztein, D. C.; Gill, M.; Lawlor, B.; Lynch, A.; Morgan, K.; Brown, K. S.; Passmore, P. A.; Craig, D.; McGuinness, B.; Todd, S.; Holmes, C.; Mann, D.; Smith, A. D.; Love, S.; Kehoe, P. G.; Hardy, J.; Mead, S.; Fox, N.; Rossor, M.; Collinge, J.; Maier, W.; Jessen, F.; Schurmann, B.; van den Bussche, H.; Heuser, I.; Kornhuber, J.; Wiltfang, J.; Dichgans, M.; Frolich, L.; Hampel, H.; Hull, M.; Rujescu, D.; Goate, A. M.; Kauwe, J. S.; Cruchaga, C.; Nowotny, P.; Morris, J. C.; Mayo, K.; Sleegers, K.; Bettens, K.; Engelborghs, S.; De Deyn, P. P.; Van Broeckhoven, C.; Livingston, G.; Bass, N. J.; Gurling, H.; McQuillin, A.; Gwilliam, R.; Deloukas, P.; Al-Chalabi, A.; Shaw, C. E.; Tsolaki, M.; Singleton, A. B.; Guerreiro, R.; Muhleisen, T. W.; Nothen, M. M.; Moebus, S.; Jockel, K. H.; Klopp, N.; Wichmann, H. E.; Carrasquillo, M. M.; Pankratz, V. S.; Younkin, S. G.; Holmans, P. A.; O'Donovan, M.; Owen, M. J.; Williams, J., Genome-wide association study identifies variants at CLU and PICALM associated with Alzheimer's disease. *Nat Genet* **2009**, *41* (10), 1088-93.

[106]   Seshadri, S.; Fitzpatrick, A. L.; Ikram, M. A.; DeStefano, A. L.; Gudnason, V.; Boada, M.; Bis, J. C.; Smith, A. V.; Carassquillo, M. M.; Lambert, J. C.; Harold, D.; Schrijvers, E. M.; Ramirez-Lorca, R.; Debette, S.; Longstreth, W. T., Jr.; Janssens, A. C.; Pankratz, V. S.; Dartigues, J. F.; Hollingworth, P.; Aspelund, T.; Hernandez, I.; Beiser, A.; Kuller, L. H.; Koudstaal, P. J.; Dickson, D. W.; Tzourio, C.; Abraham, R.; Antunez, C.; Du, Y.; Rotter, J. I.; Aulchenko, Y. S.; Harris, T. B.; Petersen, R. C.; Berr, C.; Owen, M. J.; Lopez-Arrieta, J.; Varadarajan, B. N.; Becker, J. T.; Rivadeneira, F.; Nalls, M. A.; Graff-Radford, N. R.; Campion, D.; Auerbach, S.; Rice, K.; Hofman, A.; Jonsson, P. V.; Schmidt, H.; Lathrop, M.; Mosley, T. H.; Au, R.; Psaty, B. M.; Uitterlinden, A. G.; Farrer, L. A.; Lumley, T.; Ruiz, A.; Williams, J.; Amouyel, P.; Younkin, S. G.; Wolf, P. A.; Launer, L. J.; Lopez, O. L.; van Duijn, C. M.; Breteler, M. M., Genome-wide analysis of genetic loci associated with Alzheimer disease. *JAMA* **2010**, *303* (18), 1832-40.

[107]   (a) Hollingworth, P.; Harold, D.; Sims, R.; Gerrish, A.; Lambert, J. C.; Carrasquillo, M. M.; Abraham, R.; Hamshere, M. L.; Pahwa, J. S.; Moskvina, V.; Dowzell, K.; Jones, N.; Stretton, A.; Thomas, C.; Richards, A.; Ivanov, D.; Widdowson, C.; Chapman, J.; Lovestone, S.; Powell, J.; Proitsi, P.; Lupton, M. K.; Brayne, C.; Rubinsztein, D. C.; Gill, M.; Lawlor, B.; Lynch, A.; Brown, K. S.; Passmore, P. A.; Craig, D.; McGuinness, B.;

Todd, S.; Holmes, C.; Mann, D.; Smith, A. D.; Beaumont, H.; Warden, D.; Wilcock, G.; Love, S.; Kehoe, P. G.; Hooper, N. M.; Vardy, E. R.; Hardy, J.; Mead, S.; Fox, N. C.; Rossor, M.; Collinge, J.; Maier, W.; Jessen, F.; Ruther, E.; Schurmann, B.; Heun, R.; Kolsch, H.; van den Bussche, H.; Heuser, I.; Kornhuber, J.; Wiltfang, J.; Dichgans, M.; Frolich, L.; Hampel, H.; Gallacher, J.; Hull, M.; Rujescu, D.; Giegling, I.; Goate, A. M.; Kauwe, J. S.; Cruchaga, C.; Nowotny, P.; Morris, J. C.; Mayo, K.; Sleegers, K.; Bettens, K.; Engelborghs, S.; De Deyn, P. P.; Van Broeckhoven, C.; Livingston, G.; Bass, N. J.; Gurling, H.; McQuillin, A.; Gwilliam, R.; Deloukas, P.; Al-Chalabi, A.; Shaw, C. E.; Tsolaki, M.; Singleton, A. B.; Guerreiro, R.; Muhleisen, T. W.; Nothen, M. M.; Moebus, S.; Jockel, K. H.; Klopp, N.; Wichmann, H. E.; Pankratz, V. S.; Sando, S. B.; Aasly, J. O.; Barcikowska, M.; Wszolek, Z. K.; Dickson, D. W.; Graff-Radford, N. R.; Petersen, R. C.; van Duijn, C. M.; Breteler, M. M.; Ikram, M. A.; Destefano, A. L.; Fitzpatrick, A. L.; Lopez, O.; Launer, L. J.; Seshadri, S.; Berr, C.; Campion, D.; Epelbaum, J.; Dartigues, J. F.; Tzourio, C.; Alperovitch, A.; Lathrop, M.; Feulner, T. M.; Friedrich, P.; Riehle, C.; Krawczak, M.; Schreiber, S.; Mayhaus, M.; Nicolhaus, S.; Wagenpfeil, S.; Steinberg, S.; Stefansson, H.; Stefansson, K.; Snaedal, J.; Bjornsson, S.; Jonsson, P. V.; Chouraki, V.; Genier-Boley, B.; Hiltunen, M.; Soininen, H.; Combarros, O.; Zelenika, D.; Delepine, M.; Bullido, M. J.; Pasquier, F.; Mateo, I.; Frank-Garcia, A.; Porcellini, E.; Hanon, O.; Coto, E.; Alvarez, V.; Bosco, P.; Siciliano, G.; Mancuso, M.; Panza, F.; Solfrizzi, V.; Nacmias, B.; Sorbi, S.; Bossu, P.; Piccardi, P.; Arosio, B.; Annoni, G.; Seripa, D.; Pilotto, A.; Scarpini, E.; Galimberti, D.; Brice, A.; Hannequin, D.; Licastro, F.; Jones, L.; Holmans, P. A.; Jonsson, T.; Riemenschneider, M.; Morgan, K.; Younkin, S. G.; Owen, M. J.; O'Donovan, M.; Amouyel, P.; Williams, J., Common variants at ABCA7, MS4A6A/MS4A4E, EPHA1, CD33 and CD2AP are associated with Alzheimer's disease. *Nat Genet* **2011**, *43* (5), 429-435; (b) Naj, A. C.; Jun, G.; Beecham, G. W.; Wang, L. S.; Vardarajan, B. N.; Buros, J.; Gallins, P. J.; Buxbaum, J. D.; Jarvik, G. P.; Crane, P. K.; Larson, E. B.; Bird, T. D.; Boeve, B. F.; Graff-Radford, N. R.; De Jager, P. L.; Evans, D.; Schneider, J. A.; Carrasquillo, M. M.; Ertekin-Taner, N.; Younkin, S. G.; Cruchaga, C.; Kauwe, J. S.; Nowotny, P.; Kramer, P.; Hardy, J.; Huentelman, M. J.; Myers, A. J.; Barmada, M. M.; Demirci, F. Y.; Baldwin, C. T.; Green, R. C.; Rogaeva, E.; George-Hyslop, P. S.; Arnold, S. E.; Barber, R.; Beach, T.; Bigio, E. H.; Bowen, J. D.; Boxer, A.; Burke, J. R.; Cairns, N. J.; Carlson, C. S.; Carney, R. M.; Carroll, S. L.; Chui, H. C.; Clark, D. G.; Corneveaux, J.; Cotman, C. W.; Cummings, J. L.; Decarli, C.; Dekosky, S. T.; Diaz-Arrastia, R.; Dick, M.; Dickson, D. W.; Ellis, W. G.; Faber, K. M.; Fallon, K. B.; Farlow, M. R.; Ferris, S.; Frosch, M. P.; Galasko, D. R.; Ganguli, M.; Gearing, M.; Geschwind, D. H.; Ghetti, B.; Gilbert, J. R.; Gilman, S.; Giordani, B.; Glass, J. D.; Growdon, J. H.; Hamilton, R. L.; Harrell, L. E.; Head, E.; Honig, L. S.; Hulette, C. M.; Hyman, B. T.; Jicha, G. A.; Jin, L. W.; Johnson, N.; Karlawish, J.; Karydas, A.; Kaye, J. A.; Kim, R.; Koo, E. H.; Kowall, N. W.; Lah, J. J.; Levey, A. I.; Lieberman, A. P.; Lopez, O. L.; Mack, W. J.; Marson, D. C.; Martiniuk, F.; Mash, D. C.; Masliah, E.; McCormick, W. C.; McCurry, S. M.; McDavid, A. N.; McKee, A. C.; Mesulam, M.; Miller, B. L.; Miller, C. A.; Miller, J. W.; Parisi, J. E.; Perl, D. P.; Peskind, E.; Petersen, R. C.; Poon, W. W.; Quinn, J. F.; Rajbhandary, R. A.; Raskind, M.; Reisberg, B.; Ringman, J. M.; Roberson, E. D.; Rosenberg, R. N.; Sano, M.; Schneider, L. S.; Seeley, W.; Shelanski, M. L.; Slifer, M. A.; Smith, C. D.; Sonnen, J. A.; Spina, S.; Stern, R. A.; Tanzi, R. E.; Trojanowski, J. Q.; Troncoso, J. C.; Van Deerlin, V. M.; Vinters, H. V.; Vonsattel, J. P.; Weintraub, S.; Welsh-

Bohmer, K. A.; Williamson, J.; Woltjer, R. L.; Cantwell, L. B.; Dombroski, B. A.; Beekly, D.; Lunetta, K. L.; Martin, E. R.; Kamboh, M. I.; Saykin, A. J.; Reiman, E. M.; Bennett, D. A.; Morris, J. C.; Montine, T. J.; Goate, A. M.; Blacker, D.; Tsuang, D. W.; Hakonarson, H.; Kukull, W. A.; Foroud, T. M.; Haines, J. L.; Mayeux, R.; Pericak-Vance, M. A.; Farrer, L. A.; Schellenberg, G. D., Common variants at MS4A4/MS4A6E, CD2AP, CD33 and EPHA1 are associated with late-onset Alzheimer's disease. *Nat Genet* **2011**, *43* (5), 436-41.

[108] Gelissen, I. C.; Hochgrebe, T.; Wilson, M. R.; Easterbrook-Smith, S. B.; Jessup, W.; Dean, R. T.; Brown, A. J., Apolipoprotein J (clusterin) induces cholesterol export from macrophage-foam cells: a potential anti-atherogenic function? *Biochem J* **1998**, *331 (Pt 1)*, 231-7.

[109] Barrett, P. J.; Song, Y.; Van Horn, W. D.; Hustedt, E. J.; Schafer, J. M.; Hadziselimovic, A.; Beel, A. J.; Sanders, C. R., The amyloid precursor protein has a flexible transmembrane domain and binds cholesterol. *Science* **2012**, *336* (6085), 1168-71.

[110] Klinger, S. C.; Glerup, S.; Raarup, M. K.; Mari, M. C.; Nyegaard, M.; Koster, G.; Prabakaran, T.; Nilsson, S. K.; Kjaergaard, M. M.; Bakke, O.; Nykjaer, A.; Olivecrona, G.; Petersen, C. M.; Nielsen, M. S., SorLA regulates the activity of lipoprotein lipase by intracellular trafficking. *J Cell Sci* **2011**, *124* (Pt 7), 1095-105.

[111] Chan, S. L.; Kim, W. S.; Kwok, J. B.; Hill, A. F.; Cappai, R.; Rye, K. A.; Garner, B., ATP-binding cassette transporter A7 regulates processing of amyloid precursor protein *in vitro*. *J Neurochem* **2008**, *106* (2), 793-804.

[112] (a) Shepardson, N. E.; Shankar, G. M.; Selkoe, D. J., Cholesterol level and statin use in Alzheimer disease: II. Review of human trials and recommendations. *Arch Neurol* **2011**, *68* (11), 1385-92; (b) Shepardson, N. E.; Shankar, G. M.; Selkoe, D. J., Cholesterol level and statin use in Alzheimer disease: I. Review of epidemiological and preclinical studies. *Arch Neurol* **2011**, *68* (10), 1239-44.

[113] Corder, E. H.; Saunders, A. M.; Strittmatter, W. J.; Schmechel, D. E.; Gaskell, P. C.; Small, G. W.; Roses, A. D.; Haines, J. L.; Pericak-Vance, M. A., Gene dose of apolipoprotein E type 4 allele and the risk of Alzheimer's disease in late onset families. *Science* **1993**, *261* (5123), 921-923.

[114] (a) Mahley, R. W.; Weisgraber, K. H.; Huang, Y., Apolipoprotein E4: a causative factor and therapeutic target in neuropathology, including Alzheimer's disease. *Proceedings of the National Academy of Sciences of the United States of America* **2006**, *103* (15), 5644-51; (b) Mahley, R. W.; Huang, Y., Small-molecule structure correctors target abnormal protein structure and function: structure corrector rescue of apolipoprotein E4-associated neuropathology. *Journal of medicinal chemistry* **2012**, *55* (21), 8997-9008.

[115] Tai, L. M.; Youmans, K. L.; Jungbauer, L.; Yu, C.; Ladu, M. J., Introducing Human APOE into Abeta Transgenic Mouse Models. *International journal of Alzheimer's disease* **2011**, *2011*, 810981.

[116] (a) Cramer, P. E.; Cirrito, J. R.; Wesson, D. W.; Lee, C. Y.; Karlo, J. C.; Zinn, A. E.; Casali, B. T.; Restivo, J. L.; Goebel, W. D.; James, M. J.; Brunden, K. R.; Wilson, D. A.; Landreth, G. E., ApoE-directed therapeutics rapidly clear beta-amyloid and reverse deficits in AD mouse models. *Science* **2012**, *335* (6075), 1503-6; (b) Jarvis, C. I.; Goncalves, M. B.; Clarke, E.; Dogruel, M.; Kalindjian, S. B.; Thomas, S. A.; Maden, M.; Corcoran, J. P., Retinoic acid receptor-alpha signalling antagonizes both intracellular and extracellular amyloid-beta production and prevents neuronal cell death caused by amyloid-beta. *The European journal of neuroscience* **2010**, *32* (8), 1246-55.

[117] Fukasawa, H.; Nakagomi, M.; Yamagata, N.; Katsuki, H.; Kawahara, K.; Kitaoka, K.; Miki, T.; Shudo, K., Tamibarotene: a candidate retinoid drug for Alzheimer's disease. *Biological & pharmaceutical bulletin* **2012**, *35* (8), 1206-12.

[118] Nemani, V. M.; Lu, W.; Berge, V.; Nakamura, K.; Onoa, B.; Lee, M. K.; Chaudhry, F. A.; Nicoll, R. A.; Edwards, R. H., Increased expression of alpha-synuclein reduces neurotransmitter release by inhibiting synaptic vesicle reclustering after endocytosis. *Neuron* **2010**, *65* (1), 66-79.

[119] Chandra, S.; Gallardo, G.; Fernandez-Chacon, R.; Schluter, O. M.; Sudhof, T. C., Alpha-synuclein cooperates with CSPalpha in preventing neurodegeneration. *Cell* **2005**, *123* (3), 383-96.

[120] Burre, J.; Sharma, M.; Tsetsenis, T.; Buchman, V.; Etherton, M. R.; Sudhof, T. C., Alpha-synuclein promotes SNARE-complex assembly *in vivo* and *in vitro*. *Science* **2010**, *329* (5999), 1663-7.

[121] Greten-Harrison, B.; Polydoro, M.; Morimoto-Tomita, M.; Diao, L.; Williams, A. M.; Nie, E. H.; Makani, S.; Tian, N.; Castillo, P. E.; Buchman, V. L.; Chandra, S. S., alphabetagamma-Synuclein triple knockout mice reveal age-dependent neuronal dysfunction. *Proceedings of the National Academy of Sciences of the United States of America* **2010**, *107* (45), 19573-8.

[122] Uversky, V. N., Neuropathology, biochemistry, and biophysics of alpha-synuclein aggregation. *Journal of neurochemistry* **2007**, *103* (1), 17-37.

[123] Polymeropoulos, M. H.; Lavedan, C.; Leroy, E.; Ide, S. E.; Dehejia, A.; Dutra, A.; Pike, B.; Root, H.; Rubenstein, J.; Boyer, R.; Stenroos, E. S.; Chandrasekharappa, S.; Athanassiadou, A.; Papapetropoulos, T.; Johnson, W. G.; Lazzarini, A. M.; Duvoisin, R. C.; Di Iorio, G.; Golbe, L. I.; Nussbaum, R. L., Mutation in the alpha-synuclein gene identified in families with Parkinson's disease. *Science* **1997**, *276* (5321), 2045-7.

[124] (a) Kruger, R.; Kuhn, W.; Muller, T.; Woitalla, D.; Graeber, M.; Kosel, S.; Przuntek, H.; Epplen, J. T.; Schols, L.; Riess, O., Ala30Pro mutation in the gene encoding alpha-synuclein in Parkinson's disease. *Nature genetics* **1998**, *18* (2), 106-8; (b) Singleton, A. B.; Farrer, M.; Johnson, J.; Singleton, A.; Hague, S.; Kachergus, J.; Hulihan, M.; Peuralinna, T.; Dutra, A.; Nussbaum, R.; Lincoln, S.; Crawley, A.; Hanson, M.; Maraganore, D.; Adler, C.; Cookson, M. R.; Muenter, M.; Baptista, M.; Miller, D.; Blancato, J.; Hardy, J.; Gwinn-Hardy, K., alpha-Synuclein locus triplication causes Parkinson's disease. *Science* **2003**, *302* (5646), 841.

[125] (a) OuterioLee, H. J.; Choi, C.; Lee, S. J., Membrane-bound alpha-synuclein has a high aggregation propensity and the ability to seed the aggregation of the cytosolic form. *The Journal of biological chemistry* **2002**, *277* (1), 671-8; (b) Outeiro, T. F.; Putcha, P.; Tetzlaff, J. E.; Spoelgen, R.; Koker, M.; Carvalho, F.; Hyman, B. T.; McLean, P. J., Formation of toxic oligomeric alpha-synuclein species in living cells. *PloS one* **2008**, *3* (4), e1867.

[126] Volles, M. J.; Lansbury, P. T., Jr., Zeroing in on the pathogenic form of alpha-synuclein and its mechanism of neurotoxicity in Parkinson's disease. *Biochemistry* **2003**, *42* (26), 7871-8.

[127] Li, J. Y.; Englund, E.; Holton, J. L.; Soulet, D.; Hagell, P.; Lees, A. J.; Lashley, T.; Quinn, N. P.; Rehncrona, S.; Bjorklund, A.; Widner, H.; Revesz, T.; Lindvall, O.; Brundin, P., Lewy bodies in grafted neurons in subjects with Parkinson's disease suggest host-to-graft disease propagation. *Nature medicine* **2008**, *14* (5), 501-3.

[128]　(a) Cooper, A. A.; Gitler, A. D.; Cashikar, A.; Haynes, C. M.; Hill, K. J.; Bhullar, B.; Liu, K.; Xu, K.; Strathearn, K. E.; Liu, F.; Cao, S.; Caldwell, K. A.; Caldwell, G. A.; Marsischky, G.; Kolodner, R. D.; Labaer, J.; Rochet, J. C.; Bonini, N. M.; Lindquist, S., Alpha-synuclein blocks ER-Golgi traffic and Rab1 rescues neuron loss in Parkinson's models. *Science* **2006**, *313* (5785), 324-8; (b) Mosharov, E. V.; Staal, R. G.; Bove, J.; Prou, D.; Hananiya, A.; Markov, D.; Poulsen, N.; Larsen, K. E.; Moore, C. M.; Troyer, M. D.; Edwards, R. H.; Przedborski, S.; Sulzer, D., Alpha-synuclein overexpression increases cytosolic catecholamine concentration. *The Journal of neuroscience : the official journal of the Society for Neuroscience* **2006**, *26* (36), 9304-11; (c) Cuervo, A. M.; Stefanis, L.; Fredenburg, R.; Lansbury, P. T.; Sulzer, D., Impaired degradation of mutant alpha-synuclein by chaperone-mediated autophagy. *Science* **2004**, *305* (5688), 1292-5; (d) Orth, M.; Tabrizi, S. J.; Schapira, A. H.; Cooper, J. M., Alpha-synuclein expression in HEK293 cells enhances the mitochondrial sensitivity to rotenone. *Neuroscience letters* **2003**, *351* (1), 29-32; (e) Stefanis, L.; Larsen, K. E.; Rideout, H. J.; Sulzer, D.; Greene, L. A., Expression of A53T mutant but not wild-type alpha-synuclein in PC12 cells induces alterations of the ubiquitin-dependent degradation system, loss of dopamine release, and autophagic cell death. *The Journal of neuroscience : the official journal of the Society for Neuroscience* **2001**, *21* (24), 9549-60.

[129]　Scherzer, C. R.; Grass, J. A.; Liao, Z.; Pepivani, I.; Zheng, B.; Eklund, A. C.; Ney, P. A.; Ng, J.; McGoldrick, M.; Mollenhauer, B.; Bresnick, E. H.; Schlossmacher, M. G., GATA transcription factors directly regulate the Parkinson's disease-linked gene alpha-synuclein. *Proceedings of the National Academy of Sciences of the United States of America* **2008**, *105* (31), 10907-12.

[130]　Clough, R. L.; Stefanis, L., A novel pathway for transcriptional regulation of alpha-synuclein. *FASEB journal : official publication of the Federation of American Societies for Experimental Biology* **2007**, *21* (2), 596-607.

[131]　Jowaed, A.; Schmitt, I.; Kaut, O.; Wullner, U., Methylation regulates alpha-synuclein expression and is decreased in Parkinson's disease patients' brains. *The Journal of neuroscience : the official journal of the Society for Neuroscience* **2010**, *30* (18), 6355-9.

[132]　(a) Vogiatzi, T.; Xilouri, M.; Vekrellis, K.; Stefanis, L., Wild type alpha-synuclein is degraded by chaperone-mediated autophagy and macroautophagy in neuronal cells. *The Journal of biological chemistry* **2008**, *283* (35), 23542-56; (b) Alvarez-Erviti, L.; Rodriguez-Oroz, M. C.; Cooper, J. M.; Caballero, C.; Ferrer, I.; Obeso, J. A.; Schapira, A. H., Chaperone-mediated autophagy markers in Parkinson disease brains. *Archives of neurology* **2010**, *67* (12), 1464-72.

[133]　(a) Tsvetkov, A. S.; Miller, J.; Arrasate, M.; Wong, J. S.; Pleiss, M. A.; Finkbeiner, S., A small-molecule scaffold induces autophagy in primary neurons and protects against toxicity in a Huntington disease model. *Proceedings of the National Academy of Sciences of the United States of America* **2010**, *107* (39), 16982-7; (b) Crews, L.; Spencer, B.; Desplats, P.; Patrick, C.; Paulino, A.; Rockenstein, E.; Hansen, L.; Adame, A.; Galasko, D.; Masliah, E., Selective molecular alterations in the autophagy pathway in patients with Lewy body disease and in models of alpha-synucleinopathy. *PloS one* **2010**, *5* (2), e9313; (c) Lu, J. H.; Tan, J. Q.; Durairajan, S. S.; Liu, L. F.; Zhang, Z. H.; Ma, L.; Shen, H. M.; Chan, H. Y.; Li, M., Isorhynchophylline, a natural alkaloid, promotes the degradation of alpha-synuclein in neuronal cells via inducing autophagy. *Autophagy* **2012**, *8* (1), 98-108.

[134] Sidransky, E.; Nalls, M. A.; Aasly, J. O.; Aharon-Peretz, J.; Annesi, G.; Barbosa, E. R.; Bar-Shira, A.; Berg, D.; Bras, J.; Brice, A.; Chen, C. M.; Clark, L. N.; Condroyer, C.; De Marco, E. V.; Durr, A.; Eblan, M. J.; Fahn, S.; Farrer, M. J.; Fung, H. C.; Gan-Or, Z.; Gasser, T.; Gershoni-Baruch, R.; Giladi, N.; Griffith, A.; Gurevich, T.; Januario, C.; Kropp, P.; Lang, A. E.; Lee-Chen, G. J.; Lesage, S.; Marder, K.; Mata, I. F.; Mirelman, A.; Mitsui, J.; Mizuta, I.; Nicoletti, G.; Oliveira, C.; Ottman, R.; Orr-Urtreger, A.; Pereira, L. V.; Quattrone, A.; Rogaeva, E.; Rolfs, A.; Rosenbaum, H.; Rozenberg, R.; Samii, A.; Samaddar, T.; Schulte, C.; Sharma, M.; Singleton, A.; Spitz, M.; Tan, E. K.; Tayebi, N.; Toda, T.; Troiano, A. R.; Tsuji, S.; Wittstock, M.; Wolfsberg, T. G.; Wu, Y. R.; Zabetian, C. P.; Zhao, Y.; Ziegler, S. G., Multicenter analysis of glucocerebrosidase mutations in Parkinson's disease. *The New England journal of medicine* **2009**, *361* (17), 1651-61.

[135] Goker-Alpan, O.; Giasson, B. I.; Eblan, M. J.; Nguyen, J.; Hurtig, H. I.; Lee, V. M.; Trojanowski, J. Q.; Sidransky, E., Glucocerebrosidase mutations are an important risk factor for Lewy body disorders. *Neurology* **2006**, *67* (5), 908-10.

[136] Wong, K.; Sidransky, E.; Verma, A.; Mixon, T.; Sandberg, G. D.; Wakefield, L. K.; Morrison, A.; Lwin, A.; Colegial, C.; Allman, J. M.; Schiffmann, R., Neuropathology provides clues to the pathophysiology of Gaucher disease. *Molecular genetics and metabolism* **2004**, *82* (3), 192-207.

[137] Cullen, V.; Sardi, S. P.; Ng, J.; Xu, Y. H.; Sun, Y.; Tomlinson, J. J.; Kolodziej, P.; Kahn, I.; Saftig, P.; Woulfe, J.; Rochet, J. C.; Glicksman, M. A.; Cheng, S. H.; Grabowski, G. A.; Shihabuddin, L. S.; Schlossmacher, M. G., Acid beta-glucosidase mutants linked to Gaucher disease, Parkinson disease, and Lewy body dementia alter alpha-synuclein processing. *Annals of neurology* **2011**, *69* (6), 940-53.

[138] Manning-Bog, A. B.; Schule, B.; Langston, J. W., Alpha-synuclein-glucocerebrosidase interactions in pharmacological Gaucher models: a biological link between Gaucher disease and parkinsonism. *Neurotoxicology* **2009**, *30* (6), 1127-32.

[139] BalducciSardi, S. P.; Singh, P.; Cheng, S. H.; Shihabuddin, L. S.; Schlossmacher, M. G., Mutant GBA1 expression and synucleinopathy risk: first insights from cellular and mouse models. *Neuro-degenerative diseases* **2012**, *10* (1-4), 195-202.

[140] (a) Balducci, C.; Pierguidi, L.; Persichetti, E.; Parnetti, L.; Sbaragli, M.; Tassi, C.; Orlacchio, A.; Calabresi, P.; Beccari, T.; Rossi, A., Lysosomal hydrolases in cerebrospinal fluid from subjects with Parkinson's disease. *Movement disorders : official journal of the Movement Disorder Society* **2007**, *22* (10), 1481-4; (b) Parnetti, L.; Balducci, C.; Pierguidi, L.; De Carlo, C.; Peducci, M.; D'Amore, C.; Padiglioni, C.; Mastrocola, S.; Persichetti, E.; Paciotti, S.; Bellomo, G.; Tambasco, N.; Rossi, A.; Beccari, T.; Calabresi, P., Cerebrospinal fluid beta-glucocerebrosidase activity is reduced in Dementia with Lewy Bodies. *Neurobiology of disease* **2009**, *34* (3), 484-6.

[141] Mazzulli, J. R.; Xu, Y. H.; Sun, Y.; Knight, A. L.; McLean, P. J.; Caldwell, G. A.; Sidransky, E.; Grabowski, G. A.; Krainc, D., Gaucher disease glucocerebrosidase and alpha-synuclein form a bidirectional pathogenic loop in synucleinopathies. *Cell* **2011**, *146* (1), 37-52.

[142] Sardi, S. P.; Clarke, J.; Kinnecom, C.; Tamsett, T. J.; Li, L.; Stanek, L. M.; Passini, M. A.; Grabowski, G. A.; Schlossmacher, M. G.; Sidman, R. L.; Cheng, S. H.; Shihabuddin, L. S., CNS expression of glucocerebrosidase corrects alpha-synuclein pathology and memory in a mouse model of Gaucher-related synucleinopathy. *Proceedings of the National Academy of Sciences of the United States of America* **2011**, *108* (29), 12101-6.

[143] Sun, Y.; Ran, H.; Liou, B.; Quinn, B.; Zamzow, M.; Zhang, W.; Bielawski, J.; Kitatani, K.; Setchell, K. D.; Hannun, Y. A.; Grabowski, G. A., Isofagomine *in vivo* effects in a neuronopathic Gaucher disease mouse. *PloS one* **2011**, *6* (4), e19037.

[144] (a) West, A. B.; Moore, D. J.; Biskup, S.; Bugayenko, A.; Smith, W. W.; Ross, C. A.; Dawson, V. L.; Dawson, T. M., Parkinson's disease-associated mutations in leucine-rich repeat kinase 2 augment kinase activity. *Proceedings of the National Academy of Sciences of the United States of America* **2005**, *102* (46), 16842-7; (b) Heo, H. Y.; Park, J. M.; Kim, C. H.; Han, B. S.; Kim, K. S.; Seol, W., LRRK2 enhances oxidative stress-induced neurotoxicity via its kinase activity. *Experimental cell research* **2010**, *316* (4), 649-56; (c) Smith, W. W.; Pei, Z.; Jiang, H.; Dawson, V. L.; Dawson, T. M.; Ross, C. A., Kinase activity of mutant LRRK2 mediates neuronal toxicity. *Nature neuroscience* **2006**, *9* (10), 1231-3.

[145] (a) Daniels, V.; Baekelandt, V.; Taymans, J. M., On the road to leucine-rich repeat kinase 2 signalling: evidence from cellular and *in vivo* studies. *Neuro-Signals* **2011**, *19* (1), 1-15; (b) Milosevic, J.; Schwarz, S. C.; Ogunlade, V.; Meyer, A. K.; Storch, A.; Schwarz, J., Emerging role of LRRK2 in human neural progenitor cell cycle progression, survival and differentiation. *Molecular neurodegeneration* **2009**, *4*, 25.

[146] (a) Ramonet, D.; Daher, J. P.; Lin, B. M.; Stafa, K.; Kim, J.; Banerjee, R.; Westerlund, M.; Pletnikova, O.; Glauser, L.; Yang, L.; Liu, Y.; Swing, D. A.; Beal, M. F.; Troncoso, J. C.; McCaffery, J. M.; Jenkins, N. A.; Copeland, N. G.; Galter, D.; Thomas, B.; Lee, M. K.; Dawson, T. M.; Dawson, V. L.; Moore, D. J., Dopaminergic neuronal loss, reduced neurite complexity and autophagic abnormalities in transgenic mice expressing G2019S mutant LRRK2. *PloS one* **2011**, *6* (4), e18568; (b) Li, Y.; Liu, W.; Oo, T. F.; Wang, L.; Tang, Y.; Jackson-Lewis, V.; Zhou, C.; Geghman, K.; Bogdanov, M.; Przedborski, S.; Beal, M. F.; Burke, R. E.; Li, C., Mutant LRRK2(R1441G) BAC transgenic mice recapitulate cardinal features of Parkinson's disease. *Nature neuroscience* **2009**, *12* (7), 826-8.

[147] Tong, Y.; Yamaguchi, H.; Giaime, E.; Boyle, S.; Kopan, R.; Kelleher, R. J., 3rd; Shen, J., Loss of leucine-rich repeat kinase 2 causes impairment of protein degradation pathways, accumulation of alpha-synuclein, and apoptotic cell death in aged mice. *Proceedings of the National Academy of Sciences of the United States of America* **2010**, *107* (21), 9879-84.

[148] Hinkle, K. M.; Yue, M.; Behrouz, B.; Dachsel, J. C.; Lincoln, S. J.; Bowles, E. E.; Beevers, J. E.; Dugger, B.; Winner, B.; Prots, I.; Kent, C. B.; Nishioka, K.; Lin, W. L.; Dickson, D. W.; Janus, C. J.; Farrer, M. J.; Melrose, H. L., LRRK2 knockout mice have an intact dopaminergic system but display alterations in exploratory and motor co-ordination behaviors. *Molecular neurodegeneration* **2012**, *7*, 25.

[149] Herzig, M. C.; Bidinosti, M.; Schweizer, T.; Hafner, T.; Stemmelen, C.; Weiss, A.; Danner, S.; Vidotto, N.; Stauffer, D.; Barske, C.; Mayer, F.; Schmid, P.; Rovelli, G.; van der Putten, P. H.; Shimshek, D. R., High LRRK2 levels fail to induce or exacerbate neuronal alpha-synucleinopathy in mouse brain. *PloS one* **2012**, *7* (5), e36581.

[150] (a) Chen, H.; Chan, B. K.; Drummond, J.; Estrada, A. A.; Gunzner-Toste, J.; Liu, X.; Liu, Y.; Moffat, J.; Shore, D.; Sweeney, Z. K.; Tran, T.; Wang, S.; Zhao, G.; Zhu, H.; Burdick, D. J., Discovery of selective LRRK2 inhibitors guided by computational analysis and molecular modeling. *Journal of medicinal chemistry* **2012**, *55* (11), 5536-45; (b) Reith, A. D.; Bamborough, P.; Jandu, K.; Andreotti, D.; Mensah, L.; Dossang, P.; Choi, H. G.; Deng, X.; Zhang, J.; Alessi, D. R.; Gray, N. S., GSK2578215A; a potent and highly selective 2-arylmethyloxy-5-substitutent-N-arylbenzamide LRRK2 kinase inhibitor. *Bioorganic & medicinal chemistry letters* **2012**, *22* (17), 5625-9.

[151] (a) Choi, H. G. Z., J. Deng, X. Hatcher, J. M. Patricelli, M. P. Zhao, Z. Alessi D. R. Gray, N. S. , Brain Penetrant LRRK2 Inhibitor. *Medicinal Chemistry Letters* **2012**, *3* (8), 658-662; (b) Estrada, A. A.; Liu, X.; Baker-Glenn, C.; Beresford, A.; Burdick, D. J.; Chambers, M.; Chan, B. K.; Chen, H.; Ding, X.; DiPasquale, A. G.; Dominguez, S. L.; Dotson, J.; Drummond, J.; Flagella, M.; Flynn, S.; Fuji, R.; Gill, A.; Gunzner-Toste, J.; Harris, S. F.; Heffron, T. P.; Kleinheinz, T.; Lee, D. W.; Le Pichon, C. E.; Lyssikatos, J. P.; Medhurst, A. D.; Moffat, J. G.; Mukund, S.; Nash, K.; Scearce-Levie, K.; Sheng, Z.; Shore, D. G.; Tran, T.; Trivedi, N.; Wang, S.; Zhang, S.; Zhang, X.; Zhao, G.; Zhu, H.; Sweeney, Z. K., Discovery of highly potent, selective, and brain-penetrable leucine-rich repeat kinase 2 (LRRK2) small molecule inhibitors. *Journal of medicinal chemistry* **2012**, *55* (22), 9416-33.

[152] Schapira, A. H.; Cooper, J. M.; Dexter, D.; Jenner, P.; Clark, J. B.; Marsden, C. D., Mitochondrial complex I deficiency in Parkinson's disease. *Lancet* **1989**, *1* (8649), 1269.

[153] Effects of tocopherol and deprenyl on the progression of disability in early Parkinson's disease. The Parkinson Study Group. *The New England journal of medicine* **1993**, *328* (3), 176-83.

[154] Shults, C. W.; Oakes, D.; Kieburtz, K.; Beal, M. F.; Haas, R.; Plumb, S.; Juncos, J. L.; Nutt, J.; Shoulson, I.; Carter, J.; Kompoliti, K.; Perlmutter, J. S.; Reich, S.; Stern, M.; Watts, R. L.; Kurlan, R.; Molho, E.; Harrison, M.; Lew, M.; Parkinson Study, G., Effects of coenzyme Q10 in early Parkinson disease: evidence of slowing of the functional decline. *Archives of neurology* **2002**, *59* (10), 1541-50.

[155] Murphy, M. P.; Smith, R. A., Targeting antioxidants to mitochondria by conjugation to lipophilic cations. *Annual review of pharmacology and toxicology* **2007**, *47*, 629-56.

[156] Adlam, V. J.; Harrison, J. C.; Porteous, C. M.; James, A. M.; Smith, R. A.; Murphy, M. P.; Sammut, I. A., Targeting an antioxidant to mitochondria decreases cardiac ischemia-reperfusion injury. *FASEB journal : official publication of the Federation of American Societies for Experimental Biology* **2005**, *19* (9), 1088-95.

[157] Chandran, K.; Aggarwal, D.; Migrino, R. Q.; Joseph, J.; McAllister, D.; Konorev, E. A.; Antholine, W. E.; Zielonka, J.; Srinivasan, S.; Avadhani, N. G.; Kalyanaraman, B., Doxorubicin inactivates myocardial cytochrome c oxidase in rats: cardioprotection by Mito-Q. *Biophysical journal* **2009**, *96* (4), 1388-98.

[158] Snow, B. J.; Rolfe, F. L.; Lockhart, M. M.; Frampton, C. M.; O'Sullivan, J. D.; Fung, V.; Smith, R. A.; Murphy, M. P.; Taylor, K. M.; Protect Study, G., A double-blind, placebo-controlled study to assess the mitochondria-targeted antioxidant MitoQ as a disease-modifying therapy in Parkinson's disease. *Movement disorders: official journal of the Movement Disorder Society* **2010**, *25* (11), 1670-4.

[159] (a) Devi, L.; Anandatheerthavarada, H. K., Mitochondrial trafficking of APP and alpha synuclein: Relevance to mitochondrial dysfunction in Alzheimer's and Parkinson's diseases. *Biochimica et biophysica acta* **2010**, *1802* (1), 11-9; (b) Song, D. D.; Shults, C. W.; Sisk, A.; Rockenstein, E.; Masliah, E., Enhanced substantia nigra mitochondrial pathology in human alpha-synuclein transgenic mice after treatment with MPTP. *Experimental neurology* **2004**, *186* (2), 158-72.

[160] Chinta, S. J.; Mallajosyula, J. K.; Rane, A.; Andersen, J. K., Mitochondrial alpha-synuclein accumulation impairs complex I function in dopaminergic neurons and results in increased mitophagy *in vivo. Neuroscience letters* **2010**, *486* (3), 235-9.

[161] Kamp, F.; Exner, N.; Lutz, A. K.; Wender, N.; Hegermann, J.; Brunner, B.; Nuscher, B.; Bartels, T.; Giese, A.; Beyer, K.; Eimer, S.; Winklhofer, K. F.; Haass, C., Inhibition of mitochondrial fusion by alpha-synuclein is rescued by PINK1, Parkin and DJ-1. *The EMBO journal* **2010**, *29* (20), 3571-89.

[162] Valente, E. M.; Abou-Sleiman, P. M.; Caputo, V.; Muqit, M. M.; Harvey, K.; Gispert, S.; Ali, Z.; Del Turco, D.; Bentivoglio, A. R.; Healy, D. G.; Albanese, A.; Nussbaum, R.; Gonzalez-Maldonado, R.; Deller, T.; Salvi, S.; Cortelli, P.; Gilks, W. P.; Latchman, D. S.; Harvey, R. J.; Dallapiccola, B.; Auburger, G.; Wood, N. W., Hereditary early-onset Parkinson's disease caused by mutations in PINK1. *Science* **2004**, *304* (5674), 1158-60.

[163] Weihofen, A.; Thomas, K. J.; Ostaszewski, B. L.; Cookson, M. R.; Selkoe, D. J., Pink1 forms a multiprotein complex with Miro and Milton, linking Pink1 function to mitochondrial trafficking. *Biochemistry* **2009**, *48* (9), 2045-52.

[164] (a) Clark, I. E.; Dodson, M. W.; Jiang, C.; Cao, J. H.; Huh, J. R.; Seol, J. H.; Yoo, S. J.; Hay, B. A.; Guo, M., Drosophila pink1 is required for mitochondrial function and interacts genetically with parkin. *Nature* **2006**, *441* (7097), 1162-6; (b) Sandebring, A.; Thomas, K. J.; Beilina, A.; van der Brug, M.; Cleland, M. M.; Ahmad, R.; Miller, D. W.; Zambrano, I.; Cowburn, R. F.; Behbahani, H.; Cedazo-Minguez, A.; Cookson, M. R., Mitochondrial alterations in PINK1 deficient cells are influenced by calcineurin-dependent dephosphorylation of dynamin-related protein 1. *PloS one* **2009**, *4* (5), e5701.

[165] Kondapalli, C.; Kazlauskaite, A.; Zhang, N.; Woodroof, H. I.; Campbell, D. G.; Gourlay, R.; Burchell, L.; Walden, H.; Macartney, T. J.; Deak, M.; Knebel, A.; Alessi, D. R.; Muqit, M. M., PINK1 is activated by mitochondrial membrane potential depolarization and stimulates Parkin E3 ligase activity by phosphorylating Serine 65. *Open biology* **2012**, *2* (5), 120080.

[166] Jin, S. M.; Lazarou, M.; Wang, C.; Kane, L. A.; Narendra, D. P.; Youle, R. J., Mitochondrial membrane potential regulates PINK1 import and proteolytic destabilization by PARL. *The Journal of cell biology* **2010**, *191* (5), 933-42.

[167] Takatori, S.; Ito, G.; Iwatsubo, T., Cytoplasmic localization and proteasomal degradation of N-terminally cleaved form of PINK1. *Neuroscience letters* **2008**, *430* (1), 13-7.

[168] (a) Vives-Bauza, C.; Zhou, C.; Huang, Y.; Cui, M.; de Vries, R. L.; Kim, J.; May, J.; Tocilescu, M. A.; Liu, W.; Ko, H. S.; Magrane, J.; Moore, D. J.; Dawson, V. L.; Grailhe, R.; Dawson, T. M.; Li, C.; Tieu, K.; Przedborski, S., PINK1-dependent recruitment of Parkin to mitochondria in mitophagy. *Proceedings of the National Academy of Sciences of the United States of America* **2010**, *107* (1), 378-83; (b) Narendra, D. P.; Jin, S. M.; Tanaka, A.; Suen, D. F.; Gautier, C. A.; Shen, J.; Cookson, M. R.; Youle, R. J., PINK1 is selectively stabilized on impaired mitochondria to activate Parkin. *PLoS biology* **2010**, *8* (1), e1000298.

[169] Kitada, T.; Asakawa, S.; Hattori, N.; Matsumine, H.; Yamamura, Y.; Minoshima, S.; Yokochi, M.; Mizuno, Y.; Shimizu, N., Mutations in the parkin gene cause autosomal recessive juvenile parkinsonism. *Nature* **1998**, *392* (6676), 605-8.

[170] (a) Mortiboys, H.; Thomas, K. J.; Koopman, W. J.; Klaffke, S.; Abou-Sleiman, P.; Olpin, S.; Wood, N. W.; Willems, P. H.; Smeitink, J. A.; Cookson, M. R.; Bandmann, O., Mitochondrial function and morphology are impaired in parkin-mutant fibroblasts. *Annals of neurology* **2008**, *64* (5), 555-65; (b) Grunewald, A.; Voges, L.; Rakovic, A.; Kasten, M.; Vandebona, H.; Hemmelmann, C.; Lohmann, K.; Orolicki, S.; Ramirez, A.; Schapira, A.

H.; Pramstaller, P. P.; Sue, C. M.; Klein, C., Mutant Parkin impairs mitochondrial function and morphology in human fibroblasts. *PloS one* **2010**, *5* (9), e12962.

[171] Palacino, J. J.; Sagi, D.; Goldberg, M. S.; Krauss, S.; Motz, C.; Wacker, M.; Klose, J.; Shen, J., Mitochondrial dysfunction and oxidative damage in parkin-deficient mice. *The Journal of biological chemistry* **2004**, *279* (18), 18614-22.

[172] Chan, N. C.; Salazar, A. M.; Pham, A. H.; Sweredoski, M. J.; Kolawa, N. J.; Graham, R. L.; Hess, S.; Chan, D. C., Broad activation of the ubiquitin-proteasome system by Parkin is critical for mitophagy. *Human molecular genetics* **2011**, *20* (9), 1726-37.

[173] (a) Glauser, L.; Sonnay, S.; Stafa, K.; Moore, D. J., Parkin promotes the ubiquitination and degradation of the mitochondrial fusion factor mitofusin 1. *Journal of neurochemistry* **2011**, *118* (4), 636-45; (b) Rakovic, A.; Grunewald, A.; Kottwitz, J.; Bruggemann, N.; Pramstaller, P. P.; Lohmann, K.; Klein, C., Mutations in PINK1 and Parkin impair ubiquitination of Mitofusins in human fibroblasts. *PloS one* **2011**, *6* (3), e16746.

[174] Karbowski, M.; Youle, R. J., Regulating mitochondrial outer membrane proteins by ubiquitination and proteasomal degradation. *Current opinion in cell biology* **2011**, *23* (4), 476-82.

[175] Kuroda, Y.; Mitsui, T.; Kunishige, M.; Shono, M.; Akaike, M.; Azuma, H.; Matsumoto, T., Parkin enhances mitochondrial biogenesis in proliferating cells. *Human molecular genetics* **2006**, *15* (6), 883-95.

[176] Shin, J. H.; Ko, H. S.; Kang, H.; Lee, Y.; Lee, Y. I.; Pletinkova, O.; Troconso, J. C.; Dawson, V. L.; Dawson, T. M., PARIS (ZNF746) repression of PGC-1alpha contributes to neurodegeneration in Parkinson's disease. *Cell* **2011**, *144* (5), 689-702.

[177] (a) Junn, E.; Jang, W. H.; Zhao, X.; Jeong, B. S.; Mouradian, M. M., Mitochondrial localization of DJ-1 leads to enhanced neuroprotection. *Journal of neuroscience research* **2009**, *87* (1), 123-9; (b) Thomas, K. J.; McCoy, M. K.; Blackinton, J.; Beilina, A.; van der Brug, M.; Sandebring, A.; Miller, D.; Maric, D.; Cedazo-Minguez, A.; Cookson, M. R., DJ-1 acts in parallel to the PINK1/parkin pathway to control mitochondrial function and autophagy. *Human molecular genetics* **2011**, *20* (1), 40-50.

[178] Bonifati, V.; Rizzu, P.; van Baren, M. J.; Schaap, O.; Breedveld, G. J.; Krieger, E.; Dekker, M. C.; Squitieri, F.; Ibanez, P.; Joosse, M.; van Dongen, J. W.; Vanacore, N.; van Swieten, J. C.; Brice, A.; Meco, G.; van Duijn, C. M.; Oostra, B. A.; Heutink, P., Mutations in the DJ-1 gene associated with autosomal recessive early-onset parkinsonism. *Science* **2003**, *299* (5604), 256-9.

[179] Hao, L. Y.; Giasson, B. I.; Bonini, N. M., DJ-1 is critical for mitochondrial function and rescues PINK1 loss of function. *Proceedings of the National Academy of Sciences of the United States of America* **2010**, *107* (21), 9747-52.

[180] Lev, N.; Ickowicz, D.; Melamed, E.; Offen, D., Oxidative insults induce DJ-1 upregulation and redistribution: implications for neuroprotection. *Neurotoxicology* **2008**, *29* (3), 397-405.

[181] Park, J.; Kim, S. Y.; Cha, G. H.; Lee, S. B.; Kim, S.; Chung, J., Drosophila DJ-1 mutants show oxidative stress-sensitive locomotive dysfunction. *Gene* **2005**, *361*, 133-9.

[182] Batelli, S.; Albani, D.; Rametta, R.; Polito, L.; Prato, F.; Pesaresi, M.; Negro, A.; Forloni, G., DJ-1 modulates alpha-synuclein aggregation state in a cellular model of oxidative stress: relevance for Parkinson's disease and involvement of HSP70. *PloS one* **2008**, *3* (4), e1884.

[183] Dexter, D. T.; Wells, F. R.; Agid, F.; Agid, Y.; Lees, A. J.; Jenner, P.; Marsden, C. D., Increased nigral iron content in postmortem parkinsonian brain. *Lancet* **1987**, *2* (8569), 1219-20.

[184] Sengstock, G. J.; Olanow, C. W.; Menzies, R. A.; Dunn, A. J.; Arendash, G. W., Infusion of iron into the rat substantia nigra: nigral pathology and dose-dependent loss of striatal dopaminergic markers. *Journal of neuroscience research* **1993**, *35* (1), 67-82.

[185] Gorell, J. M.; Johnson, C. C.; Rybicki, B. A.; Peterson, E. L.; Kortsha, G. X.; Brown, G. G.; Richardson, R. J., Occupational exposures to metals as risk factors for Parkinson's disease. *Neurology* **1997**, *48* (3), 650-8.

[186] Ungerstedt, U.; Ljungberg, T.; Steg, G., Behavioral, physiological, and neurochemical changes after 6-hydroxydopamine-induced degeneration of the nigro-striatal dopamine neurons. *Advances in neurology* **1974**, *5*, 421-6.

[187] Mendez-Alvarez, E.; Soto-Otero, R.; Hermida-Ameijeiras, A.; Lopez-Martin, M. E.; Labandeira-Garcia, J. L., Effect of iron and manganese on hydroxyl radical production by 6-hydroxydopamine: mediation of antioxidants. *Free radical biology & medicine* **2001**, *31* (8), 986-98.

[188] Gerlach, M.; Double, K. L.; Ben-Shachar, D.; Zecca, L.; Youdim, M. B.; Riederer, P., Neuromelanin and its interaction with iron as a potential risk factor for dopaminergic neurodegeneration underlying Parkinson's disease. *Neurotoxicity research* **2003**, *5* (1-2), 35-44.

[189] Kontoghiorghes, G. J.; Kolnagou, A.; Peng, C. T.; Shah, S. V.; Aessopos, A., Safety issues of iron chelation therapy in patients with normal range iron stores including thalassaemia, neurodegenerative, renal and infectious diseases. *Expert opinion on drug safety* **2010**, *9* (2), 201-6.

[190] (a) Shachar, D. B.; Kahana, N.; Kampel, V.; Warshawsky, A.; Youdim, M. B., Neuroprotection by a novel brain permeable iron chelator, VK-28, against 6-hydroxydopamine lession in rats. *Neuropharmacology* **2004**, *46* (2), 254-63; (b) Kaur, D.; Yantiri, F.; Rajagopalan, S.; Kumar, J.; Mo, J. Q.; Boonplueang, R.; Viswanath, V.; Jacobs, R.; Yang, L.; Beal, M. F.; DiMonte, D.; Volitaskis, I.; Ellerby, L.; Cherny, R. A.; Bush, A. I.; Andersen, J. K., Genetic or pharmacological iron chelation prevents MPTP-induced neurotoxicity *in vivo*: a novel therapy for Parkinson's disease. *Neuron* **2003**, *37* (6), 899-909; (c) Dexter, D. T.; Statton, S. A.; Whitmore, C.; Freinbichler, W.; Weinberger, P.; Tipton, K. F.; Della Corte, L.; Ward, R. J.; Crichton, R. R., Clinically available iron chelators induce neuroprotection in the 6-OHDA model of Parkinson's disease after peripheral administration. *Journal of neural transmission* **2011**, *118* (2), 223-31.

[191] Tateishi, J., Subacute myelo-optico-neuropathy: clioquinol intoxication in humans and animals. *Neuropathology : official journal of the Japanese Society of Neuropathology* **2000**, *20 Suppl*, S20-4.

[192] (a) Ritchie, C. W.; Bush, A. I.; Mackinnon, A.; Macfarlane, S.; Mastwyk, M.; MacGregor, L.; Kiers, L.; Cherny, R.; Li, Q. X.; Tammer, A.; Carrington, D.; Mavros, C.; Volitakis, I.; Xilinas, M.; Ames, D.; Davis, S.; Beyreuther, K.; Tanzi, R. E.; Masters, C. L., Metal-protein attenuation with iodochlorhydroxyquin (clioquinol) targeting Abeta amyloid deposition and toxicity in Alzheimer disease: a pilot phase 2 clinical trial. *Archives of neurology* **2003**, *60* (12), 1685-91; (b) Sampson, E.; Jenagaratnam, L.; McShane, R., Metal protein attenuating compounds for the treatment of Alzheimer's disease. *Cochrane database of systematic reviews* **2008**, (1), CD005380; (c) Sampson, E. L.; Jenagaratnam, L.; McShane, R., Metal protein attenuating compounds for the treatment of Alzheimer's dementia. *Cochrane database of systematic reviews* **2012**, *5*, CD005380.

[193] Lannfelt, L.; Blennow, K.; Zetterberg, H.; Batsman, S.; Ames, D.; Harrison, J.; Masters, C. L.; Targum, S.; Bush, A. I.; Murdoch, R.; Wilson, J.; Ritchie, C. W.; group, P. E. s., Safety, efficacy, and biomarker findings of PBT2 in targeting Abeta as a modifying therapy for Alzheimer's disease: a phase IIa, double-blind, randomised, placebo-controlled trial. *Lancet neurology* **2008**, *7* (9), 779-86.

[194] http://www.clinicaltrials.gov/ct2/show/NCT01590888.

[195] http://www.clinicaltrials.gov/ct2/show/NCT00943748.

[196] Roy, S.; Preston, J. E.; Hider, R. C.; Ma, Y. M., Glucosylated deferiprone and its brain uptake: implications for developing glucosylated hydroxypyridinone analogues intended to cross the blood-brain barrier. *Journal of medicinal chemistry* **2010**, *53* (15), 5886-9.

[197] Ward, R. J.; Dexter, D.; Florence, A.; Aouad, F.; Hider, R.; Jenner, P.; Crichton, R. R., Brain iron in the ferrocene-loaded rat: its chelation and influence on dopamine metabolism. *Biochemical pharmacology* **1995**, *49* (12), 1821-6.

[198] Singh, S.; Epemolu, R. O.; Dobbin, P. S.; Tilbrook, G. S.; Ellis, B. L.; Damani, L. A.; Hider, R. C., Urinary metabolic profiles in human and rat of 1,2-dimethyl- and 1,2-diethyl-substituted 3-hydroxypyridin-4-ones. *Drug metabolism and disposition: the biological fate of chemicals* **1992**, *20* (2), 256-61.

[199] Ma, Y.; Roy, S.; Kong, X.; Chen, Y.; Liu, D.; Hider, R. C., Design and synthesis of fluorinated iron chelators for metabolic study and brain uptake. *Journal of medicinal chemistry* **2012**, *55* (5), 2185-95.

[200] Gogoi, S.; Antonio, T.; Rajagopalan, S.; Reith, M.; Andersen, J.; Dutta, A. K., Dopamine D(2)/D(3) agonists with potent iron chelation, antioxidant and neuroprotective properties: potential implication in symptomatic and neuroprotective treatment of Parkinson's disease. *ChemMedChem* **2011**, *6* (6), 991-5.

[201] Ritz, B.; Rhodes, S. L.; Qian, L.; Schernhammer, E.; Olsen, J. H.; Friis, S., L-type calcium channel blockers and Parkinson disease in Denmark. *Annals of neurology* **2010**, *67* (5), 600-6.

[202] Ilijic, E.; Guzman, J. N.; Surmeier, D. J., The L-type channel antagonist isradipine is neuroprotective in a mouse model of Parkinson's disease. *Neurobiology of disease* **2011**, *43* (2), 364-71.

[203] Goonasekera, S. A.; Hammer, K.; Auger-Messier, M.; Bodi, I.; Chen, X.; Zhang, H.; Reiken, S.; Elrod, J. W.; Correll, R. N.; York, A. J.; Sargent, M. A.; Hofmann, F.; Moosmang, S.; Marks, A. R.; Houser, S. R.; Bers, D. M.; Molkentin, J. D., Decreased cardiac L-type Ca^{2+} channel activity induces hypertrophy and heart failure in mice. *The Journal of Clinical Investigation* **2012**, *122* (1), 280-90.

[204] Sinnegger-Brauns, M. J.; Huber, I. G.; Koschak, A.; Wild, C.; Obermair, G. J.; Einzinger, U.; Hoda, J. C.; Sartori, S. B.; Striessnig, J., Expression and 1,4-dihydropyridine-binding properties of brain L-type calcium channel isoforms. *Molecular pharmacology* **2009**, *75* (2), 407-14.

[205] Triggle, D. J., Calcium channel antagonists: clinical uses-past, present and future. *Biochemical pharmacology* **2007**, *74* (1), 1-9.

[206] Martin, J. B.; Gusella, J. F., Huntington's disease. Pathogenesis and management. *The New England journal of medicine* **1986**, *315* (20), 1267-76.

[207] Ferrante, R. J.; Kowall, N. W.; Beal, M. F.; Martin, J. B.; Bird, E. D.; Richardson, E. P., Jr., Morphologic and histochemical characteristics of a spared subset of striatal neurons in

Huntington's disease. *Journal of neuropathology and experimental neurology* **1987**, *46* (1), 12-27.

[208] (a) Aylward, E. H.; Sparks, B. F.; Field, K. M.; Yallapragada, V.; Shpritz, B. D.; Rosenblatt, A.; Brandt, J.; Gourley, L. M.; Liang, K.; Zhou, H.; Margolis, R. L.; Ross, C. A., Onset and rate of striatal atrophy in preclinical Huntington disease. *Neurology* **2004**, *63* (1), 66-72; (b) Kassubek, J.; Juengling, F. D.; Kioschies, T.; Henkel, K.; Karitzky, J.; Kramer, B.; Ecker, D.; Andrich, J.; Saft, C.; Kraus, P.; Aschoff, A. J.; Ludolph, A. C.; Landwehrmeyer, G. B., Topography of cerebral atrophy in early Huntington's disease: a voxel based morphometric MRI study. *Journal of neurology, neurosurgery, and psychiatry* **2004**, *75* (2), 213-20; (c) Ruocco, H. H.; Lopes-Cendes, I.; Li, L. M.; Santos-Silva, M.; Cendes, F., Striatal and extrastriatal atrophy in Huntington's disease and its relationship with length of the CAG repeat. *Brazilian journal of medical and biological research = Revista brasileira de pesquisas medicas e biologicas / Sociedade Brasileira de Biofisica ... [et al.]* **2006**, *39* (8), 1129-36.

[209] DiFiglia, M.; Sapp, E.; Chase, K. O.; Davies, S. W.; Bates, G. P.; Vonsattel, J. P.; Aronin, N., Aggregation of huntingtin in neuronal intranuclear inclusions and dystrophic neurites in brain. *Science* **1997**, *277* (5334), 1990-3.

[210] Arrasate, M.; Mitra, S.; Schweitzer, E. S.; Segal, M. R.; Finkbeiner, S., Inclusion body formation reduces levels of mutant huntingtin and the risk of neuronal death. *Nature* **2004**, *431* (7010), 805-10.

[211] Sanchez, I.; Mahlke, C.; Yuan, J., Pivotal role of oligomerization in expanded polyglutamine neurodegenerative disorders. *Nature* **2003**, *421* (6921), 373-9.

[212] (a) Chan, D. C., Mitochondria: dynamic organelles in disease, aging, and development. *Cell* **2006**, *125* (7), 1241-52; (b) Green, D. R.; Kroemer, G., The pathophysiology of mitochondrial cell death. *Science* **2004**, *305* (5684), 626-9; (c) Green, D. R.; Reed, J. C., Mitochondria and apoptosis. *Science* **1998**, *281* (5381), 1309-12.

[213] (a) Lin, M. T.; Beal, M. F., Mitochondrial dysfunction and oxidative stress in neurodegenerative diseases. *Nature* **2006**, *443* (7113), 787-95; (b) Mattson, M. P., Neuronal life-and-death signaling, apoptosis, and neurodegenerative disorders. *Antioxidants & redox signaling* **2006**, *8* (11-12), 1997-2006.

[214] (a) Brouillet, E.; Conde, F.; Beal, M. F.; Hantraye, P., Replicating Huntington's disease phenotype in experimental animals. *Progress in neurobiology* **1999**, *59* (5), 427-68; (b) Beal, M. F., Does impairment of energy metabolism result in excitotoxic neuronal death in neurodegenerative illnesses? *Annals of neurology* **1992**, *31* (2), 119-30.

[215] (a) McGill, J. K.; Beal, M. F., PGC-1alpha, a new therapeutic target in Huntington's disease? *Cell* **2006**, *127* (3), 465-8; (b) Cui, L.; Jeong, H.; Borovecki, F.; Parkhurst, C. N.; Tanese, N.; Krainc, D., Transcriptional repression of PGC-1alpha by mutant huntingtin leads to mitochondrial dysfunction and neurodegeneration. *Cell* **2006**, *127* (1), 59-69.

[216] Lin, J.; Wu, P. H.; Tarr, P. T.; Lindenberg, K. S.; St-Pierre, J.; Zhang, C. Y.; Mootha, V. K.; Jager, S.; Vianna, C. R.; Reznick, R. M.; Cui, L.; Manieri, M.; Donovan, M. X.; Wu, Z.; Cooper, M. P.; Fan, M. C.; Rohas, L. M.; Zavacki, A. M.; Cinti, S.; Shulman, G. I.; Lowell, B. B.; Krainc, D.; Spiegelman, B. M., Defects in adaptive energy metabolism with CNS-linked hyperactivity in PGC-1alpha null mice. *Cell* **2004**, *119* (1), 121-35.

[217] Scarpulla, R. C., Nuclear control of respiratory gene expression in mammalian cells. *Journal of cellular biochemistry* **2006**, *97* (4), 673-83.

[218] Estrada Sanchez, A. M.; Mejia-Toiber, J.; Massieu, L., Excitotoxic neuronal death and the pathogenesis of Huntington's disease. *Archives of medical research* **2008**, *39* (3), 265-76.

[219] (a) Maiuri, M. C.; Zalckvar, E.; Kimchi, A.; Kroemer, G., Self-eating and self-killing: crosstalk between autophagy and apoptosis. *Nature reviews. Molecular cell biology* **2007**, *8* (9), 741-52; (b) Mizushima, N.; Klionsky, D. J., Protein turnover via autophagy: implications for metabolism. *Annual review of nutrition* **2007**, *27*, 19-40.

[220] Hara, T.; Nakamura, K.; Matsui, M.; Yamamoto, A.; Nakahara, Y.; Suzuki-Migishima, R.; Yokoyama, M.; Mishima, K.; Saito, I.; Okano, H.; Mizushima, N., Suppression of basal autophagy in neural cells causes neurodegenerative disease in mice. *Nature* **2006**, *441* (7095), 885-9.

[221] Martinez-Vicente, M.; Talloczy, Z.; Wong, E.; Tang, G.; Koga, H.; Kaushik, S.; de Vries, R.; Arias, E.; Harris, S.; Sulzer, D.; Cuervo, A. M., Cargo recognition failure is responsible for inefficient autophagy in Huntington's disease. *Nature neuroscience* **2010**, *13* (5), 567-76.

[222] Davies, S. W.; Turmaine, M.; Cozens, B. A.; DiFiglia, M.; Sharp, A. H.; Ross, C. A.; Scherzinger, E.; Wanker, E. E.; Mangiarini, L.; Bates, G. P., Formation of neuronal intranuclear inclusions underlies the neurological dysfunction in mice transgenic for the HD mutation. *Cell* **1997**, *90* (3), 537-48.

[223] Kuemmerle, S.; Gutekunst, C. A.; Klein, A. M.; Li, X. J.; Li, S. H.; Beal, M. F.; Hersch, S. M.; Ferrante, R. J., Huntington aggregates may not predict neuronal death in Huntington's disease. *Annals of neurology* **1999**, *46* (6), 842-9.

[224] Vonsattel, J. P.; Myers, R. H.; Stevens, T. J.; Ferrante, R. J.; Bird, E. D.; Richardson, E. P., Jr., Neuropathological classification of Huntington's disease. *Journal of neuropathology and experimental neurology* **1985**, *44* (6), 559-77.

[225] Kaplan, A.; Stockwell, B. R., Therapeutic approaches to preventing cell death in Huntington disease. *Progress in neurobiology* **2012**, *99* (3), 262-80.

[226] Fecke, W.; Gianfriddo, M.; Gaviraghi, G.; Terstappen, G. C.; Heitz, F., Small molecule drug discovery for Huntington's Disease. *Drug discovery today* **2009**, *14* (9-10), 453-64.

[227] Titus, S. A.; Southall, N.; Marugan, J.; Austin, C. P.; Zheng, W., High-Throughput Multiplexed Quantitation of Protein Aggregation and Cytotoxicity in a Huntington's Disease Model. *Current chemical genomics* **2012**, *6*, 79-86.

[228] Baldo, B.; Weiss, A.; Parker, C. N.; Bibel, M.; Paganetti, P.; Kaupmann, K., A screen for enhancers of clearance identifies huntingtin as a heat shock protein 90 (Hsp90) client protein. *The Journal of biological chemistry* **2012**, *287* (2), 1406-14.

[229] Fuentealba, R. A.; Marasa, J.; Diamond, M. I.; Piwnica-Worms, D.; Weihl, C. C., An aggregation sensing reporter identifies leflunomide and teriflunomide as polyglutamine aggregate inhibitors. *Human molecular genetics* **2012**, *21* (3), 664-80.

[230] Sarkar, S.; Perlstein, E. O.; Imarisio, S.; Pineau, S.; Cordenier, A.; Maglathlin, R. L.; Webster, J. A.; Lewis, T. A.; O'Kane, C. J.; Schreiber, S. L.; Rubinsztein, D. C., Small molecules enhance autophagy and reduce toxicity in Huntington's disease models. *Nature chemical biology* **2007**, *3* (6), 331-8.

[231] Ehrnhoefer, D. E.; Duennwald, M.; Markovic, P.; Wacker, J. L.; Engemann, S.; Roark, M.; Legleiter, J.; Marsh, J. L.; Thompson, L. M.; Lindquist, S.; Muchowski, P. J.; Wanker, E. E., Green tea (-)-epigallocatechin-gallate modulates early events in huntingtin misfolding and reduces toxicity in Huntington's disease models. *Human molecular genetics* **2006**, *15* (18), 2743-51.

[232] Steffan, J. S.; Agrawal, N.; Pallos, J.; Rockabrand, E.; Trotman, L. C.; Slepko, N.; Illes, K.; Lukacsovich, T.; Zhu, Y. Z.; Cattaneo, E.; Pandolfi, P. P.; Thompson, L. M.; Marsh, J. L., SUMO modification of Huntingtin and Huntington's disease pathology. *Science* **2004**, *304* (5667), 100-4.

[233] (a) Subramaniam, S.; Sixt, K. M.; Barrow, R.; Snyder, S. H., Rhes, a striatal specific protein, mediates mutant-huntingtin cytotoxicity. *Science* **2009**, *324* (5932), 1327-30; (b) Subramaniam, S.; Mealer, R. G.; Sixt, K. M.; Barrow, R. K.; Usiello, A.; Snyder, S. H., Rhes, a physiologic regulator of sumoylation, enhances cross-sumoylation between the basic sumoylation enzymes E1 and Ubc9. *The Journal of biological chemistry* **2010**, *285* (27), 20428-32.

[234] (a) Nasir, J.; Floresco, S. B.; O'Kusky, J. R.; Diewert, V. M.; Richman, J. M.; Zeisler, J.; Borowski, A.; Marth, J. D.; Phillips, A. G.; Hayden, M. R., Targeted disruption of the Huntington's disease gene results in embryonic lethality and behavioral and morphological changes in heterozygotes. *Cell* **1995**, *81* (5), 811-23; (b) Zeitlin, S.; Liu, J. P.; Chapman, D. L.; Papaioannou, V. E.; Efstratiadis, A., Increased apoptosis and early embryonic lethality in mice nullizygous for the Huntington's disease gene homologue. *Nature genetics* **1995**, *11* (2), 155-63.

[235] Dragatsis, I.; Levine, M. S.; Zeitlin, S., Inactivation of Hdh in the brain and testis results in progressive neurodegeneration and sterility in mice. *Nature genetics* **2000**, *26* (3), 300-6.

[236] Zhang, Y.; Li, M.; Drozda, M.; Chen, M.; Ren, S.; Mejia Sanchez, R. O.; Leavitt, B. R.; Cattaneo, E.; Ferrante, R. J.; Hayden, M. R.; Friedlander, R. M., Depletion of wild-type huntingtin in mouse models of neurologic diseases. *Journal of neurochemistry* **2003**, *87* (1), 101-6.

[237] Humbert, S.; Bryson, E. A.; Cordelieres, F. P.; Connors, N. C.; Datta, S. R.; Finkbeiner, S.; Greenberg, M. E.; Saudou, F., The IGF-1/Akt pathway is neuroprotective in Huntington's disease and involves Huntingtin phosphorylation by Akt. *Developmental cell* **2002**, *2* (6), 831-7.

[238] Luthi-Carter, R.; Strand, A.; Peters, N. L.; Solano, S. M.; Hollingsworth, Z. R.; Menon, A. S.; Frey, A. S.; Spektor, B. S.; Penney, E. B.; Schilling, G.; Ross, C. A.; Borchelt, D. R.; Tapscott, S. J.; Young, A. B.; Cha, J. H.; Olson, J. M., Decreased expression of striatal signaling genes in a mouse model of Huntington's disease. *Human molecular genetics* **2000**, *9* (9), 1259-71.

[239] Kuhn, A.; Goldstein, D. R.; Hodges, A.; Strand, A. D.; Sengstag, T.; Kooperberg, C.; Becanovic, K.; Pouladi, M. A.; Sathasivam, K.; Cha, J. H.; Hannan, A. J.; Hayden, M. R.; Leavitt, B. R.; Dunnett, S. B.; Ferrante, R. J.; Albin, R.; Shelbourne, P.; Delorenzi, M.; Augood, S. J.; Faull, R. L.; Olson, J. M.; Bates, G. P.; Jones, L.; Luthi-Carter, R., Mutant huntingtin's effects on striatal gene expression in mice recapitulate changes observed in human Huntington's disease brain and do not differ with mutant huntingtin length or wild-type huntingtin dosage. *Human molecular genetics* **2007**, *16* (15), 1845-61.

[240] Cha, J. H., Transcriptional signatures in Huntington's disease. *Progress in neurobiology* **2007**, *83* (4), 228-48.

[241] Zuccato, C.; Tartari, M.; Crotti, A.; Goffredo, D.; Valenza, M.; Conti, L.; Cataudella, T.; Leavitt, B. R.; Hayden, M. R.; Timmusk, T.; Rigamonti, D.; Cattaneo, E., Huntingtin interacts with REST/NRSF to modulate the transcription of NRSE-controlled neuronal genes. *Nature genetics* **2003**, *35* (1), 76-83.

[242] (a) Dunah, A. W.; Jeong, H.; Griffin, A.; Kim, Y. M.; Standaert, D. G.; Hersch, S. M.; Mouradian, M. M.; Young, A. B.; Tanese, N.; Krainc, D., Sp1 and TAFII130 transcriptional activity disrupted in early Huntington's disease. *Science* **2002**, *296* (5576), 2238-43; (b) Steffan, J. S.; Kazantsev, A.; Spasic-Boskovic, O.; Greenwald, M.; Zhu, Y. Z.; Gohler, H.; Wanker, E. E.; Bates, G. P.; Housman, D. E.; Thompson, L. M., The Huntington's disease protein interacts with p53 and CREB-binding protein and represses transcription. *Proceedings of the National Academy of Sciences of the United States of America* **2000**, *97* (12), 6763-8.

[243] Stack, E. C.; Del Signore, S. J.; Luthi-Carter, R.; Soh, B. Y.; Goldstein, D. R.; Matson, S.; Goodrich, S.; Markey, A. L.; Cormier, K.; Hagerty, S. W.; Smith, K.; Ryu, H.; Ferrante, R. J., Modulation of nucleosome dynamics in Huntington's disease. *Human molecular genetics* **2007**, *16* (10), 1164-75.

[244] Zuccato, C.; Liber, D.; Ramos, C.; Tarditi, A.; Rigamonti, D.; Tartari, M.; Valenza, M.; Cattaneo, E., Progressive loss of BDNF in a mouse model of Huntington's disease and rescue by BDNF delivery. *Pharmacological research : the official journal of the Italian Pharmacological Society* **2005**, *52* (2), 133-9.

[245] Xie, Y.; Hayden, M. R.; Xu, B., BDNF overexpression in the forebrain rescues Huntington's disease phenotypes in YAC128 mice. *The Journal of neuroscience : the official journal of the Society for Neuroscience* **2010**, *30* (44), 14708-18.

[246] Bates, G. P. Huntington's Disease Society of America, 2009.

[247] Simmons, D. A.; Mehta, R. A.; Lauterborn, J. C.; Gall, C. M.; Lynch, G., Brief ampakine treatments slow the progression of Huntington's disease phenotypes in R6/2 mice. *Neurobiology of disease* **2011**, *41* (2), 436-44.

[248] Borrell-Pages, M.; Canals, J. M.; Cordelieres, F. P.; Parker, J. A.; Pineda, J. R.; Grange, G.; Bryson, E. A.; Guillermier, M.; Hirsch, E.; Hantraye, P.; Cheetham, M. E.; Neri, C.; Alberch, J.; Brouillet, E.; Saudou, F.; Humbert, S., Cystamine and cysteamine increase brain levels of BDNF in Huntington disease via HSJ1b and transglutaminase. *The Journal of clinical investigation* **2006**, *116* (5), 1410-24.

[249] Saydoff, J. A.; Garcia, R. A.; Browne, S. E.; Liu, L.; Sheng, J.; Brenneman, D.; Hu, Z.; Cardin, S.; Gonzalez, A.; von Borstel, R. W.; Gregorio, J.; Burr, H.; Beal, M. F., Oral uridine pro-drug PN401 is neuroprotective in the R6/2 and N171-82Q mouse models of Huntington's disease. *Neurobiology of disease* **2006**, *24* (3), 455-65.

[250] Giampa, C.; Laurenti, D.; Anzilotti, S.; Bernardi, G.; Menniti, F. S.; Fusco, F. R., Inhibition of the striatal specific phosphodiesterase PDE10A ameliorates striatal and cortical pathology in R6/2 mouse model of Huntington's disease. *PloS one* **2010**, *5* (10), e13417.

[251] (a) Askenasy, J. J., Approaching disturbed sleep in late Parkinson's Disease: first step toward a proposal for a revised UPDRS. *Parkinsonism & related disorders* **2001**, *8* (2), 123-31; (b) Bates, G. B., Harper P.S., *Huntington's disease*. Ed 3 ed.; Oxford UP: Oxford, 2002; (c) Hatfield, C. F.; Herbert, J.; van Someren, E. J.; Hodges, J. R.; Hastings, M. H., Disrupted daily activity/rest cycles in relation to daily cortisol rhythms of home-dwelling patients with early Alzheimer's dementia. *Brain : a journal of neurology* **2004**, *127* (Pt 5), 1061-74.

[252] Pace-Schott, E. F.; Hobson, J. A., The neurobiology of sleep: genetics, cellular physiology and subcortical networks. *Nature reviews. Neuroscience* **2002**, *3* (8), 591-605.

[253] van der Burg, J. M.; Bjorkqvist, M.; Brundin, P., Beyond the brain: widespread pathology in Huntington's disease. *Lancet neurology* **2009**, *8* (8), 765-74.

[254] (a) Sanberg, P. R.; Fibiger, H. C.; Mark, R. F., Body weight and dietary factors in Huntington's disease patients compared with matched controls. *The Medical journal of Australia* **1981**, *1* (8), 407-9; (b) Djousse, L.; Knowlton, B.; Cupples, L. A.; Marder, K.; Shoulson, I.; Myers, R. H., Weight loss in early stage of Huntington's disease. *Neurology* **2002**, *59* (9), 1325-30.

[255] Mochel, F.; Charles, P.; Seguin, F.; Barritault, J.; Coussieu, C.; Perin, L.; Le Bouc, Y.; Gervais, C.; Carcelain, G.; Vassault, A.; Feingold, J.; Rabier, D.; Durr, A., Early energy deficit in Huntington disease: identification of a plasma biomarker traceable during disease progression. *PloS one* **2007**, *2* (7), e647.

[256] (a) Farrer, L. A.; Meaney, F. J., An anthropometric assessment of Huntington's disease patients and families. *American journal of physical anthropology* **1985**, *67* (3), 185-94; (b) Trejo, A.; Tarrats, R. M.; Alonso, M. E.; Boll, M. C.; Ochoa, A.; Velasquez, L., Assessment of the nutrition status of patients with Huntington's disease. *Nutrition* **2004**, *20* (2), 192-6.

[257] (a) Luthi-Carter, R.; Hanson, S. A.; Strand, A. D.; Bergstrom, D. A.; Chun, W.; Peters, N. L.; Woods, A. M.; Chan, E. Y.; Kooperberg, C.; Krainc, D.; Young, A. B.; Tapscott, S. J.; Olson, J. M., Dysregulation of gene expression in the R6/2 model of polyglutamine disease: parallel changes in muscle and brain. *Human molecular genetics* **2002**, *11* (17), 1911-26; (b) Strand, A. D.; Aragaki, A. K.; Shaw, D.; Bird, T.; Holton, J.; Turner, C.; Tapscott, S. J.; Tabrizi, S. J.; Schapira, A. H.; Kooperberg, C.; Olson, J. M., Gene expression in Huntington's disease skeletal muscle: a potential biomarker. *Human molecular genetics* **2005**, *14* (13), 1863-76.

[258] (a) Arenas, J.; Campos, Y.; Ribacoba, R.; Martin, M. A.; Rubio, J. C.; Ablanedo, P.; Cabello, A., Complex I defect in muscle from patients with Huntington's disease. *Annals of neurology* **1998**, *43* (3), 397-400; (b) Ciammola, A.; Sassone, J.; Alberti, L.; Meola, G.; Mancinelli, E.; Russo, M. A.; Squitieri, F.; Silani, V., Increased apoptosis, Huntingtin inclusions and altered differentiation in muscle cell cultures from Huntington's disease subjects. *Cell death and differentiation* **2006**, *13* (12), 2068-78.

[259] Lanska, D. J.; Lanska, M. J.; Lavine, L.; Schoenberg, B. S., Conditions associated with Huntington's disease at death. A case-control study. *Archives of neurology* **1988**, *45* (8), 878-80.

[260] (a) Pattison, J. S.; Robbins, J., Protein misfolding and cardiac disease: establishing cause and effect. *Autophagy* **2008**, *4* (6), 821-3; (b) Pattison, J. S.; Sanbe, A.; Maloyan, A.; Osinska, H.; Klevitsky, R.; Robbins, J., Cardiomyocyte expression of a polyglutamine preamyloid oligomer causes heart failure. *Circulation* **2008**, *117* (21), 2743-51.

CHAPTER 5

Multifunctional Enzyme Inhibition for Neuroprotection - A Focus on MAO, NOS, and AChE Inhibitors

Jacques Joubert[a], Jacobus P. Petzer[b], Louis H.A. Prins[c], Benjamin P. Repsold[b] and Sarel F. Malan[a],*

[a]*School of Pharmacy, University of the Western Cape, Bellville, Private Bag X17, Bellville 7535, South Africa;* [b]*Pharmaceutical Chemistry, North-West University, Private Bag X6001, Potchefstroom, 2520, South Africa and* [c]*Department of Pharmacy, University of Namibia, Windhoek, Namibia*

Abstract: Neurodegenerative disorders are known to be multifactorial in nature and current research focus has moved from a 'one-drug-one-target approach' to that of drugs which are able to act at various relevant biological targets. These drugs are designed to address more than one etiological target, thereby increasing therapeutic effect and patient compliance and may lower the likelihood of encountering unwanted side-effects. Monoamine oxidase (MAO), nitric oxide synthase (NOS), and acetylcholinesterase (AChE) are enzymes that have long been associated as potential targets for neurodegenerative disorders, including Alzheimer's disease and Parkinson's disease. The selective inhibition of the abovementioned enzymes and other relevant CNS targets may provide promising strategies in the development of multifunctional neuro-protective therapeutic agents for the treatment/prevention of neurodegenerative disorders.

Keywords: Acetylcholinesterase, Alzheimer's and Parkinson's disease, drug design, monoamine oxidase, multi-target-directed ligands, neurodegenerative disorders, nitric oxide synthase.

INTRODUCTION

Disease-modifying approaches have become an important focus of research as they interrupt the early pathologic events implicated in the onset of the disease, hampering the neurotoxic cascade in the case of neurodegenerative diseases [1, 2].

***Address correspondence to Sarel F. Malan:** School of Pharmacy, University of the Western Cape, Private Bag X17, Bellville 7535, South Africa; Tel: +27 21959 3190; Fax: +27 21959 1588; E-mail: sfmalan@uwc.ac.za

Atta-ur-Rahman / Muhammad Iqbal Choudhary (Eds.)
10.1016/B978-0-12-803959-5.50005-2

Neurodegenerative disorders are not the result of a single cause or processes but are of a multifactorial nature and most likely involve several mechanistic pathways [3]. It is hypothesised that genetic, environmental and endogenous factors may be involved in neurodegenerative disorders such as, Alzheimer's disease (AD) and Parkinson's disease (PD) and a series of general pathways have been identified that may be involved in different pathogenic cascades. These include protein misfolding and aggregation, excitotoxicity, oxidative stress and free radical formation, metal dyshomeostasis, mitochondrial dysfunction and phosphorylation impairment. Several of these seem to occur simultaneously, leading to the demise of key neuronal cells [4]. With this in mind, it is evident that the paradigm of targeting a single disease factor may not be an effective treatment strategy for neurodegenerative diseases.

To date, multiple targeting has been pursued in the clinical setting through the polypharmaceutical approach, that is, a combination of therapeutic agents that act independently on different etiological targets. This strategy has already proven to be successful in the treatment of similarly complex diseases, such as cancer, HIV, and hypertension [4, 5]. For neurodegenerative disorders, it has been suggested that therapy will be a cocktail of drugs having various mechanisms of action and activity on more than one relevant neurotherapeutic target [6]. Combination therapy is, however, a difficult and complicated avenue for drug development because of the possibility of drug–drug interactions and further complexities encountered when combining drug entities that have potentially different degrees of bioavailability, pharmacokinetic profiles, metabolism, and toxicity [5].

For these reasons the design of single drug molecules that act on multiple, but specific etiological targets of a particular disease, may be of value – especially in neurodegenerative disorders. The possibility of a single drug acting at several sites in the neurotoxic cascade has been suggested in several studies [5-7]. The advantages associated with this strategy includes a lower likelihood of encountering unwanted side-effects, the possibility to 'design out' any side-effects and simplification of pharmacokinetic considerations [6]. The fact that a single compound may be able to hit multiple targets has lead to a shift from single to multi-target-directed ligands (MTDLs) [6-8]. Bearing in mind the complexity of AD and other neurodegenerative diseases, these MTDLs should be more adept at facing the complexity of neurodegenerative diseases than the classic 'one molecule, one target' concept [8-10]. Designing multifunctional agents is, however, a major challenge since the structural requirements for activity at a specific target may be very stringent and modification of the structure to enhance activity at another relevant target frequently may lead to a loss of the original pharmacological activity [6, 8].

In neurodegenerative disorders such as AD and PD, monoamine oxidase (MAO), nitric oxide synthase (NOS), and acetylcholine esterase (AChE) are enzymes that have long been acknowledged as potential targets for therapeutic drug design. MAO enzyme inhibitors, particularly inhibitors of the MAO-B isoform are used in PD and because of the relatively broad inhibitor specificities of the MAO-B enzyme, the possibility exists to design single molecular entities which combine MAO-B inhibition with activities at other relevant CNS targets [11-13]. Such activities include but are not limited to acetylcholine esterase inhibition, antagonism of adenosine receptors, modulation of apoptosis, inhibition of oxidative stress and inflammation. Overexpression of nitric oxide synthase (NOS) or more specifically, the neuronal NOS (nNOS) isoform, and subsequent overproduction of NO, has been described in various neurological diseases [14, 15]. In the treatment/prevention of neurodegenerative disorders, nNOS inhibition alone will not be an effective treatment option because of the multitude of mechanisms that may lead to neurodegeneration. For this reason additional therapeutic options have been identified for the development of dual/multiple acting nNOS inhibitors and these include modulation of the *N*-methyl-D-aspartate receptor and excitotoxicity, acetylcholine esterase (AChE) inhibition, MAO-B inhibition and attenuation of oxidative stress. AChE is the main enzyme responsible for the hydrolysis of acetylcholine into choline and acetic acid and inhibition of this enzyme has therefore become a key target for neurodegenerative disorders, especially AD [16]. Current research on AChE is focusing on the design of multifunctional neuroprotective agents targeting AChE and other CNS targets including MAO-B, amyloid-beta, β-secretase and modulation of oxidative stress.

The multifunctional but selective inhibition of the abovementioned enzymes and other relevant CNS targets may thus provide promising strategies in the development of therapeutic agents for the treatment of neurodegenerative disorders and selected examples thereof will be discussed in this chapter.

MONOAMINE OXIDASE – ISOFORMS AND FUNCTION

The monoamine oxidases (MAOs), particularly type-B MAO, have attracted widespread attention as targets for the development of neuroprotective therapies [17]. These intracellular enzymes are mitochondrial bound and dependent on the flavin adenine dinucleotide (FAD) cofactor for their functions [18]. The MAOs consist of two distinct isoforms, MAO-A and MAO-B. Although MAO-A and -B are products of separate genes, they share approximately 70% sequence identity at the amino acid level [19]. The crystallographic structures of MAO-A and -B show that the amino acid residues comprising the active sites and their relative geometries are highly conserved with only 6 of the 16 active site amino acid residues differing between the 2 isozymes [20-22]. The human forms of MAO-A and -B consist of 529 and 520 amino acid residues, respectively. In both isoforms, the FAD cofactor is covalently bound to the enzymes *via* a cysteinyl residue (Cys-406 and Cys-397 in MAO-A and -B, respectively) [23, 24]. The physiological function of the MAOs is to catalyse the α-carbon oxidation of a variety of aminyl substrates (Figs. **1** and **2**). In spite of their structural and functional similarities,

MAO-A and -B exhibit different substrate specificities. Most notably, MAO-A exhibit a high degree of substrate selectivity for the neurotransmitter serotonin, while MAO-B selectively metabolises extraneous amines such as benzylamine and β-phenylethylamine (Fig. **2**). Dopamine, epinephrine and norepinephrine, as well as the dietary amine, tyramine, are considered to be substrates for both isozymes [11].

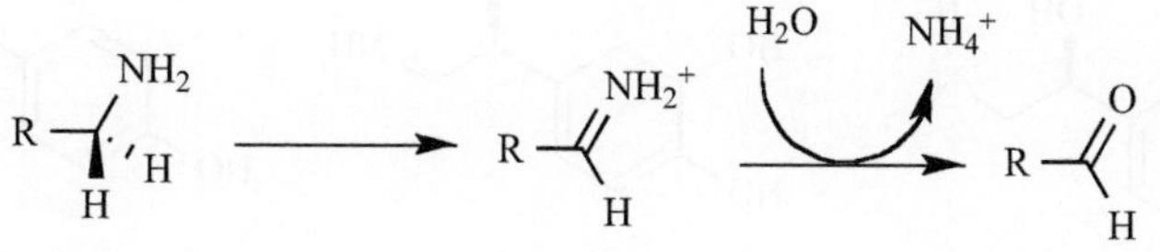

Figure 1: The oxidation of aminyl substrates by MAO with subsequent hydrolysis of the iminiumyl product.

Since the MAOs metabolise neurotransmitter amines, their main function is to terminate the actions of these amines in central and peripheral tissues. In addition, the MAO isozymes may also protect neurons from stimulation by false neurotransmitters. Among these are benzylamine and β-phenylethylamine, which are trace amines derived from the diet [11, 25]. Endogenous β-phenylethylamine is also a metabolic product of phenylalanine. These amines are metabolised to a large extent by the MAO-B found in brain microvessels, which results in the restriction of their entry into the brain [26]. The MAO enzymes also serve as metabolic barriers in the microvessels of the intestines where intestinal MAO-A catabolises tyramine, which is found in certain foods. Tyramine is an indirectly-acting sympathomimetic amine that induces the release of norepinephrine from peripheral neurons [27] and in the presence of excessive amounts of tyramine in the circulation, this may lead to a severe hypertensive crisis. The metabolic breakdown of tyramine by intestinal MAO-A reduces the amount of this amine that enters the systemic circulation and thus prevents the tyramine-associated adverse effects [27]. It should be noted that MAO-B also catabolises tyramine, but since gut microvessels are devoid of MAO-B, this protective function is entirely dependent on the action of MAO-A.

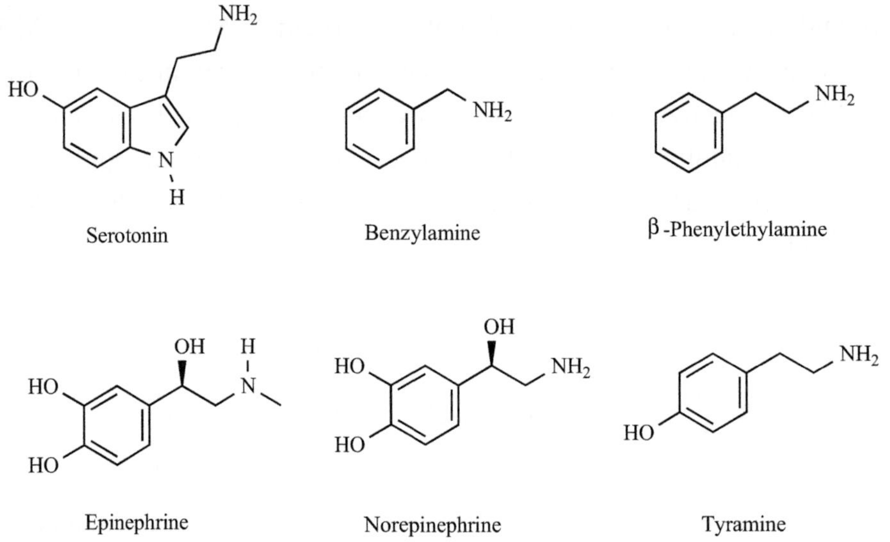

Figure 2: MAO substrates mentioned in the text.

Although MAO activity is found in most mammalian tissues, the two isoforms are expressed in a tissue-specific manner. MAO-B is the major form expressed in human liver tissue [28] while MAO-A is the main form in human placental [29] and intestinal tissues [30]. Although differently distributed, both isoforms are present in the human brain [31, 32]. In both human and subhuman primate brains, MAO-B is present in higher concentrations and exhibits a higher degree of activity [33-35]. Of particular significance is the observation that MAO-B is the dominant isoform in the human basal ganglia [34, 36] and that MAO-B activity as well as density [34, 37, 38] exhibits age-associated increases in most brain areas, including the basal ganglia. Since MAO-B is located in the glial cells, this increased activity may be attributed to glial cell proliferation, a process associated with increasing age [31, 39, 40]. In contrast, MAO-A activity remains unchanged with age [33].

Monoamine Oxidase and Parkinson's Disease

Since MAO-A and -B catabolises neurotransmitter amines, they have been targeted for the therapy of neuropsychiatric and neurodegenerative disorders [11, 12, 41]. MAO-A inhibitors prolong the central action of serotonin and are used in the treatment of anxiety disorder and depressive illness [11, 42] while MAO-B inhibitors are employed in the symptomatic therapy of Parkinson's disease [43]. The beneficial effects of MAO-B inhibitors in Parkinson's disease are thought to rely on blocking the MAO-B-catalysed metabolism of dopamine in the basal ganglia. This conserves depleted dopamine stores and prolongs the physiological action of dopamine. MAO-B inhibitors also may enhance dopamine levels derived from levodopa, the metabolic precursor of dopamine. For example, the MAO-B inhibitor, (R)-deprenyl, has been shown to enhance the elevation of dopamine levels in the striatum of primates after levodopa treatment [44, 45]. This elevation is associated with a reduction in the oxidative metabolism of dopamine [45]. Accordingly, MAO-B inhibitors are recommended as adjuvant therapy to levodopa in Parkinson's disease [43]; and in early Parkinson's disease, combination treatment with MAO-B inhibitors may allow for a reduction in the doses of levodopa and dopamine agonists. In addition, MAO-B inhibitors may also delay the emergence of disabilities that require the initiation of levodopa therapy [46, 47]. The modulation of central β-phenylethylamine levels by MAO-B inhibitors may also explain their beneficial effects in Parkinson's disease. Since β-phenylethylamine is a releaser of dopamine as well as an inhibitor of active dopamine uptake [48], blocking its metabolism results in an increase in striatal extracellular dopamine levels. The central levels of β-phenylethylamine, normally present in only traceable amounts in the central nervous system (CNS), may be enhanced several thousand-fold by the administration of MAO-B inhibitors.

Although MAO-B inhibitors are used clinically primarily for their symptomatic effects, they may also protect against the neurodegenerative processes implicated in Parkinson's disease. This is thought to occur by blocking the formation of dopanal and H_2O_2 (Fig. **3**), the metabolic by-products of dopamine metabolism by MAO-B [49-52]. In the catalytic cycle of MAO-B, the hydrolysis of the iminiumyl intermediate yields an aldehyde product, while reoxidation of the FAD cofactor requires the reduction of O_2 to H_2O_2 (Fig. **3**). One mole of each of these species is formed per mole of monoamine substrate oxidized. These by-products are considered to be neurotoxic under certain conditions and may accelerate the neurodegenerative process. For example, dopanal has been associated

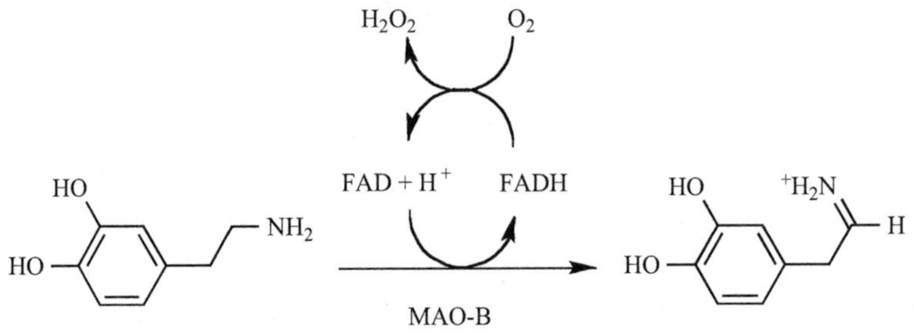

Figure 3: The oxidation of dopamine by MAO with simultaneous reduction of the FAD cofactor of MAO. The FAD is reoxidised by reaction with O_2 with the ultimate formation of H_2O_2.

with the aggregation of α-synuclein, a process which is associated with the pathogenesis of Parkinson's disease [53]. Furthermore, aldehydic products such as dopanal, may react with exocyclic amino groups of nucleosides, and *N*-terminal and lysine ϵ-amino groups of proteins [17]. Normally aldehydes are rapidly inactivated by centrally located aldehyde dehydrogenase (ADH) [49-51], but in the substantia nigra of Parkinson's disease patients, the expression of ADH may, however, be reduced. This suggests that reduced levels of ADH in the CNS may allow for the accumulation of aldehydic species derived from the action of MAO-B [54]. H_2O_2, in turn, may lead to oxidative damage by reacting in the Fenton reaction with ferrous ion to generate the highly reactive hydroxyl radical [17]. The hydroxyl radical damages virtually all types of biomolecules including proteins, DNA, lipids, carbohydrates and amino acids. This reaction may be of particular relevance when considering that the brain also exhibits an age-dependent increase in iron content. Normally H_2O_2 is deactivated in the brain by glutathione peroxidase. In the parkinsonian brain, the levels of glutathione, the electron donor for the reduction of H_2O_2 by glutathione peroxidase, may be lowered [55].

Another mechanism by which H_2O_2 may contribute to neurodegeneration is by promoting apoptotic signaling events [56]. Considering that MAO-B activity in the CNS increases with age [34, 37, 38], inhibition of the MAO-B-catalysed formation of toxic by-products in the aged parkinsonian brain is of particular significance. Inhibitors of MAO-B therefore may exert a neuroprotective effect by stoichiometrically decreasing MAO-B catalysed aldehyde and H_2O_2 production in the brain.

In the United States, (R)-deprenyl and rasagiline, are approved as adjunct therapy to levodopa, or as monotherapy in Parkinson's disease. These inhibitors are both mechanism-based irreversible MAO-B inhibitors [43]. Lazabemide, a reversible MAO-B inhibitor, has also been shown to delay the need for levodopa in early untreated Parkinson's disease patients [57]. The benefits conferred by lazabemide were similar to those observed after 1 year of (R)-deprenyl treatment [57, 58]. Phase III trials have shown that another reversible inhibitor, safinamide, results in significant improvements of motor scores when co-administered with dopamine agonist drugs [59].

MAO-B Inhibitors as Multifunctional Agents

As outlined in the introduction, Parkinson's disease is of a multifactorial nature and most likely involves several mechanistic pathways [3]. By conferring both symptomatic and neuroprotective effects, MAO-B inhibitors are useful in the treatment of Parkinson's disease. Where the goal is to prevent neurodegeneration in Parkinson's disease, it may be useful to combine activities at these pathways with inhibition of MAO-B in a single molecular entity [8]. In this respect, combining additional relevant pharmacological activities with MAO-B inhibition is especially attractive since MAO-B exhibits relatively broad inhibitor specificity and the pharmacophore for MAO-B inhibition is relatively simple. In general, there exists a relatively large degree of tolerance for structural modification of MAO-B inhibitors, and modifications made to the structure of an inhibitor to confer additional pharmacological activities are less likely to be associated with a loss of MAO-B inhibition potency.

In essence, the pharmacophore for MAO-B inhibition consists of two features, a hydrophilic feature and a lipophilic motif at the opposite end of the inhibitor. The lipophilic feature is frequently satisfied by an aromatic group containing a halogen substituent, while both hydrogen bond acceptors and donors map to the hydrophilic feature. This proposed pharmacophore model is shown in Fig. **4** with the MAO-B inhibitors, safinamide, 7-(3-chlorobenzyloxy)-4-formylcoumarin [60], (*E*)-8-(3-chlorostyryl) caffeine (CSC) and 8-benzyloxycaffeine [61-63] mapped.

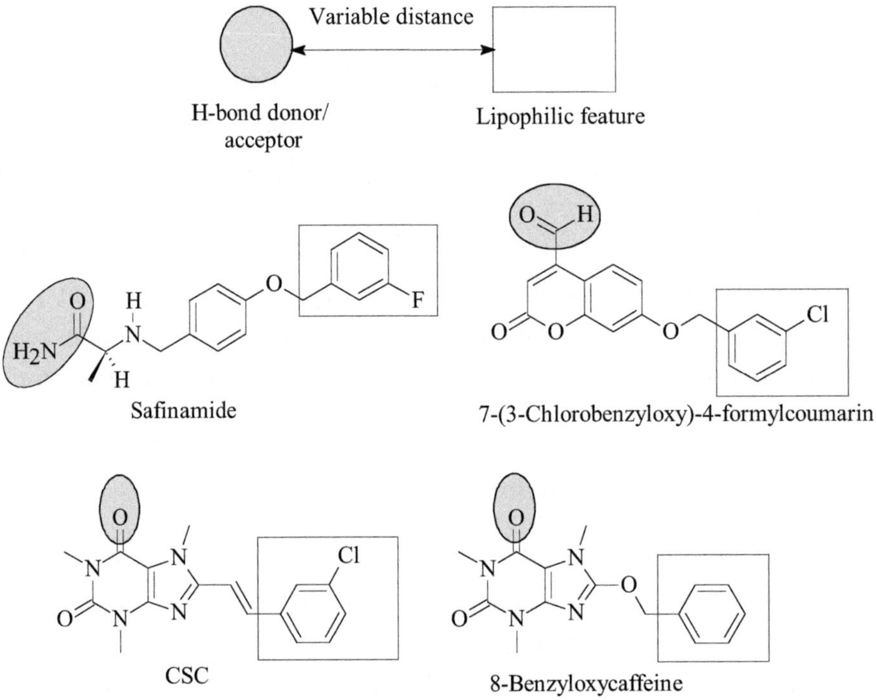

Figure 4: The pharmacophore of MAO-B inhibition with selected inhibitors mapped.

The features of this proposed pharmacophore is apparent when considering the architecture of the MAO-B active site that is reported to consist of an entrance cavity which leads to a larger substrate cavity [60]. Though the active site of MAO-B, in particular the entrance cavity, is mostly hydrophobic, a small hydrophilic region exists in the substrate cavity in front of the *re* face of the FAD isoalloxazine ring. This area is occupied by highly conserved water molecules and is also the site where the amine functional group of a substrate is predicted to bind [22]. The polar part of the MAO-B active site corresponds to the hydrophilic feature of the pharmacophore model and hydrogen bond donor and acceptor moieties of inhibitors are placed here. For example, the crystallographic structures of safinamide and 7-(3-chlorobenzyloxy)-4-formylcoumarin in complex with human MAO-B show that the propanamidyl moiety of safinamide acts as a hydrogen bond donor, and the formyl group of the coumarin as a hydrogen bond acceptor in the MAO-B substrate cavity [60]. Similarly, modeling studies suggest that a carbonyl oxygen of the caffeine rings of CSC and 8-benzyloxycaffeine acts as a hydrogen bond acceptor in the MAO-B substrate cavity [63]. The terminal phenyl rings of the inhibitors, in turn, bind within the entrance cavity of the enzyme and the lipophilic feature of the pharmacophore corresponds to this cavity. Also noteworthy is that tolerance exists for the distance between the two pharmacophore features. For example, for a series of oxycaffeines, it was found that linkers consisting of either 2 of 4 atoms separating the caffeine and the phenyl ring are appropriate for MAO-B inhibition (Fig. **5**) [64]. From this analysis, it is apparent that a wide variety of structures may satisfy the required features for binding to MAO-B. The significance of this for the design of multifunctional agents is that structures originally designed for activity at other targets are frequently found to also bind to MAO-B. Examples of compounds exhibiting this behavior are CSC and pioglitazone [61, 65, 66]. Both these compounds are potent MAO-B inhibitors, although they were originally designed for activity at adenosine A_{2A} receptors and nuclear receptor peroxisome proliferator-activated receptor gamma (PPAR-γ), respectively [61, 65].

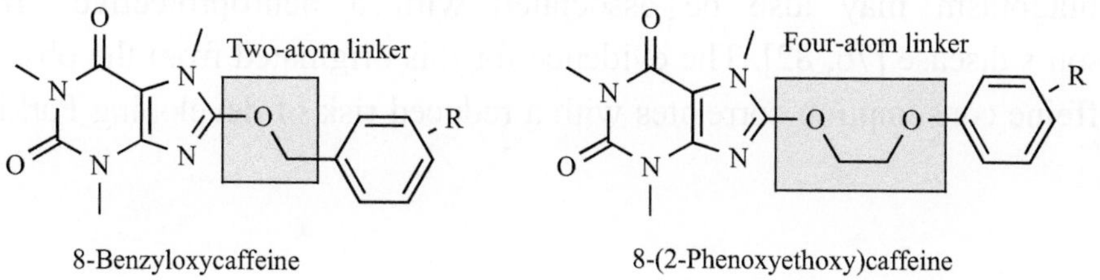

Figure 5: The structures of 8-benzyloxycaffeine and 8-(2-phenoxyethoxy)caffeine.

MAO-B Inhibition and Antagonism of Adenosine Receptors

Adenosine A_{2A} receptor antagonists are a relatively new class of agents for the symptomatic treatment of Parkinsom's disease [67-69]. A_{2A} receptors are highly enriched in the striatum and are located with the dopamine D_2 receptors on the GABAergic striatopallidal neurons [70-72]. Antagonism of the A_{2A} receptor at these neurons potentiates dopamine mediated neurotransmission *via* D_2 receptors and diminishes the effects of striatal dopamine depletion in Parkinson's disease [68, 73-75]. Another advantage of A_{2A} antagonists in Parkinson's disease is that they may reduce levodopa-induced dyskinesias. For example, in MPTP-lesioned primates, the A_{2A} antagonist KW-6002 (Fig. **6**) prevents the development of apomorphine-induced dyskinesias [76], while clinical trials showed that KW-6002 potentiates the motor benefits of a reduced dose of levodopa and at the same time produced only approximately half the amount of dyskinesias [77]. The antidyskinetic effects of A_{2A} antagonists are especially relevant in the light of the observation that the therapeutic benefits of A_{2A} antagonists are additive to those of levodopa and dopamine agonists, and it may therefore be possible to reduce the dose of the dopaminergic drugs and the severity of dyskinesias [78-81].

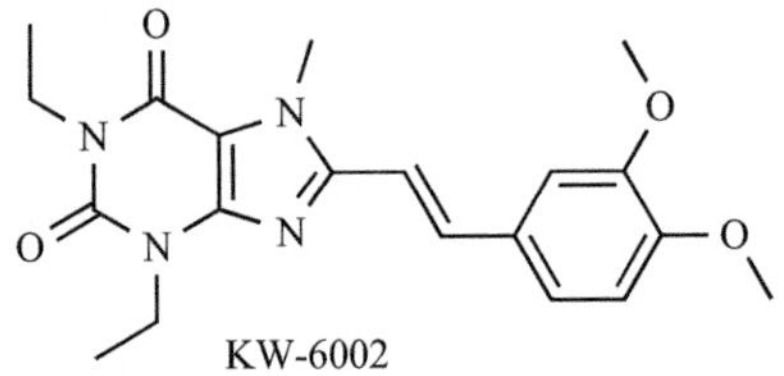

KW-6002

Figure 6: The A_{2A} antagonist KW-6002.

A_{2A} antagonism may also be associated with a neuroprotective effect in Parkinson's disease [76, 82]. The evidence for this originated from the observation that caffeine consumption correlates with a reduced risk of developing Parkinson's disease [83, 84]. This effect of caffeine, a non-selective A_1/A_{2A} antagonist, has been linked to its blockade of A_{2A} receptors since a number of selective A_{2A} antagonists, and not A_1 antagonists, protect against neurodegenerative processes in animal models [62, 85]. The mechanism by which caffeine and A_{2A} antagonists protect against neurodegeneration is, however, unclear.

Based on the relevance of both MAO-B inhibition and A_{2A} receptor antagonism in Parkinson's disease, compounds that possess both these activities would be of value. CSC is such a structure and had be used as reference A_{2A} antagonist for several years when it was discovered that it protected experimental animals against the neurodegenerative effects of 1-methyl-4-phenyl-1,2,3,6-tetrahydropyridine (MPTP) [61]. As part of an investigation to determine the mechanism by which CSC exert this protective effect, it was discovered that in addition to being a potent and selective A_{2A} antagonist, it also acted as a highly potent, competitive and reversible inhibitor of MAO-B [61]. CSC was found to inhibit MAO-B with a K_i value of 80.6 – 128 nM [61, 62, 86], This K_i value of CSC is comparable to its K_i value reported for the antagonism of A_{2A} receptors (K_i = 36–54 nM) [66, 87]. This profile is especially favorable since a compound acting at different protein targets should ideally modulate their functions at comparable concentrations. Other caffeine derived compounds have also been shown to inhibit MAO-B [86]. Unfortunately, structural modifications that improve A_{2A} antagonism have the opposite effect on MAO-B inhibition potency. Most notable, ethyl substitution at C1 and C3 of the caffeine ring reduces MAO-B inhibition potency compared to methyl substitution, while 1,3-diethyl substitution of the caffeine ring leads, in general, to enhanced A_{2A} antagonism [86, 88]. An example of this behavior is observed with KW-6002, a 1,3-diethyl-substituted caffeine derived structure with potent A_{2A} antagonistic properties (K_i = 2.2 nM), but comparatively weak MAO-B inhibition (K_i = 28 μM) [88, 89].

MAO-B and the Modulation of Oxidative Stress

Oxidative stress is thought to play a pivotal role in the mechanism of neurodegeneration in Parkinson's disease [9]. Besides directly causing damage to biomolecules in neuronal cells, oxidative stress may also lead to the misfolding of proteins, such as α-synuclein, a process associated with neurotoxicity [90]. MAO-

B inhibitors may possess neuroprotective effects in Parkinson's disease by blocking MAO-B-catalysed H_2O_2 formation, a pro-oxidant converted to the hydroxyl radical in the presence of ferrous ion [17]. Iron appears to have a role in neurodegenerative processes since in several neurodegenerative disorders, including Parkinson's disease, excessive amounts of iron are present at the injured neuronal sites [91]. Furthermore, iron chelation protects animals against 6-hydroxydopamine- (6-OHDA) and MPTP-induced neurotoxicity [92-94]. In this regard, both iron chelators, desferal and VK-28, exhibit these protective properties [95]. The detrimental effects of free iron (non-ferritin bound Fe^{2+} and Fe^{3+}) may not only depend on the reaction of ferrous ion with H_2O_2 to yield the hydroxyl radical, but free iron may be responsible for the aggregation of α-synuclein, a key contributor to degeneration in Parkinson's disease [90]. The consideration that both MAO-B activity and iron concentrations increase in the brain with age, thus increasing both components of the Fenton reaction, provides a further rational for iron chelation in Parkinson's disease [34, 37, 38, 96, 97]. Chelation of intracellular iron is expected to reduce the free iron pool and diminish the conversion of H_2O_2 to the hydroxyl radical.

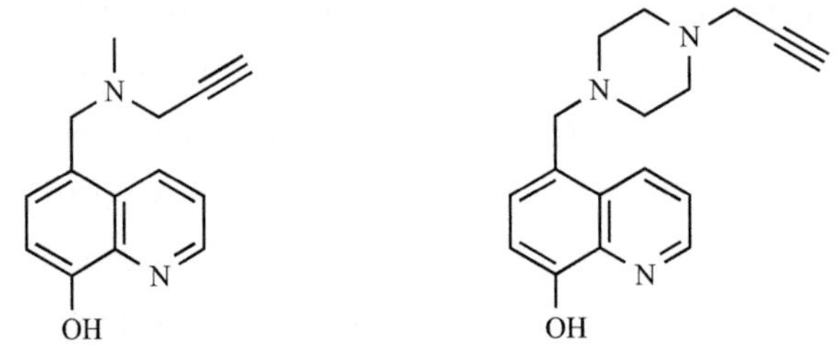

M30

MAO-B IC_{50} = 0.057 μM
MAO-A IC_{50} = 0.037 μM
LPO IC_{50} = 16 μM

HLA20

MAO-B IC_{50} = 110 μM
MAO-A IC_{50} = >200 μM
LPO IC_{50} = 12 μM

Figure 7: The structures of M30 and HLA20.

Multifunctional compounds combining MAO-B inhibition with antioxidant properties may therefore be of value in Parkinson's disease. Bifunctional compounds such as M30 and HLA20 (Fig. **7**) may be particularly effective in decreasing the generation of the hydroxyl radical by blocking MAO-B-mediated H_2O_2 formation and by sequestering intracellular iron. The structures of M30 and HLA20 were derived from the 8-hydroxyquinoline pharmacophore, which is responsible for their chelating activities [92, 93]. The ability of M30 (IC_{50} = 16 μM) to inhibit iron-dependent lipid peroxidation (LPO) is similar to that of desferal (IC_{50} = 8 μM) and like desferal, M30 also exhibits neuroprotective properties in animal models of Parkinson's disease [92]. M30 and HLA20 also contain the propargylamine functional group, which is responsible for their potent and selective *in vivo* mechanism-based inactivation of the MAO enzyme. The *in vitro* potency of M30 (IC_{50} = 0.057 μM) is comparable to that of the well-known mechanism-based MAO-B inhibitor, (*R*)-deprenyl (IC_{50} = 0.079 μM) [92, 93]. M30 is reported to also inhibit MAO-A *in vitro* (IC_{50} = 0.037 μM) with a potency comparable to that of the MAO-A inactivator, clorgyline (IC_{50} = 0.0026 μM) [92, 98]. Interestingly, in the *in vivo* setting, M30 displays selective inhibition of MAO-A and -B in the brain compared to the inhibition of these enzymes in the peripheral tissues [92]. From a safety point of view, this is highly advantageous since peripheral inhibition of MAO-A may lead to a severe hypertensive response.

MAO-A/B Mixed Inhibitors

MAO-A metabolises both serotonin and norepinephrine, two monoamines implicated in depressive illness [11] and the antidepressive effect of MAO-A inhibitors is associated with blocking the central breakdown of these monoamines. Since a significant proportion of Parkinson's disease patients exhibit signs of depression, the inhibition of MAO-A in these patients may be beneficial [100]. Besides offering a potential antidepressant action in Parkinson's disease, MAO-A inhibitors may also provide a symptomatic benefit by reducing MAO-A-catalysed oxidation of dopamine [11, 45]. Although MAO-B is present in higher concentrations than MAO-A in the human basal ganglia [34, 36], MAO-A inhibitors have also been shown to enhance dopamine levels in this region. For example, clorgyline, a selective irreversible inhibitor of MAO-A, elevates dopamine levels in the striatum of primates treated with levodopa to a similar degree than the elevation obtained with (R)-deprenyl, a selective irreversible inhibitor of MAO-B [45]. In order to conserve dopamine in the basal ganglia, mixed MAO-A/B inhibitors may therefore be more efficacious than selective inhibitors [17].

Unfortunately MAO-A inhibitors may lead to serious adverse effects when combined with certain drugs and food. These include hypertensive crisis [99-102] and serotonin toxicity (ST) [103, 104]. Because of these potential adverse effects, MAO-A inhibitors are used less frequently in the treatment of depression than the SSRIs and the tricyclic antidepressants, and are considered less suitable to enhance dopaminergic neurotransmission in the parkinsonian brain than MAO-B inhibitors. Recently developed reversible inhibitors of MAO-A, such as moclobemide, however are considered safer than the irreversible MAO-A type inhibitors [105] and is essentially free form the tyramine reaction and ST occurs only when an excessive dose of the 5-hydroxytryptaminergic agent have been used [106]. These observations suggest that, when designing MAO-A selective inhibitors or MAO-A/B mixed inhibitors, the compounds should interact reversibly with MAO-A. Another advantage of a reversible mode inhibition is that, following withdrawal of the drug, enzyme activity is recovered quickly upon elimination of the drug from the tissues. In contrast, after terminating treatment with irreversible MAO inhibitors, recovery of enzyme activity may require several weeks since the turnover rate for the biosynthesis of MAO in the human brain may be as much as 40 days [107].

Considering the high degree of similarity between the active sites of MAO-A and -B, it is surprising that few reversible MAO inhibitors that are equipotent at both enzymes have been reported, and none of these have reached the clinical market or are undergoing clinical trials [20, 21]. The most notable difference between the active sites of MAO-A and -B is that the MAO-A active site consists of a single cavity, while the active site of MAO-B consists of an entrance cavity which leads to the substrate cavity. The side chain of the Ile-199 residue appears to act as a gate between the entrance and substrate cavities of MAO-B as the small volume of the side chain of Ile-199 allows for this residue to partially rotate out of the active site, allowing sufficient space for larger inhibitors to bind in an extended conformation [108]. In MAO-A, the corresponding residue is Phe-208 and the increased size of the phenyl ring of Phe-208 prevents it from rotating into an alternative conformation. As a result, larger inhibitors are, in general, restricted from binding to MAO-A [20]. This, in part, may explain the challenge in designing reversible MAO-A/B mixed inhibitors. MAO-A/B mixed inhibitors constitute, for the most part, the older irreversible inhibitors and include compounds such as phenelzine, tranylcypromine and iproniazid (Fig. **8**). These compounds have, however, fallen into disrepute because of their potential adverse effects.

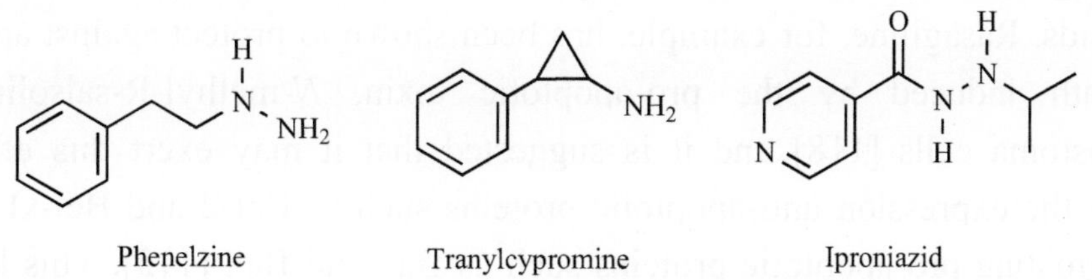

Phenelzine   Tranylcypromine   Iproniazid

Figure 8: The structures of phenelzine, tranylcypromine and iproniazid.

From the above it is clear that the development of mixed MAO-A/B inhibitors with high potencies at both enzymes could provide beneficial treatment options in neurodegenerative diseases. Such drugs should preferably possess a reversible mode of binding.

MAO-B and the Modulation of Apoptosis

Apoptosis has been demonstrated to participate in neuronal development as well as some forms of neuronal injury. In Parkinson's disease, the apoptotic process most likely occurs as an end-stage process and anti-apoptotic agents, therefore, may protect against neurodegeneration regardless of the initial cause. For example, signs of apoptosis [109], upregulation of the antiapoptotic protein Bcl-2 [110] and increased levels of the pro-apoptotic protein Bax [111] have been found in the substantia nigra of Parkinson's disease patients. In addition, caspase-3 and caspase-8 may act as effectors of apoptotic cell death in the brains of Parkinson's disease patients [112, 113].

The finding that propargylamine containing compounds activate anti-apoptotic pathways provides the opportunity of combining an anti-apoptotic effect with a MAO inhibitory effect in a single compound [114, 115]. Examples of such compounds are the irreversible MAO-B inhibitors, (*R*)-deprenyl and rasagiline. The propargylamine moiety is responsible of the mechanism-based inhibition by these compounds and, following oxidation by MAO, this moiety forms a covalent linkage to the $N(5)$ of the FAD cofactor to yield a flavocyanine adduct [116, 117]. Although the mechanism of action is still unclear, the propargylamine moiety is also thought to be responsible for the anti-apoptotic effects of this class of compounds. Rasagiline, for example, has been shown to protect against apoptotic cell death induced by the pro-apoptotic toxin, *N*-methyl-R-salsolinol, in neuroblastoma cells [118] and it is suggested that it may exert this effect by inducing the expression anti-apoptotic proteins such as Bcl-2 and Bcl-xL, while downregulating pro-apoptotic proteins such as Bax and Bad [115]. This leads to the stabilisation of the mitochondrial permeability transition pore, thus reducing the swelling of mitochondria and the decline in mitochondrial membrane potential as well as the release of cytochrome c [119]. The anti-apoptotic mechanism of propargylamines, however, involves various steps of the apoptotic cascade and also includes the suppression of the activation of caspase 3 and poly (ADP-ribose) polymerase-1 (PARP-1) as well as the prevention of the nuclear translocation of glyceraldehyde-3-phosphate dehydrogenase (GADPH) [115, 118, 120, 121]. Propargylamines such as rasagiline may also exert their neurprotective actions *via* the activation of the protein kinase C (PKC)-dependent mitogen-activated protein (MAP) kinase pathway and by inducing the expression of brain-derived neurotrophic factor (BDNF) and glial-derived neurotrophic factor (GDNF) [11, 122]. It is important to note that the anti-apoptotic effects of propargylamine containing compounds are not dependent upon the inhibition of MAO, since propargylamines devoid of MAO inhibitory activity, such as the S enantiomer of rasagiline and *N*-propargylamine, also exhibit these effects [123]. The anti-apoptotic effects of this class of compounds, therefore, seem to be an intrinsic property of the propargylamine moiety.

MAO-B and Acetylcholinesterase (AChE) Inhibition

It is widely acknowledged that the cognitive decline associated with Alzheimer's disease is related to the loss of cholinergic and glutamatergic neurons. However, behavioural change may not only be related to the severity of cholinergic loss but may also be as a result of alterations in the serotoninergic and noradrenergic systems [8]. The noradrenergic deficits of Alzheimer's disease are linked to depression [124] whereas the serotoninergic deficits are linked to depression and psychosis [125].

Keeping this in mind, drugs that increase the activity of biogenic amines while simultaneously acting on the cholinergic system, may be a promising approach for Alzheimer's disease treatment. A series of compounds incorporating a carbamate moiety and based on the pharmacophores of selegiline/(*R*)-deprenyl and rasagiline were synthesised and tested as inhibitors of MAO-B and AChE [122]. One of these, ladostigil (**1**; Fig. **9**), was selected for further investigation. It was found to be an inhibitor of both AChE and butyrylcholinesterase (BChE), and exhibited cognitive effects in animal models comparable to those of rivastigmine or galantamine, two AChE inhibitors already used for Alzheimer's disease treatment.

After chronic treatment for 1 – 8 weeks with ladostigil, both MAO isoforms were inhibited in the brain with very little inhibition of the enzyme in gut or liver. Ladostigil also presented with neuroprotective activity in cultures of neuronal cells and anti-apoptotic activity identical to that of rasagiline [122]. This compound is currently in phase II clinical trials for Alzheimer's disease and diffuse Lewy body disease and will eventually be tested for Parkinson's disease as well.

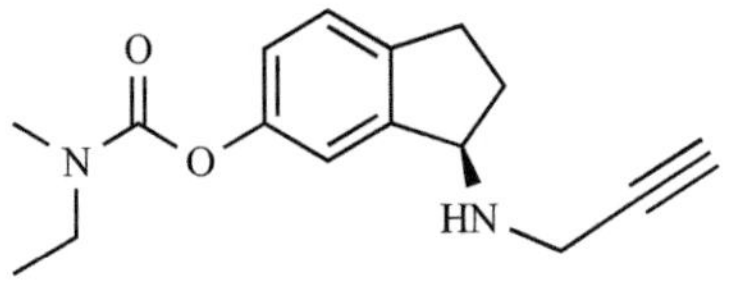

(1)

Figure 9: Ladostigil (**1**), a dual MAO and AChE inhibitor.

In an earlier study, dual MAO-AChE inhibitors were designed by decorating a tricyclic 1,2,3,4-tetrahydrocyclopent[*b*]indole carbamate scaffold, similar to the known AChE inhibitor physostigmine, with a propargylamine moiety [126]. The synthetic iminic intermediates (**2-5**, Fig. **10**) were found to inhibit both MAO-A and AChE. The results showed that the compounds exhibited moderate (**2** and **3**) to good (**4** and **5**) dual inhibitory activities. These compounds showed reversible MAO inhibition and was more selective towards MAO-A. The most exciting information obtained from structure-affinity and structure-selectivity relationship analyses performed during this study was that the R_2 substituent *ortho* to the carbamate moiety affected both enzymatic activities, leading to the multi-active compounds **2-5** [126]. However, further development of these compounds was stopped because of the low oral activity and poor brain penetration found during *ex vivo* studies on brain tissues.

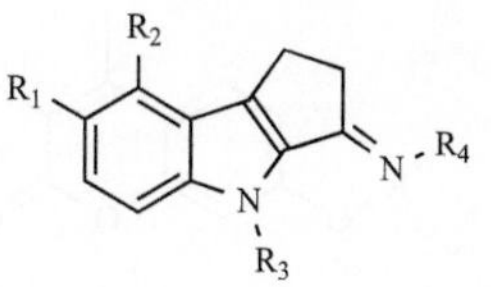

(2)	(3)	(4)	(5)
R^1 = NHMe; R_2 = H	R^1 = NHMe; R_2 = H	R^1 = NHMe; R_2 = Br	R^1 = NMe; R_2 = H
R_3 = R_4 = Me	R_3 = Me; R_4 = Et	R_3 = Me; R_4 = Et	R_3 = R_4 = Me
AChE Ki = 3.36 μM	AChE Ki = 5.68 μM	AChE Ki = 0.14 μM	AChE Ki = 0.060 μM
MAO-A IC_{50} = 13.6 μM	MAO-A IC_{50} = 3.74 μM	MAO-A IC_{50} = 0.21 μM	MAO-A IC_{50} = 0.54 μM
MAO-B IC_{50} = > 1000 μM	MAO-B IC_{50} = > 1000 μM	MAO-B IC_{50} = > 1000 μM	MAO-B IC_{50} = 51.0 μM

Figure 10: Tricyclic derivatives with dual MAO-AChE inhibitory activity.

In light of the *in vitro* AChE inhibitory activity (IC_{50} = 0.36 μM) showed by ensaculin (**6**, Fig. **11**) [127], a compound bearing a coumarin skeleton, and the potent MAO inhibition exhibited by a series of 7-substituted coumarin derivatives [128]. Bruhlmann and coworkers (2001) screened the latter series to determine their ability to inhibit AChE [129]. It was found that all of the 7-substituted coumarin derivatives (**7-10**) inhibited AChE with K_i values in the medium to low micromolar range (3-100 μM; Fig. **12**). They also presented with IC_{50} values for MAO inhibition that were in the micromolar to low nanomolar range. The benzyloxy group attached to C7 of the coumarin ring was shown to be the main molecular determinant for the high MAO-B affinity and the absence thereof reduced the AChE inhibition (see **7**; Fig. **12**).

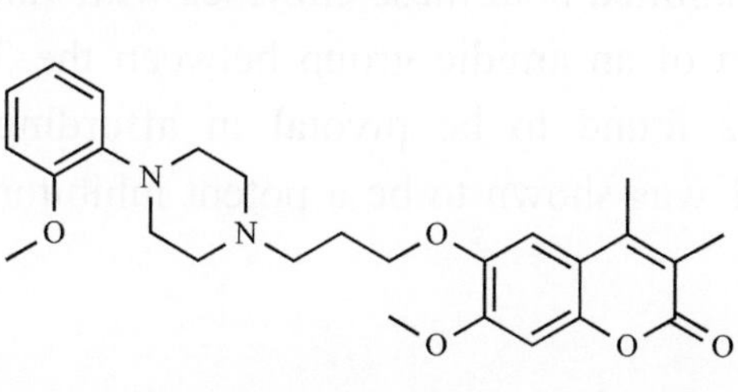

(6)

AChE IC_{50} = 0.36 μM

Figure 11: Ensaculin (**6**), a potent AChE inhibitor.

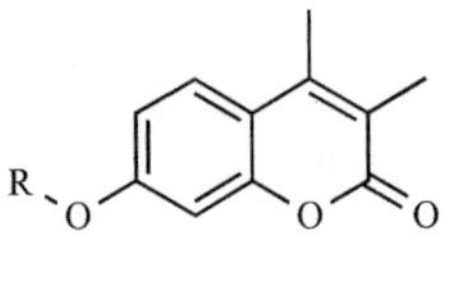

(7)	(8)	(9)	(10)
R = H	R = Benzyl	R = 3-F-benzyl	R = 3-Cl-benzyl
AChE Ki = 100 μM	AChE Ki = 4.07 μM	AChE Ki = 7.76 μM	AChE Ki = 3.39 μM
MAO-A IC_{50} = 35%	MAO-A IC_{50} = 0.69 μM	MAO-A IC_{50} = 0.58 μM	MAO-A IC_{50} = 1.12 μM
MAO-B IC_{50} = 18%	MAO-B IC_{50} = 4.37 μM	MAO-B IC_{50} = 2.82 μM	MAO-B IC_{50} = 3.31 μM

Figure 12: MAO and AChE inhibition data for coumarin derivatives **7-10**.

The series of 7-substituted coumarin derivatives was indicated to have a non-competitive mechanism of inhibiting AChE. This makes these compounds of great interest in the Alzheimer's disease context, since the kinetic mechanism may be ascribed to the interaction of the inhibitor at the peripheral anionic subsite (PAS), which is the structural motif involved in the promotion of aggregation of beta amyloid peptides (Aβ) to form amyloid plaques. Compound **10** (Fig. **12**) proved to be a potent and selective MAO-B inhibitor, also exhibiting the highest inhibition potency for AChE in the entire series [129].

Drawing inspiration from this study, Viña and coworkers (2012) designed and synthesized 3-benzamide coumarin derivatives (represented by compound **11**; Fig. **13**) and evaluated them for their ability to inhibit both MAO-B and AChE [130]. The structures inhibited both these enzymes with values in the micromolar range. The introduction of an amidic group between the 3-substituted aryl ring and the coumarin was found to be pivotal in affording this dual inhibitory potency. Compound **11** was shown to be a potent inhibitor of both MAO-B and AChE.

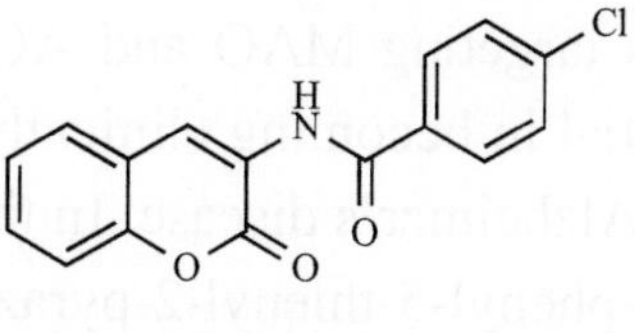

(11)

AChE IC$_{50}$ = 18.71 µM

MAO-B IC$_{50}$ = 1.95 µM

Figure 13: 3-Substituted coumarin derivative (**11**) with dual MAO-AChE inhibitory activity.

In an attempt to develop new multiple-acting drugs with a disease-modifying effect, Uçar and coworkers evaluated a series of racemic 1-*N*-substituted thiocarbamoyl-3-phenyl-5-thienyl-2-pyrazoline derivatives for their ability to inhibit ChE activity. These compounds had previously been indicated to be selective, reversible MAO-B inhibitors *in vitro* [131]. Two promising compounds were identified (**12-13**, Fig. **14**) that inhibited MAO-A/B and AChE in the micromolar range. Comparison of these two compounds revealed that AChE affinity was decreased by increasing the size of the R$_1$ substituent of the thiocarbamoyl exocyclic nitrogen [132]. The most promising derivative was **12** as it presented with a strong AChE affinity (IC$_{50}$ = 0.090 µM) and a moderate MAO-B inhibition potency (IC$_{50}$ = 43.0 µM) and selectivity. This observation gave some SAR insights, since the optimal substitution of the phenyl ring and the thioureidic moiety was represented by a *p*-methoxy and a methyl group, respectively.

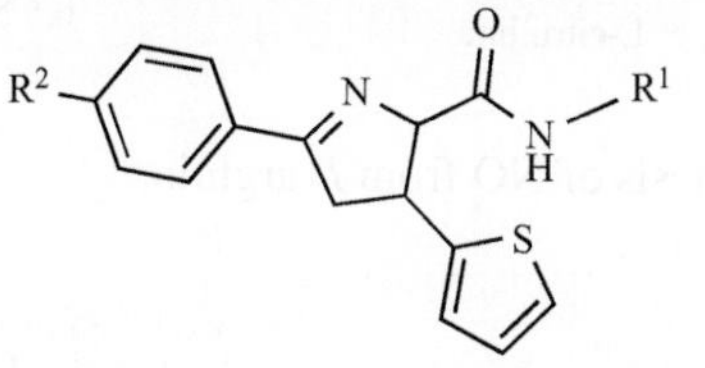

(12)	**(13)**
R^1 = Me; R^2 = OMe	R^1 = Et; R^2 = OMe
AChE IC$_{50}$ = 0.090 µM	AChE IC$_{50}$ = 2.55 µM
MAO-A IC$_{50}$ = 450 µM	MAO-A IC$_{50}$ = 421 µM
MAO-B IC$_{50}$ = 43 µM	MAO-B IC$_{50}$ = 22 µM

Figure 14: Thiocarbamoyl-3-phenyl-5-thienyl-2-pyrazoline derivatives (**12** and **13**) studied as dual MAO-AChE inhibitors.

Multi-target directed ligands targeting MAO and AChE could be of great value and have considerable potential in becoming clinically relevant candidates for the treatment and prevention of Alzheimer's disease. In this respect coumarin, indole-like and thiocarbamoyl-3-phenyl-5-thienyl-2-pyrazoline derivatives are of significant importance.

NITRIC OXIDE SYNTHASE (NOS) – ISOFORMS AND STRUCTURE

Evidence suggests that not only reactive oxygen species but also nitric oxide (NO) may play a role in oxidative damage in neurodegenerative disorders [14, 133]. NO is a cell-to-cell biological signaling molecule which controls various events in cardiovascular, immune, and nervous system physiology and is formed *via* the nitric oxide synthase (NOS) catalysed conversion of *L*-arginine to *L*-citrulline, a process that leads to the formation of the NO radical (Fig. **15**) [133, 134].

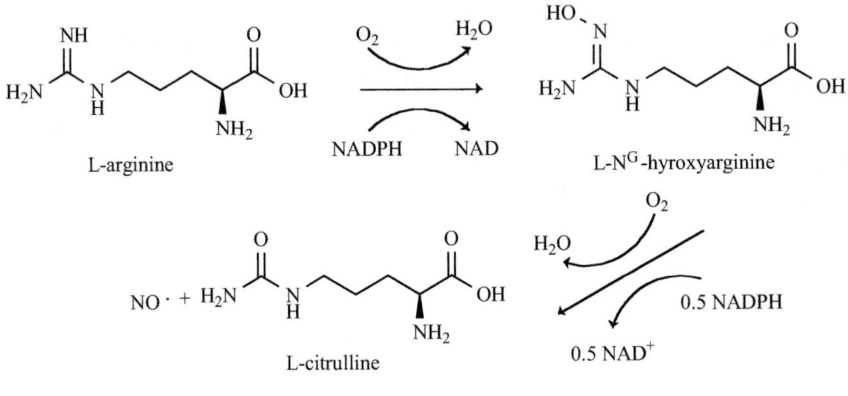

Figure 15: The catalytic biosynthesis of NO from *L*-arginine.

Three NOS isoforms are expressed in mammals from distinct genes with different subcellular localisation, catalytic properties, regulation and inhibitor sensitivity

[135]. These include two constitutive calcium/calmodulin-dependent forms of NOS: Neuronal nitric oxide (nNOS) whose activity was first identified in neurons and is involved in neurotransmission and long-term potentiation; and endothelial nitric oxide synthase (eNOS) first identified in endothelial cells which is involved in the regulation of smooth muscle relaxation and vascular tone. These two isoforms are physiologically activated by steroid hormones or neurotransmitters such as dopamine, NO, glycine and glutamate which increases intracellular calcium concentrations [136, 137]. In contrast, the inducible form of nitric oxide synthase, iNOS, is calcium independent and is expressed in a broad range of cell types. iNOS is important in the immune system's defense against pathogens and tumor cells [136] and is induced after stimulation with cytokines and exposure to microbial products [137].

All three NOS isoforms are active as homodimers that consist of two fused enzyme monomers, each containing an oxygenase domain with binding sites for NADPH, flavin adenine dinucleotide (FAD) and flavin mononucleotide (FMN), and a cytochrome reductase domain with binding sites for the substrate *L*-arginine, heme and the redox cofactor, tetrahydrobiopterin (BH_4). The NOS isoforms are the only known enzymes that require these co-factors including, NADPH, FAD, FMN, heme, BH_4 and calmodulin to be activated [138].

The *C*-terminal reductase and *N*-terminal oxygenase domains are linked together in the middle of the protein *via* a calmodulin-binding domain (Fig. **16**) [138]. Binding of calmodulin appears to act as a molecular switch that enables electron flow from the flavin prosthetic groups in the oxygenase domain to heme in the reductase domain. This facilitates the conversion of O_2 and *L*-arginine to NO and *L*-citrulline (Figs. **15** and **16**) [139]. The oxygenase domain of each NOS isoform also contains a BH_4 prosthetic group, which is required for the efficient generation of NO. Unlike other enzymes where BH_4 is used as a source of reducing equivalents and is recycled by dihydrobiopterin reductase, BH_4 activates heme-bound O_2 by donating a single electron, which is then recaptured to enable NO release [140].

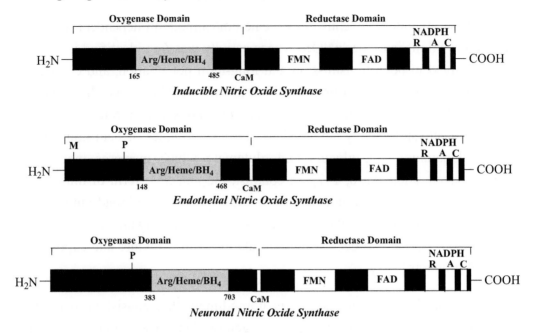

Figure 16: The different NOS isoform domains – A schematic representing the similarity of the primary amino acid sequence of human NOS isozymes. Reductase and oxygenase domains are separated by the calmodulin binding site (CaM). The reductase domain contains binding sites for FMN and FAD as well as several consensus sites for NADPH. The oxygenase domain of eNOS and nNOS contains a phosphorylation site (P). All NOS isoforms contain a region in the oxygenase domain that includes the binding sites for *L*-arginine (Arg), heme and BH₄ (adapted with permission from Kervin *et al.*, 1995) [141, 142].

Although NO controls several physiological functions, a number of disease states are associated with the overproduction of NO, making the NOS pathway an attractive target for the development of therapeutics [133, 143]. NO in its free radical form may also contribute to oxidative damage and molecular and tissue destruction. Overstimulation of individual NOS isoforms, especially nNOS and iNOS, and subsequent overproduction of NO plays a role in several disorders, including septic shock, arthiritis [144], diabetes, ischemia-reperfusion injury, pain

[134], and various neurodegenerative diseases [14]. In contrast, NO produced by eNOS in endothelial cells, has mainly a physiological role, such as maintaining normal blood pressure and flow [145, 146] and the inhibition thereof leads to unwanted effects such as hypertension and increased atherogenesis [147].

NOS and Neurodegenerative Disorders

Neuronal nitric oxide synthase (nNOS) facilitates the generation of NO in the central nervous system and is essential for normal neurological functioning [147]. Recent research has implicated the overexpression of nNOS and subsequent overproduction of NO in the neurodegeneration that is associated with Alzheimer's disease and Parkinson's disease [14]. Since nNOS plays a critical role in the production of neuronal NO, it is considered to be a promising neuroprotective therapeutic target [15]. The simultaneous inhibition of nNOS and the other NOS isoforms such as iNOS and eNOS is however undesirable, as iNOS and eNOS are responsible for maintaining crucial physiological functions [148]. Additionally, recent research has linked the inhibition of iNOS with a higher probability of developing Alzheimer's disease [149]. The selective inhibition of nNOS, however, provides a promising strategy in developing therapeutics for the treatment of neurodegenerative disorders. Typical inhibitors of nNOS include amongst others, *L*-NNA, *L*-NIO and ARL17477 (Fig. **17**) [150-153]. However, none of these earlier drugs exhibited distinct isoform selectivity. The reason for lack of selectivity could be that the immediate active sites of all three the NOS isoenzymes are identical (see Fig. **16**). A compound that could be anchored into the active site of nNOS and extend out to reach the second sphere of amino acid residues might therefor result in improved selectivity towards nNOS. Although intense research efforts have been devoted to the design and development of small molecules to selectively inhibit the activity of nNOS none have yet reached the market as a treatment strategy for neurodegenerative disorders.

NOS Inhibitors as Multifunctional Agents

Several new classes of drug-like molecules for selective inhibition of nNOS and activity on other neuroprotective targets including *N*-methyl-D-apartate (NMDA) receptor antagonism, inhibition of the MAO-B enzyme and compounds containing dual nNOS and antioxidant activities have been reported in the last few years and might find application in the future treatment of neurodegenerative disorders.

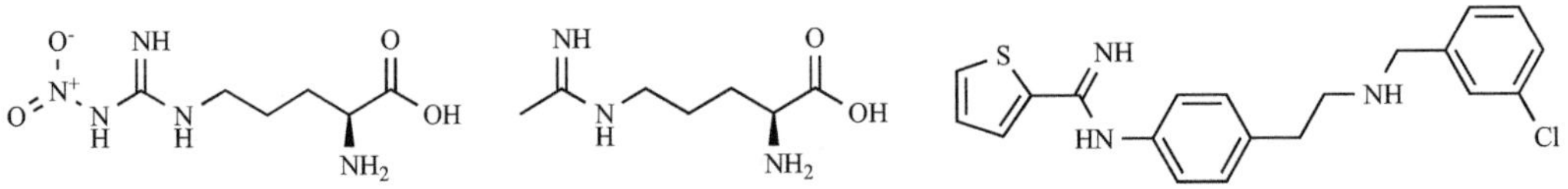

L-NNA

nNOS IC$_{50}$ = 0.29 μM
iNOS IC$_{50}$ = 3.1 μM
eNOS IC$_{50}$ = 0.35 μM

L-NIO

nNOS IC$_{50}$ = 0.1 μM
iNOS IC$_{50}$ = 60 μM
eNOS IC$_{50}$ = 12 μM

ARL17477

nNOS IC$_{50}$ = 0.07 μM
iNOS IC$_{50}$ = 0.33 μM
eNOS IC$_{50}$ = 1.6 μM

Figure 17: Typical structures of nNOS inhibitors described in literature.

NOS Inhibition and Modulation of Oxidative Stress

Both reactive oxygen species (*e.g.*, O$_2^-$) and reactive nitrogen species (*e.g.* NO·) are involved in neurotoxic processes, such as oxidation and inflammation, and contribute to neuronal death [154-156]. These species can act independently, but can also interact with other reactive species – NO· reacts rapidly with O$_2^-$ to form the highly reactive peroxynitrite (ONOO$^-$) with harmful effects on neuronal cells [157]. Research has shown that during the development of neurodegenerative disorders the formation of NO is increased and significant neuroprotective effects have been reported after treatment with certain nNOS inhibitors [158-160]. Moreover, the pre-treatment with antioxidants have also shown to reduce neuronal death in animal model studies [161]. Thus, a possible strategy for the treatment of neurodegenerative disorders would be the administration of a drug that possesses both antioxidant and selective nNOS inhibitory activities [158]. For these reasons, inhibitors of nNOS with potent antioxidant properties have become an area of research interest.

Auvin *et al.*, described the synthesis of a series of compounds where a nNOS pharmacophore is linked to an antioxidant fragment (represented by compounds **14** and **15**; Fig. **18**) [162]. The nNOS function consisted of the potent and selective nNOS pharmacophore, phenyl-2-thiophenecarboximidamide, connected *via* linkers differing in chain length to an antioxidant function consisting of a substituted phenolic moiety. Substituted phenols have long been known in literature as potent free radical scavengers [163]. These novel compounds were able to inhibit the nNOS enzyme with moderate selectivity over eNOS and iNOS. They also showed excellent antioxidant potency, which was assessed through their ability to inhibit Fe^{2+} induced lipid peroxidation (LPO) in rat brain microsomes. Among this series, compound **15**, a propofol derivative, displayed the most favorable dual potencies against nNOS (IC_{50} = 0.12 μM) and lipid peroxidation (IC_{50} = 0.4 μM; Fig. **18**). This further shows that nNOS was able to accommodate very bulky antioxidant groups such as di-tert-butyl phenol (compounds **14**) or di-iso-propyl phenol (compound **15**) in its active site. This study affirms and demonstrates the synthetic feasibility of dual nNOS/LPO inhibitors with inhibitory capacities in the low micromolar range. Moreover, compound **15** exhibits potent inhibitory properties against nNOS and lipid peroxidation while exhibiting acceptable e/nNOS (67.5) and i/nNOS (>150) selectivity [163].

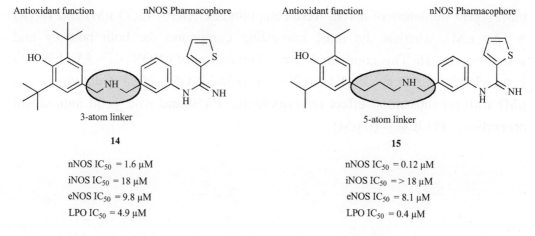

Antioxidant function	nNOS Pharmacophore		Antioxidant function	nNOS Pharmacophore

3-atom linker   5-atom linker

14   **15**

nNOS IC_{50} = 1.6 μM   nNOS IC_{50} = 0.12 μM

iNOS IC_{50} = 18 μM   iNOS IC_{50} = > 18 μM

eNOS IC_{50} = 9.8 μM   eNOS IC_{50} = 8.1 μM

LPO IC_{50} = 4.9 μM   LPO IC_{50} = 0.4 μM

Figure 18: Dual acting nNOS and LPO inhibitors (compounds **14** and **15**).

Chabrier *et al.*, has also combined the two strategies into a single molecule to protect neurons from neurodegeneration. The molecule, BN 80933, combines the known potent antioxidant activity of Trolox and a selective nNOS inhibitor pharmacophore, thiopheneamidine, *via* a piperidine linker (Fig. **19**) [164]. BN 80933 was able to selectivity inhibit nNOS ($Ki = 0.92$ µM) relative to eNOS ($Ki =$ 111 µM) and iNOS (>300 µM) in an *in vitro* enzymatic assay. This compound also exhibited the ability to inhibit iron-dependant LPO in a rat brain membrane preparation with an IC_{50} value of 0.29 µM. Interestingly, BN 80933 was more potent at inhibiting LPO ($IC_{50} = 0.29$ µM) than the antioxidant Trolox ($IC_{50} =$ 39.4 µM). This molecule was also able to protect neurons in various *in vivo* neurodegenerative animal based studies [165] and may be an useful lead compound or therapeutic agent for the treatment of neurodegenerative disorders that involve both NO and ROS.

Imidazole and simple derivatives thereof have been reported in literature as inhibitors of various NOS isoforms. Based on these findings Salerno *et al.*, (2012) developed a series of imidazole derivatives which were evaluated for their ability to act as selective nNOS inhibitors with potent antioxidant properties (Fig. **20**). Some of these compounds (**16-18**) displayed good capacity to scavenge free radicals and the ability to reduce lipid peroxidation [166]. These derivatives were also able to selectively inhibit nNOS over eNOS and iNOS. The most potent inhibitors of nNOS were compounds **17** ($Ki = 13.75$ µM) and **18** ($Ki = 12.5$ µM), whereas the most interesting compound for both potency and selectivity was **16**. This compound showed inhibition of nNOS ($Ki = 85$ µM) with good selectivity over eNOS ($Ki = > 2000$ µM) and iNOS (only 4 % activity at 500 µM) with no substantial effect on cytochrome P450 and with good antioxidant properties (LPO $IC_{50} = 10$ µM).

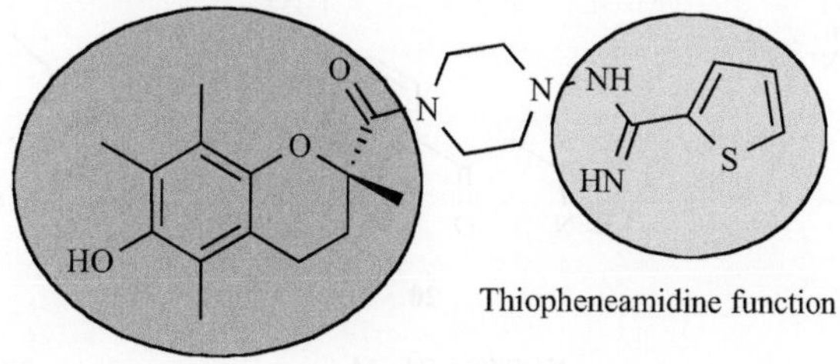

Trolox function

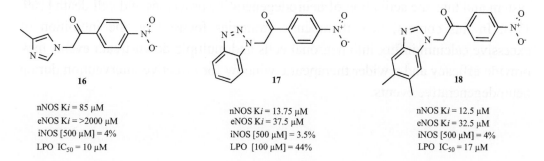

Thiopheneamidine function

BN 80933

nNOS IC$_{50}$ = 0.92 μM

iNOS IC$_{50}$ = 111 μM

eNOS IC$_{50}$ = >300 μM

LPO IC$_{50}$ = 0.29 μM

Figure 19: The chemical structure of BN 80933, a dual selective nNOS and LPO inhibitor.

In addition to the above, another series of imidazole derivatives were described as dual acting selective nNOS inhibitors endowed with potent antioxidant activity (Fig. **21**) [167]. Compounds **19–21** showed highly selective nNOS inhibition over eNOS and iNOS (Fig. **21**). The compounds also exhibited the ability to scavenge free radicals and showed effective inhibition of LPO. These imidazole derivatives (Figs. **20** and **21**) represent chemical structures which can be easily modified to improve the observed antioxidant properties and to provide new neuroprotective therapeutic strategies focused on multiple downstream events in neuro-degeneration.

16

nNOS K*i* = 85 μM

eNOS K*i* = >2000 μM

iNOS [500 μM] = 4%

LPO IC$_{50}$ = 10 μM

17

nNOS K*i* = 13.75 μM

eNOS K*i* = 37.5 μM

iNOS [500 μM] = 3.5%

LPO [100 μM] = 44%

18

nNOS K*i* = 12.5 μM

eNOS K*i* = 32.5 μM

iNOS [500 μM] = 4%

LPO IC$_{50}$ = 17 μM

Figure 20: Imidazole derivatives as selective nNOS inhibitors with potent antioxidant properties.

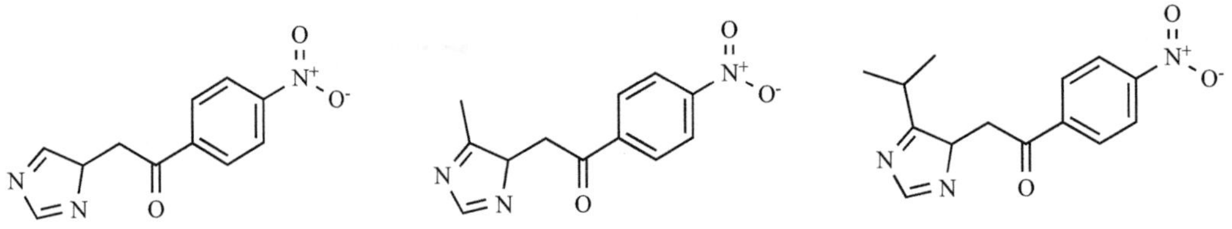

19	**20**	**21**
nNOS Ki = 22.5 µM	nNOS Ki = 74 µM	nNOS Ki = 22.5 µM
eNOS Ki = >5000 µM	eNOS Ki = >5000 µM	eNOS Ki = 550 µM
iNOS [500 µM] = 31.62%	iNOS Ki = 1050 µM	iNOS Ki = 1300 µM
LPO [50 µM] = 60%	LPO [50 µM] = 65%	LPO [50 µM] = 90%

Figure 21: Imidazole derivatives with selective nNOS inhibitory and antioxidant activity.

NOS Inhibition and Modulation of Calcium Homeostasis

Excitotoxicity is a term used to describe neuronal death due to the over stimulation of *N*-methyl-D-aspartate (NMDA) receptors, an event that causes an excessive influx of calcium into neuronal cells [168]. Calcium overloading of neuronal cells can cause the activation of calcium-dependent signals to enzymes such as phospholipases and proteases, as well as oxidative stress through ROS and RNS, culminating in cell death [169-172]. Calcium may also enter through voltage-dependent calcium channels (VDCC) and these channels are also implicated in calcium overload and mitrochondrial disruption [169, 173]. Calcium overload further stimulates the overexpression of calcium dependant nNOS, the enzyme responsible for catalysing the formation of NO in the central nervous system and thus the activation of neurodegenerative processes and cell death [169, 174]. In this context, new therapeutic strategies focused on the inhibition of excessive calcium influx into neuronal cells and multiple downstream events may provide efficacy and a wider therapeutic window for effective intervention during neurodegenerative events.

An example of this approach is the design of dual acting selective nNOS inhibitors and NMDA receptor antagonists by Neuraxon. Several indole structures, represented by compound **22** in Fig. **22** [174-179], were evaluated for NOS activity with the hemoglobin capture assay [180] using recombinant rat, bovine and murine NOS enzymes expressed in *E. coli*. Results from these screens indicated selectivity of the compounds towards nNOS inhibition. The indole derivatives were further evaluated on recombinant human iNOS, human eNOS and human nNOS produced in Baculovirus-infected Sf9 cells, using a radiometric method measuring [^{3}H]*L*-citrulline production, and were reported to possess selective inhibition of nNOS. The compounds also showed the ability to inhibit calcium flux through the NMDA receptor channel. Especially compound **22** (Fig. **22**) demonstrated a 90% inhibition of calcium flux through the NMDA receptor channel at a concentration of 50 μM and was able to selectively inhibit human nNOS (IC_{50} = 1.2 μM) over human eNOS (IC_{50} = 15 μM) and iNOS (IC_{50} = 60 μM). Further studies by Neuraxon indicated that these indole compounds were effective neuroprotective agents in various *in vitro* and *in vivo* neuroprotective studies.

A recent study by Joubert *et al.* (2011) described the synthesis of a series of fluorescent polycyclic cage derivatives as potential multifunctional neuroprotective agents [181]. The fluorescent ligands used in this study resembles the structure of 7-nitroindazole (7-NI; Fig. **24**), a selective nNOS inhibitor (IC_{50} = 0.71 μM; Handy *et al.*, 1995), which was directly conjugated to polycyclic cage structures for effective blood-brain barrier permeability [182] and to potentially increase the selective inhibition of nNOS over the other closely related isoforms. These novel structures were found to be effective inhibitors (IC_{50} values between 0.29 μM - 11 μM) of NOS function in crude rat brain homogenates. The documented calcium channel modulation observed for selected polycyclic cage structures, such as amantadine and pentacycloundecane amines [183-185] prompted further evaluation of these fluorescent polycyclic cage compounds for VGCC inhibition, NMDA receptor antagonism and the ability to scavenge detrimental oxygen and nitrogen free radicals [186]. Three heterocyclic adamantane-amines (**23**, **24** and **25**; Fig. **23**) were identified with a high degree of inhibitory activity on VGCC, NMDA receptors and NOS. They also demonstrated free radical scavenging activity in the DPPH$^{\cdot+}$ and ABTS^{+} assays. The potential of increased dopamine neuro-transmission was also postulated due to the amantadine moiety present in these structures [182, 187]. An additional application of these structures is that they may be utilised in the development of fluorescent displacement and other pharmacological studies [186, 188] to provide critical information about neurodegenerative processes.

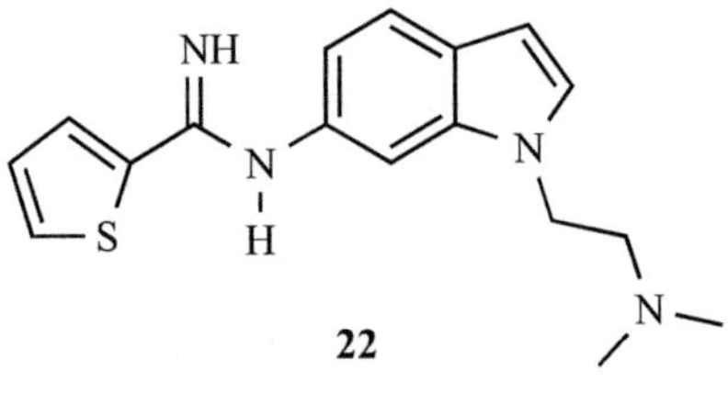

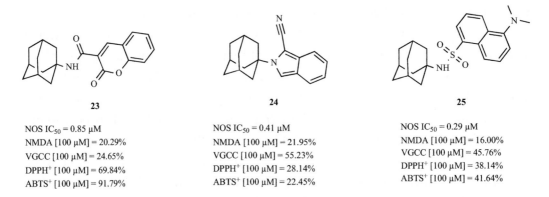

(human) nNOS IC$_{50}$ = 1.2 μM

(rat) nNOS IC$_{50}$ = 8.8 μM

NMDA [50 μM] = 90%

Figure 22: The indole derivative (**22**) designed by Neuraxon as a dual acting nNOS inhibitor and NMDA receptor antagonist.

23

NOS IC$_{50}$ = 0.85 μM

NMDA [100 μM] = 20.29%

VGCC [100 μM] = 24.65%

DPPH^{+} [100 μM] = 69.84%

ABTS^{+} [100 μM] = 91.79%

24

NOS IC$_{50}$ = 0.41 μM

NMDA [100 μM] = 21.95%

VGCC [100 μM] = 55.23%

DPPH^{+} [100 μM] = 28.14%

ABTS^{+} [100 μM] = 22.45%

25

NOS IC$_{50}$ = 0.29 μM

NMDA [100 μM] = 16.00%

VGCC [100 μM] = 45.76%

DPPH^{+} [100 μM] = 38.14%

ABTS^{+} [100 μM] = 41.64%

Figure 23: Fluorescent polycyclic compounds with potential multifunctional neuro-protective activity.

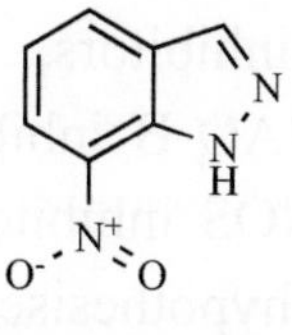

7-Nitroindazole (7-NI)

(human) MAO-B IC_{50} = >50 μM

(rat) nNOS IC_{50} = 0.71 μM

Figure 24: 7-Nitroindazole, a dual MAO-B and nNOS inhibitor.

NOS and MAO-B Inhibition

The well known nNOS inhibitor, 7-nitroindazole (7-NI, Fig. **24**) [189], has previously been indicated to have neuroprotective activity in 1-methyl-4-phenyl-1,2,3,6-tetrahydropyridin (MPTP) animal models of Parkinson's disease [190]. To gain insight into how 7-nitroindazole affords this neuroprotection, it was evaluated for its ability to impede different steps in the biochemical pathways leading to MPTP neurotoxicity. It was concluded that, apart from the well-documented nNOS inhibition (IC_{50} = 0.71 μM) and human MAO-B inhibition (IC_{50} = > 50 μM), the neuroprotective effect arises from its antioxidant activity exerted through the ability to scavenge hydroxyl radicals (•OH) [191]. Regioisomeric 5- and 6-nitroindazole can also be considered promising neuroprotective agents, since they were indicated to inhibit human MAO-B with IC_{50} values of 2.5 μM and 6.8 μM, respectively [190]. In addition to this they have similar hydroxyl radical trapping ability and a weaker *in vitro* inhibitory potency towards rat cerebellar NOS (IC_{50} = 47.3 μM for 5-nitroindazole and IC_{50} = 31.6 μM for 6-nitroindazole) [150].

In an attempt to develop multi-target directed ligands, specifically targeting nNOS and MAO-B, Prins and coworkers (2009) designed and synthesised a series of pteridine analogues and evaluated their ability to inhibit both these enzymes (Fig. **25**) [192]. The rationale for designing these structures was that firstly, pteridine-2,4-dione is structurally related to the essential NOS cofactor, tetrahydrobiopterin and several pterin analogues had been developed to target this binding site [193]; and secondly, that xanthine structures had been shown to be potent and selective MAO-B inhibitors, with (*E*)-8-(3-chlorostyryl)caffeine serving as the model xanthine MAO-B inhibitor [194, 195]. Compound **26** did however not display significant NOS inhibitory activity during its assessment in rat brain homogenate and it was hypothesised that the styryl phenyl might have sterically hindered the binding of the said compound to NOS. During *in vitro* MAO-B assays a potent and reversible MAO-B inhibitor was identified (**26**, Fig. **25**) bearing a 6-styryl-pteridine-2,4-dione structural framework. The iminic synthetic precursor (**27**) to the target pteridine compound (**26**) was less potent as a MAO-B inhibitor, but showed slightly better activity against NOS [192].

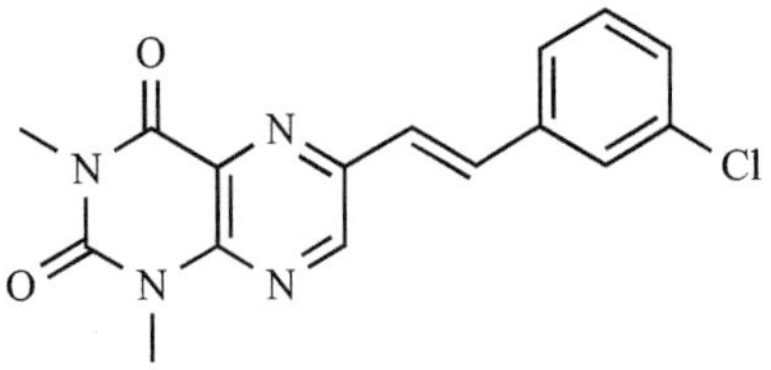

(**26**)

MAO-B Ki = 0.181 μM
NOS IC$_{50}$ = 1057 μM

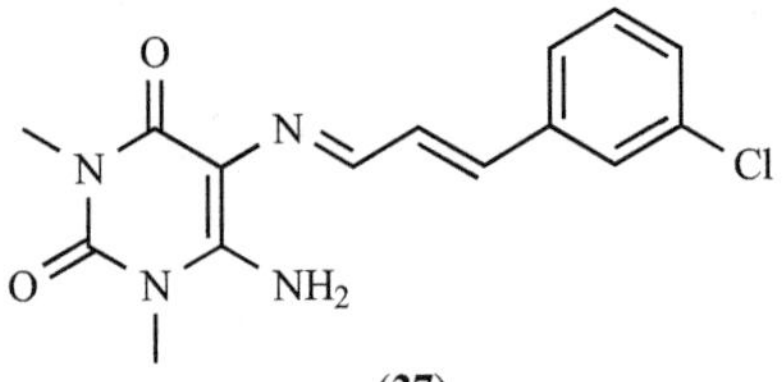

(**27**)

MAO-B Ki = 0.348 μM
NOS IC$_{50}$ = 557 μM

Figure 25: Pteridine-2,4-dione analogues designed for dual inhibition of NOS and MAO-B.

The indole nucleus has been indicated to be a very useful scaffold for developing potent NOS and MAO inhibitors [196-198]. One very good example of the above is a series of indolylmethylamine derivatives that was shown to be irreversible MAO inhibitors [199]. Of specific importance is the observation that bulky C5 substituents installed on the indole nucleus generally increased potency and selectivity for the MAO-B isoform. To this end, at least two research groups [200, 201] have indicated that the introduction of a benzyloxy group at C5 drastically improved the selectivity towards MAO-B. PF9601N (Fig. **26**) was shown to have a neuroprotective effect in different models of Parkinson's disease [202, 203].

PF9601N and its metabolite FA72 exhibited similar antioxidant activity in linoleic acid and in brain homogenate models. This effect might be as a result of a direct interaction with peroxynitrate anion (ONOO⁻) rather than an ability to scavenge free radicals. In addition to this, PF9601N and FA72 inhibited nNOS with IC_{50} values of 187 μM and 192 μM, respectively.

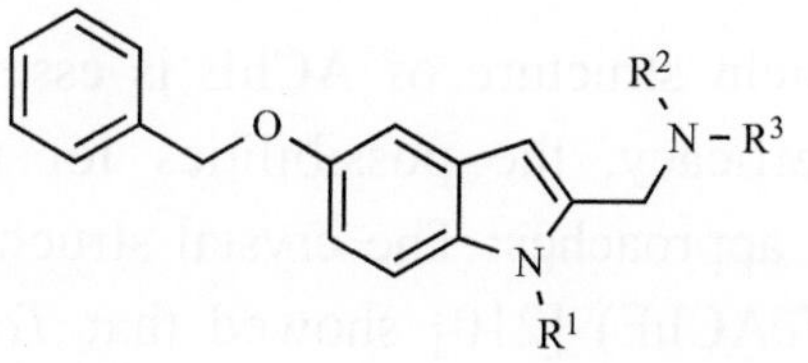

PF9601N

$R^1 = R^2 = H; R^3 = -CH_2CCH$

nNOS IC_{50} = 187 μM

LA oxidation IC_{50} = 82.8 μM

(rat liver) MAO-A IC_{50} = 1250 μM

(rat liver) MAO-B IC_{50} = 22 μM

FA72

$R^1 = R^2 = H; R^3 = -CH_2CCH$

nNOS IC50 = 187 μM

LA oxidation IC_{50} = 82.8 μM

(rat liver) MAO-A IC_{50} = 1250 μM

(rat liver) MAO-B IC_{50} = 22 μM

Figure 26: PF9601N, a dual nNOS and MAO inhibitor, and its metabolite FA72.

ACETYLCHOLINESTERASE – STRUCTURE AND FUNCTION

Acetylcholinesterase (AChE) is responsible for the termination of impulse transmission at cholinergic synapses by rapid hydrolysis of acetylcholine (ACh) into acetate and choline (Fig. **27**) [16]. Acetylcholinesterase inhibitors (AChEIs) reduces acetylcholinesterase induced destruction of ACh in the synaptic cleft, increases the intra synaptic residence time of ACh, and facilitates interaction between ACh and the postsynaptic cholinergic receptor, thus contributing to their clinical benefits. AChEIs have been reported to be effective on cognitive performance in Alzheimer's Disease (AD) and diffuse Lewy body disease (DLBD) and on cognitive and behavioural symptoms in dementia associated with PD [204-209]. Recent studies further suggest that ChEIs may reduce amyloid precursor protein (APP) processing and provide some degree of neuroprotection [16]. For these reasons AChE has become one of the most important therapeutic targets for the treatment of neurodegenerative disorders, especially AD.

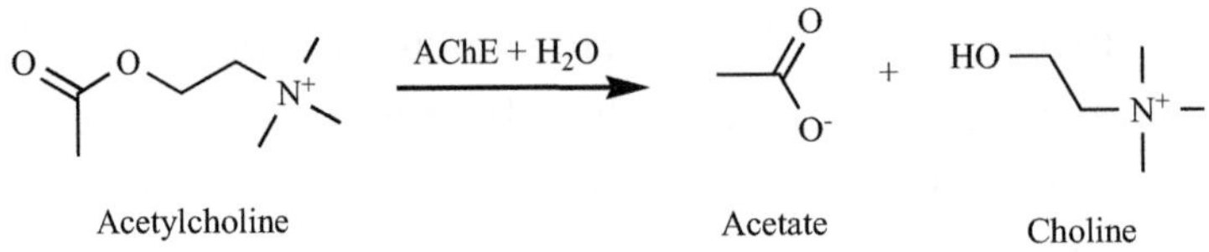

Acetylcholine   Acetate   Choline

Figure 27: Enzymatic hydrolysis of ACh by AChE.

Knowledge of the protein structure of AChE is essential for understanding its remarkable catalytic efficacy, the possibilities for rational drug design, and developing therapeutic approaches. The crystal structure of *Torpedo californica* acetylcholinesterase (*Tc*AChE) [210] showed that *Tc*AChE belongs to the α/β hydrolase family of enzymes [211, 212]. The catalytic site, which is responsible for the enzymes' rapid catalytic activity, contains the active site for binding of Ach. This site is buried close to the bottom of a deep and narrow gorge which is lined by as many as 14 conserved aromatic residues, comprising up to 70% of its surface, with the amino acids Trp84 and Phe330 contributing to the so-called catalytic anionic site (CAS), and the amino acids Tyr70, Tyr121 and Trp279 to the peripheral anionic site (PAS). The active site is buried so that the substrate ACh will be surrounded by the protein, permitting multiple substrate/enzyme interactions and in turn, creating a more effective transition state [213].

Numerous AChEIs have been developed for use in the clinical setting against AD. These include compounds like tacrine (Cognex®), donepezil (Aricept®), rivastigmine (Exelon®) and galantamine (Razadyne®) (Fig. **28**). Tacrine was the first of these drugs approved in 1993, but its use has been abandoned because of a high incidence of side effects including hepatotoxicity [214, 215]. Rivastigmine and galantamine are currently approved for mild to moderate AD, while donepezil is used for severe AD. Most of these AChEI drugs provide symptomatic relieve of AD but do not address the underlying neurodegeneration in a significant manner [216, 217].

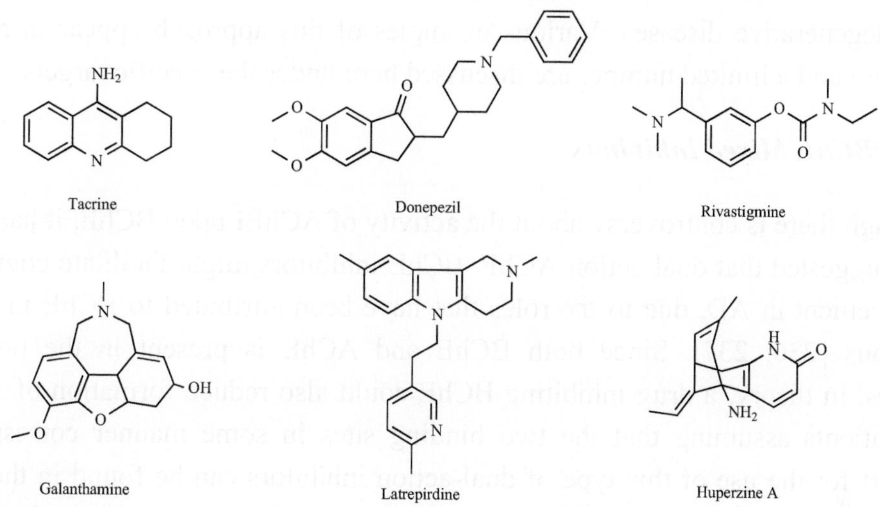

Tacrine   Donepezil   Rivastigmine

Galanthamine   Latrepirdine   Huperzine A

Figure 28: Structures of AChEIs described in literature as clinical agents used for the treatment of AD.

Acetylcholinesterase Inhibitors as Multifunctional Agents

Despite the failure of tacrine as a therapeutic drug, it has been widely used as a scaffold for the development of new multifunctional agents with additional biological properties other than AChEI [218-222]. Rivastigmine and Galantamine, though initially applied for their AChE effects, are now also known to respectively inhibit BChE and stimulate nicotinic ACh receptors to release ACh [223]. Galantamine has further been reported to modulate NMDA receptor function [224] and exert neuroprotection by inhibiting apoptosis [225, 226] and oxidative stress [227]. Another marketed AChEI of note, latrepirdine (Dimebon®), has been described to also inhibit BChE, *N*-methyl-D-apartate (NMDA) receptor activity and apoptosis [228-230]. Huperzine A (Fig. **28**), an extract of *Huperzia serrata*, exhibit potent AChE inhibition [231-233]. Further studies have revealed that huperzine A may shift APP metabolism toward the non-amyloidogenic α-secretase pathway [234] and that this compound may have the ability to reduce glutamate-induced cytotoxicity by antagonising NMDA receptors [235].

Combining therapeutic and protective effects in a single molecule with AChE inhibitors as basis could thus contribute significantly towards the treatment of neurodegenerative diseases. Various examples of this approach appear in recent literature and a limited number are discussed here under the specific targets.

AChE/BChE Mixed Inhibitors

Although there is controversy about the activity of AChEI upon BChE, it has now been suggested that dual action AChE–BChE inhibitors might facilitate cognitive improvement in AD, due to the roles that have been attributed to BChE in brain functions [236, 237]. Since both BChE and AChE is present in the neuritic plaques, in theory, a drug inhibiting BChE could also reduce formation of Aβ in AD patients assuming that the two binding sites in some manner correspond. Support for the use of this type of dual-action inhibitors can be found in the fact that brain AChE activity continuously declines but BChE activity continuously increases in AD progression [238].

Fallarero and co-workers (2008) confirmed that compound **28** (Fig. **29**), a coumarin derivative, displays AChE inhibition (IC$_{50}$ = 4.97 μM) [239]. Computational studies revealed that this compound primarily binds to this enzyme at the CAS, while a secondary binding interaction is also present at the PAS. This compound was also able to effectively inhibit BChE with an IC$_{50}$ value of 4.56 μM.

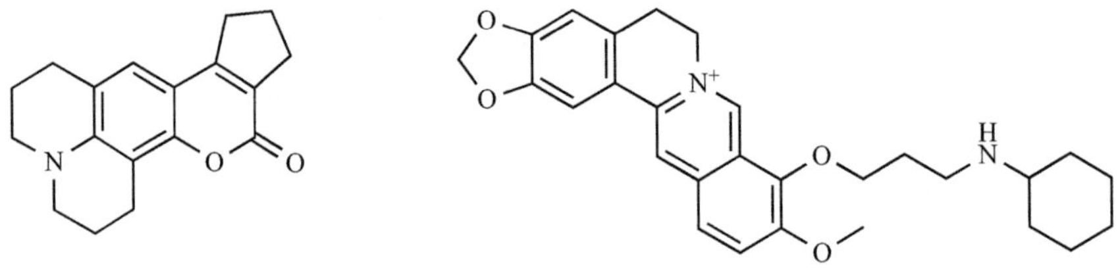

(**28**)

AChE IC$_{50}$ = 4.97 μM
BChE IC$_{50}$ = 4.56 μM

(**29**)

AChE IC$_{50}$ = 0.02 μM
BChE IC$_{50}$ = 1.35 μM

Figure 29: Structure and activities of the dual AChE/BChE inhibitors compounds **28** and **29**.

Huang *et al.*, (2010) described a series of berberine derivatives with substituted amino groups linked at the ether group in position 9 using different carbon spacers as AChEI and BChEI [240]. The derivative with the cyclohexylamino group linked to berberine *via* a three carbon spacer (**29**; Fig. **29**), was the most potent inhibitor with an IC$_{50}$ of 0.020 μM for AChE and 1.35μM for BChE. Molecular modelling simulations illustrated the presence of the compound in both the CAS and the PAS regions of the active sites of these enzymes.

Tacrine, one of the prototype AChEI is also a potent inhibitor of BChEI. Selectivity for AChE or BChE can be obtained through structural manipulation, but most structures containing this or similar scaffolds are seldom devoid of inhibition of the other, as will become evident in the following paragraphs.

AChE Inhibition and Inflammation

Systemic pro-inflammatory cytokines are depleted by the administration of AChEIs and this reduces both central and peripheral inflammation [241]. AChEIs thus seem to suppress inflammation *via* the cholinergic anti-inflammatory pathway, a mechanism by which the vagus nerve of the CNS regulates the production and release of tumour-necrotic factors and other cytokines [242]. In order to target neuro-inflammatory disorders, such as AD, PD and stroke, NSAID-AChEI conjugates have been explored [243]. These compounds seem to provide dual activity against neuro-inflammatory disorders by activating the cholinergic anti-inflammatory pathway and inhibiting cyclooxygenase (COX) enzymes [244].

Young *et al.* (2010) designed and studied two classes of these drugs, an ester carbonate and an ester series, of non-competitive AChEIs which in turn function as NSAID prodrugs (Fig. **30**) that target neuro-inflammation by releasing an NSAID *in vivo* and activating the cholinergic anti-inflammatory pathway *via* cholinergic up-regulation [245]. The most active AChEI was compound **30** which contains the NSAID diclofenac. The release of diclofenac from this prodrug was established at a half live of 357 minutes. Compound **31** was completely hydrolysed after 10 min, but showed no AChE inhibitory activity [245].

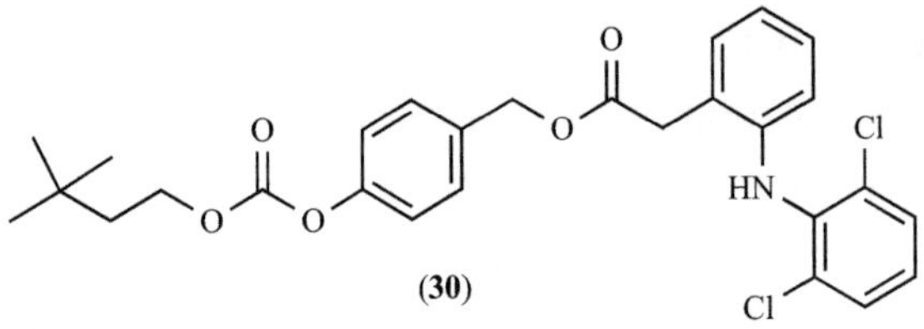

(30)

(31)

AChE IC$_{50}$ = 0.025 µM

T$_{1/2}$ = 357 min

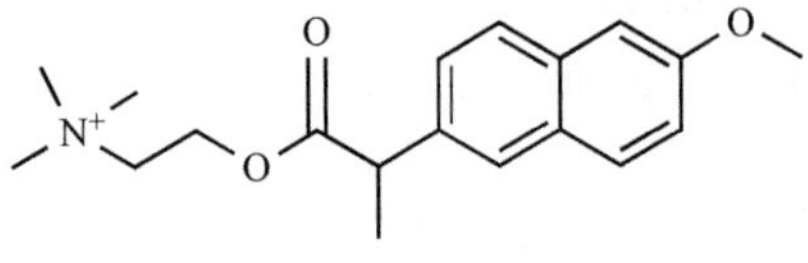

AChE IC$_{50}$ = >1000 µM

T$_{1/2}$ = 10 min

Figure 30: Structures of selected NSAID-AChEI prodrugs.

AChE Inhibition and β-Amyloid (Aβ) Modulation

Various Aβ-targeted therapeutic strategies are being pursued for the potential treatment of AD and include: (i) Modulation of Aβ production; (ii) enhancement of Aβ degradation and, (iii) inhibition of Aβ aggregation [246]. The non-cholinergic role of AChE include its binding to Aβ and the consequent promotion of its assembly and deposition into insoluble fibrils by means of a protein-protein interaction. AChE is integrated into the amyloid aggregates and forms stable complexes that, besides exerting a fibrillogenic effect, boost Aβ fibrils' neurotoxicity [247, 248]. Dual binding AChEIs that interact simultaneously with both the CAS and PAS sites of the AChE enzyme, has emerged as potential compounds to pursue a disease-modifying approach [249]. This appears to be a very promising therapeutic strategy, as dual-site AChEIs not only stimulate the cholinergic system, but also inhibit the production and/or the aggregation of Aβ promoted by AChE [250]. Many promising compounds have been developed to inhibit both AChE and AChE-mediated Aβ aggregation [251-253].

Utilising the dimethoxybenzophenone skeleton similar to that of donepezil, compound **32** (Fig. **31**) [254] was found to inhibit AChE with an IC$_{50}$ value of

0.46 µM. This compound was however not able to inhibit AChE-induced Aβ aggregation. This inability may be related to the fact that compound **32** could not properly come into contact with Trp286, a key residue in the PAS region of human AChE involved in the pro-aggregating action of AChE. As AChEIs able to interact with Trp286 might have the ability to inhibit Aß aggregation, a series of AChEIs were designed by utilising the benzophenone scaffold and *N,N'*-methylbenzylamino groups, aimed at reaching the PAS and interacting with Trp286. Different amino terminal side chains were incorporated in order to mimic the diethylmethylammonium alkyl moiety of the pure PAS ligand propidium which inhibits AChE-induced Aβ aggregation by 82% at 100 µM [255]. Compounds **33** and **34** (Fig. **31**) with a propoxy and a hexyloxy amino terminal side chain respectively, revealed good activity against human AChE and also have the ability to inhibit AChE-induced Aβ aggregation [256].

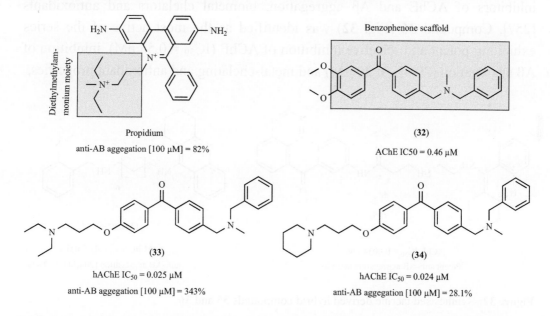

Figure 31: Propidium, lead compound **32** and the novel benzophenone derivatives **33** and **34**.

Tang and co-workers (2011) developed a series of dual binding site AChEI and inhibitors of AChE-induced and self-induced Aβ aggregation [221]. Hybrids of oxoisoaporphine-tacrine congeners were designed to simultaneously interact with both the CAS and PAS of AChE and their ability to inhibit AChE, AChE-induced and self-induced Aβ aggregation was evaluated. These compounds consist of a unit of tacrine, which occupies the CAS of AChE and the 1-azabenzanthrone moiety whose position along the enzyme gorge and the peripheral site can be modulated by a suitable tether connecting the moieties [221]. In this series compound **35** was the most potent AChEI (Fig. **32**). Compound **36** was also able to block self induced Aβ aggregation by 79% and AChE-induced Aβ aggregation by 83% in the biological assays used. BChE inhibition in the same order of that observed for AChE, which is likely to be relevant in the treatment of AD patients [221, 242] was also observed for these structures.

In an effort to identify novel multifunctional drug candidates for AD, Mao and co-workers (2012) designed a series of tacrine derived hybrid molecules active as inhibitors of AChE and Aβ aggregation, biometal chelators and antioxidants [257]. Compound **36** (Fig. **32**) was identified as the most active of the series exhibiting potent and selective inhibition of AChE (IC$_{50}$ = 0.55 nM), inhibition of Aβ aggregation (39.4% at 20 μM) and metal-chelating and antioxidant properties.

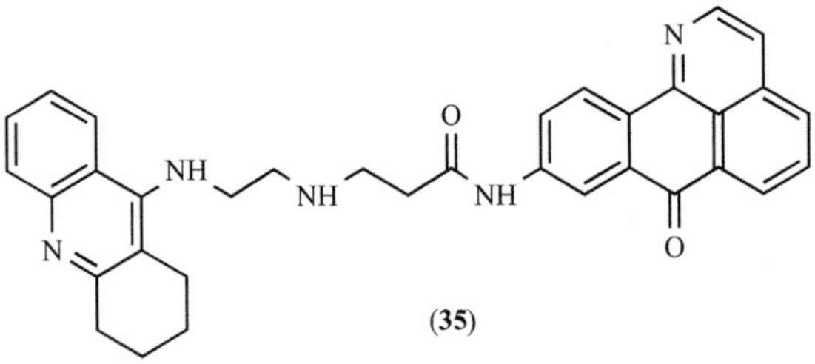

(35)

AChE IC$_{50}$ = 0.0034 μM
Potent anti-AB aggregation actiivity

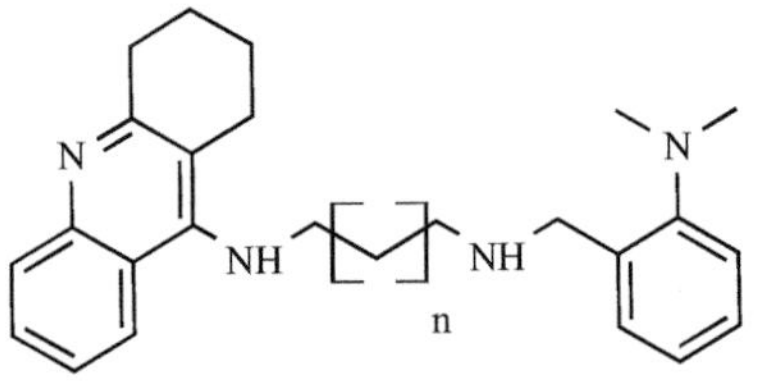

(36); n = 7

AChE IC$_{50}$ = 0.55 nM
anti-AB aggregation [20 μM] = 39.4%

Figure 32: Synthesised tacrine derived hybrid compounds **35** and **36**.

Recently, Yan *et al.*, (2012) designed a series of isaindigotone-tacrine hybrid derivatives as AChEIs and inhibitors of AChE induced Aβ aggregation [218]. Chlorine substitution in position 6 of the tacrine moiety, which was previously shown to increase the binding of these hybrid molecules with AChE [219, 220] was retained. Replacement of the ether terminal moiety with an*N*-phenylalkanamide terminal structure was found to increase hydrogen bond interaction while a five-membered aliphatic ring ring, resulting in increased planarity of the aromatic core was preferred [218, 258]. Compound **37** consisting of a five-membered aliphatic ring and an amide terminal motif was found to be the most active compound (Fig. **33**).

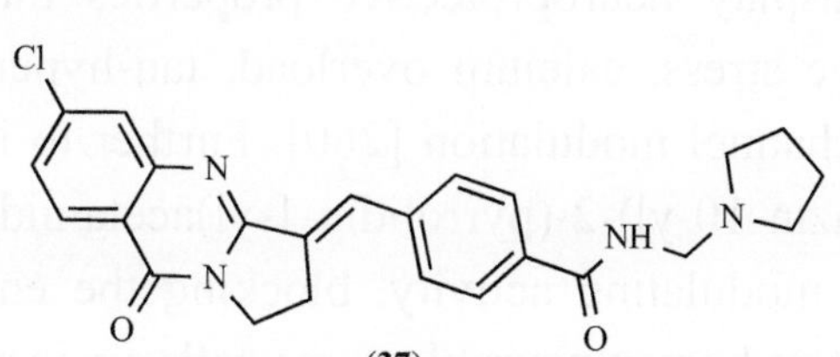

(37)

AChE IC$_{50}$ = 41 μM

Potent anti-AB aggregation activity

Figure 33: Structure of the isaindigotone-tacrine hybrid derivative (**37**), an AChEI and potent inhibitor of Aß aggregation.

AChE/BChE Inhibition and Calcium Channel Modulation

Increased intracellular Ca^{2+} and oxidative stress generated in response to Aβ enhances glutamate mediated neurotoxicity *in vitro*, suggesting that Aβ may increase NMDA responses leading to glutamate excitotoxicity through excessive influx of Ca^{2+} through the NMDA receptor channel and other calcium channels [259]. Uncompetitive NMDA receptor antagonists with fast off-rate thus represent promising neuroprotective drug candidates.

To this effect, Luo and co-workers (2012), reported that the bis-tacrine, named B3C, a dimeric AChEI (Fig. **34**) and GABA$_A$ receptor antagonist, acts as an uncompetitive antagonist or blocker of NMDA receptors (IC$_{50}$ = 0.52 μM) [222].

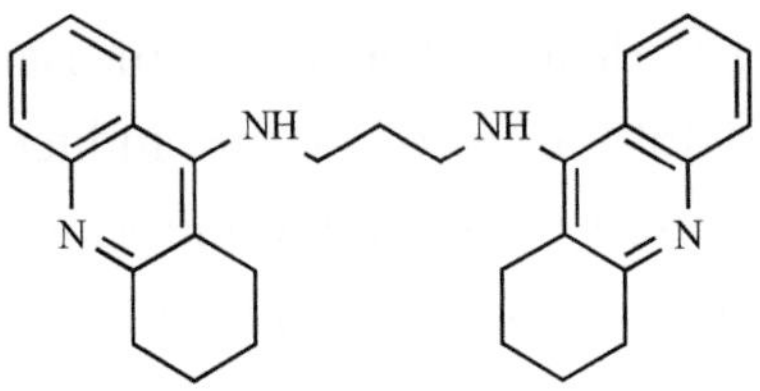

B3C

AChEI and GABA$_a$ receptor antagonist

NMDA IC$_{50}$ = 0.52 µM

Figure 34: B3C as dual acting AChEIs and NMDA receptor antagonists.

González-Muñoz *et al.*, (2011) described a series of *N*-acylamino-phenothiazines that display neuroprotective properties through modulation of AChE/BChE, oxidative stress, calcium overload, tau-hyperphosphorylation, Aβ toxicity and calcium channel modulation [260]. Further to its ChE effects, *N*-(3-Chloro-10*H*-phenothiazin-10-yl)-2-(pyrrolidin-1-yl)acetamide (**38**; Fig. **35**) showed Ca^{2+} channel modulating activity, blocking the entry of this cation by 29% at 10 µM. It protects human neuroblastoma cells against several toxic insults, such as calcium overload induced by an L-type agonist (45% at 1 µM), tau-hyperphosphorylation induced by okadaic acid (29% at 3 µM), and Aβ toxicity (91% at 0.3 µM) [260].

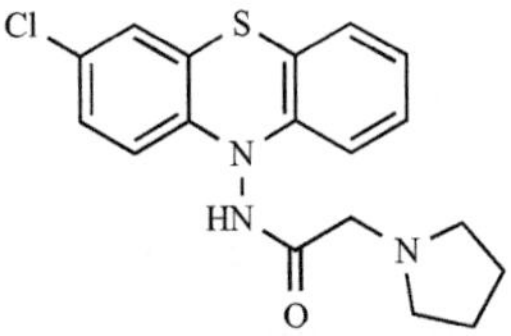

(38)

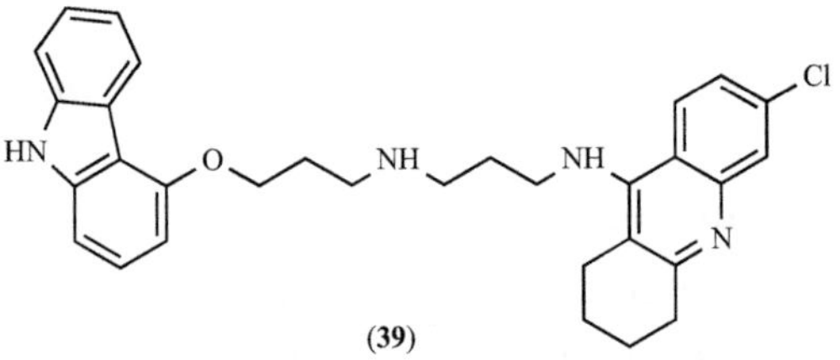

(39)

Both compounds exhibit potent AChEI and calcium channel modulating activity combined
with numerous other multifunctional activities

Figure 35: The chemical structures of multifunctional AChEI derivatives represented by compounds **38** and **39**.

Another study by Rosini and co-workers (2008) combined the cholinergic activity of AChEIs, offering symptomatic relief based on the structure of tacrine, with the neuroprotective action of NMDA receptor antagonism [219]. Compound **39**, a tacrine derived hybrid molecule (Fig. **35**) was able to inhibit AChE activity (IC_{50} =2.01 µM), block *in vitro* Aβ self-aggregation (36%), inhibit AChE-induced Aβ aggregation (57.7%), antagonize NMDA receptors (IC_{50} = 303 µM) and reduce oxidative stress.

AChE/BChE and β-Secretase (BACE-1)

AChE is one of the most important drug targets for symptomatic treatment of AD and the enzyme β-secretase (BACE-1) is a crucial factor of Aβ formation in the pathogenesis of AD. Combining these activities in one molecule led to the synthesis of chemically diverse structures with dual or multiple biological profiles including AChE, BACE, Aβ aggregation inhibiting and antioxidant activity (*e.g.*, Memoquin; Fig. **37**) [8]. Based on this assumption Zhu and co-workers (2009) designed and evaluated a series of inhibitors that inhibit both AChE and BACE-1 [261]. Compound **40** (Fig. **36**) exhibits excellent dual enzyme inhibitory potency (BACE-1: IC_{50} = 0.57 µM; AChEI: IC_{50} = 1.83 µM). This compound also showed inhibitory effects on Aβ production of amyloid precursor protein (APP) transfected in HEK293 cells (IC_{50} = 98.7 nM) and protected against H_2O_2-induced PC12 cell injury [261].

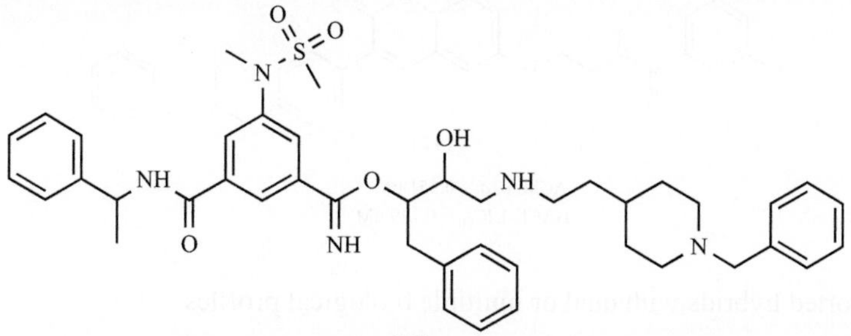

(**40**)

AChE IC_{50} = 1.83 µM
BACE-1 IC_{50} = 0.567 µM
AB production IC_{50} = 98.7 nM

Figure 36: Structure and activity of compound **40**.

Piazzi and co-workers (2008) were the first group who reported the design and evaluation of a series of coumarin-memoquin derived hybrid molecules substituted at the 6 and 7 position with halophenylalkylamidic functions as dual inhibitors (**41-42**; Fig. **37**) of AChE and BACE-1 [262]. These compounds were able to inhibit both these enzymes with sub-micro affinity. Compounds **41** and **42** were the most potent inhibitors of AChE (0.181 µM and 0.551 µM, respectively) and BACE-1 (0.114 µM and 0.149 µM, respectively).

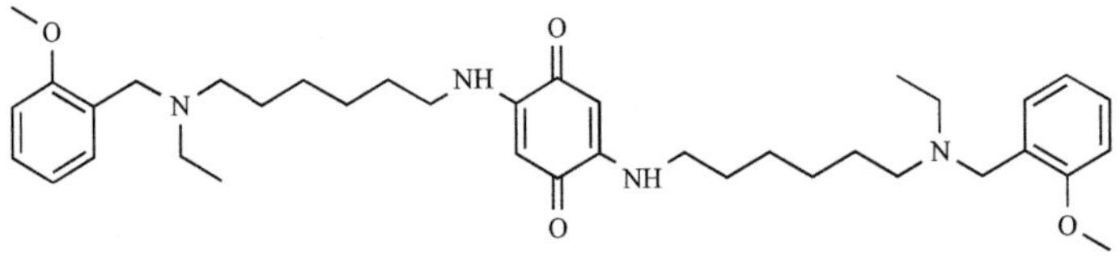

Memoquin

AChE IC_{50} = 1.55 µM
BACE-1 [3 µM] = >80%
AB aggregation [100 µM] = 87%

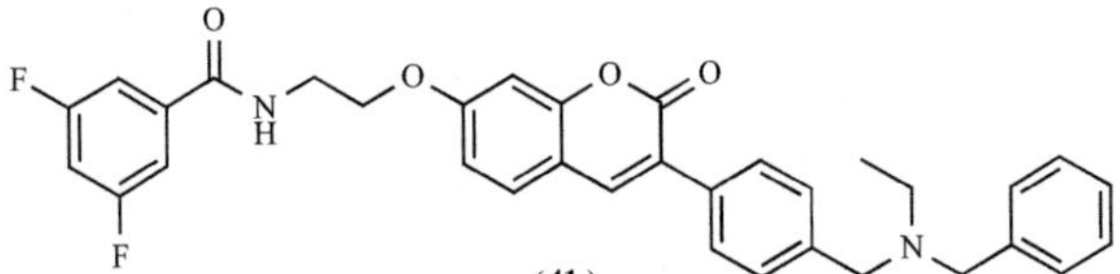

(**41**)

AChE IC_{50} = 0.181 µM
BACE-1 IC_{50} = 0.150 µM

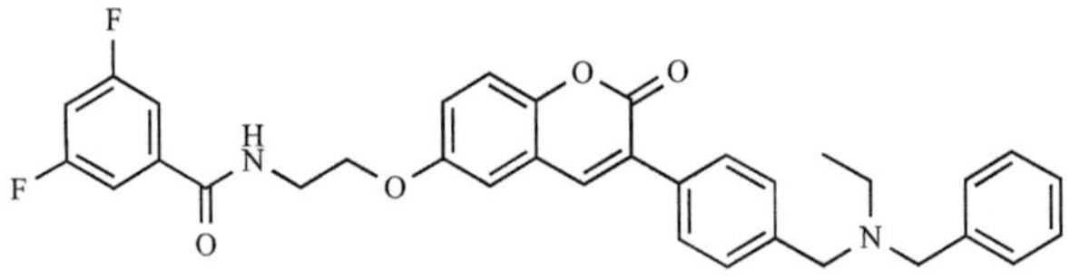

(**42**)

AChE IC_{50} = 0.551 µM
BACE-1 IC_{50} = 0.149 µM

Figure 37. Reported hybrids with dual or multiple biological profiles.

Rizzo *et al.*, (2011) reported the application of the bis(7)-tacrine concept and the design, synthesis and biological evaluation of a series of compounds that in addition to their cholinesterase activity were capable of inhibiting the non-enzymatic function of AChE, Aß-aggregation and showed remarkable activity against BACE-1 (Fig. **38**) [263]. Combination of the 9-amino-1, 2,3,4-tetrahydroacridine core with an indenoquinoline and a 7-Cl-indenoquinolineafforded the most active heterodimers (**43-44**). From the results they postulated that the four-ring heterocycles were well suited for stacking interactions with the aromatic residues at the PAS (Trp286 of hAChE). Importantly, the 7-Cl-indenoquinoline derivatives presented with selectivity towards AChE compared to BChE. The indenoquinoline homodimer (**44**) was a potent BChE inhibitor (IC_{50} = ±28.6 nM), a fact explained in that the BChE site can better tolerate the binding of larger compounds than AChE [263].

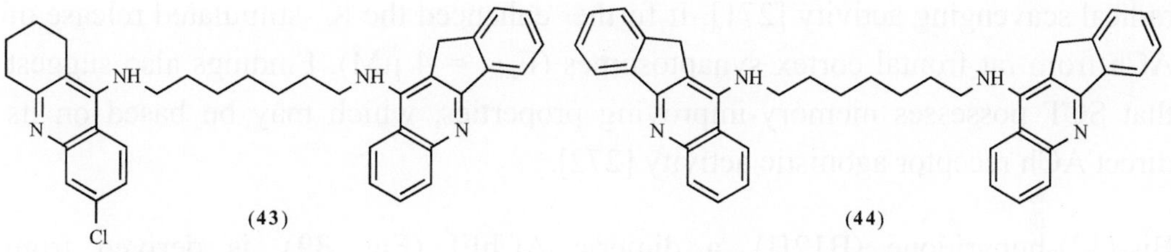

(**43**)

(human) AChE IC_{50} = 1.05 nM
(human) BChE IC_{50} = 63.7 nM
BACE1 IC_{50} = 1 µM
anti-AB aggregation [100 µM] = 46.1%

(**44**)

(human) AChE IC_{50} = 258 nM
(human) BChE IC_{50} = 28.6 nM
BACE1 IC_{50} = 0.4 µM
anti-AB aggregation [100 µM] = 30.2%

Figure 38: Bis(7)-tacrine hybrid molecules with dual AChEI and BACE1 inhibitory activity.

AChE and Other Neuroprotective Targets

Scopoletin (SCT), a coumarin analoque (Fig. **39**), was identified as an AChEI (IC_{50}= 135 µM). It was designed by employing a structure-based pharmacophore model constructed from interaction of AChEI galantamine to the AChE enzyme [264, 265]. At low concentrations SCT

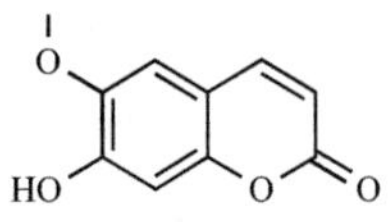

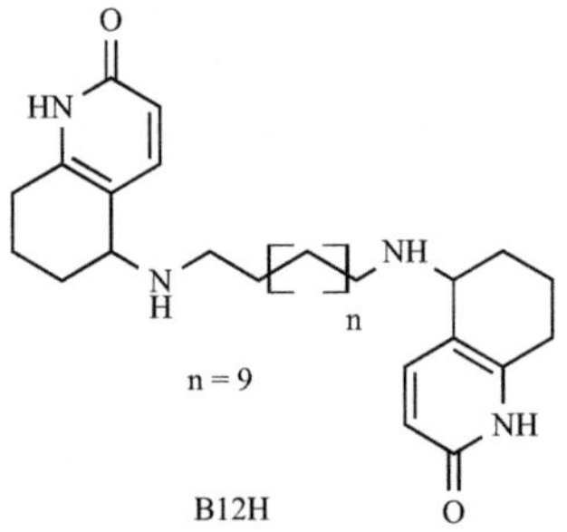

Figure 39: Multifunctional coumarin derivative SCT and dimeric Bis-(12)-hupyridone (B12H).

augmented brain ACh in the rat brain [264] which may suggest the presence of additional properties. SCT was found to have a multitarget profile which includes anti-inflammatory activity [266, 267], antiproliferative activity, inhibition of iNOS [268, 269], MAO inhibition [259], anti-antioxidant activity [270] and radical scavenging activity [271]. It further enhanced the K^+-stimulated release of ACh from rat frontal cortex synaptosomes (E_{max} = 4 μM). Findings also suggest that SCT possesses memory-improving properties, which may be based on its direct ACh receptor agonistic activity [272].

Bis-(12)-hupyridone (B12H), a dimeric AChEI (Fig. **39**), is derived from huperzine A [273] and has twice the potency of huperzine A in *in vitro* rat brain AChE inhibition [274]. After injection of this compound it easily crosses the mouse BBB, rescues scopolamine-induced learning and memory impairments in rats [274, 275], prevents H_2O_2-induced apoptosis *via* regulating vascular endothelia growth factor receptor-2 (VEGFR-2)/Akt/Glycogen synthase kinase 3β (GSK3β) signaling pathway in cerebellar granule neurons (CGNs) [276] and promotes neuronal differentiation from regulating α7 nicotinic ACh receptor [277]. Zhao and co-workers (2011) further investigated the *in vitro* and *in vivo*

NMDA receptor antagonism and neuroprotective effects of B12H and found that 0.1 nM – 1 µM prevented glutamate-induced apoptosis in a concentration- and time-dependent manner in primary rat cerebellar granule neurons [278]. Receptor-ligand binding analysis revealed a K_i value of 7.7 µM for the NMDA receptorat the MK-801 site. B12H also significantly attenuated ischemia-induced apoptosis and in a 2-hour middle cerebral artery occlusion rat model it improved neurological behavior impairment, decreased cerebral infarct volume, cerebral edema and neuronal apoptosis [278].

CONCLUSION AND FUTURE DEVELOPMENTS

The incidence of neurodegenerative disorders such as PD and AD has seen a significant increase in the global population. Despite this increase, there is no current single drug therapy that has been developed and brought to market which is able to prevent or reverse these diseases. The drug development paradigm for these diseases has recently shifted from specific single target drugs to dual- and/or multitarget-directed drugs and from treatment to protection or inhibition of progression and treatment. Understanding the multifaceted etiology and pathology of neurodegenerative diseases, these dual or mutitarget drugs should be more effective in inhibiting the progression and for the treatment of neurodegenerative diseases, than the classic 'one molecule, one target' concept [8-10]. Increasing evidence, as discussed in this chapter, from *in vitro* and *in vivo* studies suggests that multifunctional neuroprotective agents can be successfully developed, and potentially introduced to the market and used in the clinic to modify the progress of neurodegenerative diseases. Compounds such as rasagiline and selegeline, which is currently used to treat the symptoms of PD, have also shown to act as multifunctional agents able to modify neurodegenerative disease progression. By connecting such observations with experiments designed to uncover the structural requirements for targeting multiple pathological processes, researchers can now effectively develop multifunctional neuroprotective agents.

Despite major advances over the past few years, the present state of investigation of new multifunctional neuroprotective agents is not without problems. One of these relates to the almost exclusive reliance on rat or mouse models for *in vivo* efficacy studies, which may not directly correlate to efficacy in humans. Since experiments in non-human primates are expensive and becoming ethically questionable, alternative models will need to be validated for more accurate candidate assessment, to firmly establish the mechanism of action and therapeutic potential of experimental compounds. Success in the discovery of multifunctional neuroprotective agents will further clearly depend on advances of our understanding of the molecular mechanism of action of these diseases. Many of the compounds described in literature will also never progress beyond the experimental stage due to insufficient efficacy, selectivity or prohibitive toxicity in animal studies or because of pharmacokinetic properties unsuitable for orally-administered agents that would have good patient compliance. As a result the challenge in the development of therapeutically acceptable multifunctional neuroprotective compounds remain to design potent and selective (for the relevant targets) agents with high oral bioavailability, a low side effect profile and ability to reach its site of action. An additional complication in this case is the fact that modification of a structure to enhance activity at one target, frequently leads to a loss of activity on the other intended target(s) [8].

In the literature on drug design for neurodegenerative disease, the key strategies remain attennuation of oxidative stress and inhibition of monoamine oxidase and acetylcholine esterase and various compounds with multiple actions in the degenerative cascade have been developed around this. Various structures with potent multifunctional activities have been developed and described and clinical development into the drug pipeline is in progress in many instances. If research on multifunctional neuroprotective compounds continues for another decade at anything like the current level, it will surely produce answers for many of the existing questions, and may well result in the development of novel drug therapies to offer new hope for the treatment, if not cure, of many neurodegenerative diseases. Many more examples for the design of dual acting and multitarget directed drugs and even approaches can be found in recent literature and those included in this review has been limited to provide as broad picture of this approach for neurodegenerative diseases as possible within the scope of this chapter. Undoubtedly the targets focussed on here and included in this currenly favoured approach will form part of the future basis of research in this field.

ACKNOWLEDGEMENTS

Declared none.

CONFLICT OF INTEREST

The authors confirm that this chapter contents have no conflict of interest.

REFERENCES

[1]   McLean, C.A.; Cherny, R.A.; Fraser, F.W.; Fuller, S.J.; Smith, M.J.; Beyreuther, K.; Bush, A.I.; Masters, C.L. Soluble pool of Abeta amyloid as a determinant of severity of neurodegeneration in Alzheimer's disease. *Ann. Neurol.*, **1999**, *46*, 860-866.

[2]   Naslund, J.; Haroutunian, V.; Mohs, R.; Davis, K.L.; Davies, P.; Greengard, P.; Buxbaum, J.D. Correlation between elevated levels of amyloid beta-peptide in the brain and cognitive decline. *JAMA*, **2000**, *283*, 1571-1577.

[3]   Dauer, W.; Przedborski, S. Parkinson's disease: mechanisms and models. *Neuron*, **2003**, *39*(6), 889-909.

[4]   Jellinger, K.A. General aspects of neurodegeneration. *J. Neural. Transm.*, **2003**, *65*, 101.

[5]   Smid, P.; Coolen, H.K.A.C.; Keizer, H. G.; Van Hes, R.; De Moes, J.P.; Den Hartog, A.P.; Stork, B.; Plekkenpol, R.H.; Niemann, L.C.; Stroomer, C.N.J.; Tulp, M.T.M.; Van Stuivenberg, H.H.; McCreary, A.C.; Hesselink, M.B.; Herremans, A.H.J.; Kruse, C.G. Synthesis, structure-activity relationships, and biological properties of 1-heteroaryl-4-[omega-(1H-indol-3-yl)alkyl]piperazines, novel potential antipsychotics combining potent dopamine D2 receptor antagonism with potent serotonin reuptake inhibition. *J. Med. Chem.*, **2005**, *48*, 6855.

[6]   Van der Schyf, C.J.; Geldenhuys, W.J.; Youdim, M.B. Multifunctional drugs with different CNS targets for neuropsychiatric disorders. *J. Neurochem.*, **2006**, *99*, 1033-1048.

[7]   Morphy, R.; Rankovic, Z. Designed multiple ligands. An emerging drug discovery paradigm. *J. Med. Chem.*, **2005**, *48*, 6523-6543.

[8]   Cavalli, A.; Bolognesi, M.L.; Minarini, A.; Rosini, M.; Tumiatti, V.; Recanatini, M.; Melchiorre, C. Multi-target-directed ligands to combat neurodegenerative diseases. *J. Med. Chem.*, **2008**, *51*, 347-372.

[9]   Youdim, M.B.; Buccafusco, J.J. Multi-functional drugs for various CNS targets in the treatment of neurodegenerative disorders. *Trends Pharmacol. Sci.*, **2005**, *26*, 27-35.

[10]   Marlatt, M.W.; Webber, K.M.; Moreira, P.I.; Lee, H.G.; Casadesus, G.; Honda, K.; Zhu, X.; Perry, G.; Smith, M. A. Therapeutic opportunities in Alzheimer disease: one for all or all for one? *Curr. Med. Chem.*, **2005**, *12*, 1137-1147.

[11]   Youdim, M.B.; Edmondson, D.; Tipton, K.F. The therapeutic potential of monoamine oxidase inhibitors. *Nat. Rev. Neurosci.*, **2006**, *7*, 295-309.

[12]   Youdim, M.B.; Weinstock, M. Therapeutic applications of selective and non-selective inhibitors of monoamine oxidase A and B that do not cause significant tyramine potentiation. *Neurotoxicology*, **2004**, *25*, 243-250.

[13]   Riederer, P.; Lachenmayer, L.; Laux, G. Clinical applications of MAO-inhibitors. *Curr. Med. Chem.*, **2004**, *11*, 2033-2043.

[14]   Moncada, S.; Palmer, R.M.; Higgs, E.A. Nitric oxide: physiology, pathophysiology, and pharmacology. *Pharmacol. Rev.*, **1991**, *43*, 109-142.

[15]   Southan G.J.; Szabó, C. Selective pharmacological inhibition of distinct nitric oxide synthase isoforms. *Biochem. Pharmacol.*, **1996**, *51*, 383-394.

[16]   Mori, E.; Hashimoto, M.; Krishnan, K. R.; Doraiswamy, P. M. What constitutes clinical evidence for neuroprotection in Alzheimer disease: support for the cholinesterase inhibitors? *Alzheimer. Dis. Assoc. Disord.*, **2006**, *20*, S19-26.

[17]   Youdim, M.B.; Bakhle, Y.S. Monoamine oxidase: isoforms and inhibitors in Parkinson's disease and depressive illness. *Br. J. Pharmacol.*, **2006**, *147*, S287-S296.

[18]   Edmondson, D.E.; Binda, C.; Mattevi, A. The FAD binding sites of human monoamine oxidases A and B. *Neurotoxicology*, **2004**, *25*, 63-72.

[19]   Shih, J.C.; Chen. K.; Ridd. M.J. Monoamine oxidase: from genes to behavior. *Annu. Rev. Neurosci.*, **1999**, *22*, 197-217.

[20]   Son, S.Y.; Ma, J.; Kondou, Y.; Yoshimura, M.; Yamashita, E.; Tsukihara, T. Structure of human monoamine oxidase A at 2.2-A resolution: the control of opening the entry for substrates/inhibitors. *Proc. Natl. Acad. Sci. USA*, **2008**, *105*(15), 5739-5744.

[21]   Binda, C.; Newton-Vinson, P.; Hubálek, F.; Edmondson, D.E., Mattevi, A. Structure of human monoamine oxidase B, a drug target for the treatment of neurological disorders. *Nat. Struct. Biol.*, **2002**, *9*(1), 22-26.

[22]   Edmondson, D.E.; Binda, C.; Wang, J.; Upadhyay, A.K.; Mattevi, A. Molecular and mechanistic properties of the membrane-bound mitochondrial monoamine oxidases. *Biochemistry*, **2009**, *48*(20), 4220-4230.

[23]   Walker, W.H.; Kearney, E.B.; Seng, R.L.; Singer, T.P. The covalently-bound flavin of hepatic monoamine oxidase. 2. Identification and properties of cysteinyl riboflavin. *Eur. J. Biochem.*, **1971**, *24*(2), 328-331.

[24]   Kearney, E.B.; Salach, J.I.; Walker, W.H.; Seng, R.L.; Kenney, W.; Zeszotek, E.; Singer, T.P. The covalently-bound flavin of hepatic monoamine oxidase. 1. Isolation and sequence of a flavin peptide and evidence for binding at the 8alpha position. *Eur. J. Biochem.*, **1971**, *24*(2), 321-327.

[25]   Boulton, A.A. Phenylethylaminergic modulation of catecholaminergic neurotransmission. Prog. Neuropsychopharmacol. *Biol. Psychiatry.*, **1991**, *15*(2), 139-156.

[26]   Lasbennes, F.; Sercombe, R.; Seylaz, J. Monoamine oxidase activity in brain microvessels determined using natural and artificial substrates: relevance to the blood-brain barrier. *J. Cereb. Blood Flow Metab.*, **1983**, *3*(4), 521-528.

[27]   Da Prada, M.; Zürcher, G.; Wüthrich, I.; Haefely, W.E. On tyramine, food, beverages and the reversible MAO inhibitor moclobemide. *J. Neural Transm. Suppl.*, **1988**, *26*, 31-56.

[28]   Inoue, H.; Castagnoli, K.; Van Der Schyf, C.; Mabic, S.; Igarashi, K.; Castagnoli, N., Jr. Species-dependent differences in monoamine oxidase A and B-catalysed oxidation of various C4 substituted 1-methyl-4-phenyl-1,2,3, 6-tetrahydropyridinyl derivatives. *J. Pharmacol. Exp. Ther.*, **1999**, *291*(2), 856-864.

[29]   Weyler, W.; Salach, J.I. Purification and properties of mitochondrial monoamine oxidase type A from human placenta. *J. Biol. Chem.*, **1985**, *260*(24), 13199-13207.

[30]   Saura, J.; Nadal, E.; Van den Berg, B.; Vila, M,; Bombi, J.A.; Mahy, N. Localization of monoamine oxidases in human peripheral tissues. *Life Sci.*, **1996**, *59*(16),1341-1349.

[31]   Westlund, K.N.; Denney, R.M.; Kochersperger, L.M.; Rose, R.M.; Abell, C.W. Distinct monoamine oxidase A and B populations in primate brain. *Science*, **1985**, *230*(4722), 181-183.

[32]   Thorpe, L.W.; Westlund, K.N.; Kochersperger, L.M.; Abell, C.W.; Denney, R.M. Immunocytochemical localization of monoamine oxidases A and B in human peripheral tissues and brain. *J. Histochem. Cytochem.*, **1987**, *35*(1), 23-32.

[33]   Fowler, C.J.; Wiberg, A.; Oreland, L.; Marcusson, J.; Winblad, B. The effect of age on the activity and molecular properties of human brain monoamine oxidase. *J. Neural Transm.*, **1980**, *49*(1-2),1-20.

[34]   Kalaria, R.N.; Mitchell, M.J.; Harik, S.I. Monoamine oxidases of the human brain and liver. *Brain*, **1988**, *111*, 1441-1451.

[35]   Riachi, N.J.; Harik, S.I. Monoamine oxidases of the brains and livers of macaque and cercopithecus monkeys. *Exp. Neurol.*, **1992**, *115*(2), 212-217.

[36]   Collins, G.G.; Sandler, M.; Williams, E.D.; Youdim, M.B. Multiple forms of human brain mitochondrial monoamine oxidase. *Nature.*, **1970**, *225*(5235), 817-820.

[37]   Nicotra, A.; Pierucci, F.; Parvez, H.; Senatori, O. Monoamine oxidase expression during development and aging. *Neurotoxicology*, **2004**, *25*(1-2), 155-165.

[38]   Fowler, J.S.; Volkow, N.D.; Wang, G.J.; Logan, J.; Pappas, N.; Shea, C.; MacGregor, R. Age-related increases in brain monoamine oxidase B in living healthy human subjects. *Neurobiol. Aging*, **1997**, *18*(4), 431-435.

[39]   Levitt, P.; Pintar, J.E.; Breakefield, X.O. Immunocytochemical demonstration of monoamine oxidase B in brain astrocytes and serotonergic neurons. *Proc. Natl. Acad. Sci. USA*, **1982**, *79*(20), 6385-6389.

[40]   Fowler, J.S.; Logan, J.; Volkow, N.D.; Wang, G.J.; MacGregor, R.R.; Ding. Y.S. Monoamine oxidase: radiotracer development and human studies. *Methods*, **2002**, *27*(3), 263-277.

[41]   Riederer, P.; Lachenmayer, L.; Laux, G. Clinical applications of MAO-inhibitors. *Curr. Med. Chem.*, **2004**, *11*(15), 2033-2043.

[42]   Zisook, S.; Braff, D.L.; Click, M.A. Monoamine oxidase inhibitors in the treatment of atypical depression. *J. Clin. Psychopharmacol.*, **1985**, *5*(3), 131-137.

[43]   Fernandez, H.H.; Chen, J.J. Monoamine oxidase-B inhibition in the treatment of Parkinson's disease. *Pharmacotherapy*, **2007**, *27*(12 Pt 2), 174S-185S.

[44]   Finberg, J.P.; Wang, J.; Bankiewicz, K.; Harvey-White, J.; Kopin, I.J.; Goldstein, D.S. Increased striatal dopamine production from *L*-DOPA following selective inhibition of monoamine oxidase B by R(+)-*N*-propargyl-1-aminoindan (rasagiline) in the monkey. J. *Neural Transm. Suppl.*, **1998**, *52*, 279-285.

[45]   Di Monte, D.A.; DeLanney, L.E.; Irwin, I.; Royland, J.E.; Chan, P.; Jakowec, M.W.; Langston, J.W. Monoamine oxidase-dependent metabolism of dopamine in the striatum and substantia nigra of L-DOPA-treated monkeys. *Brain Res.*, **1996**, *738*(1), 53-59.

[46]   Shoulson, I.; Oakes, D.; Fahn, S.; Lang, A.; Langston, J.W.; LeWitt, P.; Olanow, C.W.; Penney, J.B.; Tanner, C.; Kieburtz, K.; Rudolph, A.; Parkinson Study Group. Impact of sustained deprenyl (selegiline) in levodopa-treated Parkinson's disease: a randomized placebo-controlled extension of the deprenyl and tocopherol antioxidative therapy of parkinsonism trial. *Ann. Neurol.*, **2002**, *51*(5), 604-612.

[47]   Pålhagen, S.; Heinonen, E.H.; Hägglund, J.; Kaugesaar, T.; Kontants, H.; Mäki-Ikola, O.; Palm, R.; Turunen, J. Selegiline delays the onset of disability in de novo parkinsonian patients. Swedish Parkinson Study Group. *Neurology*, **1998**, *51*(2), 520-525.

[48]   Finberg, J.P.; Lamensdorf, I.; Armoni, T. Modification of dopamine release by selective inhibitors of MAO-B. *Neurobiology (Bp)*, **2000**, *8*(2), 137-142.

[49]   Gesi, M.; Santinami, A.; Ruffoli, R.; Conti, G.; Fornai F. Novel aspects of dopamine oxidative metabolism (confounding outcomes take place of certainties). *Pharmacol. Toxicol.*, **2001**, *89*, 217-224.

[50]   Fornai, F.; Giorgi, F.S.; Bassi, L.; Ferrucci, M.; Alessandrì, M.G.; Corsini, G.U. Modulation of dihydroxyphenylacetaldehyde extracellular levels *in vivo* in the rat striatum after different kinds of pharmacological treatment. *Brain Res.*, **2000**, *861*, 126-134.

[51]   Marchitti, S.A.; Deitrich, R.A.; Vasiliou, V. Neurotoxicity and metabolism of the catecholamine-derived 3,4-dihydroxyphenylacetaldehyde and 3,4-dihydroxyphenylglycolaldehyde: the role of aldehyde dehydrogenase. *Pharmacol. Rev.*, **2007**, *59*, 125-150.

[52]   Lamensdorf, I.; Eisenhofer, G.; Harvey-White, J.; Nechustan, A.; Kirk, K.; Kopin, I.J.; 3,4-Dihydroxyphenylacetaldehyde potentiates the toxic effects of metabolic stress in PC12 cells. *Brain Res.*, **2000**, *868*(2), 191-201.

[53]   Burke, W.J.; Kumar, V.B.; Pandey, N.; Panneton, W.M.; Gan, Q.; Franko, M.W.; O'Dell, M.; Li, S.W.; Pan, Y.; Chung, H.D.; Galvin, J.E. Aggregation of alpha-synuclein by DOPAL, the monoamine oxidase metabolite of dopamine. *Acta Neuropathol.*, **2008**, *115*(2), 193-203.

[54]   Grünblatt, E.; Mandel, S.; Jacob-Hirsch, J.; Zeligson, S.; Amariglo, N.; Rechavi, G.; Li, J.; Ravid, R.; Roggendorf, W.; Riederer, P.; Youdim, M.B. Gene expression profiling of parkinsonian substantia nigra pars compacta; alterations in ubiquitin-proteasome, heat shock protein, iron and oxidative stress regulated proteins, cell adhesion/cellular matrix and vesicle trafficking genes. *J. Neural Transm.*, **2004**, *111*(12), 1543-1573.

[55]   Riederer, P.; Sofic, E.; Rausch, W.D.; Schmidt, B.; Reynolds, G.P.; Jellinger, K.; Youdim, M.B. Transition metals, ferritin, glutathione, and ascorbic acid in parkinsonian brains. *J. Neurochem.*, **1989**, *52*(2), 515-520.

[56]   Mallajosyula, J.K.; Kaur, D.; Chinta, S.J.; Rajagopalan, S.; Rane, A.; Nicholls, D.G.; Di Monte, D.A.; Macarthur, H.; Andersen, J.K. MAO-B elevation in mouse brain astrocytes results in Parkinson's pathology. *PLoS One*, **2008**, *3*(2), e1616.

[57]   The Parkinson study group. Effect of lazabemide on the progression of disability in early Parkinson's disease. *Ann. Neurol.*, **1996**, *40*(1), 99-107.

[58]   Pålhagen, S.; Heinonen, E.; Hägglund, J.; Kaugesaar, T.; Mäki-Ikola, O.; Palm, R. Swedish Parkinson Study Group. Selegiline slows the progression of the symptoms of Parkinson disease. *Neurology*, **2006**, *66*(8), 1200-1206.

[59]   Stocchi, F.; Vacca, L.; Grassini, P.; De Pandis, M.F.; Battaglia, G.; Cattaneo, C.; Fariello, R.G. Symptom relief in Parkinson disease by safinamide: Biochemical and clinical evidence of efficacy beyond MAO-B inhibition. *Neurology*, **2006**, *67*(7 Suppl 2), S24-S29.

[60]   Binda, C.; Wang, J.; Pisani, L.; Caccia, C.; Carotti, A.; Salvati, P.; Edmondson, D.E.; Mattevi, A. Structures of human monoamine oxidase B complexes with selective noncovalent inhibitors: safinamide and coumarin analogs. *J. Med. Chem.*, **2007**, *50*(23), 5848-5852.

[61]   Chen, J.F.; Steyn, S.; Staal, R.; Petzer, J.P.; Xu, K.; Van Der Schyf, C.J.; Castagnoli, K.; Sonsalla, P.K.; Castagnoli, N., Jr.; Schwarzschild, M.A. 8-(3-Chlorostyryl)caffeine may attenuate MPTP neurotoxicity through dual actions of monoamine oxidase inhibition and A_{2A} receptor antagonism. *J. Biol. Chem.*, **2002**, *277*(39), 36040-36044.

[62]   Chen, J.F.; Xu, K.; Petzer, J.P.; Staal, R.; Xu, Y.H.; Beilstein, M.; Sonsalla, P.K.; Castagnoli, K.; Castagnoli, N., Jr.; Schwarzschild, M.A. Neuroprotection by caffeine and

A$_{2A}$ adenosine receptor inactivation in a model of Parkinson's disease. *J. Neurosci.*, **2001**, *21*(10), RC143.

[63]   Strydom, B.; Malan, S.F.; Castagnoli, N., Jr.; Bergh, J.J.; Petzer, J.P. Inhibition of monoamine oxidase by 8-benzyloxycaffeine analogues. *Bioorg. Med. Chem.*, **2010**, *18*(3), 1018-1028.

[64]   Strydom, B.; Bergh, J.J.; Petzer, J.P. 8-Aryl- and alkyloxycaffeine analogues as inhibitors of monoamine oxidase. *Eur. J. Med. Chem.*, **2011**, *46*(8), 3474-3485.

[65]   Binda, C.; Aldeco, M.; Geldenhuys, W.J.; Tortorici, M.; Mattevi, A.; Edmondson, D.E. Molecular insights into human monoamine oxidase B inhibition by the glitazone anti-diabetes drugs. *ACS Med. Chem. Lett.*, **2011**, *3*(1), 39-42.

[66]   Jacobson, K.A.; Gallo-Rodriguez, C.; Melman, N.; Fischer, B.; Maillard, M.; Van Bergen, A.; Van Galen, P.J.; Karton, Y. Structure-activity relationships of 8-styrylxanthines as A2-selective adenosine antagonists. *J. Med. Chem.*, **1993**, *36*(10), 1333-1342.

[67]   Xu, K.; Bastia, E.; Schwarzschild, M. Therapeutic potential of adenosine A$_{2A}$ receptor antagonists in Parkinson's disease. *Pharmacol. Ther.*, **2005**, *105*, 267-310.

[68]   Pinna, A.; Wardas, J.; Simola, N.; Morelli, M. New therapies for the treatment of Parkinson's disease: adenosine A$_{2A}$ receptor antagonists. *Life Sci.*, **2005**, *77*, 3259-3267.

[69]   Morelli, M.; Di Paolo, T.; Wardas, J.; Calon, F.; Xiao, D.; Schwarzschild, M.A. Role of adenosine A$_{2A}$ receptors in parkinsonian motor impairment and l-DOPA-induced motor complications. *Prog. Neurobiol.*, **2007**, *83*, 293-309.

[70]   Jarvis, M.F.; Williams, M. Direct autoradiographic localization of adenosine A2 receptors in the rat brain using the A2-selective agonist, [^{3}H]CGS 21680. *Eur. J. Pharmacol.*, **1989**, *168*, 243-246.

[71]   Schiffmann, S.N.; Jacobs, O.; Vanderhaeghen, J.J. Striatal restricted adenosine A2 receptor (RDC8) is expressed by enkephalin but not by substance P neurons: an in situ hybridization histochemistry study. *J. Neurochem.*, **1991**, *57*, 1062-1067.

[72]   Fink, J.S.; Weaver, D.R.; Rivkees, S.A.; Peterfreund, R.A.; Pollack, A.E.; Adler, E.M.; Reppert, S.M. Molecular cloning of the rat A2 adenosine receptor: selective co-expression with D2 dopamine receptors in rat striatum. *Brain Res. Mol. Brain Res.*, **1992**, *14*(3), 186-195.

[73]   Ferré, S.; Fredholm, B.B.; Morelli, M.; Popoli, P.; Fuxe, K. Adenosine-dopamine receptor-receptor interactions as an integrative mechanism in the basal ganglia. *Trends Neurosci.*, **1997**, *20*, 482-487.

[74]   Ferré, S.; Von Euler, G.; Johansson, B.; Fredholm, B.B.; Fuxe, K. Stimulation of high-affinity adenosine A2 receptors decreases the affinity of dopamine D2 receptors in rat striatal membranes. *Proc. Natl. Acad. Sci. USA*, **1991**, *88*, 7238-7241.

[75]   Ferré, S.; O'Connor, W.T.; Fuxe, K.; Ungerstedt, U. The striopallidal neuron: a main locus for adenosine-dopamine interactions in the brain. *J. Neurosci.*, **1993**, *13*, 5402-5406.

[76]   Bibbiani, F.; Oh, J.D.; Petzer, J.P.; Castagnoli, N., Jr.; Chen, J.F.; Schwarzschild, M.A.; Chase, T.N. A$_{2A}$ antagonist prevents dopamine agonist-induced motor complications in animal models of Parkinson's disease. *Exp. Neurol.*, **2003**, *184*(1), 285-294.

[77]   Bara-Jimenez, W.; Sherzai, A.; Dimitrova, T.; Favit, A.; Bibbiani, F.; Gillespie, M.; Morris, M.J.; Mouradian, M.M.; Chase, T.N. Adenosine A$_{2A}$ receptor antagonist treatment of Parkinson's disease. *Neurology*, **2003**, *61*(3), 293-296.

[78]   Lundblad, M.; Vaudano, E.; Cenci, M.A. Cellular and behavioural effects of the adenosine A$_{2A}$ receptor antagonist KW-6002 in a rat model of l-DOPA-induced dyskinesia. *J. Neurochem.*, **2003**, *84*, 1398-1410.

[79]   Fenu, S.; Pinna, A.; Ongini, E.; Morelli, M. Adenosine A$_{2A}$ receptor antagonism potentiates L-DOPA-induced turning behaviour and c-fos expression in 6-hydroxydopamine-lesioned rats. *Eur. J. Pharmacol.*, **1997**, *321*, 143-147.

[80]   Kanda, T.; Jackson, M.J.; Smithm L.A.; Pearce, R.K.; Nakamura, J.; Kase, H.; Kuwana, Y.; Jenner, P. Adenosine A$_{2A}$ antagonist: a novel antiparkinsonian agent that does not provoke dyskinesia in parkinsonian monkeys. *Ann. Neurol.*, **1998**, *43*(4), 507-513.

[81]   Grondin, R.; Bédard, P.J.; Hadj Tahar, A.; Grégoire, L.; Mori, A.; Kase, H. Antiparkinsonian effect of a new selective adenosine A$_{2A}$ receptor antagonist in MPTP-treated monkeys. *Neurology*, **1999**, *52*, 1673-1677.

[82]   Schwarzschild, M.A.; Agnati, L.; Fuxe, K.; Chen, J.F.; Morelli, M. Targeting adenosine A$_{2A}$ receptors in Parkinson's disease. *Trends Neurosci.*, **2006**, *29*, 647-654.

[83]   Ascherio, A.; Zhang, S.M.; Hernán, M.A.; Kawachi, I.; Colditz, G.A.; Speizer, F.E.; Willett, W.C. Prospective study of caffeine consumption and risk of Parkinson's disease in men and women. *Ann. Neurol.*, **2001**, *50*(1), 56-63.

[84]   Ross, G.W.; Abbott, R.D.; Petrovitch, H.; Morens, D.M.; Grandinetti, A.; Tung, K.H.; Tanner, C.M.; Masaki, K.H.; Blanchette, P.L.; Curb, J.D.; Popper, J.S.; White, L.R. Association of coffee and caffeine intake with the risk of Parkinson disease. *JAMA*, **2000**, *283*(20), 2674-2679.

[85]   Ikeda, K.; Kurokawa, M.; Aoyama, S.; Kuwana, Y. Neuroprotection by adenosine A$_{2A}$ receptor blockade in experimental models of Parkinson's disease. *J. Neurochem.*, **2002**, *80*, 262-270.

[86]   Pretorius, J.; Malan, S.F.; Castagnoli, N., Jr.; Bergh, J.J.; Petzer, J.P. Dual inhibition of monoamine oxidase B and antagonism of the adenosine A$_{2A}$ receptor by (E,E)-8-(4-phenylbutadien-1-yl)caffeine analogues. *Bioorg. Med. Chem.*, **2008**, *16*(18), 8676-8684.

[87]   Müller, C.E.; Geis, U.; Hipp, J.; Schobert, U.; Frobenius, W.; Pawłowski, M.; Suzuki, F.; Sandoval-Ramírez, J. Synthesis and structure-activity relationships of 3,7-dimethyl-1-propargylxanthine derivatives, A$_{2A}$-selective adenosine receptor antagonists. *J. Med. Chem.*, **1997**, *40*(26), 4396-4405.

[88]   Petzer, J.P.; Steyn, S.; Castagnoli, K.P.; Chen, J.F.; Schwarzschild, M.A.; Van der Schyf, C.J.; Castagnoli, N. Inhibition of monoamine oxidase B by selective adenosine A_{2A} receptor antagonists. *Bioorg. Med. Chem.*, **2003**, *11*(7), 1299-1310.

[89]   Shimada, J.; Koike, N.; Nonaka, H.; Shiozaki, S.; Yanagawa, K.; Kanda, T.; Kobayashi, H.; Ichimura, M.; Nakamura, J.; Kase, H.; Suzuki, F. Adenosine A_{2A} antagonists with potent anti-cataleptic activity. *Bioorg. Med. Chem. Lett.*, **1997**, *18*, 2349-2352.

[90]   Giasson, B.I.; Duda, J.E.; Murray, I.V.; Chen, Q.; Souza, J.M.; Hurtig, H.I.; Ischiropoulos, H.; Trojanowski, J.Q.; Lee, V.M. Oxidative damage linked to neurodegeneration by selective alpha-synuclein nitration in synucleinopathy lesions. *Science*, **2000**, *290*(5493), 985-589.

[91]   Zecca, L.; Youdim, M.B.; Riederer, P.; Connor, J.R.; Crichton, R.R. Iron, brain ageing and neurodegenerative disorders. *Nat. Rev. Neurosci.*, **2004**, *5*(11), 863-873.

[92]   Gal, S.; Zheng, H.; Fridkin, M.; Youdim, M.B. Novel multifunctional neuroprotective iron chelator-monoamine oxidase inhibitor drugs for neurodegenerative diseases. *In vivo* selective brain monoamine oxidase inhibition and prevention of MPTP-induced striatal dopamine depletion. *J. Neurochem.*, **2005**, *95*(1), 79-88.

[93]   Zheng, H.; Weiner, L.M.; Bar-Am, O.; Epsztejn, S.; Cabantchik, Z.I.; Warshawsky, A.; Youdim, M.B.; Fridkin, M. Design, synthesis, and evaluation of novel bifunctional iron-chelators as potential agents for neuroprotection in Alzheimer's, Parkinson's, and other neurodegenerative diseases. *Bioorg. Med. Chem.*, **2005**, *13*(3), 773-783.

[94]   Dexter, D.T.; Statton, S.A.; Whitmore, C.; Freinbichler, W.; Weinberger, P.; Tipton, K.F.; Della Corte, L.; Ward, R.J.; Crichton, R.R. Clinically available iron chelators induce neuroprotection in the 6-OHDA model of Parkinson's disease after peripheral administration. *J. Neural Transm.*, **2011**, *118*(2), 223-231.

[95]   Youdim, M.B.; Stephenson, G.; Ben Shachar, D. Ironing iron out in Parkinson's disease and other neurodegenerative diseases with iron chelators: a lesson from 6-hydroxydopamine and iron chelators, desferal and VK-28. *Ann. N. Y. Acad. Sci. USA*, **2004**, *1012*, 306-325.

[96]   Sofic, E.; Paulus, W.; Jellinger, K.; Riederer, P.; Youdim, M.B. Selective increase of iron in substantia nigra zona compacta of parkinsonian brains. *J. Neurochem.*, **1991**, *56*(3), 978-982.

[97]   Dexter, D.T.; Wells, F.R.; Lees, A.J.; Agid, F.; Agid, Y.; Jenner, P.; Marsden, C.D. Increased nigral iron content and alterations in other metal ions occurring in brain in Parkinson's disease. *J. Neurochem.*, **1989**, *52*(6), 1830-1836.

[98]   Petzer, A.; Harvey, B.H.; Wegener, G.; Petzer, J.P. Azure B, a metabolite of methylene blue, is a high-potency, reversible inhibitor of monoamine oxidase. *Toxicol. Appl. Pharmacol.*, **2012**, *258*(3), 403-409.

[99]   Brown, C.; Taniguchi, G.; Yip, K. The monoamine oxidase inhibitor-tyramine interaction. *J. Clin. Pharmacol.*, **1989**, *29*(6), 529-532.

[100]   Costa, F.H.; Rosso, A.L.; Maultasch, H.; Nicaretta, D.H.; Vincent, M.B. Depression in Parkinson's disease: diagnosis and treatment. *Arq. Neuropsiquiatr.*, **2012**, *70*(8), 617-620.

[101]   Flockhart, D.A. Dietary restrictions and drug interactions with monoamine oxidaseinhibitors: an update. *J. Clin. Psychiatry.*, **2012**, *73*, Suppl 1:17-24.

[102]   Zesiewicz, T.A.; Hauser, R.A. Monoamine oxidase inhibitors. In Parkinson's disease: diagnosis and clinical management; Factor, S.A., Weiner, W.J., Eds.; Demos Medical Publishing: New York, **2002**; pp 365 378.

[103]   Ramsay, R.R.; Dunford, C.; Gillman, P.K. Methylene blue and serotonin toxicity: inhibition of monoamine oxidase A (MAO A) confirms a theoretical prediction. *Br. J. Pharmacol.*, **2007**, *152*(6), 946-951.

[104]   Stanford, S.C.; Stanford, B.J.; Gillman, P.K. Risk of severe serotonin toxicity following co-administration of methylene blue and serotonin reuptake inhibitors: an update on a case report of post-operative delirium. *J. Psychopharmacol.*, **2010**, *24*, 1433-1438.

[105]   Bonnet, U. Moclobemide: therapeutic use and clinical studies. *CNS Drug. Rev.*, **2003**, *9*(1), 97-140.

[106]   Wu, M.L.; Deng, J.F. Fatal serotonin toxicity caused by moclobemide and fluoxetine overdose. *Chang. Gung. Med. J.*, **2011**, *34*(6), 644-649.

[107]   Fowler, J.S.; Volkow, N.D.; Logan, J.; Wang, G.J.; MacGregor, R.R.; Schlyer, D.; Wolf, A.P.; Pappas, N.; Alexoff, D.; Shea, C.; Dorflinger, E.; Yoo, K.; Fazzini, E.; Patlak, C. Slow recovery of human brain MAO B after L-deprenyl (Selegeline) withdrawal. *Synapse*, **1994**, *18*(2), 86-93.

[108]   Hubálek, F.; Binda, C.; Khalil, A.; Li, M.; Mattevi, A.; Castagnoli, N.; Edmondson, D.E. Demonstration of isoleucine 199 as a structural determinant for the selective inhibition of human monoamine oxidase B by specific reversible inhibitors. *J. Biol. Chem.*, **2005**, *280*(16), 15761-15766.

[109]   Anglade, P.; Vyas, S.; Javoy-Agid, F.; Herrero, M.T.; Michel, P.P.; Marquez, J.; Mouatt-Prigent, A.; Ruberg, M.; Hirsch, E.C.; Agid, Y. Apoptosis and autophagy in nigral neurons of patients with Parkinson's disease. *Histol. Histopathol.*, **1997**, *12*(1), 25-31.

[110]   Marshall, K.; Daniel, S.E.; Cairns, N.; Jenner, P.; Halliwell, B. Upregulation of the anti-apoptotic protein Bcl-2 may be an early event in neurodegeneration: studies on Parkinson's and incidental Lewy body disease. *Biochem. Biophys. Res. Commun.*, **1997**, *240*, 84-87.

[111]   Hartmann, A.; Michel, P.P.; Troadec, J.D.; Mouatt-Prigent, A.; Faucheux, B.A.; Ruberg, M.; Agid, Y.; Hirsch, E.C. Is Bax a mitochondrial mediator in apoptotic death of dopaminergic neurons in Parkinson's disease? *J. Neurochem.*, **2001**, *76*(6), 1785-1793.

[112]   Mogi, M.; Togari, A.; Kondo, T.; Mizuno, Y.; Komure, O.; Kuno, S.; Ichinose, H.; Nagatsu, T. Caspase activities and tumor necrosis factor receptor R1 (p55) level are elevated in the substantia nigra from parkinsonian brain. *J. Neural Transm.*, **2000**, *107*(3), 335-341.

[113] Hartmann, A.; Troadec, J.D.; Hunot, S.; Kikly, K.; Faucheux, B.A.; Mouatt-Prigent, A.; Ruberg, M.; Agid, Y.; Hirsch E.C. Caspase-8 is an effector in apoptotic death of dopaminergic neurons in Parkinson's disease, but pathway inhibition results in neuronal necrosis. *J. Neurosci.*, **2001**, *21*(7), 2247-2255.

[114] Maruyama, W.; Akao, Y.; Carrillo, M.C.; Kitani, K.; Youdim, M.B.; Naoi, M. Neuroprotection by propargylamines in Parkinson's disease: suppression of apoptosis and induction of prosurvival genes. *Neurotoxicol. Teratol.*, **2002**, *24*(5), 675-682.

[115] Blandini, F. Neuroprotection by rasagiline: A new therapeutic approach to Parkinson's disease? *CNS Drug Rev.*, **2005**, *11*(2), 183-194.

[116] Edmondson, D.E.; Mattevi, A.; Binda, C.; Li, M.; Hubálek, F. Structure and mechanism of monoamine oxidase. *Curr. Med. Chem.*, **2004**, *11*(15), 1983-1993.

[117] Binda, C.; Hubálek, F.; Li, M.; Herzig, Y.; Sterling, J.; Edmondson, D.E.; Mattevi, A. Binding of rasagiline-related inhibitors to human monoamine oxidases. a kinetic and crystallographic analysis. *J. Med. Chem.*, **2005**, *48*(26), 8148-8154.

[118] Maruyama, W.; Akao, Y.; Youdim, M.B.; Naoi, M. Neurotoxins induce apoptosis in dopamine neurons: protection by N-propargylamine-1(R)- and (S)-aminoindan, rasagiline and TV1022. *J. Neural Transm. Suppl.*, **2000**, (60), 171-186.

[119] Lemasters, J.J.; Nieminen, A.L.; Qian, T.; Trost, L.C.; Elmore, S.P.; Nishimura, Y.; Crowe, R.A.; Cascio, W.E.; Bradham, C.A.; Brenner, D.A.; Herman, B. The mitochondrial permeability transition in cell death: a common mechanism in necrosis, apoptosis and autophagy. *Biochim. Biophys. Acta*, **1998**, *1366*(1-2), 177-196.

[120] Bar-Am, O.; Yogev-Falach, M.; Amit, T.; Sagi, Y.; Youdim, M.B. Regulation of protein kinase C by the anti-Parkinson drug, MAO-B inhibitor, rasagiline and its derivatives, *in vivo*. *J. Neurochem.*, **2004**, *89*(5), 1119-1125.

[121] Youdim, M.B.; Weinstock, M. Molecular basis of neuroprotective activities of rasagiline and the anti-Alzheimer drug TV3326 [(N-propargyl-(3R)aminoindan-5-YL)-ethyl methyl carbamate]. *Cell. Mol. Neurobiol.*, **2001**, *21*(6), 555-573.

[122] Mandel, S.; Weinreb, O.; Amit, T.; Youdim, M.B. Mechanism of neuroprotective action of the anti-Parkinson drug rasagiline and its derivatives. *Brain Res. Brain Res. Rev.*, **2005**, *48*(2), 379-387.

[123] Youdim, M.B.; Wadia, A.; Tatton, W.; Weinstock, M. The anti-Parkinson drug rasagiline and its cholinesterase inhibitor derivatives exert neuroprotection unrelated to MAO inhibition in cell culture and *in vivo*. *Ann. N. Y. Acad. Sci. USA*, **2001**, *939*, 450-458.

[124] Gottfries, C.G.; Neurochemical aspects on aging and diseases with cognitive impairment, *J. Neurosci. Res.*, **2009**, *27*, 541-547.

[125] Meltzer, C.C.; Smith, G.; Dekosky, S.T.; Pollock, B.G.; Mathis, C.A.; Moore, R.Y.; Kupfer, D.J.; Reynolds, C.F. Serotonin in aging, late-life depression, and Alzheimer's disease: The emerging role of functional imaging', *Neuropsychopharmacology*, **1998**, *18*, 407-430.

[126] Fink, D.M.; Palermo, M.G.; Bores, G.M.; Huger, F.P.; Kurys, B.E.; Merriman, M.C.; Olsen, G.E.; Petko, W.; O'Malley, G.J. Imino 1,2,3,4-tetrahydrocyclopent[b]indole carbamates as dual inhibitors of acetylcholine esterase and monoamine oxidase. *Bioorg. Med. Chem. Lett.*, **1996**, *6*, 625-630.

[127] Hilgert, M.; Nöldner, M.; Chatterjee, S.S.; Klein, J. KA-672 inhibits rat brain acetylcholinesterase *in vitro* but not *in vivo*. *Neurosci. Lett.*, **1999**, *263*, 193-196.

[128] Gnerre, C.; Catto, M.; Leonetti, F.; Weber, P.; Carrupt, P.A.; Altomare, C.; Carotti, A.; Testa, B. Inhibition of monoamine oxidases by functionalized coumarin derivatives: biological activities, QSARs, and 3D-QSARs. *J. Med. Chem.*, **2000**, *43*, 4747-4758.

[129] Bruhlmann, C.; Ooms, F.; Carrupt, P.A.; Testa, B.; Catto, M.; Leonetti, F.; Altomare, C.; Carotti, A. Coumarin derivatives as dual inhibitors of acetylcholinesterase and monoamine oxidase. *J. Med. Chem.*, **2001**, *44*, 3195-3198.

[130] Viña, D.; Matos, M.J.; Yáñez, M.; Santana, L.; Uriarte, E. 3-Substituted coumarins as dual inhibitors of AchE and MAO for the treatment of Alzheimer's disease. *Med. Chem. Comm.*, **2012**, *3*, 213-218.

[131] Gökhan, N.; Yesilada, A.; Uçar, G.; Erol, K.; Bilgin, AA. 1-*N*-Substituted thiocarbamoyl-3-phenyl-5-thienyl-2-pyrazolines: synthesis and evaluation as MAO inhibitors, *Archiv der Pharmazie.*, **2003**, *336*, 362-371.

[132] Uçar, G.; Gökhan, N.; Yesilada, A.; Bilgin, A.A. 1-*N*-Substituted thiocarbamoyl-3-phenyl-5-thienyl-2-pyrazolines: A novel cholinesterase and selective monoamine oxidase B inhibitors for the treatment of Parkinson's and Alzheimer's diseases. *Neurosci. Lett.*, **2005**, *382*, 327-331.

[133] Martin, I.N.; Woodward, J.J.; Winter, M.B.; Beeson, W.T.; Marletta, M.A. Design and synthesis of C5 Methylated *L*-arginine analogues as active site probes for nitric oxide synthase. *J. Am. Chem. Soc.*, **2007**, *129*, 12563-12570.

[134] Dawson, V.L.; Dawson, T.M.; Lomdon, E.D.; Bredt, D.S.; Snyder, S.H. Nitric oxide mediates glutamate neurotoxicicty in primary cortical cultures. *Proc. Natl. Acad. Sci.*, **1991**, *88*, 6368-6371.

[135] Knowles, R.G.; Moncada, S. Nitric oxide synthases in mammals. *Biochem. J.* **1994**, *298*, 249-258.

[136] Cary, S.P.L.; Winger, J.A.; Derbyshire, E.R.; Marletta, M.A. Nitric oxide signaling: no longer simply on or off. *Trends. Biochem. Sci.*, **2006**, *31*, 231-239.

[137] Hoffman, B,B.; Taylor P. Goodman and Gilman's. The pharmacological basis of therapeutics. Chapter 6: Neurotransmission. New York: McGraw-Hill. **2001**, 148p.

[138] Roman, L.J.; Mattasek, P.; Masters, B.S.S. Intrinsic and extrinsic modulation of nitric oxide synthase activity. *Chem. Rev.,* **2002**, *102*, 1179-1189.

[139] Marletta, M.A. Nitric oxide synthase structure and mechanism. *J. Biol. Chem.*, **1993**, *268*, 12231-12234.

[140] Steur, D.J. Mammalian nitric oxide synthase. Biochimica et Biophysica ACTA/General Subjects, **1999**, *1411*, 217-230.

[141] Alderton, W.K.; Cooper, C.E.; Knowles, R.G. Nitric oxide synthases: structure function and inhibition. *Biochem. J.*, **2001**, *357*, 593-615.

[142] Kerwin, J.F.; Lancaster, J.R.; Feldamn, P.L. Nitric Oxide: A new paradigm for second messengers. *J. Med. Chem.*, **1995**, *38*, 4343-4362.

[143] Joubert, J.; Malan, S.F. Nitric Oxide Synthase (NOS) inhibitors: a patent review. *Exp. Opin. Ther. Pat.*, **2011**, *21*, 537-560.

[144] Boughton, S.N.K.; Tinker, A.C. Inhibitors of nitric oxide synthase in inflammatory arthritis. *IDrugs*, **1998**, *1*, 321-333.

[145] Endres, M.; Laufs, U.; Liao, J.K.; Moskowitz, M.A. Targeting eNOS for stroke protection. *Trends. Neurosci.*, **2004**, *27*, 283-289.

[146] Braam, B.; Verhaar, M.C. Understanding eNOS for pharmacological modulation of endothelial function: a translational view. *Curr. Pharm.*, **2007**, *13*, 1727-1740.

[147] Larson, A.A.; Kovacs, K.J.; Cooper, J.C.; Kitto, K.F. Transient changes in the synthesis of nitric oxide result in long-term as well as short-term changes in acetic acid-induced writhing in mice. *Pain*, **2000**, *86*, 103-111.

[148] Vallance, P.; Leiper, J. Blocking NO Synthesis: how, where and why? *Nat. Rev. Drug. Discovery*, **2002**, *1*, 939-950.

[149] Wilcock, D.M.; Lewis, M.R.; Van Nostrand, W.E.; Davis, J.; Previti, M.L.; Gharkholonarehe, N.; Vitek, M.P.; Colton, C.A. Progression of amyloid pathology to Alzheimer's disease pathology in an amyloid precursor protein transgenic mouse model by removal of nitric oxide synthase 2. *J. Neurosci.*, **2008**, *28*, 1537-1545.

[150] Babbedge, R.C.; Bland-Ward, P.A.; Hart, S.L.; Moore, P.K. Inhibition of rat cerebellar nitric oxide synthase by 7-nitro indazole and related substituted indazoles. *Br. J. Pharmcol.* **1993**, *1*, 225-228.

[151] Babu, B.R.; Griffith, O.W.; 1998. Design of the isoform-selective inhibitors of nitric oxide synthase. *Curr. Opin. Chem.l Biol.*, **1998**, *2*, 491-500.

[152] Pekiner, C.; Kelicen, P.; Uma, S.; Miwa, I. Two nitric oxide synthase inhibitors: pyridoxal aminoguanidine and 8-quinolinecarboxylic hydrazide selectively inhibit basal but not agoniststimulated release of nitric oxide in rat aorta. *Pharmacol. Res.*, **2002**, *46*, 317-320.

[153] Zhang, Z.G.; Reif, D.; MacDonald, W.X.; Tang, W.X.; Kamp, D.K.; Gentile, R.J.; Shakespeare, W.C.; Murray, R.J.; Chopp, M. ARL 17477, a potent and selective neuronal

NOS inhibitor decreases infarct volume after transient middle cerebral artery occlusion in rats. *J. Cereb. Blood. Flow. Metab.*, **1996**, *16*, 599-604.

[154] Chan. P.H. Role of oxidants in ischemic brain damage. *Stroke*, **1996**, *27*, 1124-1129.

[155] Samdani, A.F.; Dawson, T.M.; Dawson, V.L. Nitric oxide synthase in models of focal ischemia. *Stroke.*, **1997**, *28*, 1283-1288.

[156] Hall, E.D.; Braughler, J.M. Central nervous system trauma and stroke. II. Physiological and pharmacological evidence for involvement of oxygen radicals and lipid peroxidation. *Free Radic. Biol. Med.*, **1986**, *6*, 303-313.

[157] Beckman, J.S.; Beckman, T.W.; Chen, J.; Marshall, P.A.; Freeman, B.A. Apparent hydroxyl radical production by peroxynitrite: implications for endothelial injury from nitric oxide and superoxide. *Proc. Natl. Acad. Sci. USA*, **1990**, *87*, 1620-1624.

[158] Sorrenti, V.; Di Giacomo, C.; Campisi, A.; Perez-Polo, J.R.; Vanella, A. Nitric oxide synthetase activity in cerebral post-ischemic reperfusion and effects of L-NG-nitroarginine and 7-nitroindazole on the survival, *Neurochem. Res.*, **1999**, *24*, 861-866.

[159] Escott, K.J.; Beech, J.S.; Haga, K.K.; Williams, S.C.R.; Meldrum, B.S.; Bath, P.M.W. Cerebroprotective effect of the nitric oxide synthase inhibitors, 1-(2-trifluoromethyl-phenyl)imidazole and 7-nitro indazole, after transient focal cerebral ischemia in the rat. *J. Cereb. Blood. Flow. Metab.*, **1998**, *18*, 281-287.

[160] Chalimoniuk, M.; Strosznajder, J. NMDA receptor-dependent nitric oxide and cGMP synthesis in brain hemispheres and cerebellum during reperfusion after transient forebrain ischemia in Gerbils: effect of 7-nitroindazole. *J. Neurosci. Res.*, **1998**, *54*, 681-690.

[161] Sorrenti, V.; Di Giacomo, C.; Renis, M.; Russo, A.; La Delfa, C.; Perez-Polo, J.R.; Vanella, A. Lipid peroxidation in rat cerebral cortex during post-ischemic reperfusion: effect of drugs with different molecular mechanisms. *Drugs. Exp. Clin. Res.*, **1994**, 20, 185-189.

[162] Auvin, S.; Auguet, M.; Navet, E.; Harnett, J.J; Viossat, I.; Schulz, J.; Bigg, D.; Chabrier. P.E. Novel Inhibitors of Neuronal Nitric Oxide Synthase with Potent Antioxidant Properties. *Bioorg. Med. Chem. Lett.*, **2003**, *13*, 209-212.

[163] Sgaragli, G. P.; Valoti, M.; Gorelli, B.; Fusi, F.; Palmi, M.; Mantovani, P. Calcium antagonist and antiperoxidant properties of some hindered phenols. *Br. J. Pharmacol.*, **1993**, *110*, 369.

[164] Chabrier, P.E.; Auguet, M.; Spinnewyn, B.; Auvin, S.; Cornet, S.; Demerle-Pallardy, C.; Guilmard-Favre, C.; Marin, J.G.; Pignol, B.; Gillard-Roubert, V.; Roussillot-Charnet, C.; Schulz, J.; Viossat, I.; Bigg, D.; Moncada, S. BN 80933, a dual inhibitor of neuronal nitric oxide synthase and lipid peroxidation: a promising neuroprotective strategy. *Proc. Natl. Acad. Sci. USA*, **1999**, *96*, 10824-10829.

[165] Ding-Zhou, L.; Marchand-Verrecchia, C.; Palmier, B.; Croci, N.; Chabrier, P.E.; Plotkine, M.; Margaill, I. Neuroprotective effects of (S)-N-[4-[4-[(3,4-dihydro6-hydroxy-2,5,7,8-tetramethyl-2H-1-benzopyran-2-yl)carbonyl]-1-piperazinyl]phenyl]-2-thiophenecarboximid-amide (BN 80933), an inhibitor of neuronal nitric-oxide synthase and an antioxidant, in model of transient focal cerebral ischemia in mice. *J. Pharmacol. Exp. Ther.*, **2003**, *306*, 588-594.

[166] Salerno, L.; Modica, M.N.; Romeo, G.; Pittalà, V.; Siracusa, M.A.; Amato, M.E.; Acquaviva, R.; Di Giacomo, C.; Sorrenti, V. Novel inhibitors of nitric oxide synthase with antioxidant properties. *Eur. J. Med. Chem.*, **2012**, *49*, 118-126.

[167] Sorrenti, V.; Salerno, L.; Di Giacomo, C.; Acquaviva, R.; Siracusa, M.A. Vanella, A. Imidazole derivatives as antioxidants and selective inhibitors of nNOS. *Nitric. Oxide*, **2006**, *14*, 45-50.

[168] Kemp, J.A.; Mckernan, R.M. NMDA receptor pathways as drug targets. *Nat. Neurosci.*, **2002**, *5*, 1039-1042.

[169] Mattson, M.P. Excitotoxic and excitoprotective mechanisms: abundant targets for the prevention and treatment of neurodegenerative disorders. *Neuromol. Med.*, **2003**, *3*, 65-94.

[170] Meldrum, B.; Garthwaite, J. Excitatory amino acid neurotoxicity and neurodegenerative disease. *Trends. Pharmacol. Sci.*, **1990**, *11*, 379-387.

[171] Alexi, T.; Borlogan, C.V.; Faull, R.L.M.; Williams, C.E.; Clark, R.G.; Gluckman, P.D.; Huges, P.E. Neuroprotective strategies for basal ganglia degeneration: Parkinson's and Huntington's diseases. *Prog. Neurobiol.*, **2000**, *60*, 409-470.

[172] Lipton, P. Ischemic cell death in brain neurons. *Physiol. Rev.*, **1991**, *79*, 1431-568.

[173] Cano-Abad, M.F.; Villarroya, M.; Garcia, A.G.; Gabilan, N.H.; Lopez, M.G. Calcium entry through *L*-type calcium channels causes mitochondrial disruption and chromaffin cell death. *J. Biol. Chem.*, **2001**, *276*, 39695-39704.

[174] Moncada, S.; Palmer, R.M.J.; Higgs, E.A. Biosynthesis of nitric oxide from L-arginine. A pathway for the regulation of cell function and communication. *Biochem. Pharmacol.*, **1989**, *38*, 1709-1715.

[175] Neuraxon, Inc. Substituted indole compounds having NOS inhibitory activity. US20060258721. 2006

[176] Neuraxon, Inc. Substituted indole compounds having NOS inhibitory activity. WO2007063418. 2007

[177] Neuraxon, Inc. 1,5 and 3,6-substituted indole compounds having NOS inhibitory activity. WO2007118314. 2007

[178] Neuraxon, Inc. 3,5-Substituted indole compounds having NOS and norepinephrine reuptake inhibitory activity. WO2009062318. 2009

[179] Neuraxon, Inc. Indole compounds and methods for treating visceral pain. WO2009062319. 2009.

[180] Hevel, J.M.; Marletta, M.A. Nitric oxide synthase assays. *Methods Enzymol.*, **1994**, *133*, 250-258.

[181] Joubert, J.; Van Dyk, S.; Green, I.R.; Malan, S.F. Synthesis and evaluation of fluorescent heterocyclic aminoadamantanes as multifunctional neuroprotective agents. *Bioorg. Med. Chem.*, **2011**, *19*, 3935-3944.

[182] Joubert, J.; Geldenhuys, W.J.; Van der Schyf, C.J.; Oliver, D.W.; Kruger, H.; Govender, T.; Malan. S.F. Polycyclic cage structures as lipophilic scaffolds for neuro-active drugs. *ChemMedChem.*, **2012**, *7*, 375-384.

[183] Van der Schyf, C.J.; Squier, G.J.; Coetzee, W.A. Characterization of NGP 1-01, an aromatic polycyclic amine, as a calcium antagonist. *Pharmacol. Res. Com.*, **1986**, *18*, 407-417.

[184] Malan, S.F., Van der Walt, J.J.; Van der Schyf, C.J. Structure-activity relationships of polycyclic aromatic amines with calcium channel blocking activity. *Archiv der Pharmazie.*, **2000**, *333*, 10-16.

[185] Malan, S.F., Dyason, K. Wagenaar, B., Van der Walt, J.J.; Van der Schyf, C.J. The structure and ion channel activity of 6-benzylamino-3-hydroxyhexacyclo [6.5.0.0(3,7).0(4,12).0(5,10).0(9,13)]tridecane. *Archiv der Pharmazie.*, **2003**, *336*, 127-133.

[186] Joubert, J. ; Van Dyk, S. ; Green, I.R. ; Malan. S.F. Synthesis, evaluation and application of polycyclic fluorescent analogues as *N*-methyl-D-aspartate receptor and voltage gated calcium channel ligands. *Eur. J. Med. Chem.*, **2011**, *46*, 5010-5020.

[187] Van Der Schyf, C.J.; Youdim, M.B. Multifunctional drugs as neurotherapeutics. *Neurother.*, **2009**, *6*, 1-201.

[188] Joubert, J.; Van Dyk, S.; Malan, S.F. Fluorescent polycyclic ligands for nitric oxide synthase (NOS) inhibition. *Bioorg. Med. Chem.*, **2008**, *16*, 8952-8958.

[189] Xu, K.Y.; Huso, D.L.; Dawson, T.M.; Bredt, D.S.; Becker, L.C. Nitric oxide synthase in cardiac sarcoplasmic reticulum. *Proc. Natl. Acad. Sci. USA*, **1991**, *96*, 657-662.

[190] Herraiz, T.; Aran, V.J.; Guillen, H. Nitroindazole compounds inhibit the oxidative activation of 1-methyl-4-phenyl-1,2,3,6-tetrahydropyridin (MPTP) neurotoxin to neurotoxic pyridinium cations by human monoamine oxidase (MAO). *Free. Rad. Res.*, **2009**, *43*, 975-984.

[191] Thomas, B, Saravanan, K.S., Mohanakumar, K.P. *In vitro* and *in vivo* evidences that antioxidant action contributes to the neuroprotective effects of the neuronal nitric oxide synthase and monoamine oxidase-B inhibitor, 7-nitroindazole. *Neurochem. Int.*, **2008**, *52*, 990-1001.

[192] Prins, L.H.A.; Petzer, J.P.; Malan, S.F. Synthesis and *in vitro* evaluation of pteridine analogues as monoamine oxidase B and nitric oxide synthase inhibitors. *Bioorg. Med. Chem.*, **2009**, *17*, 7523-7530.

[193] Crane, B.R.; Arvai, A.S.; Ghosh, D.K.; Wu, C.; Getzoff, E.D.; Stuehr, D.J.; Tainer, J.A. Structure of nitric oxide synthase oxygenase dimer with pterin and substrate. *Science*, **1998**, *279*, 2121-2126.

[194] Chen, J.F.; Steyn, S.; Staal, R.; Petzer, J.P.; Xu, K.; Van der Schyf, C.J.; Castagnoli, K.; Sonsalla, P.K.; Castagnioli, N.; Schwarzschild, M.A. 8-(3-chlorostyryl)caffeine may attenuate MPTP neurotoxicity through dual actions of monoamine oxidase inhibition and A_{2A} receptor antagonism. *J. Biol. Chem.*, **2002**, *277*, 36040-36044.

[195] Vlok, N.; Malan, S.F.; Castagnoli, N.; Bergh, J.J.; Petzer, J.P. Inhibition of monoamine oxidase B by analogues of the adenosine A_{2A} receptor antagonist (E)-8-(3-chlorostyryl) caffeine (CSC). *Bioorg. Med. Chem.,* **2006**, *14*, 3512-3521.

[196] Balsa, D.; Fernández-Álvarez, E.; Tipton, K.F.; Unzeta, M. Inhibition of MAO by substituted tryptamine analogues. *J. Neur. Trans. Supp.*, **1991**, *32*, 103-105.

[197] Balsa, D.; Fernández-Álvarez, E.; Tipton, K.F.; Unzeta, M. Monoamine oxidase inhibitory potencies and selectivities of 2-[*N*-(2-propynyl)-aminomethyl]-1-methyl indole derivatives. Biochem. *Soc. Trans.*, **1991**, *19*, 215-218.

[198] Fernández-García, C.; Marco, J.L.; Fernández-Álvarez, E. Acetylenic and allenic derivatives of 2-(5-methoxy-1-methylindolyl) alkylamines: Synthesis and evaluation as selective inhibitors of the monoamine oxidases A and B. *Eur. J. Med. Chem.*, **1992**, *27*, 909-918.

[199] Morón, J.A.; Campillo, M.; Pérez, V.; Unzeta, M.; Pardo, L. Molecular determinants of MAO selectivity in a series of indolylmethylamine derivatives: Biological activities, 3D-QSAR/CoMFA analysis, and computational simulation of ligand recognition. *J. Med. Chem.*, **2000**, *43*, 1684-1691.

[200] Pérez, V.; Marco, J.L.; Fernández-Álvarez, E.; Unzeta, M. Relevance of a benzyloxy group in 2-indolyl methylamines in the selective MAO-B inhibition, *Br. J. Pharmacol.*, **1991**, *127*, 869-876.

[201] Bellik, L.; Dragoni, S.; Pessina, F.; Sanz, E.; Unzeta, M.; Valoti, M. Antioxidant properties of PF9601N, a novel MAO-B inhibitor: assessment of its ability to interact with reactive nitrose species. *Acta. Biochim. Pol.*, **2010**, *57*, 235-239.

[202] Pérez, V.; Unzeta, M. PF9601N [*N*-(2-propynyl)-2-(5-benzyloxyindolyl)methylamine], a new MAO-B inhibitor, attenuates MPTP induced depletion of striatal dopamine levels in C57/BL mice. *Neurochem. Internat.*, **2003**, *42*, 221-229.

[203] Sanz, E.; Quintana, A.; Battaglia, V.; Toninello, A.; Hidalgo, J.; Ambrosio, S.; Valoti, M.; Marco, J.L.; Tipton, K.F.; Unzeta, M. Anti-apoptotic effect of MAO-B inhibitor PF9601N [*N*-(2-propynyl)-2-(5-benzyloxyindolyl)methylamine] is mediated by p53 pathway inhibition in MPP$^+$-treated SH-SY5Y human dopaminergic cells. *J. Neurochem.*, **2008**, *105*, 2404-2417.

[204] Aarsland, D.; Hutchinson, M.; Larsen, J.P. Cognitive, psychiatric and motor response to galantamine in Parkinson's disease with dementia. *Int. J. Geriatr. Psychiatry*, **2003**, *18*, 937-941.

[205] Aarsland, D.; Laake, K.; Larsen, J.P.; Janvin, C. Donepezil for cognitive impairment in Parkinson's disease: a randomised controlled study. *J. Neurol. Neurosurg. Psychiatry*, **2002**, *72*, 708-712.

[206] Giladi, N.; Shabtai, H.; Gurevich, T.; Benbunan, B.; Anca, M.; Korczyn, A.D. Rivastigmine (Exelon) for dementia in patients with Parkinson's disease. *Acta. Neurol. Scand.*, **2003**, *108*, 368-373.

[207] Leroi, I.; Brandt, J.; Reich, S. G.; Lyketsos, C.G.; Grill, S.; Thompson, R.; Marsh, L. Randomized placebo-controlled trial of donepezil in cognitive impairment in Parkinson's disease. *Int. J. Geriatr. Psychiatry*, **2004**, *19*, 1-8.

[208] Ravina, B.; Putt, M.; Siderowf, A.; Farrar, J.T.; Gillespie, M.; Crawley, A.; Fernandez, H.H.; Trieschmann, M.M.; Reichwein, S.; Simuni, T. Donepezil for dementia in Parkinson's disease: a randomised, double blind, placebo controlled, crossover study. *J. Neurol. Neurosurg. Psychiatry*, **2005**, *76*, 934-939.

[209] Linazasoro, G.; Lasa, A.; Van Blercom, N. Efficacy and safety of donepezil in the treatment of executive dysfunction in Parkinson disease: a pilot study. *Clin. Neuropharmacol*, **2005.**, *28*, 176-178.

[210] Sussman, J.L.; Harel, M.; Frolow, F.; Oefner, C.; Goldman, A.; Toker, L.; Silman, I. Atomic structure of acetylcholinesterase from Torpedo californica: a prototypic acetylcholine-binding protein. *Science*, **1991**, *253*, 872-879.

[211] Ollis, D.L.; Cheah, E.; Cygler, M.; Dijkstra, B.; Frolow, F.; Franken, S.M.; Harel, M.; Remington, S.J.; Silman, I.; Schrag, J. The alpha/beta hydrolase fold. *Protein Eng.*, **1992**, *5*, 197-211.

[212] Cygler, M.; Schrag, J.D.; Sussman, J.L.; Harel, M.; Silman, I.; Gentry, M.K.; Doctor, B.P. Relationship between sequence conservation and three-dimensional structure in a large family of esterases, lipases, and related proteins. *Protein Sci.*, **1993**, *2*, 366-382.

[213] M. Harel, D.M. Quinn, H.K. Nair, I. Silman, J.L. Sussman, The X-ray structure of a transition state analog complex reveals the molecular origins of the catalytic power and substrate specificity of acetylcholinesterase. *J. Am. Chem. Soc.*, **1996**, *118*, 2340–2346.

[214] Knapp, M.J.; Knopman, D.S.; Solomon, P.R.; Pendlebury, W.W.; Davis, C.S.; Gracon, S.I. A 30-week randomized controlled trial of high-dose tacrine in patients with Alzheimer's disease. The Tacrine Study Group. *JAMA*, **1994**, *271*, 985-991.

[215] Watkins, P.B.; Zimmerman, H.J.; Knapp, M.J.; Gracon, S.I.; Lewis, K.W. Hepatotoxic effects of tacrine administration in patients with Alzheimer's disease. *JAMA*, **1994**, *271*, 992-998.

[216] Kasa, P.; Rakonczay, Z.; Gulya, K. The cholinergic system in Alzheimer's disease. *Prog. Neurobiol*, **1997**, *52*, 511-535.

[217] Gualtieri, F.; Dei, S.; Manetti, D.; Romanelli, M.N.; Scapecchi, S.; Teodori, E. The medicinal chemistry of Alzheimer's and Alzheimer-like diseases with emphasis on the cholinergic hypothesis. *Farmaco.*, **1995**, *50*, 489-503.

[218] Yan, J.W.; Li, Y.P.; Ye, W.J.; Chen, S.B.; Hou, J.Q.; Tan, J.H.; Ou, T.M.; Li, D.; Gu, L.Q.; Huang, Z.S. Design, synthesis and evaluation of isaindigotone derivatives as dual inhibitors for acetylcholinesterase and amyloid beta aggregation. *Bioorg. Med. Chem.*, **2012**, *20*, 2527-2534.

[219] Rosini, M.; Simoni, E.; Bartolini, M.; Cavalli, A.; Ceccarini, L.; Pascu, N.; McClymont, D.W.; Tarozzi, A.; Bolognesi, M.L.; Minarini, A.; Tumiatti, V.; Andrisano, V.; Mellor, I.R.; Melchiorre, C. Inhibition of acetylcholinesterase, beta-amyloid aggregation, and NMDA receptors in Alzheimer's disease: a promising direction for the multi-target-directed ligands gold rush. *J. Med. Chem.*, **2008**, *51*, 4381-4384.

[220] Camps, P.; Formosa, X.; Galdeano, C.; Munoz-Torrero, D.; Ramirez, L.; Gomez, E.; Isambert, N.; Lavilla, R.; Badia, A.; Clos, M. V.; Bartolini, M.; Mancini, F.; Andrisano, V.; Arce, M. P.; Rodriguez-Franco, M. I.; Huertas, O.; Dafni, T.; Luque, F. J. Pyrano[3,2-c]quinoline-6-chlorotacrine hybrids as a novel family of acetylcholinesterase- and beta-amyloid-directed anti-Alzheimer compounds. *J. Med. Chem.*, **2009**, *52*, 5365-5379.

[221] Tang, H.; Zhao, L.Z.; Zhao, H. T.; Huang, S.L.; Zhong, S.M.; Qin, J.K.; Chen, Z.F.; Huang, Z.S.; Liang, H. Hybrids of oxoisoaporphine-tacrine congeners: novel acetylcholinesterase and acetylcholinesterase-induced beta-amyloid aggregation inhibitors. *Eur. J. Med. Chem.*, **2011**, *46*, 4970-4979.

[222] Luo, J.; Li, W.; Zhao, Y.; Fu, H.; Ma, D.L.; Tang, J.; Li, C.; Peoples, R.W.; Li, F.; Wang, Q.; Huang, P.; Xia, J.; Pang, Y.; Han, Y. Pathologically activated neuroprotection *via* uncompetitive blockade of *N*-methyl-D-aspartate receptors with fast off-rate by novel multifunctional dimer bis(propyl)-cognitin. *J. Biol. Chem.*, **2010**, *285*, 19947-19958.

[223] Lanctôt, K.L.; Rajaram, R.D.; Herrmann, N. Therapy for Alzheimer's Disease: How Effective are Current Treatments? Ther. *Adv. Neurol. Disord.*, **2009**, *2*(3), 163-80.

[224] Moriguchi, S.; Marszalec, W.; Zhao, X.; Yeh, J.Z.; Narahashi, T. Mechanism of action of galantamine on *N*-methyl-D-aspartate receptors in rat cortical neurons. *J. Pharmacol. Exp. Ther.*, **2004**, *310*(3), 933-42.

[225] Arias, E.; Alés, E.; Gabilan, N.H.; Cano-Abad, M.F.; Villarroya, M.; García, A.G.; López, M.G. Galantamine prevents apoptosis induced by beta-amyloid and thapsigargin: involvement of nicotinic acetylcholine receptors. *Neuropharm.*, **2004**, *46*(1), 103-14.

[226] Lorrio, S.; Sobrado, M.; Arias, E.; Roda, J.M.; García, A.G.; López, M.G. Galantamine postischemia provides neuroprotection and memory recovery against transient global cerebral ischemia in gerbils. *J. Pharmacol. Exp. Ther.*, **2007**, *322*(2), 591-599.

[227] Ezoulin, M.J.; Ombetta, J.E.; Dutertre-Catella, H.; Warnet, J.M.; Massicot, F. Antioxidative properties of galantamine on neuronal damage induced by hydrogen peroxide in SK-N-SH cells. *Neurotoxicology*, **2008**, *29*(2), 270-277.

[228] Bachurin, S.; Bukatina, E.; Lermontova, N.; Tkachenko, S.; Afanasiev, A.; Grigoriev, V.; Grigorieva, I.; Ivanov, Y.; Sablin, S.; Zefirov, N. Antihistamine agent Dimebon as a novel neuroprotector and a cognition enhancer. *Ann. N. Y. Acad. Sci.*, **2001**, *939*, 425-435.

[229]   Bachurin, S.O.; Shevtsova, E.P.; Kireeva, E.G.; Oxenkrug, G.F.; Sablin, S.O. Mitochondria as a target for neurotoxins and neuroprotective agents. *Ann. N. Y. Acad. Sci.*, **2003**, *993*, 334-44; discussion 345-9.

[230]   Grigorev, V.V.; Dranyi, O.A.; Bachurin, S.O. Comparative study of action mechanisms of dimebon and memantine on AMPA- and NMDA-subtypes glutamate receptors in rat cerebral neurons. *Bull. Exp. Biol. Med.*, **2003**, *136*, 474-477.

[231]   Wang, Y.E.; Yue, D.X.; Tang, X.C. Anti-cholinesterase activity of huperzine A. Zhongguo. Yao. Li. Xue. Bao.,**1986**, *7*, 110-113.

[232]   Ashani, Y.; Peggins, J. O.; Doctor, B.P. Mechanism of inhibition of cholinesterases by huperzine A. *Biochem. Biophys. Res. Commun.*, **1992**, *184*, 719-726.

[233]   Saxena, A.; Qian, N.; Kovach, I.M.; Kozikowski, A.P.; Pang, Y.P.; Vellom, D.C.; Radic, Z.; Quinn, D.; Taylor, P.; Doctor, B. P. Identification of amino acid residues involved in the binding of Huperzine A to cholinesterases. *Protein Sci.*, **1994**, *3*, 1770-1778.

[234]   Peng, Y.; Lee, D.Y.; Jiang, L.; Ma, Z.; Schachter, S.C.; Lemere, C.A. Huperzine A regulates amyloid precursor protein processing *via* protein kinase C and mitogen-activated protein kinase pathways in neuroblastoma SK-N-SH cells over-expressing wild type human amyloid precursor protein 695. *Neuroscience*, **2007**, 150, 386-395.

[235]   Ved, H. S.; Koenig, M. L.; Dave, J. R.; Doctor, B. P. Huperzine A, a potential therapeutic agent for dementia, reduces neuronal cell death caused by glutamate. *Neuroreport*, **1997**, *8*, 963-968.

[236]   Decker, M.; Kraus, B.; Heilmann, J. Design, synthesis and pharmacological evaluation of hybrid molecules out of quinazolinimines and lipoic acid lead to highly potent and selective butyrylcholinesterase inhibitors with antioxidant properties. *Bioorg. Med. Chem.*, **2008**, *16*, 4252-4261.

[237]   Kamal, M.A.; Klein, P.; Luo, W.; Li, Y.; Holloway, H.W.; Tweedie, D.; Greig, N.H. Kinetics of human serum butyrylcholinesterase inhibition by a novel experimental Alzheimer therapeutic, dihydrobenzodioxepine cymserine. *Neurochem. Res.*, **2008**, *33*, 745-753.

[238]   Raina, P.; Santaguida, P.; Ismaila, A.; Patterson, C.; Cowan, D.; Levine, M.; Booker, L.; Oremus, M. Effectiveness of cholinesterase inhibitors and memantine for treating dementia: evidence review for a clinical practice guideline. *Ann. Intern. Med.*, **2008**, *148*, 379-397.

[239]   Fallarero, A.; Oinonen, P.; Gupta, S.; Blom, P.; Galkin, A.; Mohan, C.G.; Vuorela, P.M. Inhibition of acetylcholinesterase by coumarins: the case of coumarin 106. *Pharmacol. Res.*, **2008**, *58*, 215-221.

[240]   Huang, L.; Luo, Z.; He, F.; Lu, J.; Li, X. Synthesis and biological evaluation of a new series of berberine derivatives as dual inhibitors of acetylcholinesterase and butyrylcholinesterase. *Bioorg. Med. Chem.*, **2010**, *18*, 4475-4484.

[241]   Pavlov, V.A.; Parrish, W.R.; Rosas-Ballina, M.; Ochani, M.; Puerta, M.; Ochani, K.; Chavan, S.; Al-Abed, Y.; Tracey, K.J. Brain acetylcholinesterase activity controls systemic

cytokine levels through the cholinergic anti-inflammatory pathway. *Brain. Behav. Immun.*, **2009**, *23*, 41-45.

[242] Terry, A.V.; Buccafusco, J.J. The cholinergic hypothesis of age and Alzheimer's disease-related cognitive deficits: recent challenges and their implications for novel drug development. *J. Pharmacol. Exp. Ther.*, **2003**, *306*, 821-827.

[243] Nizri, E.; Adani, R.; Meshulam, H.; Amitai, G.; Brenner, T. Bifunctional compounds eliciting both anti-inflammatory and cholinergic activity as potential drugs for neuroinflammatory impairments. *Neurosci. Lett.*, **2005**, *376*, 46-50.

[244] Laskin, J.D.; Fabio, K.; Lacey, C.J.; Young, S.; Mohanta, P.; Guillon, C.; Heindel, N.D.; Huang, M.T.; Heck, D.E. Unique Dual-Action Therapeutics, U. S. Patent Pending, PCT/US2009/005961, filed November 3, 2009.

[245] Young, S.; Fabio, K.; Guillon, C.; Mohanta, P.; Halton, T.A.; Heck, D.E.; Flowers, R.A.; Laskin, J.D.; Heindel, N.D. Peripheral site acetylcholinesterase inhibitors targeting both inflammation and cholinergic dysfunction. *Bioorg. Med. Chem. Lett.*, **2010**, *20*, 2987-2990.

[246] Hardy, J.; Selkoe, D.J. The amyloid hypothesis of Alzheimer's disease: progress and problems on the road to therapeutics. *Science*, **2002**, *297*, 353-356.

[247] Alvarez, A.; Alarcon, R.; Opazo, C.; Campos, E. O.; Munoz, F.J.; Calderon, F.H.; Dajas, F.; Gentry, M.K.; Doctor, B.P.; De Mello, F.G.; Inestrosa, N.C. Stable complexes involving acetylcholinesterase and amyloid-beta peptide change the biochemical properties of the enzyme and increase the neurotoxicity of Alzheimer's fibrils. *J. Neurosci.*, **1998**, *18*, 3213-3223.

[248] Alvarez, A.; Opazo, C.; Alarcon, R.; Garrido, J.; Inestrosa, N.C. Acetylcholinesterase promotes the aggregation of amyloid-beta-peptide fragments by forming a complex with the growing fibrils. *J. Mol. Biol.*, **1997**, *272*, 348-361.

[249] Munoz-Torrero, D. Acetylcholinesterase inhibitors as disease-modifying therapies for Alzheimer's disease. *Curr. Med. Chem.*, **2008**, *15*, 2433-2455.

[250] Munoz-Muriedas, J.; Lopez, J.M.; Orozco, M.; Luque, F.J. Molecular modelling approaches to the design of acetylcholinesterase inhibitors: new challenges for the treatment of Alzheimer's disease. *Curr. Pharm. Des.*, **2004**, *10*, 3131-3140.

[251] Rizzo, S.; Riviere, C.; Piazzi, L.; Bisi, A.; Gobbi, S.; Bartolini, M.; Andrisano, V.; Morroni, F.; Tarozzi, A.; Monti, J. P.; Rampa, A. Benzofuran-based hybrid compounds for the inhibition of cholinesterase activity, beta amyloid aggregation, and abeta neurotoxicity. *J. Med. Chem.*, **2008**, *51*, 2883-2886.

[252] Fernandez-Bachiller, M.I.; Perez, C.; Campillo, N.E.; Paez, J.A.; Gonzalez-Munoz, G.C.; Usan, P.; Garcia-Palomero, E.; Lopez, M.G.; Villarroya, M.; Garcia, A.G.; Martinez, A.; Rodriguez-Franco, M.I. Tacrine-melatonin hybrids as multifunctional agents for

Alzheimer's disease, with cholinergic, antioxidant, and neuroprotective properties. *ChemMedChem*, **2009**, *4*, 828-841.

[253]   Li, Y.P.; Ning, F.X.; Yang, M.B.; Li, Y.C.; Nie, M.H.; Ou, T.M.; Tan, J.H.; Huang, S.L.; Li, D.; Gu, L.Q.; Huang, Z.S. Syntheses and characterization of novel oxoisoaporphine derivatives as dual inhibitors for cholinesterases and amyloid beta aggregation. *Eur. J. Med. Chem.*, **2011**, *46*, 1572-1581.

[254]   Belluti, F.; Piazzi, L.; Bisi, A.; Gobbi, S.; Bartolini, M.; Cavalli, A.; Valenti, P.; Rampa, A. Design, synthesis, and evaluation of benzophenone derivatives as novel acetylcholinesterase inhibitors. *Eur. J. Med. Chem.*, **2009**, *44*, 1341-1348.

[255]   Bartolini, M.; Bertucci, C.; Cavrini, V.; Andrisano, V. beta-Amyloid aggregation induced by human acetylcholinesterase: inhibition studies. *Biochem. Pharmacol.*, **2003**, *65*, 407-416.

[256]   Belluti, F.; Bartolini, M.; Bottegoni, G.; Bisi, A.; Cavalli, A.; Andrisano, V.; Rampa, A. Benzophenone-based derivatives: a novel series of potent and selective dual inhibitors of acetylcholinesterase and acetylcholinesterase-induced beta-amyloid aggregation. *Eur. J. Med. Chem.*, **2011**, *46*, 1682-1693.

[257]   Mao, F.; Huang, L.; Luo, Z.; Liu, A.; Lu, C.; Xie, Z.; Li, X. O-Hydroxyl- or o-amino benzylamine-tacrine hybrids: Multifunctional biometals chelators, antioxidants, and inhibitors of cholinesterase activity and amyloid-β aggregation. *Bioorg. Med. Chem.*, **2012**, *20*, 5884-5892.

[258]   Hou, J. Q.; Tan, J. H.; Wang, X. X.; Chen, S. B.; Huang, S. Y.; Yan, J. W.; Chen, S. H.; Ou, T. M.; Luo, H. B.; Li, D.; Gu, L. Q.; Huang, Z. S. Impact of planarity of unfused aromatic molecules on G-quadruplex binding: learning from isaindigotone derivatives. *Org. Biomol. Chem.*, **2011**, *9*, 6422-6436.

[259]   Lipton, S. A. Paradigm shift in neuroprotection by NMDA receptor blockade: memantine and beyond. *Nat. Rev. Drug Discov.*, **2006**, *5*, 160-170.

[260]   González-Muñoz, G. C.; Arce, M. P.; Lopez, B.; Perez, C.; Villarroya, M.; Lopez, M. G.; Garcia, A. G.; Conde, S.; Rodriguez-Franco, M. I. Old phenothiazine and dibenzothiadiazepine derivatives for tomorrow's neuroprotective therapies against neurodegenerative diseases. *Eur. J. Med. Chem.*, **2010**, *45*, 6152-6158.

[261]   Zhu, Y.; Xiao, K.; Ma, L.; Xiong, B.; Fu, Y.; Yu, H.; Wang, W.; Wang, X.; Hu, D.; Peng, H.; Li, J.; Gong, Q.; Chai, Q.; Tang, X.; Zhang, H.; Li, J.; Shen, J. Design, synthesis and biological evaluation of novel dual inhibitors of acetylcholinesterase and beta-secretase. *Bioorg. Med. Chem.*, **2009**, *17*, 1600-1613.

[262]   Piazzi, L.; Cavalli, A.; Colizzi, F.; Belluti, F.; Bartolini, M. Multi-target-directed coumarin derivatives: hAChE and BACE1 inhibitors as potential anti-Alzheimer compounds. *Bioorg. Med. Chem. Lett.*, **2008**, *18*, 423–426.

[263]   Rizzo, S.; Bisi, A.; Bartolini, M.; Mancini, F.; Belluti, F.; Gobbi, S.; Andrisano, V.; Rampa, A. Multi-target strategy to address Alzheimer's disease: Design, synthesis and biological evaluation of new tacrine-based dimers. *Eur. J. Med. Chem.*, **2011**, *46*, 4336-4343.

[264] Rollinger, J.M.; Hornick, A.; Langer, T.; Stuppner, H.; Prast, H. Acetylcholinesterase inhibitory activity of scopolin and scopoletin discovered by virtual screening of natural products. *J. Med. Chem.*, **2004**, *47*, 6248-6254.

[265] Stuppner, H.; Langer, T.; Prast, H.; Rollinger, J.M.; Wolber, G. Use of coumarin derivatives. EP04009431.0. 2005

[266] Muschietti, L.; Gorzalczany, S.; Ferraro, G.; Acevedo, C.; Martino, V. Phenolic compounds with anti-inflammatory activity from Eupatorium buniifolium. *Planta Med.*, **2001**, *67*, 743-744.

[267] Calixto, J.B.; Otuki, M.F.; Santos, A.R. Anti-inflammatory compounds of plant origin. Part I. Action on arachidonic acid pathway, nitric oxide and nuclear factor kappa B (NF-kappaB). *Planta. Med.*, **2003**, *69*, 973-983.

[268] Kim, N.Y.; Pae, H.O.; Ko, Y.S.; Yoo, J.C.; Choi, B.M.; Jun, C.D.; Chung, H.T.; Inagaki, M.; Higuchi, R.; Kim, Y.C. *In vitro* inducible nitric oxide synthesis inhibitory active constituents from Fraxinus rhynchophylla. *Planta. Med.*, **1999**, *65*, 656-658.

[269] Kang, T.H.; Pae, H.O.; Jeong, S.J.; Yoo, J.C.; Choi, B.M.; Jun, C.D.; Chung, H.T.; Miyamoto, T.; Higuchi, R.; Kim, Y.C. Scopoletin: an inducible nitric oxide synthesis inhibitory active constituent from Artemisia feddei. *Planta. Med.*, **1999**, *65*, 400-403.

[270] Shaw, C.Y.; Chen, C.H.; Hsu, C.C.; Chen, C.C.; Tsai, Y.C. Antioxidant properties of scopoletin isolated from Sinomonium acutum. *Phytother. Res.*, **2003**, *17*, 823-825.

[271] Toda, S. Inhibitory effects of phenylpropanoid metabolites on copper-induced protein oxidative modification of mice brain homogenate, *in vitro*. Biol. *Trace Elem. Res.*, **2002**, *85*, 183-188.

[272] Hornick, A.; Lieb, A.; Vo, N.P.; Rollinger, J.M.; Stuppner, H.; Prast, H. The coumarin scopoletin potentiates acetylcholine release from synaptosomes, amplifies hippocampal long-term potentiation and ameliorates anticholinergic- and age-impaired memory. *Neuroscience.*, **2011**, *197*, 280-292.

[273] Carlier, P.R.; Du, D.M.; Han, Y.F.; Liu, J.; Perola, E.; Williams, I.D.; Pang, Y.P. Dimerization of an inactive fragment of Huperzine A produces a drug with twice the potency of the patural product. *Angew. Chem. Int. Ed Engl.*, **2000**, *39*, 1775-1777.

[274] Li, W.; Xue, J.; Niu, C.; Fu, H.; Lam, C.S.; Luo, J.; Chan, H.H.; Xue, H.; Kan, K.K.; Lee, N.T.; Li, C.; Pang, Y.; Li, M.; Tsim, K.W.; Jiang, H.; Chen, K.; Li, X.; Han, Y. Synergistic neuroprotection by bis(7)-tacrine *via* concurrent blockade of *N*-methyl-D-aspartate receptors and neuronal nitric-oxide synthase. *Mol. Pharmacol.*, **2007**, *71*, 1258-1267.

[275] Yu, H.; Li, W.M.; Kan, K.K.; Ho, J.M.; Carlier, P.R.; Pang, Y.P.; Gu, Z.M.; Zhong, Z.; Chan, K.; Wang, Y.T.; Han, Y.F. The physicochemical properties and the *in vivo* AChE

inhibition of two potential anti-Alzheimer agents, bis(12)-hupyridone and bis(7)-tacrine. *J. Pharm. Biomed. Anal.*, **2008**, *46*, 75-81.

[276] Cui, W.; Li, W.; Zhao, Y.; Mak, S.; Gao, Y.; Luo, J.; Zhang, H.; Liu, Y.; Carlier, P.R.; Rong, J.; Han, Y. Preventing H_2O_2-induced apoptosis in cerebellar granule neurons by regulating the VEGFR-2/Akt signaling pathway using a novel dimeric antiacetylcholinesterase bis(12)-hupyridone. *Brain Res.,* **2011**, *1394*, 14-23.

[277] Cui, W.; Cui, G.Z.; Li, W.; Zhang, Z.; Hu, S.; Mak, S.; Zhang, H.; Carlier, P.R.; Choi, C.L.; Wong, Y.T.; Lee, S.M.; Han, Y. Bis(12)-hupyridone, a novel multifunctional dimer, promotes neuronal differentiation more potently than its monomeric natural analog huperzine A possibly through alpha7 nAChR. *Brain. Res.*, **2011**, *1401*, 10-17.

[278] Zhao, Y.; Dou, J.; Luo, J.; Li, W.; Chan, H.H.; Cui, W.; Zhang, H.; Han, R.; Carlier, P.R.; Zhang, X.; Han, Y. Neuroprotection against excitotoxic and ischemic insults by bis(12)-hupyridone, a novel anti-acetylcholinesterase dimer, possibly *via* acting on multiple targets. *Brain Res.,* **2011**, *1421*, 100-109.

CHAPTER 6

Specific Cholinesterase Inhibitors: A Potential Tool to Assist in Management of Alzheimer Disease

Nigel H. Grieg[1], Mohammad A. Kamal[2,*], Nasimudeen R. Jabir[2], Shams Tabrez[2], Faizul H. Nasim[3], Adel M. Abuzenadah[2] and Gjumrakch Aliev[4,5]

[1]*Drug Design & Development Section, Laboratory of Neurosciences, Intramural Research Program, National Institute on Aging, National Institutes of Health, Baltimore, MD 21224, USA;* [2]*King Fahd Medical Research Center, King Abdulaziz University, P. O. Box 80216, Jeddah 21589, Saudi Arabia;* [3]*Department of Chemistry, BJ Campus, The Islamia University of Bahawalpur, Bahawalpur, Pakistan;* [4]*Department of Health Science and Healthcare Administration, University of Atlanta, Atlanta, GA, USA, and* [5]*GALLY International Biomedical Research Consulting LLC, San Antonio, TX, USA*

Abstract: Accompanying the gradual rise in the average age of the population of most industrialized countries is a regrettable escalation in individuals afflicted with progressive neurodegenerative disorders, epitomized by Alzheimer's disease (AD). The development of effective new treatment strategies for AD has therefore become one of the most critical challenges in current neuroscience. Cholinesterase inhibitors (ChEIs) remain the primary therapeutic strategy for AD, and act by amplifying residual cholinergic activity, a neurotransmitter system central in cognitive processing that is reported to be depleted in the AD brain. With the recent failure of current drug classes focused towards the molecular events known to underpin AD, including the generation of amyloid-β peptide (Aβ) containing plaques and neurofibrillary tangles (hyper-phosphorylated tau). The development of new generation of cholinergic drugs has been accomplished to take advantage of the known modulatory action of the cholinergic system on Aβ, tau production as well as the maintenance synapses, which are known to be lost in AD. Following upon the development of acetylcholinesterase inhibitors (AChE-Is), phenserine, that additionally possessed amyloid precursor protein (APP) synthesis inhibitory actions to lower the generation of Aβ. Selective butyrylcholinesterase inhibitors (BuChE-Is), cymserine analogues, have been developed on the same chemical backbone during further anti-AD research advancement. The above mentioned inhibitors retain actions on APP as well as Aβ and amplify central cholinergic actions without the classical dose-limiting adverse effect profile; therefore, these current BuChE-Is are now moving into AD clinical trials.

---

*Address correspondence to Mohammad A. Kamal: Metabolomics and Enzymology Unit, Fundamental and Applied Biology, King Fahd Medical Research Center, King Abdulaziz University, P. O. Box 80216, Jeddah 21589, Saudi Arabia; Tel: +612-98644812; Fax: +15016368847; E-mail: meu.fabg@hotmail.com

Atta-ur-Rahman / Muhammad Iqbal Choudhary (Eds.)
10.1016/B978-0-12-803959-5.50006-4

Keywords: Acetylcholinesterase, acetylcholinesterase inhibitors, Alzheimer disease, amyloid-β peptide, amyloid precursor protein, butyrylcholinesterase inhibitors, cholinesterases, clinical trials, cymserine, dementia, glial cells, inhibitors, kinetic analysis, muscle disorders, neurodegenerative disorders, neurotransmitters, synaptic cleft, tau protein.

INTRODUCTION

A significant segment of career of several scientists' in the field of anti-Alzheimer's research is spending in elucidating the interaction of various anti-Alzheimer agents on the key enzyme acetylcholinesterase (AChE), which is fundamental to nervous system function and life. There are various challenging tasks in the development of the novel drug such as in case of phenserine from laboratory to the clinic by Nigel H Greig. Specifically, the essential elucidation of the kinetics of particular agent's interaction with AChE or butyrylcholinesterase (BuChE) become the base in its application and approval for an Investigational New Drug Application by the United States Government Food and Drug Administration (US-FDA) to allow the use of the compound in humans to assess its utility in the treatment of Alzheimer's dementia. It is worthwhile, to mention in a similar manner, Greig's and his colleagues studied in the development of novel agents such as cymserine and analogues, to test the radically new hypothesis that selective BuChE inhibition is of value in the therapy of Alzheimer's disease (AD). Drugs of this class are currently in preclinical development and are expected to translate into the clinic soon. There are several research articles available in the literature related with kinetic analysis for such type of new agents such as tetrahydro-furobenzofuran cymserine, a potent BuChE inhibitor and experimental Alzheimer drug candidate [1]. Mostly these types of agents are focused on syntheses of specific carbamates related to physostigmine to figure out their anticholinesterase activities and action on amyloid precursor protein (APP). According to one study, administration of AChE inhibitors increases the ratio of APP forms in platelets of patients with AD, suggesting a potential effect of AChE inhibitors (AChE-Is) on APP trafficking or processing in a peripheral cell [2]. Overall, the conclusive progress of inhibitors of cholinesterases (ChEIs) involved in the cognitive responses, *i.e.* AD management is still under progress due to various factors such as variability in ChEIs treatment.

ALZHEIMER'S DISEASE

Consequent to improvements in preventative, diagnostic and medical treatment for cardiovascular and oncological diseases, the mean age of most industrialized countries humans, continues to steadily climb. Sadly, allied with this rise in life span there has been a steady increase in the number of individuals afflicted with age-related debilitating neurodegenerative disorders, in particular AD. Indeed, in excess of 20 million people worldwide are afflicted with AD. Neurodegenerative diseases are not only amongst the most common causes of death but are currently among the most debilitating illnesses and force an enormous strain, not only on the afflicted and their families, but on both social and healthcare budgets throughout the world. In case of the AD brain, neural dysfunction allied to the induction of numerous biological cascades that cause the classical pathological hallmarks of disease. This is characterized by the presence of (i) amyloid deposits, extra-neuronally, that contain a small toxic cleavage product of APP processing, (ii) neurofibrillary tangles, intra-neuronally, that comprise of hyperphosphorylated tau and, (iii) a dramatic synaptic loss. Of the neurotransmitters involved, the cholinergic system is the earliest and most affected, and hence has received the greatest attention with regard to drug design and development. The primary therapeutic strategy, to date, involves the use of ChEIs to amplify remaining cholinergic activity, which improve cognition and global performance and reduce psychiatric and behavioral disturbances.

AD, although an irreversible and progressive disorder, is currently treated with palliative, symptomatic therapy; primarily with AChE-Is to amplify remaining cholinergic activity. The first ChEIs approved for AD treatment was, tacrine (Cognex) in 1993, but it was proved to be relatively short-acting, unselective between the two ChE enzymes, and associated with a high incidence of reversible hepatotoxicity. In contrast, the second generation ChEIs, donepezil hydrochloride (Aricept), rivastigmine (Exelon) and galatamine (Reminyl), approved by US-FDA in the year 1996, 2000 and 2002 respectively, remain in current use, are better tolerated and have been found useful in treating mild to moderate AD patients. However, these improvements are unfortunately, quite modest. In fact, recent clinical trials have triggered considerable controversy concerning the relevance and the benefit of this class of drugs. Such concern has instigated two avenues of research. One to develop a new class of ChEIs with activity beyond the modest symptomatic one associated with initial agents, and another to optimize current agents based on a greater understanding of enzyme/inhibitor interactions.

New agents that can affect disease progression are sorely needed. Cholinergic loss is the single most replicated neurotransmitter deficiency in AD and has led to the use of AChE-Is as the mainstay of treatment [1]. AChE-Is, however, induce dose-limiting adverse effects [1]. Inhibition of brain BuChE represents a new drug target for AD treatment. Current studies indicated that selective BuChE-Is elevate acetylcholine (ACh) in brain, which augment long-term potentiation and improve cognitive performance in rodents without the classic adverse effects of AChE-Is. BuChE-Is thereby represent a new strategy to ameliorate AD, particularly since AChE activity is depleted in AD brain, in line with ACh levels, whereas BuChE activity is elevated [1]. The BuChE to AChE ratio hence changes dramatically in cortical regions affected by AD from 0.2 up to as much as 11. This altered ratio in AD brain likely modifies the normal role of BuChE in the nervous system of hydrolyzing, excess ACh to depleting already reduced levels of the neurotransmitter.

TYPES OF CHOLINESTERASES

Cholinesterases (ChEs) are the hydrolases that are responsible for the hydrolysis of choline esters. These are classified into two types: acetylcholinesterase (acetylcholine-acetylhydrolase; AChE; EC 3.1.1.7) and butyrylcholinesterase (acetylcholine-acylhydrolase; EC 3.1.1.8; BuChE). The AChE and BuChE can be distinguished on the basis of their affinity for, or reactivity with, various selective inhibitors such as BW- 284C1 for AChE, and ethopropazine, iso-OMPA and bambuterol for BuChE. However, the most specific distinguishing test between ChEs and other less specific esterases is inhibition with physostigmine (eserine). Those esterases, which are not inhibited by 10^{-5} M eserine are not referred to as ChEs [3].

HISTORY OF AChE

The physiological role of ACh was first realized by Dale (1914) during his physiological experiments with plant extracts [4]. The enzymatic hydrolysis of

ACh in nerves was also suggested by Dale, who credited this reaction with a functional role in nerve transmission. A chemical step involved in the transmission of electric signals between nerves and muscles at neuromuscular junctions was demonstrated by Loewi in 1921 and Loewi and Navratil in the year 1926 [5]. The AChE was first isolated by extraction from the electric organ of *Torpedo marmorata* in 1938. Moreover, the electric organs of *Electrophorus electricus* and *Torpedo marmorata* are rich sources of this enzyme [6].

DISTRIBUTION OF AChE IN THE BIOLOGICAL SYSTEMS

AChE is located at neuromuscular junctions and linked with both presynaptic and postsynaptic membranes of the cholinergic synapses [7]. In the nervous system of vertebrates, both the acetyl and the butyryl type ChEs have been reported. The AChE is also found at the surface of the mammalian erythrocytes [8], in cisternae of the endoplasmic reticulum and on plasma membrane of cholinergic neurons. Although some non-cholinergic neurons also contain this enzyme, many of these are cholinoceptive cells. Placenta and some vascular tissues have also been reported to contain AChE [9].

The AChE activity has been detected mainly in membrane-bound fractions in the frontal cortex of autopsied control or Alzheimer brain as well as rat cerebral cortex. However, the distribution of AChE among various membrane fractions differs between control and Alzheimer brains [10]. The distribution of AChE in human basal forebrain has been detected using light microscopic radio-autography and histo-chemical techniques. Moreover, AChE enzymatic activity has been reported to be present in the islands of calleja, putamen, medial septum, diagonal band nucleus and nucleus basalis of Meynert [11].

AChE is also found in lymphocyte [12], platelets [13], snake venom [14] in migrating neurocrest cells [15], bone marrow [16], spleen [17], kidney [18], lung [19], cerebrospinal fluid [20], muscle [21] sarcoplasmic reticulum [22] and in some cell lines of various tissues [23]. The enzyme has also been detected in the thymus cells [24], rat liver plasma membrane [25], camel blood [26] and early myotendinous junction [27]. Existence of AChE in Myelo-proliferative Leukemia Virus-transformed megakaryo-cytic cell lines has been proven using Northern blot analysis [28]. In addition, *Pseudomonas aerujinosa* A-16 strain also possesses the AChE activity [29].

BIOLOGY OF AChE

The higher catalytic efficiency of AChE than BuChE, the sensitivity, facility of spectrophotometric [30], radiometric assays [31], as well as the quality of histochemical staining methods [32] have made this enzyme a convenient development and differentiation marker for studying its molecular structure, cell biology, interactions with inhibitors and drugs, and the localization of this enzyme in various tissues. Its hydrolytic action takes place with maximal speed, with well matched activation energies and other energetic parameters such as entropy, enthalpy and free energy change for its each chemical step. Serine, Glutamic acid and Histidine have been reported to be involved in the catalytic process and the activation of serine takes place through a charge relay system [33].

FUNCTIONS OF AChE

AChE terminates the action of ACh at post-synaptic membrane at the neuromuscular junctions. Actually, neural signals pass from cell to cell at synapses, which can be either electrical (gap junctions) or chemical. At a chemical synapse, the depolarization of the presynaptic membrane by an action potential opens voltage gated Ca^{2+} channels, allowing an influx of Ca^{2+} to trigger exocytic release of neurotransmitter (ACh) from synaptic vesicles. The neurotransmitter diffuses across the synaptic cleft and binds to its specific receptor protein called "ACh receptor" that shifts abruptly from a closed to an open state and then stays open, with the ligand bound, for a randomly variable length of time, averaging about one millisecond or even less, depending upon the temperature and the species. In the open conformation the channel is indiscriminately permeable to small cations, including sodium, potassium and calcium ions (Na^+, K^+, Ca^{2+}), but it is impermeable to anions such as chloride ion [34].

The ACh, neurotransmitter is rapidly eliminated from the synaptic cleft by diffusion, by enzymatic degradation, or by re-uptake into nerve terminals or glial cells. The AChE is responsible for this said enzymatic degradation of ACh into choline, acetate ion and hydrogen ion. Physiologically ACh is very important component of nervous system. It is synthesized from choline and acetyl-CoA by the choline acetyltransferase at nerve endings and acts to transmit impulses from nerve to muscle fiber as described earlier. AChE destroys the ACh after the impulse transmission has been mediated, so that additional impulses may be transmitted, if required. Otherwise, the nerve would remain electrically charged and further conduction would not be possible. Each AChE molecule can hydrolyze up to 10 molecules of ACh per millisecond so that the entire transmitter is eliminated from the synaptic cleft within a few hundred microseconds (μ-sec) after its release from the nerve terminal [35].

BINDING SITES OF AChE

The AChE is a conformationally plastic enzyme and is comprised of following five main domains. Firstly, an anionic locus, which is used for binding the substrate within AChE molecule. This site determines specificity with respect to the substrate alcohol moiety *i.e.*, it serves as choline-binding pocket. Second, an esteratic locus, comprised of the active site serine and histidine involved in catalysis. This site is the actual catalytic machinery and is similar to the catalytic part of the other serine hydrolases such as chymotrypsin [36].

A third site is a hydrophobic region that is contiguous with or near the esteratic and anionic loci and is important in binding aryl (hydrophobic) substrates and active site ligands (aromatic cations). The fourth site is an allosteric site which is located in the regulatory subunit but physically remote from the active site of AChE [37]. The allosteric properties of AChE are due to this site [38]. This fourth domain in the enzyme binds cationic ligands such as gallamine, d-tubocurarine, decamethonium and other noncompetitive and mixed type inhibitors. When ligands bind with this site, conformation of the active site is altered very frequently. Due to this allosteric site, complex reaction dynamics and active site conformation dynamics take place, which is a hallmark of this enzyme. A characteristic property of AChE is its inhibition by high substrate concentration [39] and it has been suggested that peripheral anionic site is involved in this type of inhibition, which changes the active centre allosterically [36]. The regulatory role of this site occurs after binding of either reversible non-competitive or mixed-type inhibitor or second molecule of ACh (excess of substrate). The precise mechanism of action of the said site in the regulation of AChE activity is still ambiguous. This site has clearly been identified and distinguished from choline-binding pocket of the active centre by the use of fluorescent material such as propidium [40]. The last fifth domain is an acidic site of the AChE.

MOLECULAR MECHANISM FOR AChE ACTION

The mechanism of ACh hydrolysis by AChE *in vivo* can be divide into two steps, acetylation and deacetylation.

ACETYLATION PROCESS

The acetylation process involves formation of reversible Michaelis complex between AChE and ACh or ASCh (AChE-ACh, AChE-ASCh). This Michaelis complex serves as the covalent intermediate in ACh hydrolysis system. The AChE-ACh or AChE-ASCh complex eventually breaks down to form the acetylated enzyme (AChEa) and first product (P1) of the hydrolysis *i.e.*, choline (ChOH) in case of AChE-ACh or thiocholine (SChOH) in case of AChE-ASCh complex.

ACh AS A SUBSTRATE

The spatial arrangement of acidic and basic groups of the enzyme occurs in such a way that electrostatic attractions develop between several atoms. It has been postulated that there is an electrostatic interaction between the following atoms: i) hydrogen atom of the acidic group of AChE and alcoholic oxygen of the ACh; and ii) alcoholic oxygen of the serine and the carbonyl carbon of the ACh [41]. Along with these important electrostatic interactions, there are some other types of binding forces involved at different parts of the molecules. Wilson (1952) has shown that the methyl groups on the nitrogen of the ACh also have some role in the binding process [42].

DEACETYLATION PROCESS

Deacetylation involves breakdown of AChEa to form the free AChE and the second product (P2; acetate). The pH study proved that both the acidic and basic groups are involved in the deacylation process in both protonated and unprotonated stages.

The role of imidazole nitrogen atom of histidine in the esteratic site of the AChE is to attract a proton from the water molecule, leaving the hydroxide ion available for attack on the carbonyl carbon. At the same time the hydrogen atom of the acidic site may be transferred to the serine oxygen atom. The various bonds breaking and forming takes place simultaneously. The imidazole group of histidine serves as a general base and a proton shown at the esteratic site may involve in the catalysis, perhaps by transfer to make choline the leaving group. Overall, AChE catalysis occurs *via* an acyl-enzyme mechanism that involves nucleophilic (serine) and general acid-base (histidine) elements.

REVERSIBLE INHIBITORS OF AChE

A wide variety of chemical agents can inactivate AChE, reversibly or irreversibly and are used as insecticides *via* disrupting cholinergic transmission or in clinic to potentiate neuromuscular transmission when it is failing as a consequence of some disease [43].

Substituted ammonium ions, especially tertiary or quaternary ammonium ions have been reported as reversible inhibitors of AChE because of their ability to bind at the anionic site of the AChE [36, 44]. Clinically useful quaternary ammonium compounds such as edrophonium, eserine (physostigmine) and neostigmine (prostigmine) form a covalent intermediate in which a part of the drug molecule remains attached for a longer period of time to the essential serine group at active site of the enzyme. The inhibition of AChE activity lasts from seconds to days depending on the stability of this acyl intermediate.

The anionic site appears to be hydrophobic as the reversible inhibitors composed of large hydrocarbon chain or aromatic ring show higher binding potency for this site than the shorter hydrocarbon chain homologous member of their series [45]. Thus, phenyltrimethyl ammonium, N-methylpyridinium, N-methylquinalinium, isoquinolinium and N-methylacridinium ions are increasingly reported to be potent inhibitors of the AChE.

Tacrine, (used for the treatment of cholinergic deficits in dementia of the Alzheimer's type patients) was considered as irreversible inhibitor in very early days, however, Dawson (1990) proved it to be a reversible inhibitor of AChE [46]. Its ability to inhibit AChE was also evaluated by using camel retinal AChE. The non-significant change of the percent inhibition of AChE by tacrine with respect to various pre-incubation periods showed that it acts as a reversible inhibitor for AChE. Further analysis of the results indicated that the nature of the inhibition was linear mixed type [47]. The effect of tacrine on turnover kinetics of camel retinal AChE has also been investigated. The turnover number (K_{cat}) and specificity constant (K_{sp}) have been found to be 62.1 min^{-1} and 9.92 x 10^5 $(M.min)^{-1}$ in the control system while the values for both parameters were lowered in the tacrine treated systems [48]. The K_i value and mode of inhibition of some inhibitors of the AChE have been summarized in Table **1**.

Therapeutically used anti-ChEs inactivate both AChE as well as BuChE but with different degree of inhibition. Physostigmine, a naturally occurring alkaloid was the first anticholinesterase used in therapeutics. Puromycin, cycloheximide and acetoxyclycloheximide interfere with cholinergic synaptic transmission in sympathetic ganglion *in vitro* at concentration estimated to be effective in blocking memory [49].

IRREVERSIBLE INHIBITORS OF AChE

Irreversible inhibitors of AChE include long lasting organophosphates (OPs) like the "nerve gas" diisopropyl-fluorophosphate (DFP), methanesulfonates and carbamates. During the reaction of these inhibitors with AChE, the enzyme acts as a nucleophile and the hydroxyl group of serine in the esteratic site of the active center of AChE gets phosphorylated [50]. This could be explained as follows:

The esters of carboxylic acid and acyl halides of substituted carbamic, phosphoric and sulfonic acids resist the hydrolytic step of the normal reaction (Kn) and thus permanently inhibit the enzyme [51].

The organophosphates act as hemi-substrate for AChE, specially phosphorylating the active site serine, just as the natural substrate acylates it. Since the rate of hydrolysis of the phosphoryl or phosphonyl enzyme is very much slower than de-acylation, OPs are effectively irreversible AChE inhibitors. The acetyl enzyme reacts with water in 0.1 ms while the phosphorylated AChE reacts very slowly with water. If the reaction with water was rapid, these OPs would be the substrates of AChE. The dimethyl phosphoryl enzyme reacts with water in an hour or more than an hour [52].

Table 1. K_i Values and Nature of the Inhibition of Some Inhibitors of AChE

Inhibitor	Origin of AChE	K_i (mM)	Mode of Inhibition	References
Anisomycin	C. E.	-	l.m.(c+n)	[49]
Cycloheximide	C. retina	3.50	l.m.(c+n)	[74]
Decamethonium	Venom	0.0140	-	[75]
Gallamine	T. E.	0.015	-	[76]
Gallamine	Chicken	0.740	-	[76]
Gallamine	C. retina	0.160	c	[77]
Hexamethonium	Ch. Brain	0.242	-	[75]
Hexamethonium	T. E.	0.084	-	[75]
Hexamethonium	C. retina	0.30	p.m.(c+n)	[78]
Lannate	C. retina	0.000143	m(c+n)	[79]
Lidocaine	H. E.	-	m	[80]
Lignocaine	Rat brain	-	c	[81]
Lignocaine	Rat E.	-	n	[81]
Malathion	C. E.	K_i (ppm)	u	[82]
Mercuric Cl-	H. E.	0.0063	c	[83]
Procaine	Venom	0.133	l.m.(c+n)	[14d]
Procaine	Pigeon Brain	0.149	l.m.(c+n)	[84]
Procaine	H. E.	-	c	[80]
Tacrine	C. retina	0.000068	l.m.(c+n)	[85]
Tacrine	B. retina	0.0000045	p.m(c+n)	[86]
Tetracaine	H. E.	0.127	m(c+n)	[80]
Tetracaine	Venom	0.0072	n	[14d]
Tetracaine	Venom	0.12	m	[14d]
Sevin	C. retina	0.0062	l.m(u+c)	[87]

B., bovine; C. camel; c. californica; c, competitive; Ch., chicken; E., electrophorus; e. electricus; E. erythrocyte; H., human; l., linear; m., mixed; n, noncompetitive; p, partial; T. torpedo; u, uncompetitive.

Methanesulfonyl fluoride, (3-hydroxyphenyl) trimethyl-ammonium iodide methanesulfonate, 1-methyl-3-hydroxy-pridinium iodide methanesulfonate, 1-1-pentamethylene bis 3-hydroxypyridinium iodide have been reported as irreversible inhibitors of *Electrophorus electricus* AChE [53]. The methanesulfonyl fluoride binds to the serine-OH group in the esteratic sub-site of the AChE, while some quaternary ammonium ligands accelerate the rate of sulfonylation [54]. The influence of tetra-alkyl ammonium ions, $(CnH2+1)4N^+$ where n = 1-4 on the sulfonylation of cobra venom AChE by methane-, ethane-, propane-, and butanesulfonyl chloride has been studied and it was observed that these ions accelerate sulfonylation of the AChE. The acceleration was related to conformational changes near the esteratic site of the AChE, caused by the tetra-alkylammonium ions, which binds to the anionic sub-site of the catalytic center of the AChE [55].

The OPs when react with AChE, form phosphorylated AChE which undergoes a rapid dealkylation of the phosphorylated AChE complex. After the pinacolyl moiety is removed from the soman bound-enzyme, the phosphorylated enzyme can no longer be reactivated by oximes *etc*. This reaction is called aging. The human AChE has 1.3 minute half time for aging with soman, *i.e.* soman ages very fast [56]. The causal therapy of this intoxication consists of reactivation of the inhibited AChE by removing the phosphoryl moiety. A large number of pyridinium oximes such as P2, TMB4, obidoxime (Hagedorn oximes) have been synthesized for reactivation purpose [57]. Good nucleophiles such as NH_4OH react with phosphoryl enzyme much more rapidly than water to dephosphorylate the enzyme. This process is called reactivation.

The reaction of the organophosphate inhibitors with AChE can be blocked by reversible anionic site inhibitors such as tetramethylammonium, 3-hydroxyphenyl trimethyl-ammonium ions [58], tubocurarine, gallamine, methyl-pyridinium, decamethonium, hexamethonium [59], atropine [60] and suxamethonium ions [61]. The bis-quaternary compounds such as gallamine, pancuronium, curare, succinyldicholine and decamethonium bind to the peripheral anionic site of the AChE and thus indirectly influence the active site of the AChE.

CLINICAL ASPECT OF AChE

Regulation of ACh turnover, concentration as well as the activity of AChE in synaptic junctions, neurons and non-neuronal cells are of great interest for many neuro-scientists because of their role in a number of neurologic disorders such as AD, senile dementia, myasthenia gravies, Pick's disease, Down's syndrome and organophosphorous poisoning [62]. The different molecular forms and different modified activities of ChE have also been reported in certain pathological states such as human primary brain tumors, Guillain-Barre syndrome and Hirschsprung's disease [63].

The muscle diseases have also received considerable attention according to the scientific literature and it has been reported that muscular dystrophies and myopathies [64], motor end plate disease [65] and muscular dysgenesis [66] do have some relationship with changes in AChE activity. Investigation of the AChE in biopsies from patients with nerve and muscle disorders suggests a lack of localization and of regulation of the AChE synthesis [67].

A reference range for erythrocyte AChE in 46 mentally normal subjects aged 65 years or over has been reported [68]. Such type of reports on the relationship between changes in AChE levels and mental illness (dementia and depression) in the elderly subjects, suggest that determination may be useful in screening elderly subjects for these illnesses as well as being helpful in diagnostic matter. The intravenous administration of the monoclonal antibody is a novel model of cholinergic autoimmunity and it may have relevance for human neurological disorders of unknown etiology. These antibodies react with brain AChE and cause prolonged depression of plasma AChE without changing BuChE, lactic acid dehydrogenase and hematocrit value [69]. In case of AD, the optimum pH and sensitivity towards selective ChE-Is such as BW28C51 and iso-OMPA had changed for AChE present in the plaques and tangles of AD [70].

Production of auto antibodies that change the normal behavior of the AChE due to inhibition of human erythrocyte AChE has been reported in case of patients suffering from amyotrophic lateral sclerosis. These auto antibodies prevent the conformational changes of the AChE at break point in the Arrhenius plot (33°C), which normally occurs in control AChE [71]. Measurements of enzymatic activity of AChE in erythrocyte from insulin-dependent diabetic patients have provided conflicting results. A number of researchers reported that the AChE activity significantly increased in such patients than in normal subjects [72], while another study has demonstrated statistically significant (P < 0.001) declines in the AChE activity [73].

BACKGROUND OF BuChE

BuChE is a serine hydrolase enzyme, primarily expressed and secreted by glial cells within the human brain. It is also expressed in neuronal somata and their proximal dendrites, such as in amygdala, hippocampus and thalamus of brain areas affected with AD [1]. It has been reported that 10 to 15% of cholinergic neurons in the human hippocampus and amygdala express BuChE in their cell bodies and proximal dendrites, instead of AChE [88]. BuChE is not physiologically essential for humans and butyrylcholine is the best substrate for BuChE. Although specific function of BuChE has not been reported as yet, its wide distribution in the nervous system points towards its involvement in the neural functions and in neurodegenerative diseases [88b, 89].

BuChE efficiently catalyses the hydrolysis of numerous endogenous substances, including choline esters [88b, 89b]. It has wider substrate specificity and interacts with a broader range of inhibitors compared to AChE. Specific neurons use BuChE rather than AChE to cleave pre-synaptic ACh.

In the healthy human brain, AChE and BuChE are present in the ratio of 4:1. However, in the brains of AD patients AChE activity can decline by upto 45% during disease progression, reflecting the disappearance of neurons and axons to which it is associated, while BuChE activity can be elevated by upto 2-fold [90], thereby altering this ratio considerably.

BuChE has important role to play in the AD. The activity of AChE decreases as the disease progresses but the activity of BuChE remains unaffected or increases sometimes [89b]. In advanced stages of AD, BuChE regulates central cholinergic transmission that take place in the brain. In the diseased AD brain, BuChE hydrolyses the already depleted levels of ACh [91].

Selective BuChE inhibition has shown potential clinical value in ameliorating the symptoms of AD. Restoration of ACh levels by BuChE inhibition seems to take place without apparent adverse effects [91]. In one study by Greig *et al.* (2005), selective inhibition of BuChE by cymserine derivative has additionally augment ACh levels, increase cognitive function and decrease amyloid deposits in aged rats [91].

BuChE INHIBITORS

The AChE and BuChE differ structurally, genetically and in their substrate specificities and sensitivities to a wide range of inhibitors. Recent research indicates that selective BuChE inhibition approach could help increase in ACh levels and reduce the formation of abnormal amyloid found in AD [92]. Therefore, potent and highly selective BuChE inhibitors and/or dual AChE–BuChE inhibitors are currently being synthesized as an innovative treatment approach for AD treatment.

In one study, Grieg *et al.* (2005) reported that centrally mediated inhibition of a homologous ChE form, BuChE, may achieve the cognitive improvements associated with current AChE-Is without the classic dose limiting actions [91].

Recently discovered inhibitors acting on the hydrolase activity of BuChE include cymserine analogues [91], heterobivalent tacrine derivatives [93], quinazolinimines [94], phenothiazines [95], benzofurans [96] and isosorbide-based compounds [97].

The most potent and highly selective BuChE inhibitor reported so far belongs to the group of isosorbide-based compounds and displays an IC_{50} value of 0.15 nM with a high selectivity of over 60,000 times towards BuChE [97b]. The isosorbide-based molecules as well as the potent and selective cymserine analogues [91] are carbamates, with scaffolds inspired by the classical ChEIs physostigmine and rivastigmine.

Selective BuChE inhibition has been shown to elevate cortical extracellular ACh levels in rats in a manner similar to that achieved by selective AChE inhibition using donepezil or by dual AChE/BuChE inhibition using rivastigmine [90].

From all reported BuChE inhibitors synthesized till date, only two cymserine analog [91] and benfozuran-based [96] have been proved to act on Aβ fibril formation in the micromolar range concentration. This viewpoint was fortified by animal studies which indicates selective BuChE-Is. Cymserine analogues, elevate brain levels of ACh in AChE+/+ and -/- animals, augment LTP in brain slice preparations and improve cognitive performance of aged rodents without adverse actions [91]. However, the development of novel hybrid inhibitors which could acts on both the hydrolase activity of BuChE and Aβ fibrilogenesis is an underexplored strategy and have higher chances of success in AD pharmacotherapy [98].

In our effort to produce a potent and BuChE-selective inhibitor as a candidate to test the hypothesis that BuChE-Is would be efficacious and better tolerated than AChE-Is in AD, we have also synthesized a novel compound, tetrahydro-furobenzofuran cymserine (THFBFC) [99]. To characterize the quantitative interaction of THFBFC with human BuChE, we performed innovative enzyme kinetic analysis which gave us IC_{50} value, together with specific new kinetic constants, such as KT_{50}, $KT_{1/2}$, RI, ^{o}KRT, $^{o}Pmax$, KPT and $PT_{1/2}$, to define target concentrations for clinical translation. Additional classical kinetic parameters, including K_i, K_m or K_s, K_{cat} or V_{max} and V_{mi} were also determined. From our study, we concluded THFBFC a potent competitive inhibitor of human BuChE and like its isomer dihydrobenzodioxepine cymserine, is a potentially interesting AD drug candidate [1].

Recently, highly selective, reversible, central nervous system (CNS)-penetrable BuChE inhibitors, N^{1}-phenethyl-norcymserine (PEC) and its analogues, have been developed and are permitting elucidation of the physiological role of brain BuChE [90].

In addition to elevating brain ACh, PEC has been shown to augment long-term potentiation (LTP; a molecular correlate of learning), improve cognitive performance in aged healthy rats, and lower brain levels of $A\beta_{1-40}$ and $A\beta_{1-42}$ in transgenic mice over expressing human $A\beta$ [91]. However, the effect of selective BuChE inhibition on $A\beta$-induced cognitive dysfunction remains unclear [90].

In one study by Karlsson *et al.* (2012), a collection of small aromatic compounds were assayed for inhibition of BuChE's hydrolysis activity [98]. It comprises mostly new (not yet reported) synthetic compounds belonging to different structural scaffolds, partly based on natural products. They found that three diaryl imidazoles could act as hybrid inhibitors of both BuChE hydrolase activity and $A\beta$ fibril formation. Moreover, lead optimization generated a bifunctional molecule with a higher potency on both activities. They suggested that diaryl imidazoles-based hybrid inhibitors could evolve into novel inspirational leads for the use in AD treatment and help assess the basic functions of BuChE.

ACKNOWLEDGEMENTS

The author gratefully acknowledged research facility provided by King Fahd Medical Research Center, King Abdulaziz University, Kingdom of Saudi Arabia.

CONFLICT OF INTEREST

The authors confirm that this chapter contents have no conflict of interest.

REFERENCES

[1]   Kamal, M. A.; Klein, P.; Luo, W.; Li, Y.; Holloway, H. W.; Tweedie, D.; Greig, N. H. *Neurochem. Res.*, **2008**, *33* (*5*), 745-753.

[2]   Borroni, B.; Colciaghi, F.; Pastorino, L.; Pettenati, C.; Cottini, E.; Rozzini, L.; Monastero, R.; Lenzi, G. L.; Cattabeni, F.; Di Luca, M.; Padovani, A. *Arch. Neurol.*, **2001**, *58* (*3*), 442-446.

[3]   Augustinsson, K. B., In *Methods of Biochemical analysis*; David, G., Ed.; Interscience Publishers Inc.: New York., **1957**.

[4]   Dale, H. H. *J Pharmacol. Exp. Ther.*, **1914**, *6* (*2*), 147-190.

[5]   (a) Loewi, O. *Pflügers Archiv European Journal of Physiology*, **1924**, *204* (*1*), 629-640; (b) Loewi, O.; Navratil, E. *Pflugers. Arch.*, **1926**, *214*, 689-696.

[6]   Nachmansohn, D.; Lederer, E. *Bull. Soc. Chim. Biol.*, **1939**, *21*, 797-781.

[7]   (a) Schwarzacher, H. G. *Biblphie. Anat., Basle*, **1961**, *2*, 255; (b) Koelle, G. B. *J. Pharm. Pharmac.*, **1962**, *14*, 65-90.

[8]   (a) Galehr, O.; Plattner, F. *Pflugers Arch*, **1982**, *218*, 488-505; (b) Miki, A.; Mizoguti, H. *Histochemistry.*, **1982**, *76 (3)*, 303-314.

[9]   Simone, C.; Derewlany, L. O.; Oskamp, M.; Johnson, D.; Knie, B.; Koren, G. *J. Lab. Clin. Med.*, **1994**, *123 (3)*, 400-406.

[10] Nakamura, S.; Kawashima, S.; Nakano, S.; Tsuji, T.; Araki, W. *J. Neural. Transn. Supplementum*, **1990**, *30*, 13-23.

[11] Szigethy, E.; Quirion, R.; Beaudet, A. *The J. Comp. Neurol.*, **1990**, *297 (4)*, 487-498.

[12] Szelényi, J. G.; Bartha, E.; Hollán, S. R. *Brit. J. Haematol.* , **1982**, *50 (2)*, 241-245.

[13] Barber, A. J.; Jamieson, G. A. *J. Biol. Chem.*, **1970** *245*, 6357.

[14] (a) Kumar, V.; Elliott, W. B. *Prep. Biochem.*, **1973**, *3 (6)*, 569-582; (b) AlJafari, A. A.; Kamal, M. A.; Duhaiman, A. S.; Alhomida, A. S. *Mol. Cell. Biochem.*, **1995**, *151 (1)*, 21-26; (c) Duhaiman, A. S.; Alhomida, A. S.; Rabbani, N.; Kamal, M. A.; al-Jafari, A. A. *Biochimie.*, **1996**, *78 (1)*, 46-50; (d) al-Jafari, A. A.; Kamal, M. A.; Duhaiman, A. S.; Alhomida, A. S. *Journal of Enz. Inhib.*, **1996**, *11 (2)*, 123-134.

[15] Cochard, P.; Coltey, P. *Dev. Biol.*, **1983**, *98 (1)*, 221-238.

[16] (a) Yonemura, Y.; Kawakita, M.; Fujimoto, K.; Sakaguchi, M.; Kusuyama, T.; Hirose, J.; Kato, K.; Takatsuki, K. *Int. J. Cell Cloning*, **1992**, *10 (1)*, 18-27; (b) Lazewska, M.; Tabarowski, Z.; Dabrowski, Z. *Toxicol. Ind. Health.*, **1993**, *9 (4)*, 617-622.

[17] Bellinger, D. L.; Lorton, D.; Hamill, R. W.; Felten, S. Y.; Felten, D. L. *Brain Behav. Immun.*, **1993**, *7 (3)*, 191-204.

[18] (a) Arora, S. K.; Kaur, G. *Biochemistry and Molecular Biology International*, **1993**, *31 (3)*, 413-420; (b) Tanaka, Y.; Suzuki, A. *J. Pharm. Pharmacol.*, **1994**, *46 (3)*, 235-239.

[19] Tavakoli, R.; Buvry, A.; Le Gall, G.; Barbet, J. P.; Houssin, D.; Lockhart, A.; Frossard, N. *Am. J. Physiol.*, **1992**, *262 (3 Pt 1)*, L322-326.

[20] Pedro, B. J.; Senti, M.; Nogues, X.; Rubies, P. J.; Roquer, J.; D'olhaberriague, L.; Olive, J. *Stroke.*, **1993**, *24 (9)*, 1416-1417.

[21] Cabezas-Herrera, J.; Campoy, F. J.; Vidal, C. J. *J. Neurosci. Res.*, **1994**, *38 (5)*, 505-514.

[22] Campoy, F. J.; Cabezas-Herrera, J.; Vidal, C. J. *J. Neurosci. Res.*, **1992**, *33 (4)*, 568-578.

[23] Ehrich, M. *Clin. Exp. Pharmacol. Physiol.*, **1995**, *22 (4)*, 291-292.

[24] Topilko, A.; Caillou, B. *Blood.*, **1985**, *66 (4)*, 891-895.

[25] Kamal, M. A.; Al-Ali, N.; Al-Jafari, A. A., Acetylcholineterase activity in rat liver plasma membranes. In *20th IUPAC Symp. on the Chem. of Nat. Prod*, llinois, 1996.

[26] Al-Jafari, A. A.; M.A., K. *Cell. Biochem. Funct.*, **1994**, *12*, 26-35.

[27] Wake, K. *Cell Tissue Res.*, **1976**, *173 (3)*, 383-400.

[28] Fischer, M.; Ittah, A.; Liefer, I.; Gorecki, M. *Cell. Mol. Neurobiol.*, **1993**, *13 (1)*, 25-38.

[29] Tani, Y.; Nagasawa, T.; Sugisaki, H.; Ogata, K. *Agr. Biol. Chem*, **1975**, *39 (6)*, 1287-1294.

[30] Ellman, G. L.; Courtney, K. D.; Andres, V., Jr.; Feather-Stone, R. M. *Biochemical Pharmacology*, **1961**, *7*, 88-95.

[31] Johnson, C. D.; Russell, R. L. *Anal. Biochem.*, **1975**, *64 (1)*, 229-238.

[32] Karnovsky, M. J.; Roots, L. *J. Histochem. Cytochem.*, **1964**, *12*, 219-221.

[33] Massoulié, J.; Pezzementi, L.; Bon, S.; Krejci, E.; Vallette, F. M. *Prog. Neurobiol.*, **1993**, *41 (1)*, 31-91.

[34] Lester, H. A. *Sci. Am.*, **1977**, *236 (2)*, 106-118.

[35] (a) Massoulié, J.; Bon, S. *Annu. Rev.Neurosci.*, **1982**, *5*, 57-106; (b) Taylor, P.; Schumacher, M.; MacPhee-Quigley, K.; Friedmann, T.; Taylor, S. *Trends. Neurosci.*, **1987**, *10*, 93-95.

[36] Rosenberry, T. L. I. V. E., , , In *Advances in Enzymology*; Meister, A. J., Ed.; John Wiley & Sons: New York, **1975**; pp. 103-218.

[37] Changeux, J. P. *Mol. Pharmacol.*, **1966**, *2*, 369-392.

[38] Mabood, S. F. *Pak. J. Sci. Ind. Res.*, **1981**, *24*, 127-136.

[39] Quinn, D. M. *Chem. Rev.*, **1987**, *87*, 955-979.

[40] Taylor, P.; Lappi, S. *Biochem.*, **1975**, *14* (*9*), 1989-1997. .

[41] Krupka, R. M.; Laidler, K. J. *J. Med. Chem.*, **1961**, *83*, 1458-1460.

[42] Wilson, I. B. *J. Biol. Chem.*, **1952**, *197*, 215-218.

[43] Taylor, P., In *In The Pharma-coligical Basis of Therapeutics*; Gilman, A. G.; Goodman, L. S.; Gilman, A., Eds.; MacMillan Co: New York, **1980**; pp. 100-119.

[44] Cohen, S. G.; Chishti, S. B.; Bell, D. A.; Howard, S. I.; Salih, E.; Cohen, J. B. *Biochim. Biophys. Acta.*, **1991**, *1076* (*1*), 112-122.

[45] Belleau, B., In *Advances in Drug Res*; Harper, N. J.; Simmonds, A. B., Eds.; New York: Academic Press: London, **1967**; p 89.

[46] Dawson, R. *Neurosci. Lett.*, **1990**, *11*, 85-89.

[47] Dawson, R., Reversibility of the inhibition of acetyl-cholinesterase by tacrine. *Neurosci. Lett.*, **1990**, *11*, 85-89.

[48] al-Jafari, A. A.; Kamal, M. A. *Biochem. Mol. Biol. Int.*, **1996**, *39* (*5*), 917-922.

[49] Moss, D. E.; Fahrney, D. *J. Neurochem.*, **1976**, *26* (*6*), 1155-1157.

[50] Schwarz, M.; Glick, D.; Loewenstein, Y.; Soreq, H. *Pharmacol. Ther.*, **1995**, *67* (*2*), 283-322.

[51] Das, Y. T.; Brown, H. D.; Chattopadhyay, S. K. *Biochem. Cell Biol.*, **1987**, *65* (*9*), 798-802.

[52] (a) Hobbiger, F. *Chem. Ind.*, **1956**, 415-419; (b) Davies, D. R.; Holland, P.; Rumens, M. J. *Biochem. Pharm.*, **1966**, *15* (*11*), 1783-1789.

[53] Kitz, R.; Wilson, I. B. *J. Biol. Chem.*, **1962**, *237*, 3245-3249.

[54] Pavlic, M. R. *Arch. Biochem. Biophys.*, **1987**, *253* (*2*), 446-452.

[55] Sepp, A. V.; Iarv, I. L. *Bioorg. Khim.*, **1989**, *15* (*11*), 1499-1503.

[56] Harris, L. W.; Heyl, W. C.; Stitcher, D. L.; Broomfield, C. A. *Biochem. Pharmacol.*, **1978**, *27* (*5*), 757-761.

[57] Schoene, K. *Biochem. Pharmacol.*, **1973**, *22* (*23*), 2997-3003.

[58] Berman, H. A.; Becktel, W.; Taylor, P. *Biochemistry.*, **1981**, *20* (*16*), 4803-4810.

[59] Schoene, K. *Biochim. Biophys. Acta.*, **1978**, *525* (*2*), 468-471.

[60] Kuhnen, H.; Schrichten, A.; Schoene, K. *Arzneimittelforschung.*, **1985**, *35* (*9*), 1454-1456.

[61] Hallak, M.; Szinicz, L. *Biochem. Pharmacol.*, **1988**, *37*, 819-822.

[62] Lieske, C. N.; Gepp, R. T.; Maxwell, D. M.; Clark, J. H.; Broomfield, C. A.; Blumbergs, P.; Tseng, C. C. *J. Enz. Inhib.*, **1992**, *6* (*4*), 283-291.

[63] (a) Meier-Ruge, W.; Lutterbeck, P. M.; Herzog, B.; Morger, R.; Moser, R.; Schärli, A. *J. Pediatr. Surg.*, **1972**, *7* (*1*), 11-17; (b) Ikawa, H.; Kim, S. H.; Hendren, W. H.; Donahoe, P. K. *Arch. Surg.*, **1986**, *121* (*4*), 435-438.

[64] Rowland, L. P., In *Principal of Neural Sci*; Kandel, E. R.; Schwartz, J. H., Eds.; Elsevier-North Holland: New York, **1985**; pp. 196-208.

[65] Rieger, F.; Shelanski, M. L.; Sidman, R. L. *Exp. Neurol.*, **1983**, *79* (*2*), 299-315.

[66] Rieger, F.; Pinçon-Raymond, M.; Lombet, A.; Ponzio, G.; Lazdunski, M.; Şidman, R. L. *Dev. Biol.*, **1984**, *101* (*2*), 401-409.

[67] Rakonczay, Z. *Prog. Neurobiol.*, **1988**, *31* (*4*), 311-330.

[68]   McWilliam, C.; Wood, N.; Copeland, J. R.; Taylor, W. H. *Age. Ageing.*, **1990**, *19* (*2*), 104-106.

[69]   Brimijoin, S.; Balm, M.; Hammond, P.; Lennon, V. A. *J. Neurochem.*, **1990**, *54* (*1*), 236-241.

[70]   Geula, C.; Mesulam, M. *Brain. Res.*, **1989**, *498* (*1*), 185-189.

[71]   Sindhuphak, R.; Karlsson, E.; Conradi, S.; Ronnevi, L. O. *J. Neurol. Sci.*, **1988**, *86* (*2-3*), 195-202.

[72]   Testa, I.; Rabini, R. A.; Fumelli, P.; Bertoli, E.; Mazzanti, L. *J. Clin. Endocrinol. Metab.*, **1988**, *67* (*6*), 1129-1133.

[73]   Suhail, M.; Rizvi, S. I. *Biochem. J.*, **1989**, *259* (*3*), 897-899.

[74]   Al-Jafari, A. A.; Kamal, M. A.; Alhomida, A. S. *Cell. Biol. Toxicol.*, **1998**, *14* (*3*), 167-174.

[75]   Eichler, J.; Anselmet, A.; Sussman, J. L.; Massoulie, J.; Silman, I. *Mol. Pharm.*, **1994**, *45*, 335-340.

[76]   Eichler, J.; Kreimer, D. I.; Varon, L.; Silman, I.; Weiner, L. *J. Biol. Chem.*, **1994**, *269*, 30093-30096.

[77]   Al-Jafari, A. A. *Toxicol. Lett.*, **1997** *90*, 45-51.

[78]   Alhomida, A. S.; Kamal, M. A.; Al-Jafari, A. A. *J. Enz. Inhib.*, **1997**, *12*, 303-311.

[79]   Kamal, M. A. *Biochem. Mol. Biol. Int.*, **1997**, *43* (*5*), 1183-1193.

[80]   Spinedi, A.; Pacini, L.; Luly, P. *Biochem. J.*, **1989**, *261*, 569-573.

[81]   Haque, S. J.; Poddar, M. K. *Biochem. Pharmacol.*, **1983**, *32*, 3443-3446.

[82]   Kamal, M. A. *Mol. Biol. Int.*, **1997**, *43* (*1*), 89-97.

[83]   Al-Jafari, A. A. *Biochem. Mol. Biol. Int.*, **1995**, *36*, 1243-1253.

[84]   Al-Jafari, A. A.; A.S., D. *Cell. Biochem. Funct.*, **1994**, *12*, 206-216.

[85]   Al-Jafari, A. A.; A.S., D., Kinetics of the inhibition of acetylcholinesterase from pigeon brain by procaine hydrochloride. *Cell. Biochem. Funct.*, **1994**, *12*, 206-216.

[86]   Al-Jafari, A. A.; Kamal, M. A.; Alhomida, A. S. *J. Biochem. Mol. Toxicol.*, **1998**, *12*, 245-251.

[87]   Kamal, M. A. *Biochem. Mol. Biol. Int.*, **1997**, *42* (*2*), 235-246.

[88]   (a) Darvesh, S.; Grantham, D. L.; Hopkins, D. A. *The Journal of Comparative Neurology*, **1998**, *393* (*3*), 374-390; (b) Darvesh, S.; Hopkins, D. A.; Geula, C. *Nat. Rev. Neurosci.*, **2003**, *4* (*2*), 131-138.

[89]   (a) Greig, N. H.; Lahiri, D. K.; Sambamurti, K. *Int Psychogeriatr.*, **2002**, *14*, 77-91; (b) Giacobini, E. *Pharmacol. Res.*, **2004**, *50* (*4*), 433-440.

[90]   Furukawa-Hibi, Y.; Alkam, T.; Nitta, A.; Matsuyama, A.; Mizoguchi, H.; Suzuki, K.; Moussaoui, S.; Yu, Q.-S.; Greig, N. H.; Nagai, T.; Yamada, K. *Behav. Brain Res.*, **2011**, *225* (*1*), 222-229.

[91]   Greig, N. H.; Utsuki, T.; Ingram, D. K.; Wang, Y.; Pepeu, G.; Scali, C.; Yu, Q.-S.; Mamczarz, J.; Holloway, H. W.; Giordano, T.; Chen, D.; Furukawa, K.; Sambamurti, K.; Brossi, A.; Lahiri, D. K. *Proceedings of the National Academy of Sciences of the United States of America*, **2005**, *102* (*47*), 17213-17218.

[92]   Sekutor, M.; Mlinarić-Majerski, K.; Hrenar, T.; Tomić, S.; Primožič, I. *Bioorg. Chem.*, **2012**, *41-42*, 28-34.

[93]   Elsinghorst, P. W.; Tanarro, C. M. G.; Gütschow, M. *J. Med.Chem.*, **2006**, *49* (*25*), 7540-7544.

[94]   Decker, M.; Kraus, B.; Heilmann, J. *Bioorg. Med. Chem.*, **2008**, *16* (*8*), 4252-4261.

[95]   (a) Darvesh, S.; McDonald, R. S.; Darvesh, K. V.; Mataija, D.; Conrad, S.; Gomez, G.; Walsh, R.; Martin, E. *Bioorg. Med. Chem.*, **2007**, *15* (*19*), 6367-6378; (b) Darvesh, S.; Pottie, I. R.; Darvesh, K. V.; McDonald, R. S.; Walsh, R.; Conrad, S.; Penwell, A.; Mataija, D.; Martin, E. *Bioorg. Med. Chem.*, **2010**, *18* (*6*), 2232-2244.

[96]   Rizzo, S.; Rivière, C.; Piazzi, L.; Bisi, A.; Gobbi, S.; Bartolini, M.; Andrisano, V.; Morroni, F.; Tarozzi, A.; Monti, J.-P.; Rampa, A. *J. Med.Chem.*, **2008**, *51* (*10*), 2883-2886.

[97]   (a) Carolan, C. G.; Dillon, G. P.; Gaynor, J. M.; Reidy, S.; Ryder, S. A.; Khan, D.; Marquez, J. F.; Gilmer, J. F. *J. Med. Chem.*, **2008**, *51* (*20*), 6400-6409; (b) Carolan, C. G.; Dillon, G. P.; Khan, D.; Ryder, S. A.; Gaynor, J. M.; Reidy, S.; Marquez, J. F.; Jones, M.; Holland, V.; Gilmer, J. F. *J. Med. Chem.*, **2010**, *53* (*3*), 1190-1199; (c) Dillon, G. P.; Gaynor, J. M.; Khan, D.; Carolan, C. G.; Ryder, S. A.; Marquez, J. F.; Reidy, S.; Gilmer, J. F. *Bioorg. Med. Chem.*, **2010**, *18* (*3*), 1045-1053.

[98]   Karlsson, D.; Fallarero, A.; Brunhofer, G.; Guzik, P.; Prinz, M.; Holzgrabe, U.; Erker, T.; Vuorela, P. *Eur. J. Pharm. Sci.*, **2012**, *45* (*1-2*), 169-183.

[99]   Kamal, M. A.; Qu, X.; Yu, Q.-S.; Tweedie, D.; Holloway, H. W.; Li, Y.; Tan, Y.; Greig, N. H. *J. Neural. Transm.*, **2008**, *115* (*6*), 889-898.

CHAPTER 7

Role of Acetylcholinesterase Inhibitors and Alzheimer Disease

Zafar Saied Saify[1,*] and Nighat Sultana[2]

[1]*International Center for Chemical Sciences, H. E. J. Research Institute of Chemistry, University of Karachi, Karachi 75270, Pakistan;* [2]*Pharmaceutical Research Center, PCSIR Laboratories Complex, Shahrah-e-Dr. Salimuzzaman Siddiqui, Karachi 75280, Pakistan*

Abstract: Alzheimer's disease (AD) has a progressive neurodegenerative pathology with severe economic and social impact. There is currently no cure, although cholinesterase inhibitors provide effective temporary relief of symptoms in some patients. Nowadays, drug research and development are based on the cholinergic hypothesis that supports the cognition improvement by regulation of the synthesis and release of acetylcholine in the brain. There are only five commercial medicines Donepezil, galantamine, memantine, rivastigmine, tacrine approved for treatment of AD. Natural products have played an important alternative role in the research for new acetylcholinesterase inhibitors, as exemplified through the discovery of galantamine. AD is the dementia associated with aging, which initially targets memory and progressively destroys the functions of the brain, as the neocortex suffers neuronal, synaptic, and dendritic losses. The whole phenomena proliferate due to deposition of amyloid plaques. The goal of the present work is to analyse the search on drugs for the treatment and prevention of AD in the light of Acetyl cholinesterase activity. It is based on systematisation of the data on biochemical and structural similarities in the interaction between physiologically active compounds and their biological targets related to the development of such pathologies.

Keywords: Acetylcholinesterase, acetyl cholinesterase and disease dementia, adamantyl derivative, Alzheimer's disease, 7-Azaindole, butyrylcholinesterase, dementia of natural product, drugs available for senile, enzyme and dementia, enzyme inhibition and alzheimer, galantamine, loss of memory and plaque formation, mechanism of dementia drugs, microtubule, napthyl derivative, nitrophenacyl derivative, physostigmine, regulation of memory by effective bioactive agents, synthetic drugs and alzheimer disease, synthesis, solving the mystery of dementia.

***Address correspondence to Zafar Saied Saify:** International Center for Chemical Sciences, H. E. J. Research Institute of Chemistry, University of Karachi, Karachi 75270, Pakistan; Tel: 03332333291; E-mail: zssaify@gmail.com

INTRODUCTION

Acetylcholinesterase (AChE) has proven to be the most viable therapeutic target for symptomatic improvement in AD because cholinergic deficit is a consistent and early finding in AD. Inhibition of AChE was considered to be achievable as a therapeutic target because of proven efficacy of inhibition of peripheral AChE as a treatment for myasthenia gravis (MG) proving that the approach was feasible. However, selective inhibition of the central nervous system (CNS) AChE initially proved to be daunting. Before tacrine, physostigmine, the classic AChE inhibitor (AChEI) was investigated as a treatment for AD. Physostigmine was subsequently abandoned because of poor tolerability. Four drugs are currently available for AD treatment: galantamine, rivastigmine, donepezil, and memantine. The major AD therapeutics available on the market are AChE inhibitors, such as tacrine and donepezil.

Acetylcholinesterase (AChE) Inhibitors

AChEI are the first and the most developed group of drugs proposed for Alzheimer treatment. According to the classical concepts of the impact of AChEI on neurotransmission, the main effect of AChEI is thought to be associated with the increase of both the duration of action and concentration of acetylcholine (ACh) neurotransmitter in the synaptic cleft that result in a potentiation of the activation of cholinergic receptors decreased in AD-type pathology [1, 2].

Natural products have been used for medicinal purposes for a long time. The effort to develop natural products as potential therapeutics and advances in extraction and isolation techniques led to the development of 63% of the natural product-derived drugs from 1981–2006.

The 7-azaindole nucleus has proven to be an important model for working on its own synthesis and synthesizing its different analogues. In past indole derivatives have attracted the interest of scientists for their potential biological activities [3-13]. We have previously reported a number of novel azaindole compounds [14-16]. We report in this chapter the AChE inhibitory properties of the different compounds, and their mechanism of AChE inhibition.

Mechanisms of Action

The cholinergic theory of alzheimer's disease proposes that the selective loss of cholinergic neurons in AD results in a deficit of acetylcholine in specific regions of the brain that mediate learning and memory functions [17, 18]. The primary approach for treating AD has, therefore, aimed at raising the acetylcholine levels in the brain by using acetylcholinesterase inhibitors (AChEIs) such as donepezil, galantamine or rivastigmine [19]. On the other hand, calcium over load seems to be the main factor initiating the processes leading to cell death. Several lines of evidences showed that calcium dysfunction, involved in the pathogenesis of Alzheimer's disease [20]. Augments β amyloid-(Aβ) formation, [21] and tau hyper phosphorylation [22] calcium entry through the L-subtype of high- voltage activated calcium channels causes both calcium over load and mitochondrial disruption, which activate the apoptotic cascade and cell death [23]. Hence, blocking the entrance of calcium through calcium channels could be a good strategy to prevent cell death.

In U.S.A alone Alzheimer's disease related expenses surpass \$100 billion every year or \$ 195,000 per patient [24, 25]. Recently Alzheimer's disease affecting more than 35 million people through out the world [26]. Its two feature abnormalities are plaques deposits of a protein fragment called beta amyloids and tangle, twisted strands of another protein tau [27]. Alzheimer's disease can affect different people in different ways [28], but the most common indication pattern begins with gradually worsening difficulty in identification fresh information. AChE catalyzes the rapid hydrolysis of the neurotransmitter acetylcholine in the cholinergic synapses [29]. This enzyme is an attractive target for rational drug design and discovery of mechanism-based inhibitors for the treatment of many neurodegenerative disorders such as Alzheimer's disease and myasthenia gravis [30, 31]. the dual inhibitor of AChE and BChE, rivastigmine, showed potential therapeutic benefits in AD and related dementias [32, 33].

The active site of torpedo californica AChE (which shares more than 70% sequence homology with the human enzyme) is buried near the bottom of a 20A° deep narrow gorge that penetrates halfway into the enzyme and widens out close to its base. This gorge is lined by 14 conserved aromatic amino acids and hence frequently called the 'aromatic gorge' [34]. The recently solved crystal structure of the BChE enzyme [35] was shown to be very similar to that of torpedo californica AChE. However, several aromatic amino acids, lining the gorge of the AChE, have been replaced in BChE by hydrophobic analogues. Due to this, the BChE gorge becomes relatively larger and capable of accommodating bulkier and relatively nonpolar ligands. Since the discovery of physostigmine [36] as the first natural AChE inhibitor, only few other natural inhibitors have attracted the attention of neuropharmacologists [37]. A list of AChE- inhibitors are shown in Table **1**.

SYNTHESIS OF ACETYLCHOLINE (ACh)

Acetylcholine synthesis initiated with the reaction of choline with acetate Fig. (**1**). It is synthesized in neurons. Acetate is activated by an enzymes coenzyme A and it becomes Acetyl Co-enzyme A. The Acetyl Co-enzyme A and choline reacts and the reaction are catabolized by enzymes choline transferase. The Acetylcholine is taken up into synaptic vesicles by vascular transporter.

Acetylcholine must rapidly remove from the synapse if repolarization is occurring Fig. (**2**). The removal occurs by way of hydrolysis of Acetylcholine to chloine and acetate and the reaction is catalyzed by acetylcholine esterase [38].

Acetylcholine receptors have been divided into two main types on the basis of three pharmacological properties. Muscarinic receptors and Nicotinic receptors. Nicotinic acetylcholine receptors (nAChRs) are subtypes of acetylcholine-operated receptors and are members of the superfamily of ligand-gated ion channel receptors [39]. Pentameric combinations of multiple α and β subunits lead to the large number of nicotinic receptors in the brain, complicating their studies. The nicotinic acetylcholine receptors have been played a role in variety of physiological and pathological function. The large number of neuronal nicotinic acetylcholine receptors present at neurons throughout the CNS where they are involved in a number of processes connected to cognitive function learning and memory, motor control and analgesia [40], additionally in concentration, stimulation and sensory discernment. Areas of the brain classically considered centers of memory formation and storage are rich in nAChRs. Nicotine and other

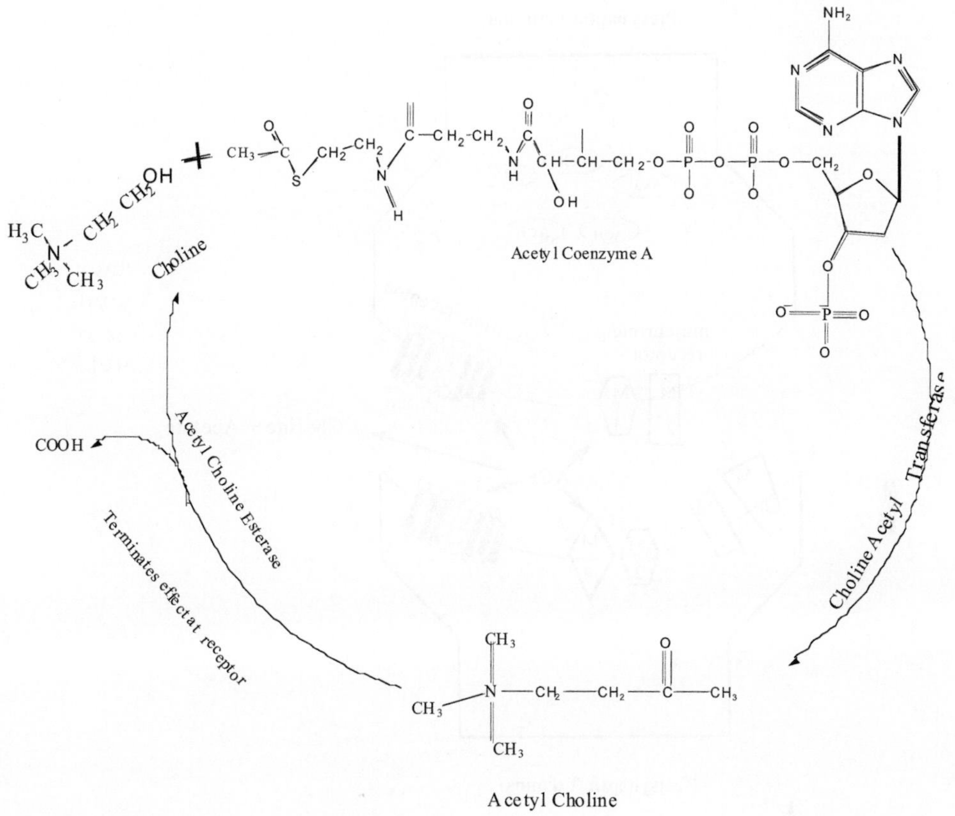

Figure 1: Biosynthesis of acetylcholine.

nAChR agonists enhance learning and memory in animals, while lesions in areas rich in nAChRs impair memory formation. Similarly significant to the overall input of nicotinic acetylcholine receptors to cholinergic neurotransmission are the roles of presynpatic and preterminal nicotinic acetylcholine receptors as autoreceptors and hetroreceptors regulating the synaptic discharge of acetylcholine and other neurotransmitters including dopamine, norepinephrine, serotonin, glutamate and γ-amino-butyric acid. Because of their modulatory influences on these neurotransmitter systems, neuronal nicotinic acetylcholine receptors have been projected as potential beneficial targets for the management of pain epilepsy and a wide variety of neuron degenerative and psychiatric syndrome such as AD, Parkinson's disease, Schizopherenia, anxiety, depression and the treatment of smoking cessation [41, 42].

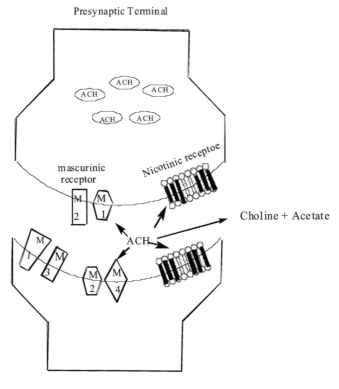

Figure 2: Cholinergic synapse.

As a part of our program to discover novel analogues of 7-azaindole (1H-pyrrolo [2,3-b] pyridine having useful biological activities, some derivatives have been synthesized and evaluated for their acetylcholinesterase and butyryl cholinesterase activity. Galanthamine was used as reference drug.

In the bio-screening of the compounds for AChE inhibition, (nitrophenacyl derivative) 7-[2-(4-nitrophenyl)-2-oxo-ethyl]-1H-pyrrolo[2,3-b]pyridinium;bromide and 7-[2-(3-nitrophenyl)-2-oxo-ethyl]-1H-pyrrolo[2,3-b]pyridinium bromide were found most active with almost same IC_{50} values 0.957 μM and 0.975 μM respectively. Napthyl derivative, 7-(2-naphthalen-2-yl-2-oxo-ethyl]-1H-pyrrolo [2,3-b]pyridinium bromide with IC_{50} value 1.34 μM was the next to the nitro derivative while adamantyl derivative 7-(2-adamantan-1-yl-2-oxo-ethyl]-1H-pyrrolo[2,3-b]pyridinium bromide (**3**) showed lowest activity among all the compound with IC_{50} value 11.60 μM.

Significant results of the synthesized compounds for AChE inhibitory activity can be explained by correlating the structures of synthesizes molecules of 7-azaindole with physostigmine.

Physostigmine is one of the clinically tested AChE inhibitor [159]. Azaindole derivatives are closely related to physostigmine and can be compared for their use in the treatment of alzheimer's disease. Fig. (**3**). showing active binding site regions [155] on the enzyme surface for acetylcholine, explaining possible three major binding areas like ionic, esteratic and possible hydrophobic region. Physostigmine (known potent ligand) and synthesized compounds showing the presence of same common essential features necessary for binding Figs. (**4-7**).

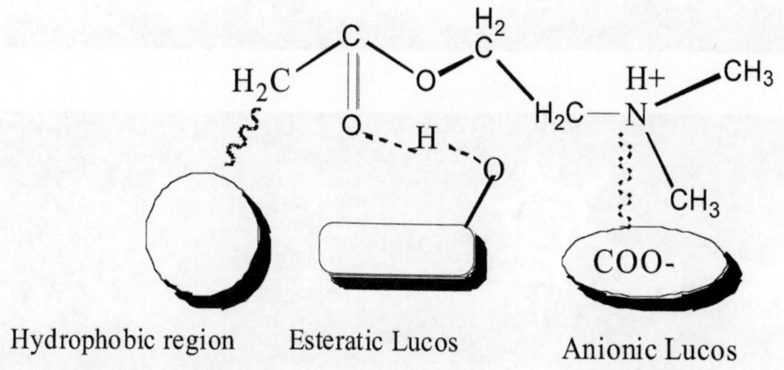

Hydrophobic region   Esteratic Lucos   Anionic Lucos

Figure 3: Acetylcholine with active binding site.

Hydrophobic region   Esteratic Lucos   Anionic Lucos

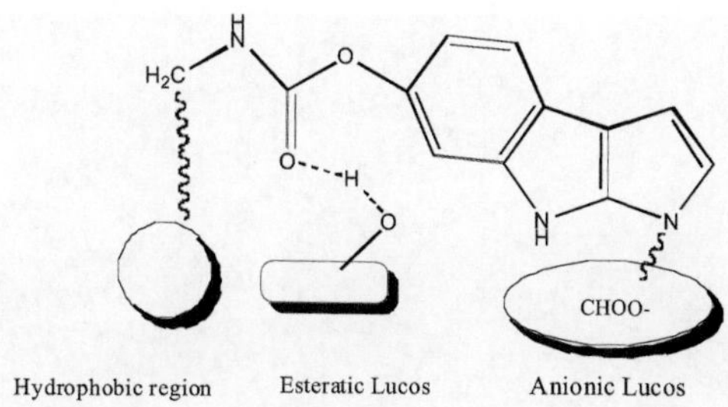

Figure 4: Physostigmine with active binding site.

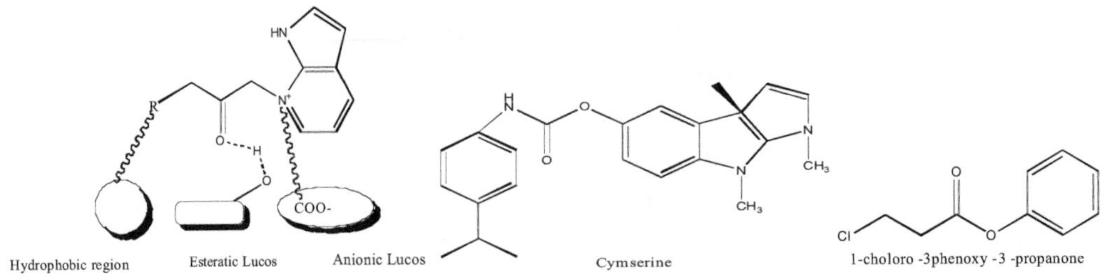

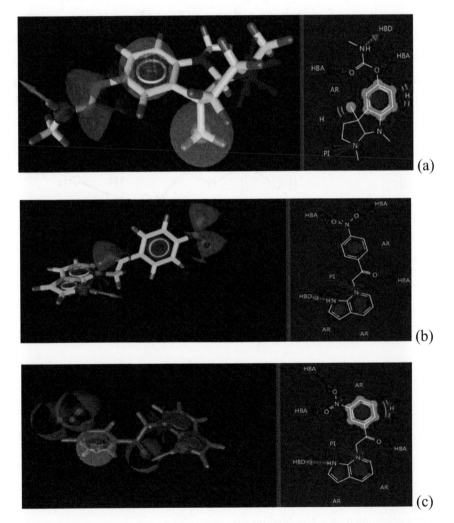

Figure 5: Synthesized derivative of 7 Azaindole. R = phenacyl, naphthyl and adamantly.

Figure 6: Showing 2D and 3D pictures Pharmacophoric region (**a**) Physostigmine (**b**) compound **4** (**c**) compound **5**.

Figure 7: showing 3D Picture showing Shared features in Physostigmine, compound **4** and compound **5**. Red = HBA :Hydrogen Bond Acceptor, Green = HBD: Hydrogen Bond Donor, Blue =AR: Aromatic Ring, Yellow=H: Hydrophobic region.

Fig. (**7**). displaying shared pharmacophoric regions of Physostigmine and compound 7-[2-(4-nitrophenyl)-2-oxo-ethyl]-1H-pyrrolo[2,3-b]pyridinium bromide and 7-[2-(3-nitrophenyl)-2-oxo-ethyl]-1H-pyrrolo[2,3-b]pyridinium bromide [149-152], which are important for binding with receptor. Excellent match of these regions can be taken as a good explanation of pronounced effect.

Fig. (**8**). representing the same binding regions in cymserine and derivative **2** and **3**. These 3 structures were checked to show the shared pharmacophoric regions showing excellent match of the active areas present and which may be involved in good binding with BChE enzymes.

Fig. (**9**) showing shared pharmacophoric regions of Cymserine and compound **2** and **3** which are imperative for binding with receptor. Admirable match of these areas can be taken as a good elucidation of marked effect.

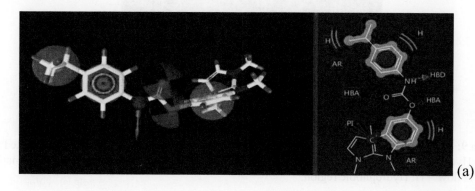

(a)

Fig. (8). contd….

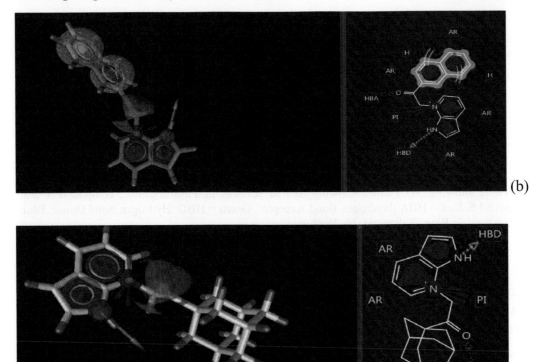

Figure 8: Showing 2D and 3D pictures Pharmacophoric region (**a**) Cymserine (**b**) compound **2** (**c**) compound **3**.

Figure 9: 3D Picture showing Shared features in cymserine, compound **2** and compound **3**. Red = HBA: Hydrogen Bond Acceptor, Green = HBD: Hydrogen Bond Donor, Blue =AR: Aromatic Ring, Yellow = H: Hydrophobic region.

Table 1:   **AChE- inhibitors**

Compound Structures with IUPAC Name	Acetylcholinesterase Activity	References
Tacrine: (1,2,3,4-tetrahydroacridin-9-amine) 	Tacrine is the potent centrally acting inhibitor of acetylcholinesterase. It is observed that when tacrine is given in combination with lecithin exhibited effect on memory performance.	[43-53]
Donipezil: (*RS*)-2-[(1-benzyl-4-piperidyl) methyl]-5,6-dimethoxy-2,3-dihydroinden-1-one 	Donepezil (Aricept) was accepted for the treatment in 1996 and is an indanone derivative of tacrine and physostigmine. It is selective inhibitor of acetyl cholinesterase in the CNS with the little effect on acetylcholine in peripheral tissues. Donepezil has long half life approximately 60 to 100 hours.	[54-58]
Rivastigmine: (*S*)-3-[1-(dimethylamino)ethyl]phenyl N-ethyl-N-methylcarbamate 	Rivastigmine is effective against acetylcholinestrase as well as butyrylcholinesterase and exhibits prolong inhibition of both the enzymes. Rivastigmine is also called as brain region selective inhibitor cholinesterase inhibitor. Its short plasma elimination half life is 1hour.	[59, 60]
Galantamine: (4a*S*,6*R*,8a*S*)- 5,6,9,10,11,12-hexahydro-3-methoxy-11-methyl-4aH-[1]benzofuro[3a,3,2-ef] benzazepin- 6-ol. 	Much research effort has been devoted to the development of anti-AD agents from natural sources. Galantamine isolated from bulbs and flowers of snowdrop *Galanthus woronowii* (Amaryllidaceae) has been approved by the FDA as an anti-AD medication due to its inhibitory effect against AChE. Galantamine, is in a class of medications called reversible acetyl cholinesterase inhibitors that is weak positive allosteric modular of nicotinic acetyl cholinesterase receptor. It is used to treat mild to moderate dementia caused by Alzheimer's disease.	[61-65]
Metrifonate: (*RS*)-Dimethyl (2,2,2-trichloro-1-hydroxyethyl)phosphonate 	Metrifonate is a long-acting irreversible ChEI that was originally used to treat schistosomiasis. Although there was a low risk of side effects with short-term use, long-term use caused respiratory paralysis and neuromuscular transmission dysfunction similar to a myasthenic crisis.	[66]
Older Cholinesterase Inhibitors		
Physostigmine: (Eserine) (3a*R*,8a*S*)-1,3a,8-trimethyl- *1*H,*2*H,*3*H,*3a*H,8H,*8a*H- pyrrolo [2,3-b] indol-5-ylN-methylcarbamate. 	The general and dominant pharmacology of physostigmine is due to a short-acting inhibition of the AChE. Physostigmine exerts a stereoselective inhibition by acting as a pseudosubstrate and transferring a carbamate residue to the enzyme's active site. Spontaneous hydrolysis regenerates the native enzyme and function.	[67-72]

Table 1. contd….

Compound Structures with IUPAC Name	Acetylcholinesterase Activity	References
Dichlorvos: 2,2-dichlorovinyl dimethyl phosphate	Dichlorvos is absorbed through all routes of exposure. Since it is an acetylcholinesterase inhibitor, its overdose symptoms are weakness, headache, tightness in chest, blurred vision, salivation, sweating, nausea, vomiting, diarrhea, and abdominal cramps.	[160]
Amiridine $C_{12}H_{16}N_2$	Amiridine has a positive effect on various aspects of the pathological process (e.g., memory disorders, speech disorders).	[161]
New Cholinesterase Inhibitors		
Phenserine: (-)-Eseroline phenylcarbamate, (-)-Phenserine, (3as,8ar)-1,3a,8-trimethyl-1,2,3,3a,8,8a-exahydropyrrolo[2,3-b]indol-5-yl phenylcarbamate	Phenserine is the phenylcarbamate derivation of physostigmine that has a dual effect: decreasing beta-amyloid precursor rotein and has a reversible AChE inhibition.	[73]
Tolserine	Tolserine (eptastigmine) only differs from phenserine at the 2'-methyl substitution on its phenylcarbamoyl moiety. Preclinical studies were initiated in 2000, and it was shown to be 200-fold more selective against AChE versus BuChE.	[74, 75]
NS2330 (Tesofensine) (1R,2R,3S)-3-(3,4-dichlorophenyl)-2-(ethoxymethyl)-8-methyl-8-azabicyclo[3.2.1]octane	Both *in vitro* and *in vivo* studies have shown that tesofensine inhibits the presynaptic uptake of the neurotransmitters serotonin, norepinephrine, and dopamine.	[76, 77]

Table 1. contd….

Compound Structures with IUPAC Name	Acetylcholinesterase Activity	References
Naturally Drive		
Huperzine: A (1R,9S,13E)- 1-Amino-13-ethylidene-11-methyl-6-azatricyclo[7.3.1.0] trideca- 2(7),3,10- trien- 5-one.	Huperzine A (HupA) is a Lycopodium alkaloid isolated from the Chinese medicinal herb. *Huperzia serrata* used for memory deficiency. *Huperzia serrata* is widely grown in China and Chinese medical tradition emphasize herbal remedies.	[78-80]
Huperzine: B 8,15-Didehydrolycodin-1(18H)-one;12-Methyl-2,3,4,4a,5,6-hexahydro-1H-5,10b-prop[1]eno-1,7-phenanthrolin-8(7H)-one.	Natural Huperzine B (HupB) is a Lycopodium alkaloid isolated from the Chinese medicinal herb *Huperzia serrata* which has been demonstrated as an effective and reversible inhibitor of AChE. HupB is less potent and selective than HupA, but it has higher therapeutic index and other positive benefits. HupB derivatives were created to be more potent than natural HupB. A novel series of 16-substituted derivatives were synthesized. 9i has the highest potency *in vitro*. The efficacy of 9i was more potent *in vitro* than HupA, galantamine, and rivastigmine, and it was equivalent to donepezil. 9c and 9i showed moderate neuroprotection against H_2O_2-induced neurotoxicity. However, given the ubiquitous availability of HupA and its higher potency, clinical development of HupB has been reportedly limited. Side effects would be expected to be similar in nature (GI) to HupA and other ChEIs.	[81]
Galangin: 3,5,7-trihydroxy-2-phenylchromen-4-one.	Guo *et al.* studied 21 different flavonoids for potential AChE inhibition properties in the brain *in vitro*. Flavonoids have been of great interest in AD research and treatment because of their free radical scavenging properties.	[82]
Cardanol: (3-[(E)-pentadec-8-enyl] derivatives).	De Paula *et al.* designed new AChEI from nonisoprenoid phenolic lipids (NIPLs) of *Anacardium occidentale*. Cardols, cardanols, anacardic acids, and methylcardols are the primary NIPL components of cashew nut-shell liquid and have been used to generate potential bioactive compounds. The derivates have structural, electrical, and hydrophobic properties that are relevant to recognition of AChE molecules.	[83]
Donepezil: $C_{24}H_{29}NO_3$ (RS)-2-[(1-benzyl-4-piperidyl)methyl]- 5,6-dimethoxy-2,3-dihydroinden-1-one.	AChE inhibitors are the only agents recommended for the treatment of cognitive decline in patients with mild to moderate alzheimers disease. Donepezil is more effective than placebo and is well tolerated in improving the major symptomd of this disease.	[84]

Table 1. contd....

Compound Structures with IUPAC Name	Acetylcholinesterase Activity	References
AP2238: 3-[4-(N-Benzyl-N-methylaminomethyl)phenyl]-6,7-dimethoxy-2H-1-benzopyran-2-one.	Rizzo *et al.* reports on a series of hybrids developed from donepezil and AP2238 in which the indanone core from donepezil is linked to the phenyl-N-methylbenzylamino moiety from AP2238. A derivate in which the phenyl-N-methylbenzylamino moiety from AP2238 was replaced by phenyl-N-ethylbenzylamino moiety from AP2243 and the idanone ring was replaced by a tetralone scaffold.	[85]
Hybrids		
Donepezil-Tacrine Hybrids	Camps *et al.* designed a novel series of donepezil-tacrine hybrids, which interact simultaneously at the peripheral, active, and midgorge binding sites of AChE [79-84].	[86-89]
Ladostigil: [(3R)-3-(prop-2-ynylamino)indan-5-yl]-N-propylcarbamate	Ladostigil is a novel anti-alzheimer's disease drug, with neuroprotective, multimodal brain-selective monoamine oxidase, and cholinesterase inhibitor properties.	[90, 91]
Aceglutamide: $C_7H_{12}N_2O_4$ (N^2-Acetylglutamine)	Aceglutamide is psychostimulant and nootropic used to improve memory and concentration. Chemically, it is the acetyl derivative of the amino acid L-glutamine.	[92]
Acetylcarnitine: ($C_9H_{17}NO_4$) *R*)-3-acetyloxy-4-trimethylammonio-butanoate.	Acetyl carnitine is best taken during a meal at 200 mg to 1000 mg, 2 to 3 times a day.	[93-103]
Aniracetam:($C_{12}H_{13}NO_3$)1-[(4-methoxybenzoyl)]-2-pyrrolidinone	Aniracetam is an ampakine and nootropic of the racetam chemical class purported to be considerably more potent than piracetam.	[104, 105]

Table 1. contd….

Compound Structures with IUPAC Name	Acetylcholinesterase Activity	References
Besipirdine ($C_{16}H_{17}N_3$)	Besipirdine is a nootropic drug developed for the treatment of alzheimer's disease acting as a blocker of voltage-sensitive sodium channels.	[106-109]
Bifemelane: ($C_{18}H_{23}NO$) *N*-methyl-4-[2-(phenylmethyl)phenoxy]butan-1-amine.	Bifemelane (Alnert, Celeport) is a pharmaceutical drug used in the treatment of senile dementia in Japan.	[110-114]
Choline alfoscerate ($C_8H_{20}NO_6P$) [(2*S*)-2,3-dihydroxypropyl]2-trimethylazaniumylethyl phosphate	L-Alpha Glycerylphosphorylcholine (Alpha GPC, choline alfoscerate) is a natural choline compound found in the brain and in milk. It is also a parasympathomimetic acetylcholine precursorwhich may have potential for the treatment of Alzheimer's disease and is used as a nootropic dietary supplement to enhance memory and cognition.	[115-117]
Domperidone 5-chloro-1-(1-[3-(2-oxo-2,3-dihydro-1*H*-benzo[*d*]imidazol-1-yl)propyl]piperidin-4-yl)-1*H*-benzo[*d*]imidazol-2(3*H*)-one	Domperidone blocks the action of dopamine. It has strong affinities for the D_2 and D_3 dopamine receptors, which are found in the chemoreceptor trigger zone, located just outside the blood brain barrier, which, among others, regulates nausea and vomiting (area postrema on the floor of the fourth ventricle and rhomboid fossa).	[118-120]
Fipexide ($C_{20}H_{21}ClN_2O_4$) 1-[4-(1,3-benzodioxol-5-ylmethyl)piperazin-1-yl]-2-(4-chlorophenoxy)ethanone	Fipexide: (Attentil, Vigilor) is a psychoactive drug of the piperazine chemical class which was developed in Italy in 1983. It was used as a nootropic drug in Italy and France, mainly for the treatment of senile dementia.	[121-123]
Idebenone: ($C_{19}H_{30}O_5$) 2-(10-hydroxydecyl)-5,6-dimethoxy-3-methyl-cyclohexa-2,5-diene-1,4-dione	Benzoquinone derivative, it is hypothesized to improve brain metabolism and protect cell membranes against lipid peroxidation. It also promotes secretion of nerve growth factor. Idebenone improved learning and memory in experiments with mice. Dosage: 90 mg/day.	[124]

Table 1. contd….

Compound Structures with IUPAC Name	Acetylcholinesterase Activity	References
Indeloxazine hydrochloride ($C_{14}H_{18}ClNO_2$) 2-(3*H*-inden-4-yloxymethyl)morpholine	Scientists studied the influence of indeloxazine hydrochloride (IH) upon photic driving responses (PDRs) elicited by a 5 Hz flickering dot pattern and red flicker stimuli. Comparisons were made from a total of 22 elderly patients. By oral IH administration, clinical improvement was found in 12 patients with cerebral arteriosclerosis (improved group).	[125]
Ispronicline: ($C_{14}H_{22}N_2O$) (2*S*,4*E*)-5-(5-isopropoxypyridin-3-yl)-*N*-methylpent-4-en-2-amine	Ispronicline (TC-1734, AZD-3480) is a drug which acts as a partial agonist at neural nicotinic acetylcholine receptors. It is subtype-selective, binding primarily to the α4β2 subtype. It has antidepressant, nootropic and neuroprotective effects, and is under development for the treatment of dementia and alzheimer's disease.	[126-130]
Leteprinim ($C_{15}H_{13}N_5O_4$) (Neotrofin, AIT-082) 4-[(3-(6-oxo-3,6-dihydro-9H-PURIN-9-YL) Propanoyl) benzoic acid	Leteprinim is a hypoxanthine derivative drug with neuroprotective and nootropic effects. It stimulates release of nerve growth factors and enhances survival of neurons in the brain.	[131-140]
Nefiracetam: ($C_{14}H_{18}N_2O_2$) N-(2,6-dimethylphenyl)-2-(2-oxopyrrolidin-1-yl)acetamide	**Nefiracetam** user patients in the study must have a caregiver and designated representative found to improved learning impairment and memory.	[139, 140]
Nizofenone: ($C_{21}H_{21}ClN_4O_3$) N-(2,6-dimethylphenyl)-2-(2-oxopyrrolidin-1-yl) acetamide	Nizofenone (Ekonal, Midafenone) is a neuroprotective drug which protects neurons from death following cerebral anoxia (interruption of oxygen supply to the brain). It might thus be useful in the treatment of acute neurological conditions such as stroke.	[141-144]

Table 1. contd….

Compound Structures with IUPAC Name	Acetylcholinesterase Activity	References
Oxiracetam (C₁₂H₂₂S₂N₂O₈) (*RS*)-2-(4-hydroxy-2-oxopyrrolidin-1-yl)acetamide	The efficacy and tolerability of the nootropic agent oxiracetam were compared in a single-blind, controlled, parallel study. The trial involved 22 men and 18 women with mild-to-moderate senile and presenile dementia of the alzheimer type.	[145]
Piracetam: (C₆H₁₀N₂O₂) *N*-(4-hydroxyphenyl)ethanamide *N*-(4-hydroxyphenyl)acetamide	Piracetam (sold under many brand names) (2-oxo-1-pyrrolidine acetamide) is a nootropic drug. it shares the same 2-oxo-pyrrolidone base structure with 2-oxo-pyrrolidine carboxylic acid (pyroglutamate). Piracetam is a cyclic derivative of GABA.	[146]
Posatirelin: (C₁₇H₂₈N₄O₄) (2*S*)-N-[(2*S*)-1-[(2*S*)-2-carbamoylpyrrolidin-1-yl]-4-methyl	Posatirelin (L-pyro-2-aminoadipyl-L-leucil-L-prolinamide) a new synthetic tripeptide with cholinergic, catecholaminergic and neurotrophic properties, was investigated in the treatment of Alzheimer's disease.	[147]
Pramiracetam: (C₁₄H₂₇N₃O₂) N-[2-(diisopropylamino)ethyl]-2-(2-oxopyrrolidin-1-yl)acetamide	Pramiracetam is a part of the class of drugs called nootropics or smart drugs. The term nootropic was originally invented by Dr. Giurgea during the 1970s. It is extracted from a greek word that means "acting on or towards the mind." These types of drugs are brain boosters and are able to promote the nervous system thus enhancing many key functions of the brain such as improving memory, intelligence and attention.	[147]
Propeniofyline: C₁₅H₂₂N₄O₃ 3-methyl-1-(5-oxohexyl)-7-propyl-3,7-dihydro-1*H*-purine-2,6-dione	Propentofylline Adenosine uptake/phosphodiesterase inhibitor (A neuroprotective agent).	[147]

Table 1. contd....

Compound Structures with IUPAC Name	Acetylcholinesterase Activity	References
Pyritinol 5,5'-[dithiobis(methylene)]bis[4-(hydroxymethyl)-2-methylpyridin-3-ol] ($C_{16}H_{20}N_2S_2O_4$)	Pyritinol is also known as pyrithioxine and pyridoxine disulfate the therapeutic effect of pyritinol is superior to placebo in patients with mild to moderate dementia of both degenerative and vascular etiology.	[148]
Rivastigmine: ($C_{14}H_{22}N_2O_2$) (S)-3-[1-(dimethylamino)ethyl]phenyl N-ethyl-N-methylcarbamate	Rivastigmine (sold under the trade name Exelon) is a parasympathomimetic or cholinergic agent for the treatment of mild to moderate dementia of the Alzheimer's type and dementia due to Parkinson's disease.	[149, 150]
Sabcomeline: ($C_{10}H_{15}N_2O$) (3R)-N-methoxyquinuclidine-3-carboximidoyl cyanide hydrochloride	Sabcomeline (Memric; SB-202,026) is a selective M_1 receptor partial agonist that was under development for the treatment of Alzheimer's disease http://en.wikipedia.org/wiki/Sabcomeline - cite_note-pmid9399977-0. It made it to phase III clinical trials before being discontinued due to poor results.	[151]
Sabeluzole: ($C_{22}H_{26}FN_3O_2S$) 1-[4-[1,3-benzothiazol-2-yl(methyl)amino]piperidin-1-yl]-3-(4-fluorophenoxy)propan-2-ol	A benzothiazol derivative Sabeluzole, shown to enhance memory function in animals and in normal volunteers. May slow clinical progression of Alzheimer's disease. It is thought to be a glutamate antagonist with modulatory effects on calcium channels. A more recent study proposes that it exerts neuroprotective effects by a novel mechanism of stabilizing the neuronal cytoskeleton.	[152, 153, 158]
Aza-indole derivatives		
7-(2-Naphthalen-2-yl-2-oxo-ethyl]-1H-pyrrolo[2,3-b]pyridinium;bromide **(1)**	Compound 1 showed Acetylcholinesterase inhibition ($IC_{50} \pm$ SEM [μM] 1.34 ± 0.05)	[154-157]

Table 1. contd....

Compound Structures with IUPAC Name	Acetylcholinesterase Activity	References
7-(2-adamantan-1-yl-2-oxo-ethyl]-1H-pyrrolo[2,3-b]pyridinium;bromide **(2)** 	Compound 2 showed Acetylcholinesterase inhibition ($IC_{50} \pm$ SEM [μM] 11.60 $\pm$ 0.173)	
7-[2-(4-nitrophenyl)-2-oxo-ethyl]-1H-pyrrolo[2,3-b]pyridinium,bromide **(3)** 	Compound 3 showed Acetylcholinesterase inhibition ($IC_{50} \pm$ SEM [μM] 0.957 $\pm$ 0.009)	
7-[2-(3-nitrophenyl)-2-oxo-ethyl]-1H-pyrrolo[2,3-b]pyridinium;bromide **(4)** 	Compound 4 showed Acetylcholinesterase inhibition($IC_{50} \pm$ SEM [μM] 0.975 $\pm$ 0.0518)	[154-157]
Compounds whose therapeutic potential in alzheimer disease is associated with their anti-inflammatory propperty		
Ibuprofin: (*RS*)-2-(4-(2-methylpropyl)phenyl) propanoic acid. 	Ibuprofin whose therapeutic potential in alzheimer disease is associated with their anti-inflammatory propperty. Ibuprofin is the first in a series of NSAIDs suggested for the therapy of AD.	[162]

Table 1. contd....

Compound Structures with IUPAC Name	Acetylcholinesterase Activity	References
Naproxen: (+)-(*S*)-2-(6-methoxynaphthalen-2-yl) propanoic acid	Naproxen is commonly used for the reduction of pain, fever, inflammation, and stiffness caused by conditions including migraine, osteoarthritis, kidney stones, rheumatoid arthritis, psoriatic arthritis, gout, ankylosing spondylitis, menstrual cramps, tendinitis, and bursitis. It is also used for the treatment of primary dysmenorrhea.	[163]
Rofecoxib: 4-(4-methylsulfonylphenyl)-3-phenyl-5H-furan-2-one.	Treatment of primary dysmenorrhea is under investigation now in phase II clinical trials in AD patients.	[164]
Dapsone:4-[(4-amino benzene)sulfonyl]aniline	Dapsone is an anti-infective agent. Anti-inflammatory effect of Dapsone may slow down the progression of AD.	[165]
Prednisone: 17,21-dihydroxypregna-1,4-diene-3,11,20-trione.	Pilot clinical trials demonstrated a low efficacy of prednisone in the treatment of AD patient.	[166]
Colchicine;N-[(7*S*)-1,2,3,10-tetramethoxy-9-oxo-5,6,7,9-tetrahydrobenzo[*a*]heptalen-7-yl]acetamide	Primary clinical trials with colchicine on AD patients demonstrated only a moderate efficacy, its clinical application is currently being questioned.	[167]
Hydroxychloroquine: (*RS*)-2-[{4-[(7-chloroquinolin-4-yl)amino]pentyl}(ethyl)amino] ethanol.	Clinical trials with this medicine (both alone and in combination with colchicine)on a limited number of AD patients have been recently initiated in the Netherlands.	[168]

Table 1. contd....

Compound Structures with IUPAC Name	Acetylcholinesterase Activity	References
GP1-1046. 3-pyridin-3-ylpropyl (2S)-1-(3,3-dimethyl-2-oxopentanoyl)pyrrolidine-2-carboxylate	The treatment of model animals with the compound GPI-1046, that structurally simulates a receptor binding domain of FK506, stimulated a partiall recovery of central cholinergic systems.	[169]
Compounds whose therapeutic potential in alzheimer disease is associated with their neurotrophic activity		
Propentophyllene: 3-methyl-1-(5-oxohexyl)-7-propyl-3,7-dihydro-1*H*-purine-2,6-dione	This compound has a broad spectrum of neuro-protective and cognition-stimulating effects.	[170]
AIT-082: $C_{15}H_{13}N_5O_4$ 4-[3-(6-oxo-3H-purin-9-yl)propanoyl amino]benzoic acid; 4-{[3-(6-oxo-3,6-dihydro-9H-purin-9-yl)propanoyl]amino}benzoic acid,	AIT-082 is currently in phase IIb/III clinical trials.	[171]
Citicoline 5'-*O*-[hydroxy({hydroxy[2-(trimethylammonio)ethoxy]phosphoryl}oxy)phosphoryl]cytidine	Citicoline may improve memory *via* its neutrophic effect. It may also regulate a blood circulation in brain vessels and stabilize mitochondria functions by the membrane stabilization. Clinical trials demonstrated a positive influence of Citicoline on cognitive functions in elderly people.	[172, 173]
Compounds affecting mitochondrial functions and ligands of GABA and monoaminergic systems as potential anti-alzheimer agents		
FCCP	It is phenyl hydrazon derivative.	[174]

Table 1. contd....

Compound Structures with IUPAC Name	Acetylcholinesterase Activity	References
Dantrolene: 1-{[5-(4-nitrophenyl)-2-furyl]methylidene amino} imidazolidine-2,4-dione.	Dantrolene, used in clinical practice as a muscle relaxant, which displayed an antiproptotic protective effect.	[175]
Valproic acid, (2-Propylpentanoic acid).	Currently Valproic acid is in phase III clinical trial.	[176]
CGP36742: 3-aminopropyl)(n-butyl)phosphinic acid.	CGP36742 displays pronounced cognition-stimulating properties in AD models.	[177]
Suritozol(C10H10FN3S): 3-(m-flourofenil)-1,4-dimetil-Δ^2-1,2,4-triazolina-5-tiona.	Suritozol agent was studied in clinical trials on AD patients.	[178]
Piracetam: (2-oxo-1-pyrrolidine acetamide).	Because of the ability of nootropic compounds, such as piracetam, to enhance cognitive functions in animal models of dementia, these agents may also be suggested as potential drugs for the treatment of AD.	[179]
Aniracetam: 1-[(4-methoxybenzoyl)]- 2-pyrrolidinone.	Because of the ability of nootropic compounds, such as Aniracetam, to enhance cognitive functions in animal models of dementia, these agents may also be suggested as potential drugs for the treatment of AD.	[179, 180]
Deprenyl (siligiline): (R)-N-methyl-N-(1-phenylpropan-2-yl)prop-1-yn-3-amine.	During the past decade, controversial data were obtained about deprenyl in AD treatment. It was found that deprenyl improved memory and cognitive functions of experimental animals.	[181, 182]

Table 1. contd….

Compound Structures with IUPAC Name	Acetylcholinesterase Activity	References
Lazabemide: *(N*-(2-aminoethyl)-5-chloro-pyridine-2-carboxamide).	Lazabemide has been proposed as a potential agent for the prevention of neurodegenerative processes in AD-type dementia.	[183]
8-OH-DPAT. 8-Hydroxy-2-(di-n-propylamino)tetralin.	8-OH-DPAT protects cells against an excitotoxic effect of NMDA and stimulates an increase of extracellular acetyl choline in the brain (in experimental animal models).	[184, 185]
Ondansetron: (*RS*)-9-methyl-3-[(2-methyl-1*H*-imidazol-1-yl)methyl]-2,3-dihydro-1*H*-carbazol-4(9*H*)-one.	Ondansetron potentiate the release of ACh in the hippocampus and cortex. However, according to recent results of double-blind clinical trials, as therapeutic effect of Ondansetron is insignificant.	[186]
Compounds whose therapeutic potential in alzheimer disease is associated with their anti-oxidant activity		
α-Tocopherol: (2*R*)-2,5,7,8-Tetramethyl-2-[(4R,8R)-(4,8,12-trimethyltridecyl)]-6-chromanol.	Neuroprotective properties of the vitamin E were studied in various experimental models of AD. α-Tocopherol tested in clinical trials on AD patients. It was revealed that vitamen E might slow down the development of AD in patients with a moderate AD.	[187, 188]
Idebenone:(2-(10-hydroxydecyl)-5,6-dimethoxy-3-methyl-cyclohexa-2,5-diene-1,4-dione).	It was revealed that synthetic analogue, Idebenone vitamen E might slow down the development of AD in patients with a moderate AD.	[189]
Raxofelast: 2-(5-Acetoxy-4,6,7-trimethyl-2,3-dihydrobenzofuran-2-yl)acetic acid.	Raxofelast (IRF10016) and MDL-74180DA: These synthetic analogues of the vitamin E are patented as effective anti-oxidants are now in preclinical studies.	[190]

Table 1. contd….

Compound Structures with IUPAC Name	Acetylcholinesterase Activity	References
MDL-74180DA: (±)-3-(4-Methyl-1-piperazinylmethyl)-2,2,4,6,7-pentamethyl-2,3-dihydro-1-benzofuran-5-ol dihydrochloride monohydrate.	Raxofelast (IRF10016) and MDL-74180DA: These synthetic analogues of the vitamin E are patented as effective anti-oxidants are now in preclinical studies.	[190]
17a-Estradiol: (17β)-estra-1,3,5(10)-triene-3,17-diol.	Recent studies demonstrated a low efficacy of the estrogen replacement therapy of mild to moderate AD. Estrogen is currently on phase III clinical trials as a potential neuroprotector that may delay the onset or the risk of AD in postmenopausal women.	[191]
17b-Estradiol.	The diversity of possible mechanisms of the effect of steroid hormone on nervous cells the homeostasis of nervous cells should be taken into account in the analysis of the neuroprotective properties of these compounds in the development of AD type pathology.	[192, 193]
Testosterone: (8R,9S,10R,13S,14S,17S)- 17-hydroxy-10,13-dimethyl- 1,2,6,7,8,9,11,12,14,15,16,17-dodecahydrocyclopenta[a]phenanthren-3-one.	Androgens are becoming a center of attention as potential anti-oxidants and neuroprotective agents.	[194-196]
J861, J811.	Synthetic drugs such as J811 display a strong anti-oxidant activity in experiments and an ability to stimulate cognitive functions on a neurotoxicological animal model of AD.	[197, 198]

Table 1. contd….

Compound Structures with IUPAC Name	Acetylcholinesterase Activity	References
Celastrol: 3-hydroxy-9β,13α-dimethyl-2-oxo-24,25,26-trinoroleana-1(10),3,5,7-tetraen-29-oic acid	Celastrol also displays an anti-inflammatory activity. This drug significantly stimulates learning, memory, and a psychomotor activity of experimental animals with a model dementia. Celastrol was recommended for clinical trials on AD patients.	[198]

CONCLUSIONS

In conclusion, cholinesterase inhibitors are the only pharmacological agents proved to be effective for the treatment of Alzheimer's disease. The importance of acetylcholine esterase is emphasized and correlated to the reported anti alzheimer compounds. One the active esterase is physostigmine, a natural product from physostigma venenosum. Various derivatives of 7-aza indole are synthesized and their efficacy is determined in order to get more effective molecule. With the inclusion of available drugs and active compounds as acetyl cholinesterase inhibitor. It can be conducted that a novel drug will be discovered in the year to come.

ACKNOWLEDGEMENTS

During the course of research work financial support is acknowledged by PSF, PCSIR, HEC, Ministry of health, ICCBS. Services of the Library ICCBS is greatly acknowledged, which was established by Prof. Atta-ur-Rahman, (F.R.S.).

CONFLICT OF INTEREST

The authors confirm that this chapter contents have no conflict of interest.

PATIENT CONFLICT

These drugs have not been tested on any patient so there is no question of any consent.

ABBREVIATIONS

ACh	=	acetylcholine
AChE	=	Acetylcholinesterase
AChEI	=	Acetylcholinesterase Inhibitors
AD	=	Alzheimer's disease
ADP	=	Adenine dinucleotide phosphate
C°A	=	Acetyl-coenzyme A
CDDO	=	2-cyano-3,12-dioxoolean-1,9-dien-28-oic acid
CDDO-Me	=	C-28 methyl ester of 2-cyano-3,12-dioxoolean-1,9-dien-28-oic acid
CNS	=	Central Nervous System
COX	=	Cyclooxygenase
Glu	=	Glutamine
Gly	=	Glycine
GSH	=	Glutathione
GSHPx	=	Glutathione peroxidase
GSHPx	=	Glutathione peroxidases
Hb	=	Hemoglobin
iNOS	=	inducible nitric oxide synthase
Loo.	=	Lipid peroxy radical

LOOH   =   Lipid hydroperoxide

LOX   =   Lipoxygenase

5-LOX   =   5-Lipoxygenase

LPS   =   Lipopolysaccharide

MMPs   =   Metalloproteinases

nAChRs   =   Nicotinic acetylcholine receptors

PG   =   Prostaglandin

ROS   =   Reactive oxygen species

SOD   =   Superoxide dismutase

TNF   =   Tumor necrosis factor

TNF   =   Tumor necrosis factor alpha

REFERENCES

[1]   Wright CL., Geula C., Mesulam MM., Protease inhibitors and indolamines selectively inhibit cholinesterase in the histopathologic structures of Alzheimer's disease. *Ann NY Acad Sci.*, **1993**, *695*, 65-88.

[2]   Giacobini E., Therapy of Alzheimer disease: Symptomatic or neuroprotective ? *J. Neural. Transm. Suppl.*, **1994**, *43*, 211-217.

[3]   Hugon B., Anizon F., Bailly C., Golsteyn RM, Pierré A., Léonce S., Hickman J., Pfeiffer B., Prudhomme M., Synthesis and biological activities of isogranulatimide analogues. *Bioorg Med. Chem.*, **2007**, *15*(17), 5965-5980.

[4]   Marminon C., Pierre A., Pfeiffer B., Perez V., Leonce S., Joubert A., Bailly C., Renard P., Hickman J., Prudhomme M., Syntheses and Antiproliferative Activities of 7-Azarebeccamycin Analogues Bearing One 7-Azaindole Moiety. *J. Med. Chem.*, **2003**, *46*(4), 609-622.

[5]   Kelly TA., McNeil DW., Rose JM., David E., Shih CK., and Grob PM., Novel Non-Nucleoside Inhibitors of Human Immunodeficiency Virus Type 1 Reverse Transcriptase. 6. 2-Indol-3-yl- and 2-azaindol-3- ldipyridodiazepinones. *J. Med. Chem.*, **1997**, *40*(15), 2430-2433.

[6]   Anthony JV., Synthesis of 1-pchlorobenzyl- 7-azaindole-3-.alpha.-piperidyl methanol as a potential antimalarial agent. *J. Med. Chem.*, **1972**, *15*(2), 149-152.

[7]   Song JJ., Reeves JT., Gallou F., Tan Z., Yee NK., Senanayake CH., Organometallic methods for the synthesis and functionalization of azaindoles. *Chem. Soc. Rev.,* **2007**, *36*(7), 1120-1132.

[8]   Lebouvier N., Pagniez F., Duflos M., LePape P., Na YM., Le Baut G and Le Borgne M., Synthesis and antifungal activities of new fluconazole analogues with azaheterocycle moiety, *Bioorg Med. Chem. Lett.,* **2007**, *17*(13), 3686-3689.

[9]   Guillard J., Decrop M., Gallay N., Espanel C., Boissier E., Herault O., Viaud-Massuard MC., Synthesis and biological evaluation of 7-azaindole derivatives, synthetic cytokinin analogues. *Bioorg. Med. Chem. Lett.,* **2007**, *17*(7), 1934-1937.

[10]   Trejo A., Arzeno H., Browner M., Chanda S., Cheng S., Comer DD., Dalrymple SA., Dunten P., Lafargue J., Lovejoy B., Freire-Moar J., Lim J, Mcintosh J., Miller J., Papp E., Reuter D., Roberts R., Sanpablo F., Saunders J., Song K., Villasenor A., Warren SD., Welch M., Weller P., Whiteley PE., Zeng L., Goldstein DM., *J. Med. Chem.*, **2003**, *23*, 46(22), 4702-13.

[11]   Marie CV., Patricia J., Marie-Laure B., Acylation of oxazolo[4,5-b]pyridine-2(3H)-ones, 2-phenyloxazolo[4,5-b]pyridines and pyrrolo-[2,3-b]pyridine-2(2H)-ones, Tetrahedron, **1997**, 53, 5159-61.

[12]   Shen TY., Ellis RL., Windholz TB., Matzuk AR., Rosegay A., Lucas S., Holly FW., Wilson AN., Non steroid anti-inflammatory agents, J. Am. Chem. Soc., 1963, 85, 488-492.

[13]   Matsuoka, Hiroharu, Maruyama, Noriaki, PCT., Int. Appl. W097 30,030 (Cl. C07 D209/08), **1996**, p.79-83.

[14]   Saify ZS., Synthesis of new carbonic anhydrase inhibitor. Pak. J. Pharm., 1982, 1(1), 83-88.

[15]   Saify ZS., Haider SM., Ahmed M., Saeed M., Khan A., Siddiqui BS., Pak. *J. Sci. & Indust. Res.,* **1994**, *37*(10), 439-441.

[16]   Saify ZS., Mozzam SM., Nisa M., Khan SA., Ahmed A., Haider S., Aryne A., Khanam M., Arshad N., Ghani M., Pak. *J. Sci. Ind. Res.*, **2009**, *52*(1), 1-7.

[17]   Scarpini E., Scheltens P., Feldman H., Treatment of Alzheimer's disease: current status and new perspectives. *Lancet Neurol*, **2003**, *2*, 539-542.

[18]   Talesa VN., Acetylcholinesterase in Alzheimer's disease, *Mech. Ageing Dev.*, **2001**, *122*, 1961-1965.

[19]   Lahiri DK., Rogers JT., Greig NH., Sambamurti K., Rationale for the development of cholinesterase inhibitors as anti-Alzheimer agents. *Curr. Pharm. Des*., **2004**, *10*, 3111-3116.

[20]   Selkoe DJ. Biochemistry of altered brain proteins in Alzheimer's disease. *Annu. Rev. Neurosci*., **1989**, *12*, 463-469.

[21]   Kruman I., Guo Q., Mattson MP., Calcium and reactive oxygen species mediate staurosporine-induced mitochondrial dysfunction and apoptosis in PC12 cells, *J. Neurosci. Res*., **1998**, *51*, 293-298.

[22]   Mattson MP., Cheng B., Davis D., Bryant K., Lieberburg I., Rydel RF., p-Amyloid peptides destabilize calcium Homeostasis and render human cortical neurons vulnerable to excitotoxicity, *J. Neurosci*., **1992**, *12*, 376-379.

[23]   Cano Abad MF., Villarroya M., García AG., Gabilán NH., López MG., Calcium entry through L-type calcium channels causes mitochondrial disruption and chromaffin cell death, *J. Biol. Chem*., **2001**, *276*, 39695.

[24]   Scott L., Goa K., Galantamine: A Review of its Use in Alzheimers Disease. *Drugs*, **2000**, *60*, 1095-1122.

[25]   Zarotosky V., Sramek JJ., Cutler NR., Galantamine hydrobromide: an agent for Alzheimer's disease, *American Journal of Health-System Pharmacists,* **2003**, *60*, 446-452.

[26]   Mount C., Downton. Alzheimer disease: progress or profit? *Nat. Med;* **2006**, *12*, 780-786.

[27]   Wang H., *Journal of Biological Chemistry*, **2003**, *278*, 5626-5632.

[28]   *www.alzheimers.org.uk* (August 2007). Retrieved on. **2008**, 02-21.

[29]   Lester HA., The response to acetylcholine, *Sci Am.*, **1977**, *236*, 100-106.

[30]   Taylor P., In: Gilman AG., Rall TW., Nies AS., Taylor P., editors. The pharmacologial basis of therapeutics. *New York, Pergamon*, **1990**, p 131-136.

[31]   Yu Q., Holloway HW., Utsuki T., Brossi A., Greig NH., Synthesis of novel phenserine-based-selective inhibitors of butyrylcholinesterase for Alzheimer's disease, *J. Med., Chem.*, **1999**, *42*, 1855-1859.

[32]   Greig NH., Lahiri DK., Sambamurti K., Butyrylcholinesterase: an important new target in Alzheimer's disease therapy, *Int Psychogeriatr*, **2002**, 14*(*Suppl 1), 77-81.

[33]   Standridge JB., *Clin. Ther,* **2004**, *26*, 615-619.

[34]   Sussman JL., Harel M., Frolow F., Oefner C., Goldman A., Toker L., Silman I., Atomic structure of cetylcholinesterase from Torpedo californica: a prototypic acetylcholine-binding protein, *Science*, **1991**, *253*, 872-879.

[35]   Nicolet Y., Lockridge O., Masson P., Fontecilla-Camps JC., Nachon FJ., A Secondary Isotope Effect Study of Equine Serum Butyrylcholinesterase-Catalyzed Hydrolysis of Acetylthiocholine, *Biol. Chem.*, **2003**, *278*, 41141-41146.

[36]   Kaur J., Zhang MQ., *Med. Chem.*, **2000**, *7*, 273-278.

[37]   Greenblatt HM., Kryger G., Lewis T., Silman I., Sussman JL., *FEBS Letters*, **1999**, *463*, 321-326.

[38]   Review of Medical Physiology by William F Ganong (22nd edition).

[39]   Romanelli D., Manetti MN., Scapecchi D., Borea S., Dei P., Bartolini AS., Ghelardini A., Gualtieri C., Guandalini F., Varani LK., *J. Med. Chem.*, **2001**, *44*, 3946-3949.

[40]   Jensen E., Frolund AA., Liljefors BT., *J. Med. Chem.*, **2005**, *44*, 4750-4757.

[41]   Coe F., Brooks JW., Wirtz PR., Bashore MC., Bianco CG., Veterlino KE., Arnold MG., Lebel EP., Fox LA., Tingley CB., Schuzl FD., Davis DW., Sands TI., Mansbach SB., Rolema, RS., 'Nneill, HO., Bioorg. Med. Chem. Lett., 2005,15, 2974-2976.

[42]   Cohen G., Bergis C., Galli OE., Jegham F., Biton SS., Leonardon B., Avenet J., Sgard P., Besnard F., Graham F., Coste D., Oblin A., Scation AB., J. Pharmacol. Exp. Ther., 2003, 306, 407-410.

[43]   Freeman SE., Dawson RM., Tacrine A Pharmacological. Review, Prog .Neurobiol, 1991, 36, 257-277.

[44]   Chatellier G., Lacomblez L., Tacrine (tetrahydroaminoacridine; THA) and Lecithine in senile dementia of the Alzheimer's type: A multicentre trail. Groupe Francais d la tetrahydroaminoacridine, BMJ,1990, 300, 495-499.

[45]   Eagger SA., Levy R., Sahakian Lancet, 1991, 337, 889-894.

[46]   Watkines PB., Zimmerman HJ., Knapp MJ., Gracon SI., Lewis KW., J. Am. Med. Assoc.,1994, 271, 992-995.

[47]   Thomson T., Zendeh B., Fischer JP., Kewtz H., Biochem. Pharmachol, 1991, 41, 139-144.

[48]   Tumiatti V., Minarini A., Bolognesi ML., Milelli A., Rosini M., Melchiorre C., "Tacrine derivatives and Alzheimer's disease," Current Medicinal Chemistry, 2010, 17(17), 1825-1838.

[49] Fern´andez-Bachiller MI., Erez CP., Gonz´alez-Mu˜noz, GC., "Novel tacrine-8-hydroxyquinoline hybrids as multifunctional agents for the treatment of Alzheimers disease, with neuroprotective, cholinergic, antioxidant, and copper complexing properties," Journal of Medicinal Chemistry, 2010, 53(13), 4927-4937.

[50] Tacrine. Drugs in Clinical Trials. Alzheimer Research Forum, http://www.alzforum. org/drg/drc/detail.asp?id=90.

[51] Kaul PN., Are drugs targeted at Alzheimer's disease useful? 1. Useful for what? J. Pharm. Pharmacol., 1962, 14, 243-247.

[52] Qizilbash N., Whitehead A., Higgins J., "Cholinesterase inhibition for Alzheimer disease: a meta-analysis of the tacrine trials". Journal of the American Medical Association, 1998, 280 (20), 1777-82.

[53] Rang HP., Dale MM., Ritter JM., Moore PK., Pharmacology (5th ed. ed.). Edinburgh: Churchill Livingstone, 2003, ISBN 978-0-443-07145-4.

[54] Shigeta M., Homma A., CNS Drug Review., 2001, 7, 353-368.

[55] Rogers SL., Doody RS., Mohs RC., Friedhoff LT., "Donepezil improves cognition and global function in Alzheimer disease: a 15-week, double-blind, placebo- controlled study," Archives of Internal Medicine, 1998, 158, no. 9, 1021-1031.

[56] Jacobson SA., Sabbagh MN., "Donepezil: potential neuroprotective and disease-modifying effects," Expert Opinion on DrugMetabolism and Toxicology, 2008, 4 (10), 1363-1369.

[57] Petersen RC., Thomas RG., Grundman M., "Vitamin E and donepezil for the treatment of mild cognitive impairment," The New England Journal of Medicine, 2005, 352(23), 2379-2388.

[58] Salloway S., Ferris S., Kluger A., "Efficacy of donepezil in mild cognitive impairment: a randomized placebo-controlled trial," Neurology, 2004, 63(4), 651-657.

[59] Gottwald MD., Rozanski RI., Exp. Opin. Investing. Drugs, 1999, 8, 1673-1682.

[60] Corey-Bloom J., Anand R., Veach J., "A randomized trial evaluating the efficacy and safety of ENA 713 (rivastigmine tartrate), a new acetylcholinesterase inhibitor, in patients with mild to moderately severe Alzheimer's disease," International Journal of Geriatric Psychopharmacology, 1998, 1, 2, 55-65,.

[61] Lilienfeld S., Galantamine--a novel cholinergic drug with a unique dual mode of action for the treatment of patients with Alzheimer's disease. CNS Drug Rev., 2002, 8,159-176.

[62] Raskind MA., Peskind ER., Truyen L., Kershaw P., Damaraju CV., The cognitive benefits of galantamine are sustained for at least 36 months: a long-term extension trial. Arch Neurol., 2004, 61(2), 252-6.

[63] Tariot PN., Solomon PR., Morris JC., Kershaw P., Lilienfeld S., Ding C., "A 5-month, randomized, placebocontrolled trial of galantamine in AD," Neurology, 2000, 54, 12, 2269-2276.

[64] Proskurnina NF., Yakovleva AP., Galanthamine in Alzheimer's Disease, J. Gen. Chem., 1952, 22, 1899-1902.

[65] Giacobini E., Long-term stabilizing effect of cholinesterase inhibitors in Alzheimer treatment. Rev Neurol, 2001, 157(10), 31-37.

[66] Opez-Arrieta JML.,´ and Schneider L., Metrifonate for Alzheimer's disease," Cochrane Database of Systematic Reviews, no. 2, Article ID CD003155, 2006, 1-40.

[67] Orhan GI., Orhan NO¨ ztekin-Subutay F., Ak., ener BS., "Contemporary anticholinesterase pharmaceuticals of natural origin and their synthetic analogues for the treatment of Alzheimer's disease," Recent Patents on CNS Drug Discovery, 2009, 4(1), 43-51.

[68]   Bartus RT., Dean RL, Beer B., Lippa AS., The cholinergic hypothesis of geriatric memory dysfunction, Science, 1997, 217,408-417.

[69]   Deutsch JA., The cholinergic synapse and the site of memory. Science, 1971, 174, 788-794.

[70]   Deutsch JA., The cholinergic synapse and the site of memory. In: Deutsch J, ed. *Physiological basis of memory.* New York: Academic Press, 1983, 367-386.

[71]   Dumery V., Derer P., Blozovski D., Enhancement of passive avoidance learning through small doses of intra-amygdaloid physostigmine in the young rat. Its relation to the development of acetylcholinesterase. Dev Psychobiol, 1988, 21, 553-565.

[72]   Murray A., Cottrell DF., Woodman MP., Cholinergic activity of intestinal muscle *in vitro* taken from horses with and without equine grass sickness. Vet Res Comm, 1994, 18,199-207.

[73]   Steinberg GM., Mednick ML., Maddox J., Rice R., J. Med. Chem. Soc., Chem., 1975, 18, 1056-1060.

[74]   Orhan G., Orhan I., Öztekin-Subutay N., Ak F., Ş ener B., "Contemporary anticholinesterase pharmaceuticals of natural origin and their synthetic analogues for the treatment of Alzheimer's disease," Recent Patents on CNS Drug Discovery, 2009, 4(1), 43-51.

[75]   Zhan ZJ., Bian HL., Wang JW., Shan WG., "Synthesis of physostigmine analogues and evaluation of their anticholinesterase activities," Bioorganic and Medicinal Chemistry Letters, 2010, 20, 5, 1532-1534.

[76]   NS2330. Drugs in Clinical Trials. Alzheimer Research Forum, http://www.alzforum.org/ drg/drc/detail.asp?id=83.

[77]   Lehr T., Staab A., Tillmann C., "Population pharmacokinetic modelling of NS2330 (tesofensine) and its major metabolite in patients with Alzheimer's disease," British Journal of Clinical Pharmacology, 2007, 64, 1, 36-48.

[78]   Wang BS., Wang H., Wei ZH., Song YY., Zhang L., Chen, HZ., "Efficacy and safety of natural acetylcholinesterase inhibitor huperzine A in the treatment of Alzheimer's disease: an updatedmeta-analysis," Journal of Neural Transmission, 2009, 116, 4, 457-465.

[79]   Bai D., "Development of huperzine A and B for treatment of Alzheimer's disease," Pure and Applied Chemistry, 2007, 79, 4, pp. 469-479.

[80]   Sabbagh MN., "Drug development for Alzheimer's disease: where are we now and where are we headed?" American Journal Geriatric Pharmacotherapy, 2009, 7, 3, 167-185.

[81]   Shi YF., Zhang HY., Wang W., "Novel 16-substituted bifunctional derivatives of huperzine B: multifunctional cholinesterase inhibitors," Acta Pharmacologica Sinica, 2009, 30, 8, 1195-1203.

[82]   Guo AJ., YXie HQ., Choi RCY., "Galangin, a flavonol derived from Rhizoma Alpiniae Officinarum, inhibits acetylcholinesterase activity *in vitro*," Chemico-Biological Interactions, 2010, vol. 187, no. 1-3, pp. 246-248.

[83]   de Paula AN., Martins JBL., dos Santos ML., "New potential AChE inhibitor candidates," European Journal of Medicinal Chemistry, 2009, 44, 9, 3754-3759.

[84]   Zheng H., Youdim MBH., Fridkin M., "Site-activated chelators targeting acetylcholinesterase and monoamine oxidase for Alzheimer's therapy," ACS Chemical Biology, 2010, 5(6), 603-610.

[85]   Rizzo S., Bartolini M., Ceccarini L., "Targeting Alzheimer's disease: novel indanone hybrids bearing a pharmacophoric fragment of AP2238," Bioorganic and Medicinal Chemistry, 2010, 18 (5), 1749-1760.

[86]   Camps P., Formosa X., Galdeano C., "Novel donepezilbased inhibitors of acetyl-and butyrylcholin butyrylcholinesterase and acetylcholinesterase-induced -amylo amyloid aggregation," Journal of Medicinal Chemistry, 2008, 51, 12, 3588-3598.

[87]   Fernández-Bachiller MI., Pérez C., González-Muñoz GC., "Novel tacrine-8-hydroxyquinoline hybrids as multifunctional agents for the treatment of Alzheimers disease, with neuroprotective, cholinergic, antioxidant, and copper-complexing properties," Journal of Medicinal Chemistry, 2010, 53(13), 4927-4937.

[88]   Ali MA., Yar MS., Hasan MZ., Ahsan MJ., Pandian S., "Design, synthesis and evaluation of novel 5,6-dimethoxy-1- oxo-2,3-dihydro-1H-2-indenyl-3,4-substituted phenyl methanone analogues," Bioorganic andMedicinal Chemistry Letters, 2009, 19 (17), pp. 5075-5077.

[89]   Korabecny J., Musilek K., Holas O., "Synthesis and *in vitro* evaluation of N-alkyl-7-methoxytacrine hydrochlorides as potential cholinesterase inhibitors in Alzheimer disease,"Bioorganic and Medicinal Chemistry Letters, 2010, 20. (20), 6093-6095.

[90]   Weinreb O., Amit T., Bar-Am O., Youdim MBH., "A novel anti-Alzheimer's disease drug, ladostigil. neuroprotective, multimodal brain-selective monoamine oxidase and cholinesterase inhibitor," International Review of Neurobiology, 2011, 100, 191-215.

[91]   Weinstock M., Luques L., Poltyrev T., Bejar C., Shoham S., "Ladostigil prevents age related glial activation and spatial memory deficits in rats," Neurobiology of Aging, 2011, 32, 6, 1069-1078.

[92]   Macdonald, F, ed., Dictionary of pharmacological agents. 1997, p. A-00018.

[93]   Barhwal K., Hota SK., Jain V., Prasad D., Singh SB., Ilavazhagan G., "Acetyl-l-carnitine (ALCAR) prevents hypobaric hypoxia-induced spatial memory impairment through extracellular related kinase-mediated nuclear factor erythroid 2-related factor 2 phosphorylation". Neuroscience, 2009,161 (2): 501-14.

[94]   Al-Majed AA., Sayed-Ahmed MM., Al-Omar FA., Al-Yahya AA., Aleisa AM., Al-Shabanah OA., "Carnitine esters prevent oxidative stress damage and energy depletion following transient forebrain ischaemia in the rat hippocampus", Clinical and Experimental Pharmacology & Physiology, 2006, 33 (8), 725-33.

[95]   Wilson AD., Hart A., Brännström T., Wiberg M., Terenghi G., "Delayed acetyl-L-carnitine administration and its effect on sensory neuronal rescue after peripheral nerve injury", Journal of Plastic, Reconstructive & Aesthetic Surgery, 2007, 60 (2), 114-8.

[96]   Samir P., Patel PG., Sullivan, TS., Lyttle, Alexander G. Rabchevsky, "Acetyl-l-carnitine ameliorates mitochondrial dysfunction following contusion spinal cord injury". Journal of Neurochemistry, 2010, 114 (1), 291-301.

[97]   Beal MF., "Bioenergetic approaches for neuroprotection in Parkinson's disease". Annals of Neurology, 2003, 53 (Suppl 3), S39-47, discussion S 47-8.

[98]   Hathcock JN., Shao A., "Risk assessment for carnitine". Regulatory Toxicology and Pharmacology, 2006, 46 (1), 23-8.

[99]   Claudio Teloken, Tulio Graziottin & Patrick E. Teloken (2007). "Oral Therapy for Peyroni's Disease". In Laurence A. Levine MD. FACS. *Peyronies Disease: A Guide to Clinical Management*. Humana Press.ISBN 978-1-58829-614-6. Retrieved 2009-06-26.

[100] Ruggenenti P., Cattaneo D., Loriga G., "Ameliorating hypertension and insulin resistance in subjects at increased cardiovascular risk, effects of acetyl-L-carnitine therapy". Hypertension, 2009, 54(3), 567-74.

[101] Zhang Z., Zhao M., Li Q., Zhao H., Wang J., Li Y., "Acetyl-l-carnitine inhibits TNF-alpha-induced insulin resistance via AMPK pathway in rat skeletal muscle cells". FEBS Letters, 2009, 583 (2), 470-4.

[102] Schroeder MA., Atherton HJ., Ball DR., Cole MA., Heather LC., Griffin JL., Clarke K., Radda GK., Tyler DJ., (August). "Real-time assessment of Krebs cycle metabolism using hyperpolarized 13C magnetic resonance spectroscopy". FASEB J., 2009, 23 (8), 2529-2538.

[103] Kotil K., Kirali M., Eras M., Bilge T., Uzun H., "Neuroprotective effects of acetyl-L-carnithine in experimental chronic compression neuropathy. A prospective, randomized and placebo-control trials.". Turk Neurosurg. 2007, 17 (2), 67-77.

[104] Nakamura K., Kurasawa M., "Anxiolytic effects of aniracetam in three different mouse models of anxiety and the underlying mechanism". Eur. J. Pharmacol. (Kanagawa, Japan). 2001, 420 (1), 33-43.

[105] Ito I., Tanabe S., Kohda A., Sugiyama H., Aniracetam reduces glutamate receptor desensitization and slows the decay of fast excitatory synaptic currents in the hippocampus, J. Physiol, 1990, 424, 533-543.

[106] Huff FJ., (January). "Preliminary evaluation of besipirdine for the treatment of Alzheimer's disease. Besipirdine Study Group". Annals of the New York Academy of Sciences, 1996, 777, 410-4.

[107] Klein JT., Davis L., Olsen GE., Wong GS., Huger FP., Smith CP., Petko WW., Cornfeldt M., Wilker JC., Blitzer RD., Landau E., Haroutunian V., Martin LL., Effland RC., "Synthesis and structure-activity relationships of N-propyl-N-(4-pyridinyl)-1H-indol-1-amine (besipirdine) and related analogs as potential therapeutic agents for Alzheimer's disease". Journal of Medicinal Chemistry, 1996, 39 (2), 570-81.

[108] Tang L., Smith CP., Huger FP., Kongsamut S., "Effects of besipirdine at the voltage-dependent sodium channel". British Journal of Pharmacology, 1995, 116 (5), 2468-72.

[109] Tang L., Kongsamut S., "Frequency-dependent inhibition of neurotransmitter release by besipirdine and HP 184". European Journal of Pharmacology, 1996, 300 (1-2), 71-4.

[110] David J., Triggle, Dictionary of Pharmacological Agents. Boca Raton., Chapman & Hall/CRC. 1996, p. 265.

[111] Kondo Y., Ogawa N., Asanuma M., Matsuura K., Nishibayashi K., Iwata E., "Preventive effects of bifemelane hydrochloride on decreased levels of muscarinic acetylcholine receptor and its mRNA in a rat model of chronic cerebral hypoperfusion". Neuroscience research, 1996, 24 (4), 409-14.

[112] Egashira T., Takayama F., Yamanaka Y., "Effects of bifemelane on muscarinic receptors and choline acetyltransferase in the brains of aged rats following chronic cerebral hypoperfusion induced by permanent occlusion of bilateral carotid arteries". Japanese journal of pharmacology, 1996, 72 (1), 57-65.

[113] Moryl E., Danysz W., Quack G., "Potential antidepressive properties of amantadine, memantine and bifemelane". Pharmacology & Toxicology, 1993, 72 (6), 394-7.

[114] Shigemitsu T., Majima Y., "Use of bifemelane hydrochloride in improving and maintaining the visual field of patients with glaucoma". Clinical therapeutics, 1996, 18(1), 106-13.

[115] De Jesus M., Moreno M., "Cognitive improvement in mild to moderate Alzheimer's dementia after treatment with the acetylcholine precursor choline alfoscerate: a multicenter, double-blind, randomized, placebo-controlled trial". Clin Ther., 2003, 25 (1): 178-93.

[116] Parnetti L., Mignini F., Tomassoni D., Traini E., "Cholinergic precursors in the treatment of cognitive impairment of vascular origin: Ineffective approaches or need for re-evaluation". Journal of the Neurological Sciences, 2007, 257 (1-2), 264-9.

[117] Barbagallo SG., barbagallo M., giordano M., "Alpha-Glycerophosphocholine in the mental recovery of cerebral ischemic attacks." An Italian multicenter clinical trial. Ann NY Acad Sci., 1994, 717, 253-69.

[118] Shindler JS., Finnerty GT., Towlson K., Dolan AL., Davies CL., Parkes JD.,"Domperidone and levodopa in Parkinson's disease". British Journal of Clinical Pharmacology, 1984, 18(6), 959-62.

[119] Rossi S., editor. Australian Medicines Handbook 2006. Adelaide: Australian Medicines Handbook, 2006.

[120] Silvers D., Kipnes M., Broadstone V., "Domperidone in the management of symptoms of diabetic gastroparesis: efficacy, tolerability, and quality-of-life outcomes in a multicenter controlled trial. DOM-USA-5 Study Group". Clinical therapeutics, 1998, 20 (3), 438-53.

[121] Missale C., Pasinetti G., Govoni S., Spano PF., Trabucchi M., Fipexide: a new drug for the regulation of dopaminergic system at the macromolecular level. Bollettino Chimico Farmaceutico, 1983, 122(2), 79-85.

[122] Bompani R., Scali G., Fipexide, an effective cognition activator in the elderly: a placebo-controlled, double-blind clinical trial. Current Medical Research and Opinion. 1986,10(2), 99-106.

[123] Guy C., Blay N., Rousset H., Fardeau V., Ollagnier M., Fever caused by fipexide, Evaluation of the national pharmacovigilance survey, Therapie. 1990, 45(5), 429-31.

[124] Liu. XJ., Wu WT., "Effects of ligustrazine, tanshinone II A, ubiquinone, and idebenone on mouse water maze performance.". Zhongguo yao li xue bao = Acta pharmacologica Sinica, 1999, 20 (11), 987-90.

[125] Takahashi T., Kataoka K., Influence of indeloxazine hydrochloride upon photic driving responses elicited by flickering dot pattern and red flicker stimuli in elderly patients. Jpn J Psychiatry Neurol, 1990, 44(4), 709-15.

[126] Gatto GJ., Bohme GA., Caldwell WS., Letchworth SR., Traina VM., Obinu MC., Laville M., Reibaud M., Pradier L., Dunbar G., Bencherif M. TC-1734: an orally active neuronal nicotinic acetylcholine receptor modulator with antidepressant, neuroprotective and long-lasting cognitive effects. CNS Drug Reviews, 2004, 10(2), 147-66.

[127] Dunbar G., Demazières A., Monreal A., Cisterni C., Metzger D., Kuchibhatla R., Luthringer R., Pharmacokinetics and safety profile of ispronicline (TC-1734), a new brain nicotinic receptor partial agonist, in young healthy male volunteers, Journal of Clinical Pharmacology, 2006, 46(7), 715-26.

[128] Lippiello P., Letchworth SR., Gatto GJ., Traina VM., Bencherif M. Ispronicline: a novel alpha 4 beta 2 nicotinic acetylcholine receptor-selective agonist with cognition-enhancing and neuroprotective properties, Journal of Molecular Neuroscience, 2006, 30(1-2),19-20.

[129] Dunbar G., Boeijinga PH., Demazières A., Cisterni C., Kuchibhatla R., Wesnes K., Luthringer R., Effects of TC-1734 (AZD3480), a selective neuronal nicotinic receptor agonist, on cognitive performance and the EEG of young healthy male volunteers. Psychopharmacology (Berlin), 2007, 191(4), 919-29.

[130] Dunbar GC., Inglis F., Kuchibhatla R., Sharma T., Tomlinson M., Wamsley J., Effect of ispronicline, a neuronal nicotinic acetylcholine receptor partial agonist, in subjects with age

associated memory impairment (AAMI), Journal of Psychopharmacology, 2007, 21(2), 171-8.

[131] Di Iorio P., Virgilio A., Giuliani P., Ballerini P., Vianale G., Middlemiss PJ., Rathbone MP., Ciccarelli R., AIT-082 is neuroprotective against kainate-induced neuronal injury in rats, Experimental Neurology, 2001,169(2), 392-9.

[132] Lahiri DK., Ge YW., Farlow MR., Effect of a memory-enhancing drug, AIT-082, on the level of synaptophysin. Annals of the New York Academy of Sciences, 2000, 903, 387-93.

[133] Yan R., Nguyen Q., Gonzaga J., Johnson M., Ritzmann RF., Taylor EM., Reversal of cycloheximide-induced memory disruption by AIT-082 (Neotrofin) is modulated by, but not dependent on, adrenal hormones. Psychopharmacology (Berlin*)*, 2003, 166(4), 400-7.

[134] Rathbone MP., Middlemiss PJ., Crocker CE., Glasky MS., Juurlink BH., Ramirez JJ., Ciccarelli R., Di Iorio P., Caciagli F., AIT-082 as a potential neuroprotective and regenerative agent in stroke and central nervous system injury. Expert Opinion in Investigational Drugs, 1999, 8(8), 1255-62.

[135] Ramirez JJ., Parakh T., George MN., Freeman L., Thomas AA., White CC., Becton A., The effects of Neotrofin on septodentate sprouting after unilateral entorhinal cortex lesions in rats. Restorative Neurology and Neuroscience, 2002, 20(1-2), 51-9.

[136] Holmes M., Maysinger D., Foerster A., Pertens E., Barlas C., Diamond J., Neotrofin, a novel purine that induces NGF-dependent nociceptive nerve sprouting but not hyperalgesia in adult rat skin. Molecular and Cellular Neuroscience, 2003, 24(3), 568-80.

[137] Jiang S., Khan MI., Middlemiss PJ., Lu Y., Werstiuk ES., Crocker CE., Ciccarelli R, Caciagli F, Rathbone MP. AIT-082 and methylprednisolone singly, but not in combination, enhance functional and histological improvement after acute spinal cord injury in rats. International Journal of Immunopathology and Pharmacology, 2004, 17(3), 353-66.

[138] Calcutt NA., Freshwater JD., Hauptmann N., Taylor EM., Mizisin AP., Protection of sensory function in diabetic rats by Neotrofin. European Journal of Pharmacology, 2006, 18, 534(1-3), 187-93.

[139] Potkin SG., Alva G., Keator D., Carreon D., Fleming K., Fallon JH., Brain metabolic effects of Neotrofin in patients with Alzheimer's disease. Brain Research. 2002, 27, 951(1), 87-95.

[140] Johnston TH., Brotchie JM., Drugs in development for Parkinson's disease. Current Opinion in Investigational Drugs, 2004, 5(7), 720-6.

[141] Ohta T., Kikuchi H., Hashi K., Kudo Y., Nizofenone administration in the acute stage following subarachnoid hemorrhage. Results of a multi-center controlled double-blind clinical study. Journal of Neurosurgery. 1986, 64(3), 420-6.

[142] Yasuda H., Izumi N., Nakanishi M., Maruyama Y., Brain protection against oxygen deficiency by nizofenone. Advances in Experimental Medicine and Biology. 1988, 222, 403-10.

[143] Yasuda H., Nakajima A., Brain protection against ischemic injury by nizofenone. Cerebrovascular and Brain Metabolism Reviews. 1993, 5(4), 264-76.

[144] Matsumoto Y., Aihara K., Kamata T., Goto N., Nizofenone, a neuroprotective drug, suppresses glutamate release and lactate accumulation. European Journal of Pharmacology, 1994, 262(1-2),157-61.

[145] Falsaperla A., Monici Preti PA., Oliani C., Selegiline versus oxiracetam in patients with Alzheimer-type dementia. Clin Ther., 1990, 12(5), 376-84.

[146] "Nootropil". NetDoctor.co.uk. 8 July 2004. Retrieved 21 September 2009.

[147] Parnetti L., Ambrosoli L., Abate G., Azzini C., Balestreri R., Bartorelli L., Bordin A, Crepaldi G., Cristianini G., Cucinotta D., Posatirelin for the treatment of late-onset Alzheimer's disease: a double-blind multicentre study vs citicoline and ascorbic acid, Acta Neurol Scand, 1995, 92(2),135-40.

[148] Hindmarch I., Coleston DM., Kerr JS., Psychopharmacological effects of pyritinol in normal volunteers. Neuropsychobiology. 1990-1991, 24(3), 159-64.

[149] Winblad B., Grossberg G., Frolich L., Farlow M., Zechner S., Nagel J., Lane R., "IDEAL: a 6-month, double-blind, placebo-controlled study of the first skin patch for Alzheimer disease". Neurology, 2007, 24, 69(4 Suppl 1), S14-22.

[150] Inglis F., "The tolerability and safety of cholinesterase inhibitors in the treatment of dementia". Int J Clin Pract. 2002,127, 45-63.

[151] Loudon JM., Bromidge SM., Brown F., "SB 202026: a novel muscarinic partial agonist with functional selectivity for M1 receptors". The Journal of Pharmacology and Experimental Therapeutics, 1997, 283 (3): 1059-68.

[152] Mohr E., Nair NP., Sampson M., Murtha S., Belanger G., Pappas B., Mendis T., Treatment of Alzheimer's disease with sabeluzole: functional and structural correlates. Clin Neuropharmacol, 1997, 20(4), 338-45.Abstract

[153] Geerts H., Nuydens R., De Jong M., Cornelissen F., Nuyens R., Wouters L., Sabeluzole stabilizes the neuronal cytoskeleton, Neurobiol Aging, 1996, 17(4), 573-81. Abstract

[154] Aldenkamp AP., Overweg J., Smakman J., Beun AM., Diepman L., Edelbroek P., Gutter T., Mulder OG, vt Slot B, Vledder B., Effect of sabeluzole (R 58,735) on memory functions in patients with epilepsy. Neuropsychobiology, 1995, 32(1), 37-44. Abstract

[155] Pauwels PJ., Van Assouw HP., Peeters L., Moeremans M., Leysen JE., Chronic treatment with sabeluzole protects cultured rat brain neurons from the neurotoxic effects of excitatory amino acids. Synapse, 1992, 12(4), 271-80. Abstract

[156] Geerts H., Nuydens R., Nuyens R., Cornelissen F., De Brabander M., Pauwels P., Janssen PA., Song YH., Mandelkow EM., Sabeluzole, a memory-enhancing molecule, increases fast axonal transport in neuronal cell cultures. Exp Neurol. 1992, 117(1), 36-43. Abstract.

[157] De Deyn PP., Van de Velde V., Verslegers W., Saerens J., Pickut BA., Clincke B., Woestenborghs R., Van Peer A., Single- dose and steady-state pharmacokinetics of sabeluzole in senile dementia of Alzheimer type patients, Eur. J. Clin Pharmacol. 1992, 43(6), 661-2. Abstract.

[158] Nikolov R., Scheller D., Nikolova M., Study on the cerebral effects of sabeluzole. Methods Find Exp. Clin. Pharmacol, 1991, 13(6), 385-90. Abstract.

[159] Davis KL., Mohs. RRC., Pfefferbaum A., Hollister LE., Kopell BS., Physostigmin, improvement of long term process in normal human, Science, 1978, 201, 272-274.

[160] Schmidt BH., Heinig R., The pharmacological basis of Metrifonate's favourable tolerability in the treatment of Alzheimer's disease. Dement Geriatr Cogn Disord, 1998, 9, 15-19.

[161] Yu V., Burov TD., Baimanov NI., Maisov, Effect of amiridine and tacrine, drugs effective in Alzheimer's disease, on synaptosomal neurotransmitter uptake, Bulletin of Experimental Biology and Medicine, 1992, 113, 4, pp 499-501.

[162] Lim GP., Yang F., Chu T., Chen P., Beech W., Teter B., Tran T., Ubeda O., Ashe KH., Frautschy SA., Cole GM., Ibuprofen suppresses plaque pathology and inflammation in a mouse model for Alzheimer's disease, J. Neurosci, 2000, 20, 5709-5714.

[163] French L., "Dysmenorrhea". Am Fam Physician, 2005, 71 (2), 285-91.

[164] Blain H., Jouzeau JY., Blain A., Terlain B., Trechot P., Touchon J., Netter P., Jeandel C., Non-steroidal anti-inflammatory drugs with selectivity for cyclooxygenase-2 I Alzheimer's disease. Rationale and perspectives, Press Med., 2000, 29, 267-273.

[165] Endoh M., Kunishita T., Tabira T., No effect of anti-leprosy drugs in the prevention of Alzheimers disease and beta-amyloid neurotoxicity, J. Neurol Sci., 1999, 165, 2830.

[166] Aisen PS., Davis Kl., Berg JD., Schafer k., Camphell k., Thomas RG., Weiner MF., Farlow MR., Sano M., Grundman M., Thal LJ., A randomized controlled trial of prednisone in Alzheimer's disease. Alzheimer's disease cooperative study, Neurology, 2000, 54, 588-593.

[167] Aisen PS., Davis KL., Anti-inflammatory therapy for Alzheimer's disease: a status report. Int. J. Geriatric Psychopharm,1997, 1, 2-5.

[168] Aisen PS., Marin DB., Brickman AM., Santoro J., Fusco M., Pilot tolerability studies of hydroxychloroquine and colchicine in Alzheimer disease. Alzheimer Dis. Assoc Disord, 2001, 15, 96-101.

[169] Hamilton GS., Steiner JP., Immunophilins, Beyong immunosuppression, J. Med. Chem., 1998, 41, 5119-5143.

[170] Noble S., Wagstaff A., Propentofylline. The Epidemiology of Use of Analgesics for Chronic Pain CNS Drugs, 1997, 8, 257-266.

[171] Lahiri DK., Ge YW., Farlow MR., Effect of a memory-enhancing drug, AIT-082, on the level of synaptophysin. Ann NY Acad Sci 2000, 903, 387-393.

[172] Farber SA., Slack BE., Blusztajn JK., Acceleration of phosphatidylcholine synthesis and break down by inhibitors of mitochondrial function in neuronal cells, A model of the membrane defect of Alzheimer disease. FASEB J 2000, 14, 2198-2206.

[173] Alvarez XA., Mouzo R., Pichel V., Perez P., Laredo M., Fernandez-Novoa L, Corzo L., Zas R., Alcaraz M., Secades JJ., Lozano R., Cacabelos R., Double-blind placebo-controlled study with citicoline in APOE genotyped Alzheimer's disease patients, Effects on cognitive performance, brain bioelectrical activity and cerebral perfusion, Methods Find Expe Clin pharmacol, 1999, 21, 633-644.

[174] Stout AK., Raphael HM., Kanterewicz BI, Klann E, Reynolds IJ. Glutamate-induced neuron death requires mitochondrial calcium uptake. Nature Neurosci, 1998, 1, 366-373.

[175] Wei H., Leeds P., Chen RW., Wei W., Leng Y., Bredesen DE., Chuang DM., Neuronal apoptosis induced by pharmacological concentrations of 3-hydroxykynurenine: Characterization and protection by dantrolene and BcL-2 overexpression, J. Neurochem, 2000, 75, 81-90.

[176] Porsteinsson AP., Tariot PN., Erb R., Cox C., Smith E., Jakimovich L., Noviasky J, Kowalski N. Holt CJ., Irvine C, Placebo-controlled study of divalproex sodium for agitation in dementia, Am J Geriatr psychiatry, 2001, 9, 58-66.

[177] Pittaluga A., Feligioni M., Ghersi C., Gemignani A., Raiteri M., Potentiation of NMDA receptor function through somatostatin release: A possible mechanism for the cognition-enhancing activity of GABA(B)receptor antagonists. Neuropharmacology, 2001, 41,301-310.

[178] Robbins DK., Hutcheson SJ., Miller TD., Green VI., Bhargava VO., Weir SJ., Pharmacokinetics of MDL 26479, a novel benzodiazepine inverse agonist, in normal volunteers, Biopharm Drug Dispos, 1997,18, 325-334.

[179] Scheuer K., Rostock A., Bartsch R., Muller WE., Piracetam improves cognitive performance by restoring neurochemical deficits of the aged rat brain, Pharmacopsychiatry, 1999, 32, 10-16.

[180] Nakamura K., Kurasawa M., Anxiolytic effects of aniracetam in three different mouse models of anxiety and the underlying mechanism, Eur J Pharmacol, 2001, 420 (1), 33-43.

[181] Riekkinen PJ., Review on the long-term efficacy and safety of selegeline in the treatment of Alzheimer's disease. Neurobiol aging 9abstract), 1998, 19, 1263.

[182] Cesura AM., Borroni E., Gottowik J., Kuhu C., Malherbe P., Martin J., Richards JG., Lazabemide for the treatment of Alzheimer's disease: Neurobiol aging 9 abstract), 1998, 19, 1263.

[183] Oosterink BJ., Korte SM., Nyakas C., Korf J., Luiten PG., Neuroprotection against N-methyl-D-aspartate-induced excitotoxicity in rat magnocellular nucleus basalois by the 5-HTIA receptor agonist 8-OH-DPAT., Eur J. Pharmacol, 1998, 358, 147-152.

[184] Katsu H., Selective 5- HTIA receptor agonist 8-OH-DPAT,locally administered into the dorsal raphe nucleus increased extracellular acetyl choline concentrations in the medial predfrontal cortex of conscious rats. Nihon Shinkei Seishin Yakurigaku Zasshi, 2001, 21,121-123.

[185] Dysken MW., Kuskowski M., Ondansetron in the treatment of cognitive decline in Alzheimer's dementia, Neurobiol aging, 1998, 10, 747.

[186] Behl C., Davis J., Cole GM., Schubert D., Vitamin E protects nerve cells from amyloid-b protein toxicity. Biochem Biophys Res Commun, 1992, 186, 944-950.

[187] Thal LJ., Clinical and preclinical information of anti-oxidants:Idebenone and vitamin E., Neurobiol Aging, 1998, 19, 20.

[188] Noble S., Wagstaff A., Propentofylline. CNS Drugs 1997, 8, 257-266.

[189] Bolkenius FN., Verne-mismer J., Wagner J., Grisar JM., Amphiphilic alpha-tocopherol analogues as inhibitors of brain lipid peroxidation. Eur. J. Pharmacol, 1996, 298, 37-43.

[190] Mulnard RA., Cotman CW., Kawas C., Van Dyck CH., Sano M., Doody R., Koss E., Pfeiffer E., Jin S., Gamst A., Grundman M., Thomas R., Thal LJ., Estrogen replacement therapy for treatment of mild to moderate Alzheimer's disease: a randomized controlled trial. Alzheimer's disease Cooperative Study. JAMA, 2000, 283,1007-1015.

[191] Dubal D., Wilson ME., Wise PM., Estradiol: a protective and trophic factor in the brain, Alzheimer Dis Rev., 1999, 4, 1-9.

[192] Xu H., Gouras GK., Greenfield JP., Vincent B., Naslund J., Mazzarelli L., Fried G., Jovanovic JN., Seeger M., Relkin NR., Liao F., Checler F., Buxbaum JD., Chait BT., Thinakaran G., Sisodia SS., Wang R., Greengard P., Gandy S., Estrogen reduces neuronal generation of Alzheimer beta- amyloid peptides. Nat. Med., 1998, 4, 447-451.

[193] Gouras GK., XU H., Gross RS., Greenfield JP., Hai B., Wang R., Greengard P., Testosterone reduces neuronal secretion of Alzheimer's b-amylioid peptides. Proc NatL Acad Sci., 2000, 97, 1202-1205.

[194] Hammond J., Le Q., Goodyer C., Gelfand M., Trifiro M., LeBlanc A., Testosterone-mediated neuroprotection through the androgen receptor in human primary neurons. J. Neurochem, 2001, 77, 1319-1326.

[195] Papasozomenos SCh., Shanavas A., Testosterone prevents the heat shock-induced overactivation of glycogen synthase kinase-3 beta but not of cyclin-dependent kinase 5 and c-Jun NH2-terminal kinase and concomitantly abolishes hyperphosphorylation of tau:Implications for Alzheimer's disease. Proc Natl Acad Sci USA, 2002, 99, 1140.

[196] Blum-Degen D., Gotz ME., Novel 17a-estradiol analogues as potent radical scavengers. In iqbal K., Swaab DF., Winblad B., Wisniweski HM., Editors. Alzheimer's disease and related disorders. Chichester, J., Wiley and sons Ltd., 1999, 671-677.

[197] Lermontova NN., P chev VK., Beznosko BK., Van'kin GL., Koroleva IV., Lukoyanova EA., MMukhina TV., Serkova TP., Bachurin SO., Effects of 17beta-estradiol and its isomer 17 alpha-estradiol on learning in rats with chronic cholinergic deficiency in the brain. Bull. Exp. Biol Med., 2000, 129, 525-527.

[198] Allison AC., Cacabelos R., Lombardi VR., Alvarez XA., Vigo C., Celastrol, a potent antioxidant and anti-inflammatory drug, as a possible treatment for Alzheimer's disease. Prog Neuropsychopharmacol Biol Psychiatry, 2001, 25, 1341-1357.

CHAPTER 8

Research Strategies Developed for the Treatment of Alzheimer's Disease. Reversible and Pseudo-Irreversible Inhibitors of Acetylcholinesterase: Structure-Activity Relationships and Drug Design

Mauricio Alcolea-Palafox[a,*], Paloma Posada-Moreno[b,c], Ismael Ortuño-Soriano[b,c], José L. Pacheco-del-Cerro[c], Carmen Martínez-Rincón[c], Dolores Rodríguez-Martínez[c] and Lara Pacheco-Cuevas[c]

[a]*Chemical-Physics Department, Chemistry Faculty, Complutense University of Madrid, Spain;* [b]*Institute for Health Research at the San Carlos Clinical Hospital (IdISSC), Madrid, Spain;* [c]*Nursing Department, Complutense University of Madrid, Spain*

Abstract: Although several research strategies have been developed in the last decades, the current therapeutic options for the treatment of Alzheimer's disease are limited to three acetylcholinesterase inhibitors: galantamine, donepezil and rivastigmine. However, they have only offered a modest improvement in memory and cognitive function. Moreover, these drugs show side effects, and relatively low bioavailability among other problems. These features limit their use in medicine and they lead to a great demand for discovering new acetylcholinesterase inhibitors. In addition to its important role in cholinergic neurotransmission, acetylcholinesterase also participates in other functions related to neuronal development, differentiation, adhesion and amyloid-β processing. Acetylcholinesterase accelerates amyloid-β aggregation and this effect is sensitive to peripheral anionic site blockers. Both features have lead to the development of dual inhibitors of both catalytic active and peripheral anionic sites. These compounds are promising disease-modifying Alzheimer's disease drug candidates. On the other hand, due to the pathological complexity of Alzheimer's disease, multifunctional molecules with two or more complementary biological activities may represent an important advance for the treatment of this disease. All these features are described in detail in the present chapter.

Keywords: Alzheimer's disease, donepezil, galantamine, huperzine, huprines, infractopicrin, inhibitors AChE, ladostigil, physostigmine derivatives, rivastigmine, tacrine, tacrine hybrid, tacripyrines and donepezil hybrids, TAK-147.

*Address correspondence to Mauricio Alcolea Palafox: Chemical-Physics Department, Chemistry Faculty, Complutense University of Madrid, Spain; Tel: 913944272; E-mail: alcolea@quim.ucm.es

Atta-ur-Rahman / Muhammad Iqbal Choudhary (Eds.)

10.1016/B978-0-12-803959-5.50008-8

1. INTRODUCTION

Alzheimer's disease (AD), the most common type of dementia worldwide, today represents a major public health issue and it is characterized by amyloid-β (Aβ) deposits, τ-protein aggregation, low levels of ACh and oxidative stress, among other events [1]. The first therapeutic strategy for the treatment of AD was mainly centered on the restoration of cholinergic functionality [2, 3]. Most of the AD treatments have been focused on the inhibition of AChE in order to enhance cholinergic neurotransmission by increasing ACh availability in the synaptic cleft. A decrease of ACh in the brain of AD patients appears to be a critical element in producing dementia [4]. Today, an acetylcholinesterase inhibitor (AChEI) is commonly used soon after the AD diagnosis [5].

Actually, the AChEIs approved by the United States Food and Drug Administration (FDA) for the symptomatic treatment of patients with mild or moderate AD include: Donepezil (a benzyl piperidine), Rivastigmine (a carbamate) and Galantamine (a tertiary alkaloid). However, they do not halt the progression of the disease or alter its final outcomes. Clinical experience has shown that AChE inhibition is a viable therapeutic approach to the palliative treatment of AD. One of the most common untoward effects of such therapy is gastrointestinal complaints resulting from stimulation of peripheral autonomic cholinergic system.

An ideal cholinesterase inhibitor (ChEI) should be well tolerated, highly selective in the brain for both the molecular enzyme forms and the specific region (cerebral cortex and hippocampus), minimal effects on the peripheral cholinergic system, and finally no organ toxicity. In addition, further compounds should be tested both *in vitro* and *in vivo* in the search for AChE molecular form-specific inhibitors for the treatment of AD [4]. Therefore, there is a great demand today in the medical community for a better cholinesterases inhibitor (ChEI).

A renewed interest in the search of AChEIs appeared when evidences suggested that AChE has an additional role in mediating the aggregation and deposition of Aβ peptide [6].

In the recent years, multifunctional compounds have been the subject of increasing attention by many investigators which have developed a number of compounds acting simultaneously of different target implicated in AD. These drugs could have more therapeutic efficacy than single-target compounds.

2. ACETYLCHOLINESTERASE

Cholinesterases (ChE) are a family of enzymes that catalyzes the hydrolysis of ACh, an essential process allowing for the restoration of the cholinergic neuron. The two types of ChE are: acetylcholinesterase (EC 3.1.1.7) and butyrylcholinesterase (BuChE; EC 3.1.1.8).

Acetylcholinesterase is one of the well-know enzyme, which plays an important role in the central nervous system (CNS). The availability of AChE crystal structures for various species with and without ligands provides a solid basis for structure-based design of novel AChE inhibitors [7].

The AChE structure has been extensively investigated since 1990s. The first experiment with X-ray was carried out on AChE in the electric eel, *Torpedo californica* (tcAChE), due to its availability [8]. The results obtained have lead to an informal model until the commercialization of human recombinant AChEs [9].

Target enzyme consists of a narrow gorge with two separate ligand binding sites: the catalytic active site (CAS) and the peripheral anionic site (PAS) Fig. (**1**) [6, 11]. The gorge itself is a narrow hydrophobic channel with a length of about 20Å, connecting the PAS to the active site [8]. It is surrounded by aromatic amino acids enabling a high selectivity for ACh. Substrate penetration is allowed by cation-π interactions between ACh quaternary ammonium atom and π electrons of phenylalanine (F), tryptophan (W) and tyrosine (Y) aromatic cores [12, 13]. These sites are described below.

The AChE catalytic active site is located at the bottom of the gorge and it contains serine (S)-histidine (H)-glutamate (E), the catalytic triad (the same in AChE and BuChE) within an esteratic site. The anionic site (also called α -anionic site) is another part of the active site, and it is close to the esteratic site, Fig. (**1**). While the esteratic site hydrolyzes the ester bond, the anionic site interacts with the acetylcholine quaternary ammonium atom and it is responsible for its correct orientation [9].

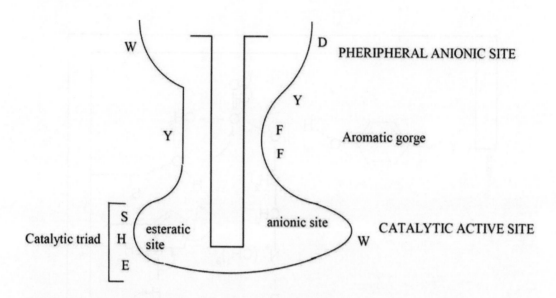

Figure 1: AChE active site with two sites of ligand binding: CAS, at the base of the gorge and PAS near the gorge entrance (This figure is modified of Botti and co-workers) [10].

In Fig. (**2**) the participation of the catalytic triad in the acetylcholine hydrolysis is shown: the primary alcohol moiety of the serine residue (catalytic triad) participates in a transesterification reaction with ACh, resulting in acetylation of the enzyme. A neighboring group, the imidazol ring (part of a histidine residue) participates and facilitates the acetyl group transfer. The resulting acetylated serine moiety is extremely labile and rapidly undergoes spontaneous hydrolytic cleavage to liberate acetate anion and to regenerate the active catalytic surface [15].

The peripheral anionic site (also called β-anionic site) is located at the active center gorge entry, to a distance of approximately 14 Å from the main active site [16]. Tryptophan, tyrosine, and aspartate (D) amino acids residues are the most significant in the PAS [17]. This PAS encompasses binding sites for allosteric ligands (activators and inhibitors) [6]. Ligand binding to the PAS affects enzymatic activity through a combination of both by steric blockade of ligands moving through the gorge, and by allosteric alteration of the catalytic triad conformation [18].

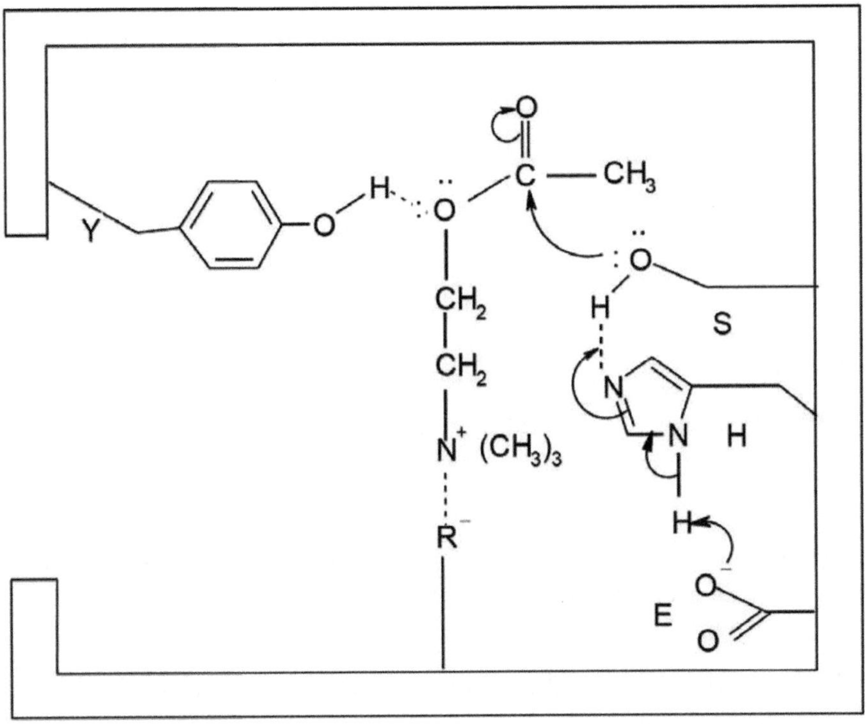

Figure 2: Catalytic active site. Electronic displacements reported in the acetylcholine hydrolysis (This figure is modificated of Delgado and co-workers) [14].

Guo and co-workers [11] described a receptor-specific scoring function for predicting binding affinities for human AChE inhibitors. This method reported a list of those residues making the most important electrostatic and Van der Waals (VdW) contributions within the main active site, gorge area, acyl binding pocket and PAS.

It is well known that AChE exists in two different forms (with identical active sites): the globular forms, consisting of monomer (G_1), dimer (G_2) and tetramer

(G$_4$), and the asymmetric forms. In the human brain, the most abundant AChE forms are G$_4$ and G$_1$ [19]. A selective loss of the membrane-associated AChE molecular form G$_4$ has been observed in the AD brain, while the G$_1$ form is relatively preserved [4].

The AChE's primary function is the rapid splitting of ACh and thus terminating cholinergic neurotransmission. However, a new noncholinergic role has been discovered for AChE, related to neuronal development, differentiation, adhesion and Aβ peptide processing. The presence of the PAS seems to be fundamental for some of its so called "non-classical action". This unique structural feature of the enzyme (PAS) may be responsible for the aggregation-promoting action of AChE. The AChE promotes amyloid fibrils assembly in the brain, by binding to Aβ through the PAS. This fact gives stable AChE-Aβ complexes that are more toxic than single Aβ peptides [20]. Amyloid β peptide interacts with the PAS inducing a conformational transition to the amyloidogenic conformation with the subsequent amyloid fibril formation [21], and consequent damage of cholinergic neurons. Several ligands that bind to the PAS have shown to retard aggregation [22]. Prefibrillar oligomers of the Aβ are recognized as potential mediators of AD pathophysiology. AChE is one of the several proteins associated with Aβ aggregation and amyloid plaque deposits. Modulation of Aβ-induced toxicity by increasing its degradation in order to reduce brain Aβ burden would be a rational strategy for the treatment and prevention of AD.

The fact that AChE accelerates Aβ aggregation, and this effect is sensitive to PAS blockers, has led to the development of dual inhibitors of both CAS and PAS. These compounds are promising modifying drug candidates because they can simultaneously improve cognition and slow the rate of Aβ-elicited neuro-degeneration [23]. The interest in these dual-site inhibitors has recently emerged.

3. INHIBITORS

Molecular modeling has been often used to design novel enzyme inhibitors. In this way the research of prior AChE inhibitor has focused on using ligand-based design methods. Molecular docking can also be very useful in characterizing ligand-receptor binding by providing predictions of the bound conformation for the ligand and a scheme for energetically ranking the ligand-receptor interaction [10, 24, 25].

Molecular modeling techniques include: the investigation of AChE-drug complexes, the ligand-binding sites calculation within the active site of the enzyme, the virtual screening toxicophorical analysis, and the estimation of pharmacokinetics properties. The toxicophoric and pharmacokinetic predictions constitute valuable tools helping the continuous search for pharmaceutical interesting molecules with low toxicity and suitable pharmacokinetic profile [26].

On the other hand, x-ray crystallographic analysis of AChE from *Torpedo californica* [8] followed by x-ray determination of the complexes of the enzyme with several structurally diverse inhibitors such as tacrine (THA) [27], donepezil (E2020) [16], galantamine [28, 29] and Huperzine A (Hup A) [30] provides crucial information with respect to orientation of these inhibitors in the active site of the enzyme.

Three classes of acetylcholinesterase inhibitors have been described: reversibles, pseudoirreversibles, and irreversibles [15]. In this chapter, we have only considered the pseudoirreversible and reversible inhibitors. Several inhibitors with synthetic and natural origins are available today in drug market. However, the side effects (due to lack of selectivity) and the relatively low bioavailability, among others factors, limit their use in medicine.

Early inhibition research was mainly focused on ligands binding in the active site. The research was later focused on finding novel ligands that bind to both sites in order to search more potent reversible inhibitors, that selectively favor the inhibition of AChE rather than the related BuChE [11]. Now, inhibition of the PAS can be considered the most promising AD treatment. The deposition of amyloid plaque in AD may be accelerated or even triggered by interaction of β-amyloid with PAS. Therefore, by interfering with PAS, the aggregation and neurotoxic effects of Aβ may be reduced. The PAS inhibitors are considered not only symptomatic drugs for AD, but also probably causative ones [23]. In Table **1** a resume of the most interesting AChEIs included here is shown.

In addition, there are studies indicating that AChEIs have the ability to affect the expression of nicotinic acetylcholine receptors (nAChRs) that play a major role in cognitive functions [31].

Table 1.   Summary of the Most Interesting AChEIs Described in the Present Chapter

Compounds	Significant Features
PSEUDOIRREVERSIBLE INHIBITORS	
Physostigmine	Prototype of AChEI. First drug investigated by the AD treatment. It has been shown to APL of nAChRs. It was discontinued after phase III clinical trials due to side effects.
Physostigmine Derivatives	
Epastigmine	Less toxic than physostigmine. Adverse hematologic effects lead to suspension of clinical trials.
Phenserine	Selective AChEI. It is a dual AChE and β-APP inhibitor. The lack of significant efficacy leads to abandon clinical trials.
Tolserine	A phenserine derivative. Preclinical studies show a highly potent inhibitor of h-AChE compared to physostigmine and phenserine.
Rivastigmine	It is an inhibitor of both AChE and BuChE. It was approved in 2000 for the treatment of mild-to-moderate AD.
Ladostigil	It is a propargylamine and rivastigmine derivative, which combines neuroprotective effects with MAO-A and –B and cholinesterase inhibitory activities. Actually in phase IIb clinical trials.
REVERSIBLE INHIBITORS	
Acridines	
Tacrine	First AChEI approved in 1993. It has been abandoned because of remarkable side effects, including hematoxicity.
Bis(7)Tacrine	Dimer of tacrine. Interesting lead compounds.
Cystamine-tacrine dimer	It has lower toxicity than bis(7) Tacrine. It acts as a radical scavenger. Potentially useful in AD treatment.
N-benzylpiperidines	
Donepezil	Approved in 1996 for the treatment of mild-to-moderate and latterly to severe AD. Selective AChEI. Side effects on the cardiovascular system are still unclear.
TAK-147	It presents high selectivity for AChE and BuChE, and it acts selectively on the CNS. Actually it is under clinical trials.
Isoquinoline Alkaloids	
Galantamine	Approved for the treatment of mild-to-moderate AD in 2007. It is an AChEI and APL of the nAChER. It has excellent pharmacological and pharmacokinetics profiles and it exhibits low side effects.
Indole Alkaloids	
Infractopicrin and 10-hydroxy derivative	They could be useful candidates for investigation as AChEI and as AD-drugs.
Other Alkaloids	
Huperzine A	It has been approved for the symptomatic AD treatment in China. In European Union and USA it is yet in clinical trials.
Huperzine B	It is less potent than HupA. It exhibits other benefits. It appears to be a promising lead compound.

Table 1. contd....

Compounds	Significant Features
REVERSIBLE INHIBITORS	
Tacrine Hybrid Derivatives Huprines	Highly potent AChEI. Huprine-tacrine heterodimers can be considered very promising compounds.
Antioxidant hybrids	
- Tacrine-8-hydroxyquinoline	They have antioxidant and neuroprotective and dual inhibition of AChE and BuChE. They are very interesting multifunctional prototypes.
- Mercapto-tacrine	Display ChE inhibition, neuroprotection and less hematotoxicity. Good candidates for the development of novel drugs for AD.
- Tacrine-4-oxo-4H-chromene	Interesting *in vitro* biological activities: inhibition of hAChE and the β-secretase, radical scavenger activity. Actually further studies are being carried out.
- Tacrine-feluric acid	One of these significantly inhibited Aβ aggregation induced by AChE and blocked the ROS. It may be a promising multifunctional drug candidate.
- Tacripyrines	Potent and selective hAChEI. Protect efficiently against free radicals. They are neuroprotective agents. Interesting new candidates for the treatment of AD
Donepezil Hybrids	Highly potent inhibitors human AChE and BuChE compounds. They have a potential disease-modifying role in the AD treatment. They could represent new templates for further optimization studies.
Hybrids MAO and Cholinesterase Inhibitors	Selective for both AChE inhibitors activity and MAO-A. They can be new, attractive and promising drugs for the AD treatment.

4. PSEUDOIRREVERSIBLE INHIBITORS

In this group are included agents with a carbamate ester moiety which is hydrolyzed by AChE, but much more slowly than ACh. It should be pointed that in stress conditions [32] they can enhance carbamate diffusion into the CNS.

These compounds interact with OH-serine in the catalytical triad of active site in essentially the same manner as ACh does, providing stable esters. The resulting carbamylated enzyme is much more stable than the acetylated enzyme. The carbamoyl moiety can be split from cholinesterase by spontaneous hydrolysis [33]. From the chemical point of view, these carbamates are in general N-alkyl and N,N-dialkyl derivatives. The natural product is physostigmine.

4.1. Physostigmine

Fig. (3), also known as eserine, is an alkaloid isolated from the seeds of the calabar bean, *Physostigmina venenosum*. It is the prototype of pseudoirreversible inhibitors of AChE and it has been the first ChEI investigated by the treatment of

AD [15]. This alkaloid exhibits equal inhibitory activity against AChE and BuChE. The inhibition of both cholinesterases is highly enentioselective [34].

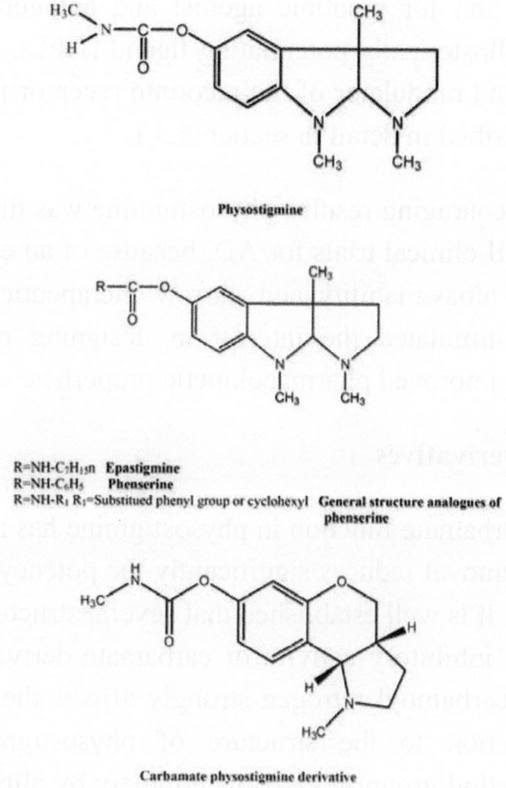

Physostigmine

R=NH-C₇H₁₅n **Epastigmine**
R=NH-C₆H₅   **Phenserine**
R=NH-R₁ R₁=Substitued phenyl group or cyclohexyl **General structure analogues of phenserine**

Carbamate physostigmine derivative

Figure 3: Pseudoirreversible inhibitors: Physostigmine, epastigmine, fenserine, analogs phenserine and carbamate physostigmine derivative.

There is evidence suggesting that in the AD brain the AChE activity remains unchanged or declined, while the BuChE activity rises in an attempt to modulate ACh levels in the cholinergic neurons. Therefore, both enzymes are likely involved in regulating ACh levels and consequently, they may represent therapeutic targets for the development of agents that, with the ability to inhibit BuChE in addition to AChE, should lead to improved clinical outcomes [35].

Physostigmine has memory enhancing effects in patients with AD [36], because it passes through the blood-brain barrier (BBB). Besides, the physostigmine has been shown to act as an agonist on nAChRs from muscle and brain, by binding to sites that are distinct from those for the natural transmitter ACh [37]. Allosteric modulators interact with the receptor through binding sites, that they are distinct from those for ACh and for nicotinic agonist and antagonist [38, 39]. Thus, physostigmine is an allosterically potentiating ligand (APL), which may act as a low potency agonist and modulator of the nicotinic receptor [40]. The interest of nicotinic APLs is described in detail in section 5.3.1.

In spite of the first encouraging results, physostigmine was discontinued after the completion of phase III clinical trials for AD, because of an effect too diminutive in duration, variable bioavailability and narrow therapeutic index [41]. These considerations have stimulated the interest in designing new physostigmine-related inhibitors with improved pharmacokinetic properties.

4.2. Physostigmine Derivatives

The removal of the carbamate function in physostigmine has no effect on potency as an APL, but this removal reduces significantly the potency of physostigmine's AChE inhibition [42]. It is well established that several structural elements concur to determinate AChE inhibitory activity of carbamate derivatives. For instance, the alkyl substituent carbamoyl nitrogen strongly affects the affinity profile [6]. One useful modification to the structure of physostigmine has been the replacement of the methyl group of carbamate moiety by aliphatic alkyl group or phenyl group. It leads to the derivatives: epastigmine and fenserine.

Epastigmine Fig. (3) (heptyl-physostigmine tartrate) is a carbamate derivative of physostigmine in which the carbamoylmethyl group has been substituted by a carbamoylheptyl group. This compound is less toxic than physostigmine [43] and it shows greater potency for BuChE [44]. *In vitro* and ex vivo results suggest that epastigmine has a long-lasting reversible brain ChE (*i.e.* AChE and BuChE) inhibitory effect. Specially, it preferentially inhibits the G_1 form of AChE [45, 46].

Clinical investigations demonstrated that epastigmine significantly improved cognitive performance. Also, pharmacokinetic studies have revealed that after oral administration it is rapidly distributed to the tissues and readily enters the CNS, where it can be expected to inhibit AChE for a prolonged period. It is generally well tolerated and the majority of adverse events (cholinergic) are mild to moderate in intensity. Although its cholinergic tolerability was found to be favorable, however the adverse hematologic (granulocytopenia) effects reported in two studies have led to the suspension of further clinical trials [47, 48].

Phenserine Fig. (**3**), phenylcarbamate derivative of physostigmine, is a selective inhibitor of AChE with minimal effect on BuChE [49]. This carbamate is a dual AChE and β-amyloid precursor protein (β-APP) inhibitor being developed to treat mild to moderate AD [50]. It has a quick absorption rate and it is less toxic than physostigmine. Although phase III clinical trials have been conducted during 2003-2004, however, further clinical trials for AD have been abandoned [41].

Results in the short-term study by Winblad and co-workers [51] have shown the phenserine potentially benefiting mild to moderate AD symptomatically, but this derivative has not addressed possible amyloid metabolic mediated effects on disease processes in AD. Therefore, phenserine may represent an important new catalog of compounds for the treatment of AD, because the inhibitors available in the market do not reduce the β-APP level [52].

Analogs of phenserine were also synthesized and their cholinesterase inhibition activities were evaluated [52]. These derivatives, Fig. (**3**), contain an electron-withdrawn substituent in each position of the phenyl group in the phenylcarbamoyl moieties, and they showed less inhibition against AChE and BuChE than phenserine. By contrast, an electron-donor substituent as the methyl/methoxyl groups, maintained or improved their AChE potencies. The results obtained clearly showed that small electron-donor substituents in *meta-* or *ortho-* position of the phenyl ring are better choices for the retention of AChE inhibition potency.

One of these analogs, Tolserine with a 2'-methyl group on the phenylcarbamoyl moiety of phenserine, proved to be a highly potent inhibitor of human AChE (hAChE) compared to physostigmine and phenserine. Preclinical studies were initiated in 2000 and they were shown to be more selective against hAChE than against BuChE [53]. Another compound, the cyclohexylcarbamate derivative exhibited extremely higher selectivity inhibiting AChE over BuChE, compared with phenserine [54].

Other **complex molecules** have been synthesized by replacing the pyrroline ring of physostigmine with a dihydropyran ring. Thus, the hexahydropyrrolo[2,3-b]indole moiety of physostigmine can be replaced by hexahydrochromeno[4,3-b]pyrrole without affecting the affinity for AChE. The carbamate physostigmine derivative, Fig. (**3**), has been found as potent as physostigmine *in vitro* against human AChE and BuChE. However, the benzene ring position isomers of the carbamate moiety are much less potent. The position of the carbamate moiety in the phenyl ring has a pivotal role in determining anticholinesterase activity [6, 15].

4.3. Rivastigmine

(ENA-713), Fig. (**4**), is other carbamate which shows an inhibitory action toward AChE, but less marked than physostigmine. However, it has superior global pharmacological profile, including a good combination of longer duration of action, good tolerability and lower toxicity. Rivastigmine was approved by FDA for the treatment of mild-to-moderate AD in 2000 [6]. Recent investigation aims at developing films capable of delivering the drug *in vivo* in a sustained manner with reduction in gastrointestinal side adverse effects [55].

In addition rivastigmine is referred as a "brain-region" selective ChEI [56] since it preferentially inhibits AChE and BuChE of hippocampus and cortex [57]. It is stronger inhibitor of AChE G_1 than G_4 [4], and it is an inhibitor of both AChE and BuChE. This inhibition improves the cognition in elderly patients with late onset AD [58]. Clinical trials have demonstrated that patients treated with this drug have not shown the widespread cortical atrophic changes in parietotemporal regions invariable reported in untreated AD patients, and which were detectable in the subgroups treated with selective acetylcholinesterase inhibitors [59, 60]. These findings are consistent with the hypothesis that inhibition of both enzymes may have neuroprotective and disease-modifying effects [61]. However, further longitudinal and long-term studies on this issue are needed.

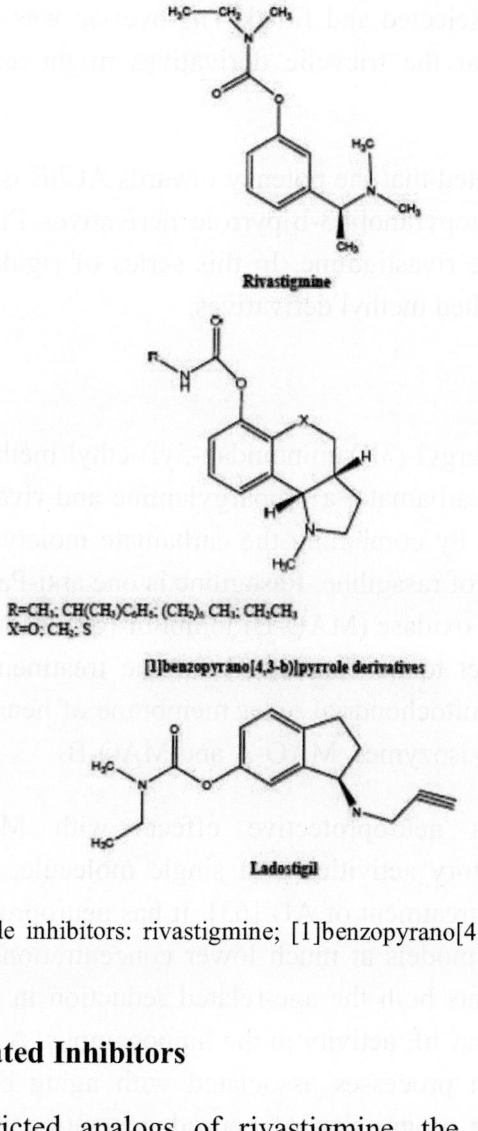

Rivastigmine

R=CH₃: CH(CH₃)C₆H₅: (CH₂)ₙ CH₃: CH₂CH₃
X=O: CH₂: S

[1]benzopyrano[4,3-b)]pyrrole derivatives

Ladostigil

Figure 4: Pseudoirreversible inhibitors: rivastigmine; [1]benzopyrano[4,3-b]pyrrole derivatives and Lagostigil.

4.4. Rivastigmine Related Inhibitors

Conformationally restricted analogs of rivastigmine, the [1]benzopyrano[4,3-b]pyrrole derivatives have been synthesized [6] by inserting the dimethylamino-ethyl-phenyl moiety of rivastigmine, Fig. (**4**), in different tricyclic systems related to carbamate physotigmine derivative, Fig. (**3**).

A superimposition between the conformation of rivastigmine and the carbamate physostigmine derivative (X=CH₂; R=CH₃), as obtained from Monte Carlo simulations, has been reported. In this study, several low energy conformations of each molecule were selected and fitted. The overlap was satisfactory with these forms confirming that the tricyclic derivatives might act as rigid analogs of rivastigmine.

This study demonstrated that the potency towards AChE is generally increased in the rigidified [1]benzopyrano[4,3-b]pyrrole derivatives Fig. (**4**) as compared to the flexible prototype rivastigmine. In this series of rigid compounds, the most potent inhibitors resulted methyl derivatives.

4.5. Ladostigil

(TV 3326), [(N-propargyl-(3R)aminoindan-5yl)-ethyl methyl carbamate] Fig. (**4**) is another synthetic carbamate, a propargylamine and rivastigmine derivative. It has been synthesized by combining the carbamate moiety of rivastigmine in the aminoindan structure of rasagiline. Rasagiline is one anti-Parkinsonian irreversible, selective monoamine oxidase (MAO-B) inhibitor [62]. Monoamine oxidase MAO is an important target to be considered for the treatment of AD. MAO is an enzyme bounded to mitochondrial outer membrane of neuronal glial and of other cells that exist as two isozymes, MAO-A and MAO-B.

Ladostigil combines neuroprotective effects with MAO-A and –B and cholinesterase inhibitory activities in a single molecule, presently in phase IIb clinical trials for the treatment of AD [63]. It has neuroprotective and antioxidant activities in cellular models at much lower concentrations than those inhibiting AChE. It also prevents both the age-related reduction in cortical AChE activity and the increase of BuChE activity in the hippocampus. A molecule that has such a selective effect on processes associated with aging could provide an ideal treatment against the progression of neurodegeneration in AD [64]. Moreover, recent finding demonstrated that the major metabolite of ladostigil, hydroxyl-1-(R)-aminoindan has also a neuroprotective activity and thus, it may contribute to the over activity of its parent compound [65, 66].

5. REVERSIBLE INHIBITORS

The currently most accepted AD therapy is the application of mild and reversible AChEI to restore ACh levels and therefore cholinergic brain activity. Reversible inhibitors combine with the substrate cation-binding site of the catalytic surface of AChE and thus they deny acetylcholine's access to this site. These compounds have a short duration of action due to the facile reversibility of their binding [15].

AChE structural complexity accounts for the large diversity of reversible inhibitors which can interact with either the active site, the peripheral site or both, and with the aromatic gorge, making use of distinct sets of interactions [9]. The chemical structures of these inhibitors are very different which fall into several classes.

5.1. Acridines

Cholinesterase inhibitors binding to the α-anionic site are a group of chemical compounds containing certain common motives: firstly, these compounds typically contain condensed aromatic cores; and secondly, they should have quaternary ammonium or nitrogen included as a heteroatom. Interesting examples are acridines and tetrahydroacridines.

5.1.1. Tacrine

Fig. (**5**) (1,2,3,4-tetrahydro-9-acridinamine) has been described in 1961 [67] as a reversible inhibitor of AChE. It binds into the α-anionic site and selectively it inhibits the G_1 form in the rat brain [4]. Also, THA inhibits BuChE, and it shows greater potency for BuChE as compared with AChE [44].

Tacrine has been the first cholinesterase inhibitor approved by the FDA for the treatment of AD in 1993 [68]. Clinical efficacy in relief of the symptoms of AD has been claimed for THA, but this positive finding is tempered by its tendency to produce hepatotoxicity. Thus, its use has been largely abandoned because of a high incidence of side effects including nausea, vomiting, dizziness, diarrheas, seizures and syncope. The serious hepatotoxicity of THA has been the main limitation for its clinical use [69]. Tacrine can cause reactive oxygen species (ROS) production stimulation and gluthatione depletion in the human liver cell. Cytotoxic studies point out that oxidative stress might be involved in THA hepatotoxicity [70].

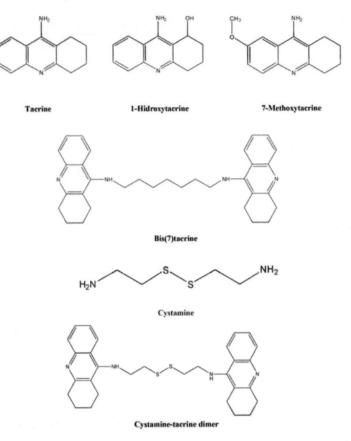

Figure 5: Reversible inhibitors. Acridines and cystamine.

5.1.2. Tacrine Derivatives

To find less toxic derivatives of tacrine, several compounds have been synthesized. The most interesting are:

1-Hidroxytacrine, Fig. (**5**) (Velnacrine) was designed in the hope that the OH group performs the glucuronidation and subsequently facilitated elimination. This compound is somewhat less potent against AChEI *in vitro* than the THA [15].

7-Methoxytacrine, Fig. (**5**) is less toxic than tacrine, and *in vitro* as *in vivo* test it proves to be superior to tacrine [71, 72]. **N-alkyl-7-methoxytacrine hydrochloride** derivatives are also of interest [73].

Bis(7)tacrine, Fig. (**5**). Several strategies have been employed to design high affinity dual inhibitors with dimeric structure. Recent studies have demonstrated that its homo- and hetero- dimers can improve and enlarge its biological profile with less side-effects [68]. Bis(7)tacrine is a heptamethylene-linked dimer of tacrine, and it has been designed taking into account that there are two AChE binding sites. It is more potent than THA in inhibition of rat AChE and it is more selective for AChE than for BuChE. The importance of ligand hydrophobicity has been indicated for effective cation-π interaction of the homodimer with PAS [15].

Bis(7)tacrine represents an interesting lead compound to design novel dual binding AChEIs. In this context, Minarini and co-workers [74] focused on the important biological properties of cystamine Fig. (**3**) as antioxidant, cyto- and neuroprotective agent [75]. These authors replaced the heptamethylene linker of bis(7)tacrine by the structure of cystamine, leading to **cystamine-tacrine dimer** Fig. (**3**), which appears to be characterized by a disulfide bridge. This dimer is able to inhibit human AChE, BuChE, self- and AChE-induced Aβ aggregation in the same range of the reference compound. It has lower toxicity than bis(7)tacrine, and it acts as a radical scavenger. All these results allowed us to consider the cystamine-tacrine dimer as potentially useful in AD treatment [74].

5.2. N-Benzylpiperidines

In this group of inhibitors the prototype is the Donepezil compound.

5.2.1. Donepezil

Fig. (**6**) was approved in 1996 for the treatment of mild-to-moderate AD and latterly it was also approved for severe AD [76]. In addition, Howard and co-workers [77] detected in double-blind, placebo-controlled trial involving patients with moderate or severe AD, significant benefits of continued donepezil therapy with respect to cognitive and functional outcomes over the course of 12 months.

Donepezil, a benzylpiperidine is a selective reversible inhibitor of AChE. Its affinity for this enzyme is greater than for BuChE [44]. The three-dimensional structure of AChE in complex with donepezil shows that donepezil interacts with both the anionic site in the bottom of the gorge and the PAS near its entrance [78]. Molecular modeling based on the crystal structure has illustrated the complementarity between this class of inhibitors and the narrow gorge of the enzyme [79].

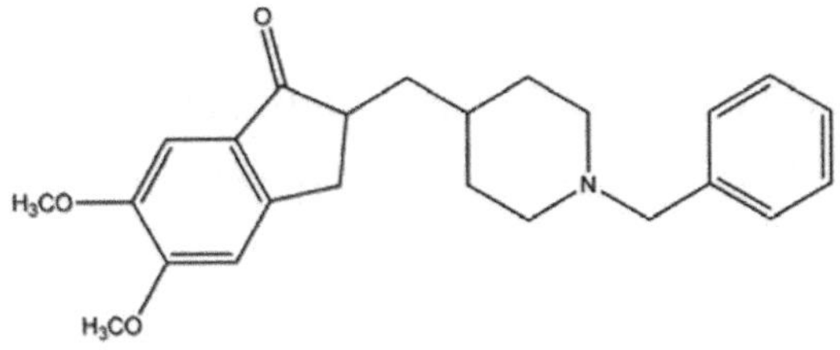

Donepezil

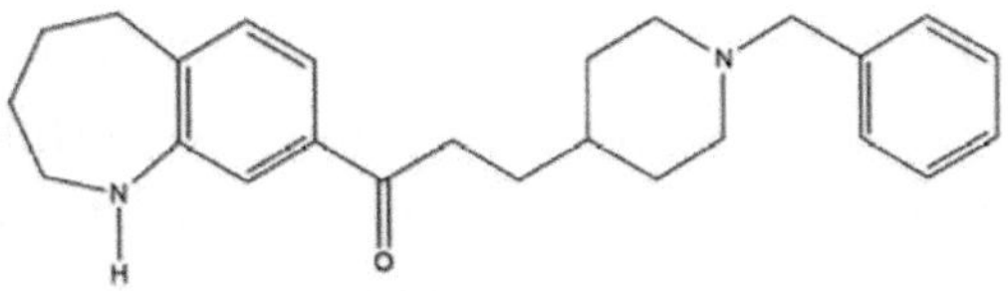

TAK-147

Figure 6: Reversible inhibitors. N-benzylpiperidines.

Donepezil is well absorbed with a relative oral bioavailability of 100% [80], with good penetration through the BBB and slow excretion [16]. There is some evidence that it may be better tolerated, with less gastrointestinal side effects, than rivastigmine or galantamine [81]. Its cholinergic adverse side effects on the cardiovascular system are still unclear. although a recent study demonstrated that donepezil is not associated with increased negative chromotropic, arrythmogenic or hypotensive effects for elderly patients with AD [82].

5.2.2. TAK-147

Based on a working hypothesis of the enzyme's active site, Ishiara and co-workers [83] designed a series of benzylamino compounds and they established several structure-activity relationships. As consequence, these studies lead to the discovery of TAK-147 Fig. (**6**).

Preclinical pharmacological and pharmacokinetic studies have shown that TAK-147 presents high selectivity for AChE and BuChE, and it acts selectively on the central nervous system. Thus, it exerts ameliorating effects on cholinergic deficits without showing excessive peripheral effects. On the basis of these and other results obtained, it is expected that TAK-147 would not only ameliorate the clinical symptoms in AD, but also that it can prevent or slow the progression of the disease [83, 84]. Actually, TAK-147 is under clinical trials as a therapeutic drug for AD.

5.3. Alkaloids

Natural sources provide a variety of structurally distinct and biologically active metabolites. Numerous plants have been used to treat neurodegenerative diseases and different neuropharmacological disorders. Plant extract is one of the major sources for discovery of new compounds with AChE inhibitory activity. We have described below several of the most interesting alkaloids.

5.3.1. Isoquinolines Alkaloids

Galantamine, Fig. (**7**), is the latest anticholinesterase drug used against AD, which has been approved for the treatment of mild-to-moderate AD in 2001 [80]. It is a tertiary alkaloid, having isoquinoline skeleton, found in the bulbs and flowers of the common snowdrop (*Galanthus nivalis*) and of other members of the *Amaryllidaceae* family. *Amaryllidaceae* alkaloids exhibited several types of pharmacological activities including on central-nervous system [85]. The difficulty of isolating galantamine from its natural source hindered its commercial use or even as a starting material for the synthesis of derivatives. Actually, there are several synthesis procedures described and it has been biosynthetically obtained [86].

Galantamine is among the very few drugs that exhibit a dual activity being both an inhibitor of AChE and an allosteric potentiator of the nicotinic response induced by ACh and by competitive agonist [42]. There is much evidence indicating that neural nAChR plays an important role in learning and memory. Moreover, the density/activity of brain nAChR is substantially reduced in AD patients compared with a control group of the same age [87]. Studies of human brain tissue collected during postmortem [88] and brain imaging studies in living AD patients [89] demonstrated specific loss of nicotinic cholinergic receptors in AD [90]. These abnormalities are closely associated with increased levels of neuritic plaques and neurofibrillary tangles in the AD patients [91]. A means to up-modulate or to potentiate the activity of nicotinic receptors in response to ACh is to use allosterically potentiating ligands. Several problems originated by administration of nicotinic cholinergic agonists can be avoided with allosteric modulators [90].

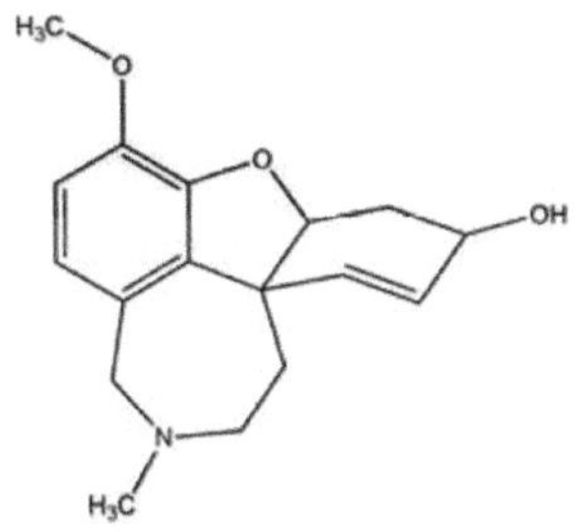

Galantamine

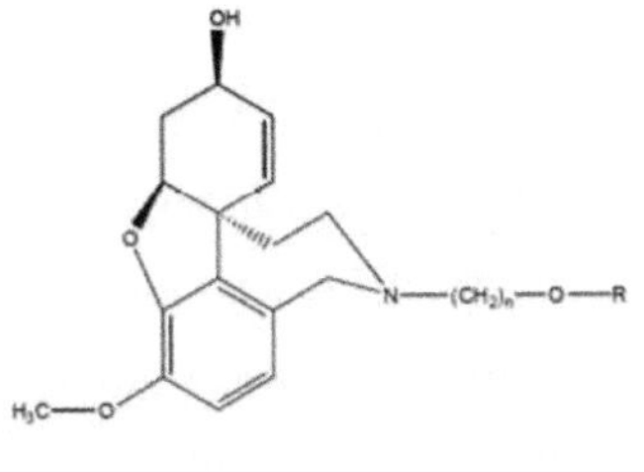

R= benzyl amino groups; n=2-12.

N-substituted galantamine derivatives

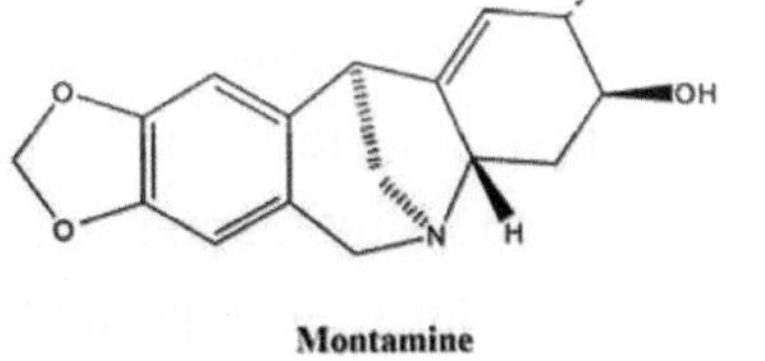

Montamine

Figure 7: Reversible inhibitors. Isoquinolines alkaloids.

Compared with conventional AChE inhibitors, galantamine produces relatively less AChE inhibition; it is less potent than physostigmine, tacrine and donepezil, but it has excellent pharmacological and pharmacokinetics profiles and it exhibits low side effects, *i.e.* less toxic [92, 93].

Galantamine Derivatives

Bartolucci and co-workers [29] determined the correct orientation and interactions of galantamine within the active site gorge AChE. The observed binding mode explains the affinities of a series of structural analogs of galantamine and it provides a rational basis for structure-based drug design aimed at developing synthetic analogs of galantamine with improved pharmacological properties.

A docking procedure has also been described which can be applied to produce models of ligand-receptor complexes for AChE and other macromolecular targets of drug design. In this study a galantamine derivative was included, which has a N-propylpiperidine substituent at the nitrogen atom instead of a methyl group. A molecular model of the complex between tcAChE and this galantamine derivative has been reported. The side chain of this ligand is predicted to extend along the enzyme active site gorge from the anionic site, at the bottom, to the peripheral anionic site, at the top [94].

Therefore, structure-activity relationship (SAR) studies reveal that substitution on the nitrogen atom of galantamine is favorable for AChE inhibitory activity. May be these substituents display interactions with the PAS. Thus, derivatives that contain a long substituent on the nitrogen are of particular importance. Such structures potentially span the whole binding cavity of tcAChE and they interact with several amino acids at the PAS, which should lead to increased potency of AChE inhibition [94].

Jia and co-workers [95] developed a new series of galantamine derivatives capable of interacting with both, the active and the peripheral sites of AChE. They have designed, synthesized and evaluated as AChE inhibitors several **N-substituted galantamine derivatives** Fig. (**7**) by selecting benzyl-amino groups and modified benzyl-amino moieties as pharmacophoric units and by incorporating them into the galantamine molecule. Besides, different lengths of the alkyl chain between galantamine and benzylamino moieties have been also explored.

Structure-activity studies showed that the potency of AChE inhibition has been mainly influenced by the function at the end of the linker, as well as, the length of the connecting units. Especially, the incorporation of a phenyl ring between the alkyl chain and the terminal nitrogen-containing moieties provided additional sites of interactions between the inhibitor and the enzyme. These results suggested that the benzyl-piperidine moiety might be a potent segment for fishing the PAS. This group binds better to the PAS of AChE than to the other moieties, which could be used in the development of novel bivalent ligands. The results of this study provided a basis for future design and development of bivalent AChE inhibitors [95].

Galantamine derivatives are predicted to improve interactions with AChE, in particular with the PAS. Several N-alkyl-phenyl substituents appear as favorable since the PAS is predominantly composed of hydrophobic aromatic residues. In addition, virtual-screening simulations have been used to select novel inhibitor candidates containing different structural scaffolds in order to find novel structural patterns with potential AChE inhibitory activity [26].

Montamine

After the discovery of galantamine, the isolation and characterization of alkaloids from *Amaryllidaceae* have increased in order to find better AChEIs. Pagliosa and co-workers [96] studied the activity of isoquinolines alkaloids isolated from *Hippeartrum* species (*Amaryllidaceae*). One of them, montamine Fig. (**5**), significantly inhibits AChE activity, although this alkaloid requires further investigations.

5.3.2. Indole Alkaloids

Indole-containing compounds play a role in diverse pharmacological actions. Various indole-structures have been discovered from plants extract and they have showed good AChE inhibitory activity. A natural source for new drugs is the fungal fruiting bodies (macromycetes). Several groups of macromycetes yield diversity of bioactive secondary metabolites and they constitute a valuable complementary source for novel lead compounds.

Geissler and co-workers [97] studied two indole alkaloids, **infractopicrin** and **10-hydroxy-infractopicrin,** Fig. (**8**), isolated from fruiting bodies of *Cortinarius infractus Berk* (*Cortinariaceae*). These alkaloids have shown inhibitory activity against AChE whilst having low cytotoxicity, and also they possess a higher selectivity than galantamine. Docking studies suggest possible reasons for the selectivity to the AChE active site versus BuChE, such as the lacking of π-π interactions in BuChE. Also *in vitro* fibril formation is positively influenced viz. reduced formation. These results, as well as the positive pharmacokinetic properties, suggest that these alkaloids could be useful candidates for further investigation as AChEI and as AD-drugs.

Recently, the AChE inhibitory activity of quinoline (biostere of the indole ring) derivatives Fig. (**8**) and β-carboline derivatives Fig. (**8**) has been studied to understand the SAR of indole structure. The simple indoles with substitutions of electron donating (methoxy and hydroxy) and electron withdrawing (carboxyl) groups on the benzene ring show a low percent of inhibitory activity in tcAChE. It implies that the structures might be too small and they are not appropriate to fit within AChE binding pocket. Adding a side chain at the pyrrole ring, β-carbolines and quinolines improves the inhibitory activity significantly. Both, β-carboline and quinoline analogs bound in the same binding site with some slightly overlapped structure. These two compound types could be used for developing AChE inhibitors in the future [98].

5.3.3. Other Alkaloids

Huperzine A, Fig. (**9**) is an alkaloid isolated from the *Huperzia serrata*. This plant has been used in Chinese traditional medicine against memory deficits, contusions, strains, swellings, *etc.* since ages [99]. The natural compound is a strong AChE inhibitor, preferentially the G_4 form [4], binding to the PAS [9]. It is three times as potent as physostigmine against AChE but it is less potent against BuChE [100]. In addition, other effects of neurodegenerative disorders have been studied such as link between huperzine and oxidative stress [101, 102] and mechanisms of neuroprotection by huperzine A [103].

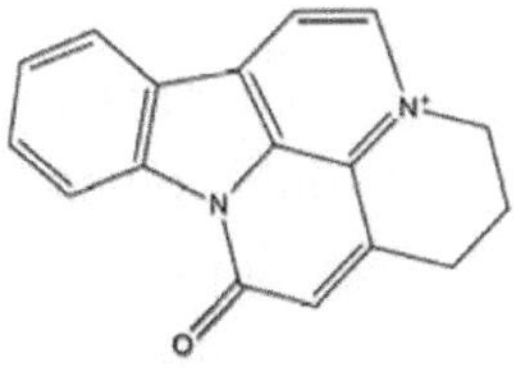

Infractopicrin 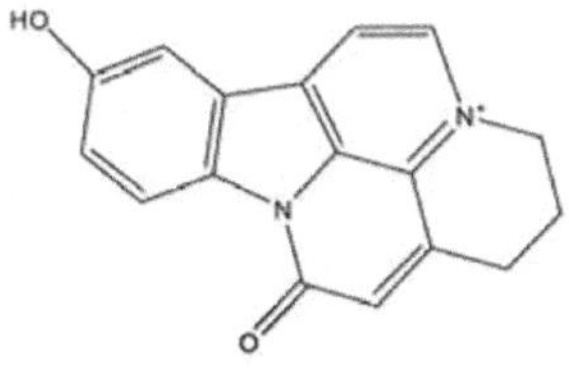   10-Hydroxy-infractopicrin

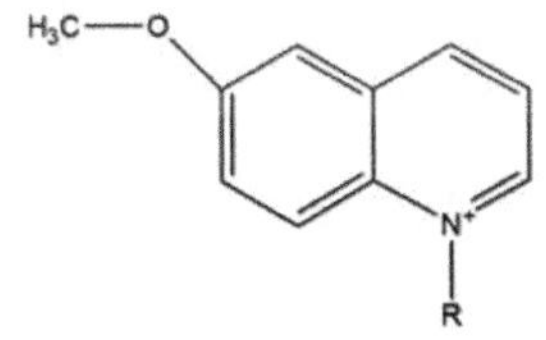

R= CH₃; CH₂ -C₆H₅

Quinoline derivatives

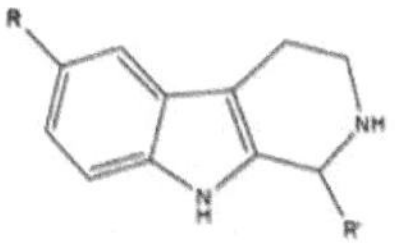

R= H; OCH₃; R'= H; COOH 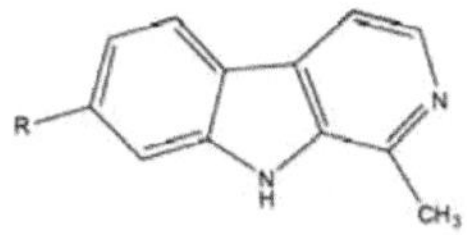   R=H; OCH₃; OH   R= OCH₃; OH

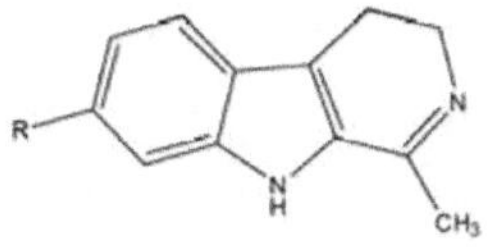

β-Carboline derivatives

Figure 8: Reversible inhibitors. Indole alkaloids.

Huperzine A is the most-promising drug candidate with potent anticholinesterase effect. Clinical trials at phase IV in China showed that this alkaloid significantly improved memory shortages in aged people with benign senescent forgetfulness and in patients with AD or vascular dementia, with minimal peripheral cholinergic side effects and no unexpected toxicity [104]. It has been approved as a new drug for the symptomatic treatment of AD in China [105]. However, in the European Union and United States it is yet in clinical trials [106].

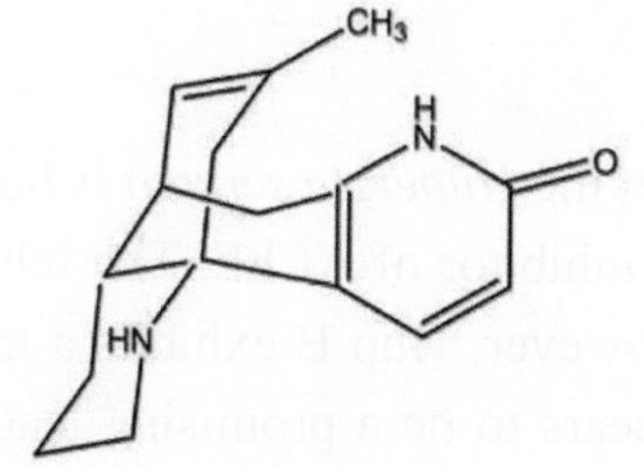

Huperzine B

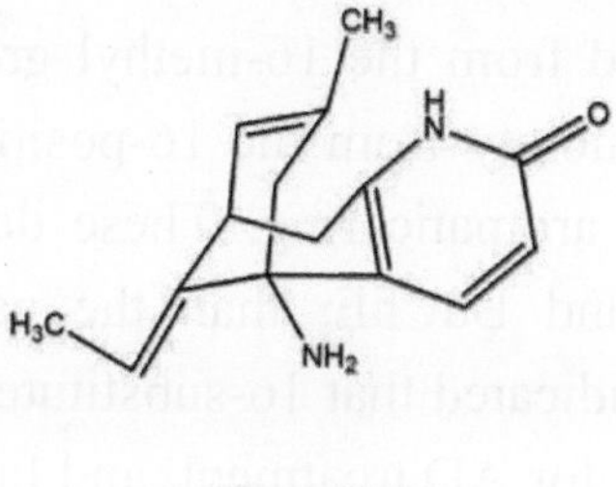

Huperzine A

Figure 9: Reversible inhibitors. Other alkaloids.

Total synthesis of the compound has been achieved [107], although the synthetic racemic mixture of huperzine A has less AChE inhibitory effects than the natural kind [108]. Several analogs and derivatives of huperzine A have been prepared and tested for their inhibitory activities towards AChE. Many of these derivatives demonstrated lower potency than the natural Hup A [109].

Other interesting derivative is bis(12)-hupyridone (B12H), a novel dimeric AChE inhibitor derived from a naturally occurring monomeric analog Hup A, which was investigated *in vitro* and *in vivo* [110]. Compared with Hup A, B12H has twice the potency in inhibiting rat brain AChE *in vitro*. Moreover, B12H shows other interesting biological activities. Although more experiments are needed, the authors indicate that B12H might provide greater therapeutic efficacy for the treatment of AD, and it might exert neuroprotection *via* acting on multiple targets [110].

Huperzine B

Other alkaloid isolated from the *Huperzia serrata* is huperzine B (Hup B), Fig. **(9)** an effective and reversible inhibitor of AChE. The inhibitory activity of Hup B is less potent than Hup A. However, Hup B exhibits a higher therapeutic index and other benefits. Thus, it appears to be a promising lead compound for developing novel multifunctional AChE inhibitors.

Shi and co-workers [111] described the synthesis, inhibitory activities and preliminary pharmacological results of 16-substituted bifunctional Hup B derivatives, which are derived from the 16-methyl group. Their main structural characteristic is that Hup B moiety from the 16-position is connected through a tether chain with a terminal aromatic ring. These derivatives are more potent inhibitors for both, AChE and BuChE, than the parent Hup B. Preliminary pharmacological evaluation indicated that 16-substituted derivatives of Hup B are potential new drug candidates for AD treatment, and further exploration is needed to evaluate their pharmacological and clinical efficacies [111].

5.4. Hybrids

Anticholinesterase drugs such as rivastigmine, donepezil and galantamine, have shown only a modest improvement in memory and cognitive function. Due to the pathological complexity of AD, it is unlikely that a unitary mechanism of action will provide a comprehensive therapeutic approach to such multifaceted neurodegenerative disease [112]. Efficient therapy is more likely to be achieved by drugs that incorporate several pharmacological effects into a single chemical entity. Thus, molecules with two or more complementary biological activities may represent an important advance in the treatment of the disease [113].

This novel therapeutic strategy in which drug candidates are designed to possess diverse pharmacological properties and to act on multitude targets has led recently to the discovery of several anti-AD drug candidates such as ladostigil, mentioned above in section 4.5, and the hybrid derivatives, that they are described below.

5.4.1. Tacrine Hybrid Derivatives

Tacrine has been extensively studied, being the first marketed product. Today, it remains as a reference structure even if the adverse troubles induced, such as hepatotoxicity and gastrointestinal disorders led to its withdrawal from the marked. The structure of tacrine has been widely used as scaffold to provide new molecules endowed with additional properties beyond simple AChE inhibition. Several tacrine hybrids of interest are described below:

Huprines

Huprines are hybrid molecules of tacrine and Hup A and they have been described as highly potent AChE inhibitors. Synthesis, SAR, and docking calculations of several huprines derivatives have been reported based on recombinant human AChE inhibitory activity [114].

One of these hybrids, huprine X (HX), Fig. (**10**) binds to AChE with high affinity. It has been shown to tightly bind to the active site [115] and it has also been suggested to interact with the PAS [116]. HX inhibits the amyloidogenic process induced by AChE *in vitro*, and moreover, it presents agonist activity on muscarinic acetylcholine receptors (mAChRs) [117] and potential allosteric activity on nAChRs [118].

While *in vitro* pharmacological characterization of HX is relatively complete, its *in vivo* effects are needed to be investigated. Thus, Hedberg and co-workers [119] studied whether HX could affect the AD-related neuropathology *in vivo* in two mouse models. Their results provide further evidence that drugs targeting AChE affect some of the fundamental processes that contribute to neurodegeneration, but whether HX might act in a disease-modifying manner in AD patients remains to be proven.

A type of huprine derivatives heterodimeric, Fig. (**10**), consists of: (i) a unit of racemic or enantiopure huprine Y Fig. (**10**), with high-affinity reversible AChEI [116]; (ii) a unit of tacrine Fig. (**3**), which is a known AChEI and with reported affinity for both the active and PAS of AChE [120], or for 6-chlorotacrine, and

(iii) a linker between (i) and (ii) of suitable length, an oligomethylene chain of 6-10 methylene groups or a 4-methyl-4-aza-heptamethylene chain.

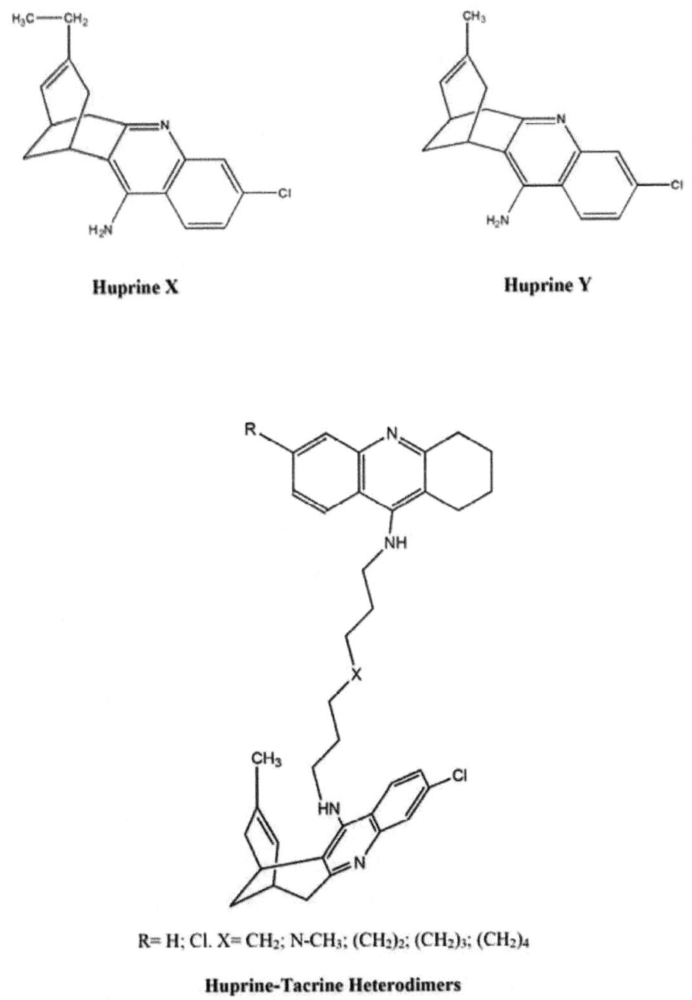

Huprine X   **Huprine Y**

R= H; Cl. X= CH₂; N-CH₃; (CH₂)₂; (CH₂)₃; (CH₂)₄

Huprine-Tacrine Heterodimers

Figure 10: Reversible inhibitors. Tacrinde hybrid derivatives.

These heterodimers have been designed to simultaneously interact with both the active and peripheral sites of AChE, and in some cases also with the aromatic residues at the midgorge of the enzyme. Their dual site binding for AChE, supported by kinetic and molecular modeling studies, gives rise to a very potent inhibition of the catalytic activity of human AChE, and moreover, to an *in vitro* neutralization of the effect of this enzyme towards Aβ aggregation. These compounds are also able to cross the BBB, as predicted in an artificial membrane model assay. Overall, huprine-tacrine heterodimers can be considered very promising lead compounds for AD [121, 122].

Tacrine-Antioxidant Hybrids

Currently, the synthesis of multifunctional compounds that combine neuroprotective effects has a special interest. In this way, a family of tacrine-antioxidant hybrids has shown hepatoprotective properties [123]. The tacrine-induced oxidative stress can be prevented by treating hepatocytes with a free radical scavenger such as vitamin E [124]. Thus, tacrine derivatives endowed with additional antioxidant properties might be beneficial by reducing its toxicity.

It has been demonstrated that oxidative damage is an event that precedes the appearance of pathological hallmarks of the AD, namely, amyloid plaques and neurofibrillary tangles [125]. During aging, the endogenous antioxidant protection system progressively decays and may be further diminished in AD. Drugs that specifically scavenger oxygen radical could be useful for either the prevention or the treatment of AD [126]. The development of tacrine derivatives endowed with additional antioxidant properties is an active field in the current AD research. Several selected compounds of this type are:

Tacrine-8-hydroxyquinoline Hybrids

New multiactive neuroprotectants with antioxidant, metal-binding properties and dual inhibition of AChE and BuChE in a simple molecule, tacrine-8-hydroxyquinoline hybrids, Fig. (**11**), have been synthesized by using moieties with well-known properties for each biological activity, such as: (i) tacrine for the inhibition of ChE through its binding to the CAS; (ii) 8-hydroxyquinoline derivative (PBT2) for its metal-chelating, neuroprotective and antioxidant properties [127] as well as for its potential interaction with the PAS; and (iii) the binding of tacrine and quinoline fragments with chains of different lengths (from 6 to 12 carbons) or with a triamine skeleton. These flexible linkers could be lodged by the enzyme cavity, allowing simultaneous interaction between the heteroaromatic fragments and both, the CAS and PAS of AChE.

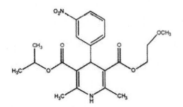

R¹=H;CH₃ R²= H;Cl
Z=(CH₂)₄; (CH₂)₅; (CH₂)₆; (CH₂)₇;
(CH₂)₈; (CH₂)₁₀;(CH₂)₂NCH₃(CH₂)₂

Tacrine-8-hydroxyquinoline hybrids

IQM – 622

n=1-4; R₁=acetyl; H R₂= H; Cl

Mercapto-Tacrine derivatives

Tacrine-4-oxo-4H-chromene hybrids

Tacrine-6-feluric acid

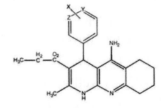

Nimodipine

Tacripyrines

Figure 11: Reversible inhibitors. Tacrine-antioxidant hybrids and nimodipine.

It is noted that PBT2 inhibits the redox-dependent formation of toxic soluble oligomers of Aβ, prevents brain Aβ deposition, and promotes its clearance [128]. Recently, PBT2 has entered in phase IIb clinical test for AD showing promised results [129].

These tacrine-8-hydroxyquinoline hybrids display interesting *in vitro* biological activities for the treatment of AD: (a) they are more potent inhibitors of human AChE and BuChE than the parent fragment tacrine; (b) they show better antioxidant properties than the aromatic portion of vitamin E, responsible for radical capture which displays neuroprotective properties against mitochondrial free radicals and; (c) they displace the PAS-specific ligand propidium from the PAS of the AChE and thus, these hybrids could be able to inhibit Aβ aggregation promoted by AChE.

Such biological properties, along with their ability to reach therapeutic targets in CNS, highlight these tacrine-PBT2 hybrids as very interesting multifunctional prototypes in the search for new disease-modifying drugs useful in the treatment of AD [130].

Recently, Antequera and co-workers [131] evaluated the efficacy of a novel tacrine-8-hydroxyquinoline hybrid, named IQM-622 Fig. (**11**), using *in vitro* and *in vivo* models. This study demonstrates the following: (a) IQM-622 has neuroprotective effects in neuronal and astrocytic cell cultures; (b) it has both antiamyloid and neuroprotective effects in a mouse model and; (c) It has proved to be a potent inhibitor of human AChE and BuChE. Moreover, IQM-622 has shown to have interesting *in vitro* biological activities for the treatment of AD including cholinergic, antioxidant, copper-complexing and neuroprotective properties. The results of this study suggest that IQM-622 holds potential for the treatment of AD-associated brain damage.

Mercapto-tacrine Hybrids

Chemical entities containing mercapto group have attracted our interest because of their multiple pharmacological actions. Due to some exogenous sulfhydryl compounds could maintain the level of intracellular glutathione, the major antioxidant in nerve cells when exposed to the ROS insults, they may also lead to neuroprotection in the CNS [132].

In this context, it has been proposed that tacrine derivatives cooperating with mercapto group in a single molecular entity may work in a synergistic manner. Thus, a series of mercapto-tacrine derivatives Fig. (11) has been designed and synthesized. These compounds displayed a synergistic pharmacological profile of long-term potentiation enhancement, ChE inhibition, neuroprotection and less hepatotoxicity. It is expected that these multifunctional compounds are more efficient to improve the memory and cognitive impairment, with less side effect, and to diminish the oxidative damage caused by free radicals. They might be good candidates for further studies directed to the development of novel drugs for AD [133].

Tacrine-4-oxo-4H-chromene Hybrids

By using fragments endowed with interesting and complementary properties for the treatments of AD, a new family of tacrine-4-oxo-4H-chromene hybrids Fig. (11) has been designed, synthesized and evaluated biologically. The flavonoid scaffold derived from 4-oxo-4H-chromene has been chosen for its antioxidant action among other activities, as well as for its potential interaction with AChE-PAS due to its aromatic character.

Flavonoids have attracted much attention in recent years because they can limit the neurodegeneration associated with a variety of neurological disorders [134]. Flavonoids mediate their effects by several routes, including their capacity to scavenge neurotoxic species, such as free radicals [135].

Among the possible structural modifications on the tacrine fragment, Fernández-Bachiller and co-workers inserted one or two chlorine atoms to study possible effects on ChE inhibition. To reach radical capture capacity, they envisaged introducing one or two phenol groups to the 4-oxo-4H-chromene fragments. In addition, they considered connecting tacrine and 4-oxo-4H-chromene fragments by alkylenediamine tethers of different lengths. The flexible linkers used could be lodged in the narrow enzymatic cavity, allowing simultaneous interaction of both the CAS and PAS of AChE.

These new hybrids showed interesting *in vitro* biological activities for the potential treatment of AD, such as inhibition of the human AChE, BuChE (being more potent than the parent inhibitor, tacrine) and the β-secretase (BACE-1), as well as radical scavenger activity, and they could be able to penetrate into the CNS. Actually, further studies are being carried out [136].

Tacrine-Feluric Acid Hybrids

Feluric acid (4-hydroxy-3-methoxycinamic acid), a bioactive component of traditional chinese medicine, has several properties such as antioxidant and anti-inflammatory effects, inhibition of Aβ fibril aggregation and hepatoprotective effects. Pi and co-workers [137] synthesized a series of hybrid compounds by linking feluric acid to tacrine as multifunctional agent. One of these hybrids, tacrine-6-feluric acid (T6FA), Fig. (**11**) has been recently evaluated. *In vitro* results demonstrated that it significantly inhibited Aβ aggregation induced by AChE and it blocked the cell death and the intracellular ROS accumulation. T6FA may be a promising multifunctional drug candidate for AD.

Tacripyrines

Tacripyrines have been designed by combining an AChEI, tacrine, with a calcium antagonist such as nimodipine Fig. (**11**). These compounds are targeted to develop multitarget therapeutic strategy to confront AD. Because 1,4-dihydropyridines (DHPs) are compounds that selectively block L-type voltage-dependent Ca^{+2} channel (VDCC), hybrid molecules that combine an AChEI and DHP, such as tacrine and nimodipine, might represent a promising approach to the AD treatment. This strategy is based on the fact that bis(7)tacrine Fig. (**5**) attenuates β-amyloid neuronal apoptosis by regulating L-type calcium channels [138].

Marco-Contelles and co-workers [139] have synthesized and evaluated a series of tacrine-DHP hybrids, named tacripyrines Fig. (**11**) that are potent and selective inhibitors of hAChE. These compounds have also been devoid of human BuChE inhibition activity, showing therefore, an extremely high selectivity. Besides inhibition of AChE and blockade of VDCC, tacripyrines protect much more efficiently against free radicals than the parent compounds, tacrine and nimodipine, that have not shown any neuroprotective effect. Tacripyrines are neuroprotective agents, also they show moderate Ca^{+2} channel blocking effect and they cross the BBB. Therefore, they can be considered as interesting new chemical entity candidates for AD treating [139]. New tacripyrines are being currently studied [140].

5.4.2. Donepezil Hybrids

A new series of donepezil-tacrine hybrids related derivatives Fig. (**12a**) has been synthesized as dual AChEIs, that could bind simultaneously to the peripheral and to the catalytic sites of the enzyme. The molecular structure of these hybrids contains: (i) the tacrine heterocyclic ring or 6-chlorotacrine, which is recognized as a catalytic site AChEI, (ii) the indanone ring (the heterocyclic present in donepezil) or the related heterocyclic, the phthalimide moiety, as responsible for the binding to the PAS of the enzyme, and (iii) the linker connecting (i) and (ii) through a different tether length.

Biological activity and molecular modeling studies have been performed in these compounds to explore their binding to the enzyme. Thus, it has been suggested that the phtalimide moiety acts as an efficient ligand for the PAS. As it has been found for the AChE inhibition, the best binding to the PAS of AChE corresponds to a tether length between the two anchoring groups (9-aminoacridine and phthalimide) of nine units [141].

The synthesis, pharmacological evaluation (AChE and BuChE inhibition and Aβ-antiaggregating effect) and molecular modeling of other class of highly potent donepezil-tacrine hybrids have also been described, Fig (**12b**). On the basis of the binding modes of donepezil [78] and tacrine [27] within tcAChE, these novel hybrids have been designed by combining the 5,6-dimethoxy-2-[(4-piperidinyl) methyl]-1-indanone moiety of donepezil with tacrine.

All these new compounds are highly potent inhibitors of bovine and human AChE and BuChE. The most potent AChEIs are those bearing an indanone system, a chloride atom at the tacrine unit and, a tether length of three methylenes. These results suggest that the novel donepezil-tacrine hybrids herein reported may have a potential disease-modifying role in the treatment of AD [142].

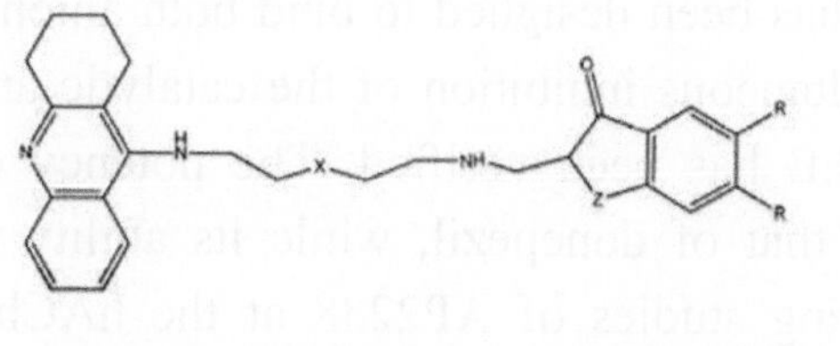

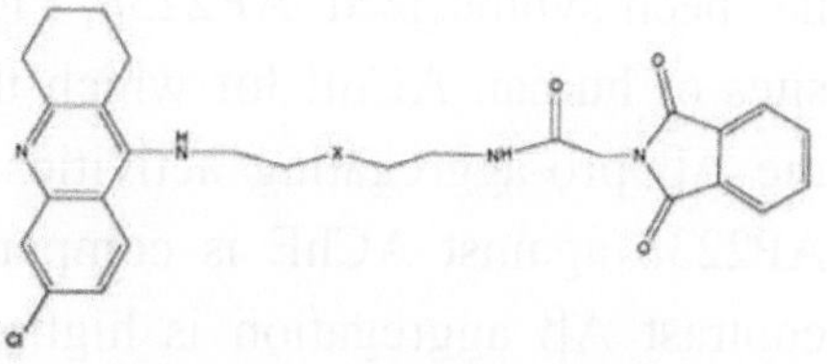

Z=H; CO.   R=OCH₃; H.   X= (CH₂)₂; (CH₂)₃   X= (CH₂)₂; (CH₂)₃; CH₃N(CH₂)₂

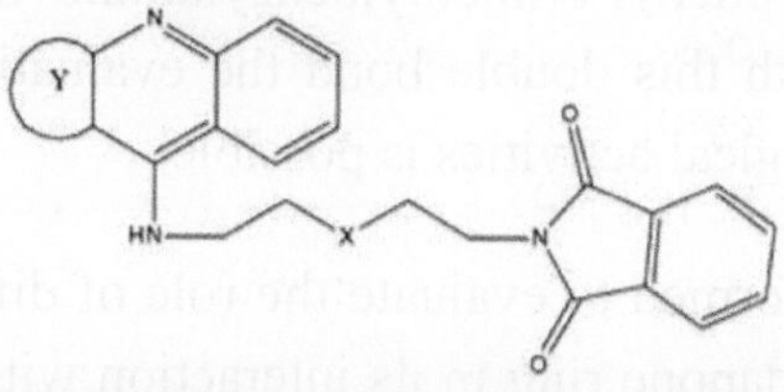

X= (CH₂)₂; (CH₂)₃; (CH₂)₄; (CH₂)₅.   Y= C₄H₄; C₄H₈

a. Donepezil-Tacrine hybrids

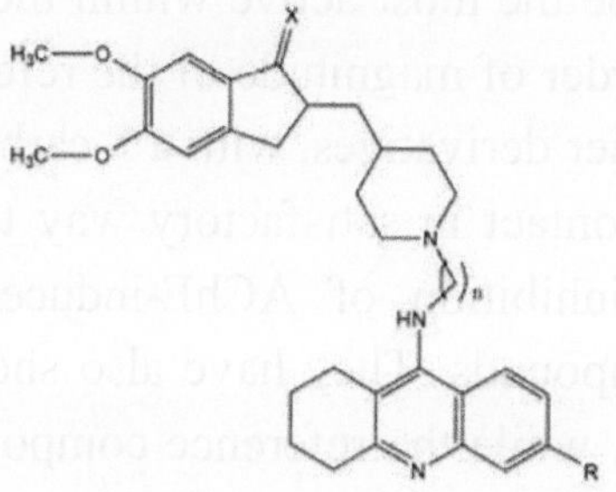

X=O;H   R=H;Cl   n=2-3

b. Donepezil-Tacrine hybrids

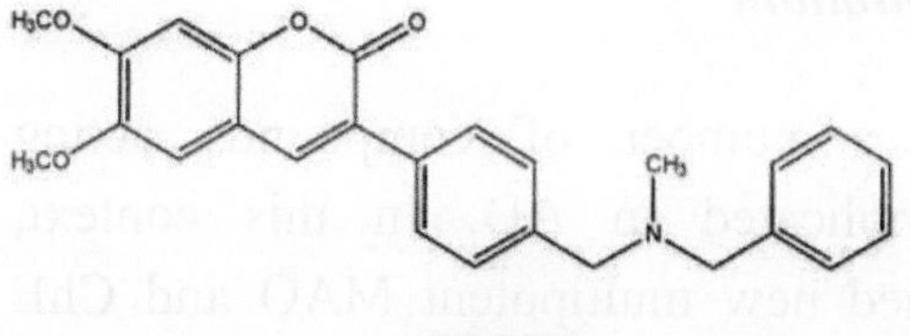

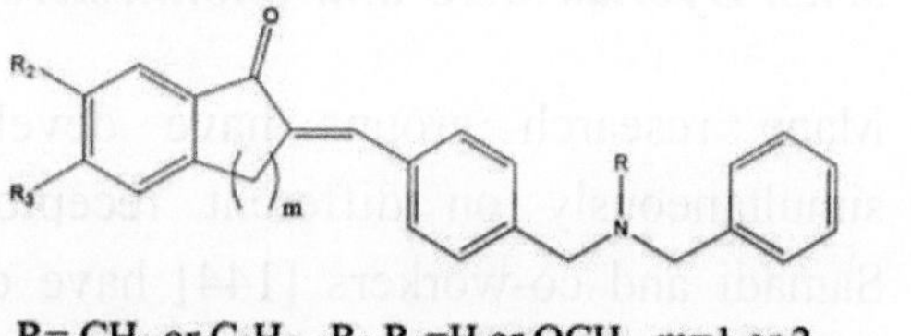

AP-2238   R= CH₃ or C₂H₅ . R₂,R₃=H or OCH₃. m=1 or 2

Donepezil- AP2238 hybrids

Figure 12: Reversible inhibitors. Donepezil hybrids and AP2238.

Other Donepezil Hybrids

A series of hybrid compounds structurally derived from donepezil and AP2238 has been synthesized. AP2238, Fig. (**12**), has been designed to bind both anionic sites of human AChE for which the simultaneous inhibition of the catalytic and the Aβ pro-aggregating activities of AChE has been verified. The potency of AP2238 against AChE is comparable to that of donepezil, while its ability to contrast Aβ aggregation is higher. Docking studies of AP2238 at the hAChE gorge have shown several interactions [143]. Thus, the indanone core from donepezil is linked to the phenyl-N-methylbenzylamino moiety from AP2238 through a double bond. With this double bond the evaluation of the decreasing linker flexibility in the biological activities is possible.

SAR studies have been performed to evaluate the role of different substituents in the position 5 or 6 of the indanone ring in its interaction with the PAS, as well as of different alkyl chains of several lengths carrying diverse amines at one end. Furthermore, the indanone ring itself has been replaced by a tetralone scaffold. Two compounds proved to be the most active within the series and their potency against AChE in the same order of magnitude of the reference compounds. These compounds together with other derivatives, with a 5-carbon alkyl chain bearing an amino moiety at one end, contact in satisfactory way to the PAS. This binding remarkably improves the inhibition of AChE-induced Aβ aggregation with respect to the reference compounds. They have also shown activity against self-aggregation of Aβ$_{42}$ peptide, while the reference compounds resulted ineffective. These three compounds mentioned could represent new templates for further optimization studies [80].

5.4.3. Hybrids MAO and Cholinesterase Inhibitors

Many research groups have developed a number of compounds acting simultaneously on different receptors implicated in AD. In this context, Samadi and co-workers [144] have designed new multipotent MAO and ChE inhibitors for the potential treatment of AD. They reported the synthesis and pharmacological evaluation of hybrids from donepezil and PF9601N, Fig. (**13**).

Last one is a well known MAO inhibitor [145], bearing N-benzyl piperidine and propargylamine moieties attached to a central pyridine or naphthyridine ring.

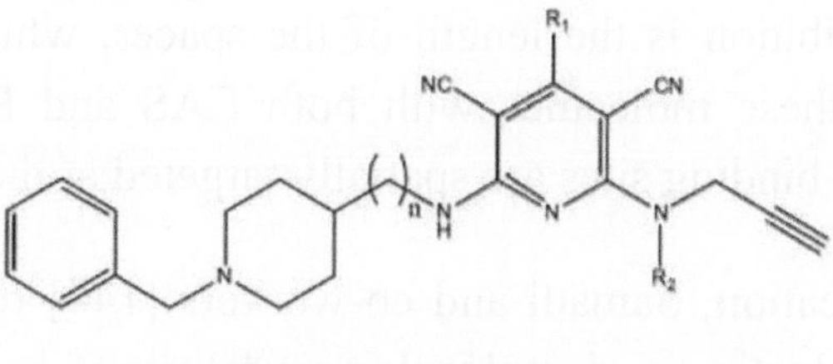

R₁=C₆H₅; H.   R₂= CH₃; H.   n=0-3

a-h

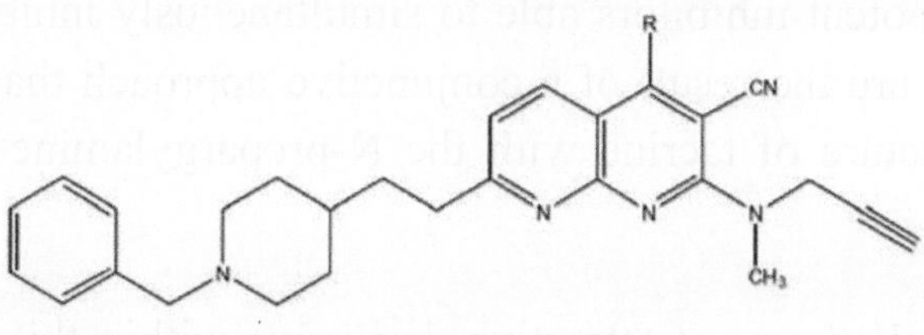

i R= C₆H₅   **j** R=H

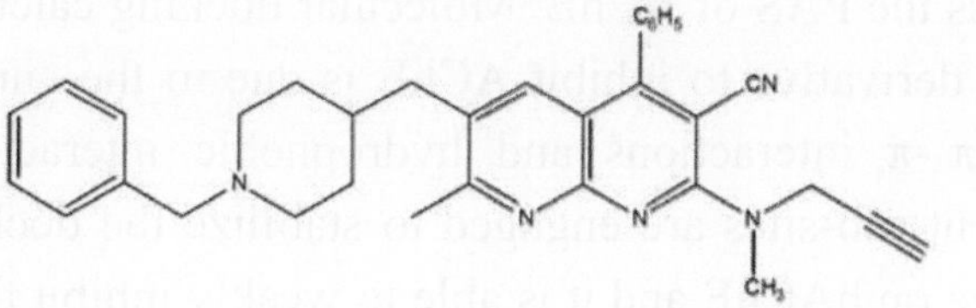

k

Hybrids from donepezil and PF9601N

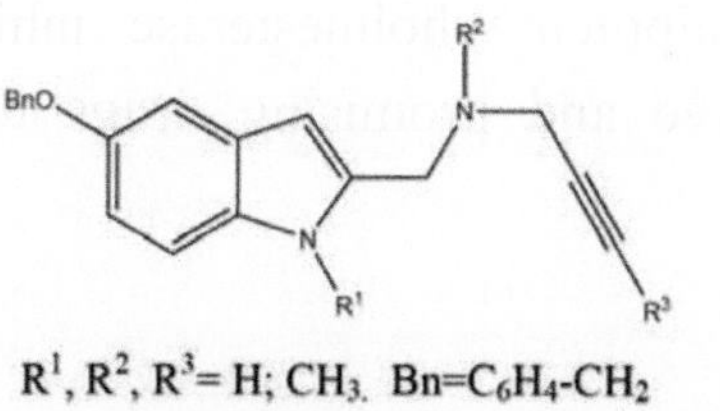

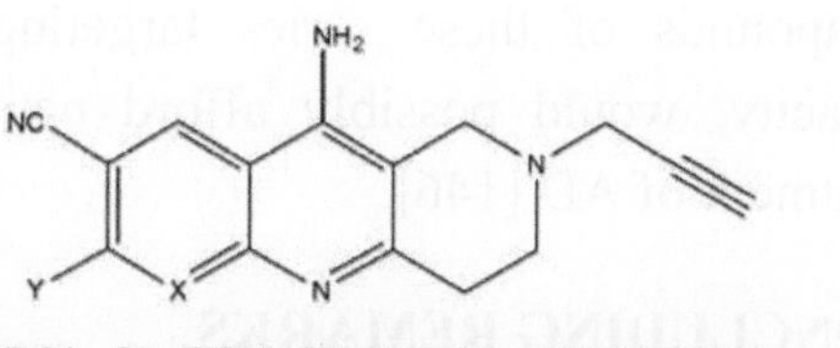

R¹, R², R³= H; CH₃.   Bn=C₆H₄-CH₂   X= CH; N.   Y= OCH₃; Cl; N(CH₃)₂; NC₄H₈;NC₅H₁₀

type I   **type II**

Heterocyclic substituted alkyl and cycloalkyl propargyl amines type I and type II

Figure 13: Reversible inhibitors. Hybrids MAO and ChEIs.

The compounds **h** (R_1=H; R_2=H; n=2) and particularly **k**, Fig. (**13**), have shown a strong and selective AChE inhibitory activity and moderate, but selective MAO-A inhibitory profile. The authors conclude that the most sensitive moiety to modulate AChE inhibition is the length of the spacer, which would control the dual interaction of these molecules with both CAS and PAS sites, improving inhibition when both binding sites are spatially targeted at the same time [144].

In a latter communication, Samadi and co-workers [144] reported the synthesis, pharmacological evaluation and molecular modeling of heterocyclic substituted alkyl and cycloalkyl propargylamine type I and type II, Fig. (**13**), that were designed as multipotent inhibitors able to simultaneously inhibit MAO and ChE. Molecules type II are the result of a conjunctive approach that combines for the first time the structure of tacrine with the N-propargylamine moiety present in PF9601N.

In molecules type II the most attractive derivative within this series is that with X=N and Y=NC_5H_{10}, which is a non-competitive inhibitor of tcAChE, that it preferentially binds the PAS of AChE. Molecular docking calculations reveal that the ability of this derivative to inhibit AChE is due to the cumulative effects of hydrogen bonds, π -π interactions and hydrophobic interactions. Amino acid residues in different sub-sites are engaged to stabilize the docked complex. This compound is active on hAChE and it is able to weakly inhibit the pro-aggregating action exerted by hAChE on amyloid. Thus, it may also be considered for further development aimed at enlarging its biological activity.

The authors suggested that a hybrid molecule resulting from the juxtaposition of compounds of these series targeting an equipotent cholinesterase inhibition capacity, would possibly afford new attractive and promising drugs for the treatment of AD [146].

CONCLUDING REMARKS

The development of effective drugs against AD is an acute clinical need. The difficult task today of developing newer AChEIs is that they will need to be more effective than those actually FDA approved.

The search for AChEIs bind to its CAS and PAS has become an area of very active research. Particularly attractive is the multi-target-directed ligand, approach based on the "one-molecule, multiple-target" paradigm. This innovative strategy has provided a number of compounds acting simultaneously on different receptors and enzymatic systems implicated in AD.

ACKNOWLEDGEMENTS

Declared none.

CONFLICT OF INTEREST

The authors confirm that this chapter contents have no conflict of interest.

ABBREVIATIONS

Aβ	=	amyloid-β
ACh	=	acetylcholine
AChE	=	acetylcholinesterase
AChEI	=	acetylcholinesterase inhibitor
AD	=	Alzheimer's disease
APLs	=	allosterically potentiating ligands
BBB	=	blood-brain barrier
B12H	=	bis(12)-hupyridone
β-APP	=	β-amyloid precursor protein
BuChE	=	butyrylcholinesterase
CAS	=	catalytic active site

ChE	=	cholinesterases
ChEI	=	cholinesterases inhibitor
CNS	=	central nervous system
D	=	aspartate
DHPs	=	1,4-dihydropyridines
E	=	glutamate
E2020	=	donepezil
F	=	phenylalanine
FDA	=	Food and Drug Administration
G_1	=	globular monomer form acetylcholinesterase
G_2	=	globular dimer form acetylcholinesterase
G_4	=	globular tetramer form acetylcholinesterase
H	=	histidine
hAChE	=	human acetylcholinesterase
Hup A	=	huperzine A
Hup B	=	huperzine B
HX	=	hupryne X
mAChRs	=	muscarinic acetylcholine receptors
MAO	=	monoamine oxidase
MAO-A	=	monoamine oxidase isozyme A

MAO-B   =   monoamine oxidase isozyme B

nAChRs =   nicotinic acetylcholine receptors

PAS   =   peripheral anionic site

PBT2   =   8-hydroxyquinoline derivative

ROS   =   reactive oxygen species

S   =   serine

SAR   =   structure-activity relationship

tcAChE =   *torpedo californica* acetylcholinesterase

T6FA   =   tacrine-6-feluric acid

THA   =   tacrine

VDCC   =   voltage-dependent Ca^{+2} channel

W   =   tryptophan

Y   =   tyrosine

REFERENCES

[1]   Querfurth, H. W.; LaFerla F.M. Alzheimer's disease. *N. Engl. J. Med.*, **2010**, *362*(4), 329-344.

[2]   Bartus, R.T. On neurodegenerative diseases, models, and treatment strategies: lessons learned and lessons forgotten a generation following the cholinergic hypothesis. *Exp. Neurol.*, **2000**, *163(2)*, 495-529.

[3]   Terry Jr, A.V.; Buccafusco, J.J. The cholinergic hypothesis of age and Alzheimer's disease-related cognitive deficits: recent challenges and their implications for novel drug development. *J. Pharmacol. Exp. Ther.*, **2003**, *306(3)*, 821-827.

[4]   Rakonczay, Z. Potencies and selectivities of inhibitors of acetylcholinesterase and its molecular forms in normal and Alzheimer's disease brain. *Acta Biol. Hung.*, **2003**, *54*(2), 183-189.

[5]   Tayeb, H.O.; Yang, H.D.; Price, B.H.; Tarazi, F.I. Pharmacotherapies for Alzheimer's disease: beyond cholinesterase inhibitors. *Pharmacol. Ther.*, **2012**, *134*(1), 8-25.

[6] Bolognesi, M.L.; Andrisano, V.; Bartolini, M.; Cavalli, A.; Minarini, A.; Recanatini, M.; Rosini, M.; Tumiatti, V.; Melchiorre, C. Heterocyclic inhibitors of AChE acylation and peripheral sites. *Farmaco*, **2005**, *60(6-7)*, 465-473.

[7] Barril, X.; Orozco, M.; Luque, F.J. Towards improved acetylcholinesterase inhibitors: a structural and computational approach. *Mini Rev. Med. Chem.*, **2001**, *1(3)*, 255-266.

[8] Sussman, J.L.; Harel, M.; Frolow, F.; Oefner, C.; Goldman, A.; Toker, L.; Silman, I. Atomic structure of acetylcholinesterase from Torpedo californica: a prototypic acetylcholine-binding protein. *Science,* **1991**, *253(5022)*, 872-879.

[9] Pohanka, M. Cholinesterases, a target of pharmacology and toxicology. *Biomed. Pap. Med. Fac. Univ. Palacky Olomouc Czech. Repub.*, **2011**, *155* (3), 219-229.

[10] Botti, S.A.; Felder, C.E.; Lifson, S.; Sussman, J.L.; Silman, I.A. A modular treatment of molecular traffic through the active site of cholinesterase. *Biophys. J.*, **1999**, *77*, 2430-2450.

[11] Guo, J.; Hurley, M. M.; Wright, J. B.; Lushington, G. H. A docking score function for estimating ligand-protein interactions: application to acetylcholinesterase inhibition. *J. Med. Chem.*, **2004**, *47(22)* , 5492-5500.

[12] Ripoll, D.R.; Faerman, C.H; Axelsen, P.H.; Silman, I.; Sussman, J. L. An electrostatic mechanism for substrate guidance down the aromatic gorge of acetylcholinesterase. *Proc. Natl. Acad. Sci. USA*. **1993**, *90(11)*, 5128-5132.

[13] Koellner, G.; Steiner, T.; Millard, C.B.; Silman, I; Susssman, J. L. A neutral molecule in a cation-binding site: specific binding of a PEG-SH to acetylcholinesterase from Torpedo californica. *J. Mol. Biol.*, **2002**, *320(4)*, 721-725.

[14] Delgado, A.; Minguillon, C.; Joglar, J. Introducción a la Química terapéutica. 2° ed.; Díaz de Santos: Madrid, 2004..

[15] Cannon, J.G. Cholinergic. In: Burger's Medicinal Chemistry and drug discovery; Abraham, A.J. Ed.; 6 th ed; John, Wiley of Sons: New Jersey, **2003**; Vol. *6*, pp. 39-108.

[16] Kryger, G.; Silman, I.; Sussman, J.L. Three-dimensional structure of a complex of E2020 with acetylcholinesterase from Torpedo californica. *J. Physiology*, **1998**, *92(3-4)*, 191-194.

[17] Johnson, G.; Moore, S.W. The peripheral anionic site of acetylcholinesterase: structure, functions and potential role in rational drug design. *Curr. Pharm. Des.*, **2006**, *12(2),* 217-225.

[18] Bourne, Y.; Taylor, P.; Radic, Z.; Marchot, P. Structural insights into ligand interactions at the acetylcholinesterase peripheral anionic site. *EMBO J.*, **2003**, *22* (1), 1-12.

[19] Rakonczay, Z.; Brimijoin, S. Monoclonal antibodies to human brain acetylcholinesterase: properties and applications. *Cell. Mol. Neurobiol.*, **1988**, *8(1)*, 85-93.

[20] Reyes, A.E.; Chacón, M.A.; Dinamarca, M.C.; Cerpa, W.; Morgan, C.; Inestrosa, N.C. Acetylcholinesterase-Abeta complexes are more toxic than Abeta fibrils in rat hippocampus: effect on rat beta- amyloid aggregation, laminin expression, reactive astrocytosis, and neuronal cell loss. *Am. J. Pathol.*, **2004**, *164(6)*, 2163-2174 .

[21] Campos, E.O.; Alvarez, A.; Inestrosa, N.C. Brain acetylcholinesterase promotes amyloid-beta-peptide aggregation but does not hydrolyze amyloid precursor protein peptides. *Neurochem. Res.*, **1998**, *23(2)*, 135-140.

[22] Cavalli, A.; Bottegoni, G.; Raco, C.; De Vivo, M.; Recanatini, M. A computational study of the binding of propidium to the peripheral anionic site of human acetylcholinesterase. *J. Med. Chem.*, **2004**, *47(16)*, 3991-3999.

[23] Castro, A.; Martínez, A. Targeting beta-amyloid pathogenesis through acetylcholinesterase inhibitors. *Curr. Pharm. Des.*, **2006**, *12(33)*, 4377-4387 .

[24]   Martín-Santamaría, S.; Muñoz-Muriedas, J.; Luque, F. J.; Gago, F. Modulation of binding strength in several classes of active site inhibitors of acetylcholinesterase studied by comparative binding energy analysis. *J. Med. Chem.*, **2004**, *47(18)*, 4471-4482.

[25]   Mizutani, M.Y.; Itai, A. Efficient method for high-throughput virtual screening based on flexible docking: discovery of novel acetylcholinesterase inhibitors. *J. Med. Chem.*, **2004**, *47(20)*, 4818-4828.

[26]   da Silva, V.B .; de Andrade, P.; Kawano, D.F.; Morais, P.A.; de Almeida, J.R.; Carvalho, I.; Taft, C.A.; da Silva, C.H. In silico design and search for acetylcholinesterase inhibitors in Alzheimer´s disease with a suitable pharmacokinetic profile and low toxicity. *Future Med. Chem.*, **2011**, 3 (8), 947-960.

[27]   Harel, M.; Schalk, I.; Ehret-Sabatier, L.; Bouet, F.; Goeldner, M.; Hirth, C.; Axelsen, P.H.; Silman, I.; Sussman, J.L. Quaternary ligand binding to aromatic residues in the active-site gorge of acetylcholinesterase. *Proc. Natl. Acad. Sci. USA.*, **1993**, *90(19), 9031-9035.

[28]   Greenblatt, H.M.; Kryger, G.; Lewis, T.; Silman, I.; Sussman, J.L. Structure of acetylcholinesterase complexed with (-)- galanthamine at 2.3 A resolution. *FEBS Lett.*, **1999**, *463*(3), 321-326.

[29]   Bartolucci, C.; Perola, E.; Pilger, C.; Fels, G.; Lamba, D. Three-dimensional structure of a complex of galanthamine (Nivalin) with acetylcholinesterase from Torpedo californica: implications for the design of new anti-Alzheimer drugs. *Proteins*, **2001**, *42(2)*, 182-191.

[30]   Raves, M.L.; Harel, M.; Pang, Y.P.; Silman, I.; Kozikowski, A.P.; Sussmann, J.L. Structure of acetylcholinesterase complexed with the nootropic alkaloid, (-)-huperzine A. *Nat. Struct. Biol.*, **1997**, *4(1)*, 57-63.

[31]   Kadir, A.; Darreh-Shori, T.; Almkvist, O.; Wall A. Langström, B.; Nordberg, A. Changes in brain ^{11}C-nicotine binding sites in patients with mild Alzheimer's disease following rivastigmine treatment as assessed by PET. *Psychopharmacology (Berl)*, **2007**, *191(4)*, 1005-1014.

[32]   Friedman, A.; Kaufer, D.; Shemer, J.; Hendler, I.; Soreq, H.; Tur-Kaspa, I. Pyridostigmine brain penetration under stress enhances neuronal excitability and induces early immediate transcriptional response. *Nat. Med.*, **1996**, *2(12)*, 1382-1385 .

[33]   Darvesh, S.; Darvesh, K.V. ;McDonald, R.S.; Mataija, D.; Walsh, R.; Mothana, S.; Lockridge, O.; Martin, E. Carbamates with differential mechanism of inhibition toward acetylcholinesterase and butyrylcholinesterase. *J. Med. Chem.*, **2008**, *51(14)*, 4200-4212.

[34]   Yu, Q.; Greig N.H.; Holloway, H.W.; Brossi, A. Synthesis and anticholinesterase activities of (3aS)-N1, N8-bisnorphenserine, (3aS)-N1, N8-bisnorphysostigmine, their antipodal isomers, and other potential metabolites of phenserine. *J. Med. Chem.*, **1998**, *41(13)*, 2371-2379 .

[35]   Giacobini, E.; Spiegel, R.; Enz, A.; Veroff, A.E.; Cutler, N.R. Inhibition of acetyl-and butyryl-cholinesterase in the cerebrospinal fluid of patients with Alzheimer´s disease by rivastigmine: correlation with cognitive benefit. *J. Neural. Transm.*, **2002**, *109(7-8)*, 1053-1065.

[36]   Volger, B.W. Alternatives in the treatment of memory loss in patients with Alzheimer's disease. *Clin. Pharm.*, **1991**, *10(6)*, 447-456.

[37]   Storch, A.; Schrattenholz, A., Cooper, J.C., Abdel Ghani, E.M., Gutbrod, O., Weber, K.H., Reinhardt, S., Lobron, C., Hermsen, B., Soskic, V. Physostigmine, galanthamine and codeine act as'noncompetitive nicotinic receptor agonists'on clonal rat pheochromocytoma cells. *Eur. J. Pharmacol.*, **1995**, *290* (3), 207-219.

[38]   Iorga, B.; Herlem, D.; Barré, E.; Guillou, C. Acetylcholine nicotinic receptors: finding the putative binding site of allosteric modulators using the "blind docking" approach. *J. Mol. Model.*, **2006**, *12*(3), 366-372.

[39]   Luttmann, E.; Ludwig, J.; Höffle-Maas, A.; Samochocki, M.; Maelicke, A.; Fels, G. Structural model for the binding sites of allosterically potentiating ligands on nicotinic acetylcholine receptors. *Chem. Med. Chem.*, **2009**, *4*(11), 1874-1882.

[40]   Svobodová, L,; Krusek, J.; Hendrych, T.; Vyskocil, F. Physostigmine modulation of acetylcholine currents in COS cells transfected with mouse muscle nicotinic receptor. *Neurosci Lett.* ,**2006**, *401* (1-2), 20-24.

[41]   Orhan, G.; Orhan, I.; Subutay -Öztekin, N.; Ak, F.; Sener, B. Contemporary anticholinesterase pharmaceuticals of natural origin and their synthetic analogues for the treatment of Alzheimer's disease. *Recent Pat. on CNS Drug Discov.*, **2009**, *4*(1), 43-51 .

[42]   Maelicke, A.; Schrattenholz, A.; Samochocki, M.; Radina, M.; Albuquerque, E.X. Allosterically potentiating ligands of nicotinic receptors as a treatment strategy for Alzheimer's disease. *Behav. Brain Res.*, **2000**, *113(1-2)*, 199-206 .

[43]   Rupniak, N.M.; Tye, S.J.; Brazell, C.; Heald, A.; Iversen, S.D.; Pagella, P.G. Reversal of cognitive impairment by heptyl physostigmine, a long-lasting cholinesterase inhibitor, in primates. *J. Neurol. Sci.*, **1992**, *107(2)*, 246-249.

[44]   Liston, D.R.; Nielsen, J.A.; Villalobos, A.; Chapin, D.; Jones, S.B..; Hubbard, S.T.; Shalaby, I.A.; Ramirez, A.; Nason, D.; White, W.F. Pharmacology of selective acetylcholinesterase inhibitors: implications for use in Alzheimer´s disease. *Eur. J. Pharmacol.*, **2004**, *486(1),* 9-17.

[45]   Ogane, N.; Giacobini, E.; Messamore, E. Preferential inhibition of acetylcholinesterase molecular forms in rat brain. *Neurochem. Res.*, **1992**, *17(5)*, 489-495.

[46]   Ogane, N.; Giacobini, E.; Struble, R. Differential inhibition of acetylcholinesterase molecular forms in normal and Alzheimer disease brain. *Brain. Res.*, **1992**, *589(2)*, 307-312.

[47]   Imbimbo, B.P.; Martelli, P.; Troetel, W.M.; Lucchelli, F.; Lucca, U.; Thal, L.J. Efficacy and safety of eptastigmine for the treatment of patients with Alzheimer's disease. *Neurology.*, **1999**, *52*(4), 700-708 .

[48]   Braida, D.; Sala, M. Eptastigmine: ten years of pharmacology, toxicology, pharmacokinetic, and clinical studies. *CNS Drug Rev.*, **2001**, *7*(4), 369-386.

[49]   Greig, N.H.; Pei, X.F.; Soncrant, T.T.; Ingram, D.K.; Brossi, A. Phenserine and ring C hetero-analogues: drugs candidates for the treatment of Alzheimer's disease *Med. Chem. Res.*, **1995**, *15(1)*, 3-31.

[50]   Shaw, K.T.; Utsuki, T.; Rogers, J.; Yu, Q.S.; Sambamurti, K.; Brossi, A.; Ge, Y.W., Lahiri, D.K.; Greig, N.H. Phenserine regulates translation of beta-amyloid precursor protein mRNA by a putative interleukin-1 responsive element, a target for drug development .*Proc. Natl. Acad. Sci. U.S.A.*, **2001**, *98(13)*, 7605-7610.

[51]   Winblad, B.; Giacobini, E.; Frölich, L.; Friedhoff, L.T.; Bruinsma, G.; Becker, R.E.; Greig, N.H. Phenserine efficacy in Alzheimer's disease. J. Alzheimers Dis., **2010**, *22(4)*, 1201-1218.

[52]   Zhan, Z.J.; Bian, H.L.; Wang, J.W.; Shan, W.G. Synthesis of physostigmine analogues and evaluation of their anticholinesterase activities. *Bioorg. Med. Chem. Lett.*, **2010**, *20(5)*, 1532-1534.

[53]   Mehta, M.; Adem, A.; Sabbagh, M. New acetylcholinesterase inhibitors for Alzheimer's disease. *Int. J. Alzheimers Dis.*, **2012**, doi:10.1155/2012/728983.

[54]   Yu, Q.S.; Atack, J.R.; Rapoport, S.I.; Brossi, A. Carbamate analogues of (-)-physostigmine: *in vitro* inhibition of acetyl- and butyrylcholinesterase. *FEBS Lett.*, **1988**, *234(1)*, 127-130 .

[55]   Kapil, R.; Dhawan, S.; Beg, S.; Singh, B. Buccoadhesive films for once-a-day administration of rivastigmine: systematic formulation development and pharmacokinetic evaluation. *Drug. Dev. Ind. Pharm.*, **2012**, doi:10.3109/03639045.2012.665926.

[56]   Williams, B.R.; Nazarians, A.; Gill, M.A. A review of rivastigmine: a reversible cholinesterase inhibitior. *Clin. Ther.*, **2003**, *25(6)*, 1634-1653.

[57]   Bullock, R. The clinical benefits of rivastigmine may reflect its dual inhibitory mode of action: an hypothesis. *Int. J. Clin. Pract.*, **2002**, *56(3)*, 206-214.

[58]   Isik, A.T.; Bozoglu, E.; Eker, D. AChE and BuChE inhibition by rivastigmin have no effect on peripheral insulin resistance in elderly patients with Alzheimer disease. *J. Nutr. Health Aging.*, **2012**, *16*(2), 139-141.

[59]   Venneri,A.; McGeown, W.J.; Shanks, M.F. Empirical evidence of neuroprotection by dual cholinesterase inhibition in Alzheimer's disease. *Neuroreport.*, **2005**, *16(2)*, 107-110 .

[60]   Venneri,A.; Lane R. Effects of cholinesterase inhibition on brain white matter volume in Alzheimer's disease. *Neuroreport.*, **2009**, *20(3)*, 285-288.

[61]   Shanks, M; Kivipelto, M.; Bullock, R.; Lane, R. Cholinesterase inhibition: is there evidence for disease-modifying effects?. *Curr. Med. Res. Opin.*, **2009**, *25(10)*, 2439-2446 .

[62]   Zheng, H.; Weiner, LM.; Bar-Am, O.; Epsztejn, S.; Cabantchik, Zl.; Warshawsky, A.; Youdim, M.B.; Fridkin, M. Design, synthesis, and evaluation of novel bifunctional iron-chelators as potential agents for neuroprotection in Alzheimer's, Parkinson's, and other neurodegenerative diseases. *Bioorg. Med. Chem.*, **2005**, *13(3)*, 773-783.

[63]   Weinreb, O.; Amit, T.; Bar-Am, O.; Youdim, MB. A novel anti- Alzheimer's disease drug, ladostigil neuroprotective, multimodal brain-selective monoamine oxidase and cholinesterase inhibitor. *Int. Rev. Neurobiol.*, **2011**, *100*, 191-215.

[64]   Weinstock, M.; Luques, L.; Poltyrev, T.; Bejar, C.; Shoham, S. Ladostigil prevents age-related glial activation and spatial memory deficits in rats. *Neurobiol. Aging.*, **2011**, *32*(6), 1069-1078.

[65]   Weinreb, O.; Amit, T.; Bar-Am, O.; Youdim, MB. Ladostigil: a novel multimodal neuroprotective drug with cholinesterase and brain-selective monoamine oxidase inhibitory activities for Alzheimer's disease treatment. *Curr. Drug. Targets.*, **2012**, *13*(4), 483-494.

[66]   Panarsky, R.; Luques, L.; Weinstock, M.; Anti-inflammatory effects of ladostigil and its metabolites in aged rat brain and in microglial cells. *J. Neuroimmune Pharmacol.*, **2012**, *7*(2), 488-498.

[67]   Heilbronn, E. Inhibition of cholinesterases by tetrahydroaminacrin. *Acta Chem. Scand.*, **1961**, *15(6)*, 1386-1390.

[68]   Tumiatti, V.; Minarini, A.; Bolognesi, M.L.; Milelli, A.; Rosini, M.; Melchiorre, C. Tacrine derivatives and Alzheimer's disease. *Curr. Med. Chem.*, **2010**, *17*(17), 1825-1838.

[69]   Watkins, P.B.; Zimmerman, H.J.; Knapp, M.J.; Gracon, S.I.; Lewis, K.W. Hepatotoxic effects of tacrine administration in patients with Alzheimer's disease. *JAMA*, **1994**, *271(13)*, 992-998.

[70]   Ezoulin, M.J.; Dong, C.Z.; Liu, Z.; Li, J.; Chen, H.Z.; Heymans, F.; Leliévre, L.; Ombetta, J.E.; Massicot, F. Study of PMS777, a new type of acetylcholinesterase inhibitor, in human

HepG2 cells. Comparison with tacrine and galanthamine on oxidative stress and mitochondrial impairment. *Toxicol. In Vitro*, **2006**, *20(6)*, 824-831 .

[71]   Bajgar, J.; Bisso, G.M.; Michalek, H. Differential inhibition of rat brain acetylcholinesterase molecular forms by 7-methoxytacrine *in vitro*. *Toxicol. Lett.*, **1995**, *80(1-3)*, 109-114.

[72]   Pohanka, M.; Kuca, K.; Kassa, J. New performance of biosensor technology for Alzheimer's disease drugs: *in vitro* comparison of tacrine and 7-methoxytacrine. *Neuro. Endocrinol. Lett.*, **2008**, *29(5)*, 755-758.

[73]   Korabecny, J.; Musilek, K.; Holas, O.; Binder, J.; Zemek, F.; Marek, J.; Pohanca, M.; Opletalova, V.; Dohnal, V.; Kuca, K.,. Synthesis and *in vitro* evaluation of N-alkyl-7-methoxytacrine hydrochlorides as potential cholinesterase inhibitors in Alzheimer disease. *Bioorg. Med. Chem. Lett.*, **2010**, *20* (20), 6093-6095.

[74]   Minarini, A.; Milelli, A.; Tumiatti, V.; Rosini, M.; Simoni, E.; Bolognesi, M.L.; Andrisano, V.; Bartolini, M.; Motori, E.; Angeloni, C.; Hrelia, S. Cystamine-tacrine dimer: a new multi-target-directed ligand as potential therapeutic agent for Alzheimer's disease treatment. *Neuropharmacology*, **2012**, *62(2)*, 997-1003.

[75]   Wood, P.L.; Khan, M.A.; Moskal, J. R. Cellular thiol pools are responsible for sequestration of cytotoxic reactive aldehydes: central role of free cysteine and cysteamine. *Brain. Res.*, **2007**, *1158*, 158-163.

[76]   Birks, J. Cholinesterase inhibitors for Alzheimer's disease. *Cochrane Database Syst. Rev.* **2006**, *(1), CD005593.*

[77]   Howard, R.; McShane, R.; Lindesay, J.; Ritchie, C.; Baldwin, A.; Barber, R.; Burns, A.; Dening, T.; Findlay, D.; Holmes, C.; Hughes, A.; Jacoby, R.; Jones, R.; Jones, R.; McKeith, I.; Macharouthu, A.; O'Brien, J.; Passmore, P.; Sheehan, B.; Juszczak, E.; Katona, C.; Hills, R.; Knapp, M.; Ballard, C.; Brown, R.; Banerjee, S.; Onions, C.; Griffin, M.; Adams, J.; Gray, R.; Johnson, T.; Bentham, P.; Phillips, P. Donepezil and Memantine for moderate-to-severe Alzheimer's disease . *N. Engl. J. Med.*, **2012**, *366(10)*, 893-903.

[78]   Kryger, G.; Silman, I.; Sussman, J.L. Structure of acetylcholinesterase complexed with E2020 (Aricept): implications for the design of new anti-Alzheimer drugs. *Structure.*,**1999**, *7(3)*, 297-307.

[79]   Villalobos, A.; Blake, J.F.; Biggers, C.K.; Butler, T.W.; Chapin, D.S.;Chen, Y.L.;Ives, J.L.; Jones, S.B.; Liston, D.R.; Nagel, A.A.; Nason, D.M.; Nielsen, J.A.; Shalaby, L.A.; White, W.F. Novel benzisoxazole derivatives as potent and selective inhibitors of acetylcholinesterase. *J. Med. Chem.*, **1994**, *37(17)*, 2721-2734.

[80]   Rizzo, S.; Bartolini, M.; Ceccarini, L.; Piazzi, L.; Gobbi, S.; Cavalli, A.; Recanatini, M.; Andrisano, V.; Rampa, A. Targeting Alzheimer's disease: novel indanone hybrids bearing a pharmacophoric fragment of AP2238. Bioorg. Med. Chem., **2010**, *18(5)*, 1749-1760.

[81]   Cummings, J.L. Use of cholinesterase inhibitors in clinical practice: evidence-based recommendations. *Am. J. Geriatr. Psychiatry*, **2003**, *11*(2), 131-145.

[82]   Isik, A.T.; Yildiz, G.B.; Bozoglu, E.; Yay, A.; Aydemir, E. Cardiac safety of donepezil in elderly patients with Alzheimer disease. *Intern. Med.*, **2012**, *51(6)*, 575-578.

[83]   Ishihara, Y.; Goto, G.; Miyamoto, M. Central selective acetylcholinesterase inhibitor with neurotrophic activity: structure-activity relationships of TAK-147 and related compounds. *Curr. Med. Chem.*, **2000**, *7(3)*, 341-354.

[84]   Hatip-Al-Khatib, I.; Iwasaki, K.; Yoshimitsu, Y.; Arai, T.; Egashira, N.; Mishima, K.; Ikeda, T.; Fujiwara, M. Effect of oral administration of zanapezil (TAK-147) for 21 days

on acetylcholine and monoamines levels in the ventral hippocampus of freely moving rats. *Br. J. Pharmacol*, **2005**, *145*(8), 1035-1044.

[85]   Mroczek, T.; Mazurek, J. Pressurized liquid extraction and anticholinesterase activity-based thin-layer cromatografy with bioautography of Amaryllidaceae alkaloids. *Anal. Chim. Acta*, **2009** , *633(2)*, 188-196.

[86]   Bastida, J.; Lavilla, R.; Viladomat, F. Chemical and biological aspects of Narcissus alkaloids. Alkaloid Chem. Biol., **2006**, *63*, 89-179.

[87]   Schroeder, H.; Lobron, C.; Wevers, A.; Maelicke, A.; Giacobini, E. Cellular acetylcholine receptor expression in the brain of patients with Alzheimer's and Parkinson's dementia. Adv. Behav. Biol., **1995**, *44,* 63-67.

[88]   Perry, E.K..; Morris, C.M.; Court, J.A.; Cheng, A.; Fairbairn, A.F.; McKeith, I.G.; Irving, D.; Brown, A.; Perry, R.H. Alteration in nicotine binding sites in Parkinson's disease, Lewy body dementia and Alzheimer's disease: possibles index of early neuropathology. *Neuroscience*, **1995**, *64(2)*, 385-395.

[89]   Nordberg, A.; Lundqvist, H.; Hartvig, P.; Lilja, A.; Langstrom, B. Kinetic analysis of regional (S) (-)11C-nicotine binding in normal and Alzheimer brains-*in vivo* assessment using positron emission tomography. *Alzheimer Dis. Assoc. Disord.*, **1995**, *9(1)*, 21-27.

[90]   Woodruff-Pak, D.S.; Vogel, R.W.3rd; Wenk, G.L. Galantamine: effect on nicotinic receptor binding, acetylcholinesterase inhibition, and learning. *PNAS*, **2001**, *98*(4), 2089-2094.

[91]   Strittmatter, W.J.; Weisgraber, K.H.,Goedert, M.; Saunders, A.M.; Huang, D.; Corder, E.H.; Dong, L. M.; Jakes, R.; Alberts, M. J.; Gilbert, J. R.; Han, S.H.; Hulette, C.; Einstein, G.; Schmechel, D. E.; Pericak-Vance, M. A.; Roses, A.D. Hypothesis: microtubule instability and paired helical filament formation in the Alzheimer disease brain are related to apolipoprotein-E genotype. *Exp. Neurol.*, **1994**, *125(2)*, 163-171 .

[92]   Mary, A.; Renko, D.Z.; Guillou, C.; Thal, C. Potent acetylcholinesterase inhibitors: design, synthesis, and structure-activity relationships of bis-interacting ligands in the galanthamine series. *Bioorg. Med. Chem.*, **1998**, *6(10)*, 1835-1850.

[93]   Tariot, P.N.; Solomon, P.R.; Morris, J.C.; Kershaw, P.; Lilienfed, S.; Ding, C. A 5-month, randomized, placebo-controlled trial of galantamine in AD. The Galantamine USA-10 Study Group. *Neurology*, **2000**, *54(12)*, 2269-2276.

[94]   Pilger, C.; Bartolucci, C.; Lamba, D.; Tropsha, A.; Fels, G. Accurate prediction of the bound conformation of galanthamine in the active site of Torpedo califórnica acetylcholinesterase using molecular docking. *J. Mol. Graph. Model*, **2001**, *19(3-4)*, 288-296, 374-378.

[95]   Jia, P.; Sheng, R.; Zhang, J.; Fang, L.; He, Q.; Yang, B.; Hu, Y. Design, synthesis and evaluation of galanthamine derivatives as acetylcholinesterase inhibitors. *Eur. J. Med. Chem.*, **2009**, *44*(2), 772-784.

[96]   Pagliosa, L.B.; Monteiro, S.C.; Silva, K.B.; de Andrade, J.P.; Dutilh, J.; Bastida, J.; Cammarota, M.; Zuanazzi, J.A. Effect of isoquinoline alkaloids from two Hippeastrum species on *in vitro* acetylcholinesterase activity. *Phytomedicine* **2010**, *17(8-9)*, 698-701.

[97]   Geissler, T.; Brandt, W.; Porzel, A.; Schlenzig, D.; Kehlen, A.; Wessjohann, L.; Arnold, N. Acetylcholinesterase inhibitors from the toadstool Cortinarius infractus. *Bioorg. Med. Chem.*, **2010**, *18*(6), 2173-2177.

[98]   Khorana, N.; Changwichit, K,; Ingkaninan, K.; Utsintong, M. Prospective acetylcholinesterase inhibitory activity of indole and its analogs. *Bioorg. Med. Chem. Lett.*, **2012**, *22(*8), 2885-2888.

[99]   Guo, B.; Xu, L.; Wei, Y.; Liu, C. Research advances of Huperzia serrata (Thunb.) Trev. *J. Med.*, **2009**, *34(16)*, 2018-2023.

[100] Kozikowski, A.P.; Xia, Y.; Reddy, E.R.; Tückmantel, W.; Hanin, I.; Tang, X.C. Synthesis of huperzine A, its analogs, and their anticholinesterase activity. *J. Org. Chem.*, **1991**, *56(15)*, 4636-4645.

[101] Little, J.T.; Walsh, S.; Aisen, P.S. An update on huperzine A as a treatment for Alzheimer's disease. *Expert. Opin. Investig. Drugs*, **2008**, *17(2)*, 209-215.

[102] Pohanka, M.; Zemek, F.; Bandouchova, H.; Pikula, J. Toxicological scoring of Alzheimer's disease drug huperzine in a guinea pig model. *Toxicol. Mech. Methods*, **2012**, *22*(3), 231-235.

[103] Yang, L.; Ye, C.Y.; Huang, X.T.; Tang, X.C.; Zhang, H.Y. Decreased accumulation of subcellular amyloid-β with improved mitochondrial function mediates the neuroprotective effect of huperzine A. *J. Alzheimers Dis.*, **2012**, *31*(1), 131-142.

[104] Zhang, Z.; Wang, X.; Chen, Q.; Shu, L.; Wang, J.; Shan, G. Clinical efficacy and safety of huperzine alpha in treatment of mild to moderate Alzheimer disease, a placebo-controlled, double-blind, randomized trial. *Zhonghua Yi Xue Za Zhi*, **2002**, *82(14)*, 941-944.

[105] Wang, B.S.; Wang, H.; Wei, Z.H.; Song, Y.Y.; Zhang, L.; Chen, H.Z. Efficacy and safety of natural acetylcholinesterase inhibitor huperzine A in the treatment of Alzheimer's disease: an updated meta-analysis. *J. Neural Transm.*, **2009**, *116(4)*, 457-465.

[106] Rafii, M.S.; Walsh, S.; Little, J.T.; Behan, K.; Reynolds, B.; Ward, C.; Jin, S.; Thomas, R.;Aisen, P.S. A phase II trial of huperzine A in mild to moderate Alzheimer disease. *Neurology,* **2011**, *76* (16), 1389-1394.

[107] Lucey, C.; Kelly, S. A.; Mann, J. A concise and convergent (formal) total synthesis of huperzine A. *Org. Biomol. Chem.*, **2007**, *5(2)*, 301-306.

[108] Zhang, H.Y.; Liang, Y.Q.; Tang, X.C.; He, X.C.; Bai, D.L. Stereoselectivities of enantiomers of huperzine A in protection against β amyloid $_{25-35}$-induced injury in PC12 and NG108-15 cells and cholinesterase inhibition in mice. *Neurosci. Lett.*, **2002**, *317(3)*, 143-146.

[109] Orhan, I.E; Orhan, G.; Gurkas, E. An overview on natural cholinesterase inhibitors- A multi-targeted drug class- and their mass production. *Mini Rev. Med. Chem.*, **2011**, *11* (10), 836-842.

[110] Zhao , Y.; Dou, J.;Luo, J; Li, W,; Chan, H.H.; Cui, W., Zhang, H.; Han, R.; Carlier, P.R.; Zhang, X.; Han, Y. Neuroprotection against excitotoxic and ischemic insults by bis (12)-hupyridone, a novel anti- acetylcholinesterase dimer, possibly *via* acting on multiple targets. *Brain Res*, **2011**, *1421,* 100-109.

[111] Shi, Y.F.; Zhang, H.Y.; Wang, W.; Fu, Y.; Xia, Y.; Tang, X.C.; Bai, D.L.; He, X. Ch. Novel 16-substituted bifunctional derivatives of huperzine B: multifunctional cholinesterase inhibitors. *Acta Pharmacol. Sin.*, **2009**, *30(8)*, 1195-1203.

[112] Bolognesi, M.L.; Cavalli, A.; Valgimigli, L.; Bartolini, M.; Rosini, M.; Andrisano, V.; Recanatini, M.; Melchiorre, C. Multi-target-directed drug design strategy: from a dual binding site acetylcholinesterase inhibitor to a trifunctional compound against Alzheimer's disease. J. Med. Chem., **2007**, *50(26)*, 6446-6449.

[113] Bolognesi, M.L., Cavalli, A.; Melchiorre, C. Memoquin: a multi-target-directed ligand as an innovative therapeutic opportunity for Alzheimer's disease. *Neurotherapeutics*, **2009**, *6(1)*, 152-162 .

[114] Ronco, C.; Sorin, G.; Nachon, F.; Foucault, R.; Jean, L.; Romieu, A.; Renard, P.Y. Synthesis and structure- activity relationship of Huprine derivatives as human acetylcholinesterase inhibitors. *Bioorg. Med. Chem.*, **2009**, *17(13)*, 4523-4536.

[115] Dvir, H.; Wong, D.M.; Harel, M.; Barril, X.; Orozco, M.; Luque, F.J.; Muñoz-Torrero, D.; Camps, P.; Rosenberry, T.L.; Silman, I.; Sussman, J.L. 3D structure of Torpedo californica acetylcholinesterase complexed with huprine X at 2.1 A resolution: kinetic and molecular dynamic correlates. *Biochemistry*, **2002**, *41(9)*, 2970-2981.

[116] Camps, P.; Cusack, B.; Mallender, W.D.; El Achab R.E.; Morral, J.; Muñoz-Torrero, D.; Rosenberry, T.L. Huprime X is a novel high-affinity inhibitor of acetylcholinesterase that is of interest for treatment of Alzheimer's disease. *Mol. Pharmacol.*, **2000**, *57(2)*, 409-417.

[117] Roman, S.; Vivas, N.M.; Badia, A.; Clos M.V. Interaction of a new potent anticholinesterasic compound (+/-) huprine X with muscarinic receptors in rat brain. *Neurosci. Lett.*, **2002**, *325(2)*, 103-106.

[118] Roman, S.; Badia, A.; Camps, P.; Clos, M.V. Potentiation effects of (+/-) huprine, X, a new acetylcholinesterase inhibitor, on nicotinic receptors in rat cortical synaptosomes. *Neuropharmacology*, **2004**, *46(1)*, 95-102.

[119] Hedberg, M.M.; Clos, M.V.; Ratia, M.; Gonzalez, D.; Lithner, C.U.; Camps, P.; Muñoz-Torrero, D.; Badia, A.; Giménez-Llort, L.; Nordberg, A. Effect of huprine X on β-amyloid, synaptophysin and α7 neuronal nicotinic acetylcholine receptors in the brain of 3xTg-AD and APPswe transgenic mice. *Neurodegener. Dis.*, **2010**, *7(6)*, 379-388.

[120] Pang, Y.P.; Quiram, P.; Jelacic, T.; Hong, F.; Brimijoin, S. Highly potent, selective, and low cost bis-tetrahydroaminacrine inhibitors of acetylcholinesterase: steps toward novel drugs for the treating Alzheimer's disease. *J. Biol. Chem.*, **1996**, *271(39)*, 23646-23649.

[121] Galdeano, C.; Viayna, E.; Sola, I.; Formosa, X.; Camps, P.; Badia, A.; Clos, M.V.; Relat, J.; Ratia, M.; Bartolini, M.; Mancini, F.; Andrisano, V.; Salmona, M.; Minguillon, C.; González-Muñoz, G.C.; Rodríguez- Franco, M.I.; Bidon-Chanal, A.; Luque, F.J.Muñoz-Torrero, D. Huprine-tacrine heterodimers as anti-amyloidogenic compounds of potential interest against Alzheimer's and prion diseases. *J. Med. Chem.*, **2012**, *55*(2), 661-669.

[122] Muñoz-Torrero, D.; Pera, M.; Relat, J.; Ratia, M.; Galdeano, C.; Viayna, E.; Sola, I.; Formosa, X.; Camps, P.; Badia, A.; Clos, M.V. Expanding the multipotent profile of huprine-tacrine heterodimers as disease-modifying anti- Alzheimer agents. *Neurodegener. Dis.*, **2012**, *10*(1-4), 96-99.

[123] Fang, L.; Appenroth, D.; Decker, M.; Kiehntopf, M.; Roegler, C.; Deufel, T.; Fleck, C.; Peng, S.; Zhang, Y.; Lehmann, J. Synthesis and biological evaluation of NO-donor-tacrine hybrids as hepatoprotective anti- Alzheimer drug candidates. *J. Med. Chem.*, **2008**, *51(4)*, 713-716.

[124] Dogterom, P.; Nagelkerke, J.F.; Mulder, G. J. Hepatotoxicity of tetrahydroaminoacridine in isolated rat hepatocytes: effect of glutathione and vitamin E. *Biochem. Pharmacol.*, **1988**, *37(12)*, 2311-2313 .

[125] Moreira, P. I.; Santos, M. S.; Oliveira, C. R.; Shenk, J. C.; Nunomura, A.; Smith, M. A.; Zhu, X.; Perry, G. Alzheimer disease and the role of free radicals in the pathogenesis of the disease. *CNS Neurol. Disord. Drug Targets*, **2008**, *7(1)*, 3-10.

[126] Lee, H.P.; Casadesus, G.; Zhu, X.; Lee, H.G.; Perry, G.; Smith, M.A.; Gustaw-Rothenberg, K.; Lerner, A. All-trans retinoic acid as a novel therapeutic strategy for Alzheimer's disease. *Expert Rev. Neurother*, **2009**, *9(11)*, 1615-1621.

[127] Wang, T.T.; Zeng, G.C.; Li, X.C.; Zeng, H.P. *In vitro* studies on the antioxidant and protective effect of 2-substituted -8-hydroxyquinoline derivatives against H(2)O(2)-induced oxidative stress in BMSCs. *Chem. Biol. Drug Des.*, **2010**, *75(2)*, 214-222.

[128] Adlard, P.A.; Bica, L.; White, A.R.; Nurjono, M.; Filiz, G.; Crouch, P.J.; Donnelly, P.S.; Cappai, R.; Finkelstein, D.I.; Bush, A.I. Metal ionophore treatment restores dendritic spine density and synaptic protein levels in a mouse model of Alzheimer's disease. *PLoS One* **2011**, *6(3)*: e17669.

[129] Faux, N.G.; Ritchie, C.W.; Gunn, A.; Rembach, A.; Tsatsanis, A.; Bedo, J.; Harrison, J.; Lannfelt, L.; Blennow, K.; Zetterberg, H.; Ingelsson, M.; Masters, C.L.; Tanzi, R.E.; Cummings, J.L.; Herd, C.M.; Bush, A.I. PBT2 rapidly improves cognition Alzheimer's disease: additional phase II analyses. *J. Alzheimers Dis.*, **2010**, *20(2)*, 509-516.

[130] Fernández-Bachiller, M.I.; Pérez, C.; González-Muñoz, G.C.; Conde, S.; López, M.G.; Villarroya, M.; García, A.G.; Rodríguez-Franco, M.I. Novel tacrine-8-hydroxyquinoline hybrids as multifunctional agents for the treatment of Alzheimer's disease, with neuroprotective, cholinergic, antioxidant, and copper-complexing properties. *J. Med. Chem.*, **2010**, *53*(13), 4927-4937.

[131] Antequera, D.; Bolos, M.; Spuch, C.; Pascual, C.; Ferrer, I.; Fernández-Bachiller, M.I.; Rodríguez-Franco, M.I.; Carro, E. Effects of a tacrine-8-hydroxyquinoline hybrid (IQM-622) on Aβ accumulation and cell death: involvement in hippocampal neuronal loss in Alzheimer's disease. *Neurobiol. Dis.*, **2012**, *46*(3), 682-691.

[132] Deneke, S.M. Thiol-based antioxidants. *Curr. Top. Cell. Regul.*, **2000**, *36*, 151-180.

[133] Wang, Y.; Guan, X.L.; Wu, P.F.; Wang, C.M.; Cao, H.; Li, L.; Guo, X.J.; Wang, F.; Xie, N.; Jiang, F.C.; Chen, J.G. Multifunctional mercapto-tacrine derivatives for treatment of age-related neurodegenerative diseases. *J. Med. Chem.*, **2012**, *55*(7), 3588-3592.

[134] Spencer, J. P. The impact of flavonoids on memory: physiological and molecular considerations. *Chem. Soc. Rev.*, **2009**, *38(4)*, 1152-1161.

[135] Spencer, J. P. Beyond antioxidants: the cellular and molecular interactions of flavonoids and how these underpin their actions on the brain. *Proc. Nutr. Soc.*, **2010**, *69(2)*, 244-260.

[136] Fernández-Bachiller, M.I.; Pérez, C.; Monjas, L.; Rademann, J.; Rodríguez-Franco, M.I. New tacrine -4-Oxo-4H-chromene hybrids as multifunctional agents for the treatment of Alzheimer's disease, with cholinergic, antioxidant, and β-amyloid-reducing properties. *J. Med. Chem.*, **2012**, *55(3)*, 1303-1317.

[137] Pi, R.; Mao, X.; Chao, X.; Cheng, Z.; Liu, M.; Duan, X.; Ye, M.; Chen, X.; Mei, Z.; Liu, P.; Li, W.; Han, Y.Tacrine-6-ferulic acid, a novel multifunctional dimer, inhibits amyloid-β-mediated Alzheimer's disease-associated pathogenesis *In vitro* and *In vivo*. *PLoS ONE*, **2012**, *7*(2), e31921.

[138] Fu, H.; Li, W.; Lao, Y.; Luo, J.; Lee, N.T.; Kan, K.K.; Tsang, H.W.; Tsim, K.W.; Pang, Y.; Li, Z.; Chang, D.C.; Li, M.; Han, Y. Bis(7)-tacrine attenuates beta amyloid-induced neuronal apoptosis by regulating L-type calcium channels. *J. Neurochem.*, **2006**, *98(5)*, 1400-1410.

[139] Marco-Contelles, J.; León, R.; de los Rios, C.; Samadi, A.; Bartolini, M.; Andrisano, V.; Huertas, O.; Barril, X; Luque, F.J.; Rodríguez-Franco, M.I.; López, B.; López, M.G.; García, A.G.; Carreiras, Mdo., C.; Villarroya, M. Tacripyrines, the first tacrine-

dihydropyridine hybrids, as multitarget-directed ligands for the treatment of Alzheimer's disease. *J. Med. Chem.*, **2009**, *52*(9), 2724-2732.

[140] Pereira, J.D.; Caricati-Neto, A.; Miranda-Ferreira, R.; Smaili, S.S.; Godinho, R.O.; de los Rios, C.; León, R.; Villaroya, M.; Samadi, A.; Marco-Contelles, J.; Jurkiewicz, N.H.; Garcia, A.G.; Jurkiewicz, A. Effect of novel tacripyrines ITH12117 and ITH12118 on rat vas deferens contractions, calcium transients and cholinesterase activity. *Eur. J. Pharmacol.*, **2011**, *660*(2-3), 411-419.

[141] Alonso, D.; Dorronsoro, I.; Rubio, L.; Muñoz, P.; García-Palomero, E.; Del Monte, M.; Bidon-Chanal, A.; Orozco, M.; Luque, F.J.; Castro, A.; Medina, M.; Martínez, A. Donepezil-tacrine hybrid related derivatives as new dual binding site inhibitors of AChE. *Bioorg. Med. Chem.*, **2005**, *13*(24), 6588-6597.

[142] Camps, P.; Formosa, X.; Galdeano, C.; Gómez, T.; Muñoz-Torrero, D.; Scarpellini, M.; Viayna, E.; Badia, A.; Clos, M.V.; Camins, A.; Pallàs, M.; Bartolini, M.; Mancini, F.; Andrisano, V.; Estelrich, J.; Lizondo, M.; Bidon-Chanal, A.; Luque, F.J. Novel donepezil-based inhibitors of acetyl-and butyrylcholinesterase and acetylcholinesterase-induced beta-amyloid aggregation. *J. Med. Chem.*, **2008**, *51*(12), 3588-3598..

[143] Piazzi, L.; Rampa, A.; Bisi, A.; Gobbi, S.; Belluti, F.; Cavalli, A.; Bartolini, M.; Andrisano, V.; Valenti, P.; Recanatini, M. 3-(4-[[Benzyl(methyl)amino]phenyl]-6,7-dimetoxy-2H-2-chromenone (AP2238) inhibits both acetylcholinesterase and acetylcholinesterase-induced beta-amyloid aggregation: a dual function lead for Alzheimer's disease therapy. *J. Med. Chem.*, **2003**, *46(12)*, 2279-2282 .

[144] Samadi, A.; Chioua, M.; Bolea, I.; De los Ríos, C.; Iriepa, I.; Moraleda, I.; Bastida, A.; Esteban, G.; Unzeta, M.; Gálvez, E.; Marco-Contelles, J. Synthesis, biological assessment and molecular modeling of new multipotent MAO and cholinesterase inhibitors as potential drugs for the treatment of Alzheimer's disease. *Eur J. Med. Chem.*, **2011**, *46(9)*, 4665-4668.

[145] Pérez, V.; Marco, J.L.; Fernández-Álvarez, E.; Unzeta, M. Relevance of benzyloxy group in 2-indolyl methylamines in the selective MAO-B inhibition. *Br. J. Pharmacol.*, **1999**, *127(4)*, 869-876.

[146] Samadi, A; De los Ríos, C.; Bolea, I.; Chioua, M.; Iriepa, I.; Moraleda, I.; Bartolini, M.; Andrisano, V.; Gálvez, E.; Valderas, C.; Unzeta, M.; Marco-Contelles, J. Multipotent MAO and cholinesterase inhibitors for the treatment of Alzheimer's disease: synthesis, pharmacological analysis and molecular modeling of heterocyclic substituted alkyl and cycloalkyl propargyl amine. *Eur. J. Med. Chem.*, **2012**, *52*, 251-262.

CHAPTER 9

Modulation of BACE1 Activity as a Potential Therapeutic Strategy for Treating Alzheimer's Disease

David William Klaver and Giuseppina Tesco[*]

Alzheimer's Disease Research Laboratory, Department of Neuroscience, Tufts University School of Medicine, Boston, MA, USA

Abstract: Inhibiting the generation of β-amyloid (Aβ) from the amyloid precursor protein (APP) by targeting the protease BACE1, the Alzheimer's disease β-secretase, is a key strategy for the development of therapeutic compounds aimed at treating Alzheimer's disease. However, progress in developing biologically active inhibitors has been slow. This is in part because BACE1 possesses a broad and open active site, which cannot be effectively inhibited by small molecules capable of penetrating the blood-brain barrier. Therefore, there is a great interest in developing modulators of BACE1 activity that are not associated with active site inhibition, rather disrupting the physiological function of the enzyme. This review will discuss the regulation of BACE1 transcriptional expression and modulation of activity by other cellular components, in particular lipids, proteins that interact directly with BACE1, and ubiquitination, as well as BACE1 immunotherapy. This review will also examine the potential of each of these as therapeutic strategies for the treatment of AD.

Keywords: Alzheimer's disease, aging, BACE1, β-amyloid, cognitive dysfunction, dementia, neurodegenerative diseases, secretases, therapies.

ALZHEIMER'S DISEASE – A BACKGROUND

Alzheimer's disease (AD) is a progressive neurodegenerative disease, primarily affecting the elderly. Clinically, AD is characterized by several clinical features, including a progressive loss of memory, which begins with degeneration of short- to medium-term memory in the earlier stages of the disease, and ultimately progressing to long-term memory loss in later stages. Behavioral changes are also

[*]Address correspondence to Giuseppina Tesco: Alzheimer's Disease Research Laboratory, Department of Neuroscience, Tufts University School of Medicine, 136 Harrison Avenue, St 328A, Boston, MA 02111, USA; Tel: 617 636 4050; Fax: 617 636 2413; E-mail: Giuseppina.Tesco@Tufts.edu

a feature observed in AD patients, with loss of inhibition, disturbances of sleep and aggression commonly described [1]. With the average age of the population ever increasing, the incidence of Alzheimer's disease amongst the population is estimated to rise drastically over the next fifty years [2]. As AD is a relatively slow progressing disease, the burden on society caused by AD is considerably high. Hence, there is increasing demand for effective treatments to combat AD.

Pathologically, AD is characterized by two classical hallmarks, neurofibrillary tangles and amyloid plaques. Neurofibrillary tangles are intraneuronal deposits of hyperphosphorylated tau protein, which is crucial for maintaining the integrity of microtubules [3-7]. It should be noted, however, that neurofibrillary tangles are not a specific feature of AD, as they are observed in several other neurodegenerative diseases, and are thought to be a secondary feature associated with AD [8-10]. The classical hallmark of AD is the presence of extracellular macromolecular proteinaceous deposits, known as amyloid plaques. The main component of these plaques is a small 38-43-amino-acid peptide known as β-amyloid (Aβ) [11-15]. It is generally accepted that the Aβ protein is the underlying neurotoxic protein responsible for the neurodegeneration observed in the AD brain, and hence a large body of research is ongoing aimed at reducing or eliminating production of Aβ as a therapeutic strategy for treating AD.

Aβ is a small 38-43 amino-acid peptide derived from a much larger precursor protein, known as the β -amyloid precursor protein (APP) [16]. APP can be proteolytically processed by sequential cleavage of APP by a class of proteases called the APP secretases. Sequential cleavage of APP by the β-secretase at the N-terminus of the Aβ domain followed by cleavage of APP at the C-terminus of the Aβ sequence by γ-secretase liberates the full-length Aβ peptide. APP can also be processed *via* an alternative route, where APP is cleaved by α-secretase within the Aβ region, which precludes formation of Aβ [17-19] (Fig. (**1**)).

BACE1 IS THE ALZHEIMER'S β-SECRETASE

Because cleavage of APP by β-secretase is the rate-limiting step in the formation of Aβ, there is considerable interest in targeting this protease pharmacologically. However, it was not until 1999 that the identity of the β -secretase was finally elucidated. Five groups almost simultaneously reported the amino-acid sequence

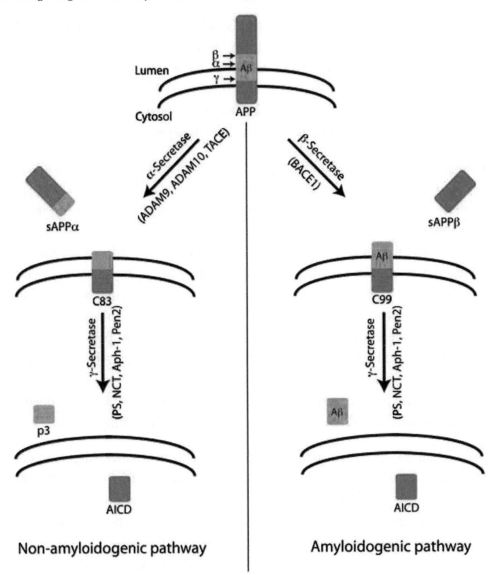

Figure 1: Processing pathways of the amyloid precursor protein (APP). The non-amyloidogenic pathway involves cleavage of APP within the Aβ region by α-secretase (ADAM family of proteins) to release the N-terminal fragment sAPPα and retain the C-terminal fragment C83. C83 can then be cleaved by the γ-secretase complex (presenilin, nicastrin, aph-1 and pen2) to liberate the p3 fragment and the APP intracellular domain (AICD). Conversely, APP can also be processed by β-secretase (BACE1) at the N-terminus of the Aβ region to produce the secreted fragment sAPPβ and the C-terminal fragment C99. C99 can then be cleaved by γ-secretase to liberate Aβ and AICD.

of the β-secretase, and named it memapsin 2, asp2 or BACE1 [20-24]. The name BACE1 (β-site APP cleaving enzyme 1) is now the commonly accepted name for the β-secretase. There is considerable evidence suggesting that BACE1 is the sole β-secretase responsible for production of Aβ from APP [25-31]. Aβ production and amyloid plaque pathology are eliminated in BACE1 knockout mice that express human APP, suggesting that BACE1 alone is responsible for the production of Aβ [25-27]. Furthermore, memory deficits and cognitive decline are rescued following BACE1 knockout in APP transgenic mice [28-31].

BACE1 is a 501 amino-acid protein, consisting of a 21-residue signal peptide, a prodomain between residues 22 and 45, and the mature protein between residues 46 and 501 [21]. BACE1 is a type I transmembrane aspartyl protease, and contains two catalytical aspartate residues, D93 and D289. Both of these aspartate residues are critical for BACE1 activity, as mutation of either of these residues abolishes BACE1 catalytic activity [20].

Like many other proteases, the prodomain of BACE1 is critical for proper folding of the catalytic domain, and is critical for autoinhibition of BACE1 activity [32]. Indeed, the zymogen does have drastically reduced protease activity compared to mature BACE1 [32]. Unlike other proteases, however, the zymogen retains some activity [32]. The prodomain has been shown through theoretical modeling to sit adjacent to the active site, however it was also shown that a flexible loop between residues 46 and 65 might also play a critical role in proBACE1 autoinhibition by blocking substrate access to the active site [33, 34]. Furthermore, progastricsin and pepsinogen, two proteases which share sequence homology with BACE1, contain a lysine residue in their prodomain which directly interacts with the catalytic residues in their respective active sites, creating a clamp-like mechanism which inhibits activity. BACE1 lacks this critical residue [33]. The flexibility of the loop, combined with the lack of a clamp residue, may provide the mechanism for why proBACE1 retains some activity.

BACE1 is a unique protease, in that it is membrane bound. BACE1 can be S-palmitoylated at cysteine residues, which targets the enzyme to cluster at sterol-rich lipid raft regions of the plasma membrane [35]. While there is some evidence showing the importance of BACE1 localization to lipid rafts in regulation of β-secretase activity, recent evidence suggests that palmitoylation and localization of BACE1 to lipid rafts does not affect the ability of BACE1 to cleave APP, both in cell culture and primary neuronal culture [36, 37].

BACE1 contains one phosphorylation site, close to the C-terminus at serine 498. Phosphorylation of BACE1 at this site does not affect endocytosis but regulates the trafficking of BACE1 between the secretory and endocytic pathway [35, 38].

TARGETING BACE1 THERAPEUTICALLY – IS IT A GOOD IDEA?

As BACE1 catalyzes the rate-limiting step in the production of Aβ, naturally it became a target of much interest for groups pursuing new and better ways to treat AD. Initial studies examining BACE1 knockout mice suggested that genetic deletion of BACE1 produced a relatively mild phenotype, suggesting that inhibiting BACE1 pharmacologically would have very few, if any, potential side effects [26, 27]. However, subsequent studies revealed abnormalities in myelination in both the central and peripheral nervous systems in BACE1 knockout mice [39-41]. This has since been attributed to perturbations in neuregulin (NRG) signaling. NRG proteins are ligands for the ErbB family of receptors, which directly stimulate myelination in both the CNS and PNS [42-44]. Both NRG1 and NRG3 are substrates for BACE1, and their proteolytic degradation is inhibitory to myelination [39-41]. A further study demonstrated that while the effect of BACE1 knockout on myelination is much more subtle in adult mice, this might be sufficient to cause disturbances in higher brain functions [45].

Since the identification of NRG1 and NRG3 as substrates of BACE1, numerous more substrates have been identified. Another such substrate for BACE1 is the $Na_v1.1$ sodium channel, a sodium channel that is a key regulator of sodium homeostasis in the CNS [46, 47]. Mutations in this channel have been linked to epilepsy [48], and indeed, BACE1 knockout mice do exhibit an epileptic-like phenotype [49]. Interestingly, epileptic-like seizures have been observed in a subset AD patients, which may be a result of excessive BACE1 activity [50]. Therefore, BACE1 inhibition may be beneficial in these patients from the perspective of reducing seizures, as well as reducing Aβ levels in the brain, however at this time, this is speculative, as there is no conclusive evidence that the seizures observed in AD patients are caused by perturbations in BACE1 activity.

More recently, several groups have identified a link between BACE1 and neuronal plasticity. Neuronal plasticity is a vital process required for brain functions that require circuit remodeling, such as learning and memory [51]. BACE1 knockout mice exhibit deficits in such brain functions [28, 52], and it is extremely well known that AD patients display deficits in such functions as well. Recent work has demonstrated that axon guidance is disrupted in the olfactory bulb of BACE1 knockout mice, an area of the brain that is known to undergo constant neurogenesis and remodeling, even into adult life [53]. Interestingly, at approximately the same time, two separate studies identified two cell-adhesion associated molecules, L1 and CHL1, as BACE1 substrates [54, 55]. Both of these proteins have been previously implicated in regulation of synaptic plasticity [56]. While it is true that the majority of axon pathfinding and synaptic connections are completed early in life, there are some neuronal populations, such as the dentate gyrus, that do undergo neurogenesis, and hence require axon guidance, throughout life. Therefore, inhibiting BACE1 may perturb the normal function of these areas of the brain, and hence produce unwanted side effects related to this.

The large body of research that has now been published on the physiological roles of BACE1 raises the very important question: is it possible to inhibit BACE1 without unwanted side effects? The answer may lie in modulation, rather than complete inhibition, of BACE1 activity. There is compelling evidence that partially reducing BACE1 levels in the brain is sufficient to have a noticeable effect on AD pathology. BACE1$^{+/-}$ heterozygous APP transgenic mice have dramatically reduced amyloid plaque burden, neuritic dystrophy and synaptic dysfunction compared with BACE1$^{+/+}$ APP transgenic mice [57, 58]. This suggests that instead of completely ablating BACE1 activity, one can merely modulate it to produce the desired effect on Aβ production with a lower risk of off-target non-specific side effects related to other BACE1 substrates.

Active Site Inhibition

After the identity of BACE1 was discovered, work began in earnest to develop inhibitors of the enzyme. Hong *et al.* were the first to solve the crystal structure of the BACE1 catalytic domain, shortly after its identification as the β-secretase, which was pivotal for the rational design of active site inhibitors [59]. Examination of the BACE1 crystal structure reveals an extremely broad and open active site, which limits the development of bioavailable active site inhibitors [59]. While there have been numerous inhibitors developed, the vast majority of these inhibitors are high molecular weight, bulky compounds, which do not cross the blood-brain barrier effectively. Furthermore, the majority of these inhibitors are peptidomimetic, yielding them susceptible to proteolytic degradation before they can reach the site of action [19]. However, there is an increasing body of work aimed at developing non-peptidomimetic inhibitors of the BACE1 active site, with some promise [60].

Very recently, several selective active site inhibitors of BACE1 were reported by research teams from the drug companies Merck, Eli Lilly and Eisai to effectively lower levels of β-secretase-derived APP processing fragments in the CSF of patients in phase I clinical trials. Phase II trials are currently ongoing (for more information, refer to http://www.alzforum.org/new/detail.asp?id=3222). While these drugs show initial promise for the treatment of AD, it remains unclear whether they will remain viable in future trials. As such, researching alternative methods of BACE1 inhibition remains highly important.

Allosteric Inhibition of BACE1

Many proteases contain binding sites at areas distal to their active sites. These sites, termed exosites, can assist in stabilizing enzyme-substrate interactions, determine specificity of proteases for their substrates, and in some cases act as allosteric regulators of protease activity. While exosites are commonly found on serine and cysteine proteases, they are rarely found on aspartyl proteases. BACE1, however, has been found to contain an exosite that binds to small peptides independent of active site occupancy [61, 62]. The binding of small peptides to this exosite inhibits proteolysis of APP by BACE1 [61, 62]. Inhibitors at the exosite may prove to have advantages over traditional active-site BACE1 inhibitors, in that they may ultimately end up to be small non-peptidomimetic inhibitors with improved bioavailability. The presence of a regulatory BACE1 exosite raises the exciting possibility that the rational design of small molecule

BACE1-specific inhibitors may still be possible, and work to develop molecules that bind to this site may be critical for the future hope of specific BACE1 inhibition.

Targeting BACE1 Gene Expression

Another strategy to modulating BACE1 activity *in vivo* is to regulate *BACE1* gene expression. The mechanisms that regulate *BACE1* gene expression are complex. Several transcription factor-binding sites have been found in the *BACE1* gene promoter, including sites for NF-κB, NFAT and PPARγ [63]. Each of these transcription factors represents a possible therapeutic target for reducing BACE1 protein levels *in vivo*. There is a growing body of evidence that BACE1 protein levels are increased in both the AD brain and cerebrospinal fluid, both from animal studies and analysis of post-mortem tissue from AD patients [64-66]. Therefore, targeting *BACE1* transcription may be beneficial in restoring BACE1 levels and activity to normal levels. Table **1** summarizes the potential therapeutic potential of targeting *BACE1* transcription that is outlined below.

Table 1.   Modulators of *BACE1* Expression

Name	Effect on *BACE1* Transcription	Comments	References
PPARγ	↓	NSAIDs activate PPARγ Decreased in AD PPARγ agonists trialed for AD	[69-77]
NF-κB	↓ in neurons ↑ in reactive astrocytes	Potential mechanism of Aβ-mediated positive feedback of Aβ production Decreased around plaques	[78-82]
NFAT	↑	Potentially involved in AD through calcium dysregulation	[86, 88]
hnRNP H	↑	Regulates *BACE1* alternative splicing	[89]
Micro RNAs	↓	Inhibit translation of BACE1 mRNA Decreased in AD	[95, 100-107]
Phospho-eIF2α	↑	Response to energy deprivation and oxidative stress. Translational control of BACE1 mRNA	[108-110]

NSAIDs and PPARγ

Perhaps the most widely studied class of compounds that alters *BACE1* gene expression is the non-steroidal anti-inflammatory drugs (NSAIDs). Inflammation has been implicated in AD for numerous years, and activation of microglia and upregulation of proinflammatory factors in the AD brain has been well described. This prompted researchers to examine any potential benefits of NSAID use on AD pathology and progression. Indeed, there may be a reduced risk of AD in people who regularly take NSAIDs [67]. Treatment with NSAIDs has been shown to significantly improve cognitive function, and reduce amyloid pathology in APP transgenic mice, indicating that NSAIDs have a direct effect on Aβ [68]. Subsequently, NSAIDs were shown to alter the expression of BACE1 through an effect on the transcription factor PPARγ. Administration of NSAIDs was shown to reverse the stimulatory effect of proinflammatory cytokines on APP and Aβ secretion, as well as *BACE1* transcription, in neuroblastoma cells [69]. Furthermore, administration of a PPARγ antagonist blocked the effect, indicating a central role for PPARγ in mediating the effect of NSAIDs. PPARγ agonists were also shown to mimic the effect of NSAIDs [69].

In transgenic mice, levels of PPARγ were shown to be markedly decreased [70]. Treatment of these mice with the NSAID ibuprofen or the PPARγ agonist pioglitazone significantly upregulated PPARγ, and decreased both BACE1 and Aβ levels [70]. When the authors examined the brains of AD patients, they found that PPARγ levels were significantly lower in the AD brain compared to controls, and this was concurrent with increased BACE1 levels [70]. Furthermore, using gel shift mobility assays, the binding of PPARγ to its sequence within the BACE1 promoter was found to be significantly decreased in the AD brain [70]. Similarly, in another study, PGC-1α, a co-activator of PPARγ, has been shown to negatively regulate *BACE1* promoter activity and BACE1 levels, and levels of PGC-1α are markedly decreased in the AD brain [71]. Taken together, these data indicate that dysregulation of PPARγ-dependent transcription of *BACE1* may be a causative factor in the pathogenesis of AD, and presents a possible attractive therapeutic target for treating AD.

The evidence that PPARγ may be involved in AD progression suggests that NSAIDs and PPARγ agonists, both drugs that are commonly prescribed for other conditions, may be useful for the treatment of AD. Pioglitazone and rosiglitazone, PPARγ agonists commonly prescribed for the treatment of diabetes mellitus type 2, have reached clinical trials for treatment of AD, with mixed results. While some small studies have claimed that PPARγ agonists improve cognition in mild AD patients [72-74], other, larger studies have shown them to have little to no benefit on cognitive decline [75-77].

NF-κB

Another transcription factor that has received considerable attention is NF-κB. There is evidence that NF-κB can act as both a repressor and activator of *BACE1* gene transcription. Through luciferase reporter assays, NF-κB was found to be a repressor of *BACE1* transcription in neuronal-like differentiated SH-SY5Y or PC12 cells [78, 79]. The finding that addition of an NF-κB antagonist increased *BACE1* transcription further strengthened the evidence that NF-κB was a repressor of *BACE1* transcription [79]. Interestingly, NF-κB immunoreactivity has been shown to be reduced surrounding plaques in the AD brain, suggesting a potential causative role for NF-κB in the production of Aβ, through a loss of suppression of *BACE1* transcription in neurons [80].

However, Bourne *et al.* published evidence that activation of NF-κB actually enhances *BACE1* transcription in reactive astrocytes [79]. By studying the effects of mutating the NF-κB binding site on the rat BACE1 promoter, the authors found that under physiological conditions, NF-κB suppressed *BACE1* transcription, which is consistent with the observations that under normal conditions, astrocytes are immunonegative for BACE1. However, upon activation of astrocytic cultures with TNF-α, NF-κB actually enhanced transcription of *BACE1*. Interestingly, this NF-κB-dependent upregulation of BACE1 was mimicked by treatment of both neuronal and astrocytic cell lines with Aβ peptides [79]. Given that reactive astrocytes that are immunopositive for BACE1 are found around plaques [81], this suggests a mechanism by which Aβ stimulates NF-κB-mediated transcription of *BACE1*, which in turn increases Aβ production, driving the progression of the disease.

Further evidence for an activator role of NF-κB in *BACE1* transcription has stemmed from work investigating the link between the receptor for advanced glycation end products (RAGE) and the potential increased risk of developing AD in patients suffering from diabetes mellitus. By inducing a diabetic condition in rats *via* treatment with streptozotocin, a concurrent upregulation of RAGE, NF-κB and BACE1 in the brain has been demonstrated [82]. Treatment of neuroblastoma cells with two different advanced glycation end products (AGEs), pentosidine and GLAP, which are ligands for RAGE, significantly increased BACE1 protein levels, and the authors then demonstrated that this effect was mediated by an activation of NF-κB activity [82]. Intriguingly, in the cortex of AD patients, levels of both pentosidine and GLAP are dramatically increased [82], and there is some evidence that Aβ binds to RAGE [83-85], suggesting a further link between RAGE, NF-κB activation and AD progression. Further work should be conducted to determine whether inhibition or stimulation of the NF-κB pathway would be the more effective route to potentially treat AD.

NFAT

Cho *et al.*, reported a binding site for the transcription factor NFAT within the BACE1 promoter in 2008 [86]. NFAT is a transcription factor critical for induction of transcription of various cytokines in immune responses, and is activated by calcineurin, a calcium-binding protein [87]. Given that inflammation and calcium dysregulation are both strong candidates for mechanisms behind AD pathogenesis, NFAT may be central to the disease process. This theory is strengthened by the observation that disrupting intracellular calcium levels using ionomycin activated NFAT, and increased *BACE1* transcription. Interestingly, treatment of cells with Aβ mimicked this effect, and induced NFAT translocation from the cytosol into the nucleus [86].

There is some evidence that RAGE can also regulate *BACE1* expression *via* NFAT [88]. Adenoviral-mediated overexpression of full-length RAGE increased deposition of Aβ-positive plaques, Aβ generation, BACE1 levels and *BACE1* promoter activity [86]. The authors thus demonstrated that this effect was mediated *via* activation of NFAT1. Treatment with RAGE ligands mimicked this effect, whereas co-infection with soluble RAGE, which is an antagonist of RAGE activity, inhibited both NFAT activation and *BACE1* transcription mediated by full length RAGE [88].

hnRNP H

A recent paper examined the higher-order structure of the *BACE1* gene in an attempt to decipher the regulation of *BACE1* alternative splicing [89]. It is known that BACE1 can undergo significant alternative splicing, yielding proteins 476, 457, 455, 432 and 127 amino acids in length, along with the major 501 amino acid form of the enzyme [90]. It has been shown that the 501 amino-acid form of BACE1 is the only form that has significant proteolytic activity; therefore a strategy whereby the alternative splicing of the *BACE1* gene is modulated may be a novel route to modulating BACE1 activity [90].

The authors reported that the *BACE1* gene contains several G-rich sequences, which can form a G-quadruplex structure and recruit the splicing regulator hnRNP H [89]. Recruitment of hnRNP H promoted the production of the proteolytically active 501 amino-acid form of BACE1, and silencing hnRNP H by shRNA reduced the amount of active, full-length BACE1 produced, and reduced Aβ production concurrently [89].

MicroRNAs

Several studies have suggested that BACE1 mRNA levels remain unchanged in AD brains [65, 91-93], whereas levels of BACE1 protein have been found to be elevated [64, 65, 92, 94-99], suggesting that regulation of BACE1 levels occurs post-transcriptionally. This prompted researchers to examine the role of microRNAs (miRNAs) in regulation of BACE1 levels. miRNAs are small, non-coding RNAs that bind to the 3'-untranslated region (3'-UTR) of their target mRNAs, and regulate the expression of target genes post-transcriptionally either by degradation of, or inhibition of translation of their respective target mRNAs. The field of miRNA research is still relatively new, however they have been shown to regulate the expression of genes related to many diverse physiological functions (for reviews of the role of miRNAs in neurodevelopment and neurodegeneration, see [100-102]).

Several miRNAs that are expressed in the brain have been implicated in regulation of BACE1 levels. Wang *et al.* first demonstrated that the miRNA miR-107 regulates *BACE1* promoter activity in cell culture, and that expression of miR-107 is decreased in AD brains [103, 104]. Increased levels of BACE1 were observed to coincide with the loss of miR-107, suggesting a causative role for miR-107 loss in increasing BACE1 activity in the brain [104]. A further examination of AD cases showed a significant correlation between decreased miR-107 levels and neuritic plaque load [103]. Around the same time, Hébert *et al.* published a study implicating another cluster of miRNAs, miR-29a/b-1, in regulation of BACE1 [95]. The authors found that loss of miR-29a/b-1 correlated well with increased BACE1 levels in sporadic AD brains, and that miR-29a/b-1 expression was coregulated with BACE1 expression in the developing brain, indicating a central role for miR-29a/b-1 in suppressing BACE1 expression *in vivo* [95]. It should be noted, however, that only a subset of AD patients (approximately one third of patients studied) had increased BACE1 levels, indicating that this mechanism may be only one of several associated with development of AD [95]. Another study reported that another member of the miR-29 family, miR-29c, can also suppress BACE1 expression in cell culture, and that BACE1 protein levels, but not mRNA levels, are significantly decreased in a mouse line overexpressing miR-29c [105]. Similarly, the miRNAs miR-298 and miR-328 have been shown to suppress BACE1 expression in both cell culture and primary neuronal cultures through a synergistic mechanism [106]. Additionally, a study published by Fang *et al.* demonstrated that miR-124 can influence BACE1 expression, and that inhibition of miR-124 increases BACE1 levels, and enhances neuronal cell death [107].

Taken together, these studies indicate that BACE1 expression is under complex control both pre- and post-transcriptionally. Further, there is merit to investigating the therapeutic potential of targeting BACE1 transcription to treat AD. However, transcription factors regulate the expression of many genes that are critical for proper cellular function, which increases the likelihood of off-target effects significantly. Therefore, an extremely thorough and robust examination of safety and tolerance would need to be conducted for any drugs that modulate transcription.

eIF2α

Through both PET imaging studies and post-mortem brain tissue analysis, a close link between impaired energy metabolism in the brain and AD has been postulated. Additionally, inhibition of energy metabolism leads to an increase in both BACE1 protein levels, and Aβ. It has been suggested that this may be the result of stress-induced changes in translational control of BACE1 mRNA.

In support of this theory, O'Connor and colleagues demonstrated that energy deprivation induces phosphorylation of the translation initiation factor eIF2α, a well-known marker of stress-induced translational control [108]. Further, direct induction of eIF2α phosphorylation increases BACE1 levels in cell culture, while inhibition of eIF2α phosphorylation decreases BACE1 levels. Moreover, selective inhibition of eIF2α dephosphorylation markedly increases BACE1 levels to a similar extent to that of energy deprivation. The authors demonstrated that this effect on BACE1 levels occurs post-transcriptionally, as no changes in BACE1 mRNA were observed following energy deprivation [108].

Importantly, several studies have shown that levels of phosphorylated eIF2α are increased in both transgenic mouse and AD patient brains, which correlates with both increased levels of BACE1 protein and amyloid load [108-110]. Taken together, these studies suggest a potential *in vivo* relevance for eIF2α phosphorylation in translational regulation of BACE1 levels, and hence modulating eIF2α phosphorylation may be a useful target for therapeutic intervention for the disease.

Targeting Interactions of BACE1 with Other Cellular Components

BACE1 activity has been shown to be regulated by its interaction with multiple cellular components, and these interactions regulate varying aspects of BACE1 biology. These interactions may present novel therapeutic targets for modulating BACE1 activity *in vivo*, either through directly modulating activity of BACE1, or altering BACE1 trafficking and localization. Table **2** summarizes the body of work that will be discussed in this section.

Targeting BACE1-Lipid Interactions

Interactions of BACE1 with lipids within the plasma membrane have been well documented. As BACE1 is a membrane-bound protease, and APP is also a transmembrane protein, this raised the possibility that BACE1 activity could be regulated by interactions with the lipid bilayer. However, BACE1 that is not palmitoylated, and hence is not targeted to lipid rafts, cleaves APP just as much as palmitoylated BACE1, suggestingthat localization of BACE1 to lipid rafts is not required for optimal BACE1 activity [36, 37].

Table 2.   Interactors with BACE1 and their Effect on BACE1 Activity

Name	Effect on BACE1 Activity	Comments	References
Lipid Rafts	↔	Potentially useful for targeting inhibitors to BACE1	[36-37, 111-112]
Sphingosine-1-phosphate	↑	Sphingosine kinase inhibitors potentially useful	[113]
Glycosaminoglycans/ proteoglycans	↑ ↓	At low concentrations *in vitro*, possible *in vivo*? At high concentrations *in vitro*, possibly *in vivo*? Also decreases BACE1 protein levels	[34, 132, 138-139] [34, 131-132, 140]
Reticulons	↓	Arrests trafficking of BACE1 to active compartments	[141-143, 146]
GGAs	↓	Regulates trafficking of BACE1 from the endosome to lysosome to be degraded	[99, 148-149, 153-158

While lipid rafts may not directly regulate BACE1 activity, they may prove useful in the development of specific BACE1 inhibitors. Theoretically, anchoring a BACE1 inhibitor to the plasma membrane could increase its potency by enriching it in membrane-rich fractions, such as the endosome, where BACE1 proteolytic activity is highest. Rajendran *et al.* showed this experimentally, where anchoring a BACE1 inhibitor to the plasma membrane through sterol moieties increased the potency of the inhibitor substantially compared to free, unbound inhibitor in cellular assays of BACE1 activity [111]. Since this study, several membrane-linked BACE1 inhibitors have been developed, however none have as yet been tested in animal models or AD clinical trials [112]. Nonetheless, membrane-anchored compounds may represent a promising future direction for the design of BACE1 inhibitors.

Changes in the metabolism of the lipid sphingosine have also been implicated in the pathogenesis of AD. Recently, a study by Takasugi *et al.* reported that sphingosine-1-phosphate, a metabolite of sphingosine, markedly increased BACE1 activity [113]. Sphingosine-1-phosphate is a biologically active GPCR-linked signaling lipid, which is produced by phosphorylation of sphingosine by two kinases, sphingosine kinase 1 and sphingosine kinase 2 (SphKI/SphKII) [113]. Addition of SphK inhibitors markedly reduced β-secretase activity in mouse neuroblastoma cells as measured by levels of both β-secretase related APP cleavage products and Aβ, an effect that was mimicked by both siRNA knockdown of SphKII and overexpression of sphingosine-1-phosphate degrading enzymes [113]. Moreover, overexpression of SphKII increased β-secretase activity in these cells. When a SphK inhibitor was directly injected into the brains of eight-week-old wild type mice, the levels of Aβ were modestly lowered in the injected hemisphere compared to the uninjected hemisphere of the brain. Oral administration of the same inhibitor to A7 transgenic mice, which express APP with two mutations associated with familial forms of AD, also decreased levels of Aβ in the brain [113]. Intriguingly, the study also found that treatment of mouse neuroblastoma cells with fibrillar $A\beta_{42}$ increased SphKII activity, suggesting a direct relationship between Aβ production, SphKII, and BACE1 activity. Furthermore, a trend towards slightly decreased levels of SphKII in postmortem AD brain tissue, yet greater SphKII enzymatic activity was observed, suggesting a positive feedback mechanism where elevated Aβ levels increase SphKII activity, that in turn increases sphingosine-1-phosphate, which stimulates BACE1 activity, producing more Aβ [113].

If this turns out to be the case in AD brains, SphK inhibitors may be beneficial to treat AD, however one must be cautious of potential side effects of SphK inhibition. SphK has been implicated in numerous critical processes in the body including the immune response [114, 115], ubiquitination pathways [116-118] and histone deacetylase activity [119]. Any potential side effects of SphK inhibition need to be evaluated in future studies.

Targeting Glycosaminoglycans and Proteoglycans

The role of glycosaminoglycans (GAGs) and proteoglycans (PGs) in the pathogenesis of AD has been studied extensively for over two decades. GAGs are long, linear polysaccharides that can be covalently attached to a proteinaceous core to form PGs [34, 120]. GAGs and PGs have been shown to regulate many functions, both in the CNS and the periphery. Several studies conducted in the 1980s and 1990s demonstrated that both heparin sulfate proteoglycans (HSPGs) and chondroitin sulfate proteoglycans (CSPGs) are constituents of both diffuse and neuritic plaques in the AD brain [121-125]. Since this initial discovery, PGs have been found in deposits associated with other amyloidoses, including Parkinson's disease [126], familial amyloidotic polyneuropathy [127, 128], familial British dementia [129], familial Danish dementia [129], the prion diseases Creutzfeld-Jakob disease and Gerstmann-Sträussler-Scheinker syndrome [130], dementia with Lewy bodies [126] and progressive supranuclear palsy [126]. This body of evidence suggests that PGs may be involved in a common mechanism behind many neurodegenerative diseases.

There is increasing evidence that GAGs and PGs can regulate BACE1 activity. Using *in vitro* assays, several studies have shown that GAGs can influence BACE1 activity. Scholefield *et al.* showed that the GAG heparan sulfate inhibited BACE1 in a dose-dependent manner [131]. Conversely, GAGs were more recently shown to be stimulatory to BACE1 activity at lower concentrations, and inhibitory to activity at higher concentrations [34, 131, 132]. Small, low molecular weight GAG analogues were shown to effectively inhibit the stimulatory effect of GAGs on BACE1 [34]. Importantly, the stimulatory effect was found to be dependent on the presence of the prodomain on BACE1, indicating that the stimulatory and inhibitory binding sites may be independent of each other [132]. Given that the BACE1 zymogen retains activity, and that APP processing has been shown to be present in compartments where proBACE1 localizes [131, 133-137], this stimulation of activity by GAGs may be relevant therapeutically.

In cell culture, Leveugle *et al.* first reported that exogenous addition of the GAG heparin could increase secretion of sAPPβ, suggesting a specific effect on β-secretase processing of APP [138]. Interestingly, secretion of full length APP was also increased following treatment with heparin, and this effect could be blocked by low molecular weight heparin mimetics [138, 139]. Conversely, heparin, and a low molecular weight heparin analogue, enoxaparin, have also been shown to inhibit BACE1 processing of APP through a dose-dependent decrease in sAPPβ secretion as well as C99 production [140]. However, the effect of GAGs on BACE1 appears to be more complex than merely an effect on activity, as heparin and enoxaparin were found to decrease BACE1 protein levels in a dose-dependent manner. Further this effect was not specific for BACE1, as levels of ADAM10, the most likely α -secretase candidate, were also decreased following GAG administration, along with α-secretase related APP processing products, indicating that administration of GAGs interferes with general processing of APP [140]. Given that α-secretase related APP processing products such as sAPPα have been shown to be neuroprotective, along with the fact that GAGs and PGs are intimately involved in many process both in the CNS and periphery, administering GAG mimetics as BACE1 modulators may produce a whole slew of side effects. More work needs to be done to pinpoint the exact role GAGs and PGs play in APP metabolism, and to determine the tolerability of GAG mimetics as therapeutic compounds.

Reticulon Proteins

Through coimmunoprecipitation screening of BACE1 binding partners, members of the reticulon family of proteins were found to bind to BACE1 [141, 142]. RTNs bind to BACE1 *via* a conserved three amino-acid QID domain [143]. One RTN family member in particular, RTN3, has been the subject of much scrutiny due to the fact that it colocalizes with BACE1 in neurons [141]. Analysis of the cellular distribution of RTN3 in the brain reveals that RTN3 expression is localized to pyramidal neurons in the cerebral cortex, an expression pattern similar to that of BACE1 [144, 145]. Overexpression of RTN3 results in a marked decrease in Aβ production in HEK293 cells [141, 142], and RTN3 was shown to decrease BACE1 activity directly using an *in vitro* assay of BACE1 activity [141]. Moreover, RNAi interference of RTN3 expression resulted in a significant increase in Aβ production in these cells [141]. RTN3 was then shown to function to modulate BACE1 activity by directly inhibiting the interaction between BACE1 and APP [141]. These effects were subsequently confirmed in neuronal cells as well [142]. Overexpression of RTN3 in APP/PS1 transgenic mice under a murine prion promoter results in decreased plaque formation and decreased β -secretase processing of APP [146]. The mechanism behind these observations was subsequently shown to be an effect on BACE1 trafficking, demonstrated by the accumulation of BACE1 in the ER and Golgi, where activity of BACE1 is substantially reduced compared to the endosome [146]. This suggests that RTN3 acts to either facilitate retrograde trafficking of BACE1 from endosomal compartments to the ER and Golgi, or arrest the trafficking of BACE1 out of the

ER and Golgi. As the primary sub-cellular sites of RTN3 localization consist of the ER and Golgi, this suggests that RTN3 is most likely acting to arrest BACE1 trafficking [146].

Several studies have endeavored to address the question of what happens to reticulon proteins in the AD brain. These studies have shown that total RTN3 levels are unchanged in the AD brain compared to non-demented controls. Interestingly, however, the levels of monomeric RTN3 are markedly decreased in the AD brain, with accumulations of RTN3 observed in dystrophic neurites [146, 147]. Furthermore, aggregation of RTN3 leads to a decreased interaction with BACE1; thereby reducing it's ability to modulate BACE1 activity [143]. This indicates that the formation of RTN3-positive dystrophic neurites, and thus liberation of BACE1 activity, may be a pathological event in the development of AD.

Based on the emerging evidence that RTN proteins regulate BACE1 trafficking and activity both in cell culture and *in vivo*, combined with the fact that the binding sites of RTN proteins for BACE1 have been mapped out, this could be useful in developing small-molecule compounds which mimic the effect of RTN proteins, thereby reducing BACE1 activity.

GGA Proteins

Alongside RTN proteins, another family of proteins, called the Golgi-localized γ-ear-containing ARF-binding (GGA) proteins, has also been shown to regulate BACE1 activity and trafficking. First described as BACE1-binding proteins by He *et al.*, GGA proteins bind to BACE1 *via* their VHS (VPS-27, Hrs and STAM) domain, which binds to specific DXXLL dileucine motifs within their target proteins [148, 149]. BACE1 contains such a dileucine motif at the C-terminus, at positions 499 and 500 [148]. This motif has been shown to be critical in regulating trafficking of BACE1 from the cell surface to the endosome. Deletion of the dileucine motif, or mutagenesis of either L499 or L500 results in retention of BACE1 at the cell surface, as well as increases the total amount of BACE1, suggesting a role of GGA proteins in regulating the degradation of BACE1 [150, 151]. This theory was further strengthened by the finding that mutagenesis of

L499/L500 results in reduced accumulation of BACE1 in lysosomes, where it is normally degraded, following lysosomal hydrolase inhibition [152]. Furthermore, GGA-BACE1 interactions appear to be mediated by ubiquitination of BACE1 at K501, a key modification that targets BACE1 for degradation [153]. Additionally, depletion of GGA3 results in a stabilization of BACE1 levels, an accumulation of BACE1 within endosomal compartments and enhancement of β-secretase activity, whereas overexpression of GGA3 protein produces the opposite effect [99, 154-157]. Importantly, BACE1 levels are also markedly increased in the brains of GGA-null mice, indicating a physiological role for GGAs in regulation of BACE1 [158]. This provides compelling evidence that GGA proteins regulate trafficking of BACE1 from the endosome to the lysosome, enhancing its degradation, and negatively regulate BACE1 activity.

Interestingly, GGA3 is a substrate for caspase-3, a key molecule in apoptotic pathways. Following induction of apoptosis in cells by staurosporine or etoposide, GGA3 is degraded by caspase-3, which results in stabilization of BACE1 levels and activity [99]. This effect is also observed in mice following both cerebral ischemia and controlled cortical impact traumatic brain injury [99, 158]. This could be a mechanism behind the observation that previous traumatic brain injury has been linked to an increased risk in the future development of AD [159]. It also raises the possibility that perturbations in GGA3 may be an underlying cause for AD pathogenesis. Indeed, GGA3 levels are markedly reduced in the AD brain, suggesting a loss of regulation of BACE1 trafficking and degradation mediated by GGA3 in AD [99]. Based on this evidence, therapeutically mimicking the effect of, or increasing the levels of GGA3 in the brain may be a new strategy for the treatment of AD.

Targeting BACE1 Ubiquitination and Degradation

The preceding strategies for modulating BACE1 activity all focus on arresting the synthesis and trafficking of BACE1. Another strategy may be to escalate the degradation of BACE1; by influencing it's trafficking towards the proteasome or lysosome, and increasing it's degradation, this may be an effective way to reduce BACE1 processing of APP in the AD brain.

Ubiquitination of proteins is the most characterized protein degradation signal. Addition of ubiquitin chains to lysine residues on proteins targets these proteins for degradation. Proteins can be shuttled down different degradation pathways depending on the type of ubiquitin chain attached to them. Ubiquitin is added to a protein as a monomer, however, given ubiquitin has seven lysine residues of its own, additional ubiquitin molecules can be added to yield polyubiquitin chains. Elongation of these chains can occur at any of the seven lysine residues, however the most studied are K48- and K63-linked polyubiquitin chains. K48-linked polyubiquitin chains mainly signal proteins to be degraded *via* the proteasomal pathway, whereas K63-linked polyubiquitination plays a role in endocytosis and sorting of proteins into endosomes and lysosomes [160].

Ubiquitin is added to protein targets through the actions of three separate classes of proteins. Firstly, ubiquitin is attached to the E1 ubiquitin-activating enzyme. Secondly, ubiquitin is transferred from E1 to an E2 ubiquitin-conjugating enzyme, of which several dozen are known. Finally, ubiquitin is transferred from the E2 enzyme to the target protein *via* the action of an E3 ubiquitin ligase. To date, several hundred different E3 ligases are known, and they provide the substrate-recognition specificity of the reaction. Ubiquitin can also be cleaved off ubiquitinated cargo by the action of a class of proteases called deubiquitinases [160].

It is now known that BACE1 is ubiquitinated at the C-terminus at lysine 501, and that this ubiquitination state signals BACE1 to be degraded [153, 161]. There is still some conjecture as to whether BACE1 is primarily degraded *via* the proteasomal pathway, or is degraded within lysosomes, both pathways that degrade ubiquitinated cargo. The first study to identify that BACE1 was ubiquitinated provided evidence that degradation was mediated through the proteasomal pathway, as an inhibitor of the proteasome, lactacystin, increased both BACE1 levels and C99 and Aβ production [161]. However, the authors did not investigate the effect of inhibition of other degradation pathways on BACE1 levels or activity, and as such left the question open as to whether other pathways may contribute to BACE1 degradation.

Since this initial paper was published, several studies have reported conflicting evidence about which pathway BACE1 is degraded. While the initial study showed that inhibiting the proteasomal pathway increases BACE1 levels, others have shown little to no evidence that the proteasome plays a role in BACE1 degradation. Rather, an emerging view is that BACE1 is degraded primarily through the lysosomal pathway. Koh *et al.* showed that inhibiting the proteasome by both lactacystin and MG132 had little to no effect on BACE1 levels, however inhibition of lysosomal proteases by chloroquine and NH_4Cl drastically increased levels of both ectopically and endogenously expressed BACE1 in a variety of cell types, including primary cortical neurons [152]. Furthermore treatment with chloroquine and NH_4Cl caused BACE1 to accumulate in LAMP-2 positive compartments, a specific marker for lysosomes [152]. Additional evidence was presented when Kang *et al.* analyzed the specific type of ubiquitin chains that are attached to BACE1 through systematic mutation of the various lysine residues present in ubiquitin [153]. BACE1 was found to be mainly monoubiquitinated and K63-linked polyubiquitinated; further strengthening the case that BACE1 is shuttled down the endosomal/lysosomal pathway, and is degraded in lysosomes [153].

The main avenue for therapeutic intervention related to BACE1 ubiquitination lies with targeting the specific E3 ligases and deubiquitinases that regulate this activity, as these proteins confer the specificity for BACE1 ubiquitination.

Perhaps the most promising candidate in this regard is the ubiquitin carboxyl-terminal hydrolase L1 (Uch-L1). Uch-L1 can function as both an ubiquitin ligase, as well as a deubiquitinase [162, 163]. This dual function is dependent on the self-association state of the protein. Monomeric Uch-L1 acts as a deubiquitinase, whereas homodimeric Uch-L1 acts as an ubiquitin ligase [163]. Uch-L1 has been shown to regulate BACE1 levels both in cell culture and transgenic AD animal models [164]. In HEK cells, inhibition of Uch-L1 markedly increases BACE1 protein levels, and overexpression of Uch-L1 increases its degradation. The consequences of this increased degradation are decreases in both C99 and Aβ production in these cells [164]. In the *gad* mouse, which is a Uch-L1-null mouse line, BACE1, along with C99 and Aβ, have been shown to be markedly upregulated, further strengthening the evidence for a role of Uch-L1 in degradation of BACE1, and indicating that Uch-L1 acts as a ubiquitin ligase in this regard [164].

Interestingly, a recent study reported that $A\beta_{42}$ downregulates Uch-L1 expression, and increases BACE1 levels [165]. NF-κB activation was shown to be critical for this effect [165]. Similarly, in the brains of patients who had sporadic AD, decreased Uch-L1 levels were observed, which correlated with increased BACE1 levels [165]. This further strengthens the theory that $A\beta$ creates a positive feedback loop, which acts to dysregulate BACE1 activity, which thus stimulates further production of $A\beta$ and drives the degenerative processes of the disease.

However, one must remain cautious when considering targeting ubiquitination pathways, as there is a high chance of detrimental off-target effects owing to the universal nature of ubiquitination in the regulation of protein levels.

BACE1 Immunotherapy

Treating AD *via* immunotherapy has been a long-standing idea that has received considerable attention. Immunotherapy employs the use of monoclonal antibodies to bind to and promote clearance of proteins associated with disease progression. This approach is already in use for the treatment of several different diseases, including cancer and immune diseases.

While the majority of immunotherapeutic research has been focused on the development of antibodies raised against $A\beta$, human trials have been less than successful, largely due to undesirable, serious side effects (for a review, see [166]). An alternative immunotherapeutic strategy that has emerged recently involves employing antibodies targeted against BACE1.

Immunotherapy provides some significant advantages over conventional small-molecule inhibitors. Molecules aimed at disrupting interactions of BACE1 to other adaptor proteins will always provide off-target effects relating to interfering with the adaptor protein's normal physiological function. Off-target effects are also a common issue when dealing with active site inhibition, as these molecules tend to inhibit closely related enzymes, such as cathepsin D or BACE2 in the case of BACE1. Antibodies offer a more specific route to inhibition of BACE1, as well as an increased half-life in the body compared to standard small-molecule inhibitors.

Zhou and colleagues first published proof of this concept in 2011, when they developed an inhibitory monoclonal antibody, named 1A11, that binds to a domain that maps to loops D and F of BACE1 [167]. This binding domain confers high specificity to BACE1, as these loops are unique for BACE1 compared to other aspartyl proteases [167]. 1A11 was found to potently inhibit BACE1 in the sub-nanomolar range in an *in vitro* assay of protease activity. The authors then tested the ability of 1A11 to inhibit BACE1-mediated processing of APP. 1A11 potently inhibited C99 and Aβ production in primary cortical cultures from wild type mice, indicating that the antibody was able to enter and inhibit activity of BACE1 in intact cells [167]. Importantly, when 1A11 was stereotactically injected into the hippocampus and cortex, BACE1-mediated processing of APP was markedly decreased, indicating that this antibody can regulate BACE1 activity *in vivo* [167]. However, the therapeutic relevance of this remains yet to be determined, as treatment was not *via* a feasible route of administration for a therapeutic agent, and as such its bioavailability was not assessed. Nonetheless, this study injected some much-needed vigor into BACE1 inhibitory research.

Atwal *et al.* reported the development of another potentially therapeutic BACE1 antibody in the same year [168]. Using phage display screening, they were able to develop a highly specific BACE1 antibody that potently inhibited BACE1 activity, yet displayed no activity against both cathepsin D and BACE2. Mapping of the binding domain of the inhibitor revealed that it binds to an exosite on BACE1, as it did not compete with the BACE1 active site inhibitor OM99-2 for binding to BACE1 [168]. While the data from initial cell culture experiments proved promising, when the antibody was administered to transgenic mice, it failed to lower Aβ levels in the brain at low doses, while it was potent at reducing plasma Aβ levels. At a high dose of 100 mg/kg, levels of Aβ in the brain were reduced several days after administration, indicating that the antibody does have therapeutic properties [168].

The main pitfall of immunotherapy in the brain is the low blood-brain barrier penetrance of antibodies. Atwal *et al.* demonstrated this in mice by measuring the concentration of their anti-BACE1 antibody in the plasma and the brain following a single dose of 100 mg/kg. They showed that while the plasma concentration of their anti-BACE1 antibody was very high, only approximately 0.1% of the antibody actually penetrated into the brain. While the concentration of antibody measured in the brain was sufficient to exert a beneficial effect on BACE1 activity, the dose required to achieve this effect is impractical for therapy [168]. In an elegant study, they addressed this issue by producing a bispecific antibody, with one end containing a low-affinity transferrin receptor epitope, and the other their anti-BACE1 antibody [169]. Transferrin receptor is expressed highly by endothelial cells that comprise the blood-brain barrier, and antibodies targeted against transferrin receptor have been shown to cross the blood-brain barrier (for a review, see [170]). It was hoped that by using a conjugated transferrin receptor, uptake of the antibody into the brain would be improved. Furthermore, by lowering the affinity of the antibody for transferrin receptor, release of the antibody into the brain would also be improved. Indeed, the bispecific antibody demonstrated a much-improved penetrance into the brain compared to that of the monospecific anti-BACE1 antibody used in the previous study, and release of the antibody into the brain was shown to be effective [169]. Additionally, the bispecific antibody successfully reduced brain Aβ levels by approximately 50%, indicating that the antibody was physiologically active once it reached the brain [169]. This study has significant implications not only for the development of anti-BACE1 antibodies, but also for other immunotherapeutic compounds targeted to the brain, such as anti-amyloid antibodies that have been in trials for several years.

The data from these studies provide an intriguing avenue of research for the development of AD therapeutics. While more work needs to be done to determine the effects of chronic dosing of these antibodies, as these studies only examined the immediate, acute effects of single dosing, it will be interesting to see how these antibodies perform in both animal behavioral and human trials, if they get that far.

Modulation of Calcium

Calcium dysregulation has been proposed as a key pathogenic process underpinning AD [171, 172]. There is evidence that both Aβ can elicit a significant influx of calcium into cells, and that increasing intracellular calcium increases Aβ production, again providing evidence that a feedback mechanism regulating Aβ production exists, which drives neurodegeneration and hence AD progression. This increase in cytosolic calcium has been proposed as the mechanism through which Aβ exerts its neurotoxic effects (for a review, see [171, 172]). Therefore, the restoration of calcium homeostasis may be another avenue of therapy that could be useful to treat AD.

There is emerging evidence that changes in cytosolic calcium can regulate various aspects of BACE1 biology, from modulation of *BACE1* expression to direct regulation of BACE1 activity. As noted earlier, administration of ionomycin, a calcium ionophore, elicited a significant increase in *BACE1* transcription through an NFAT-dependent mechanism [86]. Additionally, the calcium-binding phosphatase calcineurin blocked this activity through inhibiting the nuclear translocation of NFAT [86]. Furthermore, this effect was mediated through RAGE, which is known to interact with calcium-binding proteins [88]. Similarly, the calcium-dependent protease calpain has been demonstrated to increase BACE1 levels both in cell culture and in animal models of AD, and that overexpression of the calpain inhibitor calpastatin inhibited this effect, reduced plaque load and improved behavioral performance in AD transgenic mice [173]. Furthermore, overexpression of calpastatin inhibited Aβ-induced upregulation of BACE1 levels in cell culture [173]. Interestingly, calcium has also been shown to directly influence BACE1 activity through a direct interaction with the enzyme [174]. Conversely, increasing intracellular calcium using ionomycin decreased maturation of BACE1, through inhibition of the calcium-dependent, furin-like protease activity that cleaves the propeptide from BACE1, indicating that reduction of calcium pharmacologically may increase maturation of BACE1 from the less-active zymogen to the fully active mature form [175].

More work will need to be conducted on assessing the therapeutic potential of calcium channel blockers for the treatment of AD. While several clinical trials testing the usefulness of calcium channel blockers in AD have been disappointing, this strategy should not be ruled out completely (for a review, see [176]).

CONCLUDING REMARKS

With an aging population, the incidence of AD is on the rise, and hence the need for an effective therapy to combat the disease is ever growing. While the identity of BACE1 as the Alzheimer's β-secretase has been known for over a decade, there are still no compounds targeting BACE1 that have been approved for treating AD. While initial results *in vitro* and in animal AD models appear promising, poor pharmacokinetic properties, bioavailability and efficacy of active site-directed compounds in human trials have often stymied progress in the development of BACE1 inhibitors for several years.

These pitfalls have forced researchers to look at other strategies to modulate BACE1 activity in the brain. This review has focused on several of these potential avenues, and discussed the potential for therapeutic intervention. While modulation of BACE1 activity *via* an indirect mechanism presents a greater potential for off-target side effects, one must weigh the benefits against the risks. Given the majority of off-target effects of BACE1 inhibition or knockout have thus far been identified as disturbances in processes mainly involved in development, it is unknown whether inhibiting BACE1 activity in aged patients would present any great risk of unacceptable side effects. Further, recent developments in the design of non-active site directed, yet specific, BACE1 modulators present an exciting new avenue for therapeutic research. Therefore, BACE1 should remain an attractive target for therapeutic intervention.

Given AD is a largely heterogeneous disease, different patients may respond to different treatments. Therefore, it may be necessary to combine BACE1 inhibition with other therapeutic compounds, such as anti-amyloid therapies. It would be interesting to assess whether combining therapies produces a beneficial outcome.

ACKNOWLEDGEMENTS

This work was supported by Award Number R01AG033016 and R01AG025952 from the National Institute On Aging (to GT).

CONFLICT OF INTEREST

The authors confirm that this chapter contents have no conflict of interest.

REFERENCES

[1]   Hori, K.; Oda, T.; Asaoka, T.; Yoshida, M.; Watanabe, S.; Oyamada, R.; Tominaga, I.; Inada, T., First episodes of behavioral symptoms in Alzheimer's disease patients at age 90

and over, and early-onset Alzheimer's disease: comparison with senile dementia of Alzheimer's type. *Psychiatry Clin Neurosci* **2005**, *59* (6), 730-5.

[2]   Alzheimer's Association, 2009 Alzheimer's disease facts and figures. *Alzheimer's dement* **2009**, *5* (3), 234-70.

[3]   Delacourte, A.; Defossez, A., Alzheimer's disease: Tau proteins, the promoting factors of microtubule assembly, are major components of paired helical filaments. *J Neurol Sci* **1986**, *76* (2-3), 173-86.

[4]   Grundke-Iqbal, I.; Iqbal, K.; Quinlan, M.; Tung, Y. C.; Zaidi, M. S.; Wisniewski, H. M., Microtubule-associated protein tau. A component of Alzheimer paired helical filaments. *J Biol Chem* **1986**, *261* (13), 6084-9.

[5]   Grundke-Iqbal, I.; Iqbal, K.; Tung, Y. C.; Quinlan, M.; Wisniewski, H. M.; Binder, L. I., Abnormal phosphorylation of the microtubule-associated protein tau (tau) in Alzheimer cytoskeletal pathology. *Proc Natl Acad Sci USA* **1986**, *83* (13), 4913-7.

[6]   Ihara, Y.; Nukina, N.; Miura, R.; Ogawara, M., Phosphorylated tau protein is integrated into paired helical filaments in Alzheimer's disease. *J Biochem (Tokyo)* **1986**, *99* (6), 1807-10.

[7]   Nukina, N.; Ihara, Y., One of the antigenic determinants of paired helical filaments is related to tau protein. *J Biochem (Tokyo)* **1986**, *99* (5), 1541-4.

[8]   Farlow, M. R.; Cummings, J., A modern hypothesis: The distinct pathologies of dementia associated with Parkinson's disease *vs.* Alzheimer's disease. *Dement Geriatr Cogn Disord* **2008**, *25* (4), 301-8.

[9]   Burack, M. A.; Hartlein, J.; Flores, H. P.; Taylor-Reinwald, L.; Perlmutter, J. S.; Cairns, N. J., *In vivo* amyloid imaging in autopsy-confirmed Parkinson disease with dementia. *Neurology* **2010**, *74* (1), 77-84.

[10]   Ghetti, B.; Tagliavini, F.; Masters, C. L.; Beyreuther, K.; Giaccone, G.; Verga, L.; Farlow, M. R.; Conneally, P. M.; Dlouhy, S. R.; Azzarelli, B.; Bugiani, O., Gerstmann-Straussler-Scheinker disease. II. Neurofibrillary tangles and plaques with PrP-amyloid coexist in an affected family. *Neurology* **1989**, *39* (11), 1453-61.

[11]   Glenner, G. G.; Wong, C. W., Alzheimer's disease: initial report of the purification and characterization of a novel cerebrovascular amyloid protein. *Biochem Biophys Res Commun* **1984**, *120* (3), 885-90.

[12]   Masters, C. L.; Simms, G.; Weinman, N. A.; Multhaup, G.; McDonald, B. L.; Beyreuther, K., Amyloid plaque core protein in Alzheimer disease and Down syndrome. *Proc Natl Acad Sci USA* **1985**, *82* (12), 4245-9.

[13]   Selkoe, D. J.; Abraham, C. R.; Podlisny, M. B.; Duffy, L. K., Isolation of low-molecular-weight proteins from amyloid plaque fibers in Alzheimer's disease. *J Neurochem* **1986**, *46* (6), 1820-34.

[14]   Shoji, M.; Golde, T. E.; Ghiso, J.; Cheung, T. T.; Estus, S.; Shaffer, L. M.; Cai, X. D.; McKay, D. M.; Tintner, R.; Frangione, B.; Younkin, S. G., Production of the Alzheimer amyloid beta protein by normal proteolytic processing. *Science* **1992**, *258* (5079), 126-9.

[15]   Iwatsubo, T.; Odaka, A.; Suzuki, N.; Mizusawa, H.; Nukina, N.; Ihara, Y., Visualization of A beta 42(43) and A beta 40 in senile plaques with end-specific A beta monoclonals: evidence that an initially deposited species is A beta 42(43). *Neuron* **1994**, *13* (1), 45-53.

[16]   Kang, J.; Lemaire, H. G.; Unterbeck, A.; Salbaum, J. M.; Masters, C. L.; Grzeschik, K. H.; Multhaup, G.; Beyreuther, K.; Muller-Hill, B., The precursor of Alzheimer's disease amyloid A4 protein resembles a cell-surface receptor. *Nature* **1987**, *325* (6106), 733-6.

[17] Nunan, J.; Small, D. H., Regulation of APP cleavage by alpha-, beta- and gamma-secretases. *FEBS Lett***2000**, *483* (1), 6-10.

[18] Nunan, J.; Small, D. H., Proteolytic processing of the amyloid-beta protein precursor of Alzheimer's disease. *Essays Biochem* **2002**, *38*, 37-49.

[19] Klaver, D. W.; Wilce, M. C. J.; Cui, H.; Hung, A. C.; Gasperini, R.; Foa, L.; Small, D. H., Is BACE1 a suitable therapeutic target for the treatment of Alzheimer's disease? Current strategies and future directions. *Biol Chem* **2010**, *391* (8), 849-59.

[20] Hussain, I.; Powell, D.; Howlett, D. R.; Tew, D. G.; Meek, T. D.; Chapman, C.; Gloger, I. S.; Murphy, K. E.; Southan, C. D.; Ryan, D. M.; Smith, T. S.; Simmons, D. L.; Walsh, F. S.; Dingwall, C.; Christie, G., Identification of a novel aspartic protease (Asp 2) as beta-secretase. *Mol Cell Neurosci* **1999**, *14* (6), 419-27.

[21] Vassar, R.; Bennett, B. D.; Babu-Khan, S.; Kahn, S.; Mendiaz, E. A.; Denis, P.; Teplow, D. B.; Ross, S.; Amarante, P.; Loeloff, R.; Luo, Y.; Fisher, S.; Fuller, J.; Edenson, S.; Lile, J.; Jarosinski, M. A.; Biere, A. L.; Curran, E.; Burgess, T.; Louis, J. C.; Collins, F.; Treanor, J.; Rogers, G.; Citron, M., Beta-secretase cleavage of Alzheimer's amyloid precursor protein by the transmembrane aspartic protease BACE. *Science* **1999**, *286* (5440), 735-41.

[22] Yan, R.; Bienkowski, M. J.; Shuck, M. E.; Miao, H.; Tory, M. C.; Pauley, A. M.; Brashier, J. R.; Stratman, N. C.; Mathews, W. R.; Buhl, A. E.; Carter, D. B.; Tomasselli, A. G.; Parodi, L. A.; Heinrikson, R. L.; Gurney, M. E., Membrane-anchored aspartyl protease with Alzheimer's disease beta-secretase activity. *Nature* **1999**, *402* (6761), 533-7.

[23] Sinha, S.; Anderson, J. P.; Barbour, R.; Basi, G. S.; Caccavello, R.; Davis, D.; Doan, M.; Dovey, H. F.; Frigon, N.; Hong, J.; Jacobson-Croak, K.; Jewett, N.; Keim, P.; Knops, J.; Lieberburg, I.; Power, M.; Tan, H.; Tatsuno, G.; Tung, J.; Schenk, D.; Seubert, P.; Suomensaari, S. M.; Wang, S.; Walker, D.; Zhao, J.; McConlogue, L.; John, V., Purification and cloning of amyloid precursor protein beta-secretase from human brain. *Nature* **1999**, *402* (6761), 537-40.

[24] Lin, X.; Koelsch, G.; Wu, S.; Downs, D.; Dashti, A.; Tang, J., Human aspartic protease memapsin 2 cleaves the beta-secretase site of beta-amyloid precursor protein. *Proc Natl Acad Sci USA***2000**, *97* (4), 1456-60.

[25] Cai, H.; Wang, Y.; McCarthy, D.; Wen, H.; Borchelt, D. R.; Price, D. L.; Wong, P. C., BACE1 is the major beta-secretase for generation of Abeta peptides by neurons. *Nat Neurosci* **2001**, *4* (3), 233-4.

[26] Luo, Y.; Bolon, B.; Kahn, S.; Bennett, B. D.; Babu-Khan, S.; Denis, P.; Fan, W.; Kha, H.; Zhang, J.; Gong, Y.; Martin, L.; Louis, J.-C.; Yan, Q.; Richards, W. G.; Citron, M.; Vassar, R., Mice deficient in BACE1, the Alzheimer's [beta]-secretase, have normal phenotype and abolished [beta]-amyloid generation. *Nat Neurosci* **2001**, *4* (3), 231-232.

[27] Roberds, S. L.; Anderson, J.; Basi, G.; Bienkowski, M. J.; Branstetter, D. G.; Chen, K. S.; Freedman, S.; Frigon, N. L.; Games, D.; Hu, K.; Johnson-Wood, K.; Kappenman, K. E.; Kawabe, T. T.; Kola, I.; Kuehn, R.; Lee, M.; Liu, W.; Motter, R.; Nichols, N. F.; Power, M.; Robertson, D. W.; Schenk, D.; Schoor, M.; Shopp, G. M.; Shuck, M. E.; Sinha, S.; Svensson, K. A.; Tatsuno, G.; Tintrup, H.; Wijsman, J.; Wright, S.; McConlogue, L., BACE knockout mice are healthy despite lacking the primary {beta}-secretase activity in brain: implications for Alzheimer's disease therapeutics. *Hum Mol Gen* **2001**, *10* (12), 1317-1324.

[28] Laird, F. M.; Cai, H.; Savonenko, A. V.; Farah, M. H.; He, K.; Melnikova, T.; Wen, H.; Chiang, H. C.; Xu, G.; Koliatsos, V. E.; Borchelt, D. R.; Price, D. L.; Lee, H.-K.; Wong, P.

C., BACE1, a major determinant of selective vulnerability of the brain to amyloid-beta amyloidogenesis, is essential for cognitive, emotional and synaptic functions. *J Neurosci* **2005**, *25*, 11693-11709.

[29]   Ohno, M.; Chang, L.; Tseng, W.; Oakley, H.; Citron, M.; Klein, W. L.; Vassar, R.; Disterhoft, J. F., Temporal memory deficits in Alzheimer's mouse models: rescue by genetic deletion of BACE1. *Eur J Neurosci* **2006**, *23* (1), 251-60.

[30]   Ohno, M.; Cole, S. L.; Yasvoina, M.; Zhao, J.; Citron, M.; Berry, R.; Disterhoft, J. F.; Vassar, R., BACE1 gene deletion prevents neuron loss and memory deficits in 5XFAD APP/PS1 transgenic mice. *Neurobiol Dis* **2007**, *26* (1), 134-45.

[31]   Ohno, M.; Sametsky, E. A.; Younkin, L. H.; Oakley, H.; Younkin, S. G.; Citron, M.; Vassar, R.; Disterhoft, J. F., BACE1 deficiency rescues memory deficits and cholinergic dysfunction in a mouse model of Alzheimer's disease. *Neuron* **2004**, *41* (1), 27-33.

[32]   Shi, X. P.; Chen, E.; Yin, K. C.; Na, S.; Garsky, V. M.; Lai, M. T.; Li, Y. M.; Platchek, M.; Register, R. B.; Sardana, M. K.; Tang, M. J.; Thiebeau, J.; Wood, T.; Shafer, J. A.; Gardell, S. J., The pro domain of beta-secretase does not confer strict zymogen-like properties but does assist proper folding of the protease domain. *J Biol Chem* **2001**, *276* (13), 10366-73.

[33]   Chou, K. C.; Howe, W. J., Prediction of the tertiary structure of the beta-secretase zymogen. *Biochem Biophys Res Commun* **2002**, *292*, 702-8.

[34]   Klaver, D. W.; Wilce, M. C. J.; Gasperini, R.; Freeman, C.; Juliano, J. P.; Parish, C.; Foa, L.; Aguilar, M.-I.; Small, D. H., Glycosaminoglycan-induced activation of the beta-secretase (BACE1) of Alzheimer's disease. *J Neurochem* **2010**, *112* (6), 1552-61.

[35]   Benjannet, S.; Elagoz, A.; Wickham, L.; Mamarbachi, M.; Munzer, J. S.; Basak, A.; Lazure, C.; Cromlish, J. A.; Sisodia, S.; Checler, F.; Chretien, M.; Seidah, N. G., Post-translational processing of beta-secretase (beta-amyloid-converting enzyme) and its ectodomain shedding. The pro- and transmembrane/cytosolic domains affect its cellular activity and amyloid-beta production. *J Biol Chem***2001**, *276* (14), 10879-87.

[36]   Motoki, K.; Kume, H.; Oda, A.; Tamaoka, A.; Hosaka, A.; Kametani, F.; Araki, W., Neuronal beta-amyloid generation is independent of lipid raft association of beta-secretase BACE1: analysis with a palmitoylation-deficient mutant. *Brain Behav* **2012**, *2* (3), 270-82.

[37]   Vetrivel, K. S.; Meckler, X.; Chen, Y.; Nguyen, P. D.; Seidah, N. G.; Vassar, R.; Wong, P. C.; Fukata, M.; Kounnas, M. Z.; Thinakaran, G., Alzheimer disease Abeta production in the absence of S-palmitoylation-dependent targeting of BACE1 to lipid rafts. *J Biol Chem***2009**, *284* (6), 3793-803.

[38]   Walter, J.; Fluhrer, R.; Hartung, B.; Willem, M.; Kaether, C.; Capell, A.; Lammich, S.; Multhaup, G.; Haass, C., Phosphorylation regulates intracellular trafficking of beta-secretase. *J Biol Chem* **2001**, *276* (18), 14634-41.

[39]   Hu, X.; He, W.; Diaconu, C.; Tang, X.; Kidd, G. J.; Macklin, W. B.; Trapp, B. D.; Yan, R., Genetic deletion of BACE1 in mice affects remyelination of sciatic nerves. *Faseb J* **2008**, *22* (8), 2970-80.

[40]   Hu, X.; Hicks, C. W.; He, W.; Wong, P.; Macklin, W. B.; Trapp, B. D.; Yan, R., Bace1 modulates myelination in the central and peripheral nervous system. *Nat Neurosci* **2006**, *9* (12), 1520-5.

[41]   Willem, M.; Garratt, A. N.; Novak, B.; Citron, M.; Kaufmann, S.; Rittger, A.; DeStrooper, B.; Saftig, P.; Birchmeier, C.; Haass, C., Control of peripheral nerve myelination by the beta-secretase BACE1. *Science* **2006**, *314* (5799), 664-6.

[42]   Brinkmann, B. G.; Agarwal, A.; Sereda, M. W.; Garratt, A. N.; Muller, T.; Wende, H.; Stassart, R. M.; Nawaz, S.; Humml, C.; Velanac, V.; Radyushkin, K.; Goebbels, S.; Fischer, T. M.; Franklin, R. J.; Lai, C.; Ehrenreich, H.; Birchmeier, C.; Schwab, M. H.; Nave, K. A., Neuregulin-1/ErbB signaling serves distinct functions in myelination of the peripheral and central nervous system. *Neuron* **2008**, *59* (4), 581-95.

[43]   Chen, S.; Velardez, M. O.; Warot, X.; Yu, Z.-X.; Miller, S. J.; Cros, D.; Corfas, G., Neuregulin 1-erbB signaling is necessary for normal myelination and sensory function. *J Neurosci* **2006**, *26* (12), 3079-86.

[44]   Michailov, G. V.; Sereda, M. W.; Brinkmann, B. G.; Fischer, T. M.; Haug, B.; Birchmeier, C.; Role, L.; Lai, C.; Schwab, M. H.; Nave, K.-A., Axonal neuregulin-1 regulates myelin sheath thickness. *Science* **2004**, *304* (5671), 700-3.

[45]   Treiber, H.; Hagemeyer, N.; Ehrenreich, H.; Simons, M., BACE1 in central nervous system myelination revisited. *Mol Psychiatry* **2012**, *17* (3), 237-9.

[46]   Kim, D. Y.; Carey, B. W.; Wang, H.; Ingano, L. A. M.; Binshtok, A. M.; Wertz, M. H.; Pettingell, W. H.; He, P.; Lee, V. M. Y.; Woolf, C. J.; Kovacs, D. M., BACE1 regulates voltage-gated sodium channels and neuronal activity. *Nat Cell Biol* **2007**, *9* (7), 755-64.

[47]   Wong, H.-K.; Sakurai, T.; Oyama, F.; Kaneko, K.; Wada, K.; Miyazaki, H.; Kurosawa, M.; De Strooper, B.; Saftig, P.; Nukina, N., beta Subunits of voltage-gated sodium channels are novel substrates of beta-site amyloid precursor protein-cleaving enzyme (BACE1) and gamma-secretase. *J Biol Chem* **2005**, *280* (24), 23009-17.

[48]   Mulley, J. C.; Scheffer, I. E.; Petrou, S.; Dibbens, L. M.; Berkovic, S. F.; Harkin, L. A., SCN1A mutations and epilepsy. *Hum Mutat* **2005**, *25* (6), 535-42.

[49]   Hu, X.; Zhou, X.; He, W.; Yang, J.; Xiong, W.; Wong, P.; Wilson, C. G.; Yan, R., BACE1 deficiency causes altered neuronal activity and neurodegeneration. *J Neurosci* **2010**, *30* (26), 8819-29.

[50]   Mendez, M. F.; Catanzaro, P.; Doss, R. C.; Arguello, R.; Frey, W. H., 2nd, Seizures in Alzheimer's disease: clinicopathologic study. *J Geriatr Psychiatry Neurol* **1994**, *7* (4), 230-3.

[51]   Caroni, P.; Donato, F.; Muller, D., Structural plasticity upon learning: regulation and functions. *Nat Rev Neurosci* **2012**, *13* (7), 478-90.

[52]   Harrison, S. M.; Harafter, A. J.; Hawkins, J.; Duddy, G.; Grau, E.; Pugh, P. L.; Winter, P. H.; Shilliam, C. S.; Hughes, Z. A.; Dawson, L. A.; Gonzalez, M. I.; Upton, N.; Pangalos, M. N.; Dingwall, C., BACE1 (beta-secretase) transgenic and knock-out mice: identification of neurochemical deficits and behavioral changes. *Mol Cell Neurosci* **2003**, *24*, 646-655.

[53]   Rajapaksha, T. W.; Eimer, W. A.; Bozza, T. C.; Vassar, R., The Alzheimer's beta-secretase enzyme BACE1 is required for accurate axon guidance of olfactory sensory neurons and normal glomerulus formation in the olfactory bulb. *Mol Neurodegener* **2011**, *6*, 88.

[54]   Kuhn, P. H.; Koroniak, K.; Hogl, S.; Colombo, A.; Zeitschel, U.; Willem, M.; Volbracht, C.; Schepers, U.; Imhof, A.; Hoffmeister, A.; Haass, C.; Rossner, S.; Brase, S.; Lichtenthaler, S. F., Secretome protein enrichment identifies physiological BACE1 protease substrates in neurons. *EMBO J* **2012**, *31* (14), 3157-68.

[55]   Zhou, L.; Barao, S.; Laga, M.; Bockstael, K.; Borgers, M.; Gijsen, H.; Annaert, W.; Moechars, D.; Mercken, M.; Gevaer, K.; De Strooper, B., The Neural Cell Adhesion Molecules L1 and CHL1 Are Cleaved by BACE1 Protease *in vivo. J Biol Chem* **2012**, *287* (31), 25927-40.

[56]   Dityatev, A.; Bukalo, O.; Schachner, M., Modulation of synaptic transmission and plasticity by cell adhesion and repulsion molecules. *Neuron glia biol* **2008**, *4* (3), 197-209.

[57]   Kimura, R.; Devi, L.; Ohno, M., Partial reduction of BACE1 improves synaptic plasticity, recent and remote memories in Alzheimer's disease transgenic mice. *J Neurochem* **2010**, *113* (1), 248-61.

[58]   McConlogue, L.; Buttini, M.; Anderson, J. P.; Brigham, E. F.; Chen, K. S.; Freedman, S. B.; Games, D.; Johnson-Wood, K.; Lee, M.; Zeller, M.; Liu, W.; Motter, R.; Sinha, S., Partial reduction of BACE1 has dramatic effects on Alzheimer plaque and synaptic pathology in APP Transgenic Mice. *J Biol Chem* **2007**, *282* (36), 26326-34.

[59]   Hong, L.; Koelsch, G.; Lin, X.; Wu, S.; Terzyan, S.; Ghosh, A. K.; Zhang, X. C.; Tang, J., Structure of the protease domain of memapsin 2 (beta-secretase) complexed with inhibitor. *Science* **2000**, *290* (5489), 150-3.

[60]   Silvestri, R., Boom in the development of non-peptidic beta-secretase (BACE1) inhibitors for the treatment of Alzheimer's disease. *Med Res Rev* **2009**, *29* (2), 295-338.

[61]   Kornacker, M. G.; Lai, Z.; Witmer, M.; Ma, J.; Hendrick, J.; Lee, V. G.; Riexinger, D. J.; Mapelli, C.; Metzler, W.; Copeland, R. A., An inhibitor binding pocket distinct from the catalytic active site on human beta-APP cleaving enzyme. *Biochemistry* **2005**, *44* (34), 11567-73.

[62]   Gutierrez, L. J.; Enriz, R. D.; Baldoni, H. A., Structural and thermodynamic characteristics of the exosite binding pocket on the human BACE1: a molecular modeling approach. *J Phys Chem A Mol Spectrosc Kinet Environ Gen Theory* **2010**, *114* (37), 10261-9.

[63]   Rossner, S.; Sastre, M.; Bourne, K.; Lichtenthaler, S. F., Transcriptional and translational regulation of BACE1 expression--implications for Alzheimer's disease. *Prog Neurobiol* **2006**, *79* (2), 95-111.

[64]   Yang, L.-B.; Lindholm, K.; Yan, R.; Citron, M.; Xia, W.; Yang, X.-L.; Beach, T.; Sue, L.; Wong, P.; Price, D.; Li, R.; Shen, Y., Elevated beta-secretase expression and enzymatic activity detected in sporadic Alzheimer disease. *Nat Med* **2003**, *9* (1), 3-4.

[65]   Zhao, J.; Fu, Y.; Yasvoina, M.; Shao, P.; Hitt, B.; O'Connor, T.; Logan, S.; Maus, E.; Citron, M.; Berry, R.; Binder, L.; Vassar, R., Beta-site amyloid precursor protein cleaving enzyme 1 levels become elevated in neurons around amyloid plaques: implications for Alzheimer's disease pathogenesis. *J Neurosci* **2007**, *27* (14), 3639-49.

[66]   Faghihi, M. A.; Modarresi, F.; Khalil, A. M.; Wood, D. E.; Sahagan, B. G.; Morgan, T. E.; Finch, C. E.; St Laurent, G., 3rd; Kenny, P. J.; Wahlestedt, C., Expression of a noncoding RNA is elevated in Alzheimer's disease and drives rapid feed-forward regulation of beta-secretase. *Nat Med* **2008**, *14* (7), 723-30.

[67]   Cote, S.; Carmichael, P.-H.; Verreault, R.; Lindsay, J.; Lefebvre, J.; Laurin, D., Nonsteroidal anti-inflammatory drug use and the risk of cognitive impairment and Alzheimer's disease. *Alzheimer's dement* **2012**, *8* (3), 219-26.

[68]   Trepanier, C. H.; Milgram, N. W., Neuroinflammation in Alzheimer's disease: are NSAIDs and selective COX-2 inhibitors the next line of therapy? *J Alzheimers Dis* **2010**, *21* (4), 1089-99.

[69]   Sastre, M.; Dewachter, I.; Landreth, G. E.; Willson, T. M.; Klockgether, T.; van Leuven, F.; Heneka, M. T., Nonsteroidal anti-inflammatory drugs and peroxisome proliferator-activated receptor-gamma agonists modulate immunostimulated processing of amyloid precursor protein through regulation of beta-secretase. *J Neurosci* **2003**, *23* (30), 9796-804.

[70]   Sastre, M.; Dewachter, I.; Rossner, S.; Bogdanovic, N.; Rosen, E.; Borghgraef, P.; Evert, B. O.; Dumitrescu-Ozimek, L.; Thal, D. R.; Landreth, G.; Walter, J.; Klockgether, T.; van Leuven, F.; Heneka, M. T., Nonsteroidal anti-inflammatory drugs repress beta-secretase gene promoter activity by the activation of PPARgamma. *Proc Natl Acad Sci USA* **2006**, *103* (2), 443-8.

[71]   Katsouri, L.; Parr, C.; Bogdanovic, N.; Willem, M.; Sastre, M., PPARgamma co-activator-1alpha (PGC-1alpha) reduces amyloid-beta generation through a PPARgamma-dependent mechanism. *J Alzheimers Dis* **2011**, *25* (1), 151-62.

[72]   Risner, M. E.; Saunders, A. M.; Altman, J. F. B.; Ormandy, G. C.; Craft, S.; Foley, I. M.; Zvartau-Hind, M. E.; Hosford, D. A.; Roses, A. D.; Rosiglitazone in Alzheimer's Disease Study, G., Efficacy of rosiglitazone in a genetically defined population with mild-to-moderate Alzheimer's disease. *Pharmacogenomics J* **2006**, *6* (4), 246-54.

[73]   Sato, T.; Hanyu, H.; Hirao, K.; Kanetaka, H.; Sakurai, H.; Iwamoto, T., Efficacy of PPAR-gamma agonist pioglitazone in mild Alzheimer disease. *Neurobiol Aging* **2011**, *32* (9), 1626-33.

[74]   Hanyu, H.; Sato, T.; Kiuchi, A.; Sakurai, H.; Iwamoto, T., Pioglitazone improved cognition in a pilot study on patients with Alzheimer's disease and mild cognitive impairment with diabetes mellitus. *J Am Geriatr Soc* **2009**, *57* (1), 177-9.

[75]   Miller, B. W.; Willett, K. C.; Desilets, A. R., Rosiglitazone and pioglitazone for the treatment of Alzheimer's disease. *Ann Pharmacother* **2011**, *45* (11), 1416-24.

[76]   Harrington, C.; Sawchak, S.; Chiang, C.; Davies, J.; Donovan, C.; Saunders, A. M.; Irizarry, M.; Jeter, B.; Zvartau-Hind, M.; van Dyck, C. H.; Gold, M., Rosiglitazone does not improve cognition or global function when used as adjunctive therapy to AChE inhibitors in mild-to-moderate Alzheimer's disease: two phase 3 studies. *Curr Alzheimer Res* **2011**, *8* (5), 592-606.

[77]   Gold, M.; Alderton, C.; Zvartau-Hind, M.; Egginton, S.; Saunders, A. M.; Irizarry, M.; Craft, S.; Landreth, G.; Linnamagi, U.; Sawchak, S., Rosiglitazone monotherapy in mild-to-moderate Alzheimer's disease: results from a randomized, double-blind, placebo-controlled phase III study. *Dement Geriatr Cogn Disord* **2010**, *30* (2), 131-46.

[78]   Lange-Dohna, C.; Zeitschel, U.; Gaunitz, F.; Perez-Polo, J. R.; Bigl, V.; Rossner, S., Cloning and expression of the rat BACE1 promoter. *J Neurosci Res* **2003**, *73* (1), 73-80.

[79]   Bourne, K. Z.; Ferrari, D. C.; Lange-Dohna, C.; Rossner, S.; Wood, T. G.; Perez-Polo, J. R., Differential regulation of BACE1 promoter activity by nuclear factor-kappaB in neurons and glia upon exposure to beta-amyloid peptides. *J Neurosci Res* **2007**, *85* (6), 1194-204.

[80]   Kaltschmidt, B.; Uherek, M.; Wellmann, H.; Volk, B.; Kaltschmidt, C., Inhibition of NF-kappaB potentiates amyloid beta-mediated neuronal apoptosis. *Proc Natl Acad Sci USA* **1999**, *96* (16), 9409-14.

[81]   Hartlage-Rubsamen, M.; Zeitschel, U.; Apelt, J.; Gartner, U.; Franke, H.; Stahl, T.; Gunther, A.; Schliebs, R.; Penkowa, M.; Bigl, V.; Rossner, S., Astrocytic expression of the Alzheimer's disease beta-secretase (BACE1) is stimulus-dependent. *Glia* **2003**, *41* (2), 169-79.

[82]   Guglielmotto, M.; Aragno, M.; Tamagno, E.; Vercellinatto, I.; Visentin, S.; Medana, C.; Catalano, M. G.; Smith, M. A.; Perry, G.; Danni, O.; Boccuzzi, G.; Tabaton, M., AGEs/RAGE complex upregulates BACE1 *via* NF-kappaB pathway activation. *Neurobiol Aging* **2012**, *33* (1), 196.e13-27.

[83]   Yan, S. D.; Chen, X.; Fu, J.; Chen, M.; Zhu, H.; Roher, A.; Slattery, T.; Zhao, L.; Nagashima, M.; Morser, J.; Migheli, A.; Nawroth, P.; Stern, D.; Schmidt, A. M., RAGE and amyloid-beta peptide neurotoxicity in Alzheimer's disease. *Nature* **1996**, *382* (6593), 685-91.

[84]   Arancio, O.; Zhang, H. P.; Chen, X.; Lin, C.; Trinchese, F.; Puzzo, D.; Liu, S.; Hegde, A.; Yan, S. F.; Stern, A.; Luddy, J. S.; Lue, L.-F.; Walker, D. G.; Roher, A.; Buttini, M.; Mucke, L.; Li, W.; Schmidt, A. M.; Kindy, M.; Hyslop, P. A.; Stern, D. M.; Du Yan, S. S., RAGE potentiates Abeta-induced perturbation of neuronal function in transgenic mice. *EMBO J* **2004**, *23* (20), 4096-105.

[85]   Takuma, K.; Fang, F.; Zhang, W.; Yan, S.; Fukuzaki, E.; Du, H.; Sosunov, A.; McKhann, G.; Funatsu, Y.; Nakamichi, N.; Nagai, T.; Mizoguchi, H.; Ibi, D.; Hori, O.; Ogawa, S.; Stern, D. M.; Yamada, K.; Yan, S. S., RAGE-mediated signaling contributes to intraneuronal transport of amyloid-beta and neuronal dysfunction. *Proc Natl Acad Sci USA* **2009**, *106* (47), 20021-6.

[86]   Cho, H. J.; Jin, S. M.; Youn, H. D.; Huh, K.; Mook-Jung, I., Disrupted intracellular calcium regulates BACE1 gene expression *via* nuclear factor of activated T cells 1 (NFAT 1) signaling. *Aging Cell* **2008**, *7* (2), 137-47.

[87]   Hogan, P. G.; Chen, L.; Nardone, J.; Rao, A., Transcriptional regulation by calcium, calcineurin, and NFAT. *Genes & Development* **2003**, *17* (18), 2205-32.

[88]   Cho, H. J.; Son, S. M.; Jin, S. M.; Hong, H. S.; Shin, D. H.; Kim, S. J.; Huh, K.; Mook-Jung, I., RAGE regulates BACE1 and Abeta generation *via* NFAT1 activation in Alzheimer's disease animal model. *Faseb J* **2009**, *23* (8), 2639-49.

[89]   Fisette, J.-F.; Montagna, D. R.; Mihailescu, M.-R.; Wolfe, M. S., A G-rich element forms a G-quadruplex and regulates BACE1 mRNA alternative splicing. *J Neurochem* **2012**, *121* (5), 763-73.

[90]   Mowrer, K. R.; Wolfe, M. S., Promotion of BACE1 mRNA alternative splicing reduces amyloid beta-peptide production. *J Biol Chem* **2008**, *283* (27), 18694-701.

[91]   Gatta, L. B.; Albertini, A.; Ravid, R.; Finazzi, D., Levels of beta-secretase BACE and alpha-secretase ADAM10 mRNAs in Alzheimer hippocampus. *Neuroreport* **2002**, *13* (16), 2031-3.

[92]   Holsinger, R. M. D.; McLean, C. A.; Beyreuther, K.; Masters, C. L.; Evin, G., Increased expression of the amyloid precursor beta-secretase in Alzheimer's disease. *Ann Neurol* **2002**, *51* (6), 783-6.

[93]   Yasojima, K.; McGeer, E. G.; McGeer, P. L., Relationship between beta amyloid peptide generating molecules and neprilysin in Alzheimer disease and normal brain. *Brain Res* **2001**, *919* (1), 115-21.

[94]   Fukumoto, H.; Cheung, B. S.; Hyman, B. T.; Irizarry, M. C., Beta-secretase protein and activity are increased in the neocortex in Alzheimer disease. *Arch Neurol* **2002**, *59* (9), 1381-9.

[95]   Hébert, S.; Horré, K.; Nicolaï, L.; Papadopoulou, A. S.; Mandemakers, W.; Silahtaroglu, A. N.; Kauppinen, S.; Delacourte, A.; De Strooper, B., Loss of microRNA cluster miR-29a/b-1 in sporadic Alzheimer's disease correlates with increased BACE1/beta-secretase expression. *Proc Natl Acad Sci USA* **2008**, *105* (17), 6415-20.

[96]   Holsinger, R. M. D.; Lee, J. S.; Boyd, A.; Masters, C. L.; Collins, S. J., CSF BACE1 activity is increased in CJD and Alzheimer disease *vs.* other dementias. *Neurology* **2006**, *67* (4), 710-2.

[97]   Holsinger, R. M. D.; McLean, C. A.; Collins, S. J.; Masters, C. L.; Evin, G., Increased beta-Secretase activity in cerebrospinal fluid of Alzheimer's disease subjects. *Ann Neurol* **2004**, *55* (6), 898-9.

[98]   Li, R.; Lindholm, K.; Yang, L.-B.; Yue, X.; Citron, M.; Yan, R.; Beach, T.; Sue, L.; Sabbagh, M.; Cai, H.; Wong, P.; Price, D.; Shen, Y., Amyloid beta peptide load is correlated with increased beta-secretase activity in sporadic Alzheimer's disease patients. *Proc Natl Acad Sci USA* **2004**, *101* (10), 3632-7.

[99]   Tesco, G.; Koh, Y. H.; Kang, E. L.; Cameron, A. N.; Das, S.; Sena-Esteves, M.; Hiltunen, M.; Yang, S.-H.; Zhong, Z.; Shen, Y.; Simpkins, J. W.; Tanzi, R. E., Depletion of GGA3 stabilizes BACE and enhances beta-secretase activity. *Neuron* **2007**, *54* (5), 721-37.

[100] Bian, S.; Sun, T., Functions of noncoding RNAs in neural development and neurological diseases. *Mol Neurobiol* **2011**, *44* (3), 359-73.

[101] Enciu, A.-M.; Popescu, B. O.; Gheorghisan-Galateanu, A., MicroRNAs in brain development and degeneration. *Mol Biol Rep* **2012**, *39* (3), 2243-52.

[102] Junn, E.; Mouradian, M. M., MicroRNAs in neurodegenerative diseases and their therapeutic potential. *Pharmacol Ther* **2012**, *133* (2), 142-50.

[103] Nelson, P. T.; Wang, W.-X., MiR-107 is reduced in Alzheimer's disease brain neocortex: validation study. *J Alzheimers Dis* **2010**, *21* (1), 75-9.

[104] Wang, W.-X.; Rajeev, B. W.; Stromberg, A. J.; Ren, N.; Tang, G.; Huang, Q.; Rigoutsos, I.; Nelson, P. T., The expression of microRNA miR-107 decreases early in Alzheimer's disease and may accelerate disease progression through regulation of beta-site amyloid precursor protein-cleaving enzyme 1. *J Neurosci* **2008**, *28* (5), 1213-23.

[105] Zong, Y.; Wang, H.; Dong, W.; Quan, X.; Zhu, H.; Xu, Y.; Huang, L.; Ma, C.; Qin, C., miR-29c regulates BACE1 protein expression. *Brain Res* **2011**, *1395*, 108-15.

[106] Boissonneault, V.; Plante, I.; Rivest, S.; Provost, P., MicroRNA-298 and microRNA-328 regulate expression of mouse beta-amyloid precursor protein-converting enzyme 1. *J Biol Chem* **2009**, *284* (4), 1971-81.

[107] Fang, M.; Wang, J.; Zhang, X.; Geng, Y.; Hu, Z.; Rudd, J. A.; Ling, S.; Chen, W.; Han, S., The miR-124 regulates the expression of BACE1/beta-secretase correlated with cell death in Alzheimer's disease. *Toxicol Lett* **2012**, *209* (1), 94-105.

[108] O'Connor, T.; Sadleir, K. R.; Maus, E.; Velliquette, R. A.; Zhao, J.; Cole, S. L.; Eimer, W. A.; Hitt, B.; Bembinster, L. A.; Lammich, S.; Lichtenthaler, S. F.; Hebert, S. S.; De Strooper, B.; Haass, C.; Bennett, D. A.; Vassar, R., Phosphorylation of the translation initiation factor eIF2alpha increases BACE1 levels and promotes amyloidogenesis. *Neuron* **2008**, *60* (6), 988-1009.

[109] Devi, L.; Alldred, M. J.; Ginsberg, S. D.; Ohno, M., Mechanisms underlying insulin deficiency-induced acceleration of beta-amyloidosis in a mouse model of Alzheimer's disease. *PLoS ONE* **2012**, *7* (3), e32792.

[110] Mouton-Liger, F.; Paquet, C.; Dumurgier, J.; Bouras, C.; Pradier, L.; Gray, F.; Hugon, J., Oxidative stress increases BACE1 protein levels through activation of the PKR-eIF2alpha pathway. *Biochim Biophys Acta* **2012**, *1822* (6), 885-96.

[111] Rajendran, L.; Schneider, A.; Schlechtingen, G.; Weidlich, S.; Ries, J.; Braxmeier, T.; Schwille, P.; Schulz, J. B.; Schroeder, C.; Simons, M.; Jennings, G.; Knolker, H.-J.; Simons, K., Efficient inhibition of the Alzheimer's disease beta-secretase by membrane targeting. *Science* **2008**, *320* (5875), 520-3.

[112] Schieb, H.; Weidlich, S.; Schlechtingen, G.; Linning, P.; Jennings, G.; Gruner, M.; Wiltfang, J.; Klafki, H.-W.; Knolker, H.-J., Structural design, solid-phase synthesis and activity of membrane-anchored beta-secretase inhibitors on Abeta generation from wild-type and Swedish-mutant APP. *Chemistry* **2010**, *16* (48), 14412-23.

[113] Takasugi, N.; Sasaki, T.; Suzuki, K.; Osawa, S.; Isshiki, H.; Hori, Y.; Shimada, N.; Higo, T.; Yokoshima, S.; Fukuyama, T.; Lee, V. M. Y.; Trojanowski, J. Q.; Tomita, T.; Iwatsubo, T., BACE1 activity is modulated by cell-associated sphingosine-1-phosphate. *J Neurosci* **2011**, *31* (18), 6850-7.

[114] Bode, C.; Graler, M. H., Immune regulation by sphingosine 1-phosphate and its receptors. *Arch Immunol Ther Exp (Warsz)* **2012**, *60* (1), 3-12.

[115] Spiegel, S.; Milstien, S., The outs and the ins of sphingosine-1-phosphate in immunity. *Nature Rev Immunol* **2011**, *11* (6), 403-15.

[116] Alvarez, S. E.; Harikumar, K. B.; Hait, N. C.; Allegood, J.; Strub, G. M.; Kim, E. Y.; Maceyka, M.; Jiang, H.; Luo, C.; Kordula, T.; Milstien, S.; Spiegel, S., Sphingosine-1-phosphate is a missing cofactor for the E3 ubiquitin ligase TRAF2. *Nature* **2010**, *465* (7301), 1084-8.

[117] Maeurer, C.; Holland, S.; Pierre, S.; Potstada, W.; Scholich, K., Sphingosine-1-phosphate induced mTOR-activation is mediated by the E3-ubiquitin ligase PAM. *Cell Signal* **2009**, *21* (2), 293-300.

[118] Oo, M. L.; Thangada, S.; Wu, M.-T.; Liu, C. H.; Macdonald, T. L.; Lynch, K. R.; Lin, C.-Y.; Hla, T., Immunosuppressive and anti-angiogenic sphingosine 1-phosphate receptor-1 agonists induce ubiquitinylation and proteasomal degradation of the receptor. *J Biol Chem* **2007**, *282* (12), 9082-9.

[119] Hait, N. C.; Allegood, J.; Maceyka, M.; Strub, G. M.; Harikumar, K. B.; Singh, S. K.; Luo, C.; Marmorstein, R.; Kordula, T.; Milstien, S.; Spiegel, S., Regulation of histone acetylation in the nucleus by sphingosine-1-phosphate. *Science* **2009**, *325* (5945), 1254-7.

[120] Small, D. H.; Nurcombe, V.; Reed, G.; Clarris, H.; Moir, R.; Beyreuther, K.; Masters, C. L., A heparin-binding domain in the amyloid protein precursor of Alzheimer's disease is involved in the regulation of neurite outgrowth. *J Neurosci* **1994**, *14* (4), 2117-27.

[121] Snow, A. D.; Mar, H.; Nochlin, D.; Kimata, K.; Kato, M.; Suzuki, S.; Hassell, J.; Wight, T. N., The presence of heparan sulfate proteoglycans in the neuritic plaques and congophilic angiopathy in Alzheimer's disease. *Am J Pathol* **1988**, *133* (3), 456-63.

[122] Snow, A. D.; Mar, H.; Nochlin, D.; Sekiguchi, R. T.; Kimata, K.; Koike, Y.; Wight, T. N., Early accumulation of heparan sulfate in neurons and in the beta-amyloid protein-containing lesions of Alzheimer's disease and Down's syndrome. *Am J Pathol* **1990**, *137* (5), 1253-70.

[123] Snow, A. D.; Mar, H.; Nochlin, D.; Kresse, H.; Wight, T. N., Peripheral distribution of dermatan sulfate proteoglycans (decorin) in amyloid-containing plaques and their presence in neurofibrillary tangles of Alzheimer's disease. *J Histochem Cytochem* **1992**, *40* (1), 105-13.

[124] Snow, A. D.; Sekiguchi, R.; Nochlin, D.; Kalaria, R.; Kimata, K., HSPG in diffuse plaques of hippocampus but not of cerebellum in Alzheimer's disease brain. *Am J Pathol* **1994**, *144*, 337-347.

[125] DeWitt, D. A.; Silver, J.; Canning, D. R.; Perry, G., Chondroitin sulfate proteoglycans are associated with the lesions of Alzheimer's disease. *Exp Neurol* **1993**, *121* (2), 149-52.

[126] DeWitt, D. A.; Richey, P. L.; Praprotnik, D.; Silver, J.; Perry, G., Chondroitin sulfate proteoglycans are a common component of neuronal inclusions and astrocytic reaction in neurodegenerative diseases. *Brain Res* **1994**, *656* (1), 205-9.

[127] Magnus, J. H.; Stenstad, T.; Kolset, S. O.; Husby, G., Glycosaminoglycans in extracts of cardiac amyloid fibrils from familial amyloid cardiomyopathy of Danish origin related to variant transthyretin Met 111. *Scand J Immunol* **1991**, *34* (1), 63-9.

[128] Inoue, S.; Kuroiwa, M.; Saraiva, M. J.; Guimaraes, A.; Kisilevsky, R., Ultrastructure of familial amyloid polyneuropathy amyloid fibrils: examination with high-resolution electron microscopy. *J Struct Biol* **1998**, *124* (1), 1-12.

[129] Lashley, T.; Holton, J. L.; Verbeek, M. M.; Rostagno, A.; Bojsen-Moller, M.; David, G.; van Horssen, J.; Braendgaard, H.; Plant, G.; Frangione, B.; Ghiso, J.; Revesz, T., Molecular chaperons, amyloid and preamyloid lesions in the BRI2 gene-related dementias: a morphological study. *Neuropathol Appl Neurobiol* **2006**, *32* (5), 492-504.

[130] Snow, A. D.; Wight, T. N.; Nochlin, D.; Koike, Y.; Kimata, K.; DeArmond, S. J.; Prusiner, S. B., Immunolocalization of heparan sulfate proteoglycans to the prion protein amyloid plaques of Gerstmann-Straussler syndrome, Creutzfeldt-Jakob disease and scrapie. *Lab Invest* **1990**, *63* (5), 601-11.

[131] Scholefield, Z.; Yates, E. A.; Wayne, G.; Amour, A.; McDowell, W.; Turnbull, J. E., Heparan sulfate regulates amyloid precursor protein processing by BACE1, the Alzheimer's beta-secretase. *J Cell Biol* **2003**, *163* (1), 97-107.

[132] Beckman, M.; Holsinger, R. M. D.; Small, D. H., Heparin activates beta-secretase (BACE1) of Alzheimer's disease and increases autocatalysis of the enzyme. *Biochemistry* **2006**, *45* (21), 6703-6714.

[133] Haass, C.; Lemere, C. A.; Capell, A.; Citron, M.; Seubert, P.; Schenk, D.; Lannfelt, L.; Selkoe, D. J., The Swedish mutation causes early-onset Alzheimer's disease by beta-secretase cleavage within the secretory pathway. *Nat Med* **1995**, *1* (12), 1291-6.

[134] Huse, J. T.; Liu, K.; Pijak, D. S.; Carlin, D.; Lee, V. M. Y.; Doms, R. W., Beta-secretase processing in the trans-Golgi network preferentially generates truncated amyloid species that accumulate in Alzheimer's disease brain. *J Biol Chem* **2002**, *277* (18), 16278-84.

[135] Koo, E. H.; Squazzo, S. L., Evidence that production and release of amyloid beta-protein involves the endocytic pathway. *J Biol Chem* **1994**, *269* (26), 17386-9.

[136] Perez, R. G.; Squazzo, S. L.; Koo, E. H., Enhanced release of amyloid beta-protein from codon 670/671 "Swedish" mutant beta-amyloid precursor protein occurs in both secretory and endocytic pathways. *J Biol Chem* **1996**, *271* (15), 9100-7.

[137] Thinakaran, G.; Teplow, D. B.; Siman, R.; Greenberg, B.; Sisodia, S. S., Metabolism of the "Swedish" amyloid precursor protein variant in neuro2a (N2a) cells. Evidence that cleavage at the "beta-secretase" site occurs in the golgi apparatus. *J Biol Chem* **1996**, *271* (16), 9390-7.

[138] Leveugle, B.; Ding, W.; Durkin, J. T.; Mistretta, S.; Eisle, J.; Matic, M.; Siman, R.; Greenberg, B. D.; Fillit, H. M., Heparin promotes beta-secretase cleavage of the Alzheimer's amyloid precursor protein. *Neurochem Int* **1997**, *30* (6), 543-8.

[139] Leveugle, B.; Ding, W.; Laurence, F.; Dehouck, M. P.; Scanameo, A.; Cecchelli, R.; Fillit, H., Heparin Oligosaccharides that Pass the Blood-Brain Barrier Inhibit beta-Amyloid Precursor Protein Secretion and Heparin Binding to beta-Amyloid Peptide. *J Neurochem* **1998**, *70* (2), 736-44.

[140] Cui, H.; Hung, A. C.; Klaver, D. W.; Suzuki, T.; Freeman, C.; Narkowicz, C.; Jacobson, G. A.; Small, D. H., Effects of heparin and enoxaparin on APP processing and Abeta production in primary cortical neurons from Tg2576 mice. *PLoS ONE* **2011**, *6* (7), e23007.

[141] He, W.; Lu, Y.; Qahwash, I.; Hu, X.-Y.; Chang, A.; Yan, R., Reticulon family members modulate BACE1 activity and amyloid-beta peptide generation. *Nat Med* **2004**, *10* (9), 959-65.

[142] Murayama, K. S.; Kametani, F.; Saito, S.; Kume, H.; Akiyama, H.; Araki, W., Reticulons RTN3 and RTN4-B/C interact with BACE1 and inhibit its ability to produce amyloid beta-protein. *Eur J Neurosci* **2006**, *24* (5), 1237-44.

[143] He, W.; Hu, X.; Shi, Q.; Zhou, X.; Lu, Y.; Fisher, C.; Yan, R., Mapping of interaction domains mediating binding between BACE1 and RTN/Nogo proteins. *J Mol Biol* **2006**, *363* (3), 625-34.

[144] Heath, J. E.; Siedlak, S. L.; Zhu, X.; Lee, H.-G.; Thakur, A.; Yan, R.; Perry, G.; Smith, M. A.; Castellani, R. J., Widespread distribution of reticulon-3 in various neurodegenerative diseases. *Neuropathology* **2010**, *30* (6), 574-9.

[145] Kume, H.; Konishi, Y.; Murayama, K. S.; Kametani, F.; Araki, W., Expression of reticulon 3 in Alzheimer's disease brain. *Neuropathol Appl Neurobiol* **2009**, *35* (2), 178-88.

[146] Shi, Q.; Prior, M.; He, W.; Tang, X.; Hu, X.; Yan, R., Reduced amyloid deposition in mice overexpressing RTN3 is adversely affected by preformed dystrophic neurites. *J Neurosci* **2009**, *29* (29), 9163-73.

[147] Hu, X.; Shi, Q.; Zhou, X.; He, W.; Yi, H.; Yin, X.; Gearing, M.; Levey, A.; Yan, R., Transgenic mice overexpressing reticulon 3 develop neuritic abnormalities. *EMBO J* **2007**, *26* (11), 2755-67.

[148] He, X.; Chang, W.-P.; Koelsch, G.; Tang, J., Memapsin 2 (beta-secretase) cytosolic domain binds to the VHS domains of GGA1 and GGA2: implications on the endocytosis mechanism of memapsin 2. *FEBS Lett* **2002**, *524* (1-3), 183-7.

[149] He, X.; Zhu, G.; Koelsch, G.; Rodgers, K. K.; Zhang, X. C.; Tang, J., Biochemical and structural characterization of the interaction of memapsin 2 (beta-secretase) cytosolic domain with the VHS domain of GGA proteins. *Biochemistry* **2003**, *42* (42), 12174-80.

[150] Huse, J. T.; Pijak, D. S.; Leslie, G. J.; Lee, V. M.; Doms, R. W., Maturation and endosomal targeting of beta-site amyloid precursor protein-cleaving enzyme. The Alzheimer's disease beta-secretase. *J Biol Chem* **2000**, *275* (43), 33729-37.

[151] Pastorino, L.; Ikin, A. F.; Nairn, A. C.; Pursnani, A.; Buxbaum, J. D., The carboxyl-terminus of BACE contains a sorting signal that regulates BACE trafficking but not the formation of total A(beta). *Mol Cell Neurosci* **2002**, *19* (2), 175-85.

[152] Koh, Y. H.; von Arnim, C. A. F.; Hyman, B. T.; Tanzi, R. E.; Tesco, G., BACE is degraded *via* the lysosomal pathway. *J Biol Chem* **2005**, *280* (37), 32499-504.

[153] Kang, E. L.; Cameron, A. N.; Piazza, F.; Walker, K. R.; Tesco, G., Ubiquitin regulates GGA3-mediated degradation of BACE1. *J Biol Chem* **2010**, *285* (31), 24108-19.

[154] Wahle, T.; Prager, K.; Raffler, N.; Haass, C.; Famulok, M.; Walter, J., GGA proteins regulate retrograde transport of BACE1 from endosomes to the trans-Golgi network. *Mol Cell Neurosci* **2005**, *29* (3), 453-61.

[155] Wahle, T.; Thal, D. R.; Sastre, M.; Rentmeister, A.; Bogdanovic, N.; Famulok, M.; Heneka, M. T.; Walter, J., GGA1 is expressed in the human brain and affects the generation of amyloid beta-peptide. *J Neurosci* **2006**, *26* (49), 12838-46.

[156] von Arnim, C. A. F.; Spoelgen, R.; Peltan, I. D.; Deng, M.; Courchesne, S.; Koker, M.; Matsui, T.; Kowa, H.; Lichtenthaler, S. F.; Irizarry, M. C.; Hyman, B. T., GGA1 acts as a spatial switch altering amyloid precursor protein trafficking and processing. *J Neurosci* **2006**, *26* (39), 9913-22.

[157] He, X.; Li, F.; Chang, W.-P.; Tang, J., GGA proteins mediate the recycling pathway of memapsin 2 (BACE). *J Biol Chem* **2005**, *280* (12), 11696-703.

[158] Walker, K. R.; Kang, E. L.; Whalen, M. J.; Shen, Y.; Tesco, G., Depletion of GGA1 and GGA3 Mediates Postinjury Elevation of BACE1. *J Neurosci* **2012**, *32* (30), 10423-37.

[159] Jellinger, K. A., Head injury and dementia. *Curr Opin Neurol* **2004**, *17* (6), 719-23.

[160] Lim, K.-L.; Lim, G. G. Y., K63-linked ubiquitination and neurodegeneration. *Neurobiol Dis* **2011**, *43* (1), 9-16.

[161] Qing, H.; Zhou, W.; Christensen, M. A.; Sun, X.; Tong, Y.; Song, W., Degradation of BACE by the ubiquitin-proteasome pathway. *Faseb J* **2004**, *18* (13), 1571-3.

[162] Wilkinson, K. D.; Lee, K. M.; Deshpande, S.; Duerksen-Hughes, P.; Boss, J. M.; Pohl, J., The neuron-specific protein PGP 9.5 is a ubiquitin carboxyl-terminal hydrolase. *Science* **1989**, *246* (4930), 670-3.

[163] Liu, Y.; Fallon, L.; Lashuel, H. A.; Liu, Z.; Lansbury, P. T., Jr., The UCH-L1 gene encodes two opposing enzymatic activities that affect alpha-synuclein degradation and Parkinson's disease susceptibility. *Cell* **2002**, *111* (2), 209-18.

[164] Zhang, M.; Deng, Y.; Luo, Y.; Zhang, S.; Zou, H.; Cai, F.; Wada, K.; Song, W., Control of BACE1 degradation and APP processing by ubiquitin carboxyl-terminal hydrolase L1. *J Neurochem* **2012**, *120* (6), 1129-38.

[165] Guglielmotto, M.; Monteleone, D.; Boido, M.; Piras, A.; Giliberto, L.; Borghi, R.; Vercelli, A.; Fornaro, M.; Tabaton, M.; Tamagno, E., Abeta1-42-mediated down-regulation of Uch-L1 is dependent on NF-kappaB activation and impaired BACE1 lysosomal degradation. *Aging Cell* **2012**, *23* (10), 1474-9726.

[166] Delrieu, J.; Ousset, P. J.; Caillaud, C.; Vellas, B., 'Clinical trials in Alzheimer's disease': immunotherapy approaches. *J Neurochem* **2012**, *120 Suppl 1*, 186-93.

[167] Zhou, L.; Chavez-Gutierrez, L.; Bockstael, K.; Sannerud, R.; Annaert, W.; May, P. C.; Karran, E.; De Strooper, B., Inhibition of beta-secretase *in vivo via* antibody binding to unique loops (D and F) of BACE1. *J Biol Chem* **2011**, *286* (10), 8677-87.

[168] Atwal, J. K.; Chen, Y.; Chiu, C.; Mortensen, D. L.; Meilandt, W. J.; Liu, Y.; Heise, C. E.; Hoyte, K.; Luk, W.; Lu, Y.; Peng, K.; Wu, P.; Rouge, L.; Zhang, Y.; Lazarus, R. A.; Scearce-Levie, K.; Wang, W.; Wu, Y.; Tessier-Lavigne, M.; Watts, R. J., A therapeutic antibody targeting BACE1 inhibits amyloid-beta production *in vivo*. *Sci Transl Med* **2011**, *3* (84), 84ra43.

[169] Yu, Y. J.; Zhang, Y.; Kenrick, M.; Hoyte, K.; Luk, W.; Lu, Y.; Atwal, J.; Elliott, J. M.; Prabhu, S.; Watts, R. J.; Dennis, M. S., Boosting brain uptake of a therapeutic antibody by reducing its affinity for a transcytosis target. *Sci Transl Med* **2011**, *3* (84), 84ra44.

[170] Jones, A. R.; Shusta, E. V., Blood-brain barrier transport of therapeutics *via* receptor-mediation. *Pharm Res* **2007**, *24* (9), 1759-71.

[171] Small, D. H., Dysregulation of calcium homeostasis in Alzheimer's disease. *Neurochem Res* **2009**, *34* (10), 1824-9.

[172] Small, D. H.; Gasperini, R.; Vincent, A. J.; Hung, A. C.; Foa, L., The role of Abeta-induced calcium dysregulation in the pathogenesis of Alzheimer's disease. *J Alzheimers Dis* **2009**, *16* (2), 225-33.

[173] Liang, B.; Duan, B.-Y.; Zhou, X.-P.; Gong, J.-X.; Luo, Z.-G., Calpain activation promotes BACE1 expression, amyloid precursor protein processing, and amyloid plaque formation in a transgenic mouse model of Alzheimer disease. *J Biol Chem* **2010**, *285* (36), 27737-44.

[174] Hayley, M.; Perspicace, S.; Schulthess, T.; Seelig, J., Calcium enhances the proteolytic activity of BACE1: An *in vitro* biophysical and biochemical characterization of the BACE1-calcium interaction. *Biochim Biophys Acta* **2009**, *1788* (9), 1933-8.

[175] Bennett, B. D.; Denis, P.; Haniu, M.; Teplow, D. B.; Kahn, S.; Louis, J. C.; Citron, M.; Vassar, R., A furin-like convertase mediates propeptide cleavage of BACE, the Alzheimer's beta -secretase. *J Biol Chem* **2000**, *275* (48), 37712-7.

[176] Anekonda, T. S.; Quinn, J. F., Calcium channel blocking as a therapeutic strategy for Alzheimer's disease: the case for isradipine. *Biochim Biophys Acta* **2011**, *1812* (12), 1584-90.

CHAPTER 10

BACE1 Inhibitors: Attractive Therapeutics for Alzheimer's Disease

Boris Decourt[1,*], MiMi Macias[1], Marwan Sabbagh[1] and Abdu Adem[2]

[1]Banner Sun Health Research Institute, Sun City, AZ, USA; [2]Department of Pharmacology and Therapeutics, Faculty of Medicine and Health Sciences, United Arab Emirates University, Al-Ain, UAE

Abstract: One of the neuropathological hallmarks of Alzheimer's disease (AD) is the presence of brain senile plaques made up principally of aggregated amyloid beta (Aβ) peptides. Aβ is produced during the consecutive proteolysis of the transmembrane amyloid precursor protein (APP) by β- and γ-secretases. Genetic and pharmacological manipulations have demonstrated the major β-secretase in AD that makes the initial cleavage required for synthesis of Aβ is the beta-site APP-cleaving enzyme 1 (BACE1). It is therefore very tempting to consider inhibiting BACE1 as a potential AD therapeutic intervention. Here, we review the current knowledge and the molecular and physiological challenges associated with BACE1 inhibition. We also propose alternatives to the direct targeting of CNS BACE1 to prevent AD, as well as methods to measure the therapeutic efficacy of BACE1 inhibition.

Keywords: Amyloid, Alzheimer's, APP, BACE1, bapineuzumab, beta-secretase, blood brain barrier, central nervous system, clinical trial, design, down syndrome, gamma-secretase, immunotherapy, inhibition, modulation, optimization, periphery, peptidomimetic, solanezumab, trafficking.

INTRODUCTION

Alzheimer's Disease (AD) is a progressive debilitating neurodegenerative disorder which incidence increases with age. It is estimated that the disease affects 5.5 million Americans, and 24 million people worldwide [1, 2]. These figures are expected to double every 20 years as the global population age continues to

*Address correspondence to Boris Decourt:** Banner Sun Health Research Institute, 10515 W. Santa Fe Drive, Sun City, AZ 85351, USA; Tel: 623-832-6500; Fax: 623-832-6504;
E-mail: Boris.Decourt@bannerhealth.com

increase, resulting in a significant public health burden if no disease-preventing and -modifying treatments become available in the near future.

Most often, the diagnosis of AD is sought when close relatives inform family physicians that their loved ones experience memory and behavioral dysfunctions. Patients are then administered a battery of neuro-psychological tests assessing cognitive functions, and undergo brain imaging, physical and blood exams. Expert neurologists usually make the clinical diagnosis when all test results indicate AD symptoms. However, to date, definite diagnosis is only made after specific neuropathological features are detected in the post-mortem brain. The two main features include extracellular dense core plaque deposits containing high density of aggregated amyloid beta (Aβ) peptides, and intraneuronal fibrillary tangles composed of hyperphosphorylated tau [3]. Brain inflammation and synaptic loss are also frequently observed in the AD brain [3]. The current mainstream hypothesis is that plaques and tangles progressively accumulate in the brain over several years, inducing chronic synaptic loss and neuronal death which, until recently, were often detected only after clinical symptoms appear [4]. However, in the past decades a large body of research relying on improved imaging technology (MRI, hippocampal volume, ventricular volume, *etc*) has permitted the earlier detection of brain tissue/neuronal loss [5-8]. Once optimized, these techniques will be of great importance to demonstrate therapeutic efficacy of paradigms aiming at preventing prodromal neurodegeneration.

Amyloidogenesis, the process by which amyloid beta (Aβ) is produced, results from the consecutive endoproteolysis of amyloid precursor protein (APP) by β- and γ-secretases. The enzymatic activities generate peptides of different lengths, including the main components of senile plaques Aβ40 and Aβ42 [9]. Both genetic deletion and overexpression manipulations in mice have demonstrated that transmembrane beta-site APP-cleaving enzyme 1 (BACE1) is the principal β- secretase in the brain [10-15]. Gamma-secretase was identified as a complex comprising four transmembrane proteins, *i.e.* presenilin 1/2 (PS1/2), anterior pharynx-defective 1 (APH1), nicastrin (NCT), and presenilin enhancer 2 (PEN2), that assemble into a large heteromultimer unit generating and regulating the γ-secretase activity [16-18]. Potentially competing with the amyloidogenic pathway is the non-amyloidogenic pathway in which APP is cleaved by an α-secretase enzyme in the middle of the region forming Aβ peptides (between Lys-16 and Leu-17 of Aβ), preventing the production of Aβ [19, 20]. Several candidate proteins have been suggested as α-secretases, indicating the possibility of functional redundancies in any given cell [20]. The amyloidogenic pathway is the preferred pathway in neurons and blood platelets while the non-amyloidogenic pathway is preferred in most other cells [21-23], though brain inflammation may result in glial cells synthesizing Aβ [24, 25], as well as BACE1 [26]. To balance production, Aβ peptides are catabolized by several different enzymes, the main ones being the insulin degrading enzyme (IDE) and neprilysine (NEP) [27-29]. Compared to many peripheral pathologies which mechanisms have been under investigation for decades, all the pathways described above for AD have only been identified in the past 20 years, largely resulting from the linkage of mutations in several of the above-mentioned genes to familial forms of AD. However, such genetic forms of AD account for less than 5% of the cases, and the remaining 95% are sporadic cases of unknown etiology [9].

From the chain of processing events described above, it is clear that Aβ peptide levels are tightly regulated to fulfill their physiological functions, which still are poorly understood [30, 31]. Furthermore, these processing mechanisms provide several potential therapeutic targets to prevent the accumulation of Aβ to pathological levels, such as 1- inhibit BACE1 activity [32], 2- inhibit γ-secretase activity [33, 34], 3- shift APP processing from the amyloidogenic to non-amyloidogenic pathway [35], 4- increase Aβ degradation (*e.g.* increase the activity of IDE and NEP) [36], 5- prevent the aggregation of Aβ peptides [37], 6- clearing Aβ out of the brain by modulating its blood concentration [38, 39], and 7- decrease brain inflammation that may exacerbate Aβ production [40].

Until recently, most clinical trial efforts have focused on clearing Aβ out of the brain and preventing its aggregation *via* immunotherapies, but strong challenges made these approaches quite unsuccessful [41]. For example, in August 2012 Johnson & Johnson and Pfizer announced the termination of their phase III clinical trial of bapineuzumab [42, 43]; and Eli Lilly and Company reported the failure to detect any significant effect of solanezumab in mild-to-moderate AD patients [44]. These examples emphasize the difficulty of using Aβ immunotherapies, either active or passive, to improve cognitive functions and slow down AD progression [45, 46], at least in advanced stages of the disease. However, optimism reemerged after the recent sub-analysis of solanezumab clinical trial data revealed that the mild AD group experienced 30-35% slowing of cognitive decline [47], suggesting that targeting earlier stages of AD might show better treatment efficacy. Furthermore, plasma Aβ levels increased dramatically, suggesting that the antibody did engage its target, though it is not clear yet whether this pool of Aβ was removed from the brain or not, and whether bound Aβ half-life is increased or not in the blood (*e.g.* reduced Aβ catabolism), which could partly explain the observed increase in plasma. Nonetheless, these data provide very valuable insights for future studies targeting early stages of AD, and are encouraging from a clinical perspective since it may be possible to slow down the progression of the disease by reducing soluble Aβ levels.

In addition to the direct targeting of Aβ, some clinical trials have also attempted to target its production by inhibiting γ-secretase activity. However, initial γ-secretase inhibitors induced adverse events that precluded completion of many clinical studies [48, 49], likely because of the large number of γ-secretase substrates involved in critical physiological functions, such as the well-studied Notch signaling pathway [50, 51]. Nonetheless, recent discovery of Notch-sparing γ-secretase inhibitors provides the possibility for developing inhibitory compounds specifically targeting Aβ production in the near future [52].

Because of difficulties faced during Aβ immunotherapies and γ-secretase inhibitor clinical trials, one of the most promising therapeutic intervention one could think of at this time is BACE1 inhibition/modulation. In the present review, we summarize BACE1 molecular information and inhibitor design parameters, then report the most recent advances in BACE1 inhibitor clinical trials, and later provide a few alternative strategies to central nervous system (CNS)-targeted BACE1 inhibition.

THE β-SECRETASE ACTIVITY AND BACE1 IN ALZHEIMER's

In this section, we provide only a synopsis of the β-secretase activity and BACE1 protein as numerous elegant reviews have been published elsewhere [53-55]. The β-secretase enzymatic activity was identified as the first cleavage step of APP in the production of Aβ [56]. It releases a large APP extracellular domain called sAPPβ, leaving the transmembrane and APP intracellular domains (called C99) for cleavage by γ-secretase. A major breakthrough occurred in 1999 when several laboratories independently reported that BACE1/membrane-associated aspartic protease 2(Asp2)/memapsin 2 was likely the enzyme responsible for the β - secretase activity in AD [57-60]. In the following years, the generation of several BACE1 knock-out and knock-in mouse strains crossed to APPsw overexpressing mutants (*e.g.* mice expressing the human APP transgene bearing the Swedish mutation K670N/M671L identified in familial forms of AD) strongly supported the idea that BACE1 is the major β-secretase in the brain [11, 14, 15, 61]. This hypothesis was further supported in a recent report showing that, at least *in vitro*, an APP A673T mutation protects APP from BACE1 cleavage [62], though direct evidence of BACE1 cleavage resistance, for example *via* in a cell-free assay, was not provided. However, several other proteins have been suggested to function as β-secretases in the brain. For example, cathepsins D and E have been shown to cleave APP at the β site [63, 64]. More recently, the genetic deletion of cathepsin B showed that this enzyme can cleave wild-type human APP at the β site as well [65].

Although several cathepsins may act as β-secretases, the fact that BACE1 knock-out was sufficient to prevent plaque formation in APPsw transgenic mice is strong evidence that BACE1 is the major β-secretase involved in Aβ generation in the brain. Moreover, the very recent discovery of a molecular pathway implicating BACE1 overexpression in memory deficits [66], combined with its localization around plaques in both human and mouse brains [67], also suggests that BACE1 might be involved in AD-associated cognitive deficits. Therefore, BACE1 inhibition/modulation appears as a very promising therapeutical approach for AD.

On the other hand, experiments conducted on BACE1 knock out mice (BACE1$^{-/-}$) without APP transgenic background suggest that strong inhibition of BACE1 might be deleterious. Indeed, while BACE1 knock out (BACE1$^{-/-}$)/APP bigenic models did not show any major defect initially [11, 14], detailed investigations of BACE1 complete knock out (BACE1$^{-/-}$) revealed subtle molecular and cognitive defects. Those include hypomyelination of central and peripheral nerves [68, 69], defects in remyelination [70], increased frequency of seizures [71-73], impaired long-term potentiation [74], increased schizophrenia-like behavior [75], retinal pahothology [76], and increased neonatal lethality [61]. These results suggest that complete BACE1 inhibition might induce neurological adverse effects in patients already suffering neurodegeneration. However, to date it is not clear whether most of these phenotypes result from the lack of BACE1 in adults or during embryonic and postnatal development (in the case of hypomyelination, it was concluded this results from a developmental defect [77]). Interestingly, a few experiments conducted on monogenic BACE1 heterozygous mice (BACE1$^{+/-}$) showed limited mechanism-based toxicity compared to BACE1$^{-/-}$ bigenics [12, 78]. These data provide the exciting possibility that modulating BACE1 levels/activity, for example by administrating low to moderate doses of BACE1 inhibitor(s) rather than high doses, or by administering BACE1 modulators with partial inhibitory activity even at high dose in CNS and peripheral tissues, might suffice to reduce amyloid pathology and conversion to AD while limiting adverse effects.

BACE1 INHIBITOR DESIGN

In this section, we briefly describe the requirements of and propose new ideas for developing effective and safe BACE1 inhibitors. Readers interested in the structure and pharmacological properties of currently available BACE1 inhibitors are referred to comprehensive reviews published recently [79, 80].

BACE1 is a 501 amino acid type 1 transmembrane aspartic protease from the pepsin family bearing an extracellular active–site cleft containing two aspartyl residues (Asp 92 and Asp 228) [55]. During and after synthesis in the endoplasmic reticulum, it undergoes several post-translational maturation steps for proper folding and localization, and which increase its activity [53, 55]. Each individual step, from transcription to complete maturation, represents a potential target for reducing BACE1 levels and activity [53]. However direct competition of endogenous substrates, which is the principal inhibition paradigm investigated to date, has proven very challenging [79] (see below).

A very important parameter to consider when developing enzyme inhibitors is specificity. Numerous issues have emerged with BACE1 inhibition because of the bilobal shape of its very large catalytic site, which is unique to this protein [81].

Consequently, peptidomimetic molecules, many of which derived from the APP sequence surrounding Aβ Asp+1 and Glu+11, have proven unsatisfactory with regard to specificity because most of these compounds also inhibit other aspartic proteases such as BACE2, cathepsins, pepsin, and rennin [79, 80]. To circumvent this issue, in the past decade non-peptidomimetic molecules have been developed and proven more specific for BACE1 than peptidomimetic compounds [80]. Intriguingly, to our knowledge, very few studies take into consideration the dimerization of BACE1 when developing specific inhibitors, though evidence suggest that BACE1 dimers process APP more efficiently than BACE1 monomers [82, 83]. Knowing the stoichiometry of APP cleavage by BACE1 might help generating more efficient BACE1 inhibitors. Moreover, as an effort during drug development phases, it would be interesting to compare biochemical parameters of BACE1 inhibitors in systems enriched in BACE1 dimers *vs.* monomers, though dimerization might be difficult to achieve when using recombinant, soluble proteins.

To reduce brain production of Aβ, an effective BACE1 inhibitor needs to cross the blood-brain barrier (BBB). Several strategies have been employed to achieve this goal. For example, bioengineering showed promising results in delivering peptidomimetic compounds into the brain of transgenic Alzheimer's mice Tg2576 [84], but the moderate specificity of current peptidomimetic inhibitors (see above) limits the interest of the approach. Alternatively, recent publications by Genetech suggest that manipulation of transcytosis (*e.g.* intracellular transportation *via* binding to the transferrin receptor) could facilitate brain delivery of large molecules such as antibodies [85, 86]. On the other hand, non-peptidomimetics synthesized as small molecules capable of BBB crossing are currently the most promising pharmacological compounds to inhibit brain BACE1 activity following peripheral administration [79, 80].

In addition to crossing the BBB it is important to consider the subcellular compartment(s) to be targeted by BACE1 inhibitors. In the current model, mature BACE1 is first addressed to the plasma membrane, then internalized *via* endosomes where its activity dramatically increases upon luminal acidification [55, 87, 88]. Based on crystallographic analyses, it appears that BACE1 undergoes conformational changes during its enzymatic activity cycle [89]. Thus, specific and effective inhibitors likely need to bind BACE1 not only at neutral pH during transit along the cell surface, and doing so despite conformational limitations, but also to remain bound to the enzyme after internalization and acidification. Interestingly, it has been observed that BACE1 dimers, which are more processive than monomers [83], bind to and cleave APP oligomers [90]. Thus, one could imagine an addition to the mainstream hypothesis in which BACE1 dimers would cleave a portion of APP molecules at the cell surface before further processing by γ -secretase. Such an alternative hypothesis is supported by the observation that γ-secretase is most active at neutral, rather than acidic pH [91-93], and would be consistent with the fact that the majority of Aβ is released into the extracellular space. Furthermore, BACE1 has been shown to be shed from the plasma membrane [94-96], and overexpression of soluble BACE1 increased Aβ generation *in vitro* [96]. It would be interesting to investigate whether soluble BACE1 participates in the amyloidogenic pathway *in vivo*, and whether inhibitors bind preferentially transmembrane or soluble BACE1.

Surprisingly, to date little is known about BACE1 interacting proteins and regulatory factors, though targeting these molecules might be an alternative approach to modulate BACE1 levels/activity. Those include the translation pre-initiation complex component Eukaryotic initiation factor 2 alpha (eIF2α), which regulates BACE1 translation [97]; trafficking proteins GGA1 and GGA3 involved in BACE1 cellular localization and degradation [98-100]; ARF6 which affects BACE1 endocytosis [101]; SNX6 and SNX12, which regulate BACE1 endocytic trafficking [102, 103]; and ubiquitination of BACE1 C-terminal region that affects its trafficking [104]. Given the role of each of these proteins, even limited modulation could suffice to affect BACE1 cellular production, localization and trafficking. However, more work is needed to investigate such approach since it is likely that the physiology of other important proteins maybe disturbed when targeting cellular endocytosis, trafficking, and degradation pathways *in vivo*, which could result in undesired adverse effects.

Another important step when developing BACE1 inhibitors is the testing on cellular and animal models. Cellular models provide crucial biochemical information such as the inhibition constant (K_i), dose-response, minimum effective dose, stability in solution, and eventual toxicity. While initial tests can be conducted on non-neuronal cell lines, it is often better to validate the results using neuronal cell lines and primary cortical/hippocampal neurons to assess how these post-mitotic cells respond to particular inhibitors. When biochemical parameters meet satisfactory requirements with cellular models, then compounds can be tested in animal models. Among AD animals models available [105], the preferred models to date are transgenic Alzheimer's mice which allow assessment of bioavailability, BBB crossing capabilities, effect on peripheral and brain Aβ levels, quantification of plaque deposits, and potential *in vivo* toxicity. Furthermore, animal models are used to determine the best administration route. There is currently agreement that oral administration would be the most acceptable to patients and most practical for treating large populations, while antibody-based therapies will likely require direct injection in the body by skilled professionals. To date, only a handful of BACE1 inhibitors have shown good pre-clinical efficacy and low toxicity. Several of those inhibitors are now being tested on human subjects.

HUMAN CLINICAL TRIALS

Study Design Parameters

Following the failure of almost all clinical trials aimed at treating AD in the past years, several publications have addressed potential flaws in study designs that might explain the lack of positive results [41, 106, 107]. In this section, we focus on parameters to consider when developing clinical trials for AD treatment using BACE1 inhibitors.

In our opinion, the main parameter to consider is the stage of the disease. The recently revised stages of AD by the Alzheimer's Association and National Institute of Aging (AA-NIA) integrate mild cognitive impairment (MCI) as an intermediate stage of AD [108]. It is well established that synaptic and neuronal loss are pronounced in subjects suffering mild to moderate AD [109], suggesting that prevention of subsequent Aβ synthesis and deposition is unlikely to reverse the damage that has already occured in the brain at these stages of the disease. Thus, in the past years scientists have proposed to carry out AD-preventing rather than AD-modifying clinical trials on MCI and early AD patients, which is supported by the positive data reported about mild AD subjects from the solanezumab clinical trial [47] (see above), hoping to prevent neurodegeneration. Because brain BACE1 levels are increased early in AD [67], it is possible that BACE1 inhibition will be required earlier than at the MCI stage to slow down or prevent plaque formation and reduce Aβ-induced synaptic and neuronal toxicity [37]. However, such pilot studies will need to identify individuals at risk of developing AD in the general population, *i.e.* carry out population enrichment, for example by using imaging techniques detecting amyloid plaques *in vivo*, though large population screening will increase costs dramatically. Alternatively, one could think of testing prevention therapies on Down syndrome individuals. Indeed, DS conveys a high risk for the early development of AD. Virtually all individuals with DS have the plaques and tangles characteristic of AD by the age of 40 [110], and by age 65 60-70% of individuals with DS develop overt AD. Thus, not only this population has great potential for testing prevention therapies, but DS subjects would also greatly benefit if anti-Aβ (including BACE1 inhibition) therapies are successful at preventing conversion towards AD. Also to consider are individuals from families known to develop early AD, although their number is limited for large scale studies and genetic mutations might attenuate the effect of BACE1 inhibition therapies.

A second important parameter is the choice of biomarker(s) for monitoring the efficacy of BACE1 inhibitors. Currently, there is no FDA-validated biomarker for clinical use, although brain amyloid imaging and cerebral spinal fluid (CSF) Aβ42 and tau levels are commonly used for research purposes [111-113]. Because these biomarkers are suspected to mirror the synthesis and accumulation of Aβ in the brain, as well as neuronal death (*i.e.* tau), they seem well suited to study the efficacy of BACE1 inhibitors. However, some are questioning their real theragnostic potential, *i.e.* whether they truly reflect drug-induced changes in an individual (see ref. [47] for most recent discussions). Alternatively, one could use BACE1 itself as a biomarker, for example by measuring its levels and activity in CSF [114-116]. Maybe in the near future, peripheral biomarkers will become available and be less expensive and less invasive than brain imaging and CSF collection and analyses [117-120]. The choice of marker(s) is also dependent on the duration of the study. For example, short term studies (*e.g.* lasting less than six months) will likely focus on biological marker measurements, while long-term studies could also assess stabilization or improvement of cognitive functions.

Lastly, because Aβ immunotherapy and γ -secretase inhibitors have induced adverse events that were not reported in animal models until recently [121, 122], we believe it is important to include careful monitoring of BACE1 inhibitors potential toxicity in humans. For example, since neuregulin-mediated myelination defects were detected in adult BACE1 knock-out mice [77], this parameter will likely need to be assessed in humans, as well as the potential to induce schizophrenia [75] and age-related retinal pathology [76]. Morevoer, post-mortem analysis of patients who underwent Aβ immunotherapies revealed increased brain microhemorrhages and microvascular lesions [123]. At the moment the underlying mechanism are still unclear, though some have proposed that Aβ might act as a vascular sealent and that its removal from periphery could induce hemorrhages [124, 125]. Because BACE1 inhibition is expected to decrease overall Aβ levels, integrity of CNS blood vessels should likely be controlled as well. Furthermore, because current strategies focus on peripheral administration of the inhibitor(s) but only a few BACE1 substrates have been identified [54, 126], it will be important to conduct phase I studies assessing peripheral adverse events such as blood anomalies and peripheral enzymatic levels that could indicate organ distress (*e.g.* albumin, transaminases and bilirubin for liver; and creatine kinase for heart).

Human Trials and Reported Results

As mentioned above, only a handful of BACE1 inhibitors have successfully passed pre-clinical evaluation and been tested in clinical trials (Table **1**). Below we describe the molecules for which information has been publicly released. Additional molecules from academic and private entities are being tested currently in clinical trials worldwide, but data were not available at the time we prepared this manuscript.

Historically, the first inhibitor tested was CTS-21166, a derivative of the peptidomimetic OM99-2 [84], by CoMentis, Inc. A phase I trial was conducted in 2007-2008 on healthy young adult males who received intravenous, single escalating doses (7.5-225 mg) of the compound [127]. Such regimen resulted in significant decrease in plasma Aβ levels that persisted for several days without any severe adverse event. A second phase I study using 200 mg of an oral form of the compound seemed to produce similar results [79], though no further details were revealed. No update for this molecule was found in recent literature. Thus, whether this compound crosses the BBB in humans and has significant effects on CSF and brain biomarkers, which were not reported in the clinical studies, remains to be determined.

In 2009, Eli Lilly and Company tested the first non-peptidic BACE1 inhibitor LY2811376 in a phase I study on healthy adults administered oral, single escalating doses [128]. Data indicated a significant reduction in plasma and CSF Aβ levels in the hours following administration with well tolerated adverse events [129]. Importantly, a reduction in CSF soluble APPβ (*i.e.* the extracellular segment released from APP after β-secretase cleavage) levels strongly suggested that the compound was effectively inhibiting CNS BACE1. However, a toxicology study conducted in rats and BACE1 knock-out mice revealed that, at doses beyond 30 mg/kg, LY2811376 causes cytoplasmic accumulations of autofluorescent material within the retinal epithelium, as well as neurons and glial cells in the brain [129], which was reason to discontinue development of this molecule. Subsequently, Eli Lilly continued the exploration of BACE1 inhibition by conducting pre-clinical tests with the analog LY2886721, which showed promising results in cellular, murine, and canine models [130]. In 2010 and 2011, several phase I studies using oral administration assessed the tolerability, safety, pharmacokinetics, pharmacodynamics and biodisposition of LY2886721 (Table **1**). Positive study results [131, 132], including a significant decrease in CSF Aβ levels, have led to further testing and the compound is currently in a phase I-II study on MCI and mild AD subjects (Table **1**).

In direct competition with LY2886721 is Merck's lead BACE1 inhibitor MK-8931, which was the subject of four presentations at the 2012 Alzheimer's Association International Conference (AAIC). Phase I trials conducted in Europe, USA and Japan have demonstrated a significant reduction in CSF Aβ levels of up to 90% when MK-8931 was administered in both single and multi-dose paradigms in healthy subjects [133, 134]. Although adverse effects were recorded, most seemed related to CSF collection methods rather than to the drug itself. A phase I-II trial on patients suffering mild AD, receiving oral doses ranging 12-60 mg for seven days, was conducted in the first half 2012 (Table **1**). Results are not yet available, but it will be very interesting to see whether MK-8931 is as effective in affecting CSF Aβ in AD subjects.

Table 1.   Summary of Most BACE1 Inhibitor Clinical Trials

Clinical Trial ID	Molecule	Sponsor	Phase	Subjects	Regimen*	Status
NCT-00621010	CTS2166	CoMentis [127]	I	Healthy adult males	Single escalating dose (7.5-225 mg)	Completed
NCT-00838084	LY2811376	Eli Lilly and Company [128]	I	MCI and mild AD males and females	Single escalating dose (5-90 mg)	Completed
NCT-01133405	LY2886721	Eli Lilly and Company [136]	I	Healthy adult males and females	Single escalating dose (1-200 mg)	Completed
NCT-01227252	LY2886721	Eli Lilly and Company [137]	I	Healthy adult males and females	Multiple oral doses (5-35 mg/day for 14 days)	Completed
NCT-01367262	LY2886721	Eli Lilly and Company [138]	I	Healthy adult males	Single radioactive dose (25 mg)	Completed
NCT-01534273	LY2886721	Eli Lilly and Company [139]	I	Healthy adult males and females	Single and multiple oral doses (35-140 mg/day for 14 days)	Completed
NCT-01561430	LY2886721	Eli Lilly and Company [140]	I-II	MCI and mild AD males and females	Multiple oral doses (15 or 35 mg/day for 26 weeks)	In progress
NCT-01482013	HPP584	High Point Pharmaceuticals, [141]	I	MCI and mild AD males and females	Multiple oral doses (for 28 days)	Terminated
NCT-01496170	MK8931	Merck [142]	Ib	Mild to moderate AD males and females	Single and multiple doses (12-60 mg/day for 7 days)	Completed
NCT-01294540	E2609	Eisai Inc. [143]	I	Healthy adult males and females	Single escalating dose (5-800 mg)	Completed
NCT-01511783	E2609	Eisai Inc. [144]	I	Healthy adult males and females	Multiple doses (25, 50, 400 mg/day for 14 days)	Partly completed
NCT-01600859	E2609	Eisai Inc. [145]	I	MCI males and females	Single oral dose	In progress
NCT-01463384	Minocycline	Huntington Medical Research Institutes [146]	II	Age matched cognitively normal, MCI and	Multiple oral doses (50 mg twice/day for 6 months)	In progress
NCT-01094340	Thalidomide	Banner Health / Celgene [147]	II-III	Mild to moderate AD males and females	Multiple oral dose (up to 200 mg/day for 24 weeks)	In progress

* All regimens are compared to placebo controls.

The Japanese company Eisai also presented results regarding its BACE1 inhibitor E2609 during the 2012 AAIC. A first phase I study on healthy subjects receiving single ascending oral doses (5-800 mg), revealed a dramatic decrease in plasma Aβ of up to 92% and acceptable adverse events (Table **1**). This led to conducting a second phase I study testing the effect of multiple doses, administered orally for 14 days to healthy subjects (Table **1**). The latter study showed a significant reduction in Aβ levels in both plasma and CSF compared to placebo controls [135]. In addition, this study indicated E2609 doses in the range of 50-100 mg were sufficient to decrease body fluid Aβ levels up to 80%, suggesting that chronic administration of low doses might be sufficient in preventing plaque formation if administered at early stages of AD. The concept is currently being tested in an ongoing phase I study on MCI and mild AD subjects (Table **1**).

Another promising inhibitor in pre-clinical studies was HPP854 by High Point Pharmaceuticals. The compound was tested in 2012 in a phase I study on MCI and mild AD subjects who were administered the drug orally, once daily for 28 days. The latest information available is that the trial was terminated (Table **1**), though no official comment was released regarding the reason for early termination.

In a different approach, our group is currently testing the anti-inflammatory (TNFα inhibitor), FDA-approved, anti-cancer drug thalidomide on subjects suffering mild to moderate AD (Table **1**). While the primary purpose is to reduce AD-associated brain inflammation, pre-clinical studies suggest that thalidomide also reduces brain BACE1 levels in APP23 mice (unpublished observations), partly reproducing results of TNFα receptor I deletion in this AD mouse model [148]. Intermediate data analysis of this phase II-III study, however, indicates that thalidomide is poorly tolerated in AD subjects, which is in agreement with anti-cancer studies carried out on patients of similar age [149]. In a similar fashion, the semi-synthetic antibiotic, minocycline, has been tested in AD rodent models for its anti-inflammatory properties. It was observed that minocycline prevents central neuroinflammation, tau pathology, and neuronal death [150-152]. As is the case for thalidomide analogs, brain BACE1 levels and activity also were partially reduced after minocycline treatment in an AD mouse model [151]. These pre-clinical data were the basis for initiating a six month, open label phase II clinical study, testing the potential of minocycline in MCI and AD subjects (Table **1**). It will be interesting to determine whether the anti-inflammatory paradigms decrease BACE1 and Aβ levels in human subjects since such an approach would target not only amyloidodegenesis, but also additional AD-associated neuropathological features (*i.e.* chronic brain inflammation).

ALTERNATIVE STRATEGIES

While the production of peptidomimetic and non-peptidomimetic compounds has received the most attention, additional strategies have been explored to inhibit BACE1 activity *in vivo*. Here, we present several strategies that might arise in the near future.

An alternative intervention to inhibit BACE1 activity is to use anti-BACE1 antibodies which, for example, target the catalytic site. Because of the physical size of antibodies, it is likely they would impede, by steric hindrance, the binding of endogenous substrates in the BACE1 catalytic site. A proof-of-concept study conducted *in vitro* and in Tg2576 mice showed reduced inflammation and brain Aβ levels after immunization against BACE1 [153], suggesting the viability of this approach. Using another approach, Genetech reported the development and optimization of bispecific antibodies [85, 86]. Indeed, one issue faced during most Aβ immunization trials was the poor diffusion of antibodies across the BBB (~1% of anti-Aβ antibodies). To circumvent this problem, Genetech scientists created chimeric antibodies bearing a low affinity, anti-transferrin receptor binding site on one arm and a high-affinity, anti-BACE1 binding site on the other arm. Such a combination facilitates active transport from the blood into the brain *via* transcytosis through endothelial cells forming the BBB. The anti-transferrin receptor binding site, mimicking iron transport *via* transferrin, is released into the brain interstitial space because of its low affinity to transferrin receptors. Then the chimeric antibody passively diffuses until the high affinity anti-BACE1 binding site attaches to plasma membrane-associated BACE1 protein and inhibits its activity [85, 86]. If such an approach can be translated into humans, it would represent a breakthrough for delivering hapten-specific antibodies directly in the vicinity of neurons without requiring direct injections into the brain or spinal cord.

A second method that could be used in the future is the regulation of BACE1 translation by destabilizing its mRNA. Such a procedure could employ either antisense oligonucleotides, or catalytic nucleic acids (ribozymes), or small interfering RNAs (summarized by Nawrot [154]). For example, siRNAs have been used in APP mice to knock down BACE1, which resulted in lower Aβ production and plaque number in the brain [155]. However, large scale application in humans requires perfection of delivery methods targeting the CNS, and particularly neurons. Testing of several safe and efficient delivery methods is currently underway [156, 157].

As mentioned above in regard to our thalidomide clinical study, we and others have found that powerful immunomodulators not only reduce brain inflammation, but also reduce amyloid loads in AD mouse models [158, 159]. Preliminary analyses in our laboratory strongly suggest that thalidomide-induced decrease in amyloid loads is associated with significantly lower brain BACE1 levels (unpublished observations), though it is not clear yet whether the drug directly affects BACE1 or whether the effect is secondary to reduced inflammation. Unfortunately, because thalidomide is very toxic in humans, its large scale use for AD treatment is unlikely. However, once we confirm our pre-clinical data, one could suggest testing thalidomide analogs displaying better tolerability in humans. For example, we are currently testing lenalidomide in APP23 mice and have observed similar potency as thalidomide on amyloid deposits reduction (unpublished data). Because lenalidomide is already FDA-approved for cancer treatment, it could rapidly be repurposed in phase II-III clinical studies for AD. Similarly, drugs used for other conditions might be efficient BACE1 inhibitors, such as the calcium-blocking agent bedripdil. The compound is used in cardiac patients to induce coronary vasodilation with modest peripheral effects, decrease hypertension, and selectively treat anti-arrhythmia activities by acting as a calmodulin antagonist. Bepridil was recently shown to modulate BACE1 and γ-secretase in cell lines and rodent models, and decreased Aβ levels [160, 161]. This drug has great potential for clinical trials since it is FDA-approved and because many AD subjects also suffer heart conditions.

Finally, while most laboratories aim at developing small molecule BACE1 inhibitors that can cross the BBB, some larger compounds that do not cross the BBB were also shown to be effective in reducing brain Aβ levels in rodent models [162]. The likely cause of this effect is the fact that APP, BACE1 and γ-secretase are also expressed at high levels in blood platelets that are estimated to produce up to 95% of blood Aβ [21]. Therefore, it would be interesting to test whether inhibiting peripheral BACE1 significantly reduces brain Aβ levels. Because some Aβ immunization paradigms demonstrated brain amyloid load reduction [163], peripheral BACE1 inhibition represents a plausible alternative to investigate further whether depleting blood Aβ reduces brain Aβ levels, a process referred to as the sink theory [38, 164].

CONCLUSIONS

Since its discovery over a decade ago, BACE1 has been investigated as a potential target to reduce amyloidogenesis. However, its biochemical structure and post-translational regulation, which are still poorly understood, have presented many challenges in the development of specific pharmacological inhibitors. This is reflected by the fact that hundreds of compounds have been generated to inhibit BACE1 but only two are currently being tested in phase II clinical trials (Eli Lilly's LY2886721 and Merck's MK-8931). In addition, the lack of validated biomarkers and the targeted populations (until now, AD subjects rather than MCI) have likely contributed to the failure of many AD-oriented clinical trials. Despite possible adverse events, BACE1 inhibition remains an attractive therapeutic target for AD with great potential. As additional therapeutic approaches are explored, *e.g.* reducing brain inflammation, one could anticipate that, similar to HIV treatment, combinatorial therapies targeting several features of AD will become available in a near future which will likely increase the effectiveness of AD treatments. Therefore, developing BACE1 inhibitors/modulators is of crucial importance to finding a cure for AD.

ACKNOWLEDGEMENTS

Funded by National Institute on Aging grants R01AG034155, P30AG019610-09, and 5P30AG019610-12, and by a grant from the Alzheimer's Association (NIRG-12-237512).

CONFLICT OF INTEREST

The authors confirm that this chapter contents have no conflict of interest.

REFERENCES

[1] 2012 Alzheimer's disease facts and figures. *Alzheimers Dement* **2012**, *8* (2), 131-68.

[2] Mayeux, R.; Stern, Y., Epidemiology of Alzheimer disease. *Cold Spring Harb Perspect Med* **2012**, *2* (8).

[3] Duyckaerts, C.; Delatour, B.; Potier, M. C., Classification and basic pathology of Alzheimer disease. *Acta Neuropathol* **2009**, *118* (1), 5-36.

[4] Selkoe, D. J., Toward a comprehensive theory for Alzheimer's disease. Hypothesis: Alzheimer's disease is caused by the cerebral accumulation and cytotoxicity of amyloid beta-protein. *Ann N Y Acad Sci* **2000**, *924*, 17-25.

[5] Bateman, R. J.; Xiong, C.; Benzinger, T. L.; Fagan, A. M.; Goate, A.; Fox, N. C.; Marcus, D. S.; Cairns, N. J.; Xie, X.; Blazey, T. M.; Holtzman, D. M.; Santacruz, A.; Buckles, V.; Oliver, A.; Moulder, K.; Aisen, P. S.; Ghetti, B.; Klunk, W. E.; McDade, E.; Martins, R. N.; Masters, C. L.; Mayeux, R.; Ringman, J. M.; Rossor, M. N.; Schofield, P. R.; Sperling, R. A.; Salloway, S.; Morris, J. C., Clinical and biomarker changes in dominantly inherited Alzheimer's disease. *N Engl J Med* **2012**, *367* (9), 795-804.

[6] Desikan, R. S.; Fischl, B.; Cabral, H. J.; Kemper, T. L.; Guttmann, C. R.; Blacker, D.; Hyman, B. T.; Albert, M. S.; Killiany, R. J., MRI measures of temporoparietal regions show differential rates of atrophy during prodromal AD. *Neurology* **2008**, *71* (11), 819-25.

[7] Lehmann, M.; Koedam, E. L.; Barnes, J.; Bartlett, J. W.; Barkhof, F.; Wattjes, M. P.; Schott, J. M.; Scheltens, P.; Fox, N. C., Visual ratings of atrophy in MCI: prediction of conversion and relationship with CSF biomarkers. *Neurobiol Aging* **2013**, *34* (1), 73-82.

[8] Mueller, S. G.; Schuff, N.; Yaffe, K.; Madison, C.; Miller, B.; Weiner, M. W., Hippocampal atrophy patterns in mild cognitive impairment and Alzheimer's disease. *Hum Brain Mapp* **2010**, *31* (9), 1339-47.

[9] Brouwers, N.; Sleegers, K.; Van Broeckhoven, C., Molecular genetics of Alzheimer's disease: an update. *Ann Med* **2008**, *40* (8), 562-83.

[10] Cai, H.; Wang, Y.; McCarthy, D.; Wen, H.; Borchelt, D. R.; Price, D. L.; Wong, P. C., BACE1 is the major beta-secretase for generation of Abeta peptides by neurons. *Nat Neurosci* **2001**, *4* (3), 233-4.

[11] Luo, Y.; Bolon, B.; Kahn, S.; Bennett, B. D.; Babu-Khan, S.; Denis, P.; Fan, W.; Kha, H.; Zhang, J.; Gong, Y.; Martin, L.; Louis, J. C.; Yan, Q.; Richards, W. G.; Citron, M.; Vassar, R., Mice deficient in BACE1, the Alzheimer's beta-secretase, have normal phenotype and abolished beta-amyloid generation. *Nat Neurosci* **2001**, *4* (3), 231-2.

[12] McConlogue, L.; Buttini, M.; Anderson, J. P.; Brigham, E. F.; Chen, K. S.; Freedman, S. B.; Games, D.; Johnson-Wood, K.; Lee, M.; Zeller, M.; Liu, W.; Motter, R.; Sinha, S., Partial reduction of BACE1 has dramatic effects on Alzheimer plaque and synaptic pathology in APP Transgenic Mice. *J Biol Chem* **2007**, *282* (36), 26326-34.

[13] Rockenstein, E.; Mante, M.; Alford, M.; Adame, A.; Crews, L.; Hashimoto, M.; Esposito, L.; Mucke, L.; Masliah, E., High beta-secretase activity elicits neurodegeneration in

transgenic mice despite reductions in amyloid-beta levels: implications for the treatment of Alzheimer disease. *J Biol Chem* **2005**, *280* (38), 32957-67.

[14]   Roberds, S. L.; Anderson, J.; Basi, G.; Bienkowski, M. J.; Branstetter, D. G.; Chen, K. S.; Freedman, S. B.; Frigon, N. L.; Games, D.; Hu, K.; Johnson-Wood, K.; Kappenman, K. E.; Kawabe, T. T.; Kola, I.; Kuehn, R.; Lee, M.; Liu, W.; Motter, R.; Nichols, N. F.; Power, M.; Robertson, D. W.; Schenk, D.; Schoor, M.; Shopp, G. M.; Shuck, M. E.; Sinha, S.; Svensson, K. A.; Tatsuno, G.; Tintrup, H.; Wijsman, J.; Wright, S.; McConlogue, L., BACE knockout mice are healthy despite lacking the primary beta-secretase activity in brain: implications for Alzheimer's disease therapeutics. *Hum Mol Genet* **2001**, *10* (12), 1317-24.

[15]   Ohno, M.; Sametsky, E. A.; Younkin, L. H.; Oakley, H.; Younkin, S. G.; Citron, M.; Vassar, R.; Disterhoft, J. F., BACE1 deficiency rescues memory deficits and cholinergic dysfunction in a mouse model of Alzheimer's disease. *Neuron* **2004**, *41* (1), 27-33.

[16]   St George-Hyslop, P.; Fraser, P. E., Assembly of the presenilin gamma-/epsilon-secretase complex. *J Neurochem* **2012**, *120 Suppl 1*, 84-8.

[17]   Small, D. H.; Klaver, D. W.; Foa, L., Presenilins and the gamma-secretase: still a complex problem. *Mol Brain* **2010**, *3*, 7.

[18]   Wolfe, M. S., Structure, mechanism and inhibition of gamma-secretase and presenilin-like proteases. *Biol Chem* **2010**, *391* (8), 839-47.

[19]   Lichtenthaler, S. F., Alpha-secretase in Alzheimer's disease: molecular identity, regulation and therapeutic potential. *J Neurochem* **2011**, *116* (1), 10-21.

[20]   Vingtdeux, V.; Marambaud, P., Identification and biology of alpha-secretase. *J Neurochem* **2012**, *120 Suppl 1*, 34-45.

[21]   Chen, M.; Inestrosa, N. C.; Ross, G. S.; Fernandez, H. L., Platelets are the primary source of amyloid beta-peptide in human blood. *Biochem Biophys Res Commun* **1995**, *213* (1), 96-103.

[22]   Evin, G.; Zhu, A.; Holsinger, R. M.; Masters, C. L.; Li, Q. X., Proteolytic processing of the Alzheimer's disease amyloid precursor protein in brain and platelets. *J Neurosci Res* **2003**, *74* (3), 386-92.

[23]   Li, Q. X.; Fuller, S. J.; Beyreuther, K.; Masters, C. L., The amyloid precursor protein of Alzheimer disease in human brain and blood. *J Leukoc Biol* **1999**, *66* (4), 567-74.

[24]   Pluta, R., Astroglial expression of the beta-amyloid in ischemia-reperfusion brain injury. *Ann N Y Acad Sci* **2002**, *977*, 102-8.

[25]   Burton, T.; Liang, B.; Dibrov, A.; Amara, F., Transcriptional activation and increase in expression of Alzheimer's beta-amyloid precursor protein gene is mediated by TGF-beta in normal human astrocytes. *Biochem Biophys Res Commun* **2002**, *295* (3), 702-12.

[26]   Hartlage-Rubsamen, M.; Zeitschel, U.; Apelt, J.; Gartner, U.; Franke, H.; Stahl, T.; Gunther, A.; Schliebs, R.; Penkowa, M.; Bigl, V.; Rossner, S., Astrocytic expression of the Alzheimer's disease beta-secretase (BACE1) is stimulus-dependent. *Glia* **2003**, *41* (2), 169-79.

[27]   Miners, J. S.; Baig, S.; Palmer, J.; Palmer, L. E.; Kehoe, P. G.; Love, S., Abeta-degrading enzymes in Alzheimer's disease. *Brain Pathol* **2008**, *18* (2), 240-52.

[28]   Nalivaeva, N. N.; Fisk, L. R.; Belyaev, N. D.; Turner, A. J., Amyloid-degrading enzymes as therapeutic targets in Alzheimer's disease. *Curr Alzheimer Res* **2008**, *5* (2), 212-24.

[29]   Malito, E.; Hulse, R. E.; Tang, W. J., Amyloid beta-degrading cryptidases: insulin degrading enzyme, presequence peptidase, and neprilysin. *Cell Mol Life Sci* **2008**, *65* (16), 2574-85.

[30]   Pearson, H. A.; Peers, C., Physiological roles for amyloid beta peptides. *J Physiol* **2006**, *575* (Pt 1), 5-10.

[31]   Haass, C.; Schlossmacher, M. G.; Hung, A. Y.; Vigo-Pelfrey, C.; Mellon, A.; Ostaszewski, B. L.; Lieberburg, I.; Koo, E. H.; Schenk, D.; Teplow, D. B.; *et al.*, Amyloid beta-peptide is produced by cultured cells during normal metabolism. *Nature* **1992**, *359* (6393), 322-5.

[32]   Citron, M., Beta-secretase as a target for the treatment of Alzheimer's disease. *J Neurosci Res* **2002**, *70* (3), 373-9.

[33]   Barten, D. M.; Meredith, J. E., Jr.; Zaczek, R.; Houston, J. G.; Albright, C. F., Gamma-secretase inhibitors for Alzheimer's disease: balancing efficacy and toxicity. *Drugs R D* **2006**, *7* (2), 87-97.

[34]   Evin, G.; Sernee, M. F.; Masters, C. L., Inhibition of gamma-secretase as a therapeutic intervention for Alzheimer's disease: prospects, limitations and strategies. *CNS Drugs* **2006**, *20* (5), 351-72.

[35]   Postina, R., Activation of alpha-secretase cleavage. *J Neurochem* **2012**, *120 Suppl 1*, 46-54.

[36]   Miners, J. S.; Barua, N.; Kehoe, P. G.; Gill, S.; Love, S., Abeta-degrading enzymes: potential for treatment of Alzheimer disease. *J Neuropathol Exp Neurol* **2011**, *70* (11), 944-59.

[37]   Bharadwaj, P. R.; Dubey, A. K.; Masters, C. L.; Martins, R. N.; Macreadie, I. G., Abeta aggregation and possible implications in Alzheimer's disease pathogenesis. *J Cell Mol Med* **2009**, *13* (3), 412-21.

[38]   Deane, R.; Bell, R. D.; Sagare, A.; Zlokovic, B. V., Clearance of amyloid-beta peptide across the blood-brain barrier: implication for therapies in Alzheimer's disease. *CNS Neurol Disord Drug Targets* **2009**, *8* (1), 16-30.

[39]   Gelinas, D. S.; DaSilva, K.; Fenili, D.; St George-Hyslop, P.; McLaurin, J., Immunotherapy for Alzheimer's disease. *Proc Natl Acad Sci U S A* **2004**, *101 Suppl 2*, 14657-62.

[40]   Imbimbo, B. P., The potential role of non-steroidal anti-inflammatory drugs in treating Alzheimer's disease. *Expert Opin Investig Drugs* **2004**, *13* (11), 1469-81.

[41]   Becker, R. E.; Greig, N. H., Alzheimer's disease drug development in 2008 and beyond: problems and opportunities. *Curr Alzheimer Res* **2008**, *5* (4), 346-57.

[42]   Johnson & Johnson News: Johnson & Johnson Announces Discontinuation Of Phase 3 Development of Bapineuzumab Intravenous (IV) In Mild-To-Moderate Alzheimer's Disease. http://www.jnj.com/connect/news/all/johnson-and-johnson-announces-discontinuation-of-phase-3-development-of-bapineuzumab-intravenous-iv-in-mild-to-moderate-alzheimers-disease (accessed September 25, 2012).

[43]   Pfizer News: Pfizer Announces Co-Primary Clinical Endpoints Not Met In Second Phase 3 Bapineuzumab Study In Mild-To-Moderate Alzheimer's Disease Patients Who Do Not Carry The Apoe4 Genotype. http://www.pfizer.com/news/press_releases/pfizer_press_release. jsp?guid=727 at http://pfizer.newshq.businesswire.com&source=2012&page=3 (accessed September 25, 2012).

[44]   Eli Lilly and Company News: Eli Lilly and Company Announces Top-Line Results on Solanezumab Phase 3 Clinical Trials in Patients with Alzheimer's Disease. https://investor. lilly.com/releasedetail.cfm?ReleaseID=702211 (accessed September 25, 2012).

[45] Lemere, C. A.; Masliah, E., Can Alzheimer disease be prevented by amyloid-beta immunotherapy? *Nat Rev Neurol* **2010**, *6* (2), 108-19.

[46] von Bernhardi, R., Immunotherapy in Alzheimer's disease: where do we stand? Where should we go? *J Alzheimers Dis* **2010**, *19* (2), 405-21.

[47] Alzheimer's Association News: CTAD conference report. http://www.alzforum.org/new/detail.asp?id=3312 (accessed December 05, 2012).

[48] Imbimbo, B. P.; Giardina, G. A., gamma-secretase inhibitors and modulators for the treatment of Alzheimer's disease: disappointments and hopes. *Curr Top Med Chem* **2011**, *11* (12), 1555-70.

[49] Imbimbo, B. P.; Panza, F.; Frisardi, V.; Solfrizzi, V.; D'Onofrio, G.; Logroscino, G.; Seripa, D.; Pilotto, A., Therapeutic intervention for Alzheimer's disease with gamma-secretase inhibitors: still a viable option? *Expert Opin Investig Drugs* **2011**, *20* (3), 325-41.

[50] Haapasalo, A.; Kovacs, D. M., The many substrates of presenilin/gamma-secretase. *J Alzheimers Dis* **2011**, *25* (1), 3-28.

[51] Woo, H. N.; Park, J. S.; Gwon, A. R.; Arumugam, T. V.; Jo, D. G., Alzheimer's disease and Notch signaling. *Biochem Biophys Res Commun* **2009**, *390* (4), 1093-7.

[52] Augelli-Szafran, C. E.; Wei, H. X.; Lu, D.; Zhang, J.; Gu, Y.; Yang, T.; Osenkowski, P.; Ye, W.; Wolfe, M. S., Discovery of notch-sparing gamma-secretase inhibitors. *Curr Alzheimer Res* **2010**, *7* (3), 207-9.

[53] Hunt, C. E.; Turner, A. J., Cell biology, regulation and inhibition of beta-secretase (BACE-1). *FEBS J* **2009**, *276* (7), 1845-59.

[54] Kandalepas, P. C.; Vassar, R., Identification and biology of beta-secretase. *J Neurochem* **2012**, *120 Suppl 1*, 55-61.

[55] Vassar, R.; Kovacs, D. M.; Yan, R.; Wong, P. C., The beta-secretase enzyme BACE in health and Alzheimer's disease: regulation, cell biology, function, and therapeutic potential. *J Neurosci* **2009**, *29* (41), 12787-94.

[56] Sinha, S.; Lieberburg, I., Cellular mechanisms of beta-amyloid production and secretion. *Proc Natl Acad Sci U S A* **1999**, *96* (20), 11049-53.

[57] Hussain, I.; Powell, D.; Howlett, D. R.; Tew, D. G.; Meek, T. D.; Chapman, C.; Gloger, I. S.; Murphy, K. E.; Southan, C. D.; Ryan, D. M.; Smith, T. S.; Simmons, D. L.; Walsh, F. S.; Dingwall, C.; Christie, G., Identification of a novel aspartic protease (Asp 2) as beta-secretase. *Mol Cell Neurosci* **1999**, *14* (6), 419-27.

[58] Sinha, S.; Anderson, J. P.; Barbour, R.; Basi, G. S.; Caccavello, R.; Davis, D.; Doan, M.; Dovey, H. F.; Frigon, N.; Hong, J.; Jacobson-Croak, K.; Jewett, N.; Keim, P.; Knops, J.; Lieberburg, I.; Power, M.; Tan, H.; Tatsuno, G.; Tung, J.; Schenk, D.; Seubert, P.; Suomensaari, S. M.; Wang, S.; Walker, D.; Zhao, J.; McConlogue, L.; John, V., Purification and cloning of amyloid precursor protein beta-secretase from human brain. *Nature* **1999**, *402* (6761), 537-40.

[59] Vassar, R.; Bennett, B. D.; Babu-Khan, S.; Kahn, S.; Mendiaz, E. A.; Denis, P.; Teplow, D. B.; Ross, S.; Amarante, P.; Loeloff, R.; Luo, Y.; Fisher, S.; Fuller, J.; Edenson, S.; Lile, J.; Jarosinski, M. A.; Biere, A. L.; Curran, E.; Burgess, T.; Louis, J. C.; Collins, F.; Treanor, J.; Rogers, G.; Citron, M., Beta-secretase cleavage of Alzheimer's amyloid precursor protein by the transmembrane aspartic protease BACE. *Science* **1999**, *286* (5440), 735-41.

[60] Yan, R.; Bienkowski, M. J.; Shuck, M. E.; Miao, H.; Tory, M. C.; Pauley, A. M.; Brashier, J. R.; Stratman, N. C.; Mathews, W. R.; Buhl, A. E.; Carter, D. B.; Tomasselli, A. G.;

Parodi, L. A.; Heinrikson, R. L.; Gurney, M. E., Membrane-anchored aspartyl protease with Alzheimer's disease beta-secretase activity. *Nature* **1999**, *402* (6761), 533-7.

[61]   Dominguez, D.; Tournoy, J.; Hartmann, D.; Huth, T.; Cryns, K.; Deforce, S.; Serneels, L.; Camacho, I. E.; Marjaux, E.; Craessaerts, K.; Roebroek, A. J.; Schwake, M.; D'Hooge, R.; Bach, P.; Kalinke, U.; Moechars, D.; Alzheimer, C.; Reiss, K.; Saftig, P.; De Strooper, B., Phenotypic and biochemical analyses of BACE1- and BACE2-deficient mice. *J Biol Chem* **2005**, *280* (35), 30797-806.

[62]   Jonsson, T.; Atwal, J. K.; Steinberg, S.; Snaedal, J.; Jonsson, P. V.; Bjornsson, S.; Stefansson, H.; Sulem, P.; Gudbjartsson, D.; Maloney, J.; Hoyte, K.; Gustafson, A.; Liu, Y.; Lu, Y.; Bhangale, T.; Graham, R. R.; Huttenlocher, J.; Bjornsdottir, G.; Andreassen, O. A.; Jonsson, E. G.; Palotie, A.; Behrens, T. W.; Magnusson, O. T.; Kong, A.; Thorsteinsdottir, U.; Watts, R. J.; Stefansson, K., A mutation in APP protects against Alzheimer's disease and age-related cognitive decline. *Nature* **2012**, *488* (7409), 96-9.

[63]   Gruninger-Leitch, F.; Berndt, P.; Langen, H.; Nelboeck, P.; Dobeli, H., Identification of beta-secretase-like activity using a mass spectrometry-based assay system. *Nat Biotechnol* **2000**, *18* (1), 66-70.

[64]   Schechter, I.; Ziv, E., Kinetic properties of cathepsin D and BACE 1 indicate the need to search for additional beta-secretase candidate(s). *Biol Chem* **2008**, *389* (3), 313-20.

[65]   Hook, V. Y.; Kindy, M.; Reinheckel, T.; Peters, C.; Hook, G., Genetic cathepsin B deficiency reduces beta-amyloid in transgenic mice expressing human wild-type amyloid precursor protein. *Biochem Biophys Res Commun* **2009**, *386* (2), 284-8.

[66]   Chen, Y.; Huang, X.; Zhang, Y. W.; Rockenstein, E.; Bu, G.; Golde, T. E.; Masliah, E.; Xu, H., Alzheimer's beta-Secretase (BACE1) Regulates the cAMP/PKA/CREB Pathway Independently of beta-Amyloid. *J Neurosci* **2012**, *32* (33), 11390-11395.

[67]   Zhao, J.; Fu, Y.; Yasvoina, M.; Shao, P.; Hitt, B.; O'Connor, T.; Logan, S.; Maus, E.; Citron, M.; Berry, R.; Binder, L.; Vassar, R., Beta-site amyloid precursor protein cleaving enzyme 1 levels become elevated in neurons around amyloid plaques: implications for Alzheimer's disease pathogenesis. *J Neurosci* **2007**, *27* (14), 3639-49.

[68]   Hu, X.; Hicks, C. W.; He, W.; Wong, P.; Macklin, W. B.; Trapp, B. D.; Yan, R., Bace1 modulates myelination in the central and peripheral nervous system. *Nat Neurosci* **2006**, *9* (12), 1520-5.

[69]   Willem, M.; Garratt, A. N.; Novak, B.; Citron, M.; Kaufmann, S.; Rittger, A.; DeStrooper, B.; Saftig, P.; Birchmeier, C.; Haass, C., Control of peripheral nerve myelination by the beta-secretase BACE1. *Science* **2006**, *314* (5799), 664-6.

[70]   Hu, X.; He, W.; Diaconu, C.; Tang, X.; Kidd, G. J.; Macklin, W. B.; Trapp, B. D.; Yan, R., Genetic deletion of BACE1 in mice affects remyelination of sciatic nerves. *FASEB J* **2008**, *22* (8), 2970-80.

[71]   Hitt, B. D.; Jaramillo, T. C.; Chetkovich, D. M.; Vassar, R., BACE1-/- mice exhibit seizure activity that does not correlate with sodium channel level or axonal localization. *Mol Neurodegener* **2010**, *5*, 31.

[72]   Hu, X.; Zhou, X.; He, W.; Yang, J.; Xiong, W.; Wong, P.; Wilson, C. G.; Yan, R., BACE1 deficiency causes altered neuronal activity and neurodegeneration. *J Neurosci* **2010**, *30* (26), 8819-29.

[73]   Kobayashi, D.; Zeller, M.; Cole, T.; Buttini, M.; McConlogue, L.; Sinha, S.; Freedman, S.; Morris, R. G.; Chen, K. S., BACE1 gene deletion: impact on behavioral function in a model of Alzheimer's disease. *Neurobiol Aging* **2008**, *29* (6), 861-73.

[74] Wang, H.; Song, L.; Laird, F.; Wong, P. C.; Lee, H. K., BACE1 knock-outs display deficits in activity-dependent potentiation of synaptic transmission at mossy fiber to CA3 synapses in the hippocampus. *J Neurosci* **2008**, *28* (35), 8677-81.

[75] Savonenko, A. V.; Melnikova, T.; Laird, F. M.; Stewart, K. A.; Price, D. L.; Wong, P. C., Alteration of BACE1-dependent NRG1/ErbB4 signaling and schizophrenia-like phenotypes in BACE1-null mice. *Proc Natl Acad Sci USA* **2008**, *105* (14), 5585-90.

[76] Cai, J.; Qi, X.; Kociok, N.; Skosyrski, S.; Emilio, A.; Ruan, Q.; Han, S.; Liu, L.; Chen, Z.; Bowes Rickman, C.; Golde, T.; Grant, M. B.; Saftig, P.; Serneels, L.; de Strooper, B.; Joussen, A. M.; Boulton, M. E., beta-Secretase (BACE1) inhibition causes retinal pathology by vascular dysregulation and accumulation of age pigment. *EMBO Mol Med* **2012**, *4* (9), 980-91.

[77] Sankaranarayanan, S.; Price, E. A.; Wu, G.; Crouthamel, M. C.; Shi, X. P.; Tugusheva, K.; Tyler, K. X.; Kahana, J.; Ellis, J.; Jin, L.; Steele, T.; Stachel, S.; Coburn, C.; Simon, A. J., *In vivo* beta-secretase 1 inhibition leads to brain Abeta lowering and increased alpha-secretase processing of amyloid precursor protein without effect on neuregulin-1. *J Pharmacol Exp Ther* **2008**, *324* (3), 957-69.

[78] Laird, F. M.; Cai, H.; Savonenko, A. V.; Farah, M. H.; He, K.; Melnikova, T.; Wen, H.; Chiang, H. C.; Xu, G.; Koliatsos, V. E.; Borchelt, D. R.; Price, D. L.; Lee, H. K.; Wong, P. C., BACE1, a major determinant of selective vulnerability of the brain to amyloid-beta amyloidogenesis, is essential for cognitive, emotional, and synaptic functions. *J Neurosci* **2005**, *25* (50), 11693-709.

[79] Ghosh, A. K.; Brindisi, M.; Tang, J., Developing beta-secretase inhibitors for treatment of Alzheimer's disease. *J Neurochem* **2012**, *120 Suppl 1*, 71-83.

[80] Silvestri, R., Boom in the development of non-peptidic beta-secretase (BACE1) inhibitors for the treatment of Alzheimer's disease. *Med Res Rev* **2009**, *29* (2), 295-338.

[81] Klaver, D. W.; Wilce, M. C.; Cui, H.; Hung, A. C.; Gasperini, R.; Foa, L.; Small, D. H., Is BACE1 a suitable therapeutic target for the treatment of Alzheimer's disease? Current strategies and future directions. *Biol Chem* **2010**, *391* (8), 849-59.

[82] Jin, S.; Agerman, K.; Kolmodin, K.; Gustafsson, E.; Dahlqvist, C.; Jureus, A.; Liu, G.; Falting, J.; Berg, S.; Lundkvist, J.; Lendahl, U., Evidence for dimeric BACE-mediated APP processing. *Biochem Biophys Res Commun* **2010**, *393* (1), 21-7.

[83] Westmeyer, G. G.; Willem, M.; Lichtenthaler, S. F.; Lurman, G.; Multhaup, G.; Assfalg-Machleidt, I.; Reiss, K.; Saftig, P.; Haass, C., Dimerization of beta-site beta-amyloid precursor protein-cleaving enzyme. *J Biol Chem* **2004**, *279* (51), 53205-12.

[84] Chang, W. P.; Koelsch, G.; Wong, S.; Downs, D.; Da, H.; Weerasena, V.; Gordon, B.; Devasamudram, T.; Bilcer, G.; Ghosh, A. K.; Tang, J., *In vivo* inhibition of Abeta production by memapsin 2 (beta-secretase) inhibitors. *J Neurochem* **2004**, *89* (6), 1409-16.

[85] Yu, Y. J.; Zhang, Y.; Kenrick, M.; Hoyte, K.; Luk, W.; Lu, Y.; Atwal, J.; Elliott, J. M.; Prabhu, S.; Watts, R. J.; Dennis, M. S., Boosting brain uptake of a therapeutic antibody by reducing its affinity for a transcytosis target. *Sci Transl Med* **2011**, *3* (84), 84ra44.

[86] Atwal, J. K.; Chen, Y.; Chiu, C.; Mortensen, D. L.; Meilandt, W. J.; Liu, Y.; Heise, C. E.; Hoyte, K.; Luk, W.; Lu, Y.; Peng, K.; Wu, P.; Rouge, L.; Zhang, Y.; Lazarus, R. A.; Scearce-Levie, K.; Wang, W.; Wu, Y.; Tessier-Lavigne, M.; Watts, R. J., A therapeutic antibody targeting BACE1 inhibits amyloid-beta production *in vivo*. *Sci Transl Med* **2011**, *3* (84), 84ra43.

[87]   Capell, A.; Steiner, H.; Willem, M.; Kaiser, H.; Meyer, C.; Walter, J.; Lammich, S.; Multhaup, G.; Haass, C., Maturation and pro-peptide cleavage of beta-secretase. *J Biol Chem* **2000**, *275* (40), 30849-54.

[88]   Hong, L.; He, X.; Huang, X.; Chang, W.; Tang, J., Structural features of human memapsin 2 (beta-secretase) and their biological and pathological implications. *Acta Biochim Biophys Sin (Shanghai)* **2004**, *36* (12), 787-92.

[89]   Toulokhonova, L.; Metzler, W. J.; Witmer, M. R.; Copeland, R. A.; Marcinkeviciene, J., Kinetic studies on beta-site amyloid precursor protein-cleaving enzyme (BACE). Confirmation of an iso mechanism. *J Biol Chem* **2003**, *278* (7), 4582-9.

[90]   Multhaup, G., Amyloid precursor protein and BACE function as oligomers. *Neurodegener Dis* **2006**, *3* (4-5), 270-4.

[91]   Franberg, J.; Welander, H.; Aoki, M.; Winblad, B.; Tjernberg, L. O.; Frykman, S., Rat brain gamma-secretase activity is highly influenced by detergents. *Biochemistry* **2007**, *46* (25), 7647-54.

[92]   McLendon, C.; Xin, T.; Ziani-Cherif, C.; Murphy, M. P.; Findlay, K. A.; Lewis, P. A.; Pinnix, I.; Sambamurti, K.; Wang, R.; Fauq, A.; Golde, T. E., Cell-free assays for gamma-secretase activity. *FASEB J* **2000**, *14* (15), 2383-6.

[93]   Yonemura, Y.; Futai, E.; Yagishita, S.; Suo, S.; Tomita, T.; Iwatsubo, T.; Ishiura, S., Comparison of presenilin 1 and presenilin 2 gamma-secretase activities using a yeast reconstitution system. *J Biol Chem* **2011**, *286* (52), 44569-75.

[94]   Hussain, I.; Hawkins, J.; Shikotra, A.; Riddell, D. R.; Faller, A.; Dingwall, C., Characterization of the ectodomain shedding of the beta-site amyloid precursor protein-cleaving enzyme 1 (BACE1). *J Biol Chem* **2003**, *278* (38), 36264-8.

[95]   Murayama, K. S.; Kametani, F.; Araki, W., Extracellular release of BACE1 holoproteins from human neuronal cells. *Biochem Biophys Res Commun* **2005**, *338* (2), 800-7.

[96]   Benjannet, S.; Elagoz, A.; Wickham, L.; Mamarbachi, M.; Munzer, J. S.; Basak, A.; Lazure, C.; Cromlish, J. A.; Sisodia, S.; Checler, F.; Chretien, M.; Seidah, N. G., Post-translational processing of beta-secretase (beta-amyloid-converting enzyme) and its ectodomain shedding. The pro- and transmembrane/cytosolic domains affect its cellular activity and amyloid-beta production. *J Biol Chem* **2001**, *276* (14), 10879-87.

[97]   O'Connor, T.; Sadleir, K. R.; Maus, E.; Velliquette, R. A.; Zhao, J.; Cole, S. L.; Eimer, W. A.; Hitt, B.; Bembinster, L. A.; Lammich, S.; Lichtenthaler, S. F.; Hebert, S. S.; De Strooper, B.; Haass, C.; Bennett, D. A.; Vassar, R., Phosphorylation of the translation initiation factor eIF2alpha increases BACE1 levels and promotes amyloidogenesis. *Neuron* **2008**, *60* (6), 988-1009.

[98]   Walker, K. R.; Kang, E. L.; Whalen, M. J.; Shen, Y.; Tesco, G., Depletion of GGA1 and GGA3 mediates postinjury elevation of BACE1. *J Neurosci* **2012**, *32* (30), 10423-37.

[99]   Wahle, T.; Prager, K.; Raffler, N.; Haass, C.; Famulok, M.; Walter, J., GGA proteins regulate retrograde transport of BACE1 from endosomes to the trans-Golgi network. *Mol Cell Neurosci* **2005**, *29* (3), 453-61.

[100]   Kang, E. L.; Cameron, A. N.; Piazza, F.; Walker, K. R.; Tesco, G., Ubiquitin regulates GGA3-mediated degradation of BACE1. *J Biol Chem* **2010**, *285* (31), 24108-19.

[101]   Sannerud, R.; Declerck, I.; Peric, A.; Raemaekers, T.; Menendez, G.; Zhou, L.; Veerle, B.; Coen, K.; Munck, S.; De Strooper, B.; Schiavo, G.; Annaert, W., ADP ribosylation factor 6 (ARF6) controls amyloid precursor protein (APP) processing by mediating the endosomal sorting of BACE1. *Proc Natl Acad Sci U S A* **2011**, *108* (34), E559-68.

[102] Okada, H.; Zhang, W.; Peterhoff, C.; Hwang, J. C.; Nixon, R. A.; Ryu, S. H.; Kim, T. W., Proteomic identification of sorting nexin 6 as a negative regulator of BACE1-mediated APP processing. *FASEB J* **2010**, *24* (8), 2783-94.

[103] Zhao, Y.; Wang, Y.; Yang, J.; Wang, X.; Zhang, X.; Zhang, Y. W., Sorting nexin 12 interacts with BACE1 and regulates BACE1-mediated APP processing. *Mol Neurodegener* **2012**, *7*, 30.

[104] Kang, E. L.; Biscaro, B.; Piazza, F.; Tesco, G., BACE1 Protein Endocytosis and Trafficking Are Differentially Regulated by Ubiquitination at Lysine 501 and the Di-leucine Motif in the Carboxyl Terminus. *J Biol Chem* **2012**, *287* (51), 42867-80.

[105] Van Dam, D.; De Deyn, P. P., Animal models in the drug discovery pipeline for Alzheimer's disease. *Br J Pharmacol* **2011**, *164* (4), 1285-300.

[106] Selkoe, D. J., Preventing Alzheimer's disease. *Science* **2012**, *337* (6101), 1488-92.

[107] Reiman, E. M.; Brinton, R. D.; Katz, R.; Petersen, R. C.; Negash, S.; Mungas, D.; Aisen, P. S., Considerations in the design of clinical trials for cognitive aging. *J Gerontol A Biol Sci Med Sci* **2012**, *67* (7), 766-72.

[108] Albert, M. S.; DeKosky, S. T.; Dickson, D.; Dubois, B.; Feldman, H. H.; Fox, N. C.; Gamst, A.; Holtzman, D. M.; Jagust, W. J.; Petersen, R. C.; Snyder, P. J.; Carrillo, M. C.; Thies, B.; Phelps, C. H., The diagnosis of mild cognitive impairment due to Alzheimer's disease: recommendations from the National Institute on Aging-Alzheimer's Association workgroups on diagnostic guidelines for Alzheimer's disease. *Alzheimers Dement* **2011**, *7* (3), 270-9.

[109] Tosun, D.; Schuff, N.; Truran-Sacrey, D.; Shaw, L. M.; Trojanowski, J. Q.; Aisen, P.; Peterson, R.; Weiner, M. W., Relations between brain tissue loss, CSF biomarkers, and the ApoE genetic profile: a longitudinal MRI study. *Neurobiol Aging* **2010**, *31* (8), 1340-54.

[110] Sawa, A., Neuronal cell death in Down's syndrome. *J Neural Transm Suppl* **1999**, *57*, 87-97.

[111] Humpel, C., Identifying and validating biomarkers for Alzheimer's disease. *Trends Biotechnol* **2011**, *29* (1), 26-32.

[112] Hampel, H.; Frank, R.; Broich, K.; Teipel, S. J.; Katz, R. G.; Hardy, J.; Herholz, K.; Bokde, A. L.; Jessen, F.; Hoessler, Y. C.; Sanhai, W. R.; Zetterberg, H.; Woodcock, J.; Blennow, K., Biomarkers for Alzheimer's disease: academic, industry and regulatory perspectives. *Nat Rev Drug Discov* **2010**, *9* (7), 560-74.

[113] Shaw, L. M.; Vanderstichele, H.; Knapik-Czajka, M.; Clark, C. M.; Aisen, P. S.; Petersen, R. C.; Blennow, K.; Soares, H.; Simon, A.; Lewczuk, P.; Dean, R.; Siemers, E.; Potter, W.; Lee, V. M.; Trojanowski, J. Q., Cerebrospinal fluid biomarker signature in Alzheimer's disease neuroimaging initiative subjects. *Ann Neurol* **2009**, *65* (4), 403-13.

[114] Gonzales, A.; Decourt, B.; Walker, A.; Condjella, R.; Nural, H.; Sabbagh, M. N., Development of a specific ELISA to measure BACE1 levels in human tissues. *J Neurosci Methods* **2011**, *202* (1), 70-6.

[115] Zhong, Z.; Ewers, M.; Teipel, S.; Burger, K.; Wallin, A.; Blennow, K.; He, P.; McAllister, C.; Hampel, H.; Shen, Y., Levels of beta-secretase (BACE1) in cerebrospinal fluid as a predictor of risk in mild cognitive impairment. *Arch Gen Psychiatry* **2007**, *64* (6), 718-26.

[116] Decourt, B.; Sabbagh, M. N., BACE1 as a potential biomarker for Alzheimer's disease. *J Alzheimers Dis* **2011**, *24 Suppl 2*, 53-9.

[117] Stellos, K.; Panagiota, V.; Kogel, A.; Leyhe, T.; Gawaz, M.; Laske, C., Predictive value of platelet activation for the rate of cognitive decline in Alzheimer's disease patients. *J Cereb Blood Flow Metab* **2010**, *30* (11), 1817-20.

[118] Irizarry, M. C., Biomarkers of Alzheimer disease in plasma. *NeuroRx* **2004**, *1* (2), 226-34.

[119] Oh, E. S.; Troncoso, J. C.; Fangmark Tucker, S. M., Maximizing the potential of plasma amyloid-beta as a diagnostic biomarker for Alzheimer's disease. *Neuromolecular Med* **2008**, *10* (3), 195-207.

[120] Decourt, B.; Walker, A.; Gonzales, A.; Malek-Ahmadi, M.; Liesback, C.; Davis, K. J.; Belden, C. M.; Jacobson, S. A.; Sabbagh, M. N., Can platelet BACE1 levels be used as a biomarker for Alzheimer's disease? Proof-of-concept study. *Platelets* **2012**.

[121] Yang, X.; Chen, C.; Hu, Q.; Yan, J.; Zhou, C., Gamma-secretase inhibitor (GSI1) attenuates morphological cerebral vasospasm in 24h after experimental subarachnoid hemorrhage in rats. *Neurosci Lett* **2010**, *469* (3), 385-90.

[122] Mitani, Y.; Yarimizu, J.; Saita, K.; Uchino, H.; Akashiba, H.; Shitaka, Y.; Ni, K.; Matsuoka, N., Differential effects between gamma-secretase inhibitors and modulators on cognitive function in amyloid precursor protein-transgenic and nontransgenic mice. *J Neurosci* **2012**, *32* (6), 2037-50.

[123] Boche, D.; Zotova, E.; Weller, R. O.; Love, S.; Neal, J. W.; Pickering, R. M.; Wilkinson, D.; Holmes, C.; Nicoll, J. A., Consequence of Abeta immunization on the vasculature of human Alzheimer's disease brain. *Brain* **2008**, *131* (Pt 12), 3299-310.

[124] Atwood, C. S.; Bowen, R. L.; Smith, M. A.; Perry, G., Cerebrovascular requirement for sealant, anti-coagulant and remodeling molecules that allow for the maintenance of vascular integrity and blood supply. *Brain Res Brain Res Rev* **2003**, *43* (1), 164-78.

[125] Kokjohn, T. A.; Maarouf, C. L.; Roher, A. E., Is Alzheimer's disease amyloidosis the result of a repair mechanism gone astray? *Alzheimers Dement* **2012**, *8* (6), 574-83.

[126] Kuhn, P. H.; Koroniak, K.; Hogl, S.; Colombo, A.; Zeitschel, U.; Willem, M.; Volbracht, C.; Schepers, U.; Imhof, A.; Hoffmeister, A.; Haass, C.; Rossner, S.; Brase, S.; Lichtenthaler, S. F., Secretome protein enrichment identifies physiological BACE1 protease substrates in neurons. *EMBO J* **2012**, *31* (14), 3157-68.

[127] ClinicalTrials.gov : Safety Study of CTS21166 to Treat Alzheimer Disease. http://clinicaltrials.gov/ct2/show/NCT00621010 (accessed September 25, 2012).

[128] ClinicalTrials.gov: A Safety Study of LY2811376 Single Doses in Healthy Subjects. http://clinicaltrials.gov/ct2/show/NCT00838084 (accessed September 25, 2012).

[129] May, P. C.; Dean, R. A.; Lowe, S. L.; Martenyi, F.; Sheehan, S. M.; Boggs, L. N.; Monk, S. A.; Mathes, B. M.; Mergott, D. J.; Watson, B. M.; Stout, S. L.; Timm, D. E.; Smith Labell, E.; Gonzales, C. R.; Nakano, M.; Jhee, S. S.; Yen, M.; Ereshefsky, L.; Lindstrom, T. D.; Calligaro, D. O.; Cocke, P. J.; Greg Hall, D.; Friedrich, S.; Citron, M.; Audia, J. E., Robust central reduction of amyloid-beta in humans with an orally available, non-peptidic beta-secretase inhibitor. *J Neurosci* **2011**, *31* (46), 16507-16.

[130] May, P.; Boggs, L.; Brier, R.; Calligaro, D.; Citron, M.; Day, T.; Lin, S.; Lindstrom, T.; Mergott, D.; Monk, S.; Sanchez-Felix, M. V.; Sheehan, S.; Vaught, G.; Yang, Z.; Audia, J., Preclinical characterization of LY2886721: A BACE1 inhibitor in clinical development for early Alzheimer's disease. *Alzheimers Dement* **2012**, *8* (4), P95.

[131] Martenyi, F.; Dean, R. A.; Lowe, S.; Nakano, M.; Monk, S.; Willis, B. A.; Gonzales, C.; Mergott, D.; Leslie, D.; May, P.; James, A.; Gevorkyan, H.; Jhee, S.; Ereshefsky, L.;

Citron, M., BACE inhibitor LY2886721 safety and central and peripheral PK and PD in healthy subjects (HSs). *Alzheimers Dement* **2012**, *8* (4), P583-P584.

[132]   Willis, B.; Martenyi, F.; Dean, R.; Lowe, S.; Nakano, M.; Monk, S.; Gonzales, C.; Mergott, D.; Daugherty, L.; Citron, M.; May, P., Central BACE1 inhibition by LY2886721 produces opposing effects on APP processing as reflected by cerebrospinal fluid sAPP-alpha and sAPP-beta. *Alzheimers Dement* **2012**, *8* (4), P95.

[133]   Min, K.; Forman, M.; Dockendorf, M.; Palcza, J.; Soni, P.; Ma, L.; Krishna, G.; Hodsman, P.; Masuo, K.; Tanen, M.; Wagner, J.; Troyer, M., A study to evaluate the pharmacokinetics and pharmacodynamics of single and multiple oral doses of the novel BACE inhibitor MK-8931 in Japanese subjects. *Alzheimers Dement* **2012**, *8* (4), P186.

[134]   Tseng, J.; Dockendorf, M.; Krishna, G.; Ma, L.; Palcza, J.; Leempoels, J.; Ramael, S.; Han, D.; Jhee, S.; Ereshefsky, L.; Wagner, J.; Troyer, M.; Forman, M., Safety and pharmacokinetics of the novel BACE inhibitor MK-8931 in healthy subjects following single- and Multiple-Dose administration. *Alzheimers Dement* **2012**, *8* (4), P184-P185.

[135]   Lai, R.; Albala, B.; Kaplow, J. M.; Aluri, J.; Yen, M.; Satlin, A., First-in-human study of E2609, a novel BACE1 inhibitor, demonstrates prolonged reductions in plasma beta-amyloid levels after single dosing. *Alzheimers Dement* **2012**, *8* (4), P96.

[136]   ClinicalTrials.gov: A Safety Study of LY2886721 Single Doses in Healthy Subjects. http://www.clinicaltrial.gov/ct2/show/NCT01133405 (accessed September 25, 2012).

[137]   ClinicalTrials.gov: A Safety Study of LY2886721 Multiple Doses in Healthy Subjects. http://www.clinicaltrial.gov/ct2/show/NCT01227252 (accessed September 25, 2012).

[138]   ClinicalTrials.gov: Disposition of 14C-LY2886721 Following Oral Administration in Healthy Human Subjects. http://www.clinicaltrial.gov/ct2/show/NCT01367262 (accessed September 25, 2012).

[139]   ClinicalTrials.gov: A Study of LY2886721 in Healthy Participants. http://www.clinicaltrial.gov/ct2/show/NCT01534273 (accessed September 25, 2012).

[140]   ClinicalTrials.gov: Study of LY2886721 in Mild Cognitive Impairment Due to Alzheimer's Disease or Mild Alzheimer's Disease. http://clinicaltrials.gov/ct2/show/NCT01561430 (accessed September 25, 2012).

[141]   ClinicalTrials.gov : Safety Study of HPP854 in Subjects With Mild Cognitive Impairment or a Diagnosis of Mild Alzheimer's Disease. http://www.clinicaltrial.gov/ct2/show/NCT01482013 (accessed September 25, 2012).

[142]   ClinicalTrials.gov: A Study of the Safety, Tolerability, and Pharmacodynamics of MK-8931 in Participants With Alzheimer's Disease (MK-8931-010 AM1 [P07820 AM1]). http://www.clinicaltrial.gov/ct2/show/NCT01496170 (accessed September 25, 2012).

[143]   ClinicalTrials.gov: Evaluation of the Safety, Tolerability, Pharmacokinetics, and Pharmacodynamics of E2609 in Healthy Subjects and an Elderly Cohort. http://www.clinicaltrial.gov/ct2/show/NCT01294540 (accessed September 25, 2012)).

[144]   ClinicalTrials.gov: Evaluation of the Safety, Pharmacokinetics, and Pharmacodynamics of Multiple Doses of E2609 in Healthy Subjects. http://www.clinicaltrial.gov/ct2/show/NCT01511783 (accessed September 25, 2012).

[145]   ClinicalTrials.gov: Evaluation of E2609 in Subjects With Mild Cognitive Impairment Due to Alzheimer's Disease. http://www.clinicaltrial.gov/ct2/show/NCT01600859 (accessed September 25, 2012).

[146]   ClinicalTrials.gov: Minocycline in Patients With Alzheimer's Disease. http://www.clinicaltrial.gov/ct2/show/NCT01463384 (accessed September 25, 2012)).

[147] ClinicalTrials.gov: Thalidomide for Patients With Mild to Moderate Alzheimer's Disease. http://www.clinicaltrial.gov/ct2/show/NCT01094340 (accessed September 25, 2012).

[148] He, P.; Zhong, Z.; Lindholm, K.; Berning, L.; Lee, W.; Lemere, C.; Staufenbiel, M.; Li, R.; Shen, Y., Deletion of tumor necrosis factor death receptor inhibits amyloid beta generation and prevents learning and memory deficits in Alzheimer's mice. *J Cell Biol* **2007**, *178* (5), 829-41.

[149] Palumbo, A.; Freeman, J.; Weiss, L.; Fenaux, P., The clinical safety of lenalidomide in multiple myeloma and myelodysplastic syndromes. *Expert Opin Drug Saf* **2012**, *11* (1), 107-20.

[150] Choi, Y.; Kim, H. S.; Shin, K. Y.; Kim, E. M.; Kim, M.; Park, C. H.; Jeong, Y. H.; Yoo, J.; Lee, J. P.; Chang, K. A.; Kim, S.; Suh, Y. H., Minocycline attenuates neuronal cell death and improves cognitive impairment in Alzheimer's disease models. *Neuropsychopharmacology* **2007**, *32* (11), 2393-404.

[151] Ferretti, M. T.; Allard, S.; Partridge, V.; Ducatenzeiler, A.; Cuello, A. C., Minocycline corrects early, pre-plaque neuroinflammation and inhibits BACE-1 in a transgenic model of Alzheimer's disease-like amyloid pathology. *J Neuroinflammation* **2012**, *9*, 62.

[152] Noble, W.; Garwood, C.; Stephenson, J.; Kinsey, A. M.; Hanger, D. P.; Anderton, B. H., Minocycline reduces the development of abnormal tau species in models of Alzheimer's disease. *FASEB J* **2009**, *23* (3), 739-50.

[153] Chang, W. P.; Downs, D.; Huang, X. P.; Da, H.; Fung, K. M.; Tang, J., Amyloid-beta reduction by memapsin 2 (beta-secretase) immunization. *FASEB J* **2007**, *21* (12), 3184-96.

[154] Nawrot, B., Targeting BACE with small inhibitory nucleic acids - a future for Alzheimer's disease therapy? *Acta Biochim Pol* **2004**, *51* (2), 431-44.

[155] Singer, O.; Marr, R. A.; Rockenstein, E.; Crews, L.; Coufal, N. G.; Gage, F. H.; Verma, I. M.; Masliah, E., Targeting BACE1 with siRNAs ameliorates Alzheimer disease neuropathology in a transgenic model. *Nat Neurosci* **2005**, *8* (10), 1343-9.

[156] Peng, K. A.; Masliah, E., Lentivirus-expressed siRNA vectors against Alzheimer disease. *Methods Mol Biol* **2010**, *614*, 215-24.

[157] Alvarez-Erviti, L.; Seow, Y.; Yin, H.; Betts, C.; Lakhal, S.; Wood, M. J., Delivery of siRNA to the mouse brain by systemic injection of targeted exosomes. *Nat Biotechnol* **2011**, *29* (4), 341-5.

[158] Gabbita, S. P.; Srivastava, M. K.; Eslami, P.; Johnson, M. F.; Kobritz, N. K.; Tweedie, D.; Greig, N. H.; Zemlan, F. P.; Sharma, S. P.; Harris-White, M. E., Early intervention with a small molecule inhibitor for tumor nefosis factor-alpha prevents cognitive deficits in a triple transgenic mouse model of Alzheimer's disease. *J Neuroinflammation* **2012**, *9*, 99.

[159] Tweedie, D.; Ferguson, R. A.; Fishman, K.; Frankola, K. A.; Van Praag, H.; Holloway, H. W.; Luo, W.; Li, Y.; Caracciolo, L.; Russo, I.; Barlati, S.; Ray, B.; Lahiri, D. K.; Bosetti, F.; Greig, N. H.; Rosi, S., Tumor necrosis factor-alpha synthesis inhibitor 3,6'-dithiothalidomide attenuates markers of inflammation, Alzheimer pathology and behavioral deficits in animal models of neuroinflammation and Alzheimer's disease. *J Neuroinflammation* **2012**, *9*, 106.

[160] Sarajarvi, T.; Lipsanen, A.; Makinen, P.; Peraniemi, S.; Soininen, H.; Haapasalo, A.; Jolkkonen, J.; Hiltunen, M., Bepridil decreases Abeta and calcium levels in the thalamus after middle cerebral artery occlusion in rats. *J Cell Mol Med* **2012**, *16* (11), 2754-67.

[161] Mitterreiter, S.; Page, R. M.; Kamp, F.; Hopson, J.; Winkler, E.; Ha, H. R.; Hamid, R.; Herms, J.; Mayer, T. U.; Nelson, D. J.; Steiner, H.; Stahl, T.; Zeitschel, U.; Rossner, S.;

Haass, C.; Lichtenthaler, S. F., Bepridil and amiodarone simultaneously target the Alzheimer's disease beta- and gamma-secretase *via* distinct mechanisms. *J Neurosci* **2010**, *30* (26), 8974-83.

[162] Cumming, J.; Babu, S.; Huang, Y.; Carrol, C.; Chen, X.; Favreau, L.; Greenlee, W.; Guo, T.; Kennedy, M.; Kuvelkar, R.; Le, T.; Li, G.; McHugh, N.; Orth, P.; Ozgur, L.; Parker, E.; Saionz, K.; Stamford, A.; Strickland, C.; Tadesse, D.; Voigt, J.; Zhang, L.; Zhang, Q., Piperazine sulfonamide BACE1 inhibitors: design, synthesis, and *in vivo* characterization. *Bioorg Med Chem Lett* **2010**, *20* (9), 2837-42.

[163] Boche, D.; Nicoll, J. A., The role of the immune system in clearance of Abeta from the brain. *Brain Pathol* **2008**, *18* (2), 267-78.

[164] DeMattos, R. B.; Bales, K. R.; Cummins, D. J.; Dodart, J. C.; Paul, S. M.; Holtzman, D. M., Peripheral anti-A beta antibody alters CNS and plasma A beta clearance and decreases brain A beta burden in a mouse model of Alzheimer's disease. *Proc Natl Acad Sci U S A* **2001**, *98* (15), 8850-5.

CHAPTER 11

Combining BACE1 Inhibition with Metal Chelation as Possible Therapy for Alzheimer's Disease

Maged M. Henary[1,2,3,*], **Tyler L. Dost,**[1] **Eric A. Owens**[1,2] **and Surendra Reddy Punganuru**[1]

Department of Chemistry[1]*, Center for Diagnostics and Therapeutics*[2]*, Center for Biotechnology and Drug Design*[3]*, Georgia State University, Atlanta, Georgia, 30303, USA*

Abstract: A complex disease such as Alzheimer's requires an arsenal of therapies to help combat and stabilize its advancement. Despite considerable scientific progress, current therapeutic approaches for Alzheimer's treatment offer only limited and transient benefits to patients. Therefore, in response to the molecular complexity of this disease, a new strategy has recently emerged aimed at simultaneously targeting multiple pathological processes involved in the pathogenesis cascade. β-secretase plays a critical role in β-amyloid formation, a major constituent of amyloid plaques in Alzheimer's disease (AD) brain and is likely to play a central role in the pathogenesis of this devastating neurodegenerative disorder. Thus, β-secretase is a prime drug target for the therapeutic inhibition of β-amyloid production in AD. It has been clearly established in a number of studies that metal ions are critically involved in the pathogenesis and progression of major neurological diseases (Alzheimer's, Parkinson's). Metal ion chelators have been suggested as potential therapies for diseases involving metal ion imbalance. This chapter summarizes the current therapeutic strategies based on the β-secretase inhibition, metal chelation and their combination to treat AD.

Keywords: Aβ aggregation, Alzheimer's disease, amino imidazoles, amyloid beta precursor protein, amyloid plaque, Bi-functional compounds, clinical development, cyclic hydroxyethylamines, dementia, fragment-based lead generation, high-throughput screening, metal chelators, multi-targeted drugs, neurodegenerative disorders, oxidative stress, peptidomimetics, reactive oxygen species, substrate-based design, β-secretase (BACE1), γ-secretase.

INTRODUCTION

There are numerous new cases of dementia each year around the world and the total number of these cases is approximately 7.7 million, with almost one new

---

*Address correspondence to Maged M. Henary: Department of Chemistry, Georgia State University, 100 Piedmont Ave. #315, Atlanta, GA 30303, USA; Tel: 404-413-5566; Fax: 404-413-5505; E-mail: mhenary1@gsu.edu

case every 4 seconds. Progressing age is the strongest risk factor for Alzheimer's disease, with age-specific occurrence nearly doubling every five years beyond the age of 60. The total estimated global expenditure of dementia was $604 billion in 2010 [1].

ALZHEIMER'S DISEASE

Alzheimer's disease (AD) is a neurodegenerative disease that has many signs relating to basic dementia and is increasingly becoming more predominant due to increasing average age of the global population. Stress and aging of the brain causes oxidative stress and DNA/RNA oxidation, which leads to neurotoxicity in the brain relating to Alzheimer's disease.

The exact mechanisms leading to Alzheimer disease are not completely understood, although recent studies suggest that changes in the processing and clearance of proteolytic products, such as C-terminal fragments (CTFs) and amyloid-β peptides (Aβ), of the amyloid precursor protein (APP) have a key role in the pathogenesis of Alzheimer's disease [2, 3]. The accumulation of Aβ also causes oxidative stress and will speed up the process of neurotoxicity; it also modifies the calcium that is present in the brain, which leads to tau polymerization, which is another factor leading to neurotoxicity [4]. Although the tau tangles and oxidative stress definitely lead to the disease, the Aβ protein production regulation and control of the APP cleavage are the main concerns for scientists to prevent the progression of this disease.

Cleaving of APP

The amyloid-beta precursor protein is cleaved through a process called proteolysis [5], shown below in Fig. **1**. This reaction is very common in biochemistry.

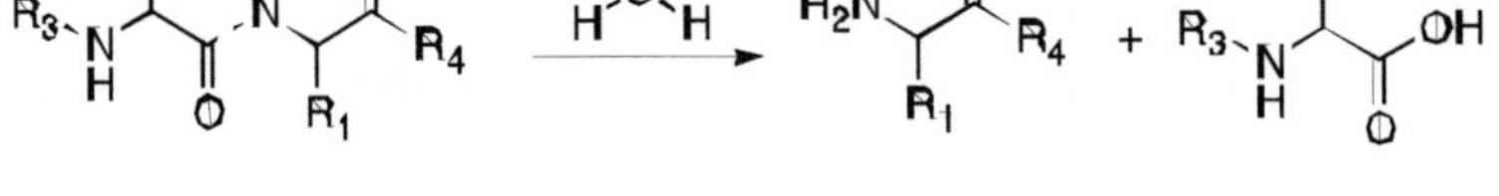

Figure 1: Proteolysis: the cleavage of Aβ protein.

When the protein comes into contact with water, a chemical reaction occurs that cleaves the protein into two pieces. The stripped bond is the scissile bond where the water splits the peptide, as shown by the dashed bond. The water attacks the inner carbonyl, forming a tetrahedral intermediate. Then a hydrogen-transfer between the water and the neighboring amine occurs, making the amine a good leaving group. This prompts the electrons from the negatively charged oxygen to collapse down and force the amine off the structure, separating the protein into two pieces.

As shown in Fig. **2** that the amyloid precursor protein is a membrane protein that passes from the exterior carboxyl-group to the interior amine functionality. Through the beta pathway of cleaving, the beta-secretase (BACE1) cleaves the part of the protein outside of the membrane, creating the beta-APP. Next, the gamma-secretase cleaves the carboxyl part of the protein, leaving just the amyloid beta precursor protein (APP), 1-42 [6].

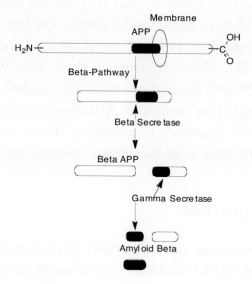

Figure 2: Cleavage pathway of APP.

BETA SECRETASE

BACE1, a solo transmembrane aspartyl-protease, was identified as the major β-secretase-like protein in 1999 [7, 8]. As previously described, the key molecule in the pathogenesis of Alzheimer's disease is the β amyloid peptide, which in either its soluble oligomeric form or in the plaque associated version leads to neurodegeneration.

By sequential proteolytic processing using β- and γ-secretases, Aβ is liberated from the membrane-spanning APP (Fig. **2**). β-Secretase activity is conferred by a transmembrane aspartyl protease, also termed BACE1 which catalyzes the rate-limiting reaction in the generation of Aβ [9, 10]. The APP cleavage by β-secretase occurs mostly in endosomes, and endocytosis of APP and β-secretase is essential for β cleavage and Aβ production [11, 12]. On the other hand, APP cleavage by α-secretase, which precludes production of the toxic Aβ peptide, happens at the plasma membrane [13]. Thus, both β- and γ-secretases are promising therapeutic targets for AD [14, 15]. γ-Secretase cleavage is less specific and produces major fragments such as 40 ($A\beta_{1-40}$), 42 ($A\beta_{1-42}$) or 43 ($A\beta_{1-43}$) amino acids contrasting to the BACE1 that specifically cleaves APP [16, 17]. Additionally, genetic ablation of γ-secretase is embryonic lethal [18]; consequently, inhibition of this protease activity could have serious side effects. On the other hand, BACE1 activity seems to be redundant, as BACE knockout mice are viable and show no harmful side effects [3]. Moreover, crossing these mice with transgenic mice expressing human APP with familial AD mutations resulted in complete absence of amyloid pathology [19]. In contrast, in Alzheimer disease, BACE1 expression and generation of APP fragments is increased. Hence, approaches that could reduce BACE1 activity have great therapeutic potential for the treatment of AD [20].

β-SECRETASE INHIBITORS

In the last decade enormous research efforts have been directed towards the development of potent BACE1 inhibitors with an aim to identify suitable AD drug candidates. There are substantial challenges in designing low-molecular-weight, brain-penetrating compounds with potent BACE1 inhibition properties. The crystal structure of BACE1 and its cell biology paved the route to explore several strategies of drug discovery in the search for potent BACE1 inhibitors, such as high-throughput screening (HTS), substrate-based design and fragment-based lead generation approaches [21].

Traditional high throughput screening methods have ended with only limited success, and most lead compounds have been derived by rational structure-based design of peptidomimetics [22-25] as presented in Fig. **3**.

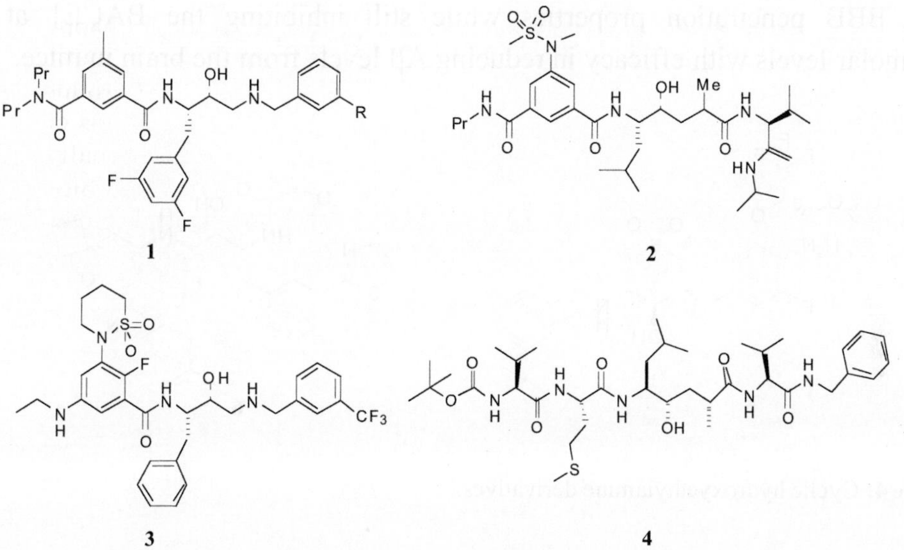

Figure 3: Compounds developed based on peptidomimetics.

Most of these compounds possess a secondary alcohol as a transition state mimic that displaces a water molecule at the active site and forms hydrogen bonding interactions with BACE1. Although some of the earlier peptidomimetic inhibitors showed potency *in vitro*, these inhibitors with multiple hydrogen bond donors and acceptors and relatively high molecular weight encountered P-glycoprotein mediated resistance to enter into the CNS [26, 27]. Consequently, most of these peptidomimetic inhibitors were not able to demonstrate significant CNS Aβ reductions in rodent pharmacodynamic models. However, compound **3** (GSK 188909) as presented in Fig. **3**, was described as an orally bioavailable β-secretase inhibitor capable of lowering brain Aβ in APP transgenic mice [28].

Recently Rueeger *et al.*, [29] reported structure-based designed cyclic hydroxyethylamine BACE1 inhibitors **5** and **6** by rational incorporation of prime- and nonprime-side fragments to a central peptidomimetic core template with and without an amide functionality (Fig. **4**). These low molecular weight inhibitors showed good BBB penetration as well as pharmacokinetic properties. Furthermore, these non-peptide compounds allowed the reduction of overall molecular weight and hydrogen bond formation ability, leading to inhibitors with good BBB penetration properties while still inhibiting the BACE1 at low nanomolar levels with efficacy in reducing Aβ levels from the brain in mice.

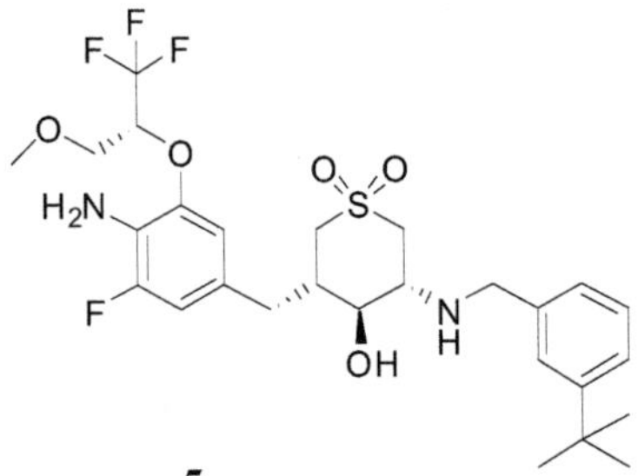

Figure 4: Cyclic hydroxyethylamine derivatives.

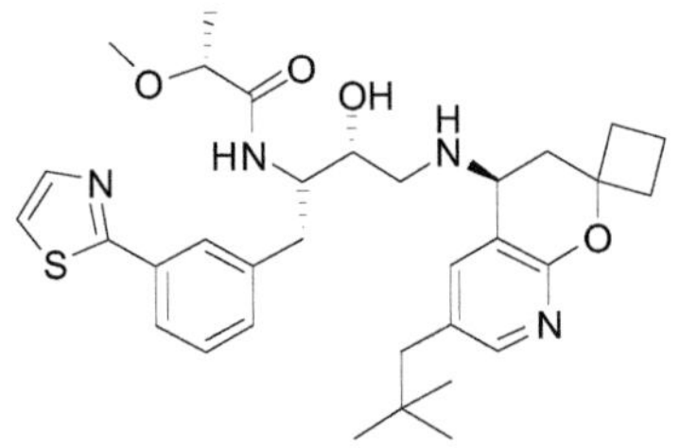

In recent years, there has been a growing interest from a number of research groups around the world in the development of non-peptidomimetic BACE1 inhibitors. Al-Tel *et al.* have developed highly selective imidazopyridines equipped with benzimidazole and/or arylimidazole as potent BACE1 inhibitors. These low molecular weight compounds demonstrated low nanomolar potency against BACE1 enzyme as measured by fluorescence resonance energy transfer and cell-based assays [30]. In addition, these potent motifs were more than 200 fold selective against the structurally similar aspartyl protease BACE2. Interestingly, a modification of the imidazopyridine **10** structure with a fluoroimidazopyridine **11** moiety resulted in an additional forty fold enhancement of the ligand's affinity (Fig. **5**). Additionally, without having peptidic nature and the less number of rotatable bonds these motifs represent a novel class of compounds to treat AD.

By using substituent-based screening of an attentive group library, Cheng *et al.* was able to identify 2-aminoquinoline **14** as an initial primary hit for BACE1 inhibition [31]. Further SAR development with the support of molecular modeling studies and X-ray structures of BACE1 co-crystallized with various ligands developed a potent compound **15**. Based on these studies they have attached a side chain on the C-3 position of the quinoline ring of **14**. This side chain extending towards deep into the P20 binding pocket of BACE1 improved the ligand's potency. Finally they were able to improve the BACE1 potency to sub-nanomolar range, over 100 fold more potent than the initial hit compound **14**. Further studies revealed that the physical properties of these lead compounds are more consistent with good blood brain barrier permeability, improved cellular activity and permeability. Compound **15** showed cellular activity at IC_{50} value of 80 nM and BACE1 of 11 nM *in vitro* against the isolated enzyme (Fig. **6**). The compound **15** was tested *in vivo* by administering to rats and demonstrated a significant reduction of Aβ levels in cerebrospinal fluid (CSF) with good pharmacokinetic profile.

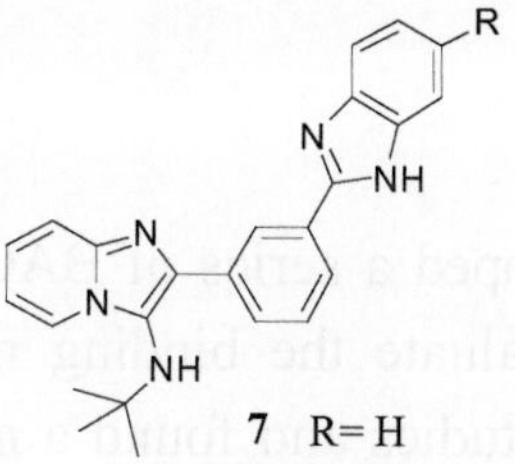

7 R= H
8 R= F
9 R= OMe

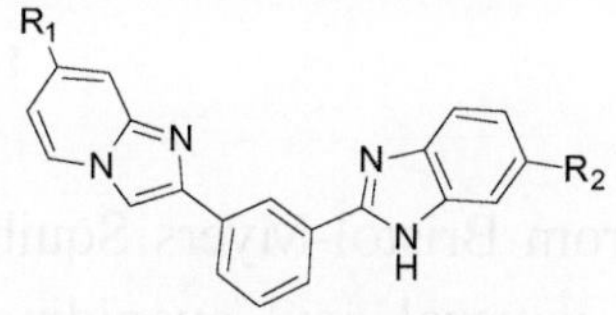

10 R₁ = H, R₂ = H
11 R₁ = H, R₂ = F
12 R₁ = H, R₂ = OMe
13 R₁ = F, R₂ = F

Figure 5: Imidazopyridine derivatives.

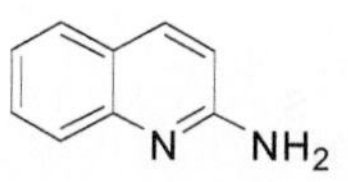

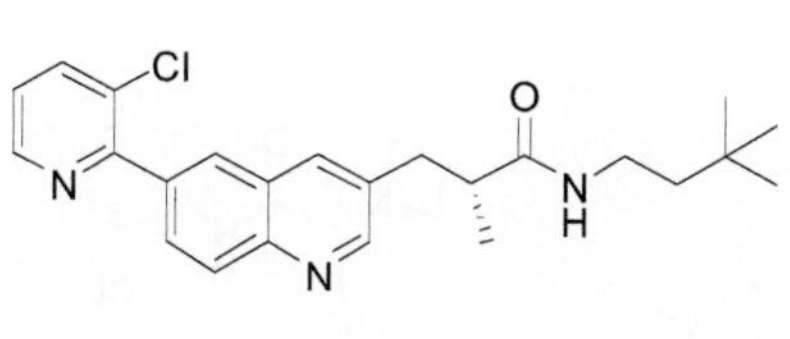

14 BACE1 Kd = 900 mM   **15** BACE1 Kd = 0.011 nM

Figure 6: Quinoline derivatives.

A structure- and property-based drug design approach was employed by Huang and co-workers to identify aminooxazoline xanthenes as potent and selective human BACE1 inhibitors [32]. After screening a library of compounds their efforts resulted in the identification of a potent, orally bioavailable and CNS penetrating compound **16**. This compound exhibited its effect against isolated BACE1 enzyme with IC_{50} of 8 nM, cellular potency at IC_{50} of 36 nM concentrations, and selectivity against the structurally similar cathepsin D aspartyl protease. Along with good cellular potency compound **16** also displays properties necessary for the desired penetration across BBB. Furthermore, they have carried out *in vivo* experiments using naive rats and these studies revealed that a single oral dose of **16** significantly reduces the CNS Aβ40.

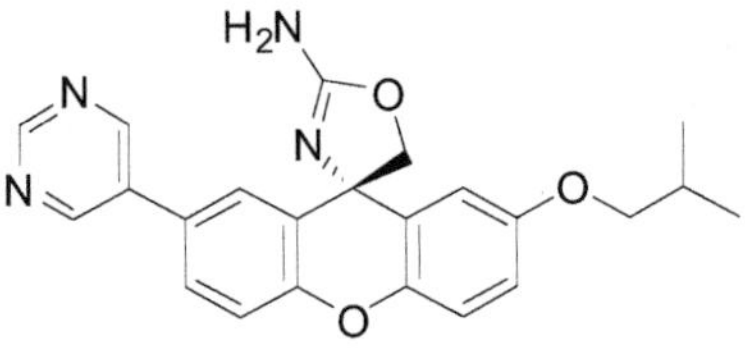

16

Researchers from Bristol-Myers Squibb developed a series of BACE1 inhibitors containing an unusual acyl guanidine. To evaluate the binding mode of these compounds they have used X-ray diffraction studies and found a number of key acyl guanidine–BACE1 interactions, such as a hydrogen bonding interactions between the side chain amide of flap residue Gln73 and the acyl guanidine carbonyl group, and a cation–π interaction between Arg235 and the isothiazole 4-methoxyphenyl substituent.

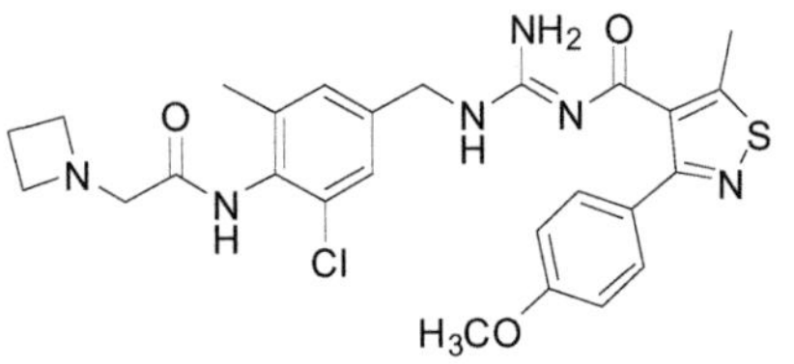

17

This acyl guanidine BACE1 inhibitor **17** exhibited its cellular activity with an IC_{50} of 5 nM. Furthermore, subcutaneous administration of **17** in rats showed good plasma exposure and a dose-dependent reduction in plasma Aβ levels.

Unfortunately, because of the Pgp-mediated resistance it exhibited poor brain exposure and significant reductions in brain Aβ levels were not obtained [33].

Gravenfors *et al.* from AstraZeneca developed Amino-2H-imidazoles **18** and **19** (Fig. 7) as a new class of BACE1 inhibitors for the treatment of Alzheimer's disease. Compound **18** was one of the most promising compounds in their report, with high potency in the cellular assay *in vitro* and a good pharmacokinetic profile *in vivo*. A concentration and time dependent decrease in Aβ40 and Aβ42 levels in plasma, brain, and CSF was observed, when guinea pigs were treated with compound **18**. A 100 μmol/kg oral dose of **18** resulted in 40-50% reduction of brain Aβ levels within 1.5 h. Based on these results amino-2H-imidazoles were evaluated as a new class of BACE1inhibitors [34, 35].

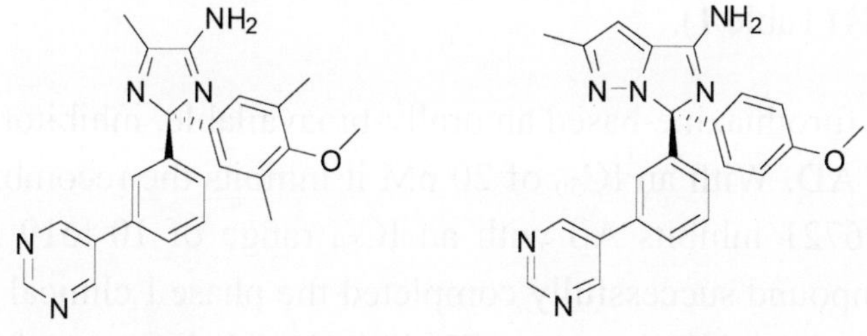

Figure 7: Amino-2H-imidazoles derivatives.

Recently, the Yonezawa group used a conformational restriction approach to develop new BACE1 inhibitors. The important key strategy in this drug discovery is the conformational restriction of sp³ hybridized carbons. By using this this strategy they have developed a four stereo isomeric cyclopropane compounds in which the ethylene linker of a known amidine-type inhibitor **20** (Fig. **8**) was replaced with chiral cyclopropane ring [36].

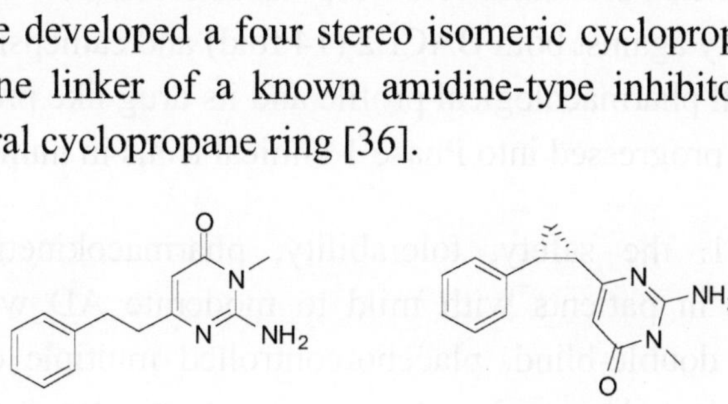

Figure 8: Stereo isomeric cyclopropane **21** based on amidine-type inhibitor **20**.

The biological evaluation of these compounds revealed that the *cis*-(1*S*,2*R*) isomer **21** exhibited the most potent BACE1 inhibitory activity among some tested compounds *in vitro*. Further they have prepared a crystalline complex of human BACE1 enzyme and **21** and analyzed with X-ray crystallography. These studies revealed its unique binding mode with BACE1 *via* CH–π interaction between the rigid cyclopropane ring and side chain of BACE1.

AGENTS IN CLINICAL DEVELOPMENT

A decade ago, BACE was considered as a very difficult and/or inflexible target, but now multiple companies, including AstraZeneca, Lilly, Merck and Takeda, have used different approaches to find molecules that inhibit BACE1. The compounds developed by Merck, AstraZeneca and Lilly have already advanced into clinical trials (Table **1**).

LY2886721 is a furothiazine-based an orally-bioavailable, inhibitor of BACE1 for the treatment of AD. With an IC_{50} of 20 nM it inhibits the recombinant hBACE1 *in vitro*. LY2886721 inhibits Aβ with an IC_{50} range of 10 to19 nM in cellular assays. This compound successfully completed the phase I clinical trial but it was unfortunately terminated during phase II clinical trials because of a liver toxicity [37, 38].

AstraZeneca Co. has identified 3-amino indene derivative AZD3839 by using fragment-based screening and structure-based design. AZD3839 exhibited the reduction of Aβ levels in mice and guinea pigs in a dose- and time-dependent manner from the plasma, brain and cerebrospinal fluid. Moreover, it demonstrated remarkable selectivity against both BACE2 (14 fold) and cathepsin D (>1000 fold). Based on this overall pharmacological profile and its drug like properties, currently AZD3839 has been progressed into Phase 1 clinical trials in humans [39].

Regarding MK8931, the safety, tolerability, pharmacokinetics and pharma-codynamic profiles in patients with mild to moderate AD were evaluated by using randomized, double-blind, placebo-controlled multiple dose study. Trial participants were selected at random to receive a particular dosage of MK-8931 medication (12, 40 and 60 mg) orally once-daily for seven days [40]. In order to monitor the efficacy of the medication, the levels of Aβ40 and Aβ42 and soluble Aβ (sAPPb) in the CSF were measured as these are known biomarkers for BACE1 activity. These studies revealed that MK-8931 exhibited a reduction in Aβ40, and in cerebrospinal fluid 57, 79 and 84 percent, respective to the dosage. The results demonstrate a dose-dependent decline of Aβ40 and MK8931 is now under phase II/III clinical trials.

TAK-070 was discovered by Takeda Pharmaceutical Co. Short-term oral administration of TAK-070 to the APP transgenic mouse model of AD resulted in the soluble Aβ brain levels and normalization of the behavioral impairments. Furthermore, an almost 60% reduction of brain Aβ levels were observed preserving the pharmacological efficacy on soluble Aβ and sAPPα levels with six-month chronic treatment with TAK-070. At present it is under phase I of clinical trials [41].

Table 1:   Selected clinical-stage β-secretase inhibitors

S. No	Name	Structure	Company	Phase of Clinical Development	Current Status
1	LY2886721		Eli Lilly and company	I/II	Terminated the Phase 2 trial because of toxicity
2	AZD3839		AstraZeneca	I	Successfully Completed Phase 1 trail
3	MK8931		Merck	II/III	Phase 2 and 3 clinical trials are underway
4	TAK-070		Takeda (University of Tokyo)	I	Phase 1 under progress

METAL CHELATORS

Because of the relatively high concentrations of zinc(II) (Zn^{2+}), copper(II) (Cu^{2+}), and iron(II) (Fe^{2+}) found in the amyloid deposits in the disease affected brains, it was hypothesized that metal-mediated chelators could help control the concentration of oxidized metals in the 40-43 polypeptides should be further investigated [42]. So, the formation of these metals, or neurotoxic reactive oxygen species (ROS) was targeted as the main concern in reducing the oxidative stress [43]. Oxidative stress is believed to enhance the cleaving of the amyloid protein through the generation of high levels of ROS and has garnered recent research interest towards controlling Alzheimer's disease. The metal chelators synthesized attempt to reduce oxidative stress and control the neurotoxicity in the brain [42]. Unlike the previous inhibitors, the metal-based compounds were smaller and were hypothesized to cross the BBB and display more compatible oral bioavailability. There are two approaches currently being researched in the design of these bifunctional compounds: direct incorporation of the metal-binding atom donors into the interacting peptide and linking the chelator to Aβ-binding fragments [44]. These metal-based compounds exhibited elevated binding affinities to the metals found in the brain and greatly reduced levels of H_2O_2, suggesting a reduction in the oxidizing conditions prevalent in Alzheimer's disease [44].

Platinum chelators have also been synthesized in order to inhibit the oxidative stress caused by BACE1 enzymes in the brain. The advancements of the previously unreported platinum chelators were helped by the known characteristics of the copper and zinc chelator properties [45]. For example, it was found that an apocyclen group attached to the Aβ recognition motifs can attract and capture Aβ-bound copper, which in turn will prevent H_2O_2 production and ultimately reduce the toxicity in the neurological cells [46]. These platinum chelators showed very high levels of inhibition in mice brains, especially when combined with cyclen [45], and this observation may be another important stepping stone in the quest for potential Alzheimer's preventions.

Despite an enormous increase in the understanding of the neuropathological and neurochemical events taking place in AD, there are still no effective drugs that can be prescribed to reverse or reduce this mental decline of affected patients with safety [47].

Combining Metal Chelation with BACE 1 Inhibition

To conquer a complex and devastating disease such as AD, more coordinated and combined strategies are required. So far, the therapeutic paradigm one-compound-one-target has failed. This could be due to the multiple pathogenic mechanisms involved in AD including Aβ aggregation to form plaques, increase in oxidative stress and the observed tau-hyperphosphorylation. In view of these complex pathogenic mechanisms, and the successful treatment of chronic diseases such as HIV or cancer, with multiple drugs having complementary mechanisms of action, the concern is growing that AD could better be treated with a single compound targeting two or more of the pathogenic mechanisms leading to neuronal death.

Recent studies have revealed that there is an increase of the levels of these metals as high as 7-fold in affected individuals in comparison to healthy individuals [48]. Based on this fact, Huang *et al.* explored metal dyshomeostasis symptoms of the Alzheimer's disease, alongside BACE-1 inhibition [49]. They have designed dual-target-directed 1,3-diphenylurea derivatives by hybridizing BACE 1 inhibitor **22** with metal chelator **23** (LR-90) (Fig. **9**). A database of compounds consisted of 1,3-diphenylurea derivatives was built and screened by the pharmacophore model (Hypo **22**) of BACE1 inhibitor.

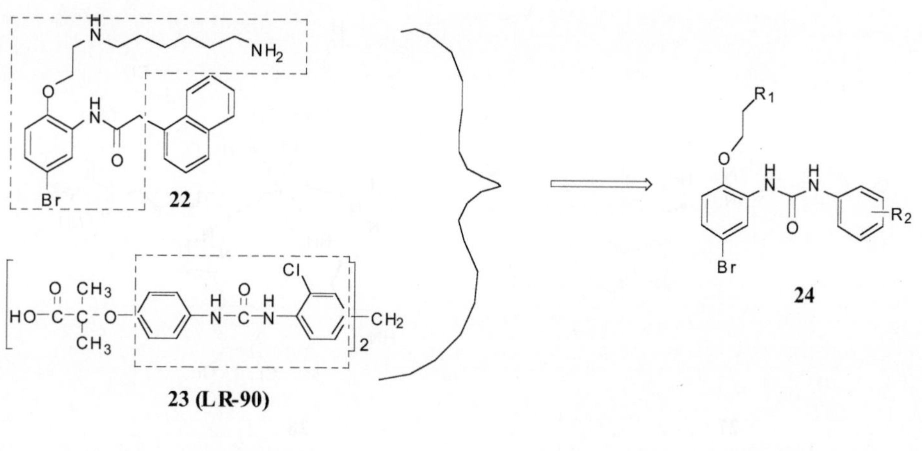

Figure 9: The basis of the 1,3-diphenylurea derivatives **24** came from the parent BACE1 inhibitor **22** and the metal chelator **23** (LR-90).

All of the synthesized compounds were able to inhibit the BACE1 activity and showed the ability to chelate metal ions which represents the multi targeted ligands' potent abilities with appropriately balanced molecular affinities.

Jiaranaikulwanitch *et al.* used tryptoline, a core structure of ochrolifuanine E **25**, which is a hit compound from virtual screening of the Thai herbal database against BACE1 was used for the design of BACE1 inhibitors [50]. The tryptoline was linked with different side chains by 1,2,3-triazole ring readily synthesized by catalytic azide-alkyne cycloaddition reactions. Among the compounds synthesized **26** showed low micromolar inhibitory activity against BACE1 and 100 times more selective to BACE1 than Cathepsin-D in enzymatic assay.

Recently, they have developed compounds based on an *in silico* design of tryptoline- and tryptamine-based BACE1 inhibitors containing additional moieties to exert multi-functionality. Among the compounds prepared, **27** showed good activity with the anti-Aβ aggregation by inhibiting the BACE1 and removing high oxidation potential through iron chelation to form complex **28** as presented in Fig. **10** [51].

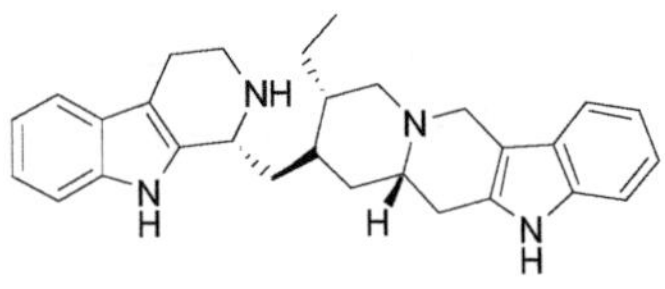

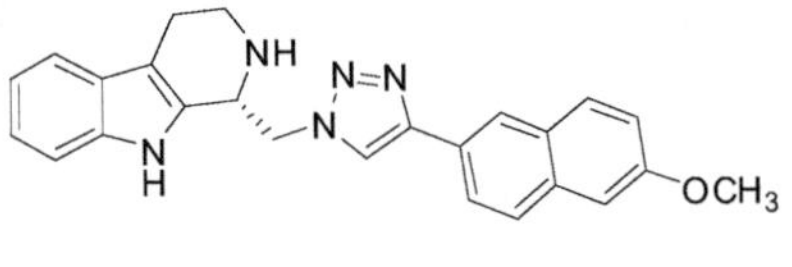

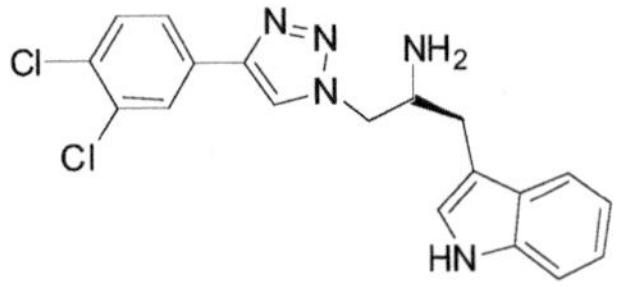

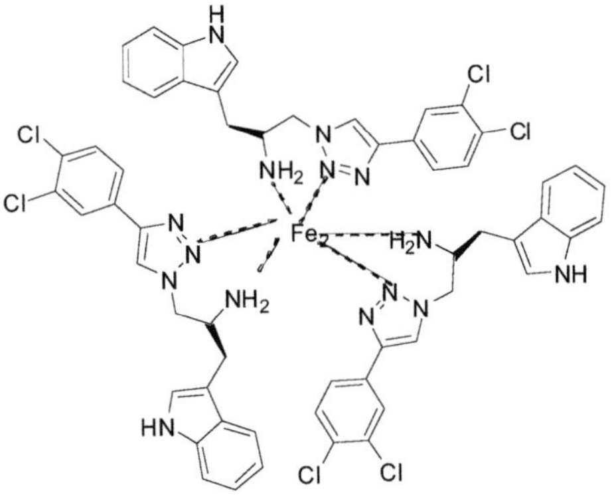

Figure 10: Compounds with BACE1 inhibition and metal chelation properties based on ochrolifuanine E.

Recently, Anuj K. Sharma, *et al.* designed and synthesized a number of bi-functional compounds (BFCs) as shown in Fig. **11**. These compounds were designed to have both Aβ and metal-binding donors incorporated into the design, and linking the two fragments. In the study, two BFCs were synthesized following the linkage approach. The effect of these compounds were evaluated and found that both **29** and **30** binds to copper and zinc ions and also has a high affinity towards Ab tangles [52].

The compounds synthesized are a combination of *N*-(2-pyridylmethyl)amine group, 2-phenylbenzothiazole, and o-vanillin groups, as shown in Fig. **11**.

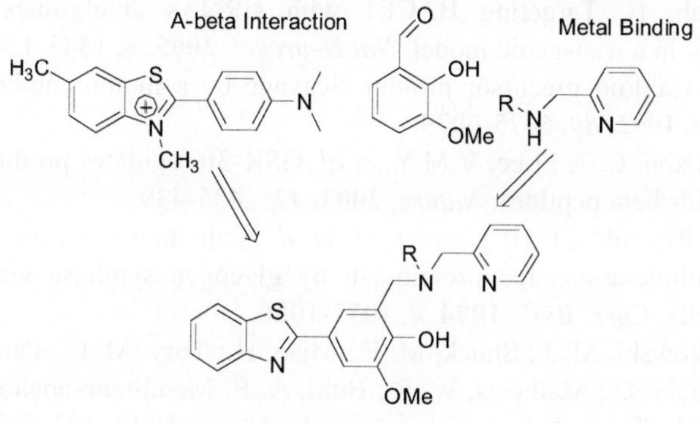

29 R = CH$_2$Py

30 R = Me

Figure 11: Inhibitors and what pieces serve for each purpose.

CONCLUSIONS

Recently, immense research efforts have been directed towards the development of multi-targeted drugs for different diseases as the one-target-one-molecule approach has not been as successful as expected in the case of complex chronic diseases. This chapter summarizes recent progress in the development of BACE1 inhibitors, metal chelators and their combination for developing new and more highly potent AD therapies.

ACKNOWLEDGEMENTS

MH would like to thank the Georgia Research Alliance grant for the support.

CONFLICT OF INTEREST

The authors confirm that this chapter contents have no conflict of interest.

REFERENCES

[1]   Alzheimer's Disease International and World Health Organization 2012 report: Dementia: a public health priority. Available at: http://www.who.int/mental_health/publications/dementia_report_2012.

[2]   Selkoe, D.J.; Schenk, D. Alzheimer's disease: Molecular understanding predicts amyloid-based therapeutics. *Annu. Rev. Pharmacol. Toxicol. 43*, **2003**, 545-584.

[3]   Singer, O.; Marr1, R. A.; Rockenstein, E.; Crews, L.; Coufal1, N. G.; Gage1, F. H.; Verma, I. M.; Masliah, E. Targeting BACE1 with siRNAs ameliorates Alzheimer disease neuropathology in a transgenic model. *Nat Neurosci.* **2005**, 8, 1343-1349.

[4]   Sisodia, S. β-Amyloid precursor protein cleavage by a membrane-bound protease. *Proc. Natl. Acad. Sci*, **1992**, *89*, 6075-6079.

[5]   Phiel, C.J.; Wilson, C. A.; Lee, V.M.Y. *et al.* GSK-3α regulates production of Alzheimer's disease amyloid-/beta peptides. *Nature*, **2003**, *423*, 435-439.

[6]   Lovestone, S; Reynolds, C.H; Latimer, D. *et al.* Alzheimer's disease-like phosphorylation of the microtubule-associated protein tau by glycogen synthase kinase-3 in transfected mammalian cells. *Curr. Biol.*, **1994**, *4*, 1077-1086.

[7]   Yan, R.; Bienkowski, M. J.; Shuck, M. E.; Miao, H.; Tory, M. C.; Pauley, A. M.; Brashier, J. R.; Stratman, N. C.; Mathews, W. R.; Buhl, A. E. Membrane-anchored aspartyl protease with Alzheimer's disease beta-secretase activity. *Nature* **1999**, *402*, 533-537.

[8]   Chamil, L.; Checler, F. BACE1 is at the crossroad of a toxic vicious cycle involving cellular stress and β-amyloid production in Alzheimer's disease. *Molecular Neurodegeneration* **2012**, *7*, 52.

[9]   Haass, C.; Selkoe, D. J. Soluble protein oligomers in neurodegeneration: lessons from the Alzheimer's amyloid β-peptide. *Nat. Rev. Mol. Cell Biol.* **2007**, *8*, 101-112.

[10]   Rajendran, L.; Schneider, A.; Schlechtingen, G.; Weidlich, S.; Ries, J.; Braxmeier, T.; Schwille, P.; Schulz, J. B.; Schroeder, C.; Simons, M.; Jennings, G.; Knölker, H-J.; Simons, K. Efficient inhibition of the Alzheimer's disease β-Secretase by membrane Targeting. *Science* **2008**, *320*, 520-523.

[11]   Small, S. A.; Gandy, S. Sorting through the cell biology of Alzheimer's disease: intracellular pathways to pathogenesis. *Neuron* **2006**, *52*, 15-31.

[12]   He, X.; Cooley, K.; Chung, C. H.; Dashti, N.; Tang, J. Apolipoprotein receptor 2 and X11 alpha/beta mediate apolipoprotein E-induced endocytosis of amyloid-beta precursor protein and beta-secretase, leading to amyloid-beta production. *J. Neurosci.* **2007**, *27*, 4052- 4060.

[13]   Kojro, E.; Fahrenholz, F. The non-amyloidogenic pathway: structure and function of alpha-secretases. *Subcell. Biochem.* **2005**, *38*, 105 - 127.

[14]   Roberson, E. D.; Mucke L.100 years and counting: prospects for defeating Alzheimer's disease. *Science* **2006**, *314*, 781- 784.

[15]   Annaert, W.; De Strooper, B. A cell biological perspective on Alzheimer's disease. *Annu. Rev. Cell Dev. Biol.* **2002**, *18*, 25-51.

[16]   Selkoe, D. *et al.* Beta-amyloid precursor protein of Alzheimer disease occurs as 110- to 135-kilodalton membrane-associated proteins in neural and nonneural tissue. *Proc. Natl. Acad. Sci.* USA **1988** *85*, 7341-7345.

[17]   Selkoe, D.J. *et al.* The role of APP processing and trafficking pathways in the formation of amyloid beta-protein. *Ann. NY Acad. Sci.* **1996**, *777*, 57-64.

[18]   Shen, J. *et al.* Skeletal and CNS deficits in Presenilin-1-deficient mice. *Cell* **1997**, *89*, 629-639.

[19]   Ohno, M. *et al.* BACE1 deficiency rescues memory deficits and cholinergic dysfunction in a mouse model of alzheimer's disease. *Neuron* **2004**, *41*, 27-33.

[20]   Vassar, R. The beta-secretase, BACE: a prime drug target for Alzheimer's disease. *J. Mol. Neurosci.* **2001**, *17*, 157-170.

[21]   Xiaoyang Luo, Riqiang Yan. Inhibition of BACE1 for therapeutic usein Alzheimer's disease *Int J Clin Exp Pathol* **2010**, *3*, 618-628.

[22]   Zhong, W.; Hitchcock, S. A.; Albrecht, B. K.; Bartberger, M. *et al.* Preparation of 2-hydroxy-1,3-diaminoalkanes including spiro substituted chroman derivatives as β-secretase modulators and their use for treatment Alzheimer's disease and related condition. Patent WO2007061670, **2007**.

[23]   Ghosh, A. K.; Kumaragurubaran, N.; Hong, L. *et al.* Design, synthesis, and X-ray structure of potent memapsin 2 (β-secretase) inhibitors with isophthalamide derivatives as the P2-P3-ligands. *J. Med. Chem.* **2007**, *50*, 2399-2407.

[24]   Maillard, M. C.; Hom, R. K.; Benson, T. E. *et al.* Design, synthesis, and crystal structure of hydroxyethyl secondary amine-based peptidomimetic inhibitors of human β-secretase. *J. Med. Chem.* **2007**, *50*, 776-781.

[25]   Ghosh, A. K.; Kumaragurubaran, N.; Tang, J. Recent developments of structure-based β-secretase inhibitors for Alzheimer's disease. *Curr. Top. Med. Chem.* **2005**, *5*, 1609-1622.

[26]   Silvestri, R. Boom in the development of non-peptidic betasecretase (BACE1) inhibitors for the treatment of Alzheimer's disease. *Med. Res. Rev.* **2009**, *29*, 295-338.

[27]   Hamada, Y.; Kiso, Y. Recent progress in the drug discovery of non-peptidic BACE1 inhibitors. *Expert Opin. Drug Discovery* **2009**, *4*, 391-416.

[28]   Hussain, I.; Hawkins, J.; Harrison, D.; Hille, C. *et al.* Oral administration of a potent and selective non-peptidic BACE-1 inhibitor decreases beta-cleavage of amyloid precursor protein and amyloid-beta production *in vivo*. *J. Neurochem.* **2007**, *100*, 802-809.

[29]   Rueeger, H.; Lueoend, R.; Rogel, O. *et al.* Discovery of Cyclic Sulfone Hydroxyethylamines as Potent and Selective β-Site APP-Cleaving Enzyme 1 (BACE1) Inhibitors: Structure-Based Design and *in Vivo* Reduction of Amyloid β-Peptides. *J. Med. Chem.* **2012**, *55*, 3364-3386.

[30]   Al-Tel, T. H.; Semreen, M. H.; Al-Qawasmeh, R. A.; Schmidt, M. F.; El-Awadi, R.; Ardah, M.; Zaarour, R.; Rao, S. N.; El-Agnaf, O. Design, Synthesis, and Qualitative Structure-Activity Evaluations of Novel *β*-Secretase Inhibitors as Potential Alzheimer's Drug Leads. *J. Med. Chem.* **2011**, *54*, 8373-8385.

[31]   Cheng, Y.; Judd, T. C.; Bartberger, M. *et al.* From Fragment Screening to *In Vivo* Efficacy: Optimization of a Series of 2-Aminoquinolines as Potent Inhibitors of Beta-Site Amyloid Precursor Protein Cleaving Enzyme 1 (BACE1). *J. Med. Chem.* **2011**, *54*, 5836-5857.

[32]   Huang, H.; La, D. S.; Cheng A. C. *et al*. Structure- and Property-Based Design of Aminooxazoline Xanthenes as Selective, Orally Efficacious, and CNS Penetrable BACE Inhibitors for the Treatment of Alzheimer's Disease. *J. Med. Chem.* **2012**, *55*, 9156-9169.

[33]   Gerritz, S. W.; Zhai, W.; Shi, S. Acyl Guanidine Inhibitors of β–Secretase (BACE-1): Optimization of a Micromolar Hit to a Nanomolar Lead *via* Iterative Solid- and Solution-Phase Library Synthesis. *J. Med. Chem.* **2012**, *55*, 9208-9223.

[34]   Gravenfors, Y.; Viklund, J.; Blid, J. et. al. New Aminoimidazoles as β–Secretase (BACE-1) Inhibitors Showing Amyloid–β (Aβ) Lowering in Brain. *J. Med. Chem.* **2012**, *55*, 9297-9311.

[35]   Ginman, T.; Viklund, J.; Malmstrom, J. *et al*. Core Refinement toward Permeable β–Secretase (BACE-1) Inhibitors with Low hERG Activity. *J. Med. Chem.* **2013**, *56*, 4181-4205.

[36]   Yonezawa, S.; Yamamoto, T.; Yamakawa, H. *et al*. Conformational Restriction Approach to β –Secretase (BACE1) Inhibitors: Effect of a Cyclopropane Ring To Induce an Alternative Binding Mode. *J. Med. Chem.* **2012**, *55*, 8838-8858.

[37]   Martenyi, F.; Dean, R. A.; Lowe, S. BACE inhibitor LY2886721 safety and central and peripheral PK and PD in healthy subjects (HSs). *Alzheimer's & Dementia* **2012**, *8*, 583-584.

[38]   Lilly Voluntarily Terminates Phase II Study for LY2886721, a Beta Secretase Inhibitor, Being Investigated as a Treatment for Alzheimer's Disease. https://investor.lilly.com/releasedetail.cfm?ReleaseID=771353

[39]   Jeppsson, F.; Eketjall, S.; Janson, J.; Karlstrom, S.; Gustavsson, S. Discovery of AZD3839, a potent and selective BACE1 inhibitor clinical candidate for the treatment of Alzheimer disease. *J Biol Chem.* **2012**, *287*, 41245-41257.

[40]   Merck Presents Findings from Phase 1b Study of Investigational BACE Inhibitor, MK-8931, in Patients with Alzheimer's Disease Merck Newsroom. Home (http://www.mercknewsroom.com)

[41]   Fukumoto, H.; Takahashi, H.; Tarui, N. *et al*. A Noncompetitive BACE1 Inhibitor TAK-070 Ameliorates Aβ Pathology and Behavioral Deficits in a Mouse Model of Alzheimer's Disease The *J. Neurosci.* **2010**, *30*, 11157-11166.

[42]   Chartier-Harlin MC; Crawford F; Houlden H; *et al*. Early-onset Alzheimer's disease caused by mutations at codon 717 of the beta-amyloid precursor protein gene. *Nature*, **1991**, 353(6347), Abstract.

[43]   Mullen, M.; Crawford, F.; Axelman, K. *et al*. A pathogenic mutation for probable Alzheimer's disease in the APP gene in the N-terminus of β-amyloid. *Nat. Gen.*, **1992**, *1*, 345-347.

[44]   Wang, X.; Wang, X.; Zhang, C. *et al*. Inhibitory action of macrocyclic platiniferous chelators on metal-induced Aβ aggregation. *Chem Sci*, **2012**, *3*, 1304-1312.

[45]   Drew, S.; Barnham, K. The Heterogeneous Nature of Cu^{2+} Interactions with Alzheimer's Amyloid-β Peptide. *Acc. Chem. Res.*, **2011**, *44(11)*, 1146-1155.

[46]   Butterfield, D.; Drake, J.; Pocernich, C. *et al*. Evidence of oxidative damage in Alzheimer's diseased brain: central role for amyloid β-peptide. *Trends Mol. Med.*, **2001**, *7(12)*, 548-554.

[47]   Hegde, M. L.; Bharathi, P.; Suram, A.; Venugopal, C.; Jagannathan, R.; Poddar, P.; Srinivas, P.; Sambamurti, K.; Rao, K. J.; Scancar, J. Messori, L.; Zecca, L.; Zatta, P. Challenges Associated with Metal Chelation Therapy in Alzheimer's Disease. *J. Alzheimers Dis.* **2009**, *17*, 457-468.

[48]   Zatta, P.; Drago, D.; Bolognin, S.; Sensi, S. L. Alzheimer's disease, metal ions and metal homeostatic therapy. *Trends Pharmacol Sci.* **2009**, *30*, 346-355.

[49]   Huang, W; Lv, D; Yu, H. *et al.* Dual-target-directed 1,3-diphenylurea derivatives: BACE1 inhibitor and metal chelator against Alzheimer's Disease. *Bioorg. Med. Chem.*, **2010**, *18*, 5610-5615.

[50]   Jiaranaikulwanitch, J.; Boonyarat, C.; Fokin V. V.; Vajragupta, O. Triazolyl tryptoline derivatives as b-secretase inhibitors. *Bioorg. Med. Chem. Lett.* **2010**, *20*, 6572-6576.

[51]   Jiaranaikulwanitch, J.; Govitrapong, P.; Fokin V. V.; Vajragupta, O. From BACE1 Inhibitor to Multifunctionality of Tryptoline and Tryptamine Triazole Derivatives for Alzheimer's Disease. *Molecules* **2012**, *17*, 8312-8333.

[52]   Sharma, A. K.; Pavlova, S. T.; Kim, J. *et al.* Bifunctional Compounds for Controlling Metal-Mediated Aggregation of the $A\beta_{42}$ Peptide. *J. Am. Chem. Soc.,* **2012**, *134*, 6625-6636.

Somatostatin Receptor-4 Agonists as Candidates for Treatment of Alzheimer's Disease

Karin E. Sandoval, Kenneth A. Witt, A. Michael Crider and Maria Kontoyianni[*]

Department of Pharmaceutical Sciences, Southern Illinois University Edwardsville, Edwardsville, Illinois 62026, USA

Abstract: Alzheimer's disease (AD) is a neurodegenerative disease characterized by a progressive loss in memory and cognitive abilities. One of the key pathologic features of AD is the accumulation of beta amyloid (Aβ). Somatostatin has been shown to regulate neuronal neprilysin activity, a key enzyme involved in Aβ catabolism. The actions of somatostatin are mediated through somatostatin receptors 1-5. The somatostatin subtype-4 receptor (sst$_4$) is expressed in key regions of the brain impacted by AD. Thus, sst$_4$ agonists may serve as disease modifying agents (*i.e.*, preventative), enhancing enzymatic activity and decreasing neurotoxic Aβ species within key brain regions of AD patients. This chapter will address the viability of such sst$_4$ agonists within the context of AD therapy, in conjunction with strategies for design, synthesis, and recognition at the macromolecular level.

Keywords: Alzheimer's disease, amyloid cascade hypothesis, beta amyloid, G protein coupled receptors, homology modeling, model building, somatostatin receptor subtype 4, somatostatin receptors, neprilysin, tau hypothesis, virtual screening.

*Address correspondence to Maria Kontoyianni: Department of Pharmaceutical Sciences, Southern Illinois University Edwardsville, Edwardsville, Illinois 62026, USA; Tel: 618-650-5166; Fax: 618-650-5145; E-mail: mkontoy@siue.edu

INTRODUCTION

Alzheimer's disease (AD) is a neurodegenerative disease clinically characterized by a progressive loss in memory and cognitive abilities. In the United States, it is estimated that one out of eight individuals over the age of 65 has AD, with an occurrence at ~45% for those over age 85 [1]. To date, all approved pharmacological treatments are strictly palliative (*i.e.*, donepezil, galantamine, rivastigmine, and memantine). Moreover, these drugs have a highly variable effect from patient to patient, as well as a relatively short window of therapeutic benefit (~6-12 months) [2]. Without the development of disease-modifying drugs (*i.e.*, preventative), the financial, societal, and emotional costs of this disease will continue to grow. To this end, our ongoing research has focused on the development and evaluation of selective somatostatin subtype-4 receptor (sst$_4$) agonists for the treatment of AD.

AD PATHOLOGY

While AD phenotypically begins with the loss of memory, it is estimated that the pathogenesis starts ~20 years before symptom onset, and is dependent on genetic and environmental variables [2]. The pathological hallmarks of the AD brain include neuritic plaques, neurofibrillary tangles, and reactive microgliosis, corresponding with a progressive loss of neurons and synapses [3]. The two principle hypotheses of AD development are the "amyloid cascade hypothesis" and the "tau and tangle hypothesis" [3, 4]. The amyloid cascade hypothesis maintains that dysregulated clearance and/or a bnormal processing of Aβ peptide initiates AD pathogenesis. Current data strongly supports this understanding, with alterations in tau and subsequent tangle formation occurring downstream [5]. Moreover, there has been further elucidation as to the neurotoxic effects of soluble Aβ-oligomers and associated contributions to AD pathogenesis [6, 7]. Thus, within the context of drug discovery and development, direct mitigation of toxic Aβ species poses the most viable manner for focused therapeutics.

Aβ is generated by proteolytic processing of amyloid precursor protein (APP) down the amyloidogenic pathway. In this pathway, APP is cleaved at the amino terminus of the Aβ sequence by β-secretase resulting in generation of soluble APPbeta (sAPPβ) and a membrane-anchored 99-amino-acid carboxyl terminal fragment-β (CTFβ). Gamma secretase then enzymatically cleaves the CTFβ fragment. Cleavage of CTFβ yields soluble Aβ peptides ranging in length from 38 to 43 amino acids and a second product, APP intracellular domain (AICD) [8]. The primary forms of Aβ within the brain are $Aβ_{1-40}$ and $Aβ_{1-42}$. $Aβ_{1-40}$ is the predominant species in the cerebral spinal fluid and plasma [9]. While the role of $Aβ_{1-40}$ continues to be investigated in AD, it is not believed to be the principal pathogenic species as it is less hydrophobic (*i.e.* lower aggregation potential), more readily cleared from the brain, and has generally been identified as being less neurotoxic than $Aβ_{1-42}$ [9, 10]. In contrast, once the soluble $Aβ_{1-42}$ monomers (~4kDa) are formed, they readily aggregate to higher molecular weight oligomers, which then can become soluble protofibrils, and ultimately assemble into insoluble fibrils which form the neuritic amyloid plaques [6, 7]. While the neuritic plaques have long been hypothesized to be the principle causative factor of AD, a number of studies have found a poor correlation between plaque load and the degree of cognitive decline [11-13]. Yet, a significant correlation has been found between the level of soluble Aβ and the severity of memory decline [11]. In recent years, increasing attention has been focused on the role of soluble $Aβ_{1-42}$ oligomers (*e.g.*, dimers, trimers, dodecamers, *etc.*) as they have been shown to correlate much better than Aβ plaque content with the extent of synaptic loss and severity of cognitive impairment [12]. While there still is a significant amount of debate as to the impact of specific $Aβ_{1-42}$ oligomeric species in AD pathogenesis, such oligomers appear to have differential neurotoxic effects based on the degree of aggregation and cellular localization (*i.e.* intracellular *vs.* extracellular content) [7, 14, 15]. Additionally, $Aβ_{1-42}$ oligomers have been shown to contribute to neurofibrillary tangle formation. In the triple transgenic mouse model of AD (3xTg-AD), Aβ deposition precedes tau pathology by several months [16], with intrahippocampal injection of an anti-Aβ oligomer antibody reducing both Aβ and tau pathology [17]. Recent evaluations have also shown that treatment of cell cultures with soluble Aβ dimers isolated from AD cortical tissue directly induces tau hyperphosphorylation and neuritic dystrophy [18]. This further supports previous studies demonstrating that soluble Aβ oligomers induce neuronal tau hyperphosphorylation [19].

Aβ DEGRADATION AND NEPRILYSIN

Clearance of Aβ from the brain is mediated by multiple processes, which can be broadly divided into transport mechanisms and enzymatic degradation. There is considerable support with alterations in Aβ transport mechanisms at the level of the cerebrovasculature (*i.e.*, blood-brain barrier; BBB) contributing to AD pathology [20, 21]. Nevertheless, any therapeutic application impacting the BBB must take into account several complex factors, including transport adaptations over time, cerebrovascular health, and the general impact of the transporter on other substances (*i.e.*, nutrients and waste products) into and out of the brain. A more direct approach, albeit with caveats of its own, is to increase Aβ enzymatic degradation. Enzymatic degradation of Aβ has received a significant deal of attention over the past decade. Several amyloid degrading enzymes (ADE) have been identified for their ability to catabolize Aβ, including neprilysin, insulin degrading enzyme (IDE), endothelin-converting enzyme (ECE-1/ECE-2), cathepsin-B, and angiotensin-converting enzyme (ACE) [22]. Of these ADEs, neprilysin appears to have the greatest potential utility to effectively mitigate toxic Aβ species.

Neprilysin (neutral endopeptidase 24.11; common lymphoblastic leukemia antigen; CD10) is a type-II transmembrane zinc metalloendopeptidase capable of cleaving a wide range of peptides [22]. Since the establishment of neprilysin as a key regulator of Aβ catabolism [23, 24], it has been heavily investigated to better understand its role in AD pathology. Neprilysin has been linked to a number of brain functions including long-term potentiation and memory [25-28], shown to be distributed throughout the brain, and importantly in regions associated with

AD pathology (*i.e.*, hippocampus and neocortex) [29-31]. Neprilysin expression and activity within the brain have been shown to decrease during aging and in the early stages of AD [32, 33]. Additionally, reductions in neprilysin levels are more pronounced in brain tissues of ApoE ε4 -positive AD patients [34-36]. Corresponding animal studies have shown that inhibition of neprilysin leads to elevation in Aβ within the brain, and is associated with enhanced formation of plaques, impaired synaptic plasticity, and cognitive decline [24, 37, 38]. In contrast, when neprilysin is overexpressed in transgenic mouse models a dramatic reduction in cerebral Aβ content with enhancement of cognition and life expectancy has been shown [39-42]. Another critical characteristic of neprilysin is that increased activity reduces both soluble and fibrillar levels of Aβ [40, 43], including the neurotoxic soluble oligomeric $A\beta_{1-42}$ species [44-46]. Yet, any approach aimed at the enhancement of neprilysin in humans needs to take into account some key factors. First, as addressed, neprilysin is capable of catabolizing a number of other physiological substrates and thus increasing its expression/ activity throughout the body has the potential to cause various side effects (*e.g.*, drop in blood-pressure). Therefore, while it may be viable to pharmacologically counter such side effects, any enhancement of neprilysin would be most practical if restricted to the brain. Secondly, direct degradation of Aβ may also pose an issue. While often ignored, Aβ is a physiological peptide whose function has not been fully elucidated and may have an important regulatory role within the brain [47-49]. Indeed, there is a growing understanding that over-suppression/ degradation of Aβ may be equally as problematic in terms of cognitive function, and that the maintenance of "normalized" (*i.e.*, homeostatic) levels of Aβ should be the focus of long-term therapeutic interventions [50]. In this regard, direct enhancement of neprilysin *via* pharmacological means may result in large fluctuations in Aβ levels, especially given the pharmacokinetic variables associated with drug uptake into the brain at any given time. However, a greater normalization of Aβ levels within the brain may be attainable *via* a secondary enhancement of neprilysin (*i.e.*, indirect enhancement of neprilysin expression/ activity through upstream regulation), reducing rapid fluctuation variables and other complications of direct Aβ degradation. Evidence supports the understanding of somatostatin regulation of neprilysin in the brain, and that a selective high affinity somatostatin receptor subtype agonist may ideally be used to focus treatment on the brain and achieve an appropriate "normal" Aβ content.

SOMATOSTATIN AND SOMATOSTATIN RECEPTORS

Somatostatin (somatotropin release inhibiting factor; SRIF) is a cyclic tetradecapeptide distributed throughout the body and involved in numerous physiological processes. There are two biologically active forms of somatostatin based upon the number of amino acids; SRIF-14 (Fig. **1**) and SRIF-28. SRIF-14 and SRIF-28 are derived from a common pre-propeptide, pre-pro-somatostatin. Depending on how pro-somatostatin is processed in different tissues, various combinations of SRIF-14 and SRIF-28 can be found. SRIF-14 is the predominant form of somatostatin found within the brain, but can also be found in other tissues in the body such as pancreatic islets, stomach, retina, peripheral nerves, and enteric neurons. In contrast, SRIF-28 is the predominant form of somatostatin found in intestinal mucosal cells. Through receptor action, somatostatin produces primarily inhibitory effects on endocrine and exocrine secretions throughout the body, regulates cellular differentiation and proliferation, and acts as a neurotransmitter/neuromodulator in the central nervous system (for full review see [51, 52]).

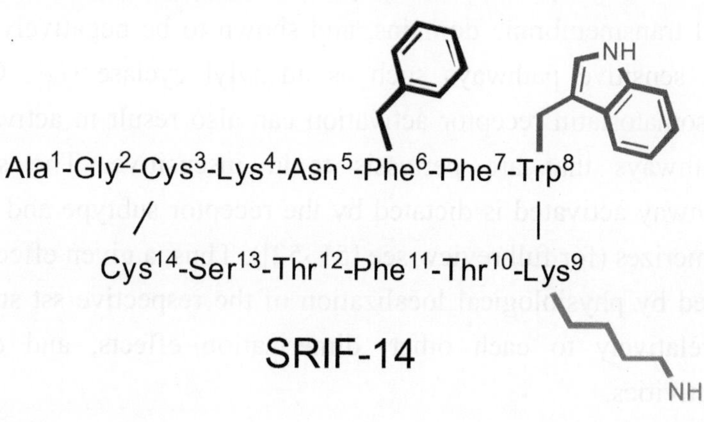

Ala^1-Gly^2-Cys^3-Lys^4-Asn^5-Phe^6-Phe^7-Trp^8

Cys^{14}-Ser^{13}-Thr^{12}-Phe^{11}-Thr^{10}-Lys^9

SRIF-14

Figure 1: Structure of SRIF-14.

Somatostatin receptors are divided into two families based on functional and structural characteristics. The SRIF-1 family consists of somatostatin receptor subtypes sst_2, sst_3, and sst_5; while the SRIF-2 family consists of the sst_1 and sst_4. The respective somatostatin receptor subtypes also include the alternatively spliced sst_{2A} and sst_{2B}. In evaluation of respective subtype expression patterns, sst_1 has been shown to be predominately expressed within the brain (*i.e.*, cortex, hypothalamus, pituitary, solitary tract, dorsal motor vagal nucleus, and ventrolateral medulla), as well as expression in the eye, pancreatic islets, and adrenals [51, 53-56]. While sst_2 has a similar expression throughout the brain, it is also expressed in the gastrointestinal tract, pancreas, microglia and adrenals [51, 53-57]. The sst_3 has been shown to be present in the brain with dense expression in the cerebellum, and also shown in microglia, spleen, kidneys, adrenals, and the liver [51, 53-55, 57, 58]. The sst_4 has been shown to be expressed in the brain, including the critical memory forming CA1 region of the hippocampus, along with presentation in the heart and moderate levels in microglia, adrenals, lungs and pancreatic islets [51, 53-55, 57, 58]. Lastly, sst_5 has a moderate degree of expression within the brain, with additional expression in the intestines, adrenals and islets [51, 53-55, 57, 58]. Delineating response of the respective receptors is complex given that they are often co-expressed/localized. When somatostatin or related peptides (*i.e.*, cortistatin) bind to the respective receptors, changes in intracellular signaling pathways occur. The somatostatin receptors are all classically defined as G protein coupled receptors (GPCR) with seven α-helical transmembrane domains, and shown to be negatively coupled to pertussis-toxin sensitive pathways such as adenylyl cyclase (*i.e.*, G_i-coupled). Nevertheless, somatostatin receptor activation can also result in activation of 2nd messenger pathways that are pertussis toxin insensitive. The specific 2nd messenger pathway activated is dictated by the receptor subtype and the manner by which it dimerizes (for full review see [51, 52]). Thus, a given effect would not only be dictated by physiological localization of the respective sst subtypes, but there ratios relatively to each other, dimerization effects, and downstream intracellular activities.

SOMATOSTATIN, NEPRILYSIN, AND AD PROGRESSION

Understanding the link between somatostatin, neprilysin, and AD pathogenesis provides the necessary insight as to why targeting somatostatin receptors for AD treatment is viable. First and foremost, somatostatin is a mediator of cognitive functioning [59, 60]. Not only is somatostatin a primary neuromodulator within the hippocampus [60], there is also strong evidence that both sst_2 and sst_4 have important impacts on learning and memory processes [61, 62]. Within the context of AD, reduced somatostatin levels within the primary brain regions associated with learning and memory (*i.e.*, cortical and hippocampal) have been shown to positively correlate with cognitive decline [63-66]. Moreover, somatostatin-immunoreactive neurons have been shown to decrease by $> 70\%$ within the cortex, compared to non-AD controls [55]. AD mouse model examinations also substantiate somatostatin's impact, showing decreases in learning and memory with somatostatin loss [67, 68]. Nevertheless, the question as to which comes first must be considered. Does the decline of somatostatin simply coincide with neuronal loss (*i.e.*, default of general brain tissue loss over disease progression) or does it precede/cause neuronal loss (*i.e.*, mediator of pathogenesis). Correspondingly, within the context of drug development, one might expect a somatostatin receptor agonist drug to be solely a cognitive enhancer (*i.e.*, palliative). However, could such a drug also be capable of mitigating disease progression (*i.e.*, disease modifying agent)? From the viewpoint of drug effect, the somatostatin receptor agonist octreotide (sst_2, sst_3, sst_5 agonist) was shown to enhance memory in AD patients [69]. While the study did not clarify long-term viability or disease mitigation potential of octreotide, it confirmed the effectiveness of a somatostatin receptor agonist to enhance memory in AD patients. Additionally, the study demonstrated that there are still functional receptors in the brain capable of inducing a beneficial effect by a somatostatin receptor agonist. Somatostatin has also been shown to be altered in patients genetically predisposed toward the development of AD. Possession of an ApoE ε4 allele, which is associated with increased incidence of AD, has been theorized to accelerate the loss of somatostatin within the brain [70-72]. Somatostatin levels have been shown to be significantly lower in AD patients carrying the ApoE ε4 allele over other groups [70]. Additionally, in both Finnish [71] and Chinese [72] population studies, ApoE ε4 positive individuals with AD were significantly more likely to be C allele carriers of the somatostatin gene single nucleotide polymorphism (SNP), rs4988514. While this is one of many such potential genetic polymorphisms, the implication is that somatostatin alterations may directly contribute/precede the actual pathogenesis. In further support of this premise, transgenic mice (PS1xAPP AD model) were shown to have significant early-age loss of somatostatin neurons and somatostatin mRNA within the hippocampus, with the absence of change in other neuronal markers (cholinergic, GABAergic, glutamatergic, or in principle cell number) [73]. So if loss of somatostatin is truly a mediator of AD, the next logical question is by what means?

The manner by which the loss of somatostatin contributes to AD pathogenesis has been evaluated, and evidence points to somatostatin's regulation of the Aβ degrading enzyme neprilysin. First, the decline in somatostatin and neprilysin content with AD progression closely correlate with each other, as well as with increase in Aβ content [27, 73, 74]. In genetically modified somatostatin-knockout mice, a significant increase in soluble $Aβ_{1-42}$ was shown over controls [27]. In the same study, somatostatin administration was reported to dose-dependently increase neprilysin activity and significantly lower soluble $Aβ_{1-42}$ in cultured neurons [27]. Second, when a somatostatin antagonist (BIM23056) or Gi-inhibitor (pertussis toxin) was co-cultured with neurons, neprilysin activity was inhibited [27], indicating a direct somatostatin receptor-mediated mechanism by which levels of Aβ could be regulated. Based on these data, it was hypothesized that decreased levels of somatostatin induces a gradual decline in neprilysin activity, resulting in an elevation of Aβ, initiating a cascade that culminates in AD [45]. Moreover, such a cascade would result in a positive feedback-loop, with the elevations of toxic Aβ species causing continued destruction and degeneration of somatostatinergic systems, accelerating AD progression [45]. Yet, these findings also suggest that use of selective somatostatin receptor agonists targeted to the brain could be utilized to break the cycle, reduce $A\beta_{1-42}$ levels, and halt disease progression.

TARGETING sst$_4$

With the knowledge that somatostatin regulates neuronal neprilysin activity *via* receptor action, and in turn reduces $A\beta$ levels in the brain, the next step was the determination of an optimal pharmacologic approach. Use of a pan-receptor somatostatin peptide-based therapeutic would in theory increase neprilysin activity. However, using any peptide as a therapeutic would not be viable given the actions of peptidases (*i.e.*, short half-life) and limited BBB penetration. Additionally, pan-somatostatin receptor based drugs would simply have too many side effects. While all somatostatin receptor subtypes are found in the brain, sst$_2$ and sst$_4$ are predominantly expressed in the cortex and hippocampus [53, 55, 58]. However, as previously addressed, the sst$_2$ is also shown to be heavily expressed in the pituitary, and is involved in hormonal regulation [53, 55, 58]. Thus, while a sst$_2$ receptor agonist would still be viable with respect to actions within AD associated brain regions, the lack of pituitary distribution of sst$_4$ makes this subtype a more attractive candidate. Moreover, sst$_4$ agonists are shown not to inhibit insulin secretion, growth hormone, or glucagon secretion [75], further reducing the potential complications. Another advantage of sst$_4$ is the reduced internalization profile following agonist treatment, in contrast to the SRIF1 receptors (sst$_2$, sst$_3$, sst$_5$) which readily internalize [51, 52], and thereby making it a viable target for continuous drug treatment. While there have been a few enzymatically stable and sst$_4$ selective agonists developed (*e.g.*, NNC 26-9100, L-803,087, J-2156, which are discussed in more detail below), to date only NNC 26-9100 has been examined within the context of AD treatment.

In terms of pharmacological viability, NNC 26-9100 is a nonpeptide drug having a >100-fold selectivity for the sst$_4$ over the other sst receptor subtypes ($K_i = 6$ nM), possesses functional activity (sst$_4$ $EC_{50} = 26$ nM), with no impact on locomotor function (rotarod test in mice) or outward toxic response at high doses

(10-100 mg/kg, *i.p.*) [76, 77]. The non-peptide characteristics of NNC 26-9100 also impart the necessary enzymatic stability [76, 77]. Pharmacokinetic evaluations of NNC 26-9100 confirm viable brain uptake *via* peripheral administration (*i.p.* and *i.v.*) [78, 79]. Evaluations of unidirectional influx-rate (Ki = 0.25 µl/g min) also identified a viable rate of brain uptake, indicative of non-saturable transport across the BBB [78]. The distribution evaluation of ^{131}I-NNC 26-9100 indicated a limited sequestration of the compound within the capillary component (~7%), and thus the calculated brain uptake was not an artifact of excessive capillary binding. Lastly, regional brain distribution of ^{131}I-NNC 26-9100 demonstrated generally uniform uptake [78], and was able to access regions associated with memory and learning decline in AD (*i.e.*, cortex, hippocampus). Thus, NNC 26-9100 was shown not only to be stable and sst_4 selective, but have the appropriate pharmacokinetic attributes for peripheral administration as well.

Evaluations of NNC 26-9100 for the treatment of AD have been principally conducted in the senescence accelerated mouse prone-8 (SAMP8) mouse strain. The SAMP8 model has distinctive age-associated learning and memory decline, with corresponding changes brain tissue changes similar to AD pathology (*i.e.*, enhanced APP, Aβ oligomers, tau, and oxidative stress) [80, 81]. By age 12 months, levels of soluble Aβ are increased ~2-fold in male SAMP8 mice [82, 83], which holds significant similarity to the 50% increase observed in human AD [84]. Initial evaluations of NNC 26-9100 in 12-month male SAMP8 mice identified increased learning and memory (20 and 200 µg, *i.p.*/daily, over 28-days) compared to age-matched vehicle treated mice [78]. No toxicity or loss in body weight observed at any given dose [78]. Subsequent brain tissue analyses also identified a significant decrease in cortical soluble $Aβ_{x-42}$ levels (20 µg dose), with no alterations in expression of the sst_4 receptor or APP [78]. The lack of change in the expression of sst_4 expression substantiates the feasibility of chronic treatments (*i.e.*, no significant receptor expression loss with long-term dosing). The lack of change in APP would also be consistent with the understanding that neprilysin does not directly impact APP processing [23]. To evaluate the effects of NNC 26-9100 effect on neprilysin activity, a single-dose i.c.v. administration study was conducted in 12-month SAMP8 mice [85]. Both learning and memory were shown to be enhanced with the NNC 26-9100 treatment (0.2 µg) over vehicle controls [85]. While the enhanced learning/memory might be accountable to direct sst_4 actions (*i.e.*, cognitive enhancer), a significant increase in cortical neprilysin activity with treatment was also found [85]. Additionally, overall expression of neprilysin did not change with NNC 26-9100 treatment [85], this result is consistent with literature identifying that neprilysin activity changes are more critical than sole expression [22, 27]. Yet interestingly, upon cellular fractionation a protein expression shift was identified with treatment, with a decrease in intracellular neprilysin expression and an increase in extracellular expression [85]. Cortical evaluation of soluble $A\beta_{1-42}$ oligomers showed a significant decrease in intracellular and extracellular trimers (~12 kDa) with NNC 26-9100 treatment compared to vehicle controls [85]. Additionally, while there was no change in the extracellular hexamer (~25 kDa) species, there was a consistent split-banding effect with NNC 26-9100 treatment [85]. Soluble $A\beta$ oligomer species are more susceptible to immediate degradation than insoluble $A\beta$, and have shown to be degraded by neprilysin [44-46]. Moreover, the $A\beta_{1-42}$ trimers have been hypothesized to be the "molecular brick" for non-fibrillar assemblies of $A\beta$ [6]. Such trimers appear to be heavily produced and secreted by primary neurons, with extracellular and intracellular concentrations shown to increase in AD transgenic mice [15]. When neprilysin-knockout mice were crossed with APP transgenic mice, there was significant increase in $A\beta$ trimers [44]. The increase in trimers positively correlated with impaired neuronal plasticity and cognitive function [44]. This demonstrated that increased levels of trimers can produce learning and memory impairments with loss of neprilysin activity. $A\beta$ trimers have also been associated with deficits in long-term potentiation [7, 86]. In regards to the hexameric $A\beta_{1-42}$ oligomer, little is known; however, an evaluation of an antibody against $A\beta_{1-42}$ oligomers in the molecular weight range of 16.5-25 kDa was shown to enhance learning and memory in SAMP8 mice [87]. Thus, the effect of NNC 26-9100 on such primary neurotoxic

$A\beta_{1-42}$ oligomers substantiates the hypothesis that an sst_4 agonist is capable of mitigating pathogenic mediators of AD.

Current data supports the hypothesis that sst_4 is a viable receptor to target for the treatment of AD. There is strong support for sst_4 agonists as disease-modifying agents, which does not negate the possibility that sst_4 agonists may also act as cognitive enhancers. Given the current state of therapeutic options, any viable approach that is not redundant with current medications would be a great benefit to AD patients. Nevertheless, continued drug development and optimization to maximize pharmacokinetic/pharmacodynamics properties are required. The ability of optimize such drugs is dependent on numerous factors, which can be further delineated *via* combination of computational and bench chemistry.

SUGAR DERIVATIVES

Peptides have extremely important functions as hormones, neurotransmitters, and enzyme substrates or inhibitors. They typically exert their effects through binding to specific receptors. A severe limitation of peptides is their rapid degradation by peptidases upon oral administration, as mentioned in the preceding discussion. Thus, development of small molecule, metabolically stable peptidomimetics has been the focus of intense research [88]. Structure-activity relationship (SAR) studies have shown that Trp8 and Lys9 are essential for pharmacological activity and that the tetrapeptide Phe7-Trp8-Lys9-Thr10 comprises the critical β-turn of SRIF. This key structural element stabilizes the secondary structure of the peptide which allows hydrogen bonding to occur between the -NH of Thr10 and the carbonyl group of Phe7 [88].

Early research on the development of the nonpeptide SRIF ligands focused on monosaccharide scaffolds (Fig. **2**) as β-turn mimetics [89, 90]. Using the β-D-glucose scaffold, side chain substitutions at positions 1 and 6 were shown to overlap with Trp8 and Lys9 of SRIF [91, 92]. The β-D-glucose analogue 1 contained the Phe7, Trp8, and Lys9 functional groups in a similar spatial orientation as octreotide; however, the compound demonstrated only weak binding in AtT-20 cells [89]. Additional studies using a L-mannose scaffold led to compound **2** with enhanced binding affinity at sst_4.

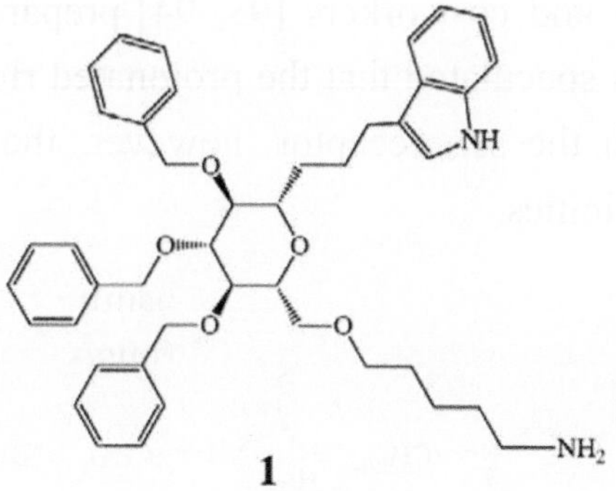

1

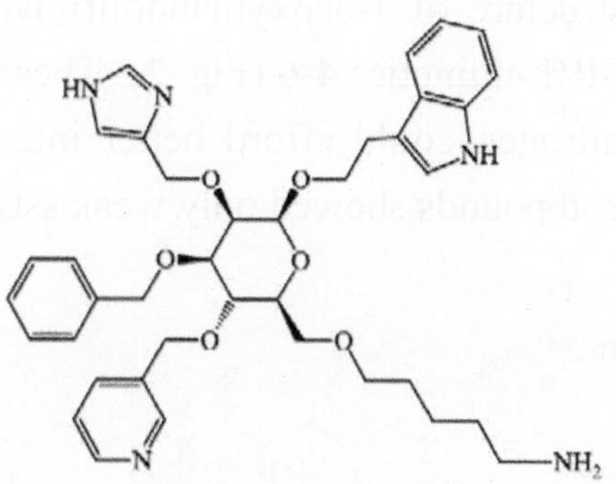

2: Ki=100 nM (sst₄)

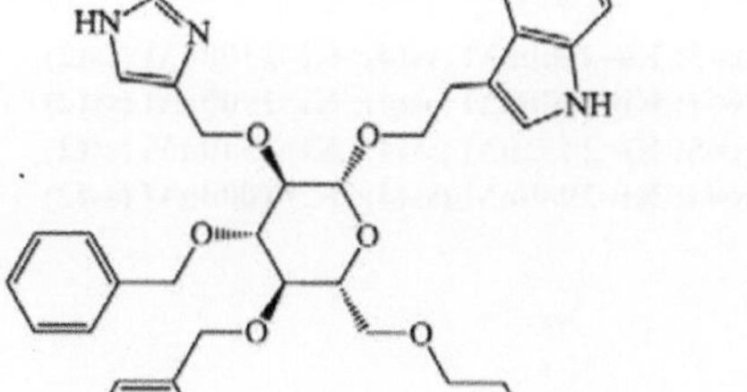

3: Ki=53 nM (sst₄)

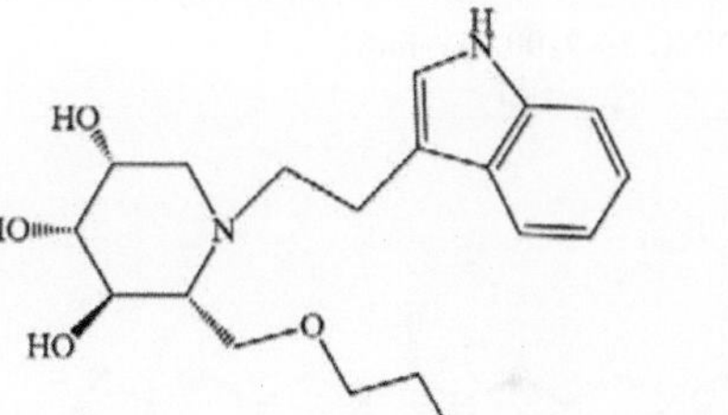

4

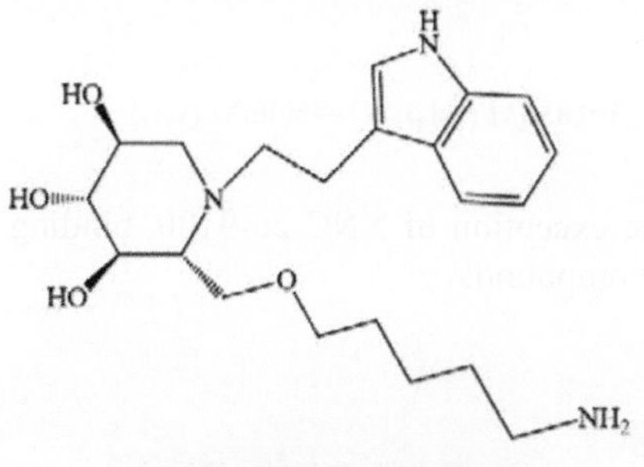

5: Ki=3200 nM (sst₄)

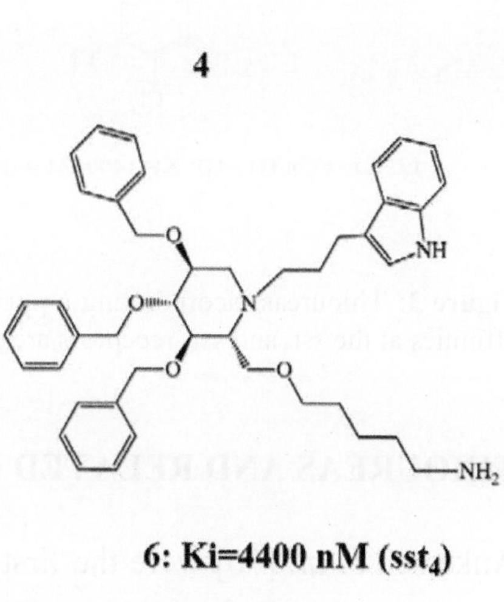

6: Ki=4400 nM (sst₄)

Figure 2: Structures of monosaccharides and iminosugars with sst_4 binding affinities.

Introduction of an (imidazol-4-yl)methyl group at C-2 and a (pyridine-3-yl)methyl moiety at C-4 gave the β-D-glucose analogue **3**. This compound had the highest affinity at sst_4 of any monosaccharide [91]. Using iminosugars based on the structure of 1-deoxymannojirimycin, Murphy and co-workers [93, 94] prepared SRIF-mimetics **4-6** (Fig. **2**). These researchers speculated that the protonated ring nitrogen could afford better interactions with the sst_4 receptor, however, these compounds showed only weak sst_4 binding affinities.

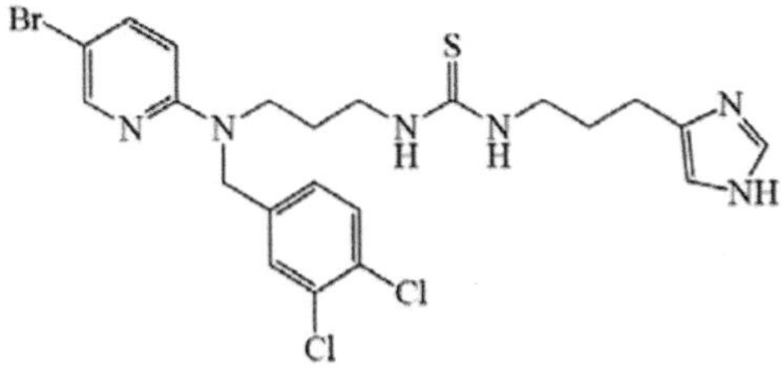

7: NNC 26-9100, Ki=6nM

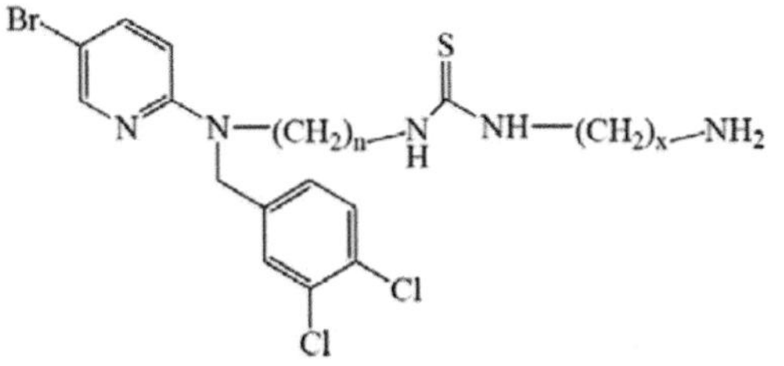

8: n=2, x=3; Ki=4200nM (sst4); Ki=2300nM (sst2)
9: n=2, x=4; Ki=2400nM (sst4); Ki=1900nM (sst2)
10: n=2, x=5; Ki=2433nM (sst4); Ki=1340nM (sst2)
11: n=3, x=4; Ki=2000nM (sst4); Ki=1000nM (sst2)

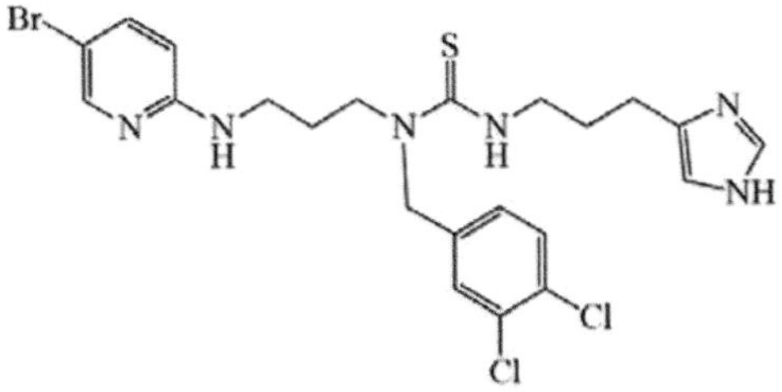

12: Ki=475nM (sst₄); Ki=1400nM (sst₂)

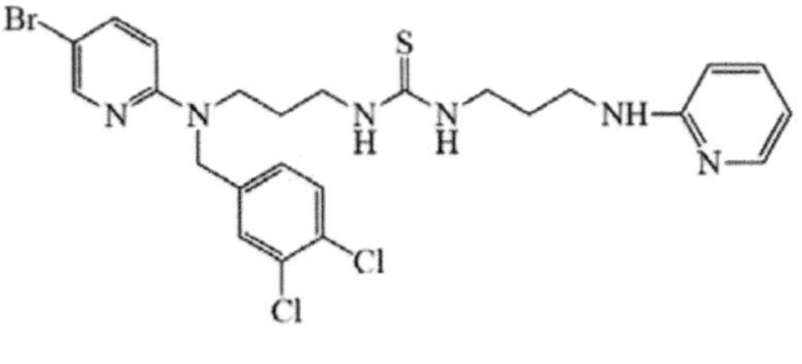

13: Ki=185nM (sst₄); Ki=>300nM (sst₂)

Figure 3: Thioureas incorporating a pyridine ring. With the exception of NNC-26-9100, binding affinities at the sst_4 and sst_2 receptors are given for all other compounds.

THIOUREAS AND RELATED COMPOUNDS

Ankersen *et al.* [76] were the first to report a nonpeptide having high affinity and selectivity at human cloned sst_4. NNC 26-9100 (compound **7**, Fig. **3**) demonstrated high affinity and selectivity at sst$_4$ compared to other ssts. This compound was shown to be a full agonist in inhibition of foskolin-stimulated cAMP accumulation [95]. Replacement of the (imidazole-4-yl)propyl group in **7** with aminoalkyl groups (8-11) led to a change in receptor binding profile. These compounds demonstrated dramatically reduced ssts binding affinity with a loss in selectivity versus sst$_2$. In fact, the primary amine analogues show slightly increased binding affinity at sst$_2$ compared to sst$_4$.

Our original hypothesis was that the pyridine ring, the 3,4-dichlorobenzyl group, and the imidazole ring in NNC 26-9100 were side-chain mimetics for Phe7, Trp8, and Lys9 [76]. This hypothesis was questioned due to the fact that substitution of aminoalkyl groups (compounds **8-11**, Fig. **3**) for the (imidazo-4-yl) propyl group led to a dramatic decrease in sst$_4$ binding affinity [96]. Movement of the 3,4-dichlorobenzyl group to the N-1 position of the thiourea moiety (compound **12**, Fig. **3**) resulted in decreased sst$_4$ binding affinity compared to NNC 26-9100 [77]. When the (imidazol-4-yl)propyl group in NNC 26-910 was replaced by a (pyrid-2-yl)amino group (compound **13**, Fig. **3**) affinity at sst$_4$ decreased, but greater selectivity versus sst$_2$ was observed [96].

Replacement of the 2-pyridyl group (Trp8 mimetic) in NNC 26-9100 with an 2-(1H-indol-3-yl)ethyl group and movement of the 3,4-dichlorobenzyl group to the N-1 position of the thiourea moiety (compound **14**, Fig. **4**) resulted in an analogue with high affinity and selectivity for sst$_4$ [97]. Substitution of a 2-phenylethyl moiety (compound **15**, Fig. **4**) for the indole group in 14 led to a significant decrease in sst$_4$ binding affinity. Unlike the case with NNC 26-9100, substitution of the (imidazole-4-yl)propyl group in 14 by an (pyrid-2-yl)amino group (compound **16**, Fig. **4**) resulted in only weak binding at sst$_4$ [97].

A major goal of our research has been to develop selectively-acting sst$_4$ agonists with potential therapeutic utility in the treatment of AD [78, 85]. Since the thiourea nucleus (**17**, Fig. **5**) has been a useful scaffold for the development of sst$_4$ ligands, we decided to incorporate this functionality into a 2-thiohydantoin ring

(**18**, Fig. **5**). The results of our binding studies revealed that the 2-thiohydantoins **19**-**24** (Fig. **5**) exhibited only moderate affinity for sst_{2A} [98]. A possible explanation for these results is that the more rigid the 2-thiohydantoin nucleus may not allow a proper orientation of the key pharmacophoric groups at sst_4.

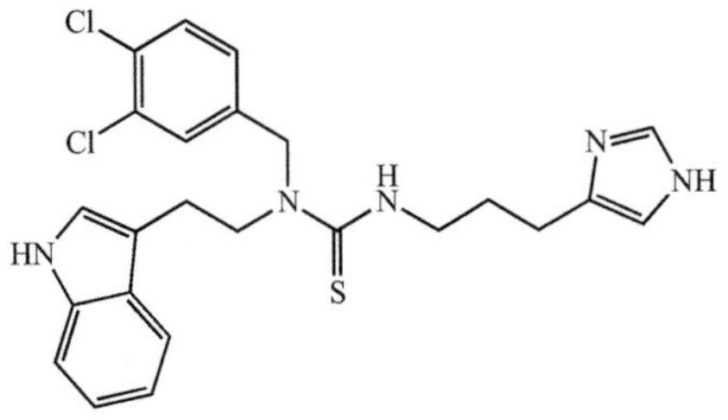

14: Ki=23nM (sst₄), Ki=2400nM (sst₂)

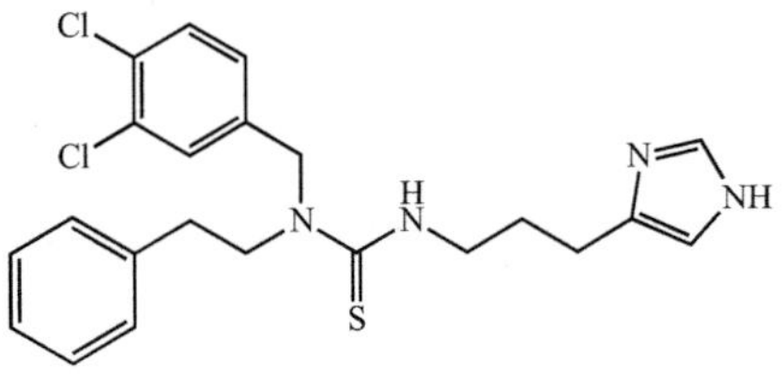

15: Ki=1088nM (sst₄), Ki=2400nM (sst₂)

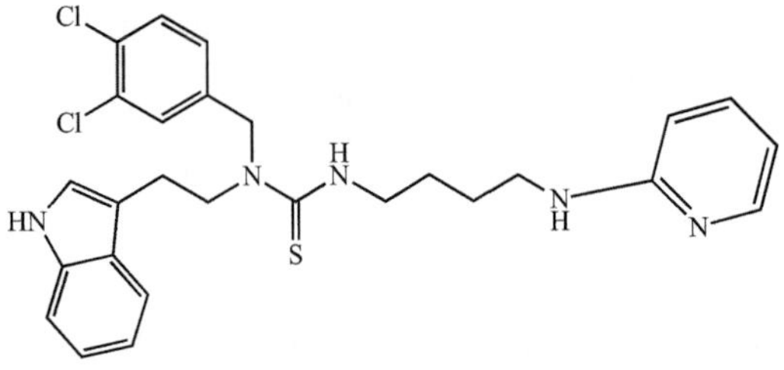

16: Ki=591nM (sst₄), Ki >5000nM (sst₂)

Figure 4: Structures of indole and phenylethyl thioureas with corresponding sst_4 and sst_2 binding affinities.

INDOLE ANALOGUES DEVELOPED AT MERCK

Soon after our report on NNC 26-9100 [76] Yang *et al.* [99] at Merck published a series of spiro[1H-indene-1,4-piperidine] derivatives (Fig. **6**) as sst agonists. Compound **25** demonstrated high binding affinity at sst_2 with >1000-fold selectivity compared to other ssts. Further studies by these investigators led to the indole analogue **26**. This compound was highly selective at sst_2 with moderate affinity for sst_4 [100]. Using combinatorial methods, the Merck group was able to discover selective agonists at all five ssts [101]. These combinatorial libraries were based on molecular modeling studies using known peptide ligands. The most active sst4 agonist from their study was the guanidine-analogue (L-803,087, **27**, Fig. **6**). L-803,087 was highly selective at sst4 in comparison with other ssts. The compound failed to exhibit actions associated with sst_2 or sst_5 including inhibition of growth hormone, insulin, and glucagon release [75].

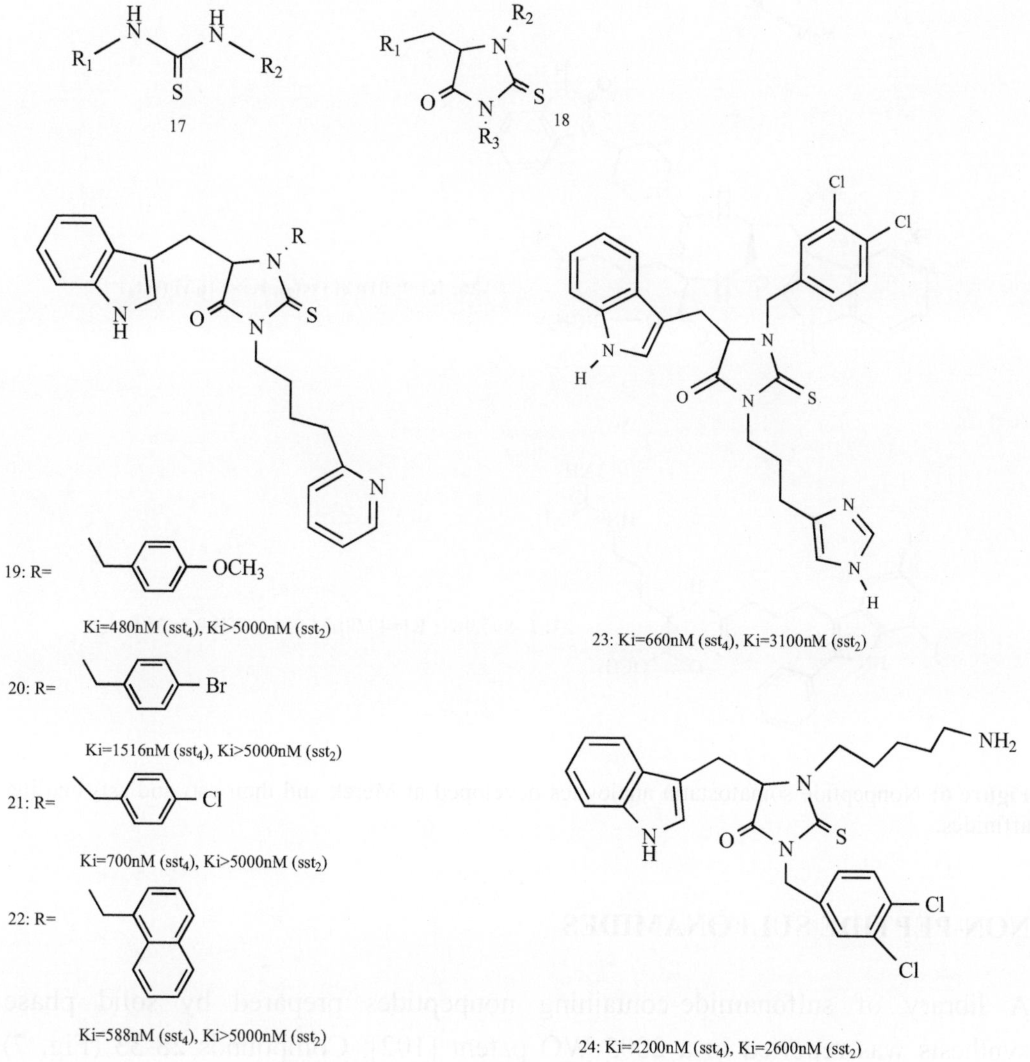

19: R=   Ki=480nM (sst4), Ki>5000nM (sst2)

20: R=   Ki=1516nM (sst4), Ki>5000nM (sst2)

21: R=   Ki=700nM (sst4), Ki>5000nM (sst2)

22: R=   Ki=588nM (sst4), Ki>5000nM (sst2)

23: Ki=660nM (sst4), Ki=3100nM (sst2)

24: Ki=2200nM (sst4), Ki=2600nM (sst2)

Figure 5: Structures of thiourea moiety (**17**), thiohydantoin nucleus (**18**), and representative thiohydantoins **19-24** with their sst4 and sst2 binding affinities.

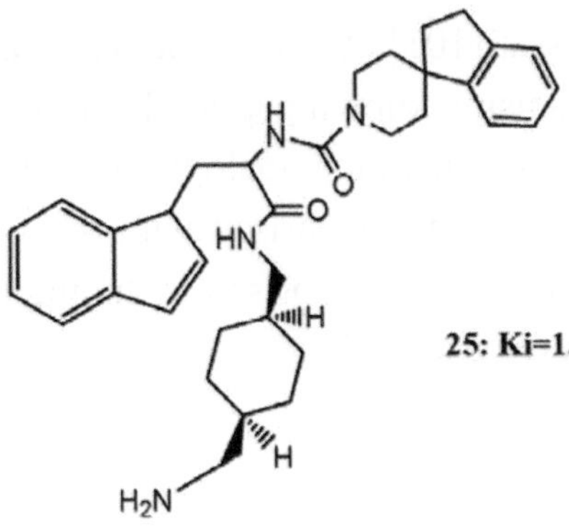

25: Ki=1.6nM (sst$_2$), Ki=2000nM (sst$_4$)

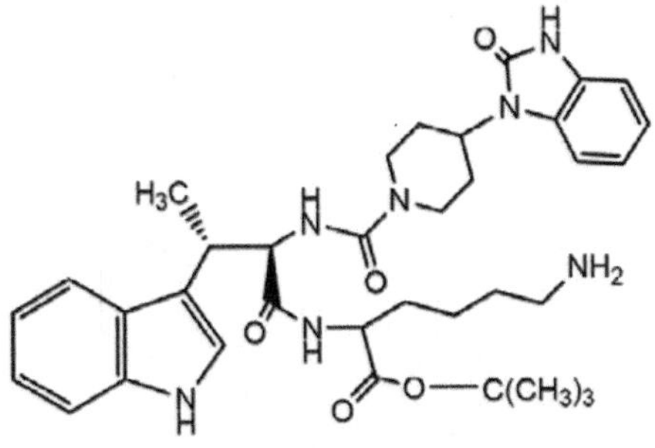

26: Ki=0.01nM (sst$_2$), Ki=84nM (sst$_4$)

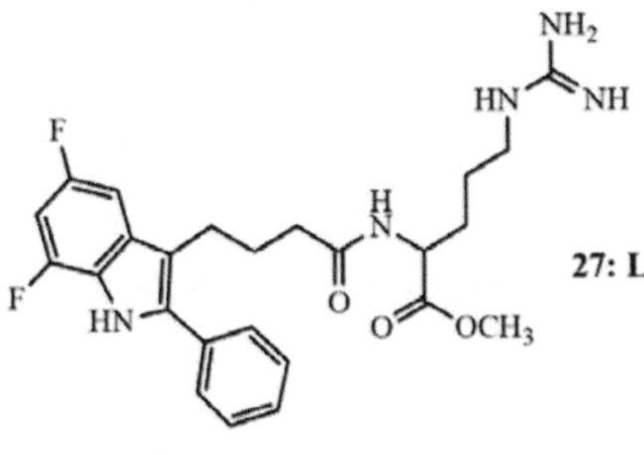

27: L-803,087; Ki=4720nM (sst$_2$), Ki=0.7nM (sst$_4$)

Figure 6: Nonpeptide somatostatin analogues developed at Merck and their sst$_4$ and sst$_2$ binding affinities.

NON-PEPTIDE SULFONAMIDES

A library of sulfonamide-containing nonpeptides prepared by solid phase synthesis was reported in a 2005 WO patent [102]. Compounds **28-33** (Fig. **7**) showed sst$_4$ binding affinities less than 5nM and were shown to be full agonists.

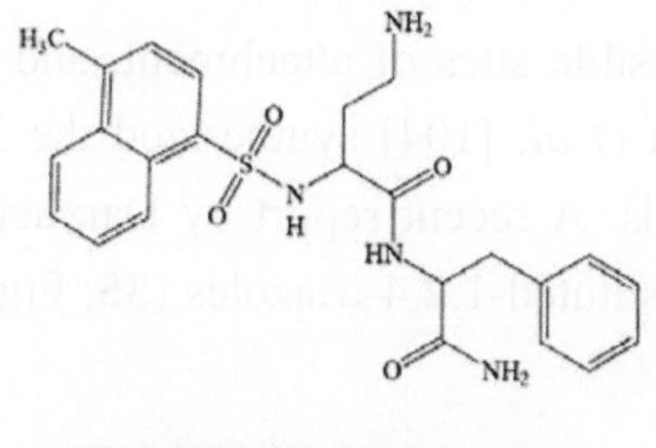

28: J-2156, Ki=1.2nM (sst$_4$)

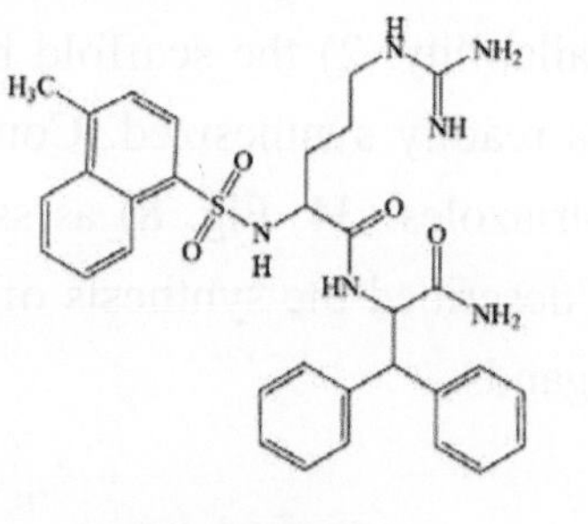

29: Ki=3.6nM (sst$_4$)

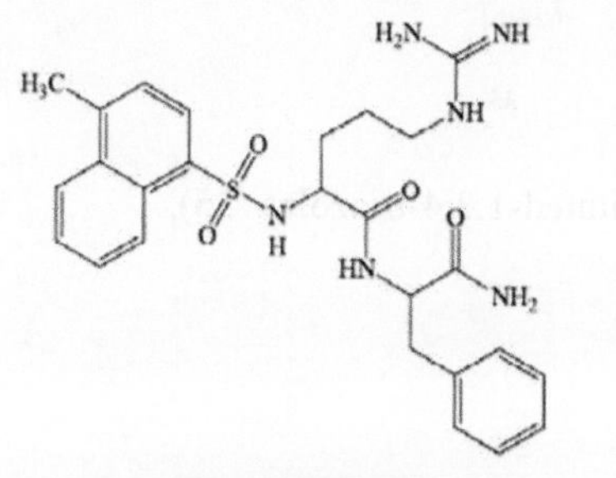

30: Ki=1.5nM (sst$_4$)

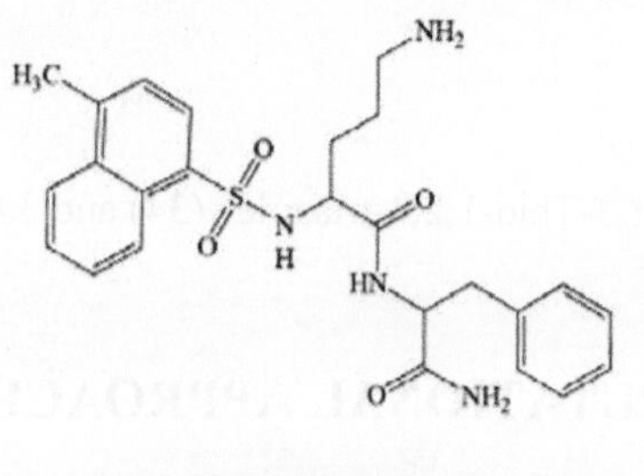

31: Ki=6.5nM (sst$_4$)

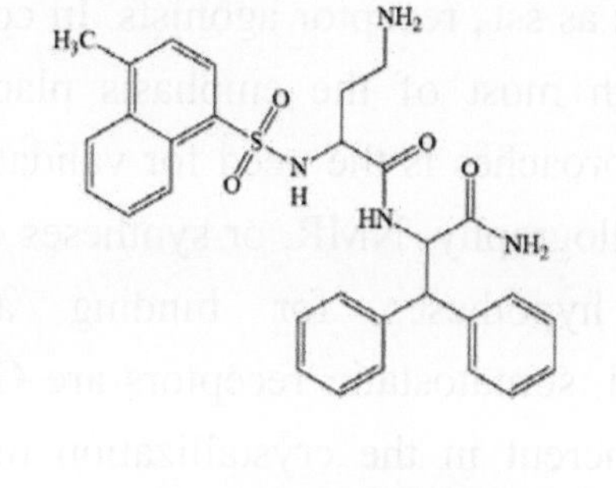

32: Ki=6.5nM (sst$_4$)

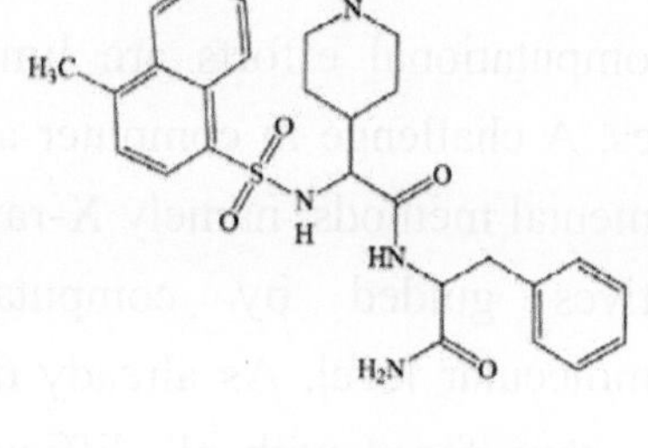

33: Ki=3.2nM (sst$_4$)

Figure 7: Structures of sulfonamide nonpeptides and their sst$_4$ binding affinities.

3,4,5-TRISUBSTITUTED-1,2,4-TRIAZOLES

The 1,2,4-triazole nucleus has been utilized as a scaffold to prepare SRIF nonpeptide ligands. This heterocycle is an excellent scaffold for several reasons:

1) the triazole ring is an amide bond isostere [103] with possibly enhanced bioavailability, 2) the scaffold has three possible sites of attachment, and 3) the ring is readily synthesized. Contour-Galcera *et al.* [104] synthesized the 3-thio-1,2,4-triazoles (**34**, Fig. **8**) as sst_2/sst_5 ligands. A recent report by Daryaei *et al.* [105] described the synthesis of 3,4,5-trisubstituted-1,2,4-triazoles (**35**, Fig. **8**) as sst_4 ligands.

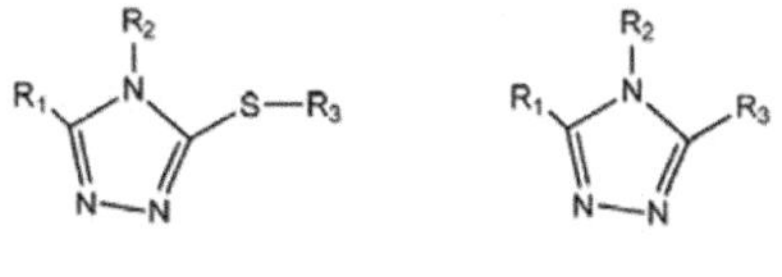

Figure 8: 3-Thio-1,2,4-triazoles (**34**) and 3,4,5-trisubstituted-1,2,4-triazoles (**35**).

COMPUTATIONAL APPROACHES

Following the preceding discussion of sst_4 ligands reported to date, it becomes apparent that extensive synthetic research has been undertaken toward peptidomimetics and small molecule ligands as sst_4 receptor agonists. In contrast, past computational efforts are limited with most of the emphasis placed on peptides. A challenge in computer-aided approaches is the need for validation by experimental methods, namely X-ray crystallography, NMR, or syntheses of new derivatives guided by computational hypotheses for binding at the macromolecular level. As already discussed, somatostatin receptors are GPCRs, and are thus faced with all difficulties inherent in the crystallization of these systems [106]. It was only in the last couple of years that GPCR advances were made toward obtaining structures of agonist-bound receptors [107-113], while previous successes were representative of antagonist-bound conformations of GPCRs [114, 115]. Because somatostatin receptors mediate the inhibitory effects of somatostatin, sst_4-selective ligands are agonists and therefore bind to the active state receptors. Consequently, the lack of available GPCR structures has become a deterrent to structure-based modeling of ssts, while peptide modeling assisted by

NMR offered a safer and more feasible alternative [116]. Still understanding somatostatin receptor subtype selectivity has been evasive, while computer-aided ligand discovery for binding to sst$_4$ is rather scarce.

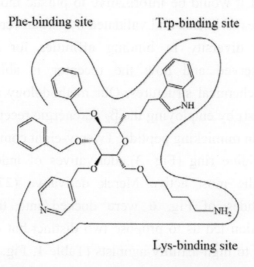

Figure 9: Proposed interactions with the receptor. Three binding sites were speculated to exist at the binding pocket of sst$_4$ [91].

The only relevant modeling attempts regarding small molecule sst$_4$ ligands date back to 2003 where various substituents on glycosides were used to probe binding to the receptor [91]. It was proposed that three sites existed in the binding pocket, that is a site binding phenylalanine which could accommodate a benzene ring, a tryptophan-binding site that could accommodate either an indole or a phenyl, but could not tolerate an heterocycle, and a lysine-binding site with which moieties such as pyridine, phenyl, and indole formed interactions (Fig. **9**). However, it was also speculated that in derivatives such as **2** (Fig. **2**), the indole and pyridine interacted with the Trp- and Lys-binding sites, respectively.

Despite recent advances in solving GPCR structures, several reviews have indicated that better templates than the existing crystal structures are needed for the majority of unavailable GPCRs in order to facilitate homology modeling and structure-building [117-119]. An interesting proposition was made by Mobarec *et al.* that the crystal structures of sst4/5 would be high-impact templates, would their 3D structures be resolved [118].

Consequently, we felt it would be informative to pursue model-building of sst4, followed by docking experiments and validated by virtual screening [120]. It was intriguing that such diversity in binding affinities for structurally similar compounds was observed and how the receptor is able to accommodate substantially diverse chemical structures. Our methodology aimed at generating an activated state of sst4 by employing the β_2 adrenergic receptor crystal structure bound with a G protein mimicking peptide. Twenty-eight compounds representing thioureas with a pyridine ring (Fig. **3**), derivatives of indole and phenylethyl thioureas (Fig. **4**), the most active Merck derivative (**27**, Fig. **6**), and all sulfonamide nonpeptides of Fig. **6** were docked into the sst4 constructed model. Visual inspection led us to propose two distinct but overlapping binding modes corresponding to high-affinity agonists (Table **1**, Fig. **10**), contrary to the low-affinity binders which lack the majority of these interactions. Virtual screening of a 996-compound library seeded with the 20 active compounds retrieved all actives when the docking algorithm Glide was employed with its internal scoring function [121, 122]. In conclusion, we were able to provide an explanation for observed binding data and narrow down the differences in interactions with either Gln243 or Asp90, in binding modes I and II, respectively.

Table 1:   Summary of interactions for proposed binding modes in the sst4/agonist complexes

sst4 Groups/Residues Involved in Binding	Binding Mode I Moieties	Binding Mode II Moieties
Hydrophobic cavity	Dichloro-benzyl and 3-bromo-pyridyl rings	Methylnaphthalene
Aromatic pocket	Imidazole	Benzene
Asp90 in helix 3	Urea	Amine
His258 in helix 7	Urea	Sulfonamide
Gln243 in helix 5	Imidazole	Amide
Trp171 backbone in extracellular loop 2	N/A	Sulfonamide

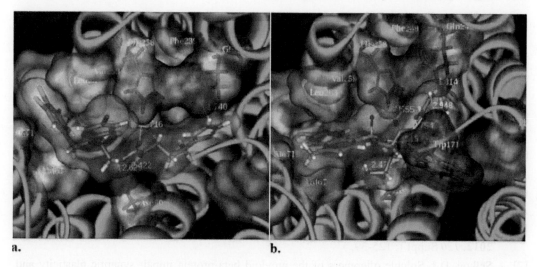

a.   b.

Figure 10: Structure-based aided studies suggested two binding modes I (panel **a**) and II (panel **b**). The urea amine-hydrogens (depicted compound **7**, Fig. **3**) interact with Asp90 of the transmembrane helix 3, while the imidazole ring hydrogen-bonds with Gln243 (**a**). In binding mode II (panel **b**), the basic amino group interacts with Asp90 (compound **28**, Fig. **7**), contrary to binding mode I where the basic imidazole interacts with Gln 243. Also depicted are hydrogens bonds between the sulfonyl and amide oxygens with His258 (see also Table **1**).

CONCLUSIONS

The sst$_4$ receptor, one of the five sst receptor types, is expressed in key regions of the brain affected by AD. In this chapter, we summarize what is known about design, synthesis, and molecular recognition of non-peptide sst$_4$ agonists, in addition to their impacts in mouse models of AD. Findings suggesting that sst$_4$ agonists may be a novel therapeutic approach for AD are discussed.

ACKNOWLEDGEMENTS

This work was supported by the National Institutes of Health National Institute on Aging (Grant: R21AG029318).

CONFLICT OF INTEREST

The authors confirm that this chapter contents have no conflict of interest.

REFERENCES

[1] 2012 Alzheimer's disease facts and figures. *Alzheimers Dement*, **2012**. *8*(2): 131-68.

[2] Mancuso, C., *et al*. Pharmacologists and Alzheimer disease therapy: to boldly go where no scientist has gone before. *Expert Opin Investig Drugs*, **2011**. *20*(9): 1243-61.

[3] Serrano-Pozo, A., *et al*. Neuropathological alterations in Alzheimer disease. *Cold Spring Harb Perspect Med*, **2011**. *1*(1): a006189.

[4] Mudher, A.; Lovestone, S. Alzheimer's disease-do tauists and baptists finally shake hands? *Trends Neurosci*, **2002**. *25*(1): 22-6.

[5] Selkoe, D.J. Resolving controversies on the path to Alzheimer's therapeutics. *Nat Med*, **2011**. *17*(9): 1060-5.

[6] Larson, M.E., Lesne, S.E. Soluble Abeta oligomer production and toxicity. *J Neurochem*, **2012**. *120* Suppl 1: 125-39.

[7] Selkoe, D.J. Soluble oligomers of the amyloid beta-protein impair synaptic plasticity and behavior. *Behav Brain Res*, **2008**. *192*(1): 106-13.

[8] Xu, X. Gamma-secretase catalyzes sequential cleavages of the AbetaPP transmembrane domain. *J Alzheimers Dis*, **2009**. *16*(2): 211-24.

[9] Gregory, G.C., Halliday, G.M. What is the dominant Abeta species in human brain tissue? A review. *Neurotox Re*s, **2005**. *7*(1-2): 29-41.

[10] Zou, K., *et al*. Amyloid beta-protein (Abeta)1-40 protects neurons from damage induced by Abeta1-42 in culture and in rat brain. *J Neurochem*, **2003**. *87*(3): 609-19.

[11] McLean, C.A., *et al*. Soluble pool of Abeta amyloid as a determinant of severity of neurodegeneration in Alzheimer's disease. *Ann Neurol*, **1999**. *46*(6): 860-6.

[12] Mucke, L., *et al*. High-level neuronal expression of abeta 1-42 in wild-type human amyloid protein precursor transgenic mice: synaptotoxicity without plaque formation. *J Neurosci*, **2000**. *20*(11): 4050-8.

[13] Terry, R.D., *et al*. Physical basis of cognitive alterations in Alzheimer's disease: synapse loss is the major correlate of cognitive impairment. *Ann Neurol*, **1991**. *30*(4): 572-80.

[14] Hung, L.W., *et al*. Amyloid-beta peptide (Abeta) neurotoxicity is modulated by the rate of peptide aggregation: Abeta dimers and trimers correlate with neurotoxicity. *J Neurosci*, **2008**. *28*(46): 11950-8.

[15] Lesne, S., *et al*. A specific amyloid-beta protein assembly in the brain impairs memory. *Nature*, **2006**. *440*(7082): 352-7.

[16] Oddo, S., *et al*. Abeta immunotherapy leads to clearance of early, but not late, hyperphosphorylated tau aggregates via the proteasome. *Neuron*, **2004**. *43*(3): 321-32.

[17] Oddo, S., *et al*. Temporal profile of amyloid-beta (Abeta) oligomerization in an in vivo model of Alzheimer disease. A link between Abeta and tau pathology. *J Biol Chem*, **2006**. *281*(3): 1599-604.

[18]   Jin, M., *et al*. Soluble amyloid beta-protein dimers isolated from Alzheimer cortex directly induce Tau hyperphosphorylation and neuritic degeneration. *Proc Natl Acad Sci U S A*, **2011**. *108*(14): 5819-24.

[19]   De Felice, F.G., *et al*. Alzheimer's disease-type neuronal tau hyperphosphorylation induced by A beta oligomers. *Neurobiol Aging*, **2008**. *29*(9): 1334-47.

[20]   Deane, R., *et al*. Clearance of amyloid-beta peptide across the blood-brain barrier: implication for therapies in Alzheimer's disease. *CNS Neurol Disord Drug Targets*, **2009**. *8*(1): 16-30.

[21]   Jeynes, B.; Provias, J. The case for blood-brain barrier dysfunction in the pathogenesis of Alzheimer's disease. *J Neurosci Res*, **2011**. *89*(1): 22-8.

[22]   Nalivaeva, N.N., *et al*. Are amyloid-degrading enzymes viable therapeutic targets in Alzheimer's disease? *J Neurochem*, **2012**. *120* Suppl 1: 167-85.

[23]   Howell, S.; Nalbantoglu, J.; Crine, P. Neutral endopeptidase can hydrolyze beta-amyloid(1-40) but shows no effect on beta-amyloid precursor protein metabolism. *Peptides*, **1995**. *16*(4): 647-52.

[24]   Iwata, N., *et al*. Identification of the major Abeta1-42-degrading catabolic pathway in brain parenchyma: suppression leads to biochemical and pathological deposition. *Nat Med*, **2000**. *6*(2): 143-50.

[25]   Meilandt, W.J., *et al*. Neprilysin overexpression inhibits plaque formation but fails to reduce pathogenic Abeta oligomers and associated cognitive deficits in human amyloid precursor protein transgenic mice. *J Neurosci*, **2009**. *29*(7): 1977-86.

[26]   Poirier, R., *et al*. Neuronal neprilysin overexpression is associated with attenuation of Abeta-related spatial memory deficit. *Neurobiol Dis*, **2006**. *24*(3): 475-483.

[27]   Saito, T., *et al*. Somatostatin regulates brain amyloid beta peptide Abeta42 through modulation of proteolytic degradation. *Nat Med*, **2005**. *11*(4): 434-9.

[28]   Zhuravin, I.A., *et al*. Epigenetic and pharmacological regulation of the amyloid-degrading enzyme neprilysin results in modulation of cognitive functions in mammals. *Dokl Biol Sci*, **2011**. *438*: 145-8.

[29]   Pollard, H., *et al*. Detailed immunoautoradiographic mapping of enkephalinase (EC 3.4.24.11) in rat central nervous system: comparison with enkephalins and substance P. *Neuroscience*, **1989**. *30*(2): 339-76.

[30]   Waksman, G., *et al*. Neuronal localization of the neutral endopeptidase 'enkephalinase' in rat brain revealed by lesions and autoradiography. *EMBO J*, **1986**. *5*(12): 3163-6.

[31]   Waksman, G., *et al*. Autoradiographic comparison of the distribution of the neutral endopeptidase "enkephalinase" and of mu and delta opioid receptors in rat brain. *Proc Natl Acad Sci U S A*, **1986**. *83*(5): 1523-7.

[32]   Yasojima, K., *et al*. Reduced neprilysin in high plaque areas of Alzheimer brain: a possible relationship to deficient degradation of beta-amyloid peptide. *Neurosci Lett*, **2001**. *297*(2): 97-100.

[33] Yasojima, K.; McGeer, E.G.; McGeer, P.L. Relationship between beta amyloid peptide generating molecules and neprilysin in Alzheimer disease and normal brain. *Brain Res*, **2001**. *919*(1): 115-21.

[34] Miners, J.S., *et al*. Decreased expression and activity of neprilysin in Alzheimer disease are associated with cerebral amyloid angiopathy. *J Neuropathol Exp Neurol*, **2006**. *65*(10): 1012-21.

[35] Wang, D.S., *et al*. Decreased neprilysin immunoreactivity in Alzheimer disease, but not in pathological aging. *J Neuropathol Exp Neurol*, **2005**. *64*(5): 378-85.

[36] Wang, S., *et al*. Expression and functional profiling of neprilysin, insulin-degrading enzyme, and endothelin-converting enzyme in prospectively studied elderly and Alzheimer's brain. *J Neurochem*, **2010**. *115*(1): 47-57.

[37] Iwata, N., *et al*. Metabolic regulation of brain Abeta by neprilysin. *Science*, **2001**. *292*(5521): 1550-2.

[38] Madani, R., *et al*. Lack of neprilysin suffices to generate murine amyloid-like deposits in the brain and behavioral deficit in vivo. *J Neurosci Res*, **2006**. 84(8): 1871-8.

[39] Hama, E., *et al*. Clearance of extracellular and cell-associated amyloid beta peptide through viral expression of neprilysin in primary neurons. *J Biochem*, **2001**. *130*(6): 721-6.

[40] Leissring, M.A., *et al*. Enhanced proteolysis of beta-amyloid in APP transgenic mice prevents plaque formation, secondary pathology, and premature death. *Neuron*, **2003**. *40*(6): 1087-93.

[41] Marr, R.A., *et al*. Neprilysin gene transfer reduces human amyloid pathology in transgenic mice. *J Neurosci*, **2003**. *23*(6): 1992-6.

[42] Spencer, B., *et al*. Long-term neprilysin gene transfer is associated with reduced levels of intracellular Abeta and behavioral improvement in APP transgenic mice. *BMC Neurosci*, **2008**. *9*: 109.

[43] Iwata, N., *et al*. Presynaptic localization of neprilysin contributes to efficient clearance of amyloid-beta peptide in mouse brain. *J Neurosci*, **2004**. *24*(4): 991-8.

[44] Huang, S.M., *et al*. Neprilysin-sensitive synapse-associated amyloid-beta peptide oligomers impair neuronal plasticity and cognitive function. *J Biol Chem*, **2006**. *281*(26): 17941-51.

[45] Iwata, N., Higuchi, M., Saido, T.C. Metabolism of amyloid-beta peptide and Alzheimer's disease. *Pharmacol Ther*, **2005**. *108*(2): 129-48.

[46] Kanemitsu, H., Tomiyama, T., Mori, H. Human neprilysin is capable of degrading amyloid beta peptide not only in the monomeric form but also the pathological oligomeric form. *Neurosci Lett*, **2003**. *350*(2): 113-6.

[47] Bailey, J.A., *et al*. Functional activity of the novel Alzheimer's amyloid beta-peptide interacting domain (AbetaID) in the APP and BACE1 promoter sequences and implications in activating apoptotic genes and in amyloidogenesis. *Gene*, **2011**. *488*(1-2): 13-22.

[48] Hardy, J. The amyloid hypothesis for Alzheimer's disease: a critical reappraisal. *J Neurochem*, **2009**. *110*(4): 1129-34.

[49]   Puzzo, D., *et al*. Endogenous amyloid-beta is necessary for hippocampal synaptic plasticity and memory. *Ann Neurol*, **2011**. *69*(5): 819-30.

[50]   Morley, J.E., Farr, S.A. Hormesis and amyloid-beta protein: physiology or pathology? *J Alzheimers Dis*, **2012**. *29*(3): 487-92.

[51]   Kumar, U., Grant, M. Somatostatin and somatostatin receptors. *Results Probl Cell Differ*, **2010**. *50*: 137-84.

[52]   Moller, L.N., *et al*. Somatostatin receptors. *Biochim Biophys Acta*, **2003**. *1616*(1): 1-84.

[53]   Bruno, J.F., *et al*. Molecular cloning and functional expression of a brain-specific somatostatin receptor. *Proc Natl Acad Sci U S A*, **1992**. *89*(23): 11151-5.

[54]   Epelbaum, J., *et al*. The neurobiology of somatostatin. *Crit Rev Neurobiol*, **1994**. *8*(1-2): 25-44.

[55]   Kumar, U. Expression of somatostatin receptor subtypes (SSTR1-5) in Alzheimer's disease brain: an immunohistochemical analysis. *Neuroscience*, **2005**. *134*(2): 525-38.

[56]   Spary, E.J. Maqbool, A. Batten, T.F. Expression and localisation of somatostatin receptor subtypes sst1-sst5 in areas of the rat medulla oblongata involved in autonomic regulation. *J Chem Neuroanat*, **2008**. *35*(1): 49-66.

[57]   Fleisher-Berkovich, S., *et al*. Distinct modulation of microglial amyloid beta phagocytosis and migration by neuropeptides (i). *J Neuroinflammation*, **2010**. *7*: 61.

[58]   Viollet, C., *et al*. Somatostatinergic systems in brain: networks and functions. *Mol Cell Endocrinol*, **2008**. *286*(1-2): 75-87.

[59]   Chen, H.X., *et al*. Long-term potentiation of excitatory synapses on neocortical somatostatin-expressing interneurons. *J Neurophysiol*, **2009**. *102*(6): 3251-9.

[60]   Tallent, M.K. Somatostatin in the dentate gyrus. *Prog Brain Res*, **2007**. *163*: 265-84.

[61]   Gastambide, F., *et al*. Cooperation between hippocampal somatostatin receptor subtypes 4 and 2: functional relevance in interactive memory systems. *Hippocampus*, **2010**. *20*(6): 745-57.

[62]   Gastambide, F., *et al*. Hippocampal SSTR4 somatostatin receptors control the selection of memory strategies. *Psychopharmacology (Berl)*, **2009**. *202*(1-3): 153-63.

[63]   Beal, M.F., *et al*. Widespread reduction of somatostatin-like immunoreactivity in the cerebral cortex in Alzheimer's disease. *Ann Neurol*, **1986**. *20*(4): 489-95.

[64]   Davies, P., Katzman, R., Terry, R.D. Reduced somatostatin-like immunoreactivity in cerebral cortex from cases of Alzheimer disease and Alzheimer senile dementa. *Nature*, **1980**. *288*(5788): 279-80.

[65]   Davis, K.L., *et al*. CSF somatostatin in Alzheimer's disease, depressed patients, and control subjects. *Biol Psychiatry*, **1988**. *24*(6): 710-2.

[66]   Dournaud, P., *et al*. Differential correlation between neurochemical deficits, neuropathology, and cognitive status in Alzheimer's disease. *Neurobiol Aging*, **1995**. *16*(5): 817-23.

[67]   Burgos-Ramos, E., *et al*. Somatostatin and Alzheimer's disease. *Mol Cell Endocrinol*, **2008**. *286*(1-2): 104-11.

[68]   Epelbaum, J., *et al*. Somatostatin, Alzheimer's disease and cognition: an old story coming of age? *Prog Neurobiol*, **2009**. *89*(2): 153-61.

[69]   Craft, S., *et al*. Enhancement of memory in Alzheimer disease with insulin and somatostatin, but not glucose. *Arch Gen Psychiatry*, **1999**. *56*(12): 1135-40.

[70]   Grouselle, D., *et al*. Loss of somatostatin-like immunoreactivity in the frontal cortex of Alzheimer patients carrying the apolipoprotein epsilon 4 allele. *Neurosci Lett*, **1998**. *255*(1): 21-4.

[71]   Vepsalainen, S., *et al*. Somatostatin genetic variants modify the risk for Alzheimer's disease among Finnish patients. *J Neurol*, **2007**. *254*(11): 1504-8.

[72]   Xue, S., Jia, L., Jia, J. Association between somatostatin gene polymorphisms and sporadic Alzheimer's disease in Chinese population. *Neurosci Lett*, **2009**. *465*(2): 181-3.

[73]   Ramos, B., *et al*. Early neuropathology of somatostatin/NPY GABAergic cells in the hippocampus of a PS1xAPP transgenic model of Alzheimer's disease. *Neurobiol Aging*, **2006**. *27*(11): 1658-72.

[74]   Hama, E., Saido, T.C. Etiology of sporadic Alzheimer's disease: somatostatin, neprilysin, and amyloid beta peptide. *Med Hypotheses*, **2005**. *65*(3): 498-500.

[75]   Rohrer, S.P., Schaeffer, J.M. Identification and characterization of subtype selective somatostatin receptor agonists. *J Physiol Paris,* **2000**. *94*(3-4): 211-5.

[76]   Ankersen, M., *et al*. Discovery of a Novel Non-Peptide Somatostatin Agonist with SST4 Selectivity. *Journal of the American Chemical Society*, **1998**. *120*(7): 1368-1373.

[77]   Crider, A.M., *et al*. Somatostatin receptor subtype 4 (sst4) ligands: Synthesis and evaluation of indol-3-yl- and 2-pyridyl-thioureas. *Letters in Drug Design & Discovery*, **2004**. *1*(1): 84-87.

[78]   Sandoval, K.E., *et al*. Chronic peripheral administration of somatostatin receptor subtype-4 agonist NNC 26-9100 enhances learning and memory in SAMP8 mice. *Eur J Pharmacol*, **2011**. *654*(1): 53-9.

[79]   Marugan, J.J., *et al*. Non-iminosugar glucocerebrosidase small molecule chaperones. *Med. Chem. Commun.* **2012**. *3*(1): 56-60.

[80]   Morley, J.E., *et al*. The senescence accelerated mouse (SAMP8) as a model for oxidative stress and Alzheimer's disease. *Biochim Biophys Acta*, **2012**. *1822*(5): 650-6.

[81]   Morley, J.E., *et al*. The SAMP8 mouse: a model to develop therapeutic interventions for Alzheimer's disease. *Curr Pharm Des*, **2012**. *18*(8): 1123-30.

[82]   Kumar, V.B., *et al*. Identification of age-dependent changes in expression of senescence-accelerated mouse (SAMP8) hippocampal proteins by expression array analysis. *Biochem Biophys Res Commun*, **2000**. *272*(3): 657-61.

[83]   Morley, J.E., *et al*. Beta-amyloid precursor polypeptide in SAMP8 mice affects learning and memory. *Peptides*, **2000**. *21*(12): 1761-7.

[84]   Rosenberg, R.N. The molecular and genetic basis of AD: the end of the beginning: the 2000 Wartenberg lecture. *Neurology*, **2000**. *54*(11): 2045-54.

[85]   Sandoval, K.E., *et al*. Somatostatin receptor subtype-4 agonist NNC 26-9100 decreases extracellular and intracellular Abeta(1-42) trimers. *Eur J Pharmacol*, **2012**.

[86]   Townsend, M., *et al*. Effects of secreted oligomers of amyloid beta-protein on hippocampal synaptic plasticity: a potent role for trimers. *J Physiol*, **2006**. *572*(Pt 2): 477-92.

[87]   Zhang, Y., *et al*. Administration of amyloid-beta42 oligomer-specific monoclonal antibody improved memory performance in SAMP8 mice. *J Alzheimers Dis*, **2011**. *23*(3): 551-61.

[88]   Liu, S., *et al*. 2-pyridylthioureas: novel nonpeptide somatostatin agonists with SST4 selectivity. *Curr Pharm Des*, **1999**. *5*(4): 255-63.

[89]   Hirschmann, R., *et al*. Nonpeptidal peptidomimetics with beta.-D-glucose scaffolding. A partial somatostatin agonist bearing a close structural relationship to a potent, selective substance P antagonist. *Journal of the American Chemical Society*, **1992**. *114*(23): 9217-9218.

[90]   Hirschmann, R., *et al*. De novo design and synthesis of somatostatin non-peptide peptidomimetics utilizing beta-D-glucose as a novel scaffolding. *Journal of the American Chemical Society*, **1993**. *115*(26): 12550-12568.

[91]   Prasad, V., *et al*. Effects of heterocyclic aromatic substituents on binding affinities at two distinct sites of somatostatin receptors. Correlation with the electrostatic potential of the substituents. *J Med Chem*, **2003**. *46*(10): 1858-69.

[92]   Angeles, A.R., *et al*. Synthesis and binding affinities of novel SRIF-mimicking beta-D-glucosides satisfying the requirement for a pi-cloud at C1. *Org Lett*, **2005**. *7*(6): 1121-4.

[93]   Gouin, S.G., Murphy, P.V. Synthesis of somatostatin mimetics based on the 1-deoxymannojirimycin scaffold. *J Org Chem*, **2005**. *70*(21): 8527-32.

[94]   Chagnault, V., Lalot, J., Murphy, P.V. Synthesis of somatostatin mimetics based on 1-deoxynojirimycin. *ChemMedChem*, **2008**. *3*(7): 1071-6.

[95]   Liu, S., *et al*. Nonpeptide somatostatin agonists with sst4 selectivity: synthesis and structure-activity relationships of thioureas. *J Med Chem*, **1998**. *41*(24): 4693-705.

[96]   Crider, A.M., Witt, K.A. Somatostatin sst4 ligands: chemistry and pharmacology. *Mini Rev Med Chem*, **2007**. *7*(3): 213-20.

[97]   Crider, A.M., Liu, S., Mahajan, S., Ankersen, M., Stidsen, C.E. Somatostatin Receptor Subtype 4 (sst 4) Ligands: Synthesis and Evaluation of Indol-3-yl-and 2-Pyridyl-thioureas. *Letters in Drug Design and Discovery*, **2004**. *1*: 84-87.

[98]   Wang, X., Mealer, D., Rodgers, L., Sandoval, K., Witt, K.A., Stidsen, C., Ankersen, M., Crider, A.M. Synthesis of 2-thiohydantoins as somatostatin subtype 4 receptor ligands. *Letters Drug Design Discovery* **2012**. *9*: 655-662.

[99]   Yang, L., *et al*. Spiro[1H-indene-1,4'-piperidine] derivatives as potent and selective non-peptide human somatostatin receptor subtype 2 (sst2) agonists. *J Med Chem*, **1998**. *41*(13): 2175-9.

[100] Yang, L., *et al*. Synthesis and biological activities of potent peptidomimetics selective for somatostatin receptor subtype 2. *Proc Natl Acad Sci U S A*, **1998**. *95*(18): 10836-41.

[101] Rohrer, S.P., *et al*. Rapid identification of subtype-selective agonists of the somatostatin receptor through combinatorial chemistry. *Science*, **1998**. 282(5389): 737-40.

[102] Tomperi, J., Engstrom, M., Wurster, S. Preparation of sulfonylamino-peptidomimetics active on the somatostatin receptor subtypes 4 (SSTR4) and 1 (SSTR1). **2005**: Oy Juvantia Pharma Ltd., Finland, WO 20055033124.

[103] Moulin, A., *et al*. Synthesis of 3,4,5-trisubstituted-1,2,4-triazoles. *Chem Rev*, **2010**. *110*(4): 1809-27.

[104] Contour-Galcera, M.O., *et al*. 3-Thio-1,2,4-triazoles, novel somatostatin sst2/sst5 agonists. *Bioorg Med Chem Lett*, **2005**. *15*(15): 3555-9.

[105] Daryaei, I., Crider, A.M., Sandoval, K., Witt, K. Synthesis of trisubstituted-1,2,4-triazoles as somatostatin subtype 4 receptor agonists in 243rd ACS Meeting **2012**: San Diego, CA.

[106] McCusker, E.C., *et al*. Heterologous GPCR expression: a bottleneck to obtaining crystal structures. *Biotechnol Prog*, **2007**. *23*(3): 540-7.

[107] Warne, T., *et al*. The structural basis for agonist and partial agonist action on a beta(1)-adrenergic receptor. *Nature*, **2011**. *469*(7329): 241-4.

[108] Lebon, G., *et al*. Agonist-bound adenosine A2A receptor structures reveal common features of GPCR activation. *Nature*, **2011**. *474*(7352): 521-5.

[109] Rasmussen, S.G., *et al*. Structure of a nanobody-stabilized active state of the beta(2) adrenoceptor. *Nature*, **2011**. *469*(7329): 175-80.

[110] Rasmussen, S.G., *et al*. Crystal structure of the beta2 adrenergic receptor-Gs protein complex. *Nature*, **2011**. *477*(7366): 549-55.

[111] Kruse, A.C., *et al*. Structure and dynamics of the M3 muscarinic acetylcholine receptor. *Nature*. *482*(7386): 552-6.

[112] Rosenbaum, D.M., *et al*. Structure and function of an irreversible agonist-beta(2) adrenoceptor complex. *Nature*, **2011**. *469*(7329): 236-40.

[113] Xu, F., *et al*. Structure of an Agonist-Bound Human A2A Adenosine Receptor. *Science*, **2011**. *332*(6027): 322-327.

[114] Jaakola, V.P., *et al*. The 2.6 angstrom crystal structure of a human A2A adenosine receptor bound to an antagonist. *Science*, **2008**. *322*(5905): 1211-7.

[115] Chien, E.Y., *et al*. Structure of the human dopamine D3 receptor in complex with a D2/D3 selective antagonist. *Science*, **2010**. *330*(6007): 1091-5.

[116] Nikiforovich, G.V., Marshall, G.R., Achilefu, S. Molecular modeling suggests conformational scaffolds specifically targeting five subtypes of somatostatin receptors. *Chem Biol Drug Des*, **2007**. *69*(3): 163-9.

[117] Congreve, M., *et al*. Progress in structure based drug design for G protein-coupled receptors. *J Med Chem*, **2011**. *54*(13): 4283-311.

[118] Mobarec, J.C., Sanchez, R., Filizola, M. Modern homology modeling of G-protein coupled receptors: which structural template to use? *J Med Chem*, **2009**. *52*(16): 5207-16.

[119] Kontoyianni, M., Liu, Z. Structure-Based Design in the GPCR Target Space. *Curr Med Chem*, **2012**. *19*(4): 544-56.

[120] Liu, Z., *et al*. A structure-based approach to understanding somatostatin receptor-4 agonism (sst4). *J Chem Inf Model*, **2012**. *52*(1): 171-86.

[121] Friesner, R. A., *et al*. Glide: a new approach for rapid, accurate docking and scoring. 1. Method and assessment of docking accuracy. *J. Med. Chem*. **2004**. *47*(7): 1739-1749.

[122] Halgren, T. A., *et al*. Glide: a new approach for rapid, accurate docking and scoring. 2. Enrichment factors in database– screening. *J. Med. Chem*. **2004**. *47*(7): 1750-1759.

CHAPTER 13

Neprilysin Inhibitors Provide Insight into its Specificity and Therapeutic Potential

Darrick Pope and Michael Cascio[*]

Department of Chemistry and Biochemistry, Duquesne University, Pittsburgh, PA 15282, USA

Abstract: Neprilysin (NEP) is one of the enzymes in the zinc-metalloendopeptidase family that displays a broad specificity in degrading small bioactive peptides. Crystal structures of seven NEP-inhibitor complexes as well as biochemical characterization of NEP activity have highlighted amino acid interactions that are crucial to the binding of various ligands. Studies also indicate that NEP is one of a select group of metalloenzymes that degrade the amyloid beta peptide (Aβ) *in vivo* and *in situ*. The accumulation of neurotoxic Aβ aggregates in the brain appears to be a causative agent in the pathophysiology of Alzheimer's Disease (AD). For this reason the enzymatic degradation of Aβ has been studied extensively, but little is currently known about the specific interactions underlying NEP degradation of Aβ. Research that pertains to these interactions may lead to critical insights for utilizing NEP inhibition of Aβ accumulation as a safe, beneficial AD therapy.

Keywords: Alzheimer's disease, amyloid beta peptide, computational modeling, metallopeptidases, neprilysin, neurodegeneration, proteolysis, substrate-binding, zinc-dependent endopeptidase.

THE BINDING POCKET OF NEPRILYSIN

Neprilysin (also known as NEP, neutral endopeptidase, EC 3.4.24.11) is a type II integral membrane protein that belongs to the M13 subfamily of zinc-dependent endopeptidases [1]. NEP and the enzymes of its family have common binding and proteolytic properties that include the promiscuous degradation of small bioactive peptides. NEP has been shown to participate in cutting off the signaling activity of enkephalin, tachykinin, and angiotensin neuropeptides in addition to clearance of amyloid beta peptide (Aβ) by degradation of these substrates [2-4]. Studies of

***Address correspondence to Michael Cascio:** Department of Chemistry and Biochemistry, Duquesne University, Pittsburgh, PA 15282, USA; Tel: 412-396-1894; Fax: 412-396-5683; E-mail: casciom@duq.edu

Atta-ur-Rahman / Muhammad Iqbal Choudhary (Eds.)

10.1016/B978-0-12-803959-5.50013-1

brain metabolism first indicated that NEP regulates neuropeptide levels after its expression on the membrane surface of nerve cells [5-8].

The extracellular proteolytic domain of NEP is believed to limit the size of substrates to less than 5 kDa (approximately 45 amino acids) for hydrolysis in a zinc-containing binding pocket (see Fig. (**1**)) [1, 9]. Clearly, a more comprehensive understanding of the NEP binding site properties and the determination of common interactions with its many substrates is essential in elucidating the relative specificity of NEP, as well as providing insights regarding its potential therapeutic utility (discussed in more depth in later sections).

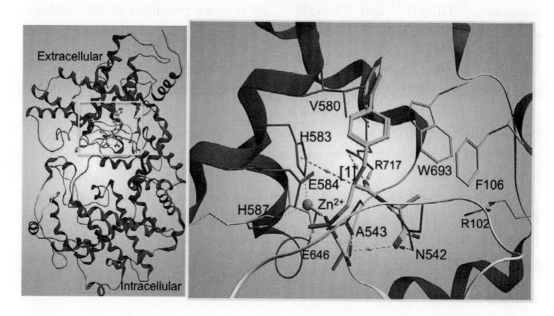

Figure 1: A cartoon representation of the NEP crystal structure with bound inhibitor 1 (cyan) (PDB 1R1H) [9]. The ribbons represent backbone loops (light blue), helices (red), turns (blue) and selected side-chains of binding (dark blue) or non-binding (gray) pocket residues are labeled in enlarged view of inset at right. The catalytic triad of H583, H587 and E646 coordinates the Zn ion (orange) in the binding pocket. Extracellular and Intracellular labels show relative orientation of this extracellular domain. Images were made using the Molecular Operating Environment (MOE) Program [10].

Zn^{2+} Coordination

Specific residues of the NEP binding pocket have been highlighted by structural and biochemical studies of NEP. Research indicates that these NEP residues participate in important binding interactions within the protease. Current knowledge of the NEP binding pocket centers on the amino acid residues surrounding its zinc (II) ion (Zn^{2+}) co-factor. Sequence alignment of zinc metallopeptidases highlighted these important residues in NEP by their conservation throughout families of mammalian and bacterial enzymes [11]. For example, two His, H, and one Glu, E, in NEP are found in a HExxH motif (where x is any amino acid) that is conserved in the NEP family and is essential to Zn^{2+}-coordinated substrate hydrolysis in both human NEP and the bacterial homologue thermolysin (TLN) [12]. X-ray crystallographic studies of seven different inhibitor-bound NEP complexes confirmed that two conserved consensus sequences, [583]HExxH[587] and [646]ExxxD[650], are in close proximity to the catalytic Zn^{2+} and each bound inhibitor [9, 13-15].

All seven structures indicate that the Zn^{2+} is coordinated near the ligand by the pocket residues H583, E646, and H587. Two of the inhibitors with phosphinic moieties (PR_2O_2H) (Fig. (**2a-b**)) have one oxygen atom located within hydrogen bonding distance of E584, which has been proposed to result from the replacement of a catalytic water molecule (Fig. (**3a**)) [9]. In addition, E584 appears to be involved in a complex hydrogen bonding network with N542 and A543 of a consensus NAFY motif (Table (**1**)) as well as R717, H711, and other residues around the binding pocket [13]. Similar Zn^{2+} coordinating and hydrogen bonding networks have been attributed to the closely related TLN enzyme based on studies of its structure [16, 17].

Binding Subsites

The orientation of NEP pocket residues around a ligand has been characterized by comparing closely related binding structures. The homologous NEP and TLN have similar ligand interactions occuring within comparable sub-regions of their respective binding pockets. These binding subsites are labeled according to convention with the letter S and a number representing their relative position, which correspond to ligand atoms that are similarly labeled using the letter P. Constituents of the S1 site in an enzyme interact with P1 atoms located on the amino (N) terminal side of a substrate's carbonyl-amide scissile bond, the bond in the peptide chain susceptible to cleavage. For substrate residues on the carboxyl (C) terminal side of the scissile bond, P1' atoms interact with enzyme residues of the S1' site. The numbers for S and P sites on either side of the scissile bond increase by one for each subsequent substrate residue. Unfortunately, all of the crystal structures of NEP, as well as those of TLN or other metallopeptidases in the NEP family, do not have bound endogenous peptide substrates, but rather are complexed with non-hydrolyzable competitive inhibitors. The NEP inhibitors are not peptides and do not contain a scissile bond, but each inhibitor contains structural similarities to peptides, including terminal N or C groups (Fig. (**2a-d**)). These peptide-like or peptidic structures in inhibitors facilitate the conventional assignment of S and P sites and a comparative analysis of inhibitor binding in the NEP crystal structures.

Table 1:   **Proposed roles for key residues identified in crystal structures of NEP complexes**

Residues	Structural Role
H583, H587, E646	The "catalytic triad" responsible for coordinating the position of Zn^{2+}
N542, A543, F544, Y545	A highly conserved NAFY motif that participates in ligand binding
E584, H711, R717	Participate in a complex hydrogen bonding network within the pocket
F106, V580, W693	Form a distinctive hydrophobic space in the S1' subsite
R102, D107, R110	Outline the far end of the S2' subsite

Binding of Inhibitors

The crystal structures of NEP complexes indicate that common structural interactions coordinate the Zn^{2+} and the inhibitors in the binding pocket [9, 15]. For inhibitor 1 (inhib1), the phosphinic group is coordinated with the Zn^{2+} and the P1 terminal N group comes into close proximity oxygen atoms from E584 and A543 of the binding pocket (Fig. (**3a**)). By this association, the E584 and A543 residues are considered to be part of the S1 site of the NEP pocket when bound to inhib1.

Conversely, E584 and A543 are considered part of the S1' site of NEP when found near the P1' amide group of other bound inhibitors [9]. The S1' site typically includes residues that surround a large, hydrophobic group from each of the inhibitors [9, 15]. An S1' specificity for these hydrophobic groups is attributed to the side-chain orientations of residues F106, V580, and W693 (Table (**1**)) [9, 13, 19].

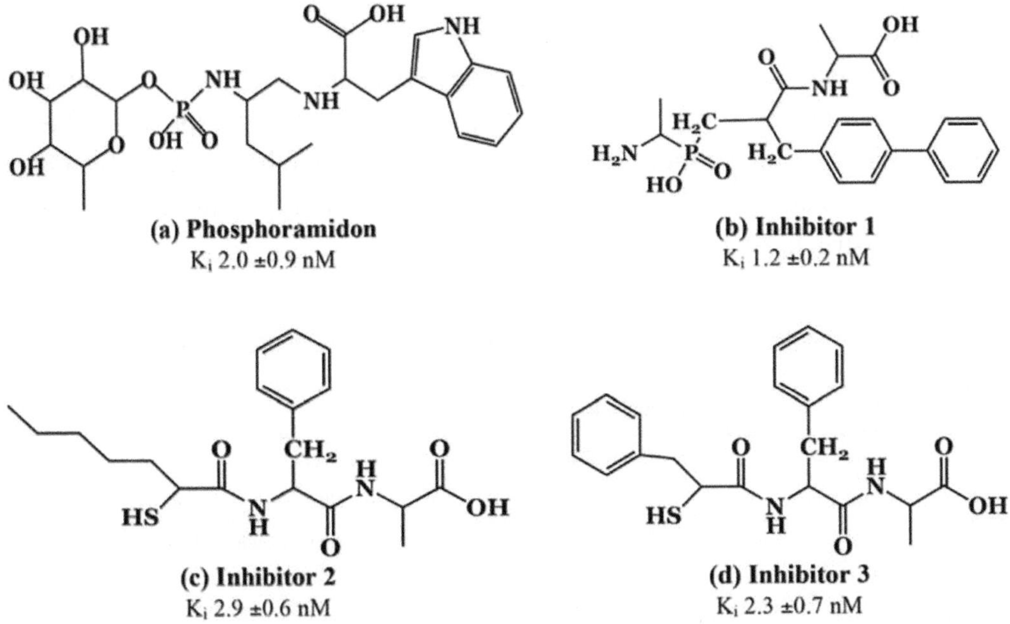

(a) **Phosphoramidon**
K_i 2.0 ±0.9 nM

(b) **Inhibitor 1**
K_i 1.2 ±0.2 nM

(c) **Inhibitor 2**
K_i 2.9 ±0.6 nM

(d) **Inhibitor 3**
K_i 2.3 ±0.7 nM

Asp-Ala-Glu-Phe-Arg-His-Asp-Ser-Gly-Tyr-Glu-Val-His-His-Gln-Lys-Leu-Val-Phe-Phe-Ala-Glu-Asp-Val-Gly-Ser-Asn-Lys-Gly-Ala-Ile-Ile-Gly-Leu-Met-Val-Gly-Gly-Val-Val-Ile-Ala

(e) **amyloid beta 42**
K_m 13.5 ±1.2 µM

Figure 2: Line structures and inhibition constants (K_i) for inhibitors crystallized with NEP [9, 13] and the Aβ amino acid sequence with its Michaelis constant (K_m) [18].

Extending away from this S1' region and towards the solvent is the large S2' subsite, outlined by the surrounding side-chains of R102, D107, and R110. Even though the charged side-chains of these residues outline the far side of the cavity, the volume of S2' site accommodates bulky side-chains with less specificity than S1' [14, 15]. The crystal structures contain no evidence of a direct binding interaction

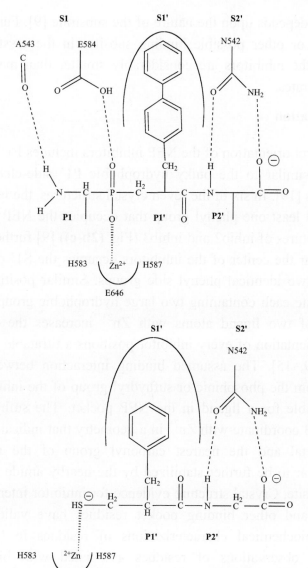

Figure 3: Line drawings of inhib1 (**a**) and thiorphan (**b**) in the binding pocket of NEP based on the respective crystal structures [9, 15]. The ligand and NEP subsites are labeled with P and S, respectively. Key binding interactions are shown by dashed lines.

between the S2' residues and any of the inhibitors. But Oefner and co-workers comment that the side-chains of R102 and R110 appear to be disordered, which stands out in "an otherwise well-defined active site" and is consistent with a binding role that depends upon the nature of the substrate [9]. Further analysis of the S2' residues or other possible binding subsites in the crystal structures is limited because the inhibitors are considerably smaller than many of the NEP polypeptide substrates.

Inhibitor Orientation

The binding pocket orientation of the NEP inhibitors includes P1' positioned side groups that are similar to the bulky hydrophobic P1' side-chains favored for peptide substrates [19]. In six of the seven crystal structures, the inhibitor P1' side group includes at least one phenyl group that occupies the NEP S1' site [9, 14, 15]. Crystal structures of inhib2 and inhib3 (Fig. (**2b-c**)) [9] further show that the phenyl group near the center of the inhibitors occupies the S1' of NEP, even in the presence of two identical phenyl side groups. Similar positioning of inhib2 and inhib3, despite each containing two large hydrophobic groups, suggests that the interaction of two ligand atoms with Zn^{2+} increases the stability of the complex. The orientation of every inhibitor positions a titratable negative charge near the Zn^{2+} [9, 15]. The assumed binding interaction between Zn^{2+} and a charged atom from the phosphinic or sufhydryl group of the inhibitors would be extremely favorable for a ligand in the NEP pocket. The sufhydryl groups of inhib2 and inhib3 coordinate with Zn^{2+} in a geometry that indicates a bond forms between the metal and the nearest carbonyl group of the inhibitors. Both interactions appear to be further stabilized by the nearby amide interacting with A543 in the S1' site. Crystal structure evidence for inhibitor interactions with the S1' site, Zn^{2+}, and other binding pocket residues have validated additional structural and biochemical characterizations of residues in NEP. Based on these congruent observations of residues associated with binding, peptidic structures have been proposed to interact with specific residues in the NEP pocket (Table (**1**)), notably F106, R102 and R717, and N542.

Roles of Individual NEP Residues in Binding

Studies involving the four NEP pocket residues mentioned above are described here in more detail. These studies found evidence of distinctive ligand binding interactions with NEP residues F106, R102, R717, and N542. The residue F106 is formally grouped with the S1' site of NEP, but crystal structures show different conformations of F106 that position the phenyl side-chain in the S2' site and reduce the volume of this second subsite [9]. A diminished S2' volume appears to account for observed reductions in NEP binding affinity for some phosphinic inhibitors. A systematic study of NEP inhibition constants (K_i) for a series of phosphinic inhibitors showed that identical, phenyl-containing P1' groups coupled with increases in the volume of the P2' substituent resulted in higher K_i, indicating an inverse relationship between binding affinity and the size of the inhibitor P2' group [20]. This trend strongly supports the hypothesis that a large P1' group induces a change in the S2' site of NEP. According to crystal structure data that includes three different phosphinic inhibitors in NEP, a large P1' phenyl group corresponds with orientation of the F106 side-chain in the S2' site [9, 13-15]. Comparison of the crystal structures for inhib1 and thiorphan (Fig. (**3a-b**)) show that the larger P1' group of inhib1 corresponds to the F106 phenyl occupying additional space in the S2' site. Changes to the NEP S2' volume based on the position of the F106 phenyl group is a rational explanation for the inverse relationship between NEP affinity and the size of P1' and P2' substituents for phosphinic inhibitors. For this reason, F106 is proposed to play a crucial role in NEP binding specificity for peptides with large P1' groups.

Site-Directed Mutagenesis

Site-directed mutagenesis studies have investigated the potential binding interactions between peptide substrates and NEP residues R102, R717, and N542. Study of R102 has indicated that the residue can bind the C-terminus of a peptide P2' residue [21, 22]. A volumetrically conservative change of NEP R102 to methionine (R102M) resulted in over 100-fold decreased binding affinity for a Phe-Ala peptide inhibitor, but only an approximately three-fold decrease in affinity for Phe-Ala derivatives without a C terminal group or a P1' carbonyl [21]. Inhibition studies with peptide substrates of NEP measured decreased binding affinity of several NEP R102 mutants for the pentapeptide [D-Ala2,Met5]-enkephalin, but not for [D-Ala2,Leu5]-enkephalin or the naturally occurring substance P and angiotensin I peptides [22]. Although the studies have insufficient evidence to propose a specific role for R102 in binding, the results are consistent with the observed orientation of R102 in the binding pocket of the NEP crystal structures.

In further agreement with the crystal structures, mutagenesis studies of NEP indicated peptide binding interactions occur with pocket residues R717 and N542. The R717M mutation of NEP dramatically decreased the enzyme binding affinity for dipeptide or polypeptide-like inhibitors, as observed by increases in inhibition values such as a 10^6–fold higher K_i for phosphoramidon [21, 23]. In contrast, the K_i of R717M-NEP did not change markedly for a modified Phe-Ala sequence containing a methyl group in place of the P1' backbone carbonyl between the residues [21]. This structural change did result in a large decrease of the Phe-Ala ligand affinity for NEP, highlighting the importance of a carbonyl group at the P1' position of the substrate for strong binding to R717. A specific binding interaction for N542 was similarly investigated because of an analogous Asn residue shown in the binding pocket of TLN. Crystal structures of TLN contain an Asn residue that appears to form two hydrogen bonds with the P2' amide and carboxylate groups of an inhibitor [24]. Comparison of the TLN sequence and pocket to NEP suggests that N542 will similarly bind the P2' of a ligand. A site-directed mutagenesis study of N542 initially tested this hypothesis with the substrate [D-Ala2,Leu5]-enkephalin and the dual TLN/NEP inhibitor thiorphan (Fig. (**3b**)) [25]. N542G and N542Q mutations led to a 14-16 fold reduction in NEP binding affinity with [D-Ala2,Leu5]-enkephalin. IC_{50} values of thiorphan were observed to increase from 5.2 for NEP to 30.2 nM for N542G. Substituting a methyl group for the proton of the P2' amide in thiorphan increased the IC_{50} for NEP from 5.2 nM to 287 nM, an approximately 60-fold change. Replacing the P2' carboxyl with a methyl group also resulted in an increase in the IC_{50} to 13.7 nM. Furthermore, methylation of the amide and carboxyl of thiorphan had an additive effect, increasing the IC_{50} with NEP to 597 nM. The evidence for the interaction of N542 with the amide and carboxylate of a thiorphan inhibitor was corroborated by subsequent crystallographic studies [15].

Summary

Structural and biochemical characterization of NEP has indicated that Zn^{2+} coordination to bound peptidic inhibitors and substrates is associated with the presence of specific residues in the S1, S1', and S2' subsites. Based on observations of ligand association to residues in those subsites, binding roles have been proposed for residues F106, R717, N542, and R102 of the NEP pocket. Researchers have used these characterizations of NEP binding pocket residues as a blueprint for predictive models of ligand binding with NEP.

COMPUTATIONAL MODELING

Computational chemistry has been used to generate predictive models for a wide variety of enzyme-substrate complexes, including matrix metalloproteinase-3, a disintegrin and metalloproteinases -9 and -10 of the zinc metallopeptidase family [26]. These peptidase models were validated with a large body of literature containing detailed structural and biochemical information about the enzymes and their respective ligands. Similarly, knowledge of NEP structure and a diverse library of inhibitors has been conducive to computational modeling of NEP ligand binding. One computational study of NEP assembled a hypothetical peptide substrate based on its compatibility with a binding pocket model (PDB 1MB7) based on the initial NEP-phosphoramidon crystal structure [27]. In the modeled substrate LATACFG, the Phe was the preferred P1' residue. The Phe side-chain is arranged in the model in the same orientation as the phenyl in the crystal structure of inhib2 and inhib3 (Fig. (**2**)). The V580 and I558 of the S1' (Table (**1**)) surround these P1' phenyl groups in a manner that was described to function like a hydrophobic lock in the binding model.

A more recent computational study predicted binding modes for a diverse set of inhibitors with potent effects on NEP and the homologous zinc metallopeptidase angiotensin-converting enzyme [28]. Crystal structures of the angiotensin peptidase with three of the selected inhibitors were used to validate computer simulations of ligand-enzyme binding. Application of those docking methods to NEP produced model structures that predicted residue interactions very similar to those supported empirically. Their inhibitor models were consistent with experimental data (Table (**1**)) and modeled novel non-covalent interactions that stabilized the binding of specific inhibitors. Theoretical binding free energies ($\Delta G°$) were also calculated using the structural information from each NEP-inhibitor model, and compared to experimental determinations.

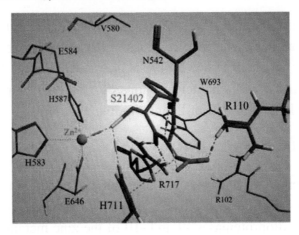

Figure 4A: A model of inhibitor S21402 docked in the binding pocket of NEP [based on Dimitropoulos *et al.* 2010]. S21402 is shown in green, Zn^{2+} in orange, and selected pocket residues in dark grey (other residues omitted for clarity).

One of the most accurate estimates of binding $\Delta G°$ was based on the model of an inhibitor designated as S21402 (Fig. (**4A**)) [29]. Using the computer model of S21402 in the NEP pocket, the binding $\Delta G°$ of the inhibitor was calculated as -11.96 kcal mol^{-1}, compared to a value of -12.41 kcal mol^{-1} calculated from experimental measurement of the K_i. The difference of -0.45 kcal mol^{-1} is well within the 2.18 kcal mol^{-1} residual standard error established for the docking method. In addition, both the structure and orientation of the model S21402 closely resembles the NEP crystal structure with the structurally similar thiorphan inhibitor (Fig. (**3b**) and (**4A-B**)).

These computational studies suggest that the interactions responsible for stabilizing any substrate in the NEP pocket may be rationally modeled. For example, one of the S21402 model's key H bonds is formed between R110 and the carboxylate in the P2' position on the inhibitor (Fig. (**4A-B**)) [28].

R110 of the S2' site is frequently predicted to form a salt bridge interaction with the carboxylate groups of other inhibitor-based models. Although there is no well-defined pattern, it appears that inhibitors with longer, peptide-like chains are frequently predicted to bond *via* a carboxylate with either R102 or R110. Previous analysis of these residues in the crystal structures and the mutagenesis studies of

R102 indicate the validity of a binding interaction between these Arg side-chains and a carboxyl from the P1' or P2' group of a substrate [9, 21, 22].

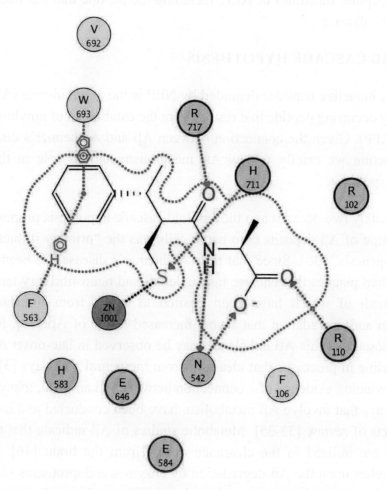

Figure 4B: Interaction map for the model of S21402 in NEP. Residues less than 4 Å from the inhibitor are shown with binding interactions represented by colored dashed lines, hydrogen/salt bridge (green) and metal-ion (purple) (Van der Waals and hydrophobic interactions are not explicitly shown). Image made with the MOE program [10].

The inhibitor models more clearly indicate how these S2' Arg residues may selectively bind carboxyl side-chains from specific substrate sequences. Computational modeling of the NEP pocket with inhibitors has accurately predicted binding free energies and NEP residue interactions using current structural and biochemical information about the NEP peptidase. This is a precedent for predicting the key interactions between the binding pocket and bioactive peptide substrates of NEP, including the peptide that has been linked to Alzheimer's disease.

AMYLOID CASCADE HYPOTHESIS

One of the bioactive peptides degraded by NEP is the amyloid-beta (Aβ) peptide, a naturally occurring peptide that results from the catabolism of amyloid precursor protein (APP). Given the connection between Aβ and Alzheimer's disease (AD), in this section we briefly review Aβ metabolism and its role in the amyloid cascade hypothesis.

Approximately two decades ago the amyloid cascade hypothesis proposed that the accumulation of Aβ deposits onto nerve cells was the "primary influence driving AD pathogenesis" [30]. Since that time Alzheimer's disease has been associated with amyloid plaques that enclose insoluble Aβ and neurofibrillary tangles of tau protein, both of which have been posited to result from imbalance in Aβ production and degradation that favors increased levels of Aβ [30]. Researchers have proposed that this Aβ imbalance may be observed in late-onset AD because of the decline in processes that clear Aβ from the neural pathways [31, 32]. Due to the convincing evidence of a connection between Aβ and AD, many studies on the pathways that involve Aβ metabolism have been conducted and have become the subjects of review [33-35]. Metabolic studies of Aβ indicate that three major pathways are utilized in the clearance of Aβ from the brain [36]. One major pathway relies upon the Aβ-degradation by enzymes endoproteases such as NEP, which is the focus of a section below. In addition to this pathway, Aβ may be eliminated from the brain by receptor-mediated absorption into the blood or through interstitial fluid drainage pathways within the walls of cerebral capillaries and arteries [36]. Absorption of Aβ into the blood is associated with low density lipoprotein receptor-related protein-1 [31], while the perivascular drainage pathways include vascular smooth muscle cells in which NEP is present [37]. Removal of Aβ from the functional brain tissue *via* perivascular drainage has been linked to toxic accumulation of the peptide and cerebral amyloid angiopathy, a disease for which NEP may also have therapeutic potential.

In spite of the attention on Aβ metabolism and elimination, none of the research has produced a viable AD therapy or removed the uncertainty surrounding the mechanism of Aβ's role in neural tangles, plaques, and neurodegeneration. On the other hand, much is known about the origin and accumulation of Aβ peptides. The name amyloid originally described protein deposits that resembled those observed for starch, with 'amyloid' simply meaning 'starch-like.' Now the term is used to refer to proteins and peptides that accumulate into a particular fibril structure in which the peptide backbone runs perpendicular to the fibril axis [38, 39]. In the case of Aβ, the assembly of these fibril structures depends upon the production of peptide monomers *via* the cleavage of APP in normal human metabolism [40]. The routine proteolytic processing of APP produces two major Aβ peptide species of 40 and 42 amino acids in length ($A\beta_{40}$ and $A\beta_{42}$, respectively). These peptide monomers associate into small polypeptides that then form large clusters called protofibrils that eventually become even larger amyloid fibrils [41, 42]. The Aβ fibrils further aggregate into fibers or bundles that are the main components of mature plaques on neuronal cells. While outside the focus of this review, readers are referred to References 33 and 39 for schematic models of fibrillization and to References 36 and 43 for photomicrographs of amyloid fibrils and plaques. While there is considerable evidence that one of these forms of Aβ plays a key role in AD pathogenesis, a debate continues over which specific species or structure is primarily responsible for the neural toxicity of Aβ. In the past this debate centered on the Aβ fibrils and plaques [43], but recent studies have supported the prefibrillar polypeptide oligomers of Aβ as the primary toxicological form [42, 44].

Evidence for the toxicity of Aβ oligomers comes from several *in vitro* studies conducted in rat, mouse, or human neuronal cell cultures [45-47]. $A\beta_{42}$ in particular was found to cause stabilization of oligomers that are toxic to primary cultures of mouse or human neurons [42, 48, 49]. Indicators of oligomeric toxicity further highlight the importance of any insight into preventing elevated concentration of Aβ in the brain. One line of research with this goal focuses on preventing the accumulation of $A\beta_{40}$ and $A\beta_{42}$, which is the most direct pathway to reduce the formation of Aβ oligomers, fibrils, and plaques. A significant area within this field centers on the enzyme NEP because of its potential to limit Aβ aggregation by degrading the preliminary peptide forms.

NEPRILYSIN AND Aβ DEGRADATION

The capability to degrade various small peptides is a feature that NEP shares with the rest of the metalloendopeptidase superfamily, which includes several members that show Aβ-degrading activities *in vitro*. The NEP homologue endothelin converting enzyme as well as insulin degrading enzyme, matrix metalloproteinase-9, and angiotensin converting enzyme are a few examples of metallopeptidases observed to have amyloid-degrading capabilities [1, 50-52]. Many of these enzymes display potential for clearing Aβ from the brain, but NEP is of special interest because of the available data on NEP structure and its efficient degradation of Aβ. Among a group of metallopeptidases sensitive to the inhibitors thiorphan and phosphoramidon, NEP was shown to degrade Aβ$_{40}$ and Aβ$_{42}$ the most rapidly and efficiently *in vitro* [53]. Significantly, NEP is one of the select few of the metallopeptidases that has been shown to cleave Aβ *in situ* or *in vivo* [54].

In vivo mouse model studies established that the proteolytic activity of NEP is a rate-limiting step in the catabolism of the Aβ peptide [55, 56]. Numerous studies have shown NEP to be highly effective in reducing levels of Aβ monomers, oligomers, and peptide deposits in the brains of mammals [55-58]. NEP activity has also been shown to increase in response to the accumulation of Aβ [59, 60]. The anti-Aβ activity of NEP led to further study of the peptidase in human brain tissue alongside symptoms of late-onset AD [50]. While the activity level of NEP was shown to increase in normal aging, further elevated levels were detected in correlation with indicators of AD progression. Increased NEP activity in AD does not support the theory that declining enzymatic degradation of Aβ is responsible for the onset of disease. Alternatively, the correlation between NEP activity and the amyloid pathogenesis associated with AD is believed by many investigators to be part of an endogenous defense against the harmful effects of Aβ [50, 61, 62]. Evidence for this theory comes from studies of rat models subjected to hypoxia to simulate neurodegenerative conditions that are believed to facilitate progression of AD [63-65]. The exposure of adult rats to prenatal and acute hypoxia led to indicators of increased APP and Aβ-production and decreased expression of NEP. Inhibition of NEP expression in the adult rats was also observed in conjunction with disrupted short-term memory following induced prenatal hypoxia [63]. In that study the use of NEP inhibitors phosphoramidon and thiorphan to inhibit the proteolytic activity in rats led to similar signs of impaired short-term memory. A follow-up study with identical rat models showed evidence that upregulation of NEP restored peptidase activity and memory function after exposure to prenatal hypoxia [64].

Research of the effects of NEP expression on Aβ pathology has also used genetic models. An *in vivo* study using transgenic mice models of human Aβ pathology reported decreases in neuronal Aβ deposits after introduction of a lentiviral vector expressing human NEP [66]. The NEP viral vector was shown to transduce human 293T cells *in vitro* to express active NEP. Addition of this NEP viral vector in the transgenic mice models of amyloid disease reduced Aβ deposition by half. This treatment also appeared to attenuate the neurodegenerative processes seen in AD, as indicated by a 16% increase in antibody reactivity with a microtubule-associated protein associated with neuronal integrity. These results were supported by another gene transfer method using the soluble catalytic domain of NEP to diminish Aβ plaque burden in mice models of genetic AD [67, 68]. Overexpression of soluble NEP in numerous other animal studies has resulted in significant decreases in levels of Aβ within the brain and improved cognitive performance [69-72]. The wealth of evidence linking Aβ accumulation, NEP, and disease-models of AD suggests that this peptidase is a promising therapeutic agent for reducing Aβ load in AD.

While the proteolytic activity of NEP has been observed to significantly reduce Aβ peptide accumulation, the enzyme has also been shown to proteolyze a multitude of other neuropeptides involved in critical physiological responses. Peptides that are known to be targeted by NEP include Met- and Leu-enkephalin, angiotensin I and II, substance P, bradykinin, neurotensin, neuropeptide Y, endothelin-1 and β-endorphin [2-4]. Excessive amounts of NEP would most likely lead to dangerous imbalances of neuropeptides in their corresponding signal pathways. Pathways that could be affected by overexpression of NEP include those that control blood pressure, pain, and other crucial physiological processes.

Increasing a patient's NEP levels may cause harmful side effects, and is a possible risk for using the peptidase in AD therapy. A solution to this potential issue is the use of site-directed mutagenesis to engineer the binding between NEP and Aβ to be more favorable than binding with other substrates. Studies of post-degradation fragments of Aβ resulting from NEP proteolysis have identified multiple cleavage sites, implying that there are several discrete binding positions or modes for the peptide in NEP [18, 73, 74]. However, little is known regarding these Aβ binding modes within NEP. Engineering an NEP peptidase for AD therapy would be problematic without detailed knowledge of the binding interactions that capture Aβ in NEP.

TOWARDS UNDERSTANDING NEP-Aβ INTERACTIONS

Direct investigation of any proposed residue interactions with Aβ would be highly constructive for mapping out the binding between Aβ and the NEP pocket. The previously mentioned methods used on inhibitors could be adapted to such an end. The solution to an X-ray crystal structure with an inhibitory compound representative of Aβ would provide high-resolution information about the binding mode of the Aβ substrate. As seen previously, crystallography work with NEP inhibitors has confirmed that specific residues in the S1 and S1' subsites play a role in binding peptidic or peptide-like structures, such as the bulky, hydrophobic P1' groups which appear to be preferred for substrates of NEP. A crystal structure of NEP containing an Aβ-like compound should provide similar confirmation of these subsite interactions and indications of additional NEP residues that participate in binding with the corresponding Aβ peptide.

Instead of compounds that represent an entire Aβ substrate, a crystallographic study could use inhibitory derivatives of the short Aβ sequences shown to be cleaved by NEP with high efficiency. This idea is inspired by a recent crystallographic study of matrix metalloproteinase-2 that produced a structure of the binding pocket occupied with an amyloid-beta precursor protein (APP) derived inhibitor [75]. In the absence of such a crystal structure for NEP, the works of Manzetti and co-workers [27] and Dimitropoulos and co-workers [28] are paradigms for using computational simulation to study the binding of NEP to Aβ. Simulations with Aβ would build on their techniques to predict the physical and energetic aspects of interactions with NEP. The information obtained from simulation of Aβ docking in NEP can then be validated by site-directed mutagenesis experiments. Independent computational and crystallographic studies of an NEP-Aβ complex would be valuable to the development of its therapeutic applications. The challenge of modeling NEP binding interactions with Aβ deserves considerable attention because of the implications for preventing the Aβ accumulation that has been linked to AD.

CONCLUDING REMARKS

Aβ peptide accumulation has been proposed to be a source of the neurodegeneration observed in AD. Recent approaches to preventing the detrimental effects of Aβ accumulation in animal models have used NEP to degrade the initial forms of the peptide. Use of NEP to prevent Aβ-related neurodegeneration comes with the inherent risk of overexpressed NEP leading to dysregulation of other biological pathways. Addressing the presumed hazards of NEP overexpression may be possible with a blueprint of the residues that hold Aβ in the binding pocket of NEP. If binding pocket residues that are of primary importance in an NEP-Aβ complex are determined, research may be able to identify NEP mutations that selectively discriminate for Aβ over other substrates, either by strengthening unique interactions with Aβ and/or weakening the interactions more crucial to binding with other substrates. The key to identifying these residues is research that provides detail about the structural compatibility of NEP for Aβ. The current abundance of information on NEP structure, binding, and proteolytic activity towards Aβ provides a basis for utilizing the NEP-Aβ interaction in development of a potential AD therapy.

ACKNOWLEDGEMENTS

We are grateful to Dr. Jeffry Madura for his helpful feedback and assistance. We also acknowledge support provided by a CURE grant from the Pennsylvania Department of Health.

CONFLICT OF INTEREST

The authors confirm that this chapter contents have no conflict of interest.

ABBREVIATIONS

Aβ	=	Amyloid beta (or beta-amyloid) peptide
AD	=	Alzheimer's disease
APP	=	Amyloid precursor protein
C	=	Carboxyl
IC_{50}	=	Half-maximal inhibitory concentration
inhib1, inhib2, inhib3	=	Inhibitor 1, inhibitor 2, or inhibitor 3, respectively
kDa	=	kilodaltons
K_i	=	Inhibition constant
MOE	=	Molecular Operating Environment
N	=	Amino
NEP	=	Neprilysin
PDB	=	Protein Database file
S or P	=	Receptor or ligand subsites, respectively
TLN	=	Thermolysin
Zn^{2+}	=	Zinc (II) cation

REFERENCES

[1]   Turner, A.J., R.E. Isaac, and D. Coates. The neprilysin (NEP) family of zinc metallo-endopeptidases: genomics and function. *Bioessays*, **2001**, *23* (3), 261-9.

[2]   Turner, A.J. and K. Tanzawa. Mammalian membrane metallopeptidases: NEP, ECE, KELL, and PEX. *FASEB J*, **1997**, *11* (5), 355-64.

[3] Roques, B.P., F. Noble, V. Dauge, M.C. Fournie-Zaluski, and A. Beaumont. Neutral endopeptidase 24.11: structure, inhibition, and experimental and clinical pharmacology. *Pharmacol Rev*, **1993**, *45* (1), 87-146.

[4] Iijima-Ando, K., S.A. Hearn, L. Granger, C. Shenton, A. Gatt, H.C. Chiang, I. Hakker, Y. Zhong, and K. Iijima. Overexpression of neprilysin reduces Alzheimer amyloid-beta 42 (A beta 42)-induced neuron loss and intraneuronal A beta 42 deposits but causes a reduction in cAMP-responsive element-binding protein-mediated transcription, age-dependent axon pathology, and premature death in Drosophila. *Journal of Biological Chemistry*, **2008**, *283* (27), 19066-19076.

[5] Barnes, K., A.J. Turner, and A.J. Kenny. An immunoelectron microscopic study of pig substantia nigra shows co-localization of endopeptidase-24.11 with substance P. *Neuroscience*, **1993**, *53* (4), 1073-82.

[6] Malfroy, B., J.P. Swerts, A. Guyon, B.P. Roques, and J.C. Schwartz. High-affinity enkephalin-degrading peptidase in brain is increased after morphine. *Nature*, **1978**, *276* (5687), 523-6.

[7] Matsas, R., I.S. Fulcher, A.J. Kenny, and A.J. Turner. Substance P and [Leu]enkephalin are hydrolyzed by an enzyme in pig caudate synaptic membranes that is identical with the endopeptidase of kidney microvilli. *Proc Natl Acad Sci U S A*, **1983**, *80* (10), 3111-5.

[8] Schwartz, J.C., S. de la Baume, B. Malfroy, G. Patey, R. Perdrisot, J.P. Swerts, M.C. Fournie-Zaluski, G. Gacel, and B.P. Roques. "Enkephalinase", a newly characterised dipeptidyl carboxypeptidase: properties and possible role in enkephalinergic transmission. *Int J Neurol*, **1980**, *14* (2-4), 195-204.

[9] Oefner, C., B.P. Roques, M.C. Fournie-Zaluski, and G.E. Dale. Structural analysis of neprilysin with various specific and potent inhibitors. *Acta Crystallographica Section D-Biological Crystallography*, **2004**, *60* (Journal Article), 392-396.

[10] Chemical Computing Group, I., *Molecular Operating Environment*. 2011, Chemical Computing Group Inc.: Montreal, QC, Canada, H3A 2R7. p. Molecular Visualization and Simulation Program.

[11] Hooper, N.M. Families of zinc metalloproteases. *FEBS Lett*, **1994**, *354* (1), 1-6.

[12] Cummins, P.M., A. Pabon, E.H. Margulies, and M.J. Glucksman. Zinc coordination and substrate catalysis within the neuropeptide processing enzyme endopeptidase EC 3.4.24.15. Identification of active site histidine and glutamate residues. *J Biol Chem*, **1999**, *274* (23), 16003-9.

[13] Oefner, C., A. D'Arcy, M. Hennig, F.K. Winkler, and G.E. Dale. Structure of human neutral endopeptidase (Neprilysin) complexed with phosphoramidon. *J Mol Biol*, **2000**, *296* (2), 341-9.

[14] Oefner, C., S. Pierau, H. Schulz, and G.E. Dale. Structural studies of a bifunctional inhibitor of neprilysin and DPP-IV. *Acta Crystallogr D Biol Crystallogr*, **2007**, *63* (Pt 9), 975-81.

[15] Dale, G.E. and C. Oefner, *Neprilysin*, in *Handbook of Metalloproteins*. 2006, John Wiley & Sons, Ltd.

[16] Weaver, L.H., W.R. Kester, and B.W. Matthews. A crystallographic study of the complex of phosphoramidon with thermolysin. A model for the presumed catalytic transition state and for the binding of extended substances. *J Mol Biol*, **1977**, *114* (1), 119-32.

[17] Kester, W.R. and B.W. Matthews. Crystallographic study of the binding of dipeptide inhibitors to thermolysin: implications for the mechanism of catalysis. *Biochemistry*, **1977**, *16* (11), 2506-16.

[18] Leissring, M.A., A. Lu, M.M. Condron, D.B. Teplow, R.L. Stein, W. Farris, and D.J. Selkoe. Kinetics of amyloid beta-protein degradation determined by novel fluorescence- and fluorescence polarization-based assays. *J Biol Chem*, **2003**, *278* (39), 37314-20.

[19] Tiraboschi, G., N. Jullian, V. Thery, S. Antonczak, M.C. Fournie-Zaluski, and B.P. Roques. A three-dimensional construction of the active site (region 507-749) of human neutral endopeptidase (EC.3.4.24.11). *Protein Eng*, **1999**, *12* (2), 141-9.

[20] Chen, H., F. Noble, A. Mothe, H. Meudal, P. Coric, S. Danascimento, B.P. Roques, P. George, and M.C. Fournie-Zaluski. Phosphinic derivatives as new dual enkephalin-degrading enzyme inhibitors: synthesis, biological properties, and antinociceptive activities. *J Med Chem*, **2000**, *43* (7), 1398-408.

[21] Beaumont, A., H. Le Moual, G. Boileau, P. Crine, and B.P. Roques. Evidence that both arginine 102 and arginine 747 are involved in substrate binding to neutral endopeptidase (EC 3.4.24.11). *J Biol Chem*, **1991**, *266* (1), 214-20.

[22] Kim, Y.A., B. Shriver, T. Quay, and L.B. Hersh. Analysis of the importance of arginine 102 in neutral endopeptidase (enkephalinase) catalysis. *J Biol Chem*, **1992**, *267* (17), 12330-5.

[23] Marie-Claire, C., E. Ruffet, S. Antonczak, A. Beaumont, M. O'Donohue, B.P. Roques, and M.C. Fournie-Zaluski. Evidence by site-directed mutagenesis that arginine 203 of thermolysin and arginine 717 of neprilysin (neutral endopeptidase) play equivalent critical roles in substrate hydrolysis and inhibitor binding. *Biochemistry*, **1997**, *36* (45), 13938-45.

[24] Roderick, S.L. and B.W. Matthews. Structure of the cobalt-dependent methionine aminopeptidase from Escherichia coli: a new type of proteolytic enzyme. *Biochemistry*, **1993**, *32* (15), 3907-12.

[25] Dion, N., H. Le Moual, M.C. Fournie-Zaluski, B.P. Roques, P. Crine, and G. Boileau. Evidence that Asn542 of neprilysin (EC 3.4.24.11) is involved in binding of the P2' residue of substrates and inhibitors. *Biochem J*, **1995**, *311 (Pt 2)*, 623-7.

[26] Manzetti, S., D.R. McCulloch, A.C. Herington, and D. van der Spoel. Modeling of enzyme-substrate complexes for the metalloproteases MMP-3, ADAM-9 and ADAM-10. *J Comput Aided Mol Des*, **2003**, *17* (9), 551-65.

[27] Manzetti, S. Computer modeling and nanosecond simulation of the enzyme-substrate complex of the common lymphoblastic leukemia antigen (neprilysin) indicates shared residues at the primary specificity pocket (S1') with matrix metalloproteases. *J Mol Model*, **2003**, *9* (5), 348-54.

[28] Dimitropoulos, N., A. Papakyriakou, G.A. Dalkas, E.D. Sturrock, and G.A. Spyroulias. A computational approach to the study of the binding mode of dual ACE/NEP inhibitors. *J Chem Inf Model*, **2010**, *50* (3), 388-96.

[29] Burrell, L.M., N.K. Farina, L.C. Balding, and C.I. Johnston. Beneficial renal and cardiac effects of vasopeptidase inhibition with S21402 in heart failure. *Hypertension*, **2000**, *36* (6), 1105-11.

[30] Hardy, J. and D.J. Selkoe. Medicine - The amyloid hypothesis of Alzheimer's disease: Progress and problems on the road to therapeutics. *Science*, **2002**, *297* (5580), 353-356.

[31] Mawuenyega, K.G., W. Sigurdson, V. Ovod, L. Munsell, T. Kasten, J.C. Morris, K.E. Yarasheski, and R.J. Bateman. Decreased clearance of CNS beta-amyloid in Alzheimer's disease. *Science*, **2010**, *330* (6012), 1774.

[32] Miners, J.S., N. Barua, P.G. Kehoe, S. Gill, and S. Love. Abeta-degrading enzymes: potential for treatment of Alzheimer disease. *J Neuropathol Exp Neurol*, **2011**, *70* (11), 944-59.

[33] Citron, M. Alzheimer's disease: strategies for disease modification. *Nature Reviews Drug Discovery*, **2010**, *9* (5), 387-398.

[34] Weiner, H.L. and D. Frenkel. Immunology and immunotherapy of Alzheimer's disease (vol 6, pg 404, 2006). *Nature Reviews Immunology*, **2006**, *6* (6), 490-490.

[35] De Strooper, B. Proteases and Proteolysis in Alzheimer Disease: A Multifactorial View on the Disease Process. *Physiological Reviews*, **2010**, *90* (2), 465-494.

[36] Carare, R.O., C.A. Hawkes, M. Jeffrey, R.N. Kalaria, and R.O. Weller. Review: cerebral amyloid angiopathy, prion angiopathy, CADASIL and the spectrum of protein elimination failure angiopathies (PEFA) in neurodegenerative disease with a focus on therapy. *Neuropathol Appl Neurobiol*, **2013**, *39* (6), 593-611.

[37] Miners, J.S., P. Kehoe, and S. Love. Neprilysin protects against cerebral amyloid angiopathy and Abeta-induced degeneration of cerebrovascular smooth muscle cells. *Brain Pathol*, **2011**, *21* (5), 594-605.

[38] Hamley, I.W. The Amyloid Beta Peptide: A Chemist's Perspective. Role in Alzheimer's and Fibrillization. *Chem Rev*, **2012.**

[39] Serpell, L.C., C.C. Blake, and P.E. Fraser. Molecular structure of a fibrillar Alzheimer's A beta fragment. *Biochemistry*, **2000**, *39* (43), 13269-75.

[40] Selkoe, D.J. and D. Schenk. Alzheimer's disease: molecular understanding predicts amyloid-based therapeutics. *Annu Rev Pharmacol Toxicol*, **2003**, *43*, 545-84.

[41] Teplow, D.B. Structural and kinetic features of amyloid beta-protein fibrillogenesis. *Amyloid*, **1998**, *5* (2), 121-42.

[42] Ahmed, M., J. Davis, D. Aucoin, T. Sato, S. Ahuja, S. Aimoto, J.I. Elliott, W.E. Van Nostrand, and S.O. Smith. Structural conversion of neurotoxic amyloid-beta(1-42) oligomers to fibrils. *Nature Structural & Molecular Biology*, **2010**, *17* (5), 561-U56.

[43] Kelly, J.W. Towards an understanding of amyloidogenesis. *Nature structural biology*, **2002**, *9* (5), 323-325.

[44] Goedert, M. and M.G. Spillantini. A century of Alzheimer's disease. *Science*, **2006**, *314* (5800), 777-81.

[45] Dahlgren, K.N., A.M. Manelli, W.B. Stine, Jr., L.K. Baker, G.A. Krafft, and M.J. LaDu. Oligomeric and fibrillar species of amyloid-beta peptides differentially affect neuronal viability. *J Biol Chem*, **2002**, *277* (35), 32046-53.

[46] Maezawa, I., P. Zimin, H. Wulff, and L.-W. Jin. Amyloid-beta protein oligomer at low nanomolar concentrations activates microglia and induces microglial neurotoxicity. *Journal of Biological Chemistry*, **2011**, *286* (5), 3693-3706.

[47]   Ono, K., M.M. Condron, and D.B. Teplow. Structure-neurotoxicity relationships of amyloid beta-protein oligomers. *Proceedings of the National Academy of Sciences of the United States of America*, **2009**, *106* (35), 14745-14750.

[48]   Hartley, D.M., D.M. Walsh, C.P. Ye, T. Diehl, S. Vasquez, P.M. Vassilev, D.B. Teplow, and D.J. Selkoe. Protofibrillar intermediates of amyloid beta-protein induce acute electrophysiological changes and progressive neurotoxicity in cortical neurons. *J Neurosci*, **1999**, *19* (20), 8876-84.

[49]   Zhang, Y., R. McLaughlin, C. Goodyer, and A. LeBlanc. Selective cytotoxicity of intracellular amyloid beta peptide1-42 through p53 and Bax in cultured primary human neurons. *J Cell Biol*, **2002**, *156* (3), 519-29.

[50]   Miners, J.S., Z. van Helmond, P.G. Kehoe, and S. Love. Changes with Age in the Activities of beta-Secretase and the A beta-Degrading Enzymes Neprilysin, Insulin-Degrading Enzyme and Angiotensin-Converting Enzyme. *Brain Pathology*, **2010**, *20* (4), 794-802.

[51]   Wang, S.Q., R. Wang, L. Chen, D.A. Bennett, D.W. Dickson, and D.S. Wang. Expression and functional profiling of neprilysin, insulin-degrading enzyme, and endothelin-converting enzyme in prospectively studied elderly and Alzheimer's brain. *Journal of neurochemistry*, **2010**, *115* (1), 47-57.

[52]   Nalivaeva, N.N., L.R. Fisk, N.D. Belyaev, and A.J. Turner. Amyloid-degrading enzymes as therapeutic targets in Alzheimer's disease. *Curr Alzheimer Res*, **2008**, *5* (2), 212-24.

[53]   Shirotani, K., S. Tsubuki, N. Iwata, Y. Takaki, W. Harigaya, K. Maruyama, S. Kiryu-Seo, H. Kiyama, H. Iwata, T. Tomita, T. Iwatsubo, and T.C. Saido. Neprilysin degrades both amyloid beta peptides 1-40 and 1-42 most rapidly and efficiently among thiorphan- and phosphoramidon-sensitive endopeptidases. *J Biol Chem*, **2001**, *276* (24), 21895-901.

[54]   Miners, J.S., S. Baig, J. Palmer, L.E. Palmer, P.G. Kehoe, and S. Love. A beta-degrading enzymes in Alzheimer's disease. *Brain Pathology*, **2008**, *18* (2), 240-252.

[55]   Iwata, N., S. Tsubuki, Y. Takaki, K. Shirotani, B. Lu, N.P. Gerard, C. Gerard, E. Hama, H.J. Lee, and T.C. Saido. Metabolic regulation of brain A beta by neprilysin. *Science*, **2001**, *292* (5521), 1550-1552.

[56]   Iwata, N., S. Tsubuki, Y. Takaki, K. Watanabe, M. Sekiguchi, E. Hosoki, M. Kawashima-Morishima, H.J. Lee, E. Hama, Y. Sekine-Aizawa, and T.C. Saido. Identification of the major A beta(1-42)-degrading catabolic pathway in brain parenchyma: Suppression leads to biochemical and pathological deposition. *Nature medicine*, **2000**, *6* (2), 143-150.

[57]   Iwata, N., H. Mizukami, K. Shirotani, Y. Takaki, S. Muramatsu, B. Lu, N.P. Gerard, C. Gerard, K. Ozawa, and T.C. Saido. Presynaptic localization of neprilysin contributes to efficient clearance of amyloid-beta peptide in mouse brain. *Journal of Neuroscience*, **2004**, *24* (4), 991-998.

[58]   El-Amouri, S.S., H. Zhu, J. Yu, R. Marr, I.M. Verma, and M.S. Kindy. Neprilysin: An enzyme candidate to slow the progression of Alzheimer's disease. *American Journal of Pathology*, **2008**, *172* (5), 1342-1354.

[59]   Miners, J.S., S. Morris, S. Love, and P.G. Kehoe. Accumulation of Insoluble Amyloid-beta in Down's Syndrome is Associated with Increased BACE-1 and Neprilysin Activities. *Journal of Alzheimer's Disease*, **2010** (1875-8908 (Electronic); 1387-2877 (Linking)).

[60]   Tampellini, D., N. Rahman, M.T. Lin, E. Capetillo-Zarate, and G.K. Gouras. Impaired beta-amyloid secretion in Alzheimer's disease pathogenesis. *J Neurosci*, **2011**, *31* (43), 15384-90.

[61]   Marr, R.A. and B.J. Spencer. NEP-like endopeptidases and Alzheimer's disease [corrected]. *Curr Alzheimer Res*, **2010**, *7* (3), 223-9.

[62]   Nalivaeva, N.N., C. Beckett, N.D. Belyaev, and A.J. Turner. Are amyloid-degrading enzymes viable therapeutic targets in Alzheimer's disease? *J Neurochem*, **2012**, *120 Suppl 1*, 167-85.

[63]   Dubrovskaia, N.M., N.N. Nalivaeva, S.A. Plesneva, A.A. Feponova, A.J. Turner, and I.A. Zhuravin. [Changes in the activity of amyloid-degrading metallopeptidases leads to disruption of memory in rats]. *Zh Vyssh Nerv Deiat Im I P Pavlova*, **2009**, *59* (5), 630-8.

[64]   Nalivaeva, N.N., N.D. Belyaev, D.I. Lewis, A.R. Pickles, N.Z. Makova, D.I. Bagrova, N.M. Dubrovskaya, S.A. Plesneva, I.A. Zhuravin, and A.J. Turner. Effect of sodium valproate administration on brain neprilysin expression and memory in rats. *J Mol Neurosci*, **2012**, *46* (3), 569-77.

[65]   Nalivaevaa, N.N., L. Fisk, E.G. Kochkina, S.A. Plesneva, I.A. Zhuravin, E. Babusikova, D. Dobrota, and A.J. Turner. Effect of hypoxia/ischemia and hypoxic preconditioning/ reperfusion on expression of some amyloid-degrading enzymes. *Ann N Y Acad Sci*, **2004**, *1035*, 21-33.

[66]   Marr, R.A., H. Guan, E. Rockenstein, M. Kindy, F.H. Gage, I. Verma, E. Masliah, and L.B. Hersh. Neprilysin regulates amyloid Beta peptide levels. *J Mol Neurosci*, **2004**, *22* (1-2), 5-11.

[67]   Hemming, M.L., M. Patterson, C. Reske-Nielsen, L. Lin, O. Isacson, and D.J. Selkoe. Reducing amyloid plaque burden *via ex vivo* gene delivery of an a beta-degrading protease: A novel therapeutic approach to Alzheimer Disease. *Plos Medicine*, **2007**, *4* (8), 1405-1416.

[68]   Li, Y., J. Wang, J. Liu, and F. Liu. A novel system for *in vivo* neprilysin gene delivery using a syringe electrode. *J Neurosci Methods*, **2010**, *193* (2), 226-31.

[69]   Guan, H., Y. Liu, A. Daily, S. Police, M.-H. Kim, S. Oddo, F.M. LaFerla, J.R. Pauly, M.P. Murphy, and L.B. Hersh. Peripherally Expressed Neprilysin Reduces Brain Amyloid Burden: A Novel Approach for Treating Alzheimer's Disease. *Journal of neuroscience research*, **2009**, *87* (6), 1462-1473.

[70]   Liu, Y., C. Studzinski, T. Beckett, H. Guan, M.A. Hersh, M.P. Murphy, R. Klein, and L.B. Hersh. Expression of Neprilysin in Skeletal Muscle Reduces Amyloid Burden in a Transgenic Mouse Model of Alzheimer Disease. *Molecular Therapy*, **2009**, *17* (8), 1381-1386.

[71]   Liu, Y., C. Studzinski, T. Beckett, M.P. Murphy, R.L. Klein, and L.B. Hersh. Circulating neprilysin clears brain amyloid. *Molecular and Cellular Neuroscience*, **2010**, *45* (2), 101-107.

[72]   Spencer, B., R.A. Marr, R. Gindi, R. Potkar, S. Michael, A. Adame, E. Rockenstein, I.M. Verma, and E. Masliah. Peripheral Delivery of a CNS Targeted, Metalo-Protease Reduces AB Toxicity in a Mouse Model of Alzheimer's Disease. *PLoS ONE*, **2011**, *6* (1), 1-12.

[73]   Hersh, L.B. and D.W. Rodgers. Neprilysin and amyloid beta peptide degradation. *Curr Alzheimer Res*, **2008**, *5* (2), 225-31.

[74]   Howell, S., J. Nalbantoglu, and P. Crine. Neutral endopeptidase can hydrolyze beta-amyloid(1-40) but shows no effect on beta-amyloid precursor protein metabolism. *Peptides*, **1995**, *16* (4), 647-52.

[75]   Hashimoto, H., T. Takeuchi, K. Komatsu, K. Miyazaki, M. Sato, and S. Higashi. Structural basis for matrix metalloproteinase-2 (MMP-2)-selective inhibitory action of beta-amyloid precursor protein-derived inhibitor. *J Biol Chem*, **2011**, *286* (38), 33236-43.

CHAPTER 14

Targeting the GSK3β/β-catenin Signaling to Treat Alzheimer′s Disease: Plausible or Utopic?

Fares Zeidán-Chuliá* and José Cláudio Fonseca Moreira

Center of Oxidative Stress Research, Department of Biochemistry, Postgraduate Program in Biological Sciences: Biochemistry, Institute of Basic Health Sciences, Federal University of Rio Grande do Sul, Porto Alegre, RS, Brazil

Abstract: Alzheimer′s disease (AD) is a neurodegenerative disorder characterized by progressive memory loss, cognitive impairment, and at the molecular level, by the presence of neurofibrillary tangles (NFTs). As opposed to degeneration, it is known that some specific regions of the brain contain neural stem cells (NSCs) able to produce neurons during adulthood. Wnt/β-catenin signaling has been described as a key pathway modulating the balance between NSC proliferation and differentiation. Wnt signaling is regulated by glycogen synthase kinase 3 (GSK3) that is constitutively active in the cells, keeps β-catenin phosphorylated on serine and threonine residues, and controls its proteosomal-mediated degradation. This raises the question whether inhibition of GSK3β activity, β-catenin stabilization, and therefore, pharmacological activation of endogenous neurogenesis would be a plausible therapeutic strategy for treating AD patients. In this chapter, we herein review the Wnt/β-catenin signaling and evaluate the strategy of inhibiting GSK3β in the disease.

Keywords: Neurofibrillary tangles, β-amyloid, drug therapy, GSK3 inhibitor, lithium, neural stem cells, diabetes, AGEs, RAGE, cognitive impairment.

INTRODUCTION

Alzheimer′s disease (AD) is a neurodegenerative disorder defined by progressive memory loss, cognitive impairment, and at the molecular level, by the presence of neurofibrillary tangles (NFTs) composed by hyper-phosphorylated forms of the microtubule-associated protein tau and insoluble β-amyloid (Aβ) plaques.

*Address correspondence to Fares Zeidán-Chuliá: Center of Oxidative Stress Research, Department of Biochemistry, Postgraduate Program in Biological Sciences: Biochemistry, Institute of Basic Health Sciences, Federal University of Rio Grande do Sul, Porto Alegre, RS, Brazil; Tel: +55 51 3308-5577; Fax: +55 51 3308-5535; E-mail: fzchulia.biomed@gmail.com

However, it is not well understood what really triggers the development of such structures that actually appear in the patient's brain at later stages of the disease.

As opposed to degeneration, certain locations of the brain such as subventricular zone (SVZ) of telencephalic lateral ventricles [1, 2] and subgranular zone (SGZ) of hippocampal dentate gyrus [3, 4] contain neural stem cells (NSCs) able to produce neurons during adulthood [5]. Several growth factors, ligands, and pathways have been proposed to regulate neural and glial progenitor cell behavior. Among other candidates, Wnt/β-catenin signaling has been described as a key pathway modulating the balance between NSC proliferation and differentiation, but it seems to have opposing roles in the developing and the adult brain [6-8]. For instance, Wnt/β-catenin signaling transiently induces cortical radial glia (RG) to proliferate during early neurogenesis but forces NSCs from SVZ to exit the cell cycle [9, 10].

Some studies are highlighting the existence of an altered and decreased neurogenesis in the brain of AD patients [11, 12]. Therefore, one could speculate whether both events, impairment of neurogenesis and generation of NFT and senile plaques (SP), might be directly or indirectly connected to the disease.

Activation of the canonical Wnt signaling pathway inhibits the activity of glycogen synthase kinase 3 (GSK3), a serine/threonine kinase which has a role in controlling diverse neuronal functions (*e.g.*, neurite outgrowth, neurotransmission, synapse formation, and neurogenesis) [13]. Interestingly, its activity has already been linked to AD-associated anomalous structures [14] since inhibition of GSK3α with therapeutic concentrations of lithium is able to stop both production and accumulation of Aβ in the brains of amyloid precursor protein (APP) over-expressing mice. Besides, GSK3β was also shown to phosphorylate tau protein of the neurofibrillary tangles [15]. Since evidences are pointing towards a deficient neurogenesis in AD patients and the possible role of GSK3 activity in the generation of AD-associated aberrant structures, the question is whether pharmacological modulation of GSK3β/β-catenin signaling may represent a real therapeutic alternative for these patients. The aim of this chapter is to critically review the pros and cons of such strategy.

THE SCENARIO: AD HALLMARKS

Nowadays, AD is the most common irreversible neurodegenerative disorder, with more than 20 million cases worldwide [16]. It is characterized by the loss of neurons and synapses, leading to cognitive impairment, loss of memory, language, reasoning, and followed by dementia. Most cases of AD are idiopathic, and advanced age, diabetes, hypertension, hypercholesterolaemia, hyperhomocysteinaemia, and inheritance of the epsilon 4 allele of the polymorphic apolipoprotein E gene are reported major risk factors, although they are not enough to cause the disease [17]. Data from postmortem brains reveal abnormal intracellular NFTs, consisting of hyperphosphorylated microtubule-binding protein tau, and extracellular senile plaques mainly composed of Aβ oligomers deposits [18]. Even though the expression pattern of NFTs correlates with the clinical onset and progression of AD [19], the relevant role for Aβ deposits has been suggested from the association between familial AD and mutations in the genes that encode APP, presenilin-1, and presenilin-2 [20], that up-regulate the general Aβ production and more specifically, the generation of a minor 42-amino acid form (Aβ42) with increased propensity for aggregation [21]. In general, Aβ can be 40 to 42 amino acids in length and it is generated by proteolytic cleavage of the transmembrane APP [22]. This is processed by two competing cellular pathways: the amyloidogenic or β-secretase-mediated route and the non-amyloidogenic or α-secretase-mediated pathway. For the generation of Aβ, APP is first cleaved by β-secretase BACE1 (β-site amyloid precursor protein cleaving enzyme 1), a membrane-bound aspartyl-protease, generating β-secreted APP (sAPPβ), and a C-terminal membrane-bound fragment, called C99 or β-CTF. C99 is further processed by γ-secretase, releasing the Aβ peptide. In addition to the amyloidogenic processing of APP involving β- and γ-secretase activity, APP can be cleaved in a nonamyloidogenic pathway by α-secretases [21, 23-25]. The excessive Aβ peptides may aggregate into toxic oligomers and induce neuronal loss by promotion of oxidative stress, lipid peroxidation, increase of intracellular Ca^{2+}, mitochondrial dysfunction, caspase pathway activation accompanied with increased macroautophagy and lysosomal ensuing apoptosis [26, 27].

Interestingly, the advanced glycation end products (AGEs) increase in the brain during normal aging and different studies support that AGEs are further increased in the brain in the presence of vascular or Alzheimer's dementia [28, 29]. As a matter of fact, several experimental evidence suggests that accumulation of N(epsilon)-(carboxymethyl)lysine protein (CML), the most abundant AGE *in vivo* [30], is greater in AD and diabetes than that observed in AD alone [29]. A study from Kim and colleagues (2009) proposed that type 1 and type 2 diabetes could actually contribute to AD but through different mechanisms: if hyperglycemia-mediated tau cleavage might be the key feature in type 2 diabetes, in case of type 1 diabetes, insulin deficiency would be the major contributing factor [31]. The close interrelation between AD, diabetes, and cognitive decline is under intense research within the scientific community [32, 33]. Certainly, the main pathophysiological problem in both type 1 and 2 diabetes mellitus is insulin resistance, which is characterized by the progressive reduction in the response of insulin-sensitive tissues to normal levels of insulin [34, 35]. Diverse studies have shown that reductions in both insulin receptor substrate 1 (IRS1) and/or 2 (IRS2) protein levels are associated with insulin resistance [36-38]. Furthermore, it has been shown that activation of GSK3β contributes to the induction of insulin resistance *via* phosphorylation of IRS1, triggering the ubiquitination and degradation of IRS1 [39].

GSK3 is an enzyme that, in physiological conditions, takes part in several cellular processes (*i.e.*, regulation of body metabolism by phosphorylation of glycogen synthase (GS) and other substrates); but it has recently gained researchers' attention by incorporating the "*GSK3 hypothesis of AD*" into the field [40]. This theory proposed that over-activity of such an enzyme has a role in the elevated production of Aβ, tau phosphorylation and microglia-associated inflammatory process and, as a final consequence, in memory impairment. All these events are typical hallmarks of AD-associated pathology. Thus, the overexpression or overactivation of GSK3 could induce a series of pathological events, most of which are common hallmarks of AD and diabetes mellitus type 2 and would represent a common cross-talk between these age-dependent diseases [41].

THE TARGET: GSK3/B-CATENIN SIGNALING IN AD

During embryonic development, adult tissue remodeling, and even tumorigenesis, cell proliferation and differentiation are two distinct (but coupled) biological processes that require appropriate changes in the metabolic status to occur; and the Wnt signaling pathway is indeed known for integrating cellular metabolism together with tissue development and function [42]. Secreted signaling proteins belonging to the Wnt family bind to specific Frizzled (Fzd) receptor complexes on the surface of target cells, triggering an intracellular signal transduction that can occur through either β-catenin-dependent (canonical Wnt-signaling) or β-catenin-independent (non-canonical Wnt-signaling) pathways. The less studied β-catenin-independent Wnt signaling pathways do not need the transcriptional activity of β-catenin and it is usually involved in cell migration and polarity [43, 44]. The more extensively studied canonical pathway regulates the ability of β-catenin protein to modulate the activation of specific target genes. In the absence of Wnt signal, β-catenin (transcriptional activator) is actively degraded in the cell by the action of a protein complex called the "destruction box". Within this complex, the axin and APC form a scaffold allowing β-catenin phosphorylation by CK1α (casein-kinase 1α) and GSK3β. Phosphorylation of cytosolic β-catenin marks it for subsequent ubiquitination-dependent proteasomal degradation [45].

Then, the lack of free β -catenin allows the DNA-binding T-cell factor/ lymphoid enhancer factor (Tcf/Lef) proteins to interact with transcriptional co-repressors to block target expression in the nucleus. On the other hand, Wnt binding to Fzd-LRP (low-density lipoprotein receptor-related protein) receptor complexes at the membrane leads to the formation of Dishevelled (Dvl)-Fzd complexes and relocalization of Axin from the destruction complex to the cell membrane. This allows β-catenin to accumulate and enter the nucleus, where it interacts with members of the Tcf/Lef family, transforms the Tcf proteins into powerful transcriptional activators by recruiting co-activator proteins that ensures efficient activation of Wnt target genes (*i.e.*, c-myc), and instructs the cell to actively proliferate and remain in an undifferentiated state [46-48]. Diverse studies highlight the critical role of Wnt pathway in controlling stem cell proliferation. It has been shown to participate in the regulation of stem cell expansion in a number of tissues such as the skin, intestine, hematopoietic, and nervous system [49-53]. In fact, when Wnt signaling pathway is blocked, progenitor cell proliferation is disrupted and causes severe reduction of hippocampus development [54, 55]. On the contrary, over-expression of Wnt signaling is known to lead to uncontrolled cell proliferation and tumorigenesis

[56]. GSK3β is inactivated by phosphorylation of serine (Ser), and its activity is increased by phosphorylation of tyrosine (Tyr); in fact, p38MAPK directly inactivates GSK3β by phosphorylating Ser389 in the C terminus of GSK3β in the brain and thymocytes, leading to an accumulation of intracellular β-catenin. Thus, GSK3β activity is a determinant of β-catenin stabilization and its accumulation in the nucleus; a required step for normal neurogenesis [57-60]. *In vitro*, GSK3 inhibition has been shown to increase the proliferation of neural progenitors in the presence of basic fibroblast growth factor (bFGF) and epidermal growth factor (EGF) with the involvement of NOTCH signaling and β-catenin stabilization. On the other hand, in the absence of growth factors, same inhibition enhances neuronal differentiation of neural progenitor cells [61]. Since the ability to generate new neurons throughout life gets compromised with age due to reduced neurogenesis and failure of newborn neurons to mature [62, 63], one could speculate whether Wnt signaling could be altered or had a role in the pathophysiology of AD. This was the aim of a study performed by He & Shen (2009), where the specific goal of their investigation was to study the neurogenic potential of glial precursor cells (GPCs) coming from AD patients, to reveal whether exogenous levels of Aβ might affect neurogenesis and the mechanism controlling such interplay [12]. To answer this question, they compared the properties of glial progenitor cells (GPCs) from the cortices of healthy controls and AD patients, showing that GPCs from patients displayed reduced renewal capability and reduced neurogenesis when compared with healthy controls. Interestingly, GPCs from patients expressed high levels of GSK3β together with an increase in phosphorylated β-catenin and a reduction in non-phosphorylated β-catenin in relation to healthy controls. Moreover, the exposure to Aβ was able to impair the ability of GPCs from healthy controls to give rise to new neurons and leading to similar changes in β-catenin signaling proteins, all together suggesting that Aβ-induced disruption of β-catenin signaling could contribute to the impairment of neurogenesis in AD progenitor cells.

THE USE OF GSK3 INHIBITORS AS THERAPEUTICAL AGENTS FOR AD PATIENTS

The serine/threonine kinase GSK3 was first isolated and purified as an enzyme able to phosphorylate and inactivate the enzyme GS [57], but subsequent purification and molecular cloning revealed two isoforms as two different gene products. Even though GSK3α (51 KDa) is slightly higher in molecular weight with an extended N-terminal glycine rich domain than GSK3β (47 KDa), both isoforms have very similar biochemical characteristics [64-66]. Under physiological conditions, insulin can stimulate glycogen synthesis by dephosphorylation of GS (GS activation) at the sites targeted by GSK3 [67], thanks to simultaneous inhibition of GSK3 activity and activation of one of the glycogen-associated forms of protein phosphatase 1 [68]. Insulin-derived GSK3α and β inhibition occurs through phosphorylation of both isoforms at Ser residues in the N-terminal lobe of the protein kinase (Ser21 for GSK3α and Ser9 for GSK3β) [66]. In opposition to inhibitory serine phosphorylation, GSK3 activity can be increased by phosphorylation of a Tyr residue, Tyr216 in GSK3β and Tyr279 in GSK3α, located in the kinase domain [69]. In contrast to most protein kinases, GSK3 is constitutively active in cells and in addition to insulin, a variety of extracellular stimuli can inhibit it such as EGF, FGF, and ligands of Wnt signaling pathway [70]. GSK3 is known to participate in numerous cellular processes (*e.g.*, glycogen metabolism, cytoskeleton regulation, intracellular vesicular transport, cell cycle progression, and apoptosis). Therefore, numerous putative substrates for this protein kinase have been identified to date and include β-catenin, axin, APC, cyclin D1, IRS1, GS, tau, and presenilin 1 (PS1) [71-78]. From those, the most studied interaction is probably the regulation of GSK3/β-catenin signaling due to the central role that GSK3 plays in the canonical Wnt signal transduction pathway, where its phosphorylation of β-catenin on key residues is required for β-catenin's cellular ubiquitination and proteasomal degradation; a signaling pathway that is well known to promote self-renewal in a number of different tissue stem cells when activated [79-82], including NSCs, and hematopoietic stem cells. Furthermore, the fact that GSK3 interacts with several neuronal proteins that have directly been related to AD raises the question whether targeting this protein kinase could be therapeutically possible. For instance, the neuron-specific microtubule-associated protein tau (main component of NFTs and known to be abnormally phosphorylated in AD) was shown to be phosphorylated by GSK3 [72]. Aβ is able to induce a significant increase in GSK3β expression *in vitro* and it is known to be elevated in AD human brains [12, 83], and it has even postulated that PS1 could act as a scaffold protein that would bring GSK3 into proximity with its substrates tau and β-catenin [70]. In that respect, the therapeutic potential of GSK3 inhibitors has become a primary focus of pharmaceutical interest. Already a decade ago, Sun and colleagues reported that lithium chloride was able to reduce in a dose-dependent manner the secreted Aβ after transient expression of APP C99 *in vitro* [84]. Lithium ions, mainly utilized as a primary mood stabilizer in the treatment of chronic patients with bipolar disorder, were demonstrated to inhibit GSK3 [85, 86]; highlighting that potential dysregulation of GSK3 may contribute to this mental disorder [87]. Bipolar disorder has been associated with cognitive dysfunction and increased risk for dementia [88, 89]. But most interestingly, it has been reported that lithium treatment reduced the prevalence of AD in patients with bipolar disorder to levels found in the general elderly population [90]. In addition to lithium-induced GSK3 inhibition, one cannot exclude the effect of additional mechanisms since lithium induces its therapeutic effects only after chronic administration, whereas direct lithium-induced inhibition of GSK3 is rapid and mild at therapeutic concentrations of 1 mM [91]. Furthermore, lithium is known to affect neuronal inositol metabolism, to inhibit adenylate cyclase, and to activate c-Jun NH2-terminal kinases (JNKs) [68, 92-95]. Perhaps, the most peculiar characteristic of lithium is its "dual" inhibition of GSK3. GSK3 catalyzes the phosphorylation of numerous protein substrates in the presence of Mg^{2+}-ATP. Lithium can directly inhibit GSK3 by acting as competitive inhibitor of Mg^{2+}, reducing the activity of this kinase [96]. Additionally, we previously commented in this chapter that GSK3 is inactivated by phosphorylation on a serine in the N-terminal domain: Ser9 in GSK3β and Ser21 in GSK3α. Once GSK3 is inactivated by serine phosphorylation, it can be re-activated by removing the phosphate from that serine through phosphatase activity. Lithium-induced indirect inhibition of GSK3 is due to a reduced effect of this phosphatase, resulting in higher levels of phosphorylated GSK3 (inactive form) [97]. A number of studies have demonstrated that Aβ production is enhanced by GSK3 and decreased by GSK3 inhibitors [84, 98]. For instance, *in vivo*, lithium treatment abolishes GSK3β-mediated Aβ increase in the brains of GSK3β transgenics and reduces plaque burden in the brains of the PDAPP (APP$_{V717F}$) transgenic mice [99]. Moreover, LiCl is able to modulate GSK3β transcription *in vitro* and *in vivo* [100]. All these studies support the potential therapeutic use of lithium in AD patients.

A number of research groups and pharmaceutical companies are searching for GSK3 inhibitory activities in compounds that have already shown other biological properties [87], such as hymenialdisine, maleimides, muscarinic agonist, paullones, thiadiazolidinones, and indirubins.

Recent studies have highlighted the relevant role of cell cycle proteins in mild cognitive impairment (MCI) and AD. The levels of key cell cycle proteins (*e.g.*, CDK2, CDK5, cyclin G1, and BRAC1) are increased in MCI brains when compared to age-matched controls [101]. Furthermore, the peptidyl-prolyl cis-trans isomerase (Pin1), a protein that plays an important role in regulating the activity of key proteins like CDK5, GSK3β, and PP2A involved in both the phosphorylation state of Tau and cell cycle, has been found to be oxidatively modified and downregulated in MCI and AD brains [102-104]. Therefore, CDK´s inhibitors may have, in theory, great potential for the treatment of AD. Indirubins, for instance, were initially identified as CDK inhibitors and they are derived from the spontaneous and non-enzymatic dimerization of isantin and indoxyl found in more than 200 species of indigo-producing plants [105]. 6-brominated indirubins [106] extracted from another natural source (Mollusk *Hexaplex trunculus*) have been shown to provide an excellent scaffold for the generation of potent kinase inhibitors selective for GSK3 (comparing to CDKs). Indirubins have been shown to be very potent inhibitors of both GSK3β and CDK5, the two major kinases involved in tau hyperphosphorylation, constituting a promising family of compounds to evaluate as therapeutic agents in AD [107].

CONCLUDING REMARKS

The demonstrations that autonomous Wnt signaling is a conserved feature of the neurogenic niche that preserves the delicate balance between NSC maintenance and differentiation [8], and how lithium is able to directly expand pools of adult hippocampal progenitors *in vitro*, inducing them to become neurons at therapeutically relevant concentrations [108], suggest a connection between deficient neurogenesis in AD patients and a possible role of GSK3 in this deficit. Furthermore, additional studies have linked GSK3 to long term potentiation (LTP) by showing that over-expression of GSK3β in mice is enough to prevent its induction [109]. Similarly, it has been reported that inhibition of GSK3β blocked long-term depression (LTD) [110]. These studies together indicate that GSK3β could be critical for the induction of memory formation, switching off LTD and allowing LTP to occur [40]. Thus, the development of new therapeutic approaches to inhibit GSK3β may facilitate not only the promotion of neurogenesis but also recovery of the memory loss by inducing LTP. As a matter of fact, the Wnt/β-catenin signaling pathway is considered a putative target for the treatment of different diseases [111], including neurodegenerative diseases; and the study from He & Shen (2009) reporting that interruption of β-catenin signaling reduces neurogenesis in AD patients support this possibility [12]. They proposed that Aβ elevates GSK3β, which in turn promotes the phosphorylation and degradation of β-catenin, downregulating the expression of proneural genes. However, their model may leave other open questions for further studies. For instance, does their data support previous work where calcium-selective Aβ channels were proposed as an entrance pathway associated with neuronal toxicity [112]? What is the mechanism by which Aβ might be able to increase the intracellular levels of GSK3? Could such increase be a result of an indirect Aβ-induced effect or does it happen as a consequence of a direct physical interaction between protein (Aβ) and enzyme (GSK3)? Is the Aβ-induced effect specific for GSK3 or does it affect other known β-catenin phosphorylators such as PKC (protein kinase C) [45]?

Despite preliminary data demonstrating that GSK3 inhibitors can reduce hyperglycemia, improve insulin sensitivity [113], and exert benefits in animal models of AD [114], there are two main concerns in conflict with the predicted usefulness in therapy: (I) GSK3 is a critical player belonging to the Wnt signaling pathway and its inhibition would mimic the activation of this route which has been linked to the development of several types of cancers in humans [115]; (II) *in vivo,* disruption of the murine GSK3β gene revealed an unexpected embryonic lethality due to massive hepatocyte apoptosis [116]. This is consistent with additional *in vitro* data where pharmacological inhibition of GSK3 in different cell lines has been shown to facilitate apoptosis triggered by different stimuli [117-119]. This means that GSK3 inhibitors would provide protection from intrinsic apoptosis signaling, induced by several stimuli that cause cell damage (*e.g.,* DNA damage, oxidative stress), but could exacerbate extrinsic apoptosis signaling (receptor-mediated). An excellent example is given by an *in vivo* study, where co-infusion of the specific GSK3 inhibitor, SB216763, corrected all responses derived from Aβ infusion, excepting for the induction of gliosis and behavioral deficits. However, SB216763 alone was associated with the induction of neurodegenerative markers and behavioral deficits, supporting the role for GSK3 hyper-activation in AD pathogenesis, but also highlighting the relevance of developing novel inhibitors that do not suppress its constitutive activity [114].

In another elegant study from 2007, Tet/DN-GSK3 mice showed increased neuronal apoptosis and impaired motor coordination, effects that were reversed when DN-GSK3 expression was shut-down and GSK3 activity came back to normal levels [120]. This proves that potential neurological toxicity induced by pharmacological inhibition of GSK3, beyond physiological levels, is plausible and must be taken into consideration. Moreover, in the heart, GSK3 seems to suppress cardiac hypertrophy [121, 122]. Then, the question is whether long-term exposure to GSK3 inhibitors could also increase the incidence of severe cardiac adverse effects because these drugs are intended to treat diabetes and/or AD patients for several years. Even though long-term use of lithium ions (generally used to treat bipolar disorder) has not been associated with an increased risk of cancer so far [123], additional pharmacological tests are required to answer these questions that still remain. Nevertheless, since the activity of the majority of GSK3 inhibitors is not restricted to GSK3, to establish a direct link between pure kinase modulation and its effects could be a difficult challenge.

ACKNOWLEDGEMENTS

We thank the Brazilian research funding agencies FAPERGS (PqG 1008860, PqG 1008857, ARD11/1893-7, PRONEX 1000274), CAPES (PROCAD 066/2007), CNPq, PROPESQ-UFRGS, and IBN-Net #01.06.0842-00 for financial support. Fares Zeidán-Chuliá acknowledges the Marie Curie Early Stage Research Training (EST) program for his previous funding at the University of Helsinki, Finland.

CONFLICT OF INTEREST

The authors confirm that this chapter contents have no conflict of interest.

ABBREVIATIONS

AD	=	Alzheimer´s disease
NFTs	=	neurofibrillary tangles
GSK3	=	glycogen synthase kinase 3
Aβ	=	β-amyloid
SVZ	=	subventricular zone
SGZ	=	subgranular zone
NSCs	=	neural stem cells
RG	=	radial glia
SP	=	senile plaques
APC	=	*adenomatous polyposis coli* protein
APP	=	amyloid precursor protein
Aβ42	=	42-amino acid form of Aβ
BACE1	=	β-site amyloid precursor protein cleaving enzyme 1
sAPPβ	=	β-secreted APP
β-CTF	=	C-terminal membrane-bound fragment
AGEs	=	advanced glycation end products
CML	=	N(epsilon)-(carboxymethyl)lysine protein
IRS1	=	insulin receptor substrate 1
IRS2	=	insulin receptor substrate 2

GS	=	glycogen synthase
Fzd	=	frizzled
CK1α	=	casein-kinase 1α
Tcf/Lef	=	T-cell factor/lymphoid enhancer factor proteins
LRP	=	low-density lipoprotein receptor-related protein
Dvl	=	dishevelled
bFGF	=	basic fibroblast growth factor
EGF	=	epidermal growth factor
GPCs	=	glial precursor cells
PS1	=	presenilin 1
JNKs	=	c-Jun NH2-terminal kinases
MCI	=	mild cognitive impairment
Pin1	=	peptidyl-prolyl cis-trans isomerase
LTP	=	long term potentiation
LTD	=	long-term depression

REFERENCES

[1]   Lois C, Alvarez-Buylla A. Proliferating subventricular zone cells in the adult mammalian forebrain can differentiate into neurons and glia. *Proc Natl Acad Sci U S A* **1993**; *90*(5): 2074-7.

[2]   Doetsch F, Garcia-Verdugo JM, Alvarez-Buylla A. Cellular composition and three-dimensional organization of the subventricular germinal zone in the adult mammalian brain. *J Neurosci* **1997**; *17*(13): 5046-61.

[3]   Kempermann G, Gage FH. Closer to neurogenesis in adult humans. Nat Med **1998**; 4(5): 555-7.

[4]   Gage FH. Mammalian neural stem cells. *Science* **2000**; *287*(5457): 1433-38.

[5]   Jin K, Galvan V. Endogenous neural stem cells in the adult brain. *J Neuroimmune Pharmacol* **2007**; *2*(3): 236-42.

[6]   Hirsch C, Campano LM, Wöhrle S, Hecht A. Canonical Wnt signaling transiently stimulates proliferation and enhances neurogenesis in neonatal neural progenitor cultures. *Exp Cell Res* **2007**; *313*(3): 572-87.

[7]   Nusse R. Wnt signaling and stem cell control. *Cell Res* **2008**; *18*(5): 523-7.

[8]   Wexler EM, Paucer A, Kornblum HI, Plamer TD, Geschwind DH. Endogenous Wnt signaling maintains neural progenitor cell potency. *Stem Cells* **2009**; *27*(5): 1130-41.

[9]   Hirabayashi Y, Itoh Y, Tabata H, *et al.* The Wnt/beta-catenin pathway directs neuronal differentiation of cortical neural precursor cells. *Development* **2004**; *131*(12): 2791-801.

[10]   Marinaro C, Pannese M, Weinandy F, *et al.* Wnt signaling has opposing roles in the developing and the adult brain that are modulated by Hipk1. *Cereb Cortex* **2012**; *22*(10): 2415-27.

[11]   Abdipranoto A, Wu S, Stayte S, Vissel B. The role of neurogenesis in neurodegenerative diseases and its implications for therapeutic development. *CNS Neurol Disord Drug Targets* **2008**; *7*(2): 187-210.

[12]   He P, Shen Y. Interruption of beta-catenin signaling reduces neurogenesis in Alzheimer's disease. *J Neurosci* **2009**; *29*(20): 6545-57.

[13]   Cole AR. GSK3 as a sensor determining cell fate in the brain. *Front Mol Neurosci* **2012**; *5*: 4.

[14]   Phiel CJ, Wilson CA, Lee VM, Klein PS. GSK-3alpha regulates production of Alzheimer's disease amyloid-beta peptides. *Nature* **2003**; *423*(6938): 435-9.

[15]   Sperber BR, Leight S, Goedert M, Lee VM. Glycogen synthase kinase-3 beta phosphorylates tau protein at multiple sites in intact cells. *Neurosci Lett* **1995**; *197*(2): 149-53.

[16]   Goedert M, Spillantini MG. A century of Alzheimer's disease. *Science* **2006**; *314*(5800): 777-81.

[17]   Salmina AB, Inzhutova AI, Malinovskaya NA, Petrova MM. Endothelial dysfunction and repair in Alzheimer-type neurodegeneration: neuronal and glial control. *J Alzheimers Dis* **2010**; *22*(1): 17-36.

[18]   Hölscher C. Development of beta-amyloid-induced neurodegeneration in Alzheimer's disease and novel neuroprotective strategies. *Rev Neurosci* **2005**; *16*(3): 181-212.

[19]   Bierer LM, Hof PR, Purohit DP, *et al.* Neocortical neurofibrillary tangles correlate with dementia severity in Alzheimer's disease. *Arch Neurol* **1995**; *52*(1): 81-8.

[20]   Hardy J, Selkoe DJ. The amyloid hypothesis of Alzheimer's disease: progress and problems on the road to therapeutics. *Science* **2002**; *297*(5580): 353-356.

[21]   Suzuki N, Cheung TT, Cai XD, *et al.* An increased percentage of long amyloid beta protein secreted by familial amyloid beta protein precursor (beta APP717) mutants. *Science* **1994**; *264*(5163): 1336-40.

[22]   Masters CL, Simms G, Weinman NA, Multhaup G, McDonald BL, Beyreuther K. Amyloid plaque core protein in Alzheimer disease and Down syndrome. *Proc Natl Acad Sci U S A* **1985**; *82*(12): 4245-9.

[23]   Lammich S, Kojro E, Postina R, *et al.* Constitutive and regulated alpha-secretase cleavage of Alzheimer's amyloid precursor protein by a disintegrin metalloprotease. *Proc Natl Acad Sci U S A* **1999**; *96*(7): 3922-7.

[24]   Ariga T, McDonald MP, Yu RK. Role of ganglioside metabolism in the pathogenesis of Alzheimer's disease--a review. *J Lipid Res* **2008**; *49*(6): 1157-75.

[25]   Rothhaar TL, Grösgen S, Haupenthal VJ, *et al.* Plasmalogens inhibit APP processing by directly affecting γ-secretase activity in Alzheimer's disease. *ScientificWorldJournal* **2012**; *2012*: 141240.

[26]   Walsh DM, Klyubin I, Fadeeva JV, *et al.* Naturally secreted oligomers of amyloid beta protein potently inhibit hippocampal long-term potentiation *in vivo*. *Nature* **2002**; *416*(6880): 535-9.

[27]   Zhao LN, Long H, Mu Y, Chew LY. The toxicity of amyloid β oligomers. *Int J Mol Sci* **2012**; *13*(6): 7303-27.

[28]   Dei R, Takeda A, Niwa H, *et al.* Lipid peroxidation and advanced glycation endproducts in the brain in normal aging and in Alzheimer's disease. *Acta Neuropathol* **2002**; *104*(2): 113-22.

[29]   Gironès X, Guimerà A, Cruz-Sánchez CZ, *et al.* N epsilon carboxymethyllysine in brain aging, diabetes mellitus, and Alzheimer's disease. *Free Radic Biol Med* **2004**; *36*(10): 1241-7.

[30]   Ikeda K, Higashi T, Sano H, *et al.* N (epsilon)-(carboxymethyl)lysine protein adduct is a major immunological epitope in proteins modified with advanced glycation end products of the Maillard reaction. *Biochemistry* **1996**; *35*(24): 8075-83.

[31]   Kim B, Backus C, Oh S, Hayes JM, Feldman EL. Increased Tau phosphorylation and cleavage inmousemodels of type 1 and type 2 diabetes. *Endocrinology* **2009**; *150*(12): 5294-301.

[32]   Accardi G, Caruso C, Colonna-Romano G, Camarda C, Monastero R, Candore G. Can Alzheimer disease be a form of type 3 diabetes? *Rejuvenation Res* **2012**; *15*(2): 217-21.

[33]   Zhong Y, Miao Y, Jia WP, Yan H, Wang BY, Jin J. Hyperinsulinemia, insulin resistance and cognitive decline in older cohort. *Biomed Environ Sci* **2012**; *25*(1): 8-14.

[34]   DeFronzo RA, Hendler R, Simonson D. Insulin resistance is a prominent feature of insulin-dependent diabetes. *Diabetes* **1982**; *31*(9): 795-801.

[35]   Donga E, van Dijk M, Hoogma RP, Corssmit EP, Romijn JA. Insulin resistance in multiple tissues in patients with type 1 diabetes mellitus on long term continuous subcutaneous insulin infusion therapy. *Diabetes Metab Res Rev* **2012**; doi: 10.1002/dmrr.2343.

[36]   Saad MJ, Araki E, Miralpeix M, Rothenberg PL, White MF, Kahn CR. Regulation of insulin receptor substrate-1 in liver and muscle of animal models of insulin resistance. *J Clin Invest* **1992**; *90*(5): 1839-49.

[37]   Lee AV, Gooch JL, Oesterreich S, Guler RL, Yee D. Insulin-like growth factor I-induced degradation of insulin receptor substrate 1 is mediated by the 26S proteasome and blocked by phosphatidylinositol 3´-kinase inhibition. *Mol Cell Biol* **2000**; *20*(5): 1489-96.

[38]   Copps KD, White MF. Regulation of insulin sensitivity by serine/threonine phosphorylation of insulin receptor substrate proteins IRS1 and IRS2. *Diabetologia* **2012**; *55*(10): 2565-82.

[39]   Leng S, Zhang W, Zheng Y, *et al.* Glycogen synthase kinase 3 beta mediates high glucose-induced ubiquitination and proteasome degradation of insulin receptor substrate 1. *J Endocrinol* **2010**; *206*(2): 171-81.

[40]   Hooper C, Killick R, Lovestone S. The GSK3 hypothesis of Alzheimer's disease. *J Neurochem* **2008**; *104*(6): 1433-1439.

[41]   Gao C, Hölscher C, Liu Y, Li L. GSK3: a key target for the development of novel treatments for type 2 diabetes mellitus and Alzheimer disease. *Rev Neurosci* **2011**; *23*(1): 1-11.

[42]   Sethi JK, Vidal-Puig A. Wnt signalling and the control of cellular metabolism. *Biochem J* **2010**; *427*(1): 1-17.

[43]   MacDonald BT, Tamai K, He X. Wnt/beta-catenin signaling: components, mechanisms, and diseases. *Dev Cell* **2009**; *17*(1): 9-26.

[44]   Clevers H, Nusse R. Wnt/β-catenin signaling and disease. *Cell* **2012**; *149*(6): 1192-205.

[45]   Chen RH, Ding WV, McCormick F. Wnt signaling to beta-catenin involves two interactive components. Glycogen synthase kinase-3beta inhibition and activation of protein kinase C. *J Biol Chem* **2000**; *275*(23): 17894-899.

[46]   Amit S, Hatzubai A, Birman Y, *et al.* Axin-mediated CKI phosphorylation of β-catenin at Ser45: a molecular switch for the Wnt pathway. *Genes Dev* **2002**; *16*(9): 1066-76.

[47]   Price MA. CKI, there's more than one: casein kinase I family members in Wnt and Hedgehog signaling. *Genes Dev* **2006**; *20*(4): 399-410.

[48]   Jin T, George Fantus I, Sun J. Wnt and beyond Wnt: multiple mechanisms control the transcriptional property of beta-catenin. *Cell Signal* **2008**; *20*(10): 1697-704.

[49]   Alonso L, Fuchs E. Stem cells in the skin: waste not, Wnt not. *Genes Dev* **2003**; *17*(10): 1189-200.

[50]   Pinto D, Gregorieff A, Begthel H, Clevers H. Canonical Wnt signals are essential for homeostasis of the intestinal epithelium. *Genes Dev* **2003**; *17*(14): 1709-13.

[51]   Reya T, Duncan AW, Ailles L, *et al.* A role for Wnt signalling in self-renewal of haematopoietic stem cells. *Nature* **2003**; *423*(6938): 409-14.

[52]   Ciani L, Salinas PC. WNTs in the vertebrate nervous system: from patterning to neuronal connectivity. *Nat Rev Neurosci* **2005**; *6*(5): 351-62.

[53]   Zeidán-Chuliá F, Noda M. "Opening" the mesenchymal stem cell tool box. *Eur J Dent* **2009**; *3*(3): 240-9.

[54]   Lee SM, Tole S, Grove E, McMahon AP. A local Wnt-3a signal is required for development of the mammalian hippocampus. *Development* **2000**; *127*(3): 457-67.

[55]   Kumar DU, Devaraj H. Expression of Wnt 3a, β-Catenin, Cyclin D1 and PCNA in Mouse Dentate Gyrus Subgranular Zone (SGZ): a Possible Role of Wnt Pathway in SGZ Neural Stem Cell Proliferation. *Folia Biol (Praha)* **2012**; *58*(3): 115-20.

[56]   Behrens J, Lustig B. The Wnt connection to tumorigenesis. *Int J Dev Biol* **2004**; *48*(5-6): 477-87.

[57]   Doble BW, Woodgett JR. GSK-3: tricks of the trade for a multi-tasking kinase. *J. Cell Sci* **2003**; *116*(Pt 7): 1175-86.

[58]   Lie DC, Colamarino SA, Song HJ, *et al. Nature* **2005**; *437*(7063): 1370-5.

[59]   Thornton TM, Pedraza-Alva G, Deng B, *et al.* Phosphorylation by p38 MAPK as an alternative pathway for GSK3beta inactivation. *Science* **2008**; *320*(5876): 667-70.

[60]   Qu Q, Sun G, Li W, *et al.* Orphan nuclear receptor TLX activates Wnt/beta-catenin signalling to stimulate neural stem cell proliferation and self-renewal. *Nat Cell Biol* **2010**; *12*(1): 31-40.

[61]   Esfandiari F, Fathi A, Gourabi H, Kiani S, Nemati S, Baharvand H. Glycogen synthase kinase-3 inhibition promotes proliferation and neuronal differentiation of human-induced pluripotent stem cell-derived neural progenitors. *Stem Cells Dev* **2012**; doi: 10.1089/scd.2011.0678.

[62]   Lazarov O, Marr R. Neurogenesis and Alzheimer's disease: at the crossroads. *Exp Neurol* **2010**; *223*(2): 267-81.

[63]   Shruster A, Melamed E, Offen D. Neurogenesis in the aged and neurodegenerative brain. *Apoptosis* **2010**; *15*(11): 1415-21.

[64]   Woodgett JR, Cohen P. Multisite phosphorylation of glycogen synthase. Molecular basis for the substrate specificity of glycogen synthase kinase-3 and casein kinase-ii (glycogen synthase kinase-5). *Biochim Biophys Acta* **1984**; *788*(3): 339-47.

[65]   Woodgett JR. Molecular cloning and expression of glycogen synthase kinase-3/factor A. *EMBO J* **1990**; *9*(8): 2431-8.

[66]   Lee J, Kim MS. The role of GSK3 in glucose homeostasis and the development of insulin resistance. *Diabetes Res Clin Pract* **2007**; *77* (Suppl 1): 49-57.

[67]   Parker PJ, Caudwell FB, Cohen P. Glycogen synthase from rabbit skeletal muscle; effect of insulin on the state of phosphorylation of the seven phosphoserine residues *in vivo*. *Eur J Biochem* **1983**; *130*(1): 227-34.

[68]   Patel S, Doble B, Woodgett JR. Glycogen synthase kinase-3 in insulin and Wnt signalling: a double-edged sword? *Biochem Soc Trans* **2004**; *32*(Pt 5): 803-8.

[69]   Sayas CL, Ariaens A, Ponsioen B, Moolenaar WH. GSK-3 is activated by the tyrosine kinase Pyk2 during LPA1-mediated neurite retraction. *Mol Biol Cell* **2006**; *17*(4): 1834-44.

[70]   Eldar-Finkelman H. Glycogen synthase kinase 3: an emerging therapeutic target. *Trends Mol Med* **2002**; *8*(3): 126-32.

[71]   Dent P, Campbell DG, Hubbard MJ, Cohen P. Multisite phosphorylation of the glycogen-binding subunit of protein phosphatase-1G by cyclic AMP-dependent protein kinase and glycogen synthase kinase-3. *FEBS Lett* **1989**; *248*(1-2): 67-72.

[72]   Hanger DP, Hughes K, Woodgett JR, Brion JP, Anderton BH. Glycogen synthase kinase-3 induces Alzheimer's disease-like phosphorylation of tau: generation of paired helical filament epitopes and neuronal localisation of the kinase. *Neurosci Lett* **1992**; *147*(1): 58-62.

[73]   Rubinfeld B, Albert I, Porfiri E, Fiol C, Munemitsu S, Polakis P. Binding of GSK3beta to the APC-beta-catenin complex and regulation of complex assembly. *Science* **1996**; *272*(5264): 1023-6.

[74]   Yost C, Torres M, Miller JR, Huang E, Kimelman D, Moon RT. The axis-inducing activity, stability, and subcellular distribution of beta-catenin is regulated in Xenopus embryos by glycogen synthase kinase 3. *Genes Dev* **1996**; *10*(12): 1443-54.

[75]   Eldar-Finkelman H, Krebs EG. Phosphorylation of insulin receptor substrate 1 by glycogen synthase kinase 3 impairs insulin action. *Proc Natl Acad Sci* U S A **1997**; *94*(18): 9660-4.

[76]   Diehl JA, Cheng M, Roussel MF, Sherr CJ. Glycogen synthase kinase-3beta regulates cyclin D1 proteolysis and subcellular localization. *Genes Dev* **1998**; *12*(22): 3499-511.

[77]   Ikeda S, Kishida S, Yamamoto H, Murai H, Koyama S, Kikuchi A. Axin, a negative regulator of the Wnt signaling pathway, forms a complex with GSK-3beta and beta-catenin and promotes GSK-3beta-dependent phosphorylation of beta-catenin. *EMBO J* **1998**; *17*(5): 1371-84.

[78]   Kirschenbaum F, Hsu SC, Cordell B, McCarthy JV. Glycogen synthase kinase-3beta regulates presenilin 1 C-terminal fragment levels. *J Biol Chem* **2001**; *276*(33): 30701-7.

[79]   Willert K, Brown JD, Danenberg E, *et al.* Wnt proteins are lipid-modified and can act as stem cell growth factors. *Nature* **2003**; *423*(6938): 448-52.

[80]   Zechner D, Fujita Y, Hülsken J, *et al.* beta-Catenin signals regulate cell growth and the balance between progenitor cell expansion and differentiation in the nervous system. *Dev Biol* **2003**; *258*(2): 406-18.

[81]   Dravid G, Ye Z, Hammond H, *et al.* Defining the role of Wnt/beta-catenin signaling in the survival, proliferation, and self-renewal of human embryonic stem cells. *Stem Cells* **2005**; *23*(10): 1489-501.

[82]   Katoh M, Katoh M. WNT signaling pathway and stem cell signaling network. *Clin Cancer Res* **2007**; *13*(14): 4042-5.

[83]   Pei JJ, Braak E, Braak H, *et al.* Distribution of active glycogen synthase kinase 3β (GSK-3β) in brains staged for Alzheimer disease neurofibrillary changes. *J Neuropathol Exp Neurol* **1999**; *58*(9): 1010-9.

[84]   Sun X, Sato S, Murayama M, Park JM, Yamaguchi H, Takashima A. Lithium inhibits amyloid secretion in COS7 cells transfected with amyloid precursor protein C100. *Neurosci Lett* **2002**; *321*(1-2): 61-4.

[85]   Jope RS. Anti-bipolar therapy: mechanism of action of lithium. *Mol Psychiatry* **1999**; *4*(2): 117-28.

[86]   Phiel CJ, Klein PS. Molecular targets of lithium action. *Annu Rev Pharmacol Toxicol* **2001**; *41*: 789-813.

[87]   Martinez A, Castro A, Dorronsoro I, Alonso M. Glycogen synthase kinase 3 (GSK-3) inhibitors as new promising drugs for diabetes, neurodegeneration, cancer, and inflammation. *Med Res Rev* **2002**; *22*(4): 373-84.

[88]   Kessing LV, Nilsson FM. Increased risk of developing dementia in patients with major affective disorders compared to patients with other medical illnesses. *J Affect Disord* **2003**; *73*(3): 261-9.

[89]   Lopes R, Fernandes L. Bipolar disorder: clinical perspectives and implications with cognitive dysfunction and dementia. *Depress Res Treat* **2012**; *2012*: 275957.

[90]   Nunes PV, Forlenza OV, Gattaz WF. Lithium and risk for Alzheimer´s disease in elderly patients with bipolar disorder. *Br J Psychiatry* **2007**; *190*: 359-60.

[91]   Jope RS. Lithium and GSK-3: one inhibitor, two inhibitory actions, multiple outcomes. *Trends Pharmacol Sci* **2003**; *24*(9): 441-3.

[92]   Berridge MJ, Downes CP, Hanley MR. Neural and developmental actions of lithium: a unifying hypothesis. *Cell* **1989**; *59*(3): 411-19.

[93]   Marmol F, Carbonell L, Cuffi ML, Forn J. Demonstration of inhibition of cyclic AMP accumulation in brain by very low concentrations of lithium in the presence of alpha-adrenoceptor blockade. *Eur J Pharmacol* **1992**; *226*(1): 93-6.

[94]   Masana MI, Bitran JA, Hsiao JK, Potter WZ. *In vivo* evidence that lithium inactivates Gi modulation of adenylate cyclase in brain. *J Neurochem* **1992**; *59*(1): 200-5.

[95]   Yuan P, Chen G, Manji HK. Lithium activates the c-Jun NH2-terminal kinases *in vitro* and in the CNS *in vivo*. *J Neurochem* **1999**; *73*(6): 2299-309.

[96]   Ryves WJ, Harwood AJ. Lithium inhibits glycogen synthase kinase-3 by competition for magnesium. *Biochem Biophys Res Commun* **2001**; *280*(3): 720-5.

[97]   Zhang F, Phiel CJ, Spece L, Gurvich N, Klein PS. Inhibitory phosphorylation of glycogen synthase kinase-3 (GSK-3) in response to lithium. Evidence for autoregulation of GSK-3. *J Biol Chem* **2003**; *278*(35): 33067-77.

[98]   Li B, Ryder J, Su Y, Zhou Y, Liu F, Ni B. FRAT1 peptide decreases Ab production in swAPP(751) cells. *FEBS Lett* **2003**; *553*(3): 347-50.

[99] Su Y, Ryder J, Li B, *et al.* Lithium, a common drug for bipolar disorder treatment, regulates amyloid-beta precursor protein processing. *Biochemistry* **2004**; *43*(22): 6899-908.

[100] Mendes CT, Mury FB, de Sá Moreira E, *et al.* Lithium reduces Gsk3b mRNA levels: implications for Alzheimer Disease. *Eur Arch Psychiatry Clin Neurosci* **2009**; *259*(1): 16-22.

[101] Keeney JT, Swomley AM, Harris JL, *et al.* Cell cycle proteins in brain in mild cognitive impairment: insights into progression to Alzheimer disease. *Neurotox Res* **2012**; *22*(3): 220-30.

[102] Flaherty DB, Soria JP, Tomasiewicz HG, Wood JG. Phosphorylation of human tau protein by microtubuleassociated kinases: GSK-3beta and cdk5 are key participants. *J Neurosci Res* **2000**; *62*(3): 463-72.

[103] Butterfield DA, Abdul HM, Opii W, *et al.* Pin1 in Alzheimer's disease. *J Neurochem* **2006**; *98*(6): 1697-706.

[104] Sultana R, Butterfield DA. Regional expression of key cell cycle proteins in brain from subjects with amnestic mild cognitive impairment. *Neurochem Res* **2007**; *32*(4-5): 655-62.

[105] Maugard T, Enaud E, Choisy P, Legoy MD. Identification of an indigo precursor from leaves of Isatis tinctoria (Woad). *Phytochemistry* **2001**; *58*(6): 897-904.

[106] Meijer L, Skaltsounis AL, Magiatis P, *et al.* GSK-3-selective inhibitors derived from Tyrian purple indirubins. *Chem Biol* **2003**; *10*(12): 1255-66.

[107] Leclerc S, Garnier M, Hoessel R, *et al.* Indirubins inhibit glycogen synthase kinase-3 beta and CDK5/p25, two protein kinases involved in abnormal tau phosphorylation in Alzheimer's disease. A property common to most cyclin-dependent kinase inhibitors? *J Biol Chem* **2001**; *276*(1): 251-60.

[108] Wexler EM, Geschwind DH, Palmer TD. Lithium regulates adult hippocampal progenitor development through canonical Wnt pathway activation. *Mol Psychiatry* **2008**; *13*(3): 285-92.

[109] Hooper C, Markevich V, Plattner F, *et al.* Glycogen synthase kinase-3 inibition is integral to long-term potentiation. *Eur J Neurosci* **2007**; *25*(1): 81-6.

[110] Peineau S, Taghibiglou C, Bradley C, *et al.* LTP inhibits LTD in the hippocampus *via* regulation of GSK3beta. *Neuron* **2007**; *53*(5): 703-17.

[111] Takahashi-Yanaga F, Sasaguri T. The Wnt/beta-catenin signaling pathway as a target in drug discovery. *J Pharmacol Sci* **2007**; *104*(4): 293-302.

[112] Jang H, Zheng J, Nussinov R. Models of beta-amyloid ion channels in the membrane suggest that channel formation in the bilayer is a dynamic process. *Biophys J* **2007**; *93*(6): 1938-49.

[113] Wagman AS, Johnson KW, Bussiere DE. Discovery and development of GSK3 inhibitors for the treatment of type 2 diabetes. *Curr Pharm Des* **2004**; *10*(10): 1105-37.

[114] Hu S, Begum AN, Jones MR, *et al.* GSK3 inhibitors show benefits in an Alzheimer's disease (AD) model of neurodegeneration but adverse effects in control animals. *Neurobiol Dis* **2009**; *33*(2): 193-206.

[115] Polakis P. The many ways of Wnt in cancer. *Curr Opin Genet Dev* **2007**; *17*(1): 45-51.

[116] Hoeflich KP, Luo J, Rubie EA, Tsao MS, Jin O, Woodgett JR. Requirement for glycogen synthase kinase-3beta in cell survival and NF-kappaB activation. *Nature* **2000**; *406*(6791): 86-90.

[117]   Beyaert R, Vanhaesebroeck B, Suffys P, Van Roy F, Fiers W. Lithium chloride potentiates tumor necrosis factor-mediated cytotoxicity *in vitro* and *in vivo*. *Proc Natl Acad Sci U S A* **1989**; *86*(23): 9494-8.

[118]   Song L, Zhou T, Jope RS. Lithium facilitates apoptotic signaling induced by activation of the Fas death domain-containing receptor. *BMC Neurosci* **2004**; *5*: 20.

[119]   Beurel E, Jope RS. The paradoxical pro- and anti-apoptotic actions of GSK3 in the intrinsic and extrinsic apoptosis signaling pathways. *Prog Neurobiol* **2006**; *79*(4): 173-89.

[120]   Gomez-Sintes R, Hernandez F, Bortolozzi A, *et al.* Neuronal apoptosis and reversible motor deficit in dominant negative GSK-3 conditional transgenic mice. *EMBO J* **2007**; *26*(11): 2743-54.

[121]   Haq S, Choukroun G, Kang ZB, *et al.* Glycogen synthase kinase 3b is a negative regulator of cardiomyocyte hypertrophy. *J Cell Biol* **2000**; *151*(1): 117-30.

[122]   Antos CL, McKinsey TA, Frey N, *et al.* Activated glycogen synthase-3 beta suppresses cardiac hypertrophy *in vivo*. *Proc Natl Acad Sci U S A* **2002**; *99*(2): 907-12.

[123]   Cohen Y, Chetrit A, Cohen Y, Sirota P, Modan B. Cancer morbidity in psychiatric patients: influence of lithium carbonate treatment. *Med Oncol* **1998**; *15*(1): 32-6.

CHAPTER 15

Targets and Small Molecules Against Tauopathies. Part 1: From Genes to Soluble, Aggregation-Prone Tau Proteins

Pierfausto Seneci[*]

Department of Chemistry, University of Milan, Via Golgi 19, I-20133 Milan, Italy

Abstract: Tau is a key dynamic regulator of microtubules in neurons. Tau-microtubule binding contributes to axonal stabilization in neurons. Its controlled-physiological weakening allows cell division and microtubule reorganization in fetal state or in mitotic neuronal cells. Tauopathies often show a decreased tau-microtubule binding and the aggregation of tau. The former leads to chronic microtubule-axonal destabilization, the latter preludes to the formation of intra-neuronal, insoluble tau deposits. Tau alternative splicing and post-translational modifications (hyper-phosphorylation, glycosylation, prolylamide bond isomerization, oxidation, *etc.*) are early events with an impact on tauopathies which may lead to disease-modifying therapeutic interventions. The most prospective therapeutic avenues targeted against these events are presented and critically discussed, selecting a single molecular target of particular relevance. Each target is presented together with its known small molecule modulators. Priority is given to mechanisms, targets and small molecules impacting on more than one tauopathy-causing event.

Keywords: Alternative splicing, Alzheimer's disease, DAPK, DYRK1A, glycosyl ation, GSK3, kinases, microtubules, neurodegeneration, O-glcnac hydrolase, peptidyl-prolyl isomerases, PIN1, post-translational modifications, TAU protein, tauopathies.

INTRODUCTION

"Tau proteins bind microtubules through the microtubule-binding domains… they are mainly expressed in neurons where they play an important role in the assembly of tubulin monomers into microtubules to constitute the neuronal microtubules network... microtubule assembly depends partially upon the phosphorylation state since phosphorylated tau proteins are less effective than non-phosphorylated tau proteins on microtubule polymerization... the most

---

*Address correspondence to Pierfausto Seneci: Department of Chemistry, University of Milan, Via Golgi 19, I-20133 Milan, Italy; Tel: +39 02 50314060; Fax: +39 02 50314075; E-mail: pierfausto.seneci@unimi.it

Atta-ur-Rahman / Muhammad Iqbal Choudhary (Eds.)

obvious pathological event in several neurodegenerative disorders is the aggregation of tau isoforms into intraneuronal filamentous inclusions. Until recently, it was thought that an abnormal phosphorylation of tau proteins was responsible for their aggregation.... data suggest that, in addition to phosphorylation, other mechanisms may be involved in the formation of pathological tau filaments".

A review dated 2000 [1] highlighted the importance of tau in maintaining neuronal shape and functionality, the relevance of tau-driven pathologies in neurodegenerative disorders, and the leading role of tau hyper-phosphorylation in such pathologies. Such statements are still valid today, but much more is known about additional molecular mechanisms leading to tau-microtubule binding disruption and to tau-based paired helical filament – PHF - formation. Such mechanistic elucidation led to the identification and validation of a number of targets which were exploited by private and public research groups in the past decade. This in turn guided the rational design, synthesis and characterization of novel disease-modifying agents against tauopathies.

Overall, we will describe the physiological and pathological role of tau in CNS, focusing on four processes (step 1 to 4, Chart **1**).

Here we will describe early tau-driven pathological alterations in the brain, referring to alternative splicing and post-translational modifications (PTMs). Rational options to restore the functions of tau in such altered conditions will be provided. Each selected alteration/repair couple will be briefly illustrated, and at least one preferred molecular target will be discussed in details as a significant therapeutic option. Targets having an additional, putative beneficial effect on other causative effects in tauopathies, such as Aβ [2] in Alzheimer's disease (AD), will be privileged.

Each molecular target will be connected to small molecules acting on such target, as leads and candidates currently under evaluation as disease-modifyers against tauopathies. Although immunotherapy can be successfully employed against tauopathies, as shown in animal models [3, 4], it will not be dealt with here.

Soluble, mis-decorated tau protein copies are the most advanced therapeutic target in the development of tauopathies which will be covered here, but not the final step in the development of tau-centered diseases. The aggregation of soluble, mis-decorated tau proteins in neurotoxic oligomers, fibrils and tangles, and the pathological consequences of tau modifications on its binding partners will be covered in a forthcoming review.

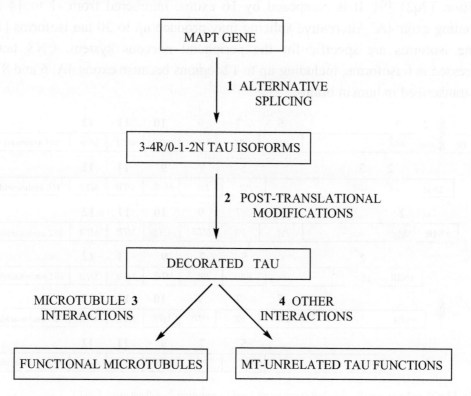

Chart 1: Tau-related physiological processes.

CNS TAU ISOFORMS: TO SPLICE OR NOT TO SPLICE, THAT IS THE QUESTION....

Tau is a highly soluble, natively unfolded protein discovered in 1975 [5]. As a microtubule-associated protein (MAP), its main role is to promote microtubule (MT) assembly. Consequently, tau ensures structural integrity to neurons in general and axons in particular, where it is predominantly expressed [6]. The almost total absence of secondary and tertiary structural elements in tau leads to a large set of low-energy inter-converting tau structures in solution [7]. A dynamic MT-tau interaction network is, thus, permanently established. Namely, MT segments can be stabilized or destabilized in a tau-dependent manner to, among other events, regulate neuritic growth and promote axonal transport [8].

The single copy human tau gene *MAPT* is located on chromosome 17 at band position 17q21 [9]. It is composed by 16 exons, numbered from -1 to 14 and including exon 4A. Alternative splicing may produce up to 30 tau isoforms [10]. Some isoforms are specific for the peripheral nervous system. CNS tau is expressed in 6 isoforms, including up to 12 regions because exons 4A, 6 and 8 are not transcribed in human brains.

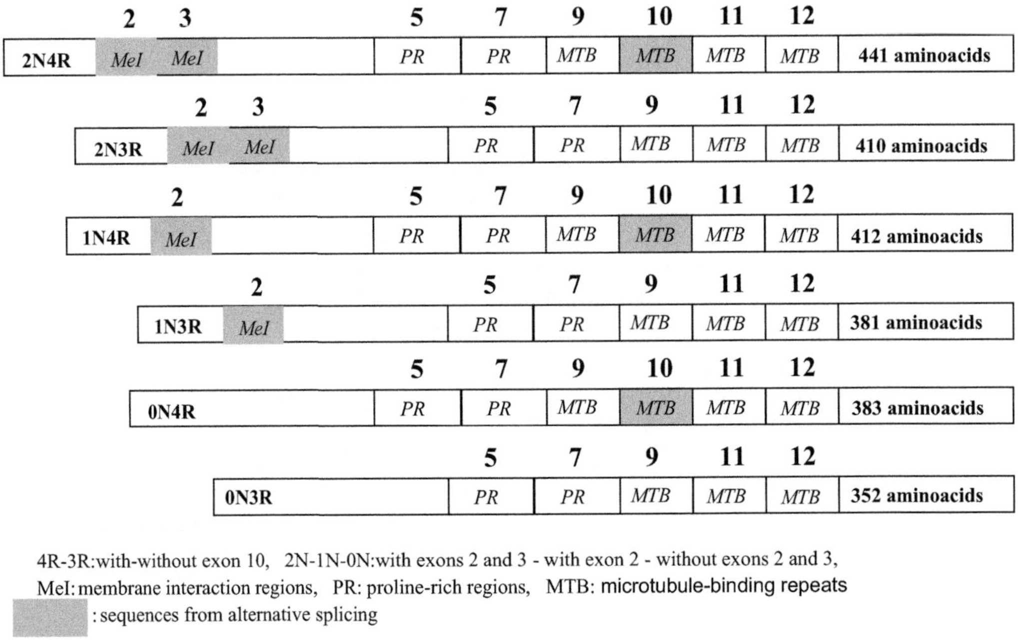

Isoform	2	3	5	7	9	10	11	12	
2N4R	MeI	MeI	PR	PR	MTB	MTB	MTB	MTB	441 aminoacids
2N3R	MeI	MeI	PR	PR	MTB		MTB	MTB	410 aminoacids
1N4R	MeI		PR	PR	MTB	MTB	MTB	MTB	412 aminoacids
1N3R	MeI		PR	PR	MTB		MTB	MTB	381 aminoacids
0N4R			PR	PR	MTB	MTB	MTB	MTB	383 aminoacids
0N3R			PR	PR	MTB		MTB	MTB	352 aminoacids

4R-3R:with-without exon 10, 2N-1N-0N:with exons 2 and 3 - with exon 2 - without exons 2 and 3,
MeI:membrane interaction regions, PR: proline-rich regions, MTB: microtubule-binding repeats
: sequences from alternative splicing

Figure 1: CNS-relevant tau isoforms.

The longer human brain tau isoform (2N4R, Fig. **1**) is composed by three domains. A basic C-terminal domain (aminoacids 244 to 441), containing four MT-binding repeats (MTB, Fig. **1**), modulates the MT-tau interactions [11]. A basic middle domain (aminoacids 151 to 243), containing two proline-rich regions (PR, Fig. **1**), contributes to MT binding and to the interaction with proteins such as actin [12]. An N-terminal domain (aminoacids 1 to 150), containing two acidic inserts (MeI, Fig. **1**), interacts with plasma membrane [13] and proteins such as Fyn [14].

Tau isoforms shown in Fig. **1** differ for the presence-absence of three sequences resulting from alternative splicing of exons 2, 3 (N-terminal domain, isoforms 2N, 1N and 0N) and 10 (C-terminal domain, isoforms 4R and 3R). These exons are shown as gray rectangles in Fig. **1**. A total of six, rather than eight isoforms, is observed because exon 3 is expressed only in presence of exon 2 [15]. The shortest, 352 aminoacids-containing tau protein is the only isoform present in fetal tau [16]. The adult tau pool in human brains encompasses the six isoforms shown in Fig. **1**.

Tau *alternative splicing* has profound implications. Exon 10 contains one of the MTB repeats, so that 4R isoforms show stronger binding to MTs than 3R isoforms. At embryonic stage, the shortest 3R isoform weakens MT-tau interactions and allows the growth of immature neurons [16]. Adult tau shows a ≈1:1 4R:3R ratio, likely representing an equilibrium between strong MT cohesion to secure neuronal integrity, and residual morphological plasticity to allow dynamic changes in the MT-tau complexes. An abnormal 4R:3R ratio in adult tau pools, regardless of which isoform family is preponderant, is invariably associated with tauopathies [17].

The transcription of N-terminal inserts corresponding to isoforms 2N, 1N and 0N has an effect on axonal membrane binding [18], and on dynactin binding [19]. Although a clear connection between tauopathies and 2N:1N:0N tau isoform ratios has not yet been elucidated, their relative abundance in adult human brain is surely connected with physiological and pathological downstream events in neuronal functions [20].

Prevention of Abnormal Tau Splicing

Alternative splicing of *MAPT* (step 1, Chart **2**) provides the physiological 3-4R/0-1-2N ratio in adult tau pools and is regulating tau functions and stability as a monomer. Abnormal splicing in tauopathies (step 1', Chart **2**) is, conversely, a significant tauopathy-promoting event. It represents a multi-targeted therapeutic objective which will be discussed further in this Section.

The human tau gene *MAPT* can cause a disease even while codifying the non-mutated, wild-type tau protein. This was proven for the pathological alteration of exon 10 splicing ratios among tau isoforms. It is also likely to happen through the splicing alteration of exon 2, but a clear-cut exon 2 splicing-disease connection has not been established yet.

Mutations in *MAPT* were first detected in 1998 as causative events for frontotemporal dementia with parkinsonism linked to chromosome 17 (FTDP-17), an inherited tauopathy [21-23]. Some missense mutations cause aminoacid variations in tau, but a number of them are silent mutations. Most silent mutations act by switching the physiological, ≈1:1 3R:4R isoform ratio towards 4R/exon 10 inclusion [24]. Eventually, they may lead to a tauopathy-like clinical outcome with an altered 3R:4R ratio in diseased brains [25]. Among such tauopathies, progressive supranuclear palsy (PSP) [26], corticobasal degeneration (CBD) [27] and argyrophilic grain disease (AGD) [28] are unbalanced towards 4R prevalence-exon 10 inclusion. Pick's disease (PiD) [29] and Down syndrome (DS) [30] show 3R prevalence-exon 10 exclusion, and accumulation of 3R aggregates in diseased brains. Thus, altering the ≈1:1 3R:4R tau isoform ratio in each direction is associated to distinct tauopathies. Opposite splicing-based therapeutic intervention strategies for 3R- and 4R-characterized tauopathies are then needed, although they could raise concerns in terms of side effects due to over-compensation of pathological splicing.

A detailed description of the mechanisms and roles for exon 10 splicing in tauopathies was recently published [31].

Molecular Targets

Stem Loop RNA Sequences

Two mechanisms of action may be imagined to restore a physiological, ≈1:1 3R:4R splicing ratio (Chart **2**). Splicing of exon 10 in tau is a complex process, that involves *cis* enhancers and silencers [32] – mostly localized in exon 10 and intron 10 – and *trans*-acting factors [33, 34].

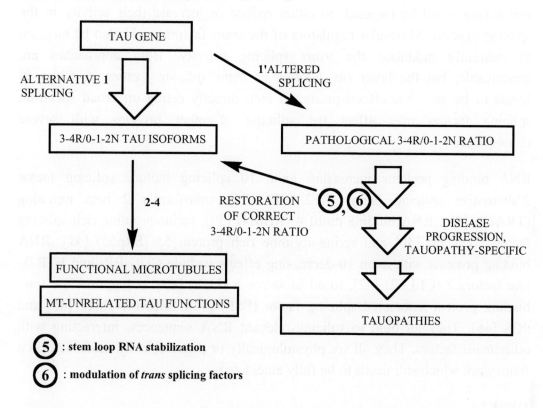

Chart 2: Alternative splicing-related pathological processes, and putative therapeutic approaches.

As to *cis* elements, the majority of silent mutations in FTDP-17 are located at the 3' end of exon 10. Sequence complementarity between the exon 10-intron 10 interface suggests the formation of a stem loop RNA structure [23, 31, 35], although its existence is questioned by some [17]. Silent MAPT mutations in this region reduce stem loop stability, make the region more accessible for interaction with the splicing machinery, and eventually increase the quantity of 4R tau isoform. *Stem loop RNA stabilization* [8, 35, 36] (**5**, Chart **2**) could be targeted to strengthen the stem loop structure, and to prevent the excess of tau 4R isoform in 4R-unbalanced tauopathies. The structure of small molecule stem loop RNA stabilizers will be shown in a forthcoming Paragraph.

As to *trans* splicing factors, past years witnessed the identification of putative molecular targets to influence the 3R:4R isoform ratio. Splicing factors themselves could be targeted, to either reduce or increase their activity in the splicing process. Molecular regulators of the *trans* factors could also be targeted to indirectly modulate the *trans* splicing activity. Both approaches are conceivable, but the latter (*modulation of trans splicing factors*, **6**, Chart **2**) seems to be less side effects-prone. In fact, directly acting on broad spectrum splicing factors may affect the splicing of other proteins with severe consequences.

RNA binding proteins increasing exon 10 splicing include splicing factor 2/alternative splicing factor (SF2/ASF) [37], transformer 2 beta homolog (TRA2β) [38], RNA binding motif 4 (RBM4) [39], serine-arginine rich splicing factor 2 (SC35) [40] and serine-arginine rich protein 55 (SRp55) [41]. RNA binding proteins with exon 10-decreasing effects include CUG-BP- and ETR-3-like factors 2 (CELF2) [42], fused in sarcoma (FUS) [43], polypyrimidine tract binding protein associated splicing factor (PSF) [44], SRp54/SFRS11 [45] and 9G8 [46]. They all bind to splicing-relevant RNA sequences, interacting with other *trans* factors. They all are physiologically or pathologically modulated in a framework which still needs to be fully elucidated.

DYRK1A

Kinases are known regulators of alternative splicing [47]. Kinase regulators for exon 10 splicing include CLKs – unknown mechanism [48] -, PKA – acting on SF2/ASF [49] and 9G8 [50] -, and GSK-3β – acting on SC35 in an Aβ-dependent cascade [51].

A stronger connection is made between *dual-specificity tyrosine phosphorylation-regulated kinase 1A (DYRK1A)* [52], tau phosphorylation and alternative exon 10 splicing. DYRK1A is the first – but not the last – tau-relevant kinase encountered in this review, that shows a pleiotropic effect on putative tauopathy-causing events. Its recent association with exon 10 splicing was fundamental to validate DYRK1A as a molecular target against tauopathies.

DYRK1A is over-expressed, in a brain region- and cell type-specific manner, in patients affected by DS, PiD and AD [53]. DYRK1A-dependent tau phosphorylation of up to 11 Ser-Thr residues is observed *in vitro*, possibly contributing to pathological hyper-phosphorylation [54]. DYRK1A is involved in priming of tau hyper-phosphorylation by GSK-3-β [55]. It inhibits the phosphatase activity of calcineurin *via* phosphorylation of the regulator of calcineurin 1 (RCAN1) protein [55]. DYRK1A-dependent tau hyper-phosphorylation is observed in a mouse model of DS [56]. Co-localization of DYRK1A is detectable in neurofibrillary tangles (NFTs) from sporadic AD patients [57], but is much higher in NFTs from DS patients [30], reflecting the varying relevance of DYRK1A in individual tauopathies.

DYRK1A is associated to nuclear speckles (sub-nuclear structures involved in splicing) and is able, when overexpressed, to promote speckle disassembly [57]. DYRK1A phosphorylates the *trans* splicing factor SF2/ASF on three DYRK1A-specific epitopes [58], causing its translocation to nuclear speckles and preventing its splicing function – inclusion of exon 10 in tau. Over-expression of DYRK1A correlates with a significant increase of 3R tau isoforms in NFTs from DS patients, and with the accelerated, AD-like neurodegeneration process of these patients. In comparison, sporadic AD patients with similar DYRK1A overexpression have a much lower 3R:4R imbalance in NFTs [59]. The link between DS and SF2/ASF-caused, DYRK1A-regulated splicing validates DYRK1A as a target for therapeutic intervention in DS.

Three other *trans* splicing factors are regulated by DYRK1A. DYRK1A exerts a 3R-increasing, negative regulation of 4R-promoting *trans* splicing factor SC35, although the exact phosphorylation sites on SC35 are unknown [60]. DYRK1A has the same negative modulation of 4R-promoting *trans* splicing factor SRp55 [61], *via* phosphorylation of residues in its proline-rich domain. Conversely, DYRK1A shows a cell type-dependent, 4R-increasing role through phosphorylation and positive regulation of the 4R-excluding *trans* splicing factor 9G8 [62].

The multi-targeted, mostly 3R-increasing regulatory role of DYRK1A in tauopathies may stem also from binding interactions with additional splicing factors. Together with its tau kinase role, and with its priming of GSK-3-β hyper-phosphorylation of tau – see Paragraph GSK-3, the regulation function of DYRK1A on alternative splicing should make it a preferred target for tauopathy-related mechanistic studies and for drug discovery projects.

Chemical Structures

Stem Loop RNA Stabilizers

The X-ray structure of the aminoglycoside neomycin (**1**, Fig. **2**) bound to the RNA sequence corresponding to the regulation sequence of tau exon 10 is available [36]. The well known tendency of aminoglycoside antibiotics to bind RNA, and their sub-optimal drug-like profile (limited bioavailability, unlikely blood-brain barrier/BBB permeability, oto- and nephrotoxicity, *etc.*) discourage their use for the development of stem loop stabilizers. Incubation with mitoxanthrone (**2a**, Fig. **2**) provides increased stability of the RNA stem loop in tau exon 10 [35]. Mitoxanthrone is the only validated hit (EC$_{50}$ = 0.89μM in the displacement of a fluorescent derivative of **1**) from a ≈110,000-membered library tested in a screening campaign. Molecular interactions between **2a** and the RNA stem loop in tau are determined, using NMR and other bioanalytical techniques [5].

A preliminary SAR among ≈20 synthesized analogues of **2a** identifies compounds **2b,c** (Fig. **2**) as the most active individuals [63]. Larger constructs, comprised of a neomycin-like part linked to a mitoxanthrone-inspired, heterocyclic moiety [64], are also able to stabilize the RNA stem loop in tau.

Mitoxanthrone-related off-target effects, and its sub-optimal drug-likeness hinders any tauopathy-related development. The discovery of drug-like stem loop stabilizers through HTS campaigns is a daunting task [35], hampering the discovery of much needed, novel chemiotypes.

DYRK1A Inhibitors

DYRK1A is a dual-specificity kinase, regulated by Tyr-321 auto-phosphorylation. Its many substrates are phosphorylated on Ser-Thr residues [65]. DYRK1A inhibitors, preventing either auto-phosphorylation or substrate phosphorylation, were reported. Their potential as treatments for AD and DS is assessed in a recent, medicinal chemistry-driven review [66].

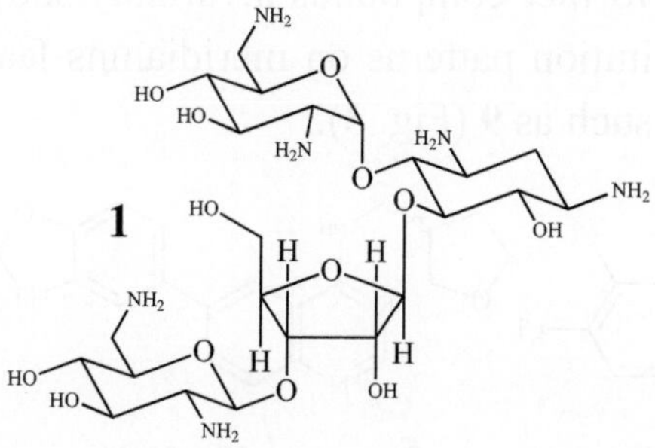

2a $R_1=R_2=OH$, $R_3=R_4=H$
      $EC_{50} = 0.89\mu M$

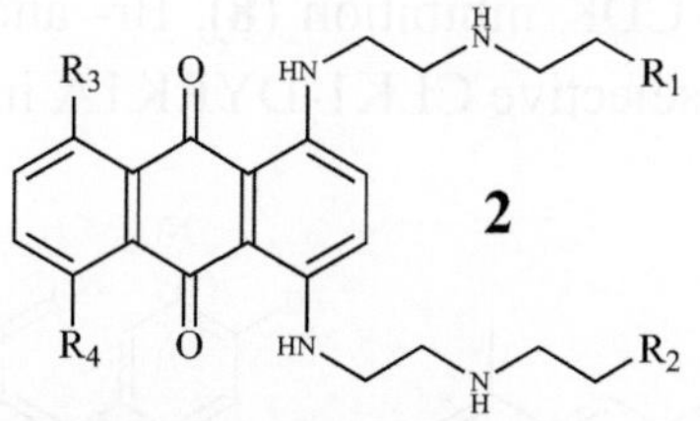

2b $R_1=R_2=NHCH_2CH_2NH_2$, $R_3=R_4=H$
      $EC_{50} = 0.13\mu M$

2c $R_1=R_2=OH$, $R_3=NHCH_2CH_2NHCH_2CH_2OH$, $R_4=OH$
      $EC_{50} = 0.80\mu M$

Figure 2: Stem loop RNA stabilizers: chemical structures.

Purvalanol A [67], rebeccamycin [68], staurosporin [68] and A-443654 [69] are nanomolar DYRK1A inhibitor. Due to their aspecificity, they have no practical use.

"Cleaner" DYRK1A inhibitors are endowed with nanomolar potency on the target and similar activities on few other kinases (see IC_{50s}, Fig. **3**). Moderately active benzocoumarin dNBC (**3**, Fig. **3**) stems from a CK2-oncology targeted project [70]. Its binding mode to DYRK1A is rationalized through modeling studies. Compound **4** inhibits both auto-phosphorylation and substrate phosphorylation by DYRK1A [71]. Poor DYRK1A inhibition for more than 20 analogues of **4** shows the limited tolerance of structural modifications for the pyrazolidine-3,5-dione scaffold. Benzodioxolane-containing CLK-DYRK1A mixed inhibitors are found between ≈50 6-arylquinazolin-6-amines profiled *in vitro* [72]. Among them, compound **5** shows a slight preference for DYRK1A, good aqueous solubility and cell permeability [72]. Compound **6** shows cell-free potency and selectivity for DYRK1A among pyridine- and pyrazine-centered, CDK5-targeted V-shaped molecules [73]. Lamellarin-inspired chromeno[3,4b]indoles are rather selective

DYRK1A inhibitors [74]. C_2-OH substitution (as in **7**) abolishes the preeminent activity on topoisomerase I shown by naturally occurring lamellarins [74]. Finally, meriolins [75] and meridianins [76] show nanomolar potency on DYRK1A in an *in vitro* screening. While the former compounds invariably show stronger CDK inhibition (**8**), Br- and I- substitution patterns on meridianins lead to more selective CLK1-DYRK1A inhibitors, such as **9** (Fig. **3**).

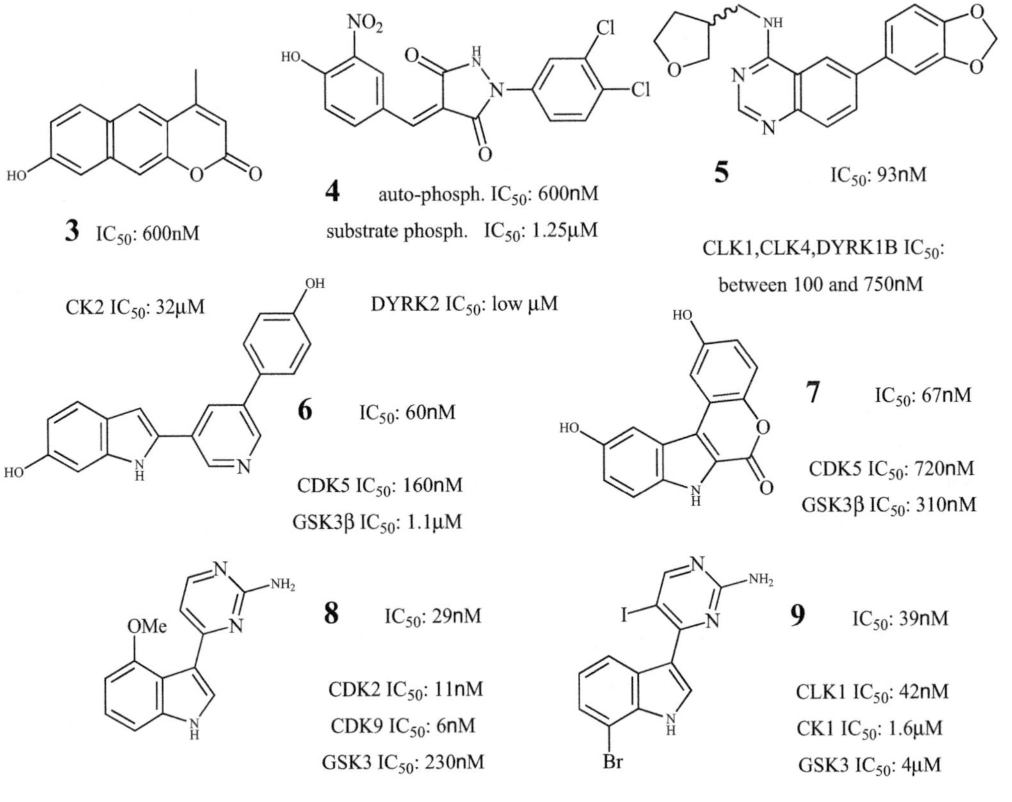

3 IC$_{50}$: 600nM

CK2 IC$_{50}$: 32μM

4 auto-phosph. IC$_{50}$: 600nM
substrate phosph. IC$_{50}$: 1.25μM

DYRK2 IC$_{50}$: low μM

5 IC$_{50}$: 93nM

CLK1,CLK4,DYRK1B IC$_{50}$:
between 100 and 750nM

6 IC$_{50}$: 60nM

CDK5 IC$_{50}$: 160nM
GSK3β IC$_{50}$: 1.1μM

7 IC$_{50}$: 67nM

CDK5 IC$_{50}$: 720nM
GSK3β IC$_{50}$: 310nM

8 IC$_{50}$: 29nM

CDK2 IC$_{50}$: 11nM
CDK9 IC$_{50}$: 6nM
GSK3 IC$_{50}$: 230nM

9 IC$_{50}$: 39nM

CLK1 IC$_{50}$: 42nM
CK1 IC$_{50}$: 1.6μM
GSK3 IC$_{50}$: 4μM

Figure 3: DYRK1A inhibitors: chemical structures, compounds **3-9**.

Three compound classes endowed with DYRK1A inhibition show cellular and *in vivo* potency in neuroprotective assays (Fig. **4**).

Harmine (**10a**, Fig. **4**) is a naturally occurring β-carboline alkaloid, characterized as a potent and inter-kinase selective DYRK1A inhibitor [69]. It acts as an ATP-competitive inhibitor [77] and has a stronger effect on substrate phosphorylation [78], rather than on Tyr321 auto-phosphorylation. Its activity on DYRK1A was confirmed on HEK293 cells, reducing the phosphorylation of the DYRK1A substrate splicing factor 3B1 (SF3B1, IC_{50}=48nM) [78]. Harmine regulates neurite formation [78], in agreement with the role of DYRK1A as a regulator of neuronal development [79]. Low µM harmine concentrations cause DYRK1A-dependent inhibition of phosphorylation at T231 and S396 tau epitopes in H4 neuroglioma cell lines [80]. Toxicity on the same cells is observed at slightly higher harmine concentrations. Harmine seems a chemically tractable, low MW scaffold for the identification of optimized DYRK1A inhibitors. Unfortunately, it also acts strongly and aspecifically on several CNS targets. In particular, its strong inhibition of monoamine oxidase A (MAO-A, IC_{50}=5nM [81]) should be eliminated to allow *in vivo* testing. Harmol **10b** and 9-ethylharmine **10c** show good potencies on DYRK1A-mediated tau phosphorylation and a preliminary SAR [80], but their cell toxicity increased proportionally with their effects on tau.

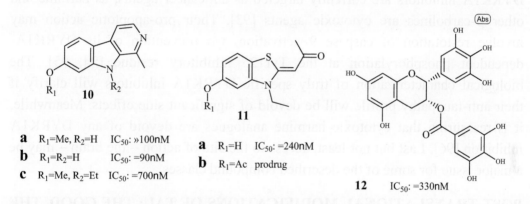

a	R_1=Me, R_2=H	IC_{50}: »100nM
b	R_1=R_2=H	IC_{50}: =90nM
c	R_1=Me, R_2=Et	IC_{50}: =700nM

a	R_1=H	IC_{50}: =240nM
b	R_1=Ac	prodrug

12　　IC_{50}: =330nM

Figure 4: DYRK1A inhibitors: chemical structures, compounds **10-12**.

Disconnecting DYRK1A inhibition from undesired CNS effects in harmine analogues may be indirectly proven by INDY (**11a**). The synthetic, ATP-competitive benzothiazole inhibitor of DYRK1A can reverse aberrant tau phosphorylation in cellular systems without any MAO-A inhibition [82]. The X-ray structure of INDY-DYRK1A and harmine-DYRK1A complexes show a similar orientation for the two inhibitors in the ATP binding site of DYRK1A [82]. A prodrug of INDY (proINDY, **11b**) shows *in vivo* efficacy – recovery of head malformation caused by DYRK1A overexpression – in a *Xenopus* model [82]. Thus, **10** and **11** should become gateways to active leads, for their *in vivo* evaluation in DYRK1A-dependent animal models of tauopathies.

Epigallocatechin gallate (ECGC, **12**, Fig. **4**) is the major catechin component in green tea. It shows beneficial effects against ischemia [83], atherosclerosis [84], cancer [85], bacterial [86] and fungal diseases [87]. Its connection to the amyloid pathway in AD is assessed [88]. ECGC is also a nanomolar, selective DYRK1A inhibitor tested on a panel of 28 kinases [89], acting as an ATP-noncompetitive inhibitor [90]. A green tea diet – and subsequent treatment with pure EGCG – rescues transgenic mice overexpressing DYRK1A from brain defects related to DYRK1A malfunctioning [91]. Similarly, a mixture of catechins from grape seed show beneficial results in another tauopathy model [92]. EGCG is well tolerated, chemically tractable, has multiple beneficial activities and seems to cross the BBB: Thus, it may represent a useful scaffold for the identification of pharmacologically cleaner, DYRK1A-inhibiting leads against tauopathies.

DYRK1A inhibitors are currently targeted as anticancer agents, as harmine and other β-carbolines are cytotoxic agents [93]. Their pro-apoptotic action may involve restoration of caspase 9 activation, *via* prevention of its DYRK1A-dependent phosphorylation at the Tyr125 inhibitory residue [94, 95]. The biological characterization of truly specific DYRK1A inhibitors will clarify if their anti-tauopathy profile will be devoid of significant side effects. Meanwhile, it is reassuring that cytotoxic harmine analogues are devoid of any DYRK1A inhibition [96]. Last but not least, access to the site of action – the brain – may be a major issue for some of the described compound classes.

POST-TRANSLATIONAL MODIFICATIONS OF TAU: THE GOOD, THE BAD AND THE UGLY

Tau is a member of the intrinsically unstructured proteins' family – (IUPs) [97]. Its lack of secondary structure and its resistance to folding stems from a high basic and polar aminoacids' content. The basic-polar nature of tau is relevant for its interaction with acidic MTs [98]. The large number of polar-basic residues also increases the occurrence of post-translational modifications (PTMs) on tau. Such crucial modifications to regulate tau and its functions (step 2, Chart **1**) were thoroughly reviewed [99].

Phosphorylation of tau is a key therapy-associated research area since early tau days [100]. It may theoretically regard up to 85 Ser, Thr and Tyr residues. Up to now, 32 residues were characterized as being phosphorylated mostly in non-diseased brains, 15 were detected phosphorylated in both physiological and pathological conditions, and 28 were found to be phosphorylated mainly in AD brains [99]. These patterns play important physiological and pathological roles through a dynamic phosphorylation balance. Such balance is kept by the interplay between tau kinases and phosphatases [101]. The large number of human kinases [102], the scarcity of selective kinase inhibitors [103], and the complexity of tauopathies [24] make it almost impossible to establish a quantitative relationship between physiological and pathological phosphorylation sites, tau kinases and phosphatases.

O-glycosylation of tau with N-acetylglucosamine (*O-GlcNAcylation*) is likely to play an essential role, as is for hundreds of substrates in physiological and pathological states [104]. Cross-talk between phosphorylation and O-GlcNacylation is well documented [105]. The latter was less recognized as a key PTM mostly due to technical difficulties in identifying O-GlcNacylated substrates [106]. O-GlcNAcylation negatively regulates tau phosphorylation [107], and consequently tau aggregation and the risk for tauopathies. O-GlcNacylation-dependent decrease of tau phosphorylation may be due to direct competition for PTM of the same tau sites, and to O-GlcNAcylation-induced conformational changes of tau. Both mechanisms lead to the inaccessibility of selected phosphorylation sites for tau kinases [105].

Proline residues often determine the folding of protein regions, due to the X-Pro amide bond assuming *cis* and *trans* conformations [108]. *Peptidyl prolyl isomerization* [109] of a single X-Pro amide may hinder (leading to a lower number of PTMs) or make more accessible to PTM enzymes (higher number of PTMs) a whole protein region, with important functional consequences [110]. The middle region of tau contains two proline-rich domains, and 21 proline residues. Indirect targeting of PTM patterns on pathogenic tau conformations *via* X-Pro amide bond isomerization was validated as a putative therapeutic approach through the use of specific antibodies [111].

Other PTMs happen downstream of O-GlcNAcylation and phosphorylation [99]. Tau acetylation [112], glycation [113], nitration [114], polyamination [115], sumoylation [116] and oxidation [117] need additional efforts to prove their therapeutic relevance and, thus, to be validated as molecular targets against tauopathies.

Re-equilibration of Abnormal PTM Patterns

O-phosphorylation, O-GlcNAcylation and peptidyl prolyl isomerization influence, either directly or indirectly, the post-translational "decoration" of tau (step 2, Chart **3**). De-regulation of PTM-introducing enzymes leads to pathological PTM patterns (step 2', Chart **3**), *i.e.,* to aggregation-prone, misfolded tau protein copies. Acting on these enzymes to promote the restoration of physiological PTM levels of tau will be thoroughly discussed below.

Tau varies its phosphorylation state depending on cellular compartment localization [118] and developmental stage, with fetal tau being hyper-phosphorylated in comparison with adult tau in humans [119]. An overall higher phosphorylation level for adult tau is a marker for tau aggregation and for an increased risk of tauopathies [120]. Remarkably, hyper-phosphorylated fetal tau is highly soluble and perfectly functional [16]. Specific tau residues are hyper-phosphorylated in human brain tissues from AD patients in comparison to human fetal tau. Site-specific, pathological hyper-phosphorylation depends on residues S202, T212, T217, T231, S396, S404 and S422 [121]. These, and a few other residues such as S214 and S262 were characterized as early, intermediate or late stage, tauopathy-specific hyper-phosphorylated epitopes [122-126] to be targeted in drug discovery efforts.

A decreased level of tau O-GlcNAcylation is found in brain tissues from AD patients. A correlation between tau hypo-GlcNAcylation and hyper-phosphorylation can be established [127]. O-GlcNAcylation and O-deGlcNAcylation on tau are carried out respectively by a single transferase and hydrolase [128]. This is an advantage for the rational design of selective O-GlcNAc modulators, when compared to the many kinases and phosphatases involved in tau phosphorylation. The large number of transferase and hydrolase substrates, though, raises concerns about side effects of O-GlcNAcylation modulators. For example, insulin resistance [129] and effects on mitotic progression and cytokinesis [130] were reported.

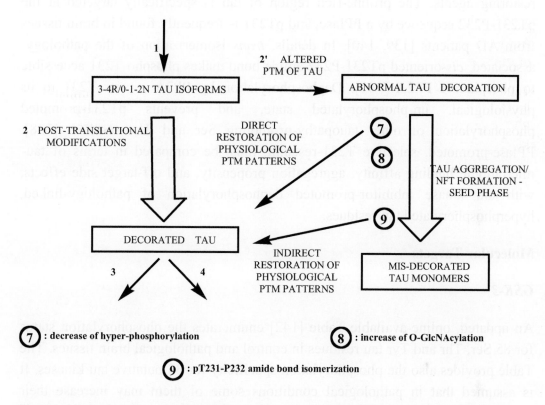

Chart 3: Misdecorated tau-related pathological processes, and putative therapeutic approaches.

Peptidyl prolyl isomerization is a safety catch mechanism, to rescue tau protein copies from a hyper-phosphorylated, aggregation-prone state. Peptidyl-prolyl isomerases (PPIases) are the corresponding enzymes, subdivided in three structural families [131]. The oldest family of FK506-binding proteins (FKBPs) [132] is composed by ≈15 proteins. FKBPs carry out a plethora of activities including – but not limited to – peptidyl prolyl isomerization [133]. Cyclophilins [134], *i.e.,* cyclosporine A-binding proteins, have ≈20 family members involved in various physiological and pathological processes [135]. FKBPs and cyclophilins are also called immunophilins, as they strongly determine the activity of the immune system [136]. A third and younger PPIase family, the parvulins, contains only three proteins which do not have any connection with the immune system

[137]. FKBPs and cyclophilins have neurodegeneration- and tau-related implications [138], but parvulins are more relevant as physiological PTM-restoring agents. The proline-rich region of tau is specifically targeted at the pT231-P232 sequence by a PPIase, and pT231 is frequently found in brain tissues from AD patients [139, 140]. In details, *trans* isomerization of the pathology-associated, *cis*-oriented pT231-P232 amide bond makes phospho-T231 accessible to phosphatase PP2A [141]. Dephosphorylation by PP2A reverts T231 to its physiological, un-phosphorylated state, and prevents pT231-promoted phosphorylation of other tauopathy-promoting Ser and Thr residues [138]. PPIase-promoted, selective T231-rescuing must be compared in terms of tau-microtubule binding affinity, aggregation propensity, and off-target side effects, with tau kinase inhibitor-promoted dephosphorylation of pathology-linked, hyperphosphorylated tau residues.

Molecular Targets

GSK-3

An updated, online-available Table [142] enumerates the phosphorylation status for 85 Ser, Thr and Tyr tau residues in control and pathological brain tissues. The Table provides also the phosphorylation specificity for 33 putative tau kinases. It is assumed that in pathological conditions some of them may increase their activity on tau, and/or that the activity of their phosphatase counterparts is reduced. Targeting a *decrease of hyper-phosphorylation* on tau (**7**, Chart **3**) is an assessed drug discovery approach against tauopathies [101]. Kinases are heavily popular as CNS targets [143], and several tau kinase inhibitors have entered clinical trials [144]. Tau kinases, *i.e.*, putative targets against hyper-phosphorylation, include previously mentioned DYRK1A [54], CDK5 [145], CK1 [146], PKA [147], PKC [148], ERK-2 [149], AMPK [150], p38α [151] and Fyn [152]. Their phosphatase counterparts include PP-1 [153], PP-2A [154], PP-2B [155] and PP-5 [156].

We focus our attention on the serine-threonine *glycogen synthase kinase 3 beta* (*GSK-3β* [157]), one among the most validated and exploited kinase target in AD and tauopathies. It is acknowledged that other kinases – the ones mentioned above, and even others – may be therapeutically as important, if not more important than GSK-3. Nevertheless, GSK-3 will be dealt with here as a single example of a validated tau kinase target.

GSK-3 is an enzyme involved in the control of glycogen metabolism [158]. GSK-3-promoted phosphorylation of substrates has also effects in apoptosis [159], cell division [160], signal transduction [161], and trafficking [162]. Aberrant GSK-3 activity is detected, and therapeutically targeted, in neurodegenerative diseases [163], diabetes [164], inflammation [165], psychiatric disorders [166] and cancer [167].

Two isoforms, GSK-3α and GSK-3β, share 85% sequence identity and possess a highly conserved catalytic domain [168]. GSK-3α null mice are viable and relatively normal [169]. GSK-3β null mice are non viable [170], showing the crucial role of this isoform. GSK-3β is predominant in the brain [171]. Its splice variant, named GSK-3β2, is further enriched in neurons [172]. GSK-3β has a higher tau phosphorylation activity compared with GSK-3α[173]. Their phosphorylation patterns – and even the patterns of splice variants GSK-3β1 and GSK-3β2 - are only partially overlapping [171, 174]. As both GSK-3α and GSK-3β are connected to AD and tauopathies [175], and available GSK-3 inhibitors hit both isoforms, we will consider GSK-3 as a single target for disease-modifying treatment of tauopathies.

GSK-3 phosphorylates *in vitro* up to 40 Ser and Thr residues [176], and is, together with CK1, the most efficient kinase in phosphorylating tau. All but one of phosphorylated residues found in control human brains, and the majority (27 out of 45) of tauopathy-specific, hyper-phosphorylated epitopes of tau found in brains from AD patients are phosphorylated *in vitro* by GSK-3 [177]. Over-expression of GSK-3 in transfected COS cells decreases electrophoretic mobility of tau, and induces its reactivity with several PHF-tau-selective monoclonal antibodies [178]. Cell lines transfected with tau and GSK-3β show a reduced tau-microtubule affinity [179]. Conversely, treatment of cultured cortical neurons with small molecule GSK-3 inhibitors in a model of tau hyper-phosphorylation induced by okadaic acid (OA, a protein phosphatase 2A inhibitor) [180, 181] reverse the OA-induced pathological tau pattern. This neuroprotective effect is observed without signs of toxicity, *i.e.,* without inhibition of essential, physiological tau phosphorylation [180]. GSK-3 also phosphorylates the Thr668 residue on the amyloid precursor protein (APP), hinting to a potential role for GSK-3 inhibitors as neuroprotective-Aβ anti-aggregant agents [182].

GSK-3 is highly active in cells under basal conditions, due also to constitutive phosphorylation on a Tyr residue in the activation loop (Tyr279 in GSK-3α, Tyr216 in GSK-3β) [183]. Its phosphorylation activity is heavily regulated by two general mechanisms. Phosphorylation of an N-terminal Ser residue (Ser21 in GSK-3α, Ser9 in GSK-3β) by several kinases inactivates GSK-3 [184]. Moreover, most of GSK-3 substrates must be primed – pre-phosphorylated elsewhere by other kinases – to be phosphorylated by GSK-3 [185]. Thus, external stimuli control the activity of GSK-3 and ensure the time- and compartment-dependent regulation of crucial CNS processes such as neurogenesis and neuronal apoptosis. The sensor role of GSK-3 in the brain [185], and the intricacies of its regulation as a tau phosphorylating enzyme by physiological and pathological stimuli [186] were recently reviewed.

In vivo validation of GSK-3 as a tauopathy-AD target is copious [187, 188]. Transgenic mice bearing conditionally expressed GSK-3β (including Ser9-mutated, more active kinases) as such [189], or crossed with tauopathies-inducing mutated tau isoforms [190], are known. They are considered suitable tauopathy models to test GSK-3 inhibitors.

Single kinase-selective inhibitors are rare. Although selectivity is often claimed, off-target kinase inhibition materialize when large panels of kinases are used to profile even clinically tested compounds [69, 191, 192]. The broad substrate specificity of GSK-3, its wide role in physiological processes and – when deregulated - its connection to various diseases raise legitimate concerns regarding the consequences of GSK-3 inhibition. Such concerns are addressed by a few considerations. More than 100 substrates are known to be phosphorylated by GSK-3 *in vitro*, but not all of them are true GSK-3 substrates [193]. The tight regulation of GSK-3 ensures that it acts at the right time and place in physiological conditions. It is conceivable to target its deregulated activity in a disease state, such as neurodegeneration [185]. In fact, small molecule inhibitorscould target disease-specific regulatory events leading to GSK-3 inactivation; and they could be delivered to specific areas of the human body by means of targeting chemical functions, or of nanoparticles, to minimize side effects. Most important, they can be tailored to avoid full inhibition of basal GSK-3 activity [185, 193]. Thus, a controlled effect on GSK-3 should be enough to abolish, or at least minimize the effects of pathological, slow hyper-phosphorylation. It should not, though, prevent GSK-3 from performing its physiological duties. The wealth of accumulated preclinical and clinical data for GSK-3 inhibitors in various therapeutic areas seems to confirm that a therapeutic window exists, although their development as chronic treatment against neurodegenerative tauopathies is a major challenge.

OGA

Tau O-GlcNAcylation is due to the action of two enzymes. OGT [194], an O-GlcNAc transferase, introduces an N-acetylglucosamine moiety on up to 11 Ser and Thr residues of tau, and on more than 500 other protein substrates. Conversely, OGA [195], an O-GlcNAc hydrolase, removes GlcNAc from the same residues/substrates in a dynamic equilibrium. Evidence exists for the involvement of dynamic O-GlcNAc cycling in cell cycle control [196], development [197], signalling [198] and trafficking [199]. The detailed mechanisms for these and other key cellular events is not yet fully elucidated. The role of O-GlcNAcylation as a nutrient sensor [200], with the OGT carbohydrate donor UDP-GlcNAc influencing O-GlcNAc cycling [201], determines a stress-related PTM pattern in response to alterations in glucose metabolism. Prolonged stress conditions may lead to deregulation of O-GlcNAcylation, and eventually to type 2 diabetes [202]. Therapeutic modulation of O-GlcNAc cycling is beneficial to cardiovascular disease [203].

O-GlcNAcylation is a common regulation mechanism in the brain and, conversely, deregulated O-GlcNAcylation is observed in neuronal disorders [204]. In AD, an overall decreased level of O-GlcNAcylation is observed together with an impaired glucose mechanism - the so-called type 3 diabetes-AD [205].

Tau O-GlcNAcylation is affected by glucose stress. Starved mice show a hypo-O-GlcNAcylation – hyper-phosphorylation pattern [206]. The same PTM pattern is observed in AD brain tissues [207, 208]. Brain-targeted deletion of OGT in mice leads to tau hyperphosphorylation and to neuronal death [197]. An increase in O-GlcNAcylation, obtained through small molecule-mediated OGA inhibition in mice, results in a residue-specific reduced level of phosporylation on tau [209]. Interestingly, a connection exists between the residue-specific effect and GSK-3 activation [209, 210]. GSK-3 activation may explain the variations observed on GSK-3-phosphorylatable epitopes Ser199, Ser202 and Ser396 in mice treated with an OGA inhibitor. A decrease in phosphorylation induced by O-GlcNAcylation, and an increase due to activation of GSK-3 *via* O-GlcNAcylation-promoted inhibition of phosphorylation on its regulatory Ser9 residue, lead to an overall negligible effect on Ser199, Ser202 and Ser396 [211]. Overall, an *increase of O-GlcNAcylation* on tau (**8**, Chart **3**) is a sound therapeutic goal.

Several papers [128, 212, 213] review the mechanism of action, the substrate specificity and the main properties of OGT and OGA. Either an increase of OGT activity, or a decrease of OGA activity could lead to an increase of tau O-GlcNAcylation. The latter is a more achievable goal, as structure-based drug design of enzyme inhibitors is well documented in literature [214]. OGA shows a high selectivity for O-GlcNAc substrates, as O-GalNAc peptidic substrates are not hydrolyzed [215]. Conversely, any peptidic sequence out of the hundreds – maybe thousands – of O-GlcNAcylated, physiological substrates is hydrolyzed by OGA. OGA belongs to the GH84 glycoside hydrolase family [216], and its substrate-assisted mechanism of action is well known [217]. The crystal structure of OGA homologues from bacteria [218, 219] is instrumental in clarifying the structural requirements for selective OGA inhibition, and in designing and synthesizing efficient small molecule OGA inhibitors [220] – see next Paragraph for structural details.

Several concerns are apparent in dealing with OGA inhibitors. First, the wide peptide substrate specificity of OGA may lead to off-target side effects. Then, OGT over-expression experiments showed detrimental effects such as disrupted glucose homeostasis and developed insulin resistance [129, 221]. Finally, insulin resistance was observed using small molecule OGA inhibitors with limited selectivity [129, 222, 223]. Recently, *in vitro* [224] and *in vivo* [225] results with a selective OGA inhibitor did not alter glucose homeostasis, and hinted to a cleaner activity profile for such molecules. Although the discrepancy between the effects of OGT expression and small molecule OGA inhibition on insulin resistance is not yet fully understood [226], the rationale for OGA inhibitors as PTM-correcting agents in neurodegeneration is strong.

Pin1

The parvulin PPIase *Pin1* (peptidyl-prolyl *cis/trans* isomerase NIMA interacting-1) [227] is endowed with activity connected with essential cellular roles. Pin1 has pSer/pThr-Pro-specific PPIase activity on protein sequences containing the phosphorylated dipeptide. The same sequence regulates key cellular events involved in physiological (transcription [228], signal transduction [229]) and pathological processes (cancer [230], neurodegeneration [231]). Ser/Thr-Pro sequences are phosphorylated by proline-directed Ser-Pro kinases (CDKs, MAPKs, GSK-3, JNKs, p38, *etc.*) [232], a large subfamily of Ser-Thr kinases. Conversely, pSer/pThr-Pro sequences are dephosphorylated by proline-directed phosphatases (PP2A, calcineurin, *etc.*) [233]. Proline-directed Ser-Pro kinases carry out their enzymatic reaction on the *trans* Ser/Thr-Pro amide isoform [234]. Significant amounts of the kinase-inaccessible *cis* Ser/Thr-Pro isomeric state may exist, depending also on the aminoacids neighboring the Ser/Thr-Pro sequence. Similarly, proline-directed Ser-Pro phosphatases dephosphorylate only *trans* pSer/pThr-Pro amide conformations [235], which are in equilibrium with their *cis* isoforms. Immunophilins isomerize unphosphorylated *cis* Ser/Thr-Pro sequences, accelerating a slow, non catalytic process [236]. Immunophilin-catalyzed *cis-trans* isomerization increase the phosphorylated/unphosphorylated ratio for a Ser/Thr-Pro-containing substrate. They promote the pSer/pThr-Pro-induced conformational changes [237] leading to essential cellular processes. Immunophilins can not isomerize *cis* pSer/pThr-Pro sequences. As their non-catalyzed *cis-trans* isomerization is even slower than for the unphosphorylated sequence [238], the concentration of *cis* pSer/pThr-Pro-containing proteins grows in time and eventually lead to pathological consequences.

Pin1 exerts PPIase activity on pSer/pThr-Pro sequences, contributing to bi-directional regulation of cellular processes involving Ser/Thr-Pro-containing proteins [239]. More than 50 Pin1 substrates are known [240], to make a complex Pin1-dependent regulation scenario. Some substrates are preferentially dephosphorylated due to their interaction with Pin1, as expected [241]. Others are preferentially phosphorylated, either by inhibition of dephosphorylation [242] or by stimulation of phosphorylation [243]. Oncoproteins [244, 245] are stabilized by the interaction with Pin1, while tumor suppressors [246] and ageing-related proteins [247] are preferentially degraded after Pin1-induced destabilization. Physiological regulation of Pin1 activity by transcription factors [248], or by phosphorylation/kinases [249, 250] is known. Pin1 expression in most tissues correlates with cell proliferation [240], and its pathological upregulation is observed in many human cancers [251].

Most neurons constantly express Pin1 at high levels [252]. Its regulatory functions are essential for neuronal survival and operational efficiency [253]. Synphilin-1 (inhibition of degradation-promotion of aggregation of α-synuclein, Parkinson disease) [254], gephyrin (modulation of the function of glycine receptors) [255] and MCL1 (prevention of neuronal apoptosis, spinal cord injury) [256] were reported are known Pin1 substrates in neurons. Pin1 also shows a putative neuroprotective role in AD through regulation of APP processing *via* dephosphorylation of the β-amyloid production-inducing pThr668 residue on APP [257, 258].

Tau is the first Pin1 substrate identified in neurons [252]. Although tau contains 15 Ser/Thr-Pro epitopes [259], Pin1 selectively targets-isomerizes the *cis* pT231-P232 amide bond on tau. pT231-P232 amide bond isomerization promotes its dephosphorylation catalyzed by *trans*-specific phosphatase PP2A [241, 260]. The pThr231 epitope is a pathology-related phosphorylated residue which appears in the early phases of tau pathologies [261]. Its appearance is likely to cause major conformational changes in tau, eventually leading to the phosphorylation of other pathological epitopes [262]. Its use as a diagnostic tool to observe the progression of AD in cerebrospinal fluid (CSF) is suggested, due to a direct correlation between increased CSF levels and AD severity [263]. Thus, *pSer/pThr-Pro cis-trans isomerization* of tau (**9**, Chart **3**) has a strong connection to tauopathies. It is targeted both to unravel the remaining scientific issues, and to discover meaningful clinical candidates.

Pin1 is significantly deregulated in neurodegeneration [231, 240, 264, 265]. Its polymorphism is linked to AD [266, 267]. Its relocation from the nucleus to the cytoplasm [252, 268], its co-localization in deposits of aggregated tau [252, 253] and in Pin1-specific granules [269, 270], and the depletion of soluble, enzymatically active Pin1 [252] are observed in pathology-affected neuronal cultures and diseased brains. A direct correlation between Pin1 deregulation and the severity-progression of the disease can be established even at asymptomatic, early stages in tauopathy-affected brains [271]. Neuroprotection by Pin1 *via* prevention of tau hyper-phosphorylation can be safely inferred, although the exact impact and role of Pin1 in tau pathology is debated [264] due to contrasting experimental observations [269, 272]. The relevance of Pin1 in AD is strengthened by its Aβ-reducing function [257, 258]. Pin1-mediated connections between tau and Aβ pathology [273], with the involvement of GSK-3β [274], underline the central role of Pin1 in both pathologies.

Knockout organisms prove that Pin1 is essential for growth in lower eukaryotes [275, 276] and viable, but neurodegeneration-relevant, in mice [253, 277]. Pin1-null mice show progressive, age-dependent motor and behavior deficits [253]. Their tau is hyper-phosphorylated, assumes abnormal conformations, and is found in insoluble filaments. Brain area-dependent neuronal loss is observed [253]. An age-dependent moderate increase of Aβ production (further increased by crossing with an APP mutation) is also observed [257]. As of today, Pin1 null mice are the only deletion transgenics causing a progressive, tau-dependent neurodegeneration. Tau transgenic mice show similar phenotypes, but require tau mutations [278], sometimes crossed with APP mutations [279] to assume a human tauopathy-like phenotype.

A Pin1-centered intervention in tauopathies should restore its functions in diseased brain areas. Small molecule Pin1 enhancers are not yet reported, and can not be easily conceived. Pin1 modulators, *i.e.,* small molecules acting on Pin1 regulators, could provide the same result. AP4 [280], a transcription factor found in brain, is a negative regulator of Pin1 expression. The observed polymorphism in the Pin1 promoter sequence leads to the abolishment of AP4-mediated suppression of Pin expression, and to a 3 years' delay in the onset of AD in a Chinese population [267]. Cellular knockdown of AP4 leads to a 2-fold increase of Pin1 [267]. Similar results may be obtained with small molecule RNA binders selective for the AT4 binding motif in the Pin1 promoter sequence.

A more druggable set of targets relies on PTMs of Pin1, and in particular on the phosphorylation of three regulatory epitopes – Ser16, Ser65 and Ser71. pSer16 is a negative regulator of Pin1 activity, probably by preventing the interaction between Pin1 and its substrates [281]. Ser16-hyper-phosphorylated Pin1 is detected in the brains of AD patients, and a tentative connection with Pin1 deregulation is proposed [282]. The identification of the kinase responsible for Ser16-phosphorylation of Pin1 should provide a novel target for the positive regulation of Pin1. Polo-like-kinase-1 (PLK1) phosphorylates Pin1 on Ser65, increasing its stability through inhibition of its ubiquitinylation [283].

Phosphorylation of Ser71 by death-associated protein kinase 1 (DAPK1) [284] inhibits the PPIase activity of Pin1 [285]. DAPK1 is a cell death-promoting kinase, together with closely related but less characterized DAPK2 and DAPK3 [284]. DAPK1 acts on caspase-dependent [286], autophagic [287], and necrotic death pathways [288]. Its involvement in neuronal death processes *in vivo*, as a DAPK1 inhibitor protects mice from ischemic neuronal injuries [289, 290]. DAPK1 is also connected to tau hyper-phosphorylation through phosphorylation-independent activation of tau Ser262-phosphorylating MARK1/2 kinases [291]. Thus, DAPK1 inhibitors should have particular value, as they interfere with two distinct tauopathy-targeted pathological mechanisms.

Pin1-targeted small molecules raise concerns, due to the involvement of Pin1 in essential cellular processes. Pin1 inhibitors are actively pursued as putative cancer treatments [292]. A positive modulator of Pin1 may disrupt the physiological regulation by Pin1 in non-neuronal cells. The lower risk of contracting cancer for AD patients, mirrored by the same risk reduction for cancer patients to develop AD, is known [293]. Pin1 may play a role in this inverse correlation, but there is no evidence of significant side effects caused by the restoration of Pin1 activity in tauopathy-like *in vitro* and *in vivo* models, with one notable exception. The P301L tau mutation found in FTDP-17 reverses the effects of Pin1 [294]. Pin1-null mice crossed with P301L tau show an attenuation of the tauopathy-like phenotype, while Pin1 over-expression in P301L mice produce a worsening-acceleration of tau hyper-phosphorylation. Observed conformational changes in P301L tau [295] may lead to an equilibrium switch for the PPIase activity of Pin1 and, consequently, to the Pin1-catalyzed enrichment of the hyper-phosphorylation-promoting *cis*-p-T231-P232 epitope. The effect of other tau mutations on Pin1 should be thoroughly evaluated, to define the tauopathy subset which should be treated with positive modulators of Pin1. Conversely, the effect of Pin1 inhibition on P301L mice models should be determined to identify a possible CNS therapeutic niche for Pin1 inhibitors.

Chemical Structures

GSK-3 Inhibitors

GSK-3 is considered a valuable therapeutic target since early '80s, and a steady flow of patents and papers dealing with hundreds of compound classes as GSK-3 inhibitors started in the last decade [296]. A provocative chemotype-based analysis, focused on 17 heterocyclic cores and on their recurrence in potent GSK-3 inhibitors, can be accessed [296]. Here we will focus on a limited number of compound families, for which extensive pharmacological characterization, and sometimes clinical study results, are reported. More details could also be found in four recent reviews [101, 144, 297, 298].

Two small metal cations have a track of records as GSK-3 inhibitors. Lithium is used as a mood stabilizing drug [299], due to its weak, millimolar affinity for GSK-3 [300]. It inhibits GSK-3 by competing with magnesium ions [301], and enhancing Ser279/216 phosphorylation [302]. Lithium is active in animal models, showing GSK-3 inhibition-compatible effects on tau phosphorylation and on other pathological, tauopathy-connected neuronal events [303, 304]. Lithium salts-based therapies ensure cognitive enhancement in humans [305-307], showing statistically significant efficacy and limited toxicity in long clinical trials. Although a lack of activity in a few AD trials using lithium salts [308], and claims that millimorar affinity for GSK-3 cannot justify the observed effects in humans

[309], the diffused, chronic use of lithium in psychiatric therapy indirectly proves the safety of weak to moderate GSK-3 inhibitors.

Zinc is naturally found in the human body as a trace element, and has an essential role in various metabolic pathways [310]. Its inhibition of GSK-3 is much stronger than lithium (IC_{50}=15μM [311]), so that lower dosages may be as effective on GSK-3. Targeted zinc-chaperoning molecules exert their cellular, therapeutic effect in models of AD also through GSK-3 inhibition [312]. One such zinc chaperone is clinically tested on AD patients [313], and will be covered later. The risk of depleting the physiological zinc pool limits its consideration as a GSK-3 inhibitor.

ATP-competitive synthetic compounds are the largest class of GSK-3 kinase inhibitors. Tetracyclic paullones [314] can be structurally optimized to minimize inhibitory potencies against close congeners CDK5 and cyclin B, as in **13a** (Fig. **5**) [315]. Alsterpaullone (**13b**) inhibits tau phosphorylation in brain slices, and tau hyperphosphorylation *in vivo* in a GSK-3-CDK5 coherent pattern [314, 316].

Bisarylmaleimides are known GSK-3 inhibitors, [317] tested *in vitro* and-or *in vivo* in neurodegeration models. Arylindolemaleimides such as SB-216763 (**14**) [318] are potent, GSK-3-selective inhibitors on a panel of 25 kinases (not including CDKs). SB-216763 is *in vivo* active in models of ischemia [319], spinal cord injury [320], and schizophrenia [321]. It shows disparate effects in models of neurodegeneration, reducing *in vivo* tau phosphorylation, [316] but showing neurodegenerative effects and cognitive deficits attributed to over-inhibition of GSK-3 [322]. 1,2,3,4-Tetrahydro[1,4]diazepino[6,7,1-hi]indol-7-yl maleimides such as **15** show strong cellular inhibition of tau phosphorylation at Ser396, and strong *in vivo* activity when administered *per os* at 10mg/Kg in a diabetes model [323, 324]. The bisindolylmaleimide BIP-35 (**16**) shows potency in an *in vivo* mouse model of mania [325]. It increases motor neuron survival in cells, and the mean survival time in an *in vivo* spinal muscular athropy (SMA) model when administered i.p. once a day at 75mg/Kg [326].

Indirubins are naturally occurring CDK inhibitors [327]. The limited GSK-3 inhibition of parent compounds is enhanced in 6-bromo, 3'-oxime indirubine (6-

BIO, **17a**) [328]. 6-BIO reduces neurite outgrowth in dorsal root ganglions [329] and decreases tau phosphorylation levels at low micromolar concentrations [330]. *In vivo* results for 6-BIO are conflicting. Significant efficacy is observed with 20mg/Kg of 6-BIO, given i.p., in a double transgenic APP/PS1 model. Spatial memory deficits are reduced and, among other biochemical parameters, tau hyperphosphorylation is attenuated [331]. Conversely, 6-BIO lacks efficacy at the same dose in a postnatal mouse model of tau hyperphosphorylation [326]. This may be due to low concentrations reached in mouse brains, or to differences between adult and post-natal mice [331]. The sub-optimal drug-like profile of 6-BIO and its low micromolar cytotoxicity shifted the focus onto more soluble, less toxic derivatives [332]. The O-acyloxime **17b** (6-BIMYEO) shows GSK-3 selectivity, decreased cytotoxicity and good cellular potency, reversing tau hyperphosphorylation and neuronal apoptosis induced by okadaic acid in neuronal cultures [333]. The O-bromoalkyl oxime **17c** (6-BIBEO) shows similar potency in neuronal cultures. It is slightly less potent on GSK-3, is selective *vs.* CDKs and has no cytotoxicity up to 100μM [334]. *In vivo* data for **17b** and **17c** are not reported yet.

Aminothiazoles include AR-A014418 (**18**, Fig. **5**), showing strong selectivity *vs.* CDKs [335]. *In vitro*, AR-A014418 prevents HIV-mediated neurotoxicity in neurons [336], prevents axon elongation in hippocampal neurons [337], shows neuroprotection in hippocampal slices [333] and prevents tau hyper-phosphorylation on GSK-3-sensitive Ser396 in an okadaic acid-induced model in SY5Y neuroblastoma cells [333]. *In vivo* behavioral studies report an AR-014418-induced anti-depressive like behavior in a mice forced swim test [338] and in a manic hyperactivity model [339]. Anti-nociceptive effects are observed in several mice models [340]. *In vivo* neurological testing shows biochemical and cognitive improvements in an amyotrophic lateral sclerosis (ALS) model [341], and reduced levels of aggregated insoluble tau in the brain stem of tau transgenic JNPL3 mice [342]. Conversely, AR-A014418 is ineffective on tau phosphorylation in a postnatal rat model [326].

Aminopyrimidine-based purine analogues **19a,b** (Fig. **6**) are picomolar GSK-3 inhibitors [343]. CHIR99021 (**19a**) inhibits neurite outgrowth in cerebellar and dorsal root ganglion neurons [344], blocks NMDA-mediated long term depression in hippocampal slices [345], enhances self renewal and pluripotency in mouse embryonic stem cells [346], enhances the level of the survival motor neuron protein in spinal muscular athropy [347] and reduces neuronal death in a cerebral ischemia model in rats [348]. CHIR98014 (**19b**) strongly reduces tau hyper-phosphorylation in neurons and in rat brains [326].

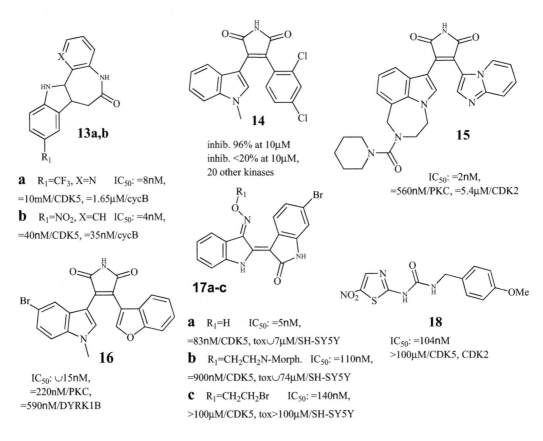

a R_1=CF$_3$, X=N   IC$_{50}$: =8nM,
=10mM/CDK5, =1.65μM/cycB

b R_1=NO$_2$, X=CH IC$_{50}$: =4nM,
=40nM/CDK5, =35nM/cycB

13a,b

14

inhib. 96% at 10μM
inhib. <20% at 10μM,
20 other kinases

15

IC$_{50}$: =2nM,
=560nM/PKC, =5.4μM/CDK2

16

IC$_{50}$: ∪15nM,
=220nM/PKC,
=590nM/DYRK1B

17a-c

a R_1=H   IC$_{50}$: =5nM,
=83nM/CDK5, tox∪7μM/SH-SY5Y

b R_1=CH$_2$CH$_2$N-Morph. IC$_{50}$: =110nM,
=900nM/CDK5, tox∪74μM/SH-SY5Y

c R_1=CH$_2$CH$_2$Br   IC$_{50}$: =140nM,
>100μM/CDK5, tox>100μM/SH-SY5Y

18

IC$_{50}$: =104nM
>100μM/CDK5, CDK2

Figure 5: GSK-3 inhibitors: chemical structures, compounds **13-18**.

Pyrazines such as **20** are potent, selective and ADMET-compliant GSK-3 inhibitors acting against Ser396 phosphorylation in cellular models [349]. Naturally occurring hymenialdisine (**21**) [350] shows low nanomolar potency against several kinases [351]. Its GSK-3 driven cellular activities include inhibition of MAP-1B in cerebellum granular cells, and inhibition of tau phosphorylation in Sf9 cells [351]. Imidazopyridine **22** [352] showed good PK properties and an effective blood glucose lowering effect when given orally at

10mg/kg. Tricyclic spiro-substituted compound **23** was designed and optimized [353, 354] by merging the structure of **19a** with a HTS hit. It showed good pharmacokinetic properties and efficacy in the oral glucose tolerance test [355]. Structurally optimized benzopyrazole **24** (Fig. **6**) [356, 357] showed efficacy in inhibiting GSK-3-dependent tau phosphorylation on Ser396/404 residues *in vitro* and *in vivo* at 10mg/kg. *In vitro* selectivity of compounds **22-24** *vs.* other kinases is unknown, as is the BBB penetration of compound classes **20-24**.

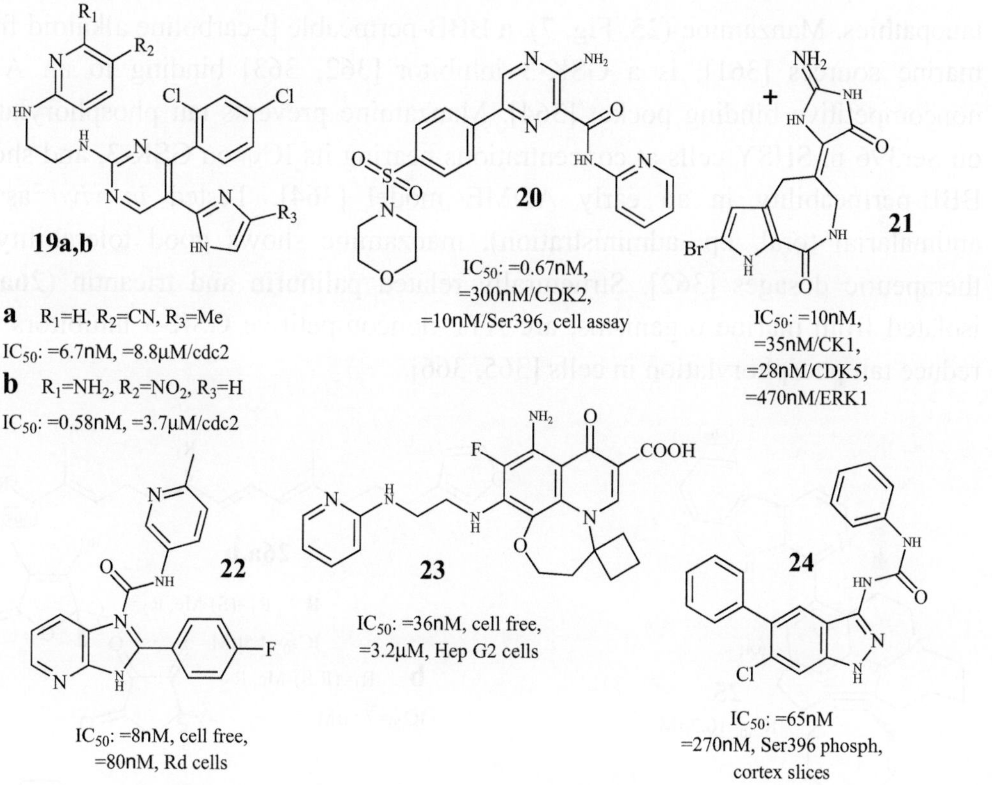

a R₁=H, R₂=CN, R₃=Me
IC₅₀: =6.7nM, =8.8μM/cdc2
b R₁=NH₂, R₂=NO₂, R₃=H
IC₅₀: =0.58nM, =3.7μM/cdc2

20
IC₅₀: =0.67nM,
=300nM/CDK2,
=10nM/Ser396, cell assay

21
IC₅₀: =10nM,
=35nM/CK1,
=28nM/CDK5,
=470nM/ERK1

22
IC₅₀: =8nM, cell free,
=80nM, Rd cells

23
IC₅₀: =36nM, cell free,
=3.2μM, Hep G2 cells

24
IC₅₀: =65nM
=270nM, Ser396 phosph,
cortex slices

Figure 6: GSK-3 inhibitors: chemical structures, compounds **19-24**.

Available *in vivo* results, coupled with several X-ray structures of the complexes between the best inhibitors and the ATP binding site of GSK-3 [342, 349, 352, 358], should have led to the clinical development of a few, fully optimized candidates. An analogue of AR-A014418 (AZD-1080, undisclosed structure) entered clinical trials for AD in 2006, only to be discontinued shortly thereafter [359]. Most likely, the low nanomolar potency of ATP-competitive GSK-3 inhibitors on a highly conserved ATP binding site may have raised concerns – and, maybe, may have caused development-threatening observations - in terms of selectivity and off-target effects [360].

Moderate GSK-3 inhibition coupled with high specificity should be enough to secure a therapeutic effect on GSK-3-driven pathologies, without affecting the physiological duties of GSK-3. ATP-noncompetitive GSK-3 inhibitors successfully fill such a profile, and currently are the focus of GSK-3 drug discovery in AD and tauopathies. Manzamine (**25**, Fig. **7**), a BBB-permeable β-carboline alkaloid from marine sources [361], is a GSK-3 inhibitor [362, 363] binding to an ATP-noncompetitive binding pocket [364]. Manzamine prevents tau phosphorylation on Ser396 in SHSY cells at concentrations nearing its IC_{50} on GSK-3, and shows BBB-permeability in an early ADME model [364]. Tested *in vivo* as an antimalarial (oral, i.p. administration), manzamine shows good tolerability at therapeutic dosages [362]. Structurally related palinurin and tricantin (**26a,b**), isolated from marine organisms, are ATP-noncompetitive GSK-3 inhibitors and reduce tau phosphorylation in cells [365, 366].

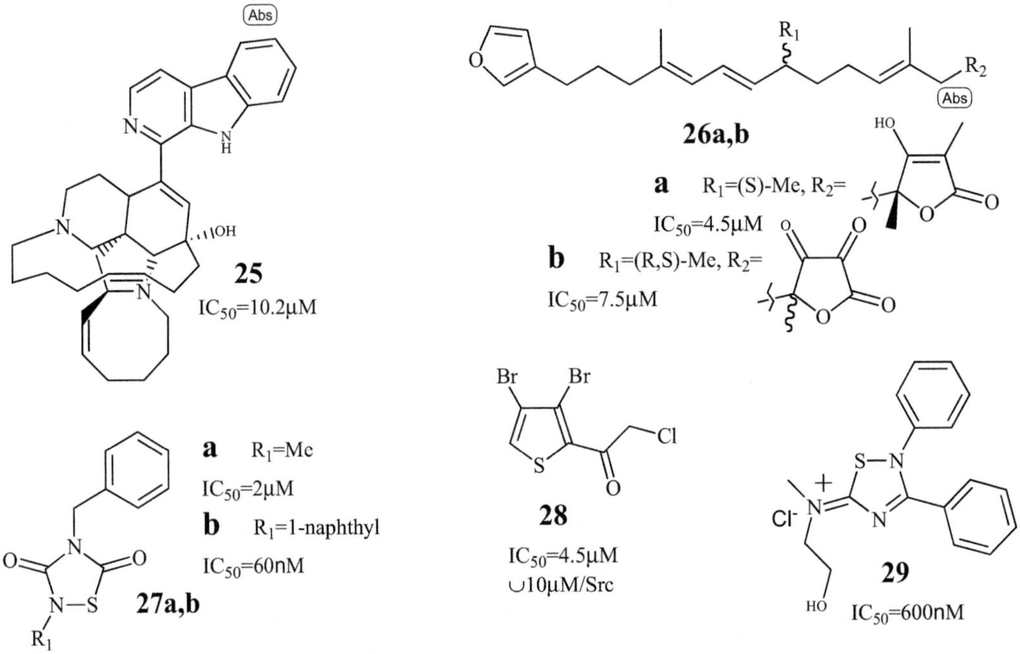

Figure 7: GSK-3 inhibitors: chemical structures, compounds **25-29**.

Thiadiazolidinones (TDZDs, **27a,b**) are the first published, most promising class of synthetic ATP-noncompetitive GSK-3 inhibitors [367]. A detailed review describes their structural optimization [368]. Several hypotheses regarding alternative binding pockets and covalent binding with Cys residues in GSK-3 are proposed [369]. First generation, methyl-substituted TDZD **27a** has *in vitro* and *in vivo* effects in studies targeting arthritis [370], asthma [371], septic shock [372], organ failure [373, 374], myocardial ischemia [375] and schizophrenia [376]. Neurological diseases targeted by **27a** include spinal cord injury [377], cerebral ischemia [378] and Parkinson [379, 380]. Compound **27a** has *in vitro* and *in vivo* GSK-3-mediated effects of on tau [380, 381].

Second generation, 1-naphthyl TDZD (tideglusib, **27b**) is developed as a treatment for AD and tauopathies [360]. It binds irreversibly to its target, without the formation of a covalent bond [382]. Tideglusib shows a stronger GSK-3 inhibition [382], better anti-inflammatory and neuroprotective effects [383] when compared to its analogues [384]. It also activates the PPARγ nuclear receptor [383]. Tideglusib is efficacious in a double APP-tau transgenic mice model [385]. A 200mg/Kg daily oral dose for 3 months provides biochemical – reduction of tau phosphorylation at GSK-3 epitopes, reduction of amyloid load – and functional efficacy – arresting neuronal loss in AD-affected brain areas [385]. Tideglusib is also active in a double APP-presenilin transgenic model [386], increasing the potent neurotrophic peptide insulin growth factor 1 (IGF-1). It also promotes endogenous hippocampal neurogenesis *in vitro* and *in vivo* through GSK-3 inhibition [387]. The good preclinical profile of tideglusib is due to its GSK-3-driven, mixed effects and to its noncompetitive/irreversible nature [382, 383, 386, 387]. Clinical trials using tideglusib show its good tolerability (Phase I study, healthy volunteers oral daily dosages between 50 to 1200mg/Kg for up to 14 days) [388]. A Phase IIa study on 30 AD patients, treated with escalating doses from 400 to 1000mg/Kg for 15 weeks, confirms the good tolerability of **27b** and give preliminary indications of efficacy in mild-to-moderate AD patients [388]. Two Phase IIb studies on AD (ARGO [389], 306 patients) and PSP (TAUROS [390], 146 patients) are approaching completion. Preliminary published results for ARGO support the continuation of the study, and final results are expected by the end of 2012 [391]. The results of TAUROS do not reach the planned endpoints, but encouraging signs of reduced brain atrophy are observed [392].

ATP-noncompetitive inhibitors include thienylhalomethylketones, such as **28** [393, 394]. These irreversible inhibitors covalently react wih the Cys199 residue of GSK-3β [395]. They reduce tau phosphorylation at several epitopes in granule cerebellar neurons [394]. 5-Imino-1,2,4-thiadiazoles (ITDZ such as **29**, Fig. **7**) are substrate-competitive GSK-3 inhibitors [396]. They cross the BBB, show anti-inflammatory and neuroprotective effects, and induce differentiation of neural stem cells into mature neurons [396].

Final results of the ongoing Phase IIb studies with tideglusib are much awaited, and should provide a clinical validation for noncompetitive GSK-3 inhibitors against tauopathies. ATP competitive GSK-3 inhibitors are to be considered therapeutically less attractive.

OGA Inhibitors

Inhibitors of the GH84 glycoside hydrolase family are targeted since several decades. Since its identification in 2001 [195], OGA has also become the target of drug discovery projects. A handful of structural chemiotypes are known as OGA inhibitors. Most are OGlcNAc mimics, as elucidation of the substrate-assisted enzymatic mechanism [217] of OGA permits their rational design-structural optimization-fine tuning. The availability of X-ray structures of human OGA homologues complexed with inhibitors [218, 219, 397-399] may provide inspiration for the rational drug design of OGA-targeted inhibitors with selectivity against other GH84 hydrolases [212, 213, 400]. Recent reviews cover in details these efforts [220, 401].

Streptozotocin (STZ, **30**, Fig. **8**) [402], a naturally occurring N-nitrosomethylamide, is used as a diabetogenic compound in pharmacology since 50 years [403] and is a weak OGA inhibitor [404, 405]. Its toxic effects on pancreatic β-cells were linked to a putative irreversible inhibition of OGA [407], due to the abundance of O-GlcNAcylation in the pancreas [406]. This theory is now proven wrong by the STZ-like toxicity on β-cells by OGA-ineffective, galactose-configured STZ [408], and by the absence of any covalent bond between bacterial OGAs and STZ in an X-ray structure of their complex [408, 409]. Thus, STZ is a poor inhibitor of OGA, and its many pharmacological effects are due to pathways other than O-GlcNAcylation.

PUGNAc (Phenyl Uretane of GlcNAc, **31a**) [410] is a glucosaminidase inhibitor with broad activity on members of GH84, GH20 [411], GH3 [412] and GH89 [413] hydrolase families. The importance of the phenyl urethane moiety for OGlcNAcase activity is indirectly confirmed by poorly active LOGNAc (Lactone Oxime of GlcNAc, **31b**) [414]. PUGNAc was used to prove the limited usefulness of STZ to study O-GlcNAcylation [415]. Its limited hydrolase electivity questions its use as a chemical tool to study the effects of OGA inhibition. Recent results with selective OGA inhibitors [224, 225, 416] seem to attribute harmful effects of PUGNAc – *i.e.,* induction of insulin resistance [129, 222, 223] and of lysosomal storage disorders [416] – to the inhibition of other hydrolases such as HexB. Over 50 *in vitro* and *in vivo* studies involving PUGNAc [220] show OGA-dependent therapeutic effects in inflammation [417], vascular diseases [418], ischemic damage [419], ataxia-telangectasia [420] and ALS [421]. The beneficial role of PUGNAc-increased OGlcNAcylation in neural stem cells [422] and in axon branching [423] is also reported.

The bacterial tetrahydroimidazopyridine nagstatin is a GH20 family-selective inhibitor [424]. Gluco-nagstatin (**32a**) shows equipotent, nanomolar binding on GH20 β-hexosaminidases and on OGA [425]. It is structurally similar to the oxazoline intermediate **33**, formed during the hydrolysis catalyzed by OGA and other GH hydrolases. A PUGNAc-nagstatin hybrid (**32b**) was synthesized [426] to combine structural features from chemiotypes **31** and **32**. Unfortunately, the hybrid is less active than both its precursors on any glycosidase.

NAG-thiazoline (**34a**) [427] is a synthetic, S-containing analogue of the enzymatic reaction intermediate **33**. It shows strong potency against OGA and similar GH20/GH84 hydrolases. 2-Methyl-extended derivatives, such as **34b** (NButGT) and **34c**, retain most of the potency of **34a** with major inter-hydrolase family selectivity improvements. In fact, steric clashes of large, OGA-compatible 2-substituents [428] prevent the binding with the active site of lysosomal β-hexosaminidases [429]. NButGT, and a few analogues, are characterized as *in vitro*-active, OGA-selective inhibitors [430]. NButGT shows a 1500 selectivity *vs.* the β-hexosaminidase HexB [431]. Its "clean" OGA inhibition profile significantly reduces the off-target effects observed using PUGNAc in cellular models and *in vivo* [225, 411, 416, 432]. NButGT is effective in cellular models of ischemia-reperfusion injuries [433]. More importantly, NButGT shows biochemical – lower γ-secretase activity, reduced Aβ accumulation, attenuation of neuroinflammation – and behavioral improvements – reduction of memory impairments - in cellular and mice models of AD [434]. It also shows putative therapeutic effects (oral route, 100mg/Kg, 3 days) in a mutant SOD1 transgenic mouse model of ALS [435].

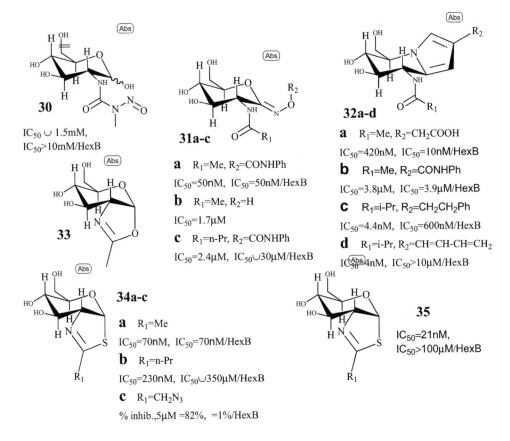

Figure 8: OGA inhibitors: chemical structures.

The structural differences in the active site of glucosaminidases are targeted within other chemical structures. Butyl-PUGNAc (**31c**) and similar analogues

[436, 437] are potent OGA inhibitors. Their selectivity compared to NButGT is significantly reduced [436], due to a different binding mode with glucosaminidases [438]. GlcNAcstatins C [439] and G [440] (respectively **32c** and **32d**) are extremely potent OGA inhibitors, with good to excellent selectivity *vs.* HexB [441].

The validation of OGlcNAcylation as a tauopathy-related mechanism, and of OGA inhibition as a therapeutic target, relies mostly upon thiamet-G (**35**, Fig. **8**) [442]. Thiamet-G is designed by taking advantage of the structural information extracted from the NButGT-bacterial OGA X-ray complex [428], and by pK_a considerations to maximize the interaction with Asp residues in the active site of OGA [442]. The replacement of a C atom with a N atom (compare **34c** with **35**) increases the pK_a of thiamet-G with respect to NButGT, and strengthens the interaction between thiamet-G and OGA – 4-fold increase in IC_{50}. Thiamet-G is more selective *vs.* HexB, and shows much higher aqueous stability than NButGT [442]. *In vitro*, thiamet-G significantly reduces tau phosphorylation in PC-12 cells at Ser 396 and Thr231. *In vivo*, when orally administered to healthy rats at 200mg/Kg, it shows penetration through the BBB and similar phosphoepitope reductions [442]. Efficacy on OGA is confirmed by immunohistochemistry, while HexB is not inhibited – HexB inhibition is observed with non-selective NAG-thiazoline [442].

Chronic treatment with thiamet-G (500mg/Kg daily for 36 weeks) shows therapeutic efficacy also on JNPL3 transgenic mice bearing the tauopathy-inducing P301L mutation [443]. Desired pharmacological – increased motor neuron count, decreased NFT count, decreased neurogenic atrophy of skeletal muscle – and behavioral effects – increased body weight, better rotarod and cage-hang performance – are observed. No signs of toxicity are apparent [443]. Processed brain samples highlight a thiamet-G-dependent increase of OGlcNAcylation. Surprisingly, no effects on tau phosphorylation are observed [443]. Nevertheless, thiamet-G reduces the aggregation propensity of tau in a phosphorylation state-independent manner [443].

An intra-cerebroventricular (icv) injection of thiamet-G in mice shows an AKT-dependent activation effect on GSK-3, leading to a complex tau phosphorylation pattern – decrease at Thr181, Thr212, Ser214, Ser262/Ser356, Ser404 and Ser409, increase at GSK-3-dependent Ser199, Ser202, Ser396 and Ser422 epitopes [209]. This observation may depend on the off-target effects from the extremely high cerebral concentrations attained after icv administration [209]. The GSK-3-targeted, detrimental effect of thiamet-G is not observed in cultured neuronal cells [209], but a cell-dependent influence of OGA inhibition on AKT-dependent GSK-3 activation is reported [211].

Thiamet-G is used *in vitro* and *in vivo* as an OGA-directed tool to study p38 MAPK signaling [444] calcium entry in cardiomyocytes [445], vascular dysfunction [446], and chondrogenic differentiation [447]. An agreement between a canadian small biotech company and Merck should lead to the clinical development of at least one thiamet-G-related molecule as a treatment for tauopathies [448].

Non sugar-like OGA inhibitors are not actively pursued. A single paper [449] discloses four hits from the screening of a 880-membered library of off-patent drugs/clinical candidates. Those hits are endowed with moderate potency against OGA, and limited selectivity *vs.* HexB. Further explorations around these chemiotypes are not yet reported. OGA-targeted screening campaigns on larger compound collections could provide additional useful hits, to be compared with existing substrate analogues.

DAPK Inhibitors

3-Aminopyridazine-based compound **36a** (MW01-26Z, Fig. **9**) [289, 450] is a moderately potent, selective DAPK1 inhibitor. Surprisingly **36a** shows neuroprotective effects *in vitro* and *in vivo* (5mg/Kg, ip), using preventative and therapeutic schedules, in mouse models of hypoxia-ischemia induced brain injury [289] and of cerebral ischemia [290]. The X-ray structure of **36a** bound to the active site of DAPK1 highlights structural features/modifications potentially suitable for potency optimization [289].

Compound **36a** stems from an HTS campaign for inhibitors of neuroinflammatory processes originated by chronically activated glial cells [451]. Structurally related, pyridazine-containing compounds from the same HTS campaign are also characterized [452, 453]. Compound **36b** (MW01-070C) [451] is *in vivo* active in a murine neuroinflammation model caused by Aβ-42 infusion [454, 455]. Minozac (**37**) [456] is a stable, orally available, non toxic, *in vivo* potent clinical candidate [457], developed by Transition Therapeutics [458] after acquisition of NeuroMedix. Both **36b** and **37** are not described as DAPK1 inhibitors, but their structural similarity with **36a** may suggest that their pharmacological profile is at least partially due to DAPK1 inhibition. Thus, their *in vitro* and *in vivo* effects on tau-centered models should also be determined.

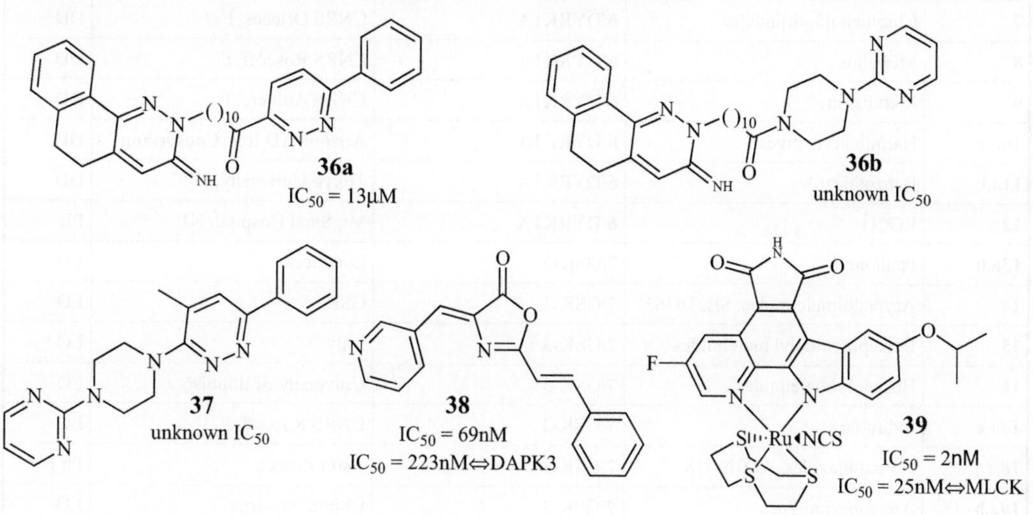

Figure 9: DAPK inhibitors: chemical structures.

Other DAPK1 inhibitors appeared in literature. The 5-oxazolinone **38** is the structurally optimized result from a virtual HTS hit [459]. It shows strong potency on closely related DAPK1 and DAPK3, and great selectivity against 20 kinases [460]. Strong potency and moderate selectivity characterize the octahedral Ru-complex **39** (Fig. **9**) [461]. Previously discussed, pleiotropic ECGC (**12**, Fig. **4**) is able to influence DAPK2-mediated neutrophil differentiation [462]. An extract from bay leaves shows DAPK1-mediated neuroprotective effects *in vitro* – SH-SY5Y cells, brain slices - and *in vivo* at 4mg/Kg, ip – middle cerebral artery occlusion [463].

Table 1: **Compounds 1-32, 34-39: chemical class, target, developing organization, development status**

Number	Chemical cpd./Class	Target	Organization	Dev. Status
1	Neomycin	5/Stem loop	-	NP*
2a-c	Mitoxanthrone analogues	5/Stem loop	Harvard	DD
3	Benzocoumarin	6/DYRK1A	University of Padua, I	DD
4	Pyrazoline-3,5-diones	6/DYRK1A	Irje University, South Korea	DD
5	6-Arylquinazolin-6-amides	6/DYRK1A	NIH	DD
6	Pyridines, pyrazines	6/DYRK1A	CNRS Orleans, F	DD
7	Chromeno[3,4b]indoles	6/DYRK1A	CNRS Orleans, F	DD
8	Meriolins	6/DYRK1A	CNRS Roscoff, F	DD
9	Meridianins	6/DYRK1A	CNRS Aubiere, F	DD
10a-c	Harmin derivatives	6/DYRK1A	Arizona AD Res. Consortium	DD
11a,b	Benzothiazoles	6/DYRK1A	Tokyo University, JP	DD
12	EGCG	6/DYRK1A	Mt. Sinai Hospital, NJ	PE
13a,b	Paullones	7/GSK-3	Lundbeck	LO
14	Arylindolmaleimides, SB216763	7/GSK-3	GSK	LO
15	Diazepinoindolyl maleimides	7/GSK-3	Lilly	LO
16	Bisindolylmaleimides	7/GSK-3	University of Illinois	LO
17a-c	Indirubins	7/GSK-3	CNRS Roscoff, F	LO
18	Aminothiazoles, Ar-014418	7/GSK-3	Astra Zeneca	Ph I
19a,b	Aminopyrimidines	7/GSK-3	Chiron/Novartis	LO
20	Pyrazines	7/GSK-3	Astra Zeneca	DD
21	Hymenialdisine	7/GSK-3	CNRS Roscoff, F	DD
22	Imidazopyridines	7/GSK-3	CrystalGenomics, South Korea	DD
23	Tryciclic spiro compounds	7/GSK-3	Kyorin Pharma, JP	DD
24	Benzopyrazoles	7/GSK-3	Sanofi-Aventis	LO
25	Manzamine	7/GSK-3, ATP-noncomp.	University of Mississippi	DD
26a,b	Palinurin, tricantin	7/GSK-3, ATP-noncomp.	CSIC Madrid, E	DD
27a,b	Thiadiazolidinones, tideglusib	7/GSK-3, ATP-noncomp.	Noscira, E	Ph II
28	Thienylhalomethyl ketones	7/GSK-3, ATP-noncomp.	CSIC Madrid, E	DD
29	5-Imino-1,2,4-thiadiazoles	7/GSK-3, ATP-noncomp.	CSIC Madrid, E	DD
30	Streptozotocin	8/OGA	-	NP*
31a-c	PUGNAc, LOGNAc, Butyl-PUGNAc	8/OGA	NIH	DD

Table 1: contd....

Number	Chemical cpd./Class	Target	Organization	Dev. Status
32a-d	GlcNAcstatins	8/OGA	University of Dundee, UK	DD
34a-c	NAG-thiazoline, NButGT	8/OGA	NIH	PE
35	Thiamet G	8/OGA	Alectos/Merck	PE
36a,b	Aminopyridazines	9/DAPK, presumed	Northwestern University, IL	LO
37	Minozac	9/DAPK, presumed	NeuroMedix	PE
38	5-Oxazolinones	9/DAPK	PharmaDesign, JP	DD
39	Ruthenium complex	9/DAPK	Marburg University	DD

Recent discoveries regarding the multi-directed putative impact of DAPK regulation of tau phosphorylation and of Pin1 activity should prompt the profiling of known DAPK inhibitors in tau-centered neurodegeneration models. Additional chemiotypes, as sources of optimized DAPK-acting leads and candidates, should also be actively pursued.

CONCLUSIONS

Even limiting our survey to the early steps leading to pathological alterations in tau and/or in tau-connected events, it is apparent how rich the target pool is. Five potentially therapeutic mechanisms were examined in details, and a single target was – maybe arbitrarily – chosen. Tens of other targets – some of which are validated and actively pursued by various labs – were neglected, mostly for reasons of space. Thirty-eight scaffolds acting on the selected targets were reported in Figs. (**2-9**), and are briefly summarized below in Table **1**. Each compound class is numbered as in Figures, and its chemical core is structurally defined; its general target (5-9, Charts **2**, **3**) and molecular target are mentioned; the developing laboratory (either public or private) is listed; and the development status – according to publicly available information – is finally provided (options: early discovery, **DD**; lead optimization, **LO**; preclinical evaluation, **PE**; clinical Phase I-II, **Ph I- Ph II**). Please note that the most advanced status for neurodegeneration-targeted experiments is listed: in other words, candidates in clinical trials for non-CNS indications with early *in vitro* characterization against tau-driven pathologies are classified as DD.

The forthcoming, final part of this survey will deal with even more therapeutic avenues. We will deeply examine the key switch between soluble, mis-decorated tau proteins and pathological tau oligomers, and the subsequent formation of insoluble aggregates. We will examine the consequences of non-functional tau proteins on their cellular partners – microtubules, and other proteins. In each case, a single target and as many as possible target-related chemiotypes will be presented and discussed.

Hopefully, once the survey will be completed, each of us may make up his or her mind, and prioritize – in terms of scientific relevance, of druggability, of market attractiveness, and so on – the described targets and molecules. I'll be so bold as to present you with my personal table of pros and cons, and I can't wait to put on paper my thoughts: hope you'll find it amusing, if not useful.

ACKNOWLEDGEMENTS

Declare none.

CONFLICT OF INTEREST

The author confirms that this chapter contents have no conflict of interest.

REFERENCES

[1]   Buee, L.; Bussie, T.; Buee-Scherrerb, V.; Delacourte, A.; Hof, P.R. Tau protein isoforms, phosphorylation and role in neurodegenerative disorders. *Brain Res. Rev.*, **2000**, *33*, 95-130.

[2]   Reitz, C. Alzheimer's disease and the amyloid cascade hypothesis: a critical review. *Int. J. Alzheimer Dis.*, **2012**, 369808.

[3]   Gu, J.; Sigurdsson, E.M. Immunotherapy for tauopathies. *J. Mol. Neurosci.*, **2011**, *45*, 690-695.

[4]   Delrieu, J.; Ousset, P.J.; Caillaud, C.; Vellas, B. "Clinical trials in Alzheimer's disease": immunotherapy approaches. *J. Neurochem.*, **2012**, *120*, 186-193.

[5]   Weingarten, M.D.; Lockwood, A.H.; Hwo, S.Y.; Kirschner, M.W. A protein factor essential for microtubule assembly. *Proc. Natl. Acad. Sci. U.S.A.*, **1975**, *72*, 1858-1862.

[6]   Hirokawa, N. Microtubule organization and dynamics dependent on microtubule-associated proteins. *Curr. Opin. Cell Biol.*, **1994**, *6*, 74-81.

[7]   Mukrasch, M.D.; Bibow, S.; Korukottu, J.; Jeganathan, S.; Biernat, J.; Griesinger, C.; Mandelkow, E.; Zweckstetter, M. Structural polymorphism of 441-residue tau at single residue resolution. *PLoS Biol.*, **2009**, *7*, 399-414.

[8]   Zheng, S.; Chen, Y.; Donahue, C.P.; Wolfe, M.S.; Varani, G. Structural basis for stabilization of the tau pre-mRNA splicing regulatory element by novantrone (mitoxantrone). *Chem. Biol.*, **2009**, *16*, 557-566.

[9]   Neve, R.L.; Harris, P.; Kosik, K.S.; Kurnit, D.M.; Donlon, T.A. Identification of cDNA clones for the human microtubule-associated protein tau and chromosomal localization of the genes for tau and microtubule-associated protein 2. *Brain Res.*, **1986**, *387*, 271- 280.

[10]   Andreadis, A. Tau gene alternative splicing: Expression pattern, regulation and modulation of function in normal brain and neurodegenerative disease. *Biochem. Biophys. Acta*, **2005**, 1739, 91-103.

[11]   Gustke, N.; Trinczek, B.; Biernat, J.; Mandelkow, E.-M.; Mandelkow, E. Domains of tau protein and interactions with microtubules. *Biochemistry*, **1994**, *33*, 9511-9522.

[12]   He, H.J.; Wang, X.S.; Pan, R.; Wang, D.L.; Liu' M.N.; He, R.Q. The proline-rich domain of tau plays a role in interactions with actin. *BMC Cell Biol.*, **2009**, *10*, 81.

[13]   Pooler, A.M.; Hanger, D.P. Functional implications of the association of tau with the plasma membrane. *Biochem. Soc. Trans.*, **2010**, *38*, 1012-1015.

[14]   Lee, G. Tau and src family tyrosine kinases. *Biochim. Biophys. Acta*, **2005**, *1739*, 323-330.

[15]   Andreadis, A.; Brown, W.M.; Kosik, K.S. Structure and novel exons of the human *T* gene. *Biochemistry*, **1992**, *31,* 10626-10633.

[16]   Hof, P.R.; Simic, G. Human fetal tau protein isoform: Possibilities for Alzheimer's disease treatment. *Int. J. Biochem. Cell Biol.*, **2012**, *44*, 1290-1294.

[17]   Crespo-Biel, N.; Theunis, C.; Van Leuven, F. Protein tau: Prime cause of synaptic and neuronal degeneration in Alzheimer's disease. *Int. J. Alzheim. Dis.*, **2012**, 251426.

[18]   Li, K.; Arikan, M.C.; Andreadis, A. Modulation of the membrane-binding domain of tau protein: splicing regulation of exon 2. *Mol. Brain Res.*, **2003**, *116*, 94-105.

[19]   Magnani, E.; Fan, J.; Gasparini, L.; Golding, M.; Williams, M.; Schiavo, G.; Goedert, M.; Amos, L.A; Spillantini, M.G. Interaction of tau protein with the dynactin complex. *EMBO J.*, **2007**, *26*, 4546-4554.

[20]   Andreadis, A. Tau splicing and the intricacies of dementia. *J. Cell. Physiol.*, **2011**, *227*, 1220-1225.

[21]   Spillantini, M.G.; Murrell, J.R.; Goedert, M.; Farlow, M.R.; Klug, A.; Ghetti, B. Mutation in the tau gene in familial multiple system tauopathy with presenile dementia. *Proc. Natl. Acad. Sci. USA*, **1998**, *95*, 7737-7741.

[22]   Poorkaj, P.; Bird, T.D.; Wijsman, E.; Nemens, E.; Garruto, R.M.; Anderson, L.; Andreadis, A.; Wiederholt, W.C.; Raskind, M.; Schellenberg, G.D. Tau is a candidate gene for chromosome 17 frontotemporal dementia. *Ann. Neurol.*, **1998**, *43*, 815-825.

[23]   Hutton, M.; Lendon, C.L.; Rizzu, P.; Baker, M.; Froelich, S.; Houlden, H.; Pickering-Brown, S.; Chakraverty, S.; Isaacs, A.; Grover, A.; Hackett, J.; Adamson, J.; Lincoln, S.; Dickson, D.; Davies, P.; Petersen, R.C.; Stevens, M.; de Graaff, E.; Wauters, E.; van Baren, J.; Hillebrand, M.; Joosse, M.; Kwon, J.M.; Nowotny, P.; Che, L.K.; Norton, J.; Morris, J.C.; Reed, L.A.; Trojanowski, J.; Basun, H.; Lannfelt, L.; Neystat, M.; Fahn, S.; Dark, F.; Tannenberg, T.; Dodd, P.R.; Hayward, N.; Kwok, J.B.J.; Schofield, P.R.; Andreadis, A.; Snowden, J.; Craufurd, D.; Neary, D.; Owen, F.; Oostra, B.A.; Hardy, J.; Goate, A.; van Swieten, J.; Mann, D.; Lynch, T.; Heutink, P. Association of missense and 5'-splice site mutations in tau with the inherited dementia FTDP-17. *Nature*, **1998**, *393*, 702-705.

[24]   Goedert, M.; Spillantini, M.G. Pathogenesis of the tauopathies. *J. Mol. Neurosci.*, **2011**, *45*, 425-431.

[25]   Ghetti, B.; Wszolek, Z.W.; Boeve, B.F.; Spina, S.; Goedert, M. Frontotemporal dementia and parkinsonism linked to chromosome 17. In: Dickson, D.; Weller, R.O. (eds.), *Neurodegeneration: the molecular pathology of dementia and movement disorders*, 2nd edn. Blackwell, Oxford, **2011**, 110-134.

[26]   Rademakers, R.; Melquist, S.; Cruts, M.; Theuns, J.; Del-Favero, J.; Poorkaj, P.; Baker, M.; Sleegers, K.; Crook, R.; De Pooter, T.; Bel Kacem, S.; Adamson, J.; Van den Bossche, D.; Van den Broeck, M.; Gass, J.; Corsmit, E.; De Rijk, P.; Thomas, N.; Engelborghs, S.; Heckman, M.; Litvan, I.; Crook, J.; De Deyn, P.P.; Dickson, D.; Schellenberg, G.D.; Van Broeckhoven, C.; Hutton, M.L. High density SNP haplotyping suggests altered regulation of Tau gene expression in progressive supranuclear palsy. *Hum. Mol. Genet.,* **2005**, *14*, 3281-3292.

[27]   Pittman, A.M.; Myers, A. J.; Abou-Sleiman, P.; Fung, H.C.; Kaleem, M.; Marlowe, L.; Duckworth, J.; Leung, D.; Williams, D.; Kilford, L.; Thomas, N.; Morris, C.M.; Dickson, D.; Wood, N.W.; Hardy, J.; Lees, A.J.; de Silva, R. Linkage disequilibrium fine mapping and haplotype association analysis of the Tau gene in progressive supranuclear palsy and corticobasal degeneration. *J. Med. Genet.,* **2005**, *42*, 837-846.

[28]   Kovacs, G.G.; Pittman, A.; Revesz, T.; Luk, C.; Lees, A.; Kiss, E.; Tariska, P.; Laszlo, L.; Molnár, K.; Molnar, M.J.; Tolnay, M.; de Silva, R. *MAPT* S305I mutation: implications for argyrophilic grain disease. *Acta Neuropathol.,* **2008**, *116*, 103-118.

[29]   Armstrong, R.A.; Cairns, N.J.; Lantos, P.L. Quantification of pathological lesions in the frontal and temporal lobe of ten patients diagnosed with Pick disease. *Acta Neuropathol.,* **1999**, *97*, 456-462.

[30]   Wegiel, J.; Dowjat, K.; Kaczmarski, W.; Kuchna, I.; Nowicki, K.; Frackowiak, J.; Mazur Kolecka, B.; Wegiel, J.; Silverman, W.P.; Reisberg, B.; de Leon, M.; Wisniewski, T.; Gong, C.-X.; Liu, F.; Adayev, T.; Chen-Hwang, M.-C.; Hwang, Y.-W. The role of overexpressed DYRK1A protein in the early onset of neurofibrillary degeneration in Down syndrome. *Acta Neuropathol.,* **2008**, *116*, 391-407.

[31]   Liu, F.; Gong, C.-X. Tau exon 10 alternative splicing and tauopathies. *Mol. Neurodegener.,* **2008**, *3*, 8.

[32]   Wang, Z.; Burge, C.B. Splicing regulation: From a parts list of regulatory elements to an integrated splicing code. *RNA*, **2008**, *14*, 802-813.

[33]   Long, J.C.; Caceres, J.F. The SR protein family of splicing factors:Master regulators of gene expression. *Biochem. J.,* **2009**, *417*, 15-27.

[34]   Martinez-Contreras, R.; Cloutier, P.; Shkreta, L.; Fisette, J.F.; Revil, T.; Chabot, B. HnRNP proteins and splicing control. *Adv. Exp. Med. Biol.,* **2007**, *623*, 123-147.

[35]   Donahue, C.P.; Ni, J.; Rozners, E.; Glicksman, M.A.; Wolfe, M.S. Identification of tau stem loop RNA stabilizers. *J. Biomol. Screening*, **2007**, *12*, 789-799.

[36]   Varani, L.; Spillantini, M.G.; Goedert, M.; Varani, G. Structural basis for recognition of the RNA major groove in the tau exon 10 splicing regulatory element by aminoglycoside antibiotics. *Nucleic Acids Res.,* **2000**, *28*, 710-719.

[37]   D'Souza, I.; Schellenberg, G.D. Arginine/serine-rich protein interaction domain-dependent modulation of a tau exon 10 splicing enhancer: altered interactions and mechanisms for functionally antagonistic FTDP-17 mutations Delta280K and N279K. *J. Biol. Chem.,* **2006**, *281*, 2460-2469.

[38]   Jiang, Z.; Tang, H.; Havlioglu, N.; Zhang, X.; Stamm, S.; Yan, R.; Wu, J.Y. Mutations in tau gene exon 10 associated with FTDP-17 alter the activity of an exonic splicing enhancer to interact with Tra2β. *J. Biol. Chem.*, **2003**, *278*, 18997-19007.

[39]   Kar, A.; Fushimi, K.; Zhou, X.; Ray, P.; Shi, C.; Chen, X.; Liu, Z.; Chen, S.; Wu, J.Y. RNA helicase p68 (DDX5) regulates *tau* exon 10 splicing by modulating a stem-loop structure at the 5' splice site. *Mol. Cell. Biol.*, **2011**, *31*, 1812-1821.

[40]   Qian, W.; Iqbal, K.; Grundke-Iqbal, I.; Gong, C.-X.; Liu, F. Splicing factor SC35 promotes tau expression through stabilization of its mRNA. *FEBS Lett.*, **2011**, *585*, 875-880.

[41]   Wang, Y.; Wang, J.; Gao, L.; Lafyatis, R.; Stamm, S.; Andreadis, A. Tau exons 2 and 10, which are misregulated in neurodegenerative diseases, are partly regulated by silencers which bind a SRp30c-SRp55 complex that either recruits or antagonizes htra2β1. *J. Biol. Chem.*, **2005**, *280*, 14230-14239.

[42]   Dhaenens, C.M.; Tran, H.; Frandemiche, M.-L.; Carpentier, C.; Schraen-Maschke, S.; Sistiaga, A.; Goicoechea, M.; Eddarkaoui, S.; Van Brussels, E.; Obriot, H.; Labudeck, A.; Gevaert, M.H.; Fernandez-Gomez, F.; Charlet-Berguerand, N.; Deramecourt, V.; Maurage, C.A.; Buee, L.; Lopez de Munain, A.; Sablonniere, B.; Caillet-Boudin, M.L.; Sergeant, N. Mis-splicing of Tau exon 10 in myotonic dystrophy type 1 is reproduced by overexpression of CELF2 but not by MBNL1 silencing. *Biochim. Biophys. Acta*, **2011**, *1812*, 732-742.

[43]   Orozco, D.; Tahirovic, S.; Rentzsch, V.; Schwenk, B.M.; Haass, C.; Edbauer, D. Loss of fused in sarcoma (FUS) promotes pathological Tau splicing. *EMBO Rep.*, **2012**, *13*, 759-764.

[44]   Ray, P.; Kar, A.; Fushimi, K.; Havlioglu, N.; Chen, X.; Wu, J.Y. PSF suppresses tau exon 10 inclusion by interacting with a stem-loop structure downstream of exon 10. *J. Mol. Neurosci.*, **2011**, *45*, 453-466.

[45]   Wu, J.Y.; Kar, A.; Kuo, D.; Yu, B.; Havlioglu, N. SRp54 (SFRS11), a regulator for tau exon 10 alternative splicing identified by an expression cloning strategy. *Mol. Cell. Biol.*, **2006**, *26*, 6739-6747.

[46]   Gao, L.; Wang, J.; Wang, Y.; Andreadis, A. SR protein 9G8 modulates splicing of tau exon 10 *via* its proximal downstream intron, a clustering region for frontotemporal dementia mutations. *Mol. Cell. Neurosci.*, **2007**, *34*, 48-58.

[47]   Stamm, S. Regulation of alternative splicing by reversible protein phosphorylation. *J. Biol. Chem.*, **2008**, *283*, 1223-1227.

[48]   Hartmann, A.M.; Rujescu, D.; Giannakouros, T.; Nikolakaki, E.; Goedert, M.; Mandelkow, E.M.; Gao, Q.S.; Andreadis, A.; Stamm, S. Regulation of alternative splicing of human tau exon 10 by phosphorylation of splicing factors. *Mol. Cell. Neurosci.*, **2001**, *18*, 80-90.

[49]   Shi, J.; Qian, W.; Yin, X.; Iqbal, K.; Grundke-Iqbal, I.; Gu, X.; Ding, F.; Gong, C.-X.; Liu, F. Cyclic AMP-dependent protein kinase regulates the alternative splicing of tau exon 10. *J. Biol. Chem.*, **2011**, *286*, 14639-14648.

[50]   Gu, J.; Shi, J.; Wu, S.; Jin, N.; Qian, W.; Zhou, J.; Grundke-Iqbal, I.; Iqbal, K.; Gong, C.-X.; Liu, F. Cyclic AMP-dependent protein kinase regulates 9G8-mediated alternative splicing of tau exon 10. *FEBS Lett.*, **2012**, *586*, 2239-2244.

[51]   Chen, K.-L.; Yuan, R.-Y.; Hu, C.-J.; Hsu, C.Y. Amyloid-β peptide alteration of tau exon-10 splicing *via* the GSK-3β-SC35 pathway. *Neurobiol. Dis.*, **2010**, *40*, 378-385.

[52]   Wegiel, J.; Gong, C.X.; Hwang, Y.W. The role of DYRK1A in neurodegenerative diseases. *FEBS J.*, **2011**, *278*, 236-245.

[53] Ferrer, I.; Barrachina, M.; Puig, B.; Martinez de Lagran, M.; Marti, E.; Avila, J.; Dierssen, M. Constitutive Dyrk1A is abnormally expressed in Alzheimer disease, Down syndrome, Pick disease, and related transgenic models. *Neurobiol. Dis.*, **2005**, *20*, 392-400.

[54] Liu, F.; Liang, Z.; Wegiel, J.; Hwang, Y.-W.; Iqbal, K.; Grundke-Iqbal, I.; Ramakrishna, N.; Gong, C.-X. Over-expression of Mnb/ Dyrk1A contributes to neurofibrillary degeneration in Down syndrome. *FASEB J.*, **2008**, *22*, 3224-3233.

[55] Jung, M.-S.; Park, J.-H.; Ryu, Y.S.; Choi, S.-H.; Yoon, S.-H.; Kwen, M.-Y.; Oh, J.Y.; Song, W.J.; Chung, S.-H. Regulation of RCAN1 protein activity by Dyrk1A protein-mediated phosphorylation. *J. Biol. Chem.*, **2011**, *286*, 40401-40412.

[56] Reeves, R.H.; Irving, N.G.; Moran, T.H.; Wohn, A.; Kitt, C.; Sisodia, S.S.; Schmidt, C.; Bronson, R.T.; Davisson, M.T. A mouse model for Down syndrome exhibits learning and behaviour deficits. *Nat. Genet.*, **1995**, *11*, 177-184.

[57] Alvarez, M.; Estivill, X.; de la Luna, S. DYRK1A accumulates in splicing speckles through a novel targeting signal and induces speckles disassembly. *J. Cell Sci.*, **2003**, *116*, 3099-3107.

[58] Shi, J.; Zhang, T.; Zhou, C.; Chohan, M.O.; Gu, X.; Wegiel, J.; Zhou, J.; Hwang, Y.-W; Iqbal, K.; Grundke-Iqbal, I.; Gong, C.-X.; Liu, F. Increased dosage of Dyrk1A alters alternative splicing factor (ASF)-regulated alternative splicing of tau in Down syndrome. *J. Biol. Chem.*, **2008**, *283*, 28660-28669.

[59] Wegiel, J.; Kaczmarski, W.; Barua, M.; Kuchna, I.; Nowicki, K.; Wang, K.-C.; Wegiel, J.; Ma, S.Y.; Frackowiak, J.; Mazur Kolecka, B.; Silverman, W.P.; Reisberg, B.; Monteiro, I.; de Leon, M.; Wisniewski, T.; Dalton, A.; Lai, F.; Hwang, Y.-W.; Adayev, T.; Liu, F.; Iqbal, K.; Grundke-Iqbal, I.; Gong, C.-X. Link between DYRK1A overexpression and several-fold enhancement of neurofibrillary degeneration with 3-repeat tau protein in Down syndrome. *J. Neuropathol. Exp. Neurol.*, **2011**, *70*, 36-50.

[60] Qian, W.; Liang, H.; Shi, J.; Jin, N.; Grundke-Iqbal, I.; Iqbal, K.; Gong, C.-X.; Liu, F. Regulation of the alternative splicing of tau exon 10 by SC35 and Dyrk1A. *Nucleic Acid Res.*, **2011**, *39*, 6161-6171.

[61] Jin, X.; Jin, N.; Gu, J.; Shi, J.; Zhou, J.; Gong, C.-X.; Iqbal, K.; Grundke-Iqbal, I.; Liu, F. Dual-specificity tyrosine-phosphorylated and regulated kinase 1A (Dyrk1A) modulates serine-arginine rich protein 55 (SRp55)-promoted tau exon 10 inclusion. *J. Biol. Chem.*, **2012**, *287*, 30497-30506.

[62] Ding, S.; Shi, J.; Qian, W.; Iqbal, K.; Grundke-Iqbal, I.; Gong, C.-X.; Liu, F. Regulation of alternative splicing of tau exon 10 by 9G8 and Dyrk1A. *Neurobiol. Aging*, **2012**, *33*, 1389-1399.

[63] Liu, Y.; Peacey, E.; Dickson, J.; Donahue, C.P.; Zheng, S.; Varani, G.; Wolfe, M.S. Mitoxanthrone analogues as ligands for a stem-loop structure of tau pre-mRNA. *J. Med. Chem.*, **2009**, *52*, 6523-6526.

[64] Lopez-Senin, P.; Gomez-Pinto, I.; Grandas, A.; Marchan, V. Identification of ligands for the tau exon 10 splicing regulatory element RNA by using dynamic combinatorial chemistry. *Chem. Eur. J.*, **2011**, *17*, 1946-1953.

[65] Becker, W.; Sippl, W. Activation, regulation, and inhibition of DYRK1A. *FEBS J.*, **2011**, *278*, 246-256.

[66] Smith, B.; Medda, F.; Gokhale, V.; Dunckley, T.; Hulme, C. Recent advances in the design, synthesis, and biological evaluation of selective DYRK1A inhibitors: a new avenue for a

disease modifying treatment of Alzheimer's? *ACS Chem. Neurosci.*, **2012**, DOI: 10.1021/cn300094k.

[67]   Frost, D.; Meechoovet, B.; Wang, T.; Gately, S.; Giorgetti, M.; Shcherbakova, I.; Dunckley, T. β-Carboline compounds, including harmine, inhibit DYRK1A and tau phosphorylation at multiple Alzheimer's disease-related sites. *PLoS One*, **2011**, *6*, e19264.

[68]   Sanchez, C.; Salas, A.P.; Brana, A.F.; Palomino, M.; Pineda-Lucena, A.; Carbajo, R.J.; Mendez, C.; Moris, F.; Sala, J.A. Generation of potent and selective kinase inhibitors by combinatorial biosynthesis of glycosylated indolocarbazoles. *Chem. Commun.*, **2009**, *27*, 4118-4120.

[69]   Bain, J.; Plater, L.; Elliott, M.; Shapiro, N.; Hastie, J.; McLauchlan, H.; Klevernic, I.; Arthur, J.S.C.; Alessi, D.R.; Cohen, P. The selectivity of protein kinase inhibitors: a further update. *Biochem. J.*, **2007**, *408*, 297-315.

[70]   Sarno, S.; Mazzorana, M.; Traynor, R.; Ruzzene, M.; Cozza, G.; Pagano, M.A.; Meggio, F.; Zagotto, G.; Battistutta, R.; Pinna, L.A. Structural features underlying the selectivity of the kinase inhibitors NBC and dNBC: Role of a nitro group that discriminates between CK2 and DYRK1A. *Cell. Mol. Life Sci.*, **2012**, *69*, 449-60.

[71]   Kyng, A.K.; Kim, N.D.; Chon, Y.S.; Jung, M.-S.; Lee, B.-J.; Kim, J.H.; Song, W.-J. QSAR analysis of pyrazolidine-3,5-diones derivatives as Dyrk1A nhibitors. *Bioorg. Med. Chem. Lett.*, **2009**, *19*, 2324-2328.

[72]   Rosenthal, A.S.; Tanega, C.; Shen, M.; Mott, B.T.; Bougie, J.M.; Nguyen, D.-T.; Misteli, T.; Auld, D.S.; Maloney, D.J.; Thomas, C.J. Potent and selective small molecule inhibitors of specific isoforms of Cdc2-like kinases (Clk) and dual specificity tyrosine-phosphorylation regulated kinases (Dyrk). *Bioorg. Med. Chem. Lett.*, **2011**, *21*, 3152-3158.

[73]   Kassis, P.; Brzeszcz, J.; Beneteau, V.; Lozach, O.; Meijer, L.; Guevel, R.L.; Guillouzo, C.; Lewiński, K.; Bourg, S.; Colliandre, L.; Routier, S.; Merour, J.-Y. Synthesis and biological evaluation of new 3-(6- hydroxyindol-2-yl)-5-(phenyl) pyridine or pyrazine V-shaped molecules as kinase inhibitors and cytotoxic agents. *Eur. J. Med. Chem.*, **2011**, *46*, 5416-5434.

[74]   Neagoie, C.; Vedrenne, E.; Buron, F.; Merour, J.-Y.; Rosca, S.; Bourg, S.; Lozach, O.; Meijer, L.; Baldeyrou, B.; Lansiaux, A.; Routier, S. Synthesis of chromeno[3,4b]indoles as lamellarin D analogues: A novel DYRK1A inhibitor class. *Eur. J. Med. Chem.*, **2012**, *49*, 379-396.

[75]   Echalier, A.; Bettayeb, K.; Ferandin, Y.; Lozach, O.; Clement, M.; Valette, A.; Liger, F.; Marquet, B.; Morris, J.C.; Endicott, J.A.; Joseph, B.; Meijer, L. Meriolins (3-pyrimidin-4-yl-7-azaindoles): synthesis, kinase inhibitory activity, cellular effects, and structure of CDK2/cyclin A/meriolin complex. *J. Med. Chem.*, **2008**, *51*, 737-751.

[76]   Giraud, F.; Alves, G.; Debiton, E.; Nauton, L.; Thery, V.; Durieu, E.; Ferandin, Y.; Lozach, O.; Meijer, L.; Anizon, F.; Pereira, E.; Moreau, P. Synthesis, protein kinase inhibitory potencies, and *in vitro* antiproliferative activities of meridianin derivatives. *J. Med. Chem.*, **2011**, *54*, 4474-4489.

[77]   Adayev, T.; Wegiel, J.; Hwang, Y.-W. Harmine is an ATP-competitive inhibitor for dual-specificity tyrosine phosphorylation-regulated kinase 1A (Dyrk1A). *Archiv. Biochem. Biophys.*, **2011**, *507*, 212-218.

[78]   Goeckler, N.; Jofre, G.; Papadopoulos, C.; Soppa, U.; Tejedor, F.J. Harmine specifically inhibits protein kinase DYRK1A and interferes with neurite formation. *FEBS J.*, **2009**, *276*, 6324-6337.

[79] Tejedor, F.J.; Haemmerle, B. MNB/DYRK1A as a multiple regulator of neuronal development. *FEBS J.*, **2011**, *278*, 223-235.

[80] Frost, D.; Meechoovet, B.; Wang, T.; Gately, S.; Giorgetti, M.; Shcherbakova, I.; Dunckley, T. β-Carboline compounds, including harmine, inhibit DYRK1A and tau phosphorylation at multiple Alzheimer's disease-related sites. *PLoS One*, **2011**, *6*, e19264.

[81] Herraiz, T.; Gonzalez, D.; Ancin-Azpilicueta, C.; Aran, V.J.; Guillen, H. Beta-carboline alkaloids in Peganum harmala and inhibition of human monoamine oxidase (MAO). *Food Chem. Toxicol.*, **2010**, *48*, 839-845.

[82] Ogawa, Y.; Nonaka, Y.; Goto, T.; Ohnishi, E.; Hiramatsu, T.; Kii, I.; Yoshida, M.; Ikura, T.; Onogi, H.; Shibuya, H.; Hosoya, T.; Ito, N.; Hagiwara, M. Development of a novel selective inhibitor of the Down syndrome-related kinase DYRK1A. *Nat. Commun.*, **2010**, *86*, 1-9.

[83] Lee, H.; Bae, J.H.; Lee, S.R. Protective effect of green tea polyphenol EGCG against neuronal damage and brain edema after unilateral cerebral ischemia in gerbils. *J. Neurosci. Res.*, **2004**, *77*, 892-900.

[84] Chyu, K.Y.; Babbidge, S.M.; Zhao, X.; Dandillaya, R.; Rietveld, A.G.; Yano, J.; Dimayuga, P.; Cercek, B.; Shah, P.K. Differential effects of green tea-derived catechin on developing versus established atherosclerosis in apolipoprotein E-null mice. *Circulation*, **2004**, *109*, 2448-2453.

[85] Leone, M.; Zhai, D.; Sareth, S.; Kitada, S.; Reed, J.C.; Pellecchia, M. Cancer prevention by tea polyphenols is linked to their direct inhibition of antiapoptotic Bcl-2-family proteins. *Cancer Res.*, **2003**, *63*, 8118-8121.

[86] Yam, T.S.; Shah, S.; Hamilton-Miller, J.M. Microbiological activity of whole and fractionated crude extracts of tea (*Camella Sinesis*), and of tea components. *FEMS Microbiol. Lett.*, **1997**, *152*, 169-174.

[87] Hamza, A.; Zhan, C.G. How can (-)-epigallocatechin gallate from green tea prevent HIV-1 infection? Mechanistic insights from computational modeling and the implication for rational design of anti-HIV-1 entry inhibitors. *J. Phys. Chem. B*, **2006**, *110*, 2910-2917.

[88] Rezai-Zadeh, K.; Shytle, D.; Sun, N.; Mori, T.; Hou, H.; Jeanniton, D.; Ehrhart, J.; Townsend, K.; Zeng, J.; Morgan, D.; Hardy, J.; Town, T.; Tan, J. Green tea epigallocatechin-3-gallate (EGCG) modulates amyloid precursor protein cleavage and reduces cerebral amyloidosis in Alzheimer transgenic mice. *J. Neurosci.*, **2005**, *25*, 8807-14.

[89] Bain, J.; McLauchlan, H.; Elliott, M.; Cohen, P. The specificities of protein kinase inhibitors: an update. *Biochem. J.*, **2003**, *371*, 199-204.

[90] Adayev, T.; Chen-Hwang, M.C.; Murakami, N.; Wegiel, J.; Hwang, Y.-W. Kinetic properties of a MNB/DYRK1A mutant suitable for the elucidation of biochemical pathways. *Biochemistry*, **2006**, *45*, 12011-12019.

[91] Guedj, F.; Sebrie, C.; Rivals, I.; Ledru, A.; Paly, E.; Bizot, J.C.; Smith, D.; Rubin, E.; Gillet, B.; Arbones, M.; Delabar, J.M. Green tea polyphenols rescue of brain deficits induced by overexpression of DYRK1A. *PLoS One*, **2009**, *4*, e4606.

[92] Santa-Maria, I.; Diaz-Ruiz, C.; Ksiezak-Reding, H.; Chen, A.; Ho, L.; Wang, J.; Pasinetti, G.M. GSPE interferes with tau aggregation *in vivo*: implication for treating tauopathy. *Neurobiol. Aging*, **2012**, *33*, 2072-2081.

[93]   Cao, R.; Chen, H.; Peng, W.; Ma, Y.; Hou, X.; Guan, H.; Liu, X.; Xu, A. Design, synthesis and *in vitro* and *in vivo* antitumor activities of novel beta-carboline derivatives. *Eur. J. Med. Chem.*, **2005**, *40*, 991–1001.

[94]   Seifert, A.; Allan, L.A.; Clarke, P.R. DYRK1A phosphorylates caspase 9 at an inhibitory site and is potently inhibited in human cells by harmine. *FEBS J.*, **2008**, *275*, 6268–6280.

[95]   Laguna, A.; Aranda, S.; Barallobre, M.J.; Barhoum, R.; Fernandez, E.; Fotaki, V.; Delabar, J.M.; de la Luna, S.; de la Villa, P.; Arbones, M.L. The protein kinase DYRK1A regulates caspase-9- mediated apoptosis during retina development. *Dev. Cell*, **2008**, *15*, 841–853.

[96]   Frederick, R.; Bruyere, C.; Vancraeynest, C.; Reniers, J.; Meinguet, C.; Pochet, L.; Backlund, A.; Masereel, B.; Kiss, R.; Wouters, J. Novel trisubstituted harmine derivatives with original *in vitro* anticancer activity. *J. Med. Chem.*, **2012**, *55*, 6489-6501.

[97]   Dunker, A.K.; Silman, I.; Uversky, V.N; Sussman, J.L. Function and structure of inherently disordered proteins. *Curr. Opin. Struct. Biol.*, **2008**, *18*, 756-764.

[98]   Goedert, M.; Spillantini, M.G.; Potier, M.C.; Ulrich, J.; Crowther, R.A. Cloning and sequencing of the cDNA encoding an isoform of microtubule-associated protein tau containing four tandem repeats: differential expression of tau protein mRNAs in human brain. *EMBO J.*, **1989**, *8*, 393-399.

[99]   Martin, L.; Latypova, X.; Terro, F. Post-translational modifications of tau protein: Implications for Alzheimer's disease. *Neurochem. Int.*, **2011**, *58*, 458-471.

[100]   Lee, V.M. Regulation of tau phosphorylation in Alzheimer's disease. *Ann. N. Y. Acad. Sci.*, **1996**, *777*, 107-113.

[101]   Mazanetz, M.P.; Fischer, P.M. Untangling tau hyperphosphorylation in drug design for neurodegenerative diseases. *Nat. Rev. Drug Discov.*, **2007**, *6*, 464-479.

[102]   Boyle, S.N.; Koleske, A.J. Dissecting kinase signaling pathways. *Drug Discov. Today*, **2007**, *12*, 717-724.

[103]   Bamborough, P.; Brown, M.J.; Christopher, J.A.; Chung, C.-w.; Mellor, G.W. Selectivity of kinase inhibitor fragments. *J. Med. Chem.*, **2011**, *54*, 5131-5143.

[104]   Hanover, J.A.; Krause, M.W.; Love, D.C Bittersweet memories: Linking metabolism to epigenetics through O-GlcNacylation. *Nat. Rev. Mol. Cell Biol.*, **2012**, *13*, 312-321.

[105]   Hart, G.W.; Slawson, C.; Ramirez-Correa, G.; Lagerlof, O. Cross talk between O-GlcNAcylation and phosphorylation: Roles in signaling, transcription, and chronic disease. *Annu. Rev. Biochem.*, **2011**, *80*, 825-858.

[106]   Wang, Z.; Udeshi, N.D.; O'Malley, M.; Shabanowitz, J.; Hunt, D.F.; Hart, G.W. Enrichment and site mapping of O-linked *N*-acetylglucosamine by a combination of chemical/enzymatic tagging, photochemical cleavage, and electron transfer dissociation mass spectrometry. *Mol. Cell. Proteomics*, **2010**, *9*, 153-60.

[107]   Lazarus, B.D.; Love, D.C; Hanover, J.A. O-GlcNAc cycling: Implications for neurodegenerative disorders. *Int. J. Biochem. Cell Biol.*, *41*, 2134-2146.

[108]   Wedemeyer, W.J.; Welker, E.; Scheraga, H.A. Proline cis-trans isomerization and protein folding. *Biochemistry*, **2002**, *41*, 14637-14644.

[109]   Zhou, X.Z.; Lu, P.-J.; Wulf, G.; Lu, K.P. Phosphorylation-dependent prolyl isomerization. A novel signaling regulatory mechanism. *Cell. Mol. Life Sci.*, **1999**, *56*, 788-806.

[110]   Shaw, P.E. Peptidyl-prolyl cis/trans isomerases and transcription: is there a twist in the tail? *EMBO Rep.*, **2007**, *8*, 40-45.

[111] Nakamura, K.; Greenwood, A.; Binder, L.; Bigio, E.H.; Denial, S.; Nicholson, L.; Zhou, X.Z.; Lu, K.P. Proline isomer-specific antibodies reveal the early pathogenic tau conformation in Alzheimer's disease. *Cell*, **2012**, *149*, 232-244.

[112] Cohen, T.J.; Guo, J.L.; Hurtado, D.E.; Kwong, L.K.; Mills, I.P.; Trojanowski, J.Q.; Lee, V.M.Y. The acetylation of tau inhibits its function and promotes pathological tau aggregation. *Nat. Commun.*, **2011**, *2*, 252.

[113] Nacharaju, P.; Ko, L.; Yen, S.H. Characterization of *in vitro* glycation sites of tau. *J. Neurochem.*, **1997**, *69*, 1709-1719.

[114] Horiguchi, T.; Uryu, K.; Giasson, B.I.; Ischiropoulos, H.; LightFoot, R.; Bellmann, C.; Richter-Landsberg, C.; Lee, V.M.; Trojanowski, J.Q. Nitration of tau protein is linked to neurodegeneration in tauopathies. *Am. J. Pathol.*, **2003**, *163*, 1021-1031.

[115] Halverson, R.A.; Lewis, J.; Frausto, S.; Hutton, M.; Muma, N.A. Tau protein is cross-linked by transglutaminase in P301L tau transgenic mice. *J. Neurosci.*, **2005**, *25*, 1226-1233.

[116] Takahashi, K.; Ishida, M.; Komano, H.; Takahashi, H. SUMO-1 immunoreactivity co-localizes with phospho-Tau in APP transgenic mice but not in mutant Tau transgenic mice. *Neurosci. Lett.*, **2008**, *441*, 90-93.

[117] Schweers, O.; Mandelkow, E.M.; Biernat, J.; Mandelkow, E. Oxidation of cysteine-322 in the repeat domain of microtubule-associated protein tau controls the *in vitro* assembly of paired helical filaments. *Proc. Natl. Acad. Sci. USA*, **1995**, *92*, 8463-8467.

[118] Riederer, B.M.; Binder, L.I. Differential distribution of tau proteins in developing rat cerebellum. *Brain Res. Bull.*, **1994**, *33*, 155-161.

[119] Brion, J.-P.; Octave, J.N.; Couck, A.M. Distribution of the phosphorylated microtubule-associated protein tau in developing cortical neurons. *Neuroscience*, **1994**, *63*, 895-909.

[120] Badiola, N.; Suarez-Calvet, M.; Lleo, A. Tau phosphorylation and aggregation as a therapeutic target in tauopathies. *CNS Neurol. Dis. Drug Targets*, **2010**, *9*, 727-740.

[121] Yu, Y.; Run, X.; Liang, Z.; Li, Y.; Liu, F.; Liu, Y.; Iqbal, K.; Grundke-Iqbal, I.; Gong, C.-X. Developmental regulation of tau phosphorylation, tau kinases, and tau phosphatases. *J. Neurochem.*, **2009**, 1480-1494.

[122] Su, B.; Wang, X.; Drew, K.L.; Perry, G.; Smith, M.A.; Zhu, X. Physiological regulation of tau phosphorylation during hibernation. *J. Neurochem.*, **2008**, *105*, 2098-2108.

[123] Dickey, C.A.; Koren, J.; Zhang, Y.-J.; Xu, Y.-F.; Jinwal, U.K.; Birnbaum, M.J.; Monks, B.; Sun, M.; Cheng, J.Q.; Patterson, C.; Bailey, R.M.; Dunmore, J.; Soresh, S.; Leon, C.; Morgan, D.; Petrucelli, L. Akt and CHIP co-regulate tau degradation through coordinated interactions. *Proc. Natl. Acad. Sci. USA*, **2008**, *105*, 3622-3627

[124] Bertrand, J.; Plouffe, V.; Senechal, P.; Leclerc, N. The pattern of human tau phosphorylation is the result of priming and feedback events in primary hippocampal neurons. *Neuroscience*, **2010**, *168*, 323-334.

[125] Ploia, C.; Antoniou, X.; Sclip, A.; Grande, V.; Cardinetti, D.; Colombo, A.; Canu, N.; Benussi, L.; Ghidoni, R.; Forloni, G.; Borsello, T. JNK plays a key role in tau hyperphosphorylation in Alzheimer's disease models. *J. Alzheim. Dis.*, **2011**, *26*, 315-329

[126] Wang, S.; Toth, M.E.; Bereczki, E.; Santha, M.; Guan, Z.-Z.; Winblad, B.; Pei, J.-J. Interplay between glycogen synthase kinase-3β and tau in the cerebellum of Hsp27 transgenic mouse. *J. Neurosci. Res.*, **2011**, *89*, 1267-1275.

[127] Liu, Y.; Liu, F.; Grundke-Iqbal, I.; Iqbal, K.; Gong, Cheng-X. Brain glucose transporters, O-GlcNAcylation and phosphorylation of tau in diabetes and Alzheimer's disease. *J. Neurochem.*, **2009**, *111*, 242-249.

[128] Gloster, T.M.; Vocadlo, D.J. Mechanism, structure, and inhibition of O-GlcNAc processing enzymes. *Curr. Signal Transduct, Ther.*, **2010**, *5*, 74-91.

[129] Yang, X.; Ongusaha, P.P.; Miles, P.D.; Havstad, J.C.; Zhang, F.; So, W.V.; Kudlow, J.E.; Michell, R.H.; Olefsky, J.M.; Field, S.J.; Evans, R.M. Phosphoinositide signalling links OGlcNAc transferase to insulin resistance. *Nature*, **2008**, *451*, 964-969.

[130] Slawson, C.; Lakshmanan, T.; Knapp, S.; Hart, G.W. A mitotic GlcNAcylation/ phosphorylation signaling complex alters the posttranslational state of the cytoskeletal protein vimentin. *Mol. Biol. Cell*, **2008**, *19*, 4130-4140.

[131] Gerard, M.; Deleersnijder, A.; Demeulemeester, J.; Debyser, Z.; Baekelandt, V. Unraveling the role of peptidyl-prolyl isomerases in neurodegeneration. *Mol. Neurobiol.*, **2011**, *44*, 13-27.

[132] Parsons, W.H.; Sigal, N.H.; Wyvratt, M.J. FK-506 - a novel immunosuppressant. *N. Y. Acad. Sci.*, **1993**, *685*, 22-36.

[133] Rulten, S.L.; Kinloch, R.A.; Tateossian, H.; Robinson, C.; Gettins, L.; Kay, J.E. The human FK506-binding proteins: characterization of human FKBP19. *Mamm. Genome*, **2006**, *17*, 322-331.

[134] Handschumacher, R.E.; Harding, M.W.; Rice, J.; Drugge, R.J.; Speicher, D.W. Cyclophilin: a specific cytosolic binding protein for cyclosporin A. *Science*, **1984**, *226*, 544-547

[135] Galat, A.; Bua, J. Molecular aspects of cyclophilins mediating therapeutic actions of their ligands. *Cell. Mol. Life Sci.*, **2010**, *67*, 3467-3488.

[136] Galat, A. Peptidylprolyl cis/trans isomerases (immunophilins): Biological diversity - targets - functions. *Curr. Top. Med. Chem.*, **2003**, *3*, 1315-1347.

[137] Mueller, J.W.; Bayer, P. Small family with key contacts: Par14 and Par17 parvulin proteins, relatives of Pin1, now emerge in biomedical research. *Perspect. Med. Chem.*, **2008**, *2*, 11-20.

[138] Koren, J.III; Jinwal, U.K.; Davey, Z.; Kiray, J.; Arulselvam, K.; Dickey, C.A. Bending tau into shape: the emerging role of peptidyl-prolyl isomerases in tauopathies. *Mol. Neurobiol.*, **2011**, *44*, 65-70.

[139] Preuss,U.; Mandelkow, E.M. Mitotic phosphorylation of tau protein in neuronal cell lines resembles phosphorylation in Alzheimer's disease. *Eur. J. Cell Biol.*, **1998**, *76*, 176-184.

[140] Augustinack, J.C.; Schneider, A.; Mandelkow, E.-M.; Hyman, B.T. Specific tau phosphorylation sites correlate with severity of neuronal cytopathology in Alzheimer's disease. *Acta Neuropathol.*, **2002**, *103*, 26-35.

[141] Rudrabhatla, P.; Pant, H.C. Role of protein phosphatase 2A in Alzheimer's disease. *Curr. Alzh. Res.*, **2011**, *8*, 623-632.

[142]. http://cnr.iop.kcl.ac.uk/hangerlab/tautable

[143] Chico, L.K.; Van Eldik, L.J.; Watterson, D. M. Targeting protein kinases in central nervous system disorders. *Nat. Rev. Drug Discov.*, **2009**, *8*, 892-909.

[144] Savage, M.J.; Gingrich, D.E. Advances in the development of kinase inhibitor therapeutics for Alzheimer's disease. *Drug Devel. Res.*, **2009**, *70*, 125-144.

[145] Cheung, Zelda H.; Ip, N.Y. Cdk5 : a multifaceted kinase in neurodegenerative diseases. *Trends Cell. Biol.*, **2012**, *22*, 169-175.

[146] Li, G.; Yin, H.; Kuret, J. Casein Kinase 1 delta phosphorylates tau and disrupts its binding to microtubules. *J. Biol. Chem.*, **2004**, *279*, 15938-15945.

[147] Zhu, B.; Zhang, L.; Creighton, J.; Alexeyev, M.; Strada, S.J.; Stevens, T. Protein kinase A phosphorylation of tau-serine 214 reorganizes microtubules and disrupts the endothelial cell barrier. *Am. J. Physiol.*, **2010**, *299*, L493-L501.

[148] Liu, S.J.; Zhang, A.H.; Li, H.L.; Wang, Q.; Deng, H.M.; Netzer, W.J.; Xu, H.; Wang, J.Z. Overactivation of glycogen synthase kinase-3 by inhibition of phosphoinositol-3 kinase and protein kinase C leads to hyperphosphorylation of tau and impairment of spatial memory. *J. Neurochem.*, **2003**, *87*, 1333-1344.

[149] Le Corre, S.; Klafki, H.W.; Plesnila, N.; Huebinger, G.; Obermeier, A.; Sahagun, H.; Monse, B.; Seneci, P.; Lewis, J.; Eriksen, J.; Zehr, C.; Yue, M.; McGowan, E.; Dickson, D.W.; Hutton, M.; Roder, H.M. An inhibitor of tau hyperphosphorylation prevents severe motor impairments in tau transgenic mice. *Proc. Natl. Acad. Sci. USA*, **2006**, *103*, 9673-9678.

[150] Cai, Z.; Yan, L.-J.; Li, K.; Quazi, S.H.; Zhao, B. Roles of AMP-activated protein kinase in Alzheimer's disease. *NeuroMol. Med.*, **2012**, *14*, 1-14.

[151] Correa, Sonia A. L.; Eales, Katherine L. The role of p38 MAPK and its substrates in neuronal plasticity and neurodegenerative disease. *J. Signal Transduct.*, **2012**, 649079.

[152] Yang, K.; Belrose, J.; Trepanier, C.H.; Lei, G.; Jackson, M.F.; MacDonald, J.F. Fyn, a potential target for Alzheimer's disease. *J. Alzheim. Dis.*, **2011**, *27*, 243-252.

[153] Rahman, A.; Grundke-Iqbal, I.; Iqbal, K. Phosphothreonine-212 of Alzheimer abnormally hyperphosphorylated tau is a preferred substrate of protein phosphatase-1. *Neurochem. Res.*, **2005**, 30, 277-287.

[154] Torrent, L.; Ferrer, I. PP2A and Alzheimer disease. *Curr. Alzheim. Res.*, **2012**, *9*, 248-256.

[155] Qian, W.; Yin, X.; Hu, W.; Shi, J.; Gu, J.; Grundke-Iqbal, I.; Iqbal, K.; Gong, C.-X.; Liu, F. Activation of protein phosphatase 2B and hyperphosphorylation of tau in Alzheimer's disease. *J. Alzheim. Dis.*, **2011**, *23*, 617-627.

[156] Liu, F.; Iqbal, K.; Grundke-Iqbal, I.; Rossie, S.; Gong, Cheng-X. Dephosphorylation of tau by protein phosphatase 5: Impairment in Alzheimer's disease. *J. Biol. Chem.*, **2005**, *280*, 1790-1796.

[157] Mondragon-Rodriguez, S.; Perry, G.; Zhu, X.; Moreira, P.I.; Williams, S. Glycogen synthase kinase 3: a point of integration in Alzheimer's disease and a therapeutic target? *Int. J. Alzheimer's Dis.*, **2012**, 276803.

[158] Cohen, P. The hormonal control of glycogen metabolism in mammalian muscle by multivalent phosphorylation. *Biochem. Soc. Trans*, **1979**, *7*, 459-480.

[159] Maurer, U.; Charvet, C.; Wagman, A.S.; Dejardin, E.; Green, D.R. Glycogen synthase kinase-3 regulates mitochondrial outer membrane permeabilization and apoptosis by destabilization of MCL-1. *Mol. Cell*, **2006**, *21*, 749-760.

[160] Happel, N.; Stoldt, S.; Schmidt, B.; Doenecke, D. M phase-specific phosphorylation of histone H1.5 at threonine 10 by GSK-3, *J. Mol. Biol.*, **2009**. *386*, 339-350.

[161] Phukan, S.; Babu, V.S.; Kannoji, A.; Hariharan, R.; Balaji, V.N. GSK-3β: role in therapeutic landscape and development of modulators. *Br. J. Pharmacol.*, **2010**, *160*, 1-19.

[162] Adachi, A.; Kano, F.; Tsuboi, T.; Fujita, M.; Maeda, Y.; Murata, M. Golgi-associated GSK-3β regulates the sorting process of post-Golgi membrane trafficking. *J. Cell Sci.*, **2010**, *123*, 3215-3225.

[163] Hooper, C.; Killick, R.; Lovestone, S. The GSK-3 hypothesis of Alzheimer's disease. *J. Neurochem.,* **2008**, *104*, 1433-1439.

[164] Eldar-Finkelman, H. Glycogen synthase kinase-3:an emerging therapeutic target. *Trends Mol. Med.,* **2002**, *8*, 126-132.

[165] Jope, R.S.; Yuskaitis, C.J.; Beurel, E. Glycogen synthase kinase-3 (GSK-3): inflammation, diseases,and therapeutics. *Neurochem. Res.,* **2007**, *32*, 577-595.

[166] Gould, T.D.; Zarate, C.A.; Manji, H.K. Glycogen synthase kinase-3: a target for novel bipolar disorder treatments. *J. Clin. Psychiatry,* **2004**, *65*, 10-21.

[167] Luo, J. Glycogen synthase kinase 3 beta (GSK-3 beta) in tumorigenesis and cancer chemotherapy. *Cancer Lett.,* **2009**, *273*, 194-200.

[168] Frame, S.; Cohen, P. GSK-3 takes centre stage more than 20 years after its discovery. *Biochem. J.,* **2001**, *359*, 1-16.

[169] MacAulay, K.; Doble, B.W.; Patel, S.; Hansotia, T.; Sinclair, E.M.; Drucker, D.J.; Nagy, A.; Woodgett, J.R. Glycogen synthase kinase 3alpha-specific regulation of murine hepatic glycogen metabolism. *Cell Metab.,* **2007**, *6*, 329-337.

[170] Hoeflich, K.P.; Luo, J.; Rubie, E.A.; Tsao, M.S.; Jin, O.; Woodgett, J.R. Requirement for glycogen synthase kinase-3beta in cell survival and NF-kappaB activation. *Nature,* **2000**, *406*, 86-90.

[171] Soutar, M.P.M.; Kim, W.-Y.; Williamson, R.; Peggie, M.; Hastie, C.J.; McLauchlan, H.; Snider, W.D.; Gordon-Weeks, P.R.; Sutherland, C. Evidence that glycogen synthase kinase-3 isoforms have distinct substrate preference in the brain. *J. Neurochem.,* **2010**, *115*, 974-983.

[172] Wood-Kaczmar, A.; Kraus, M.; Ishiguro, K.; Philpott, K.L.; Gordon-Weeks, P.R. An alternatively spliced form of glycogen synthase kinase-3beta is targeted to growing neurites and growth cones. *Mol. Cell, Neurosci.,* **2009**, *42*, 184-194.

[173] Hanger, D.P.; Hughes, K.; Woodgett, J.R.; Brion, J.P.; Anderton, B.H. Glycogen synthase kinase-3 induces Alzheimer's disease-like phosphorylation of tau: generation of paired helical filament epitopes and neuronal localisation of the kinase. *Neurosci. Lett.,* **1992**, *147*, 58-62.

[174] Saeki, K.; Machida, M.; Kinoshita, K.; Takasawa, R.; Tanuma, S. Glycogen synthase kinase-3beta2 has lower phosphorylation activity to tau than glycogen synthase kinase-3beta1. *Biol. Pharm. Bull.,* **2011**, *34*, 146-149.

[175] Hurtado, D.E.; Molina-Porcel, L.; Carroll, J.C.; MacDonald, C.; Aboagye, A.K.; Trojanowski, J.Q.; Lee, V.M.-Y. Selectively silencing GSK-3 isoforms reduces plaques and tangles in mouse models of Alzheimer's disease. *J. Neurosci.,* **2012**, *32*, 7392-7402.

[176] Hanger, D.P.; Byers, H.L.; Wray S.; Leung, K.-Y.; Saxton, M.J.; Seereeram, A.; Reynolds, C.H.; Ward, M.A.; Anderton, B.H. Novel phosphorylation sites in tau from Alzheimer brain support a role for casein kinase 1 in disease pathogenesis. *J. Biol. Chem.,* **2007**, *282*, 23645-23654.

[177] Hanger, D.P.; Noble, W. Functional implications of glycogen synthase kinase-3-mediated tau phosphorylation. *Int. J. Alzheim. Dis.,* **2011**, 352805.

[178] Lovestone, S.; Reynolds, C.H.; Latimer, D.; Davis, D.R.; Anderton, B.H.; Gallo, J.-M.; Hanger, D.; Mulot, S.; Marquardt, B.; Stabel, S.; Woodgett, J.R.; Miller, C.C.J. Alzheimer's disease-like phosphorylation of the microtubule-associated protein tau by glycogen synthase kinase-3 in transfected mammalian cells. *Curr. Biol.,* **1994**, *4*, 1077-1086.

[179] Wagner, U.; Utton, M.A.; Gallo, J.M.; Miller, C.C. Cellular phosphorylation of tau by GSK-3beta influences tau binding to microtubules and microtubule organization. *J. Cell Sci.*, **1996**, *109*, 1537-1543.

[180] Martin, L.; Page, G.; Terro, F. Tau phosphorylation and neuronal apoptosis induced by the blockade of PP2A preferentially involve GSK-3β. *Neurochem. Int.*, **2011**, *59*, 235-250.

[181] Martin, L.; Magnaudeix, A.; Wilson, C.M.; Yardin, C.; Terro, F. The new indirubin derivative inhibitors of glycogen synthase kinase-3, 6-BIDECO and 6-BIMYEO, prevent tau phosphorylation and apoptosis induced by the inhibition of protein phosphatase-2A by okadaic acid in cultured neurons. *J. Neurosci. Res.*, **2011**, *89*, 1802-1811.

[182] Aplin, A.E.; Gibb, G.M.; Jacobsen, J.S.; Gallo, J.M.; Anderson, B.H. *In vitro* phosphorylation of the cytoplasmic domain of the amyloid precursor protein by glycogen synthase kinase-3β. *J. Neurochem.*, **1996**, *67*, 699-707.

[183] Cole, A.; Frame, S.; Cohen, P. Further evidence that the tyrosine phosphorylation of glycogen synthase kinase-3 (GSK-3) in mammalian cells is an autophosphorylation event. *Biochem. J.,* **2004**, *377*, 249-255.

[184] Frame, S.; Cohen, P.; Biondi, R.M. A common phosphate binding site explains the unique substrate specificity of GSK-3 and its inactivation by phosphorylation. *Mol. Cell,* **2001**, *7*, 1321-1327.

[185] Cole, A.R. GSK-3 as a sensor determining cell fate in the brain. *Front. Molec. Neurosci.,* **2012**, *5*, 4.

[186] Medina, M.; Garrido, J.J.; Wandosell, F.G. Modulation of GSK-3 as a therapeutic strategy on tau pathologies. *Front. Molec. Neurosci.*, **2011**, *4*, 24.

[187] Gomes-Sintes, R.; Hernandez, F.; Lucas, J.J.; Avila, J. GSK-3 mouse models to study neuronal apoptosis and neurodegeneration. *Front. Molec. Neurosci.*, **2011**, *4*, 45.

[188] Noble, W.; Hanger, D.P.; Gallo, J.M. Transgenic mouse models of tauopathy in drug discovery. *CNS Neurol. Dis. Drug Targets*, **2010**, *9*, 403-428.

[189] Engel, T.; Hernandez, F.; Avila, J.; Lucas, J.J. Full reversal of Alzheimer's disease-like phenotype in a mouse model with conditional overexpression of glycogen synthase kinase-3. *J. Neurosci.,* **2006**, *26*, 5083-5090.

[190] Terwel, D.; Muyllaert, D.; Dewachter, I.; Borghgraef, P.; Croes, S.; Devijver, H.; Van Leuven, F. Amyloid activates GSK-3beta to aggravate neuronal tauopathy in bigenic mice. *Am. J. Pathol,* **2008**, *172*, 786-798.

[191] Davies, S.P.; Reddy, H.; Caivano, M.; Cohen, P. Specificity and mechanism of action of some commonly used protein kinase inhibitors. *Biochem. J.*, **2000**, *351*, 95-105.

[192] Bain, J.; McLauchlan, H.; Elliott, M.; Cohen, P. The specificities of protein kinase inhibitors: an update. *Biochem. J.*, **2003**, *371*, 199-204.

[193] Sutherland, C. What are the bona fide GSK-3 substrates? *Int. J. Alzheim. Dis.*, **2011**, 505607.

[194] Haltiwanger, R.S.; Holt, G.D.; Hart, G.W. Enzymatic addition of O-GlcNAc to nuclear and cytoplasmic proteins. Identification of a uridine diphospho-*N*-acetylglucosamine:peptide beta-*N*-acetylglucosaminyltransferase. *J. Biol. Chem.,* **1990**, *265*, 2563-2568.

[195] Comtesse, N.; Maldener, E.; Meese, E. Identification of a nuclear variant of MGEA5, a cytoplasmic hyaluronidase and a beta-*N*-acetylglucosaminidase. *Biochem. Biophys. Res. Commun.,* **2001**, *283*, 634-640.

[196] Slawson, C.; Zachara, N.E.; Vosseller, K.; Cheung, W.D.; Lane, M.D.; Hart, G.W. Perturbations in O-linked beta-N-acetylglucosamine protein modification cause severe defects in mitotic progression and cytokinesis. *J. Biol. Chem.*, **2005**, *280*, 32944-32956.

[197] O'Donnell, N.; Zachara, N.E.; Hart, G.W.; Marth, J.D. Ogt-dependent X-chromosome linked protein glycosylation is a requisite modification in somatic cell function and embryo viability. *Mol. Cell. Biol.*, **2004**, *24*, 1680-1690.

[198] Hanover, J.A.; Forsythe, M.E.; Hennessey, P.T.; Brodigan, T.M.; Love, L.C.; Ashwell, G.; Krause, M. A Caenorhabditis elegans model of insulin resistance: altered macronutrient storage and dauer formation in an OGT-1 knockout. *Proc. Natl. Acad. Sci. U.S.A.*, **2005**, *102*, 11266-11271.

[199] Guinez, C.; Morelle, W.; Michalski, J.C.; Lefebvre, T. O-GlcNAc glycosylation: a signal for the nuclear transport of cytosolic proteins? *Int. J. Biochem. Cell. Biol.*, **2005**, *37*, 765-774.

[200] Wells, L.; Vosseler, K.; Hart, G.W. A role for N-acetylglucosamine as a nutrient sensor and mediator of insulin resistance. *Cell. Mol. Life Sci.*, **2003**, *60*, 222-228.

[201] Lefebvre, T.; Dehennaut, V.; Guinez, C.; Olivier, S.; Drougat, L.; Mir, A.-M.; Mortuaire, M.; Vercoutter-Edouart, A.S.; Michalski, J.-C. Dysregulation of the nutrient/stress sensor O-GlcNAcylation is involved in the etiology of cardiovascular disorders, type-2 diabetes and Alzheimer's disease. *Biochim. Biophys. Acta*, **2010**, *1800*, 67-79.

[202] Clark, R.J.; McDonough, P.M.; Swanson, E.; Trost, S.U.; Suzuki, M.; Fukuda, M.; Dillmann, W.H. Diabetes and the accompanying hyperglycemia impairs cardiomyocyte calcium cycling through increased nuclear O-GlcNAcylation. *J. Biol. Chem.*, **2003**, *278*, 44230-44237.

[203] Jones, S.P.; Zachara, N.E.; Ngoh, G.A.; Hill, B.G.; Teshima, Y.; Bhatnagar, A.; Hart, G.W.; Marbán, E. Cardioprotection by N-acetylglucosamine linkage to cellular proteins. *Circulation*, **2008**, *117*, 1172-1182.

[204] Lefebvre, T.; Guinez, C.; Dehennaut, V.; Beseme-Dekeyser, O.; Morelle, W.; Michalski, J.C. Does O-GlcNAc play a role in neurodegenerative diseases? *Expert Rev. Proteomics*, **2005**, *2*, 265-275.

[205] Alexander, G.E.; Chen, K.; Pietrini, P.; Rapoport, S.I.; Reiman, E.M. Longitudinal PET evaluation of cerebral metabolic decline in dementia: a potential outcome measure in Alzheimer's disease treatment studies. *Am. J. Psychiatry*, **2002**, *159*, 738-45.

[206] Deng, Y.; Li, B.; Liu, Y.; Iqbal, K.; Grundke-Iqbal, I.; Gong, C.X. Dysregulation of insulin signaling, glucose transporters, O-GlcNAcylation, and phosphorylation of tau and neurofilaments in the brain: implication for Alzheimer's disease. *Am. J. Pathol.*, **2009**, *175*, 2089-98

[207] Liu, F.; Iqbal, K.; Grundke-Iqbal, I.; Hart, G.W.; Gong, C.X. O-GlcNAcylation regulates phosphorylation of tau: a mechanism involved in Alzheimer's disease. *Proc. Natl. Acad. Sci. USA*, **2004**, *101*, 10804-10809.

[208] Liu, F.; Shi, J.; Tanimukai, H.; Gu, J.; Gu, J.; Grundke-Iqbal, I.; Iqbal, K.; Gong, C.X. Reduced O-GlcNAcylation links lower brain glucose metabolism and tau pathology in Alzheimer's disease. *Brain*, **2009**, *132*, 1820-1832.

[209] Yu, Y.; Zhang, L.; Li, X.; Run, X.; Liang, Z.; Li, Y.; Liu, Y.; Lee, M.H.; Grundke-Iqbal, I.; Vocadlo, D.J.; Gong, C.-X. Differential effects of an O-GlcNAcase inhibitor on tau phosphorylation. *PLoS ONE*, **2012**, *7*, 35277.

[210]   Deng, Y.; Li, B.; Liu, Y.; Iqbal, K.; Grundke-Iqbal, I.; Gong, C.-X. Disregulation of insulin signaling, glucose transporters, O-GlcNAcylation, and phosphorylation of tau and neurofilaments in the brain. *Am. J. Pathol.*, **2009**, *175*, 2089-2098.

[211]   Shi, J.; Wu, S.; Dai, C.-l.; Li, Y.; Grundke-Iqbal, I.; Iqbal, K.; Liu, F.; Gong, C.-X. Diverse regulation of AKT and GSK-3β by O-GlcNAcylation in various types of cells. *FEBS Lett.*, **2012**, *586*, 2443-2450.

[212]   Lameira, J.; Nahum Alves, C.; Tunon, I.; Marti, S.; Moliner, V. Enzyme molecular mechanism as a starting point to design new inhibitors: a theoretical study of O-GlcNAcase. *J. Phys. Chem. B*, **2011**, *8*, 6764-6775.

[213]   Shen, D.L.; Gloster, T.M.; Yuzwa, S.A.; Vocadlo, D.J. Insights into O-linked N-Acetylglucosamine (O-GlcNAc) processing and dynamics through kinetic analysis of OGlcNAc transferase and O-GlcNACase activity on protein substrates. *J. Biol. Chem.*, **2012**, *287*, 15395-15408.

[214]   Kuntz, I.D. Structure-based drug design: past, present and future. *Solvay Pharm. Conf.*, **2009**, *9*, 23-27.

[215]   Gao, Y.; Wells, L.; Comer, F.I.; Parker, G.J.; Hart, G.W. Dynamic O-glycosylation of nuclear and cytosolic proteins: cloning and characterization of a neutral, cytosolic beta-N-acetylglucosaminidase from human brain. *J. Biol. Chem.*, **2001**, *276*, 9838-9845.

[216]   Cantarel, B.L.; Coutinho, P.M.; Rancurel, C.; Bernard, T.; Lombard, V.; Henrissat, B. The Carbohydrate-Active enZymes database (CAZy): an expert resource for glycogenomics. *Nucleic Acids Res.*, **2008**, *37*, D233-D238.

[217]   Vocadlo, D.J.; Davies, G.J. Mechanistic insights into glycosidase chemistry. *Curr. Opin. Chem. Biol.*, **2008**, *12*, 539-555.

[218]   Dennis, R.J.; Taylor, E.J.; Macauley, M.S.; Stubbs, K.A.; Turkenburg, J.P.; Hart, S.J.; Black, G.N.; Vocadlo, D.J.; Davies, G.J. Structure and mechanism of a bacterial beta-glucosaminidase having O-GlcNAcase activity. *Nat. Struct. Mol. Biol.*, **2006**, *13*, 365-371.

[219]   Ficko-Blean, E.; Boraston, A.B. Cloning, recombinant production, crystallization and preliminary X-ray diffraction studies of a family 84 glycoside hydrolase from Clostridium perfringens. *Acta Crystallogr. Sect. F, Struct. Biol. Cryst. Commun.*, **2005**, *61*, 834-836.

[220]   Macauley, M.S.; Vocadlo, D.J. Increasing O-GlcNAc levels: An overview of small-molecule inhibitors of O-GlcNAcase. *Biochim. Biophys. Acta*, **2010**, *1800*, 74-91.

[221]   McClain, D.A.; Lubas, W.A.; Cooksey, R.C.; Hazel, M.; Parker, G.J.; Love, D.C.; Hanover, J.A. Altered glycan-dependent signaling induces insulin resistance and hyperleptinemia. *Proc. Natl. Acad. Sci. USA*, **2002**, *99*, 10695-10699.

[222]   Vosseller, K.; Wells, L.; Lane, M.D.; Hart, G.W. Elevated nucleocytoplasmic glycosylation by O-GlcNAc results in insulin resistance associated with defects in Akt activation in 3T3-L1 adipocytes. *Proc. Natl. Acad. Sci. USA*, **2002**, *99*, 5313-5318.

[223]   Arias, E.B.; Kim, J.; Cartee, G.D. Prolonged incubation in PUGNAc results in increased protein O-Linked glycosylation and insulin resistance in rat skeletal muscle. *Diabetes*, **2004**, *53*, 921-930.

[224]   Macauley, M.S.; He, Y.; Gloster, T.M.; Stubbs, K.A.; Davies, G.J.; Vocadlo, D.J. Inhibition of O-GlcNAcase using a potent and cell-permeable inhibitor does not induce insulin resistance in 3T3-L1 adipocytes. *Chem. Biol.*, **2010**, *17*, 937-948.

[225]   Macauley, M.S.; Shan, X.; Yuzwa, S.A.; Gloster, T.M.; Vocadlo, D.J. (2010). Elevation of global O-GlcNAc in rodents using a selective O-GlcNAcase inhibitor does not cause insulin resistance or perturb glucohomeostasis. *Chem. Biol.*, **2010**, *17*, 949-958.

[226] Ref. 225, Supplementary Discussion.

[227] Lu, K.P.; Hanes, S.D.; Hunter, T. A human peptidylprolyl isomerase essential for regulation of mitosis. *Nature,* **1996**, *380*, 544-547.

[228] Whitmarsh, A.J.; Davis, R.J. Transcription factor AP-1 regulation by mitogenactivated protein kinase signal transduction pathways. *J. Mol. Med.,* **1996**, *74*, 589-607.

[229] Pearson, G.; Robinson, F.; Beers Gibson, T.; Xu, B.E.; Karandikar, M.; Berman, K.; Cobb, M.H. Mitogen-activated protein (MAP) kinase pathways: regulation and physiological functions. *Endocr. Rev.,* **2001**, *22*, 153C-183C.

[230] Lu, K.P. Pinning down cell signaling, cancer and Alzheimer's disease. *Trends Biochem. Sci.,* **2004**, *29*, 200-209.

[231] Balastik, M.; Lim, J.; Pastorino, L.; Lu, K.P. Pin1 in Alzheimer's disease: multiple substrates, one regulatory mechanisms? *Biophys. Biochem. Acta,* **2007**, *1772*, 422-429.

[232] Pelech, S.L. Networking with proline-directed protein kinases implicated in tau phosphorylation. *Neurobiol. Aging,* **1995**, *16*, 247-56.

[233] Rudrabhatla, P.; Albers, W.; Pant, H.C. Peptidyl-prolyl isomerase 1 regulates protein phosphatase 2A-mediated topographic phosphorylation of neurofilament proteins. *J. Neurosci.,* **2009**, *29*, 14869-14880.

[234] Weiwad, M.; Werner, A.; Rucknagel, P.; Schierhorn, A.; Kullertz, G.; Fischer, G. Catalysis of proline-directed protein phosphorylation by peptidyl-prolyl cis/trans isomerases. *J. Mol. Biol.,* **2004**, *339*, 635-646.

[235] Schelbert, B.; Rahfeld, J.-U. Phosphorylation-specific prolyl bond isomerization in eukaryotes: new insights in conformational regulation. *Recent Res. Dev. Med. Chem.,* **2001**, *1*, 1-15.

[236] Fischer, G. Peptidyl-prolyl *cis/trans* isomerases. *Angew. Chem., Int. Ed. Engl.,* **1994**, *33*, 1415-1436.

[237] Kipping, M.; Zarnt, T.; Kiessig, S.; Reimer, U.; Fischer, G.; Bayer, P. Increased backbone flexibility in threonine45-phosphorylated hirudin upon pH change. *Biochemistry,* **2001**, *40*, 7957-7963.

[238] Yaffe, M.B.; Schutkowski, M.; Shen, M.; Zhou, X.Z.; Stukenberg, P.T.; Rahfeld, J.U.; Xu, J.; Kuang, J.; Kirschner, M.W.; Fischer, G.; Cantley, L.C.; Lu, K.P. Sequence-specific and phosphorylation-dependent proline isomerization: a potential mitotic regulatory mechanism. *Science,* **1997**, *278*, 1957-1960.

[239] Wulf, G.; Finn, G.; Suizu, F.; Lu, K.P. Phosphorylation specific prolyl isomerization: is there an underlying theme? *Nat. Cell Biol.,* **2005**, *7*, 435-441.

[240] Lu, K.P.; Zhou, X.-Z. The prolyl isomerase PIN1: a pivotal new twist in phosphorylation signalling and disease. *Nat. Rev. Mol. Cell. Biol.,* **2007**, *8*, 904-916.

[241] Zhou, X.Z.; Kops, O.; Werner, A.; Lu, P.J.; Shen, M.; Stoller, G.; Kullertz, G.; Stark, M.; Fischer, G.; Lu, K.P. Pin1-dependent prolyl isomerization regulates dephosphorylation of Cdc25C and tau proteins. *Mol. Cell.,* **2000**, *6*, 873-883.

[242] Liu, W.; Youn, H.D.; Zhou, XZ.; Lu, K.P.; Liu, J.O. Binding and regulation of the transcription factor NFAT by the peptidyl prolyl cis-trans isomerase Pin1. *FEBS Lett.,* **2001**, *496*, 105-108.

[243] Rudrabhatla, P.; Zheng, Y.L.; Amin, N.D.; Kesavapany, S.; Albers, W.; Pant, H.C. Pin1-dependent prolyl isomerization modulates the stress-induced phosphorylation of high molecular weight neurofilament protein. *J. Biol. Chem.,* **2008**, *283*, 26737-26747.

[244] Ryo, A.; Nakamura, M.; Wulf, G.; Liou, Y.C.; Lu, K.P. Pin1 regulates turnover and subcellular localization of β-catenin by inhibiting its interaction with APC. *Nat. Cell Biol.*, **2001**, *3*, 793-801.

[245] Yang, J.Y.; Ding, Q.; Huo, L.; Xia, W.; Wei, Y.; Liao, Y.; Chang, C.J.; Yang, Y.; Lai, C.C.; Lee, D.F.; Yen, C.J.; Chen, Y.J.; Hsu, J.M.; Kuo, H.P.; Lin, C.Y.; Tsai, F.J.; Li, L.Y.; Tsai, C.H.; Hung, M.C. Down-regulation of myeloid cell leukemia-1 through inhibiting Erk/Pin 1 pathway by sorafenib facilitates chemosensitization in breast cancer. *Cancer Res.*, **2008**, *68*, 6109-17.

[246] Yeh, E.; Cunningham, M.; Arnold, H.; Chasse, D.; Monteith, T.; Ivaldi, G.; Hahn, W.C.; Stukenberg, P.T.; Shenolikar, S.; Uchida, T.; Counter, C.M.; Nevins, J.R.; Means, A.R.; Sears R. A signalling pathway controlling c-Myc degradation that impacts oncogenic transformation of human cells. *Nat. Cell. Biol.*, **2004**, *6*, 308-318.

[247] Lee, T.H.; Tun-Kyi, A.; Shi, R.; Lim, J.; Soohoo, C.; Finn, G.; Balastik, M.; Pastorino, L.; Wulf, G.; Zhou, X.Z.; Lu, K.P. Essential role of Pin1 in the regulation of TRF1 stability and telomere maintenance. *Nat. Cell Biol.*, **2009**, *11*, 97-105.

[248] Pulikkan, J.A.; Dengler, V.; Peer Zada, A.A.; Kawasaki, A.; Geletu, M.; Pasalic, Z.; Bohlander, S.K.; Ryo, A.; Tenen, D.G.; Behre, G. Elevated PIN1 expression by C/EBPα-p30 blocks C/EBPa-induced granulocytic differentiation through c-Jun in AML. *Leukemia*, **2010**, *24*, 914-923.

[249] Lu, K.P.; Finn, G.; Lee, T.H.; Nicholson, L.K. Prolyl cis-trans isomerization as a molecular timer. *Nat. Chem. Biol.*, **2007**, *3*, 619-629.

[250] Eckerdt, F.; Yuan, J.; Saxena, K.; Martin, B.; Kappel, S.; Lindenau, C.; Kramer, A.; Naumann, S.; Daum, S.; Fischer, G.; Dikic, I.; Kaufmann, M.; Strebhardt, K. Polo-like kinase 1-mediated phosphorylation stabilizes Pin1 by inhibiting its ubiquitination in human cells. *J. Biol. Chem.*, **2005**, *280*, 36575-36583

[251] Ryo, A.; Liou, Y.C.; Lu, K.P.; Wulf, G. Prolyl isomerase Pin1: a catalyst for oncogenesis and a potential therapeutic target in cancer. *J. Cell. Sci.*, **2003**, *116*, 773- 783.

[252] Lu, P.J.; Wulf, G.; Zhou, X.Z.; Davies, P.; Lu, K.P. The prolyl isomerase Pin1 restores the function of Alzheimer-associated phosphorylated tau protein. *Nature*, **1999**, *399*, 784-788.

[253] Liou, Y.C.; Sun, A.; Ryo, A.; Zhou, X.Z.; Yu, Z.X.; Huang, H.K.; Uchida, T.; Bronson, R.; Bing, G.; Li, X.; Hunter, T.; Lu, K.P. Role of the prolyl isomerase Pin1 in protecting against age-dependent neurodegeneration. *Nature*, **2003**, *424*, 556-561.

[254] Ryo, A.; Togo, T.; Nakai, T.; Hirai, A.; Nishi, M.; Yamaguchi, A.; Suzuki, K.; Hirayasu, Y.; Kobayashi, H.; Perrem, K.; Liou, Y.-C.; Aoki, I. Prolyl-isomerase Pin1 accumulates in lewy bodies of Parkinson disease and facilitates formation of alpha-synuclein inclusions. *J. Biol. Chem.*, **2006**, *281*, 4117-4125.

[255] Moretto Zita, M.; Marchionni, I.; Bottos, E.; Righi, M.; Del Sal, G.; Cherubini, E.; Zacchi, P. Post-phosphorylation prolyl isomerisation of gephyrin represents a mechanism to modulate glycine receptors function. *EMBO J.*, **2007**, *26*, 1761-1771.

[256] Li, Q.M.; Tep, C.; Yune, T.Y.; Zhou, X.Z.; Uchida, T.; Lu, K.P.; Yoon, S.O. Opposite regulation of oligodendrocyte apoptosis by JNK3 and Pin1 after spinal cord injury. *J. Neurosci.*, **2007**, *27*, 8395-8404.

[257] Pastorino, L.; Sun, A.; Lu, P.J.; Zhou, X.Z.; Balastik, M.; Finn, G.; Wulf, G.; Lim, J.; Li, S.H.; Li, X.; Xia, W.; Nicholson, L.K.; Lu, K.P. The prolyl isomerase Pin1 regulates amyloid precursor protein processing and amyloid-beta production. *Nature*, **2006**, *440*, 528-534.

[258] Pastorino, L.; Ma, S.L.; Balastik, M.; Huang, P.; Pandya, D.; Nicholson, L.; Lu, K.P. Alzheimer's disease-related loss of Pin1 function influences the intracellular localization and the processing of AβPP. *J. Alzheimer's Dis.*, **2012**, *30*, 277-297.

[259] Reynolds, C.H.; Betts, J.C.; Blackstock, W.P.; Nebreda, A.R.; Anderton, B.H. Phosphorylation sites on tau identified by nanoelectrospray mass spectrometry: differences *in vitro* between the mitogen-activated protein kinases ERK2, c-Jun N-terminal kinase and P38, and glycogen synthase kinase-3beta. *J. Neurochem.*, **2000**, *74*, 1587-1595.

[260] Landrieu, I.; Smet-Nocca, C.; Amniai, L.; Louis, J.V.; Wieruszeski, J.-M.; Goris, J.; Janssens, V.; Lippens, G. Molecular implication of PP2A and Pin1 in the Alzheimer's disease specific hyperphosphorylation of tau. *PLoS One*, **2011**, *6*, e21521.

[261] Hampel, H.; Buerger, K.; Kohnken, R.; Teipel, S.J.; Zinkowski, R.; Moeller, H.J.; Rapoport, S.I.; Davies, P. Tracking of Alzheimer's disease progression with cerebrospinal fluid tau protein phosphorylated at threonine 231, *Ann. Neurol.*, **2001**, *49*, 545- 546.

[262] Luna-Munoz. J.; Chavez-Macias, L.; Garcia-Sierra, F.; Mena, R. Earliest stages of tau conformational changes are related to the appearance of a sequence of specific phosphodependent tau epitopes in Alzheimer's disease. *J. Alzheimer's Dis.*, **2007**, *12*, 365-375.

[263] Kohnken, R.; Buerger, K.; Zinkowski, R.; Miller, C.; Kerkman, D.; DeBernardis, J.; Shen, J.; Moller, H.J.; Davies, P.; Hampel, H. Detection of tau phosphorylated at threonine 231 in cerebrospinal fluid of Alzheimer's disease patients. *Neurosci. Lett.*, **2000**, *287*, 187-190.

[264] Lim, J.; Lu, K.P. Pinning down phosphorylated tau and tauopathies. *Biophys. Biochem. Acta*, **2005**, *1739*, 311-322.

[265] Lee, T.H.; Pastorino, L.; Lu, K.P. Peptidyl-prolyl cis-trans isomerase Pin1 in ageing, cancer and Alzheimer disease. *Exp. Rev. Mol. Med.*, **2011**, *13*, e21.

[266] Segat, L.; Pontillo, A.; Annoni, C.; Trabattoni, D.; Vergani, C.; Clerici, M.; Arosio, B.; Crovella, S. Pin1 promoter polymorphisms are associated with Alzheimer's disease. *Neurobiol. Aging*, **2007**, *28*, 69-74.

[267] Ma, S.L.; Sang Tang, N.L.; Chi Tam, C.W.; Cheong Lui, V.W.; Wa Lam, L.C.; Kum Chiu, H.F.; Driver, J.A.; Pastorino, L.; Lu, K.P. A PIN1 polymorphism that prevents its suppression by AP4 associates with delayed onset of Alzheimer's disease. *Neurobiol. Aging*, **2012**, *33*, 804-803.

[268] Thorpe, J.R.; Morley, S.J.; Rulten, S.L. Utilizing the peptidyl-prolyl cis-trans isomerase pin1 as a probe of its phosphorylated target proteins. Examples of binding to nuclear proteins in a human kidney cell line and to tau in Alzheimer's diseased brain. *J. Histochem. Cytochem.*, **2001**, *49*, 97-108.

[269] Ramakrishnan, P.; Dickson, D.W.; Davies, P. Pin1 colocalization with phosphorylated tau in Alzheimer's disease and other tauopathies. *Neurobiol. Dis.*, **2003**, *14*, 251-264.

[270] Dakson, A.; Yokota, O.; Esiri, M.; Bigio, E.H.; Horan, M.; Pendleton, N.; Richardson, A.; Neary, D.; Snowden, J.S.; Robinson, A.; Davidson, Y.S.; Mann, D.M.A. Granular expression of prolyl-peptidyl isomerise PIN1 is a constant and specific feature of Alzheimer's disease pathology and is independent of tau, Aβ and TDP-43 pathology. *Acta Neuropathol.*, **2011**, *121*, 635-649.

[271] Wang, S.; Simon, B.P.; Bennett, D.A.; Schneider, J.A.; Malter, J.S.; Wang, D.-S. The significance of Pin1 in the development of Alzheimer's disease. *J. Alzheim. Dis.*, **2007**, *11*, 13-23.

[272] Holzer, M.; Gartner, U.; Stobe, A.; Hartig, W.; Gruschka, H.; Bruckner, M.K.; Arendt, T. Inverse association of Pin1 and tau accumulation in Alzheimer's disease hippocampus. *Acta Neuropathol.*, **2002**, *104*, 471- 481.

[273] Bulbarelli, A.; Lonati, E.; Cazzaniga, E.; Gregori, M.; Masserini, M. Pin1 affects tau phosphorylation in response to Ab oligomers. *Mol. Cell. Neurosci.*, **2009**, *42*, 75-80.

[274] Ma, S.L.; Pastorino, L.; Zhou, X.Z.; Lu, K.P. Prolyl isomerase Pin1 promotes amyloid precursor protein (APP) turnover by inhibiting glycogen synthase kinase-3β (GSK-3β) activity. *J. Biol. Chem.*, **2012**, *287*, 6969-6973.

[275] Joseph, J.D.; Daigle, S.N.; Means, A.R. PINA is essential for growth and positively influences NIMA function in *Aspergillus nidulans. J. Biol. Chem.,* **2004**, *279*, 32373-32384.

[276] Devasahayam, G.; Chaturvedi, V.; Hanes, S.D. The Ess1 prolyl isomerase is required for growth and morphogenetic switching in *Candida albicans. Genetics,* **2002**, *160*, 37-48.

[277] Liou, Y.C.; Ryo, R.; Huang, H.K.; Lu, P.J.; Bronson, R.; Fujimori, F.; Uchida, U.; Hunter, T.; Lu, K.P. Loss of Pin1 function in the mouse causes phenotypes resembling cyclin D1-null phenotypes, *Proc. Natl. Acad. Sci. U.S.A.*, **2002**, *99*, 1335-1340.

[278] Lewis, J.; McGowan, E.; Rockwood, J.; Melrose, H.; Nacharaju, P.; Van Slegtenhorst, M.; Gwinn-Hardy, K.; Murphy, M.P.; Baker, M.; Yu, X.; Duff, K.; Hardy, J.; Corral, A.; Lin, W.L.; Yen, S.H.; Dickson, D.W.; Davies, P.; Hutton, M. Neurofibrillary tangles, amyotrophy and progressive motor disturbance in mice expressing mutant (P301L) tau protein. *Nat. Genet.,* **2000**, *25,* 402-405.

[279] Oddo, S.; Caccamo, A.; Shepherd, J.D.; Murphy, M.P.; Golde, T.E.; Kayed, R.; Metherate, R.; Mattson, M.P.; Akbari, Y.; LaFerla, F.M. Triple transgenic model of Alzheimer's disease with plaques and tangles: intracellular Abeta and synaptic dysfunction. *Neuron,* **2003**, *39*, 409- 421.

[280] Yap, C.C.; Murate, M.; Kishigami, S.; Muto, Y.; Kishida, H.; Hashikawa, T.; Yano, R. Adaptor protein complex 4 (AP-4) is expressed in the central nervous system neurons and interacts with glutamate receptor delta2. *Mol. Cell. Neurosci.*, **2003**, *24*, 283-295.

[281] Lu, P.J.; Zhou, X.Z.; Liou, Y.C.; Noel, J.P.; Lu, K.P. Critical role of WW domain phosphorylation in regulating its phosphoserine-binding activity and the Pin1 function. J. *Biol. Chem.,* **2002**, *277*, 2381-2384.

[282] Ando, K.; Dourlen, P.; Sambo, A.-V.; Bretteville, A.; Belarbi, K.; Vingtdeux, V.; Eddarkaoui, S.; Drobecq, H.; Ghestem, A.; Begard, S.; Demey-Thomas, E.; Melnyk, P.; Smet, C.; Lippens, G.; Maurage, C.-A.; Caillet-Boudin, M.-L.; Verdier, Y.; Vinh, J.; Landrieu, I.; Galas, M.-C.; Blum, D.; Hamdane, M.; Sergeant, N.; Buee, L. Tau pathology modulates Pin1 post-translational modifications and may be relevant as biomarker. *Neurobiol. Aging*, **2012**, doi:10.1016/j.neurobiolaging.2012.08.004.

[283] Eckerdt, F.; Yuan,J.; Saxena, K.; Martin, B.; Kappel, S.; Lindenau, C.; Kramer, A.; Naumann, S.; Daum, S.; Fischer, G.; Dikic, I.; Kaufmann, M.; Strebhardt, K. Polo-like kinase 1-mediated phosphorylation stabilizes Pin1 by inhibiting its ubiquitination in human cells. *J. Biol. Chem.,* **2005***, 280*, 36575-36583.

[284] Bialik, S.; Kimchi, A. The death-associated protein kinases: structure, function, and beyond. *Annu. Rev. Biochem.*, **2006**,*75*, 189-210.

[285] Lee, T.H.; Chen, C.-H.; Suizu, F.; Huang, P.; Schiene-Fischer, C.; Daum, S.; Zhang, Y.J.; Goate, A.; Chen, R.-H.; Zhou, X.Z.; Lu, K.P. Death-associated protein kinase 1

phosphorylates Pin1 and inhibits its prolyl isomerase activity and cellular function. *Mol. Cell*, **2011**, *42*, 147-159.

[286] Jin, Y.; Gallagher, P.J. Anti-sense depletion of death associated protein kinase promotes apoptosis. *J. Biol. Chem.*, **2003**, *278*, 51587-51593.

[287] Inbal, B.; Bialik, S.; Sabanay, I.; Shani, G.; Kimchi, A. DAP kinase and DRP-1 mediate membrane blebbing and the formation of autophagic vesicles during programmed cell death. *J. Cell Biol.*, **2002**, *157*, 455-468.

[288] Eisenberg-Lerner, A.; Kimchi, A. DAP kinase regulates JNK signaling by binding and activating protein kinase D under oxidative stress. *Cell Death Differ.*, **2007**, *14*, 1908-1915.

[289] Velentza, A.V.; Wainwright, M.S.; Zasadzki, M.; Mirzoeva, S.; Schumacher, A.M.; Haiech, J.; Focia, P.J.; Egli, M.; Watterson, D.M. An aminopyridazine-based inhibitor of a pro-apoptotic protein kinase attenuates hypoxia-ischemia induced acute brain injury. *Bioorg. Med. Chem. Lett.*, **2003**, *13*, 3465-3470.

[290] Shamloo, M.; Soriano, L.; Wieloch, T.; Nikolich, K.; Urfer, R.; Oksenberg, D. Death-associated protein kinase is activated by dephosphorylation in response to cerebral ischemia. *J. Biol. Chem.*, **2005**, *280*, 42290-42299.

[291] Wu, P.R.; Tsai, P.I.; Chen, G.C.; Chou, H.J.; Huang, Y.P.; Chen, Y.H.; Lin, M.Y.; Kimchi, A.; Chien, C.T.; Chen, R.H. DAPK activates MARK1/2 to regulate microtubule assembly, neuronal differentiation, and tau toxicity. *Cell Death Differ.*, **2011**, *18*, 1507-1520.

[292] Theuerkorn, M.; Fischer, G.; Schiene-Fischer, C. Prolyl cis/trans isomerase signalling pathways in cancer. *Curr. Opin. Pharmacol.*, **2011**, *11*, 281-287.

[293] Roe, C.M.; Behrens, M.I.; Xiong, C.; Miller, J.P.; Morris, J.C. Alzheimer disease and cancer. *Neurology*, **2005**, *64*, 895-898.

[294] Lim, J.; Balastik, M.; Lee, T.H.; Nakamura, K.; Liou, Y.-C.; Sun, A.; Finn, G.; Pastorino, L.; Lee, V.M.-Y.; Lu, K.P. Pin1 has opposite effects on wildtype and P301L tau stability and tauopathy. *J. Clin. Invest.*, **2008**, *118*, 1877-1889.

[295] Yotsumoto, K.; Saito, T.; Asada, A.; Oikawa, T.; Kimura, T.; Uchida, C.; Ishiguro, K.; Uchida, T.; Hasegawa, M.; Hisanaga, S.-I. Effect of Pin1 or microtubule binding on dephosphorylation of FTDP-17 mutant tau. *J. Biol. Chem.*, **2009**, *284*, 16840-16847.

[296] Phukan, S.; Babu, V.S.; Kannoji, A.; Hariharan, R.; Balaji, V.N. GSK-3beta: role in therapeutic landscape and development of modulators. *Br. J. Pharmacol.*, **2010**, *160*, 1-19.

[297] Cohen, P.; Goedert, M. GSK3 inhibitors: Development and therapeutic potential. *Nat. Rev. Drug Discov.*, **2004**, *3*, 479-487.

[298] Eldar-Finkelman, H.; Martinez, A. GSK-3 inhibitors: preclinical and clinical focus on CNS. *Front. Molec. Neurosci.*, **2011**, *4*, 32.

[299] Price, L.H.; Heninger, G.R. Lithium in the treatment of mood disorders. *N. Engl. J. Med.*, **1994**, *331*, 591-598.

[300] Stambolic, V.; Ruel, L.; Woodgett, J.R. Lithium inhibits glycogen synthase kinase-3 activity and mimics wingless signaling in intact cells. *Curr.Biol.*, **1996**, *6*, 1664-1668.

[301] Klein, P.S.; Melton, D.A. A molecular mechanism for the effect of lithium on development. *Proc. Natl. Acad. Sci. U.S.A*, **1996**, *93*, 8455-8459.

[302] Zhang, F.; Phiel, C.J.; Spece, L.; Gurvich, N.; Klein, P.S. Inhibitory phosphorylation of glycogen synthase kinase-3 (GSK-3) in response to lithium. Evidence for autoregulation of GSK-3. *J. Biol. Chem.*, **2003**, *278*, 33067-33077.

[303] Perez, M.; Hernandez, F.; Lim, F.; Diaz-Nido, J.; Avila, J. Chronic lithium treatment decreases mutant tau protein aggregation in a transgenic mouse model. *J. Alzheim. Dis.,* **2003**, *5*, 301-308.

[304] Caccamo, A.; Oddo, S.; Tran, L.X.; LaFerla, F.M. Lithium reduces tau phosphorylation but not A beta or working memory deficits in a transgenic model with both plaques and tangles. *Am. J. Pathol,,* **2007**, *170*, 1669-1675.

[305] Havens, W.W.II; Cole, J. Successful treatment of dementia with lithium. *J. Clin. Psychopharm.,* **1982**, *2*, 71-72.

[306] Terao, T.; Nakano, H.; Inoue, Y.; Okamoto, T.; Nakamura, J.; Iwata, N. Lithium and dementia: a preliminary study. *Prog. Neuropsychopharmacol. Biol. Psychiatry,* **2006**, *30*, 1125-1128.

[307] Bedlack, R.S.; Maragakis, N.; Heiman-Patterson, T. Lithium may slow progression of amyotrophic lateral sclerosis, but further study is needed. *Proc. Natl. Acad. Sci. U.S.A.,* **2008**, *105*, E17; author reply E18.

[308] Hampel, H.; Ewers, M.; Burger, K.; Annas, P.; Mortberg, A.; Bogstedt, A.; Frolich, L.; Schroder, J.; Schonknecht, P.; Riepe, M.W.; Kraft, I.; Gasser, T.; Leyhe, T.; Möller, H.J.; Kurz, A.; Basun, H. Lithium trial in Alzheimer's disease: a randomized, single-blind, placebo-controlled, multicenter 10- week study. *J. Clin. Psychiatry,* **2009**, *70*, 922-931.

[309] Kremer, A.; Louis, J.V.; Jaworski, T.; Van Leuven, F. GSK-3 and Alzheimer's disease: facts and fiction… *Front. Molec. Neurosci.,* **2011**, *4*, 17.

[310] Prasad, A.S. Biochemistry of Zinc. *Plenum Press*, New York, **1993**, 291 pages.

[311] Ilouz, R.; Kaidanovich, O.; Gurwitz, D.; Eldar-Finkelman, H. Inhibition of glycogen synthase kinase-3beta by bivalent zinc ions: insight into the insulin-mimetic action of zinc. *Biochem. Biophys. Res. Commun.,* **2002**, *295*, 102-106.

[312] Crouch, P.J.; Savva, M.S.; Hung, L.W.; Donnelly, P.S.; Mot, A.I.; Parker, S.J.; Greenough, M.A.; Volitakis, I.; Adlard, P.A.; Cherny, R.A.; Masters, C.L.; Bush, A.I.; Barnham, K.I.; White, A.R. The Alzheimer's therapeutic PBT2 promotes amyloid-b degradation and GSK-3 phosphorylation *via* a metal chaperone activity. *J. Neurochem.,* **2011**, *119*, 220-230.

[313] http://alzheimersweekly.com/content/pbt2-dementia-trial-2012-editors-choice

[314] Leost, M.; Schultz, C.; Link, A.; Wu, Y.Z.; Biernat, J.; Mandelkow, E.M.; Bibb, J.A.; Snyder, G.L.; Greengard, P.; Zaharevitz, D.W.; Gussio, R.; Senderowicz, A.M.; Sausville, E.A.; Kunick, C.; Meijer, L. Paullones are potent inhibitors of glycogen synthase kinase-3beta and cyclin-dependent kinase 5/p25. *Eur. J. Biochem.,* **2000**, *267*, 5983-5994.

[315] Stukenbroch, H.; Mussmann, R.; Geese, M.; Ferandin, Y.; Lozach, O.; Lemcke, T.; Kegel, S.; Lomow, A.; Burk, U.; Dohrmann, C.; Meijer, L.; Austen, M.; Kunick, C.9-cyano- 1-azapaullone (cazpaullone), a glycogen synthase kinase-3 (GSK-3) inhibitor activating pancreatic beta cell protection and replication. *J. Med. Chem,* **2008**, *51*, 2196-2207.

[316] Selenica, M.L.; Jensen, H.S.; Larsen, A.K.; Pedersen, M.L.; Helboe, L.; Leist, M.; Lotharius, J. Efficacy of small-molecule glycogen synthase kinase-3 inhibitors in the postnatal rat model of tauhyperphosphorylation. *Br. J. Pharmacol.,* **2007**, *152*, 959-979.

[317] Bone, H.K.; Damiano, T.; Bartlett, S.; Perry, A.; Letchford, J.; Ripoll, Y.S.; Nelson, A.S.; Welham, M.J. Involvement of GSK-3 in regulation of murine embryonic stem cell self-renewal revealed by a series of bisindolylmaleimides. *Chem. Biol.,* **2009**, *16*, 15-27.

[318] Coghlan, M.P.; Culbert, A.A.; Cross, D.A.; Corcoran, S.L.; Yates, J.W.; Pearce, N.J.; Rausch, O.L.; Murphy, G.J.; Carter, P.S.; Roxbee Cox, L.; Mills, D.; Brown, M.J.; Haigh, D.; Ward, R.W.; Smith, D.G.; Murray, K.J.; Reith, A.D.; Holder, J.C. Selective small

molecule inhibitors of glycogen synthase kinase-3 modulate glycogen metabolism and gene transcription. *Chem. Biol.,* **2000**, *7*, 793-803.

[319] Valerio, A.; Bertolotti, P.; Delbarba, A.; Perego, C.; Dossena, M.; Ragni, M.; Spano, P.; Carruba, M.O.; De Simoni, M.G.; Nisoli, E. Glycogen synthase kinase-3 inhibition reduces ischemic cerebral damage, restores impaired mitochondrial biogenesis and prevents ROS production. *J. Neurochem.,* **2011**, *116*, 1148-1159.

[320] Dill, J.; Wang, H.; Zhou, F.; Li, S. Inactivation of glycogen synthase kinase 3 promotes axonal growth and recovery in the CNS. *J. Neurosci.,* **2008**, *28*, 8914-8928.

[321] Mao, Y.; Ge, X.; Frank, C.L.; Madison, J.M.; Koehler, A.N.; Doud, M.K.; Tassa, C.; Berry, E.M.; Soda, T.; Singh, K.K.; Biechele, T.; Petryshen, T.L.; Moon, R.T.; Haggarty, S.J.; Tsai, L.H. Disrupted in schizophrenia 1 regulates neuronal progenitor proliferation *via* modulation of GSK-3 beta/beta- catenin signaling. *Cell,* **2009**, *136*, 1017-1031.

[322] Hu, S.; Begum, A.N.; Jones, M.R.; Oh, M.S.; Beech, W.K.; Beech, B. H.; Yang, F.; Chen, P.; Ubeda, O.J.; Kim, P.C.; Davies, P.; Ma, Q.; Cole, G.M.; Frautschy, S.A. GSK-3 inhibitors show benefits in an Alzheimer's disease (AD) model of neurodegeneration but adverse effects in control animals. *Neurobiol. Dis.,* **2009**, *33*, 193-206.

[323] Engler, T.A.; Henry, J.R.; Malhotra, S.; Cunningham, B.; Furness, K.; Brozinick, J.; Burkholder, T.P.; Clay, M.P.; Clayton, J.; Diefenbacher, C.; Hawkins, E.; Iversen, P.W.; Li, Y.; Lindstrom, T,D.; Marquart, A.L.; McLean, J.; Mendel, D.; Misener, E.; Briere, D.; O'Toole, J.C.; Porter, W.J.; Queener, S.; Reel, J.K.; Owens, R.A.; Brier, R.A.; Eessalu, T.E.; Wagner, J.R.; Campbell, R.M.; Vaughn, R. Substituted 3-imidazo[1,2-*a*]pyridin-3-yl-4-(1,2,3,4-tetrahydro-[1,4]diazepino-[6,7,1-*hi*]indol-7-yl)pyrrole-2,5-diones as highly selective and potent inhibitors of glycogen synthase kinase-3. *J. Med. Chem.,* **2004**, *47*, 3934-3937.

[324] Engler, T.A.; Malhotra, S.; Burkholder, T.P.; Henry, J.R.; Mendel, D.; Porter, W.J.; Furness, K.; Diefenbacher, C.; Marquart, A.L.; Reel, J.K.; Li, Y.; Clayton, J.; Cunningham, B.; McLean, J.; O'Toole, J.C.; Brozinick, J.; Hawkins, E.; Misener, E.; Briere, D.; Brier, R.A.; Wagner, J.R.; Campbell, R.M.; Anderson, B.D.; Vaughn, R.; Bennett, D.B.; Meier, T.I Cook, J.I. The development of potent and selective bisarylmaleimide GSK-3 inhibitors. *Bioorg. Med. Chem. Lett.,* **2005**, *15*, 899-903.

[325] Kozikowski, A.P.; Gaisina, I.N.; Hongbin, Y.; Petukhov, P.A.; Blond, S.Y.; Fedolak, A.; Caldarone, B.; McGonigle, P. Structure-based design leads to the identification of lithium mimetics that block mania-like effects in rodents. Possible new GSK-3β therapies for bipolar disorders. *J. Am. Chem. Soc.,* **2007**, *129*, 8328–8332.

[326] Chen, P.C.; Gaisina, I.N.; El-Khodor, B.F.; Ramboz, S.; Makhortova, N.R.; Rubin, L.L.; Kozikowski, A.P. Identification of a maleimide-based glycogen synthase kinase-3 (GSK-3) inhibitor, BIP-135, that prolongs the median survival time of Δ7 SMA KO mouse model of spinal muscular atrophy. *ACS Chem. Neurosci.,* **2012**, *3*, 5-11.

[327] Hoessel, R.; Leclerc, S.; Endicott, J.; Noble, M.; Lawrie, A.; Tunnah, P.; Leost, M.; Damiens, E.; Marie, D.; Marko, D.; Niederberger, E.; Tang, W.; Eisenbrand, G.; Meijer, L. Indirubin, the active constituent of a Chinese antileukaemia medicine, inhibits cyclin-dependent kinases. *Nat. Cell Biol.,* **1999**, *1*, 60-67.

[328] Meijer, L.; Skaltsounis, A.L.; Magiatis, P.; Polychronopoulos, P.; Knockaert, M.; Leost, M.; Ryan, X.P.; Vonica, C.A.; Brivanlou, A.; Dajani, R.; Crovace, C.; Tarricone, C.; Musacchio, A.; Roe, S.M.; Pearl, L.; Greengard, P. GSK-3-selective inhibitors derived from Tyrian purple indirubins. *Chem. Biol.,* **2003**, *10*, 1255-1266.

[329] Kim, W.Y.; Zhou, F.Q.; Zhou, J.; Yokota, Y.; Wang, Y.M.; Yoshimura, T.; Kaibuchi, K.; Woodgett, J.R.; Anton, E.S.; Snider, W.D. Essential roles for GSK- 3s andGSK-3-primed substrates in neurotrophin-induced and hippocampal axon growth. *Neuron,* **2006**, *52*, 981-996.

[330] Martin, L.; Magnaudeix, A.; Esclaire, F.; Yardin, C.; Terro,F. Inhibition of glycogen synthase kinase- 3beta down regulates total tau proteins in cultured neurons and its reversal by the blockade of protein phosphatase-2A. *Brain Res.,* **2009**, *1252*, 66-75.

[331] Ding, Y.; Qiao, A.; Fan, G.H. Indirubin-3'-monoxime rescues spatial memory deficits and attenuates beta-amyloid-associated neuropathology in a mouse model of Alzheimer's disease. *Neurobiol. Dis.,* **2010**, *39*, 156-168.

[332] Vougogiannopoulou, K.; Ferandin, Y.; Bettayeb, K.; Myrianthopoulos, V.; Lozach, O.; Fan, Y.; Johnson, C.H.; Magiatis, P.; Skaltsounis, A.L.; Mikros, E.; Meijer, L. Soluble 3',6-substituted indirubins with enhanced selectivity toward glycogen synthase kinase-3 alter circadian period. *J. Med.Chem.,* **2008**, *51*, 6421-6431.

[333] Martin, L.; Magnaudeix, A.; Wilson, C.M.; Yardin, C.; Terro, F. The new indirubin derivative inhibitors of glycogen synthase kinase-3, 6-BIDECO and 6-BIMYEO, prevent tau phosphorylation and apoptosis induced by the inhibition of protein phosphatase-2A by okadaic acid in cultured neurons. *J. Neurosci. Res.,* **2011**, *89*, 1802-1811.

[334] Martin, L.; Page, G.; Terro, F. Tau phosphorylation and neuronal apoptosis induced by the blockade of PP2A preferentially involve GSK-3β. *Neurochem. Int.,* **2011**, *59*, 235-250.

[335] Bhat, R.; Xue, Y.; Berg, S.; Hellberg, S.; Ormo, M.; Nilsson, Y.; Radesater, A.C.; Jerning, E.; Markgren, P.O.; Borgegard, T.; Nylöf, M.; Giménez-Cassina, A.; Hernández, F.; Lucas, J.J.; Díaz-Nido, J.; Avila, J. Structural insights and biological effects of glycogen synthase kinase 3-specific inhibitor AR-A014418. *J. Biol. Chem.,* **2003**, *278*, 45937-45945.

[336] Nguyen, T.B.; Lucero, G.R.; Chana, G.; Hult, B.J.; Tatro, E.T.; Masliah, E.; Grant, I.; Achim, C.L.; Everall, I.P. Glycogen synthase kinase-3β (GSK-3β) inhibitors AR-A014418 and B6B3O prevent human immunodeficiency virus-mediated neurotoxicity in primary human neurons. *J. NeuroVirol.,* **2009**, *15*, 434-438.

[337] Shi, S.H.; Cheng, T.; Jan, L.Y.; Jan, Y.N. APC and GSK-3 beta are involved in mPar3 targeting to the nascent axon and establishment of neuronal polarity. *Curr. Biol.,* **2004**, *14*, 2025-2032.

[338] Gould, T.D.; Einat, H.; Bhat, R.; Manji, H.K. AR-A014418, a selective GSK-3 inhibitor, produces antidepressant-like effects in the forced swim test. *Int. J. Neuropsychopharmacol.,* **2004**, *7*, 387-390.

[339] Kalinichev, M.; Dawson, L.A. Evidence for anti manic efficacy of glycogen synthase kinase- 3 (GSK-3) inhibitors in a strain-specific model of acute mania. *Int. J. Neuropsychopharmacol.,* **2011**, *14*, 1-17.

[340] Martins, D.F.; Rosa, A.O.; Gadotti, V.M.; Mazzardo-Martins, L.; Nascimento, F.P.; Egea, J.; Lopez, M.G.; Santos, A.R.S. The antinociceptive effects of AR-A014418, a selective inhibitor of glycogen synthase kinase-3 beta, in mice. *J. Pain,* **2011**, *12*, 315-322.

[341] Koh, S.H.; Kim, Y.; Kim, H.Y.; Hwang, S.; Lee, C.H.; Kim, S.H. Inhibition of glycogen synthase kinase-3 suppresses the onset of symptoms and disease progression of G93A-SOD1 mouse model of ALS. *Exp. Neurol.,* **2007**, *205*, 336-346.

[342] Noble, W.; Planel, E.; Zehr, C.; Olm, V.; Meyerson, J.; Suleman, F.; Gaynor, K.; Wang, L.; La Francois, J.; Feinstein, B.; Burns, M.; Krishnamurthy, P.; Wen, Y.; Bhat, R.; Lewis, J.; Dickson, D.; Duff, K. Inhibition of glycogen synthase kinase-3 by lithium correlates with

reduced tauopathy and degeneration *in vivo*. *Proc. Natl. Acad. Sci. U.S.A.*, **2005**, *102*, 6990-6995.

[343] Ring, D.B.; Johnson, K.W.; Henriksen, E.J.; Nuss, J.M.; Goff, D.; Kinnick, T.R.; Ma, S.T.; Reeder, J.W.; Samuels, I.; Slabiak, T.; Wagman, A.S.; Hammond, M.E.; Harrison, S.D. Selective glycogen synthase kinase 3 inhibitors potentiate insulin activation of glucose transport and utilization *in vitro* and *in vivo*. *Diabetes*, **2003**, *52*, 588-595.

[344] Peineau, S.; Nicolas, C.S.; Bortolotto, Z.A.; Bhat, R.V.; Ryves, W.J.; Harwood, A.J.; Dournaud, P.; Fitzjohn, S.M.; Collingridge, G.L. A systematic investigation of the protein kinases involved in NMDA receptor-dependent LTD: evidence for a role of GSK-3 but not other serine/threonine kinases. *Mol.Brain*, **2009**, *2*, 22.

[345] Alabed, Y.Z.; Pool, M.; OngTone, S.; Sutherland, C.; Fournier, A.E. GSK-3 beta regulates myelin-dependent axon outgrowth inhibition through CRMP4. *J. Neurosci.*, **2011**, *30*, 5635-5643.

[346] Li, W.; Sun, W.; Zhang, Y.; Wei, W.; Ambasudhan, R.; Xia, P.; Talantova, M.; Lin, T.; Kim, J.; Wang, X.; Kim, W.R.; Lipton, S.A.; Zhang, K.; Ding, S. Rapid induction and long-term self-renewal of primitive neural precursors from human embryonic stem cells by small molecule inhibitors. *Proc. Natl. Acad. Sci. U.S.A.*, **2011**, *108*, 8299-8304.

[347] Makhortova, N.R., Hayhurst, M.; Cerqueira, A.; Sinor-Anderson, A.D.; Zhao, W.N.; Heiser, P.W.; Arvanites, A.C.; Davidow, L.S.; Waldon, Z.O.; Steen, J.A.; Lam, K.; Ngo, H.D.; Rubin, L.L. A screen for regulators of survival of motor neuron protein levels. *Nat. Chem. Biol.*, **2011**, *7*, 544-552.

[348] Kelly, S.; Zhao, H.; Hua Sun, G.; Cheng, D.; Qiao, Y.; Luo, J.; Martin, K.; Steinberg, G.K.; Harrison, S.D.; Yenari, M.A. Glycogen synthase kinase 3 beta inhibitor Chir025 reduces neuronal death resulting from oxygen-glucose deprivation, glutamate excitotoxicity, and cerebral ischemia. *Exp. Neurol.*, **2004**, *188*, 378-386.

[349] Berg, S.; Bergh, M.; Hellberg, S.; Hoedgkin, K.; Lo-Alfredsson, Y.; Soederman, P.; von Berg, S.; Weigelt, T.; Ormoe, M.; Xue, Y.; Tucker, J.; Neelissen, J.; Jerning, E.; Nilsson, Y.; Bhat, R.; Discovery of novel potent and highly selective glycogen synthase kinase-3β (GSK-3β) inhibitors for Alzheimer's disease: Design, synthesis, and characterization of pyrazines. *J. Med. Chem.*, **2012**, *55*, 9107-9119.

[350] Kitagawa, I.; Kobayashi, M.; Kitanaka, K.; Kido, M.; Kyogoku, Y. Marine natural products. XII. On the chemical constituents of the Okinawan marine sponge *Hymeniacidon aldis*. *Chem. Pharm. Bull.*, **1983**, *31*, 2321-2328.

[351] Meijer, L.; Thunnissen, A.M.; White, A.W.; Garnier, M.; Nikolic, M.; Tsai, L.H.; Walter, J.; Cleverley, K.E.; Salinas, P.C.; Wu, Y.Z.; Biernat, J.; Mandelkow, E.M.; Kim, S.H.; Pettit, G.R. Inhibition of cyclin- dependent kinases, GSK-3 beta and CK1 by hymenialdisine, a marine sponge constituent. *Chem. Biol.*, **2000**, *7*, 51-63.

[352] Lee, S.C.; Kim, H.T.; Park, C.-H.; Lee, D.Y.; Chang, H.-J.; Park, S.; Cho, J.M.; Ro, S.; Suh, Y.-G. Design, synthesis and biological evaluation of novel imidazopyridines as potential antidiabetic GSK-3β inhibitors. *Bioorg. Med. Chem. Lett.*, **2012**, *22*, 4221-4224.

[353] Cociorva,O.M.; Li, B.; Nomanbhoy, T.; Li, Q.; Nakamura, A.; Nakamura, K.; Nomura, M.; Okada, K.; Seto, S.; Yumoto, K.; Liyanage, M.; Zhang, M.C.; Aban, A.; Leen, B.; Szardenings, A.K.; Rosenblum, J.S.; Kozarich, J.W.; Kohno, Y.; Shreder, K.R. Synthesis and structure-activity relationship of 4-quinolone-3-carboxylic acid based inhibitors of glycogen synthase kinase-3β. *Bioorg. Med. Chem. Lett.*, **2011**, *21*, 5948-5951.

[354] Li, B.; Cociorva,O.M.; Nomanbhoy, T.; Li, Q.; Nakamura, A.; Nakamura, K.; Nomura, M.; Okada, K.; Yumoto, K.; Liyanage, M.; Zhang, M.C.; Aban, A.; Szardenings, A.K.; Kozarich, J.W.; Kohno, Y.; Shreder, K.R. 6-Position optimization of tricyclic 4-quinolone-based inhibitors of glycogen synthase kinase-3β: Discovery of nitrile derivatives with picomolar potency. *Bioorg. Med. Chem. Lett.*, **2012**, *22*, 5948-5951.

[355] Seto, S.; Yumoto, K.; Okada, K.; Asahina, Y.; Iwane, A.; Iwago, M.; Terasawa, R.; Shreder, K.R.; Murakami, K.; Kohno, Y. Quinolone derivatives containing strained spirocycle as orally active glycogen synthase kinase 3β (GSK-3β) inhibitors for type 2 diabetics. *Bioorg. Med. Chem.*, **2012**, *20*, 1188-1200.

[356] Lesuisse, D.; Dutruc-Rosset, G.; Tiraboschi, G.; Dreyer, M.K.; Maignan, S.; Chevalier, A.; Halley, F.; Bertrand, P.; Burgevin, M.-C.; Quarteronet, D.; Rooney, T. Rational design of potent GSK-3β inhibitors with selectivity for Cdk1 and Cdk2. *Bioorg. Med. Chem. Lett.*, **2010**, *20*, 1985-1989.

[357] Lesuisse, D.; Tiraboschi, G.; Krick, A.; Abecassis, P.-Y.; Dutruc-Rosset, G.; Babin, D.; Halley, F.; Chatreau, F.; Lachaud, S.; Chevalier, A.; Quarteronet, D.; Burgevin, M.C.; Amara, C.; Bertrand, P.; Rooney, T. Design of potent and selective GSK-3β inhibitors with acceptable safety profile and pharmacokinetics. *Bioorg. Med. Chem. Lett.*, **2010**, *20*, 2344-2348.

[358] Dajani, R.; Fraser, E.; Roe, S.M.; Young, N.; Good, V.; Dale, T.C.; Pearl, L.H. Crystal structure of glycogen synthase kinase 3 beta: structural basis for phosphate-primed substrate specificity and autoinhibition. *Cell*, **2001**, *105*, 721-732.

[359] http://www.pharmatopics.com/2011/12/astrazeneca-grants-uk-academia-free-access-to-its-patented-compounds/

[360] Martinez, A.; Gil, C.; Perez, D.I. Glycogen synthase kinase 3 inhibitors in the next horizon for Alzheimer's disease treatment. *Int. J. Alzheimer's Dis.*, **2011**, 280502.

[361] Sakai, R.; Higa, T.; Jefford, C.W.; Bernardinelli, G. Manzamine A, a novel antitumor alkaloid from a sponge. *J. Am. Chem. Soc.*, **1986**, *108*, 6404-6405.

[362] Rao, K.V.; Donia, M.S.; Peng, J.; Garcia-Palomero, E.; Alonso, D.; Martinez, A.; Medina, M.; Franzblau, S. G.; Tekwani, B.L.; Khan, S.I.; Wahyuono, S.; Willett, K. L.; Hamann, M.T. Manzamine B and E and ircinal A related alkaloids from an Indonesian Acanthostrongylophora sponge and their activity against infectious, tropical parasitic, and Alzheimer's diseases. *J. Nat. Prod.*, **2006**, *69*, 1034-1040.

[363] Hamann, M.; Alonso, D.; Martin-Aparicio, E.; Fuertes, A.; Perez-Puerto, M.J.; Castro, A.; Morales, S.; Navarro, M.L.; Del Monte-Millan, M.; Medina, M.; Pennaka, H.; Balaiah, A.; Peng, J.; Cook, J.; Wahyuono, S.; Martinez, A. Glycogen synthase kinase-3 (GSK-3) inhibitory activity and structure-activity relationship (SAR) studies of the manzamine alkaloids. Potential for Alzheimer's disease. *J. Nat. Prod.*, **2007**, *70*, 1397-1405.

[364] Peng, J.; Kudrimoti, S.; Prasanna, S.; Odde, S.; Doerksen, R.J.; Pennaka, H.K.; Choo, Y.M.; Rao, K.V.; Tekwani, B.L.; Madgula, V.; Khan, S.I.; Wang, B.; Mayer, A.M.; Jacob, M.R.; Tu, L.C.; Gertsch, J.; Hamann, M.T. Structure-activity relationship and mechanism of action studies of manzamine analogues for the control of neuroinflammation and cerebral infections. *J. Med.Chem.*, **2011**, *53*, 61-76.

[365] Alonso, D.; Martinez, A. "Marine compounds as a new source for glycogen synthase kinase 3 inhibitors,"in *Glycogen Synthase Kinase3 (GSK-3) and its Inhibitors*, **2006**. Eds A. Martinez, A. Castro, M. Medina (John Wiley & sons), 307-331.

[366] Ermondi, G.; Caron, G.; Pintos, I.G.; Gerbaldo, M.; Perez, M.; Perez, D.I.; Gandara, Z.; Martinez, A.; Gomez, G.; Fall, Y. An application of two MIFs-based tools (Volsurf+ and Pentacle) to binary QSAR: the case of a palinurin-related dataset to non-ATPcompetitive glycogen synthase kinase 3beta (GSK-3beta) inhibitors. *Eur. J. Med. Chem.*, **2011**, *46*, 860-869.

[367] Martinez, A.; Alonso, M.; Castro, A.; Perez, C.; Moreno, F.J. First non-ATP competitive glycogen synthase kinase 3 beta (GSK-3 beta) inhibitors: thiadiazolidinones (TDZD) as potential drugs for the treatment of Alzheimer's disease. *J. Med. Chem.*, **2002**, *45*, 1292-1299.

[368] Martinez, A.; Alonso, A.; Castro, A.; Dorronsoro, I. "GSK- 3 inhibitors in Alzheimer's disease: TDZDs, from the discovery toclinical trial," in *Medicinal Chemistry of Alzheimer Disease*, **2008**. Ed. A. Martinez (Kerala: Transworld Research Network), 225-253.

[369] Martinez, A.; Alonso, M.; Castro, A.; Dorronsoro, I.; Gelpí, J.L.; Luque, F.J.; Pe'rez, C.; Moreno, F.J. SAR and 3D-QSAR studies on thiadiazolidinone derivatives: exploration of structural requirements for glycogen synthase kinase 3 inhibitors. *J. Med. Chem.*, **2005**, *48*, 7103-7112.

[370] Cuzzocrea, S.; Mazzon, E.; Di Paola, R.; Muia, C.; Crisafulli, C.; Dugo, L.; Collin, M.; Britti, D.; Caputi, A.P.; Thiemermann, C. Glycogen synthase kinase-3β inhibition attenuates the degree of arthritis caused by type II collagen in the mouse. *Clin. Immunol.*, **2006**, *120*, 57-67.

[371] Bao, Z.; Lim, S.; Liao, W.; Lin, Y.; Thiemermann, C.; Leung, B.P.; Wong, W.S.F. Glycogen synthase kinase-3β inhibition attenuates asthma in mice. *Am. J. Resp. Critical Care Med.*, **2007**, *176*, 431-438.

[372] Dugo, L.; Collin, M.; Allen, D.A.; Patel, N.S.A.; Bauer, I.; Mervaala, E.M.A.; Louhelainen, M.; Foster, S.J.; Yaqoob, M.M.; Thiemermann, C. Inhibiting glycogen synthase kinase 3β in sepsis. *Novartis Foundat. Sympos.*, **2007**, *280*, 128-146.

[373] Bao, H; Ge, Y; Zhuang, S.; Dworkin, L.D.; Liu, Z.; Gong, R. Inhibition of glycogen synthase kinase-3β prevents NSAID-induced acute kidney injury. *Kidney Internat.*, **2012**, *81*, 662-673.

[374] Cuzzocrea, S.; Malleo, G.; Genovese, T.; Mazzon, E.; Esposito, E.; Muia, C.; Abdelrahman, M.; Di Paola, R.; Thiemermann, C. Effects of glycogen synthase kinase-3β inhibition on the development of cerulein-induced acute pancreatitis in mice. *Critical Care Med.*, **2007**, *35*, 2811-2821.

[375] Gao, H.-K.; Yin, Z.; Zhang, R.-Q.; Zhang, J.; Gao, F.; Wang, H.-C. GSK-3β inhibitor modulates TLR2/NF-κB signaling following myocardial ischemia-reperfusion. *Inflammation Res.*, **2009**, *58*, 377-383.

[376] Lipina, T.V.; Kaidanovich-Beilin, O.; Patel, S.; Wang, M.; Clapcote, S.J.; Liu, F.; Woodgett, J.R.; Roder, J.C. Genetic and pharmacological evidence for schizophrenia-related Disc1 interaction with GSK-3. *Synapse*, **2011**, *65*, 234-248.

[377] Cuzzocrea, S.; Genovese, T.; Mazzon, E.; Crisafulli, C.; DiPaola, R.; Muia, C.; Collin, M.; Esposito, E.; Bramanti, P.; Thiemermann, C. Glycogen synthase kinase- 3 beta inhibition reduces secondary damage in experimental spinal cord trauma. *J. Pharmacol. Exp. Ther.*, **2006**, *318*, 79-89.

[378] Collino, M.; Thiemermann, C.; Mastrocola, R.; Gallicchio, M.; Benetti, E.; Miglio, G.; Castiglia, S.; Danni, O.; Murch, O.; Dianzani, C.; Aragno, M.; Fantozzi, R. Treatment with

the glycogen synthase kinase-3β inhibitor, TDZD-8, affects transient cerebral ischemia/reperfusion injury in the rat hippocampus. *Shock*, **2008**, *30*, 299-307.

[379] Chen, G.; Bower, K.A.; Ma, C.; Fang, S.; Thiele, C.J.; Luo, J. Glycogen synthase kinase 3β (GSK3β) mediates 6-hydroxydopamine-induced neuronal death. *FASEB J.*, **2004**, *18*, 1162-1164.

[380] Duka, T.; Duka, V.; Joyce, J.N.; Sidhu, A. α-Synuclein contributes to GSK-3β-catalyzed Tau phosphorylation in Parkinson's disease models. *FASEB J.*, **2009**, *23*, 2820-2830.

[381] Maldonado, H.; Ramirez, E.; Utreras, E.; Pando, M.E.; Kettlun, A.M.; Chiong, M.; Kulkarni, A.B.; Collados, L.; Puente, J.; Cartier, Luis; Valenzuela, M.A. Inhibition of cyclin-dependent kinase 5 but not of glycogen synthase kinase 3-β prevents neurite retraction and tau hyperphosphorylation caused by secretable products of human T-cell leukemia virus type I-infected lymphocytes. *J. Neurosci. Res.*, **2011**, *89*, 1489-1498.

[382] Domínguez, J.M.; Fuertes, A.; Orozco, L.; del Monte-Milla, M.; Delgado, E.; Medina, M. Evidence for irreversible inhibition of glycogen synthase kinase-3β by tideglusib. *J. Biol. Chem.*, **2012**, *287*, 893-904.

[383] Luna-Medina, R.; Cortes-Canteli, M.; Sanchez-Galiano, S.; Morales-Garcia, J.A.; Martinez, A.; Santos, A.; Perez-Castillo, A. NP031112, a thiadiazolidinone compound, prevents inflammation and neurodegeneration under excitotoxic conditions: potential therapeutic role in brain disorders. *J. Neurosci.*, **2007**, *27*, 5766 -5776.

[384] Luna-Medina, R.; Cortes-Canteli, M.; Alonso, M.; Santos, A.; Martinez, A.; Perez-Castillo, A. Regulation of inflammatory response in neural cells *in vitro* by thiadiazolidinones derivatives through peroxisome proliferator activated receptor γ activation. *J. Biol. Chem.*, **2005**, *280*, 21453-21462.

[385] Sereno, L.; Coma, M.; Rodriguez, M.; Sanchez-Ferrer, P.; Sanchez, M.B.; Gich, I.; Agullo, J.M.; Perez, M.; Avila, J.; Guardia-Laguarta, C.; Clarimón, J.; Lleó, A.; Gómez-Isla, T. A novel GSK-beta inhibitor reduces Alzheimer's pathology and rescues neuronal loss *in vivo*. *Neurobiol. Dis.*, **2009**, *35*, 359-367.

[386] Bolos, M.; Fernandez, S.; Torres- Aleman, I. Oral administration of a GSK3 inhibitor increases brain insulin-like growth factor I levels. *J. Biol. Chem.*, **2010**, *285*, 17693-17700.

[387] Morales-Garcia, J.A.; Luna-Medina, R.; Alonso-Gil, S.; Sanz-SanCristobal, M.; Palomo, V.; Gil, C.; Santos, A.; Martinez, A.; Perez-Castillo, A. Glycogen synthase kinase 3 inhibition promotes adult hippocampal neurogenesis *in vitro* and *in vivo*. *ACS Chem. Neurosci.*, **2012**, DOI: 10.1021/cn300110c.

[388] del Ser, T.; Steinwachs, K.C.; Gertz, H.J.; Andres, M.V.; Gomez-Carrillo, B.; Medina, M.; Vericat, J.A.; Redondo, P.; Fleet, D.; Leon, T. Treatment of Alzheimer's disease with the GSK-3 inhibitor tideglusib: A pilot study. *J. Alzheim. Dis.*, **2012**, DOI: 10.3233/JAD-2012-120805.

[389] http://clinicaltrials.gov/ct2/show/NCT01350362?term=tideglusib&rank=1

[390] http://clinicaltrials.gov/ct2/show/study/NCT01049399?term=tideglusib&rank=2

[391] http://www.noscira.com/media/docs/DSMB%20ARGO_junio12_en.pdf

[392] http://www.medpagetoday.com/MeetingCoverage/AAIC/33822

[393] Conde, S.; Perez, D.I.; Martinez, A.; Perez, C.; Moreno, F.J. Thienyl and phenyl alpha-halomethyl ketones: new inhibitors of glycogen synthase kinase (GSK- 3beta) from a library of compound searching. *J. Med. Chem.*, **2003**, *46*, 4631-4633.

[394] Perez, D.I.; Conde, S.; Perez, C.; Gil, C.; Simon, D.; Wandosell, F.; Moreno, F.J.; Gelpi, J.L.; Luque, F.J.; Martinez, A. Thienyl halomethyl ketones: irreversible glycogen synthase

kinase 3 inhibitors as useful pharmacological tools. *Bioorg. Med. Chem.,* **2009**, *17,* 6914-6925.

[395]   Perez, D.I.; Palomo, V.; Perez, C.; Gil, C.; Dans, P.D.; Luque, F. J.; Conde, S.; Martinez, A. Switching reversibility to irreversibility in glycogen synthase kinase 3 inhibitors:clues for specific design of new compounds. *J. Med. Chem.,* **2011**, *54,* 4042-4056.

[396]   Palomo, V.; Perez, D.I.; Perez, C.; Morales-Garcia, J.A.; Soteras, I.; Alonso-Gil, S.; Encinas, A.; Castro, A.; Campillo, N.E.; Perez-Castillo, A.; Gil, C.; Martinez, A. 5-Imino-1,2,4-thiadiazoles: First small molecules as substrate competitive inhibitors of glycogen synthase kinase 3. *J. Med. Chem.,* **2012**, *55,* 1645-1661.

[397]   He, Y.; Bubb, A.K.; Stubbs, K.A.; Gloster, T.M.; Davies, G.J. Inhibition of a bacterial O-GlcNAcase homologue by lactone and lactam derivatives: structural, kinetic and thermodynamic analyses. *Amino Acids,* **2011**, *40,* 829-839.

[398]   Schimp, M.; Borodkin, V.S.; Gray, L.J.; van Aalten, D.M.F. Synergy of peptide and sugar in O-GlcNAcase substrate recognition. *Chem. Biol.,* **2012**, *19,* 173-178.

[399]   de Alencar, N.A.N.; Sousa, P.R.M.; Silva, J.R.A.; Lameira, J.; Nahum Alves, C.; Martí, S.; Moliner, V. Computational analysis of human OGA structure in complex with PUGNAc and NAG-thiazoline derivatives. *J. Chem. Inf. Model.,* **2012**, DOI: 10.1021/ci2006005.

[400]   Macauley, M.S.; Chan, J.; Zandberg, W.F.; He, Y.; Whitworth, G.E.; Stubbs, K.A.; Yuzwa, S.A.; Bennet, A.J.; Varki, A.; Davies, G.J.; Vocadlo, D.J. Metabolism of vertebrate amino sugars with *N*-glycolyl groups. Intracellular β-O-linked N-glycolylglucosamine (GlcNGc), UDP-GlcNGc, and the biochemical and structural rationale for the substrate tolerance of β-O-linked-β-N-acetylglucosaminidase. *J. Biol. Chem.,* **2012**, *287,* 28882-28897.

[401]   Gloster, T.M.; Vocadlo, D.J. Mechanism, structure, and inhibition of O-GlcNAc-processing enzymes. *Curr. Signal Transduct. Ther.,* **2010**, *5,* 74-91.

[402]   Herr, R.R.; Jahnke, H.K.; Argoudelis, A.D. The structure of streptozotocin. *J. Am. Chem. Soc.,* **1967**, *89,* 4808-4809.

[403]   Junod, A.; Lambert, A.E.; Orci, L.; Pictet, R.; Gonet, A.E.; Renold, A.E. Studies of the diabetogenic action of streptozotocin. *Proc. Soc. Exp. Biol.Med.,* **1967**, *126,* 201-205.

[404]   Konrad, R.J.; Mikolaenko, I; Tolar, J.F.; Liu, K.; Kudlow, J.E. The potential mechanism of the diabetogenic action of streptozotocin: inhibition of pancreatic beta-cell OGlcNAc-selective N-acetyl-beta-D-glucosaminidase. *Biochem. J.,* **2001**, *356,* 31-41.

[405]   Macauley, M.S.; Whitworth, G.E.; Debowski, A.W.; Chin, D.; Vocadlo, D.J. OGlcNAcase uses substrate-assisted catalysis: kinetic analysis and development of highly selective mechanism-inspired inhibitors. *J. Biol. Chem.,* **2005**, *280,* 25313-25322.

[406]   Roos, M.D.; Xie, W.; Su, K.; Clark, J.A.; Yang, X.; Chin, E.; Paterson, A.J.; Kudlow, K.E. Streptozotocin, an analog of N-acetylglucosamine, blocks the removal of O-GlcNAc from intracellular proteins. *Proc. Assoc. Am. Physicians,* **1998**, *110,* 422-432.

[407]   Liu, K.; Paterson, A.J.; Konrad, R.J.; Parlow, A.F.; Jimi, S.; Roh, M.; Chin Jr., E.; Kudlow, K.E. Streptozotocin, an O-GlcNAcase inhibitor, blunts insulin and growth hormone secretion. *Mol. Cell. Endocrinol.,* **2002**, *194,* 135-146.

[408]   Pathak, S.; Dorfmueller, H.C.; Borodkin, V.S.; van Aalten, D.M. Chemical dissection of the link between streptozotocin, O-GlcNAc, and pancreatic cell death. *Chem. Biol.,* **2008**, *15,* 799-807.

[409]   He, Y.; Martinez-Fleites, C.; Bubb, A.; Gloster, T.M.; Davies, G.J. Structural insight into the mechanism of streptozotocin inhibition of O-GlcNAcase. *Carbohydr. Res.,* **2008**, *340,* 627-631.

[410] Beer, D.; Maloisel, J.L.; Rast, D.M.; Vasella, A. Synthesis of 2-acetamido-2-deoxy-D-gluconhydroximolactone- and chitobionhydroximolactone-derived N-phenylcarbamates, potential inhibitors of β-N-acetylglucosaminidase. *Helv. Chim. Acta*, **1990**, *73*, 1918.

[411] Stubbs, K.A.; Macauley, M.S.; Vocadlo, D.J. A selective inhibitor Gal-PUGNAc of human lysosomal beta-hexosaminidases modulates levels of the ganglioside GM2 in neuroblastoma cells. *Angew Chem., Int. Ed.*, **2009**, *48*, 1300-1303.

[412] Stubbs, K.A.; Balcewich, M.; Mark, B.L.; Vocadlo, D.J. Small molecule inhibitors of a glycoside hydrolase attenuate inducible AmpC-mediated beta-lactam resistance. *J. Biol. Chem.*, **2007**, *282*, 21382-21391.

[413] Ficko-Blean, E.; Stubbs, K.A.; Nemirovsky, O.; Vocadlo, D.J.; Boraston, A.B. Structural and mechanistic insight into the basis of mucopolysaccharidosis IIIB. *Proc. Natl. Acad. Sci. USA*, **2008**, *105*, 6560-6565.

[414] Godknecht, A.; Honegger, T.G. Isolation, characterization, and localization of a sperm-bound N-acetylglucosaminidase that is indispensable for fertilization in the ascidian, Phallusia mammillata. *Devel. Biol.*, **1991**, *143*, 398-407.

[415] Gao, Y.; Parker, G.J.; Hart, G.W. Streptozotocin-induced beta-cell death is independent of its inhibition of O-GlcNAcase in pancreatic Min6 cells. *Arch. Biochem. Biophys.*, **2000**, *383*, 296-302.

[416] Mehdy, A.; Morelle, W.; Rosnoblet, C.; Legrand, D.; Lefebvre, T.; Duvet, S.; Foulquier, F. PUGNAc treatment leads to an unusual accumulation of free oligosaccharides in CHO cells. *J. Biochem.*, **2012**, *151*, 439-446.

[417] Xing, D; Gong, K; Feng, W; Nozell, S.E.; Chen, Y.-F.; Chatham, J.C.; Oparil, S. O-GlcNAc modification of NFκB p65 inhibits TNF-α-induced inflammatory mediator expression in rat aortic smooth muscle cells. *PLoS One*, **2011**, *6*, e24021.

[418] Xing, D.; Feng, W.; Not, L.G.; Miller, A.P.; Zhang, Y.; Chen, Y.-F.; Majid-Hassan, E.; Chatham, J.C.; Oparil, S. Increased protein O-GlcNAc modification inhibits inflammatory and neointimal responses to acute endoluminal arterial injury. *Am. J. Physiol.*, **2008**, *295*, H335-H342.

[419] Liu, J.; Marchase, R.B.; Chatham, J.C. Increased O-GlcNAc levels during reperfusion lead to improved functional recovery and reduced calpain proteolysis. *Am. J. Physiol.*, **2007**, *293*, H1391-H1399.

[420] Miura, Y.; Sakurai, Y.; Endo, T. O-GlcNAc modification affects the ATM-mediated DNA damage response. *Biochim. Biophys. Acta*, **2012**, *1820*, 1678-1685.

[421] Luedemann, N.; Clement, A.; Hans, V.H.; Leschik, J.; Behl, C.; Brandt, R. O-Glycosylation of the tail domain of neurofilament protein M in human neurons and in spinal cord tissue of a rat model of Amyotrophic Lateral Sclerosis (ALS). *J. Biol. Chem.*, **2005**, *280*, 31648-31658.

[422] Yanagisawa, M.; Yu, R.K. O-linked β-N-acetylglucosaminylation in mouse embryonic neural precursor cells. *J. Neurosci. Res.*, **2009**, *87*, 3535-3545.

[423] Francisco, H.; Kollins, K.; Varghis, N.; Vocadlo, D.; Vosseller, K.; Gallo, G. O-GlcNAc post-translational modifications regulate the entry of neurons into an axon branching program. *Devel. Neurobiol.*, **2009**, *69*, 162-173.

[424] Aoyama, T.; Naganawa, H.; Suda, H.; Uotani, K.; Aoyagi, T.; Takeuchi, T. The structure of nagstatin, a new inhibitor of N-acetyl-beta-D-glucosaminidase. *J. Antibiot.*, **1992**, *45*, 1557-1558.

[425] Terinek, M.; Vasella, A. Synthesis of N-acetylglucosamine-derived nagstatin analogues and their evaluation as glycosidase inhibitors. *Helv. Chim. Acta*, **2005**, *88*, 10-22.

[426] Shanmugasundaram, B.; Debowski, A.W.; Dennis, R.J.; Davies, G.J.; Vocadlo, D.J.; Vasella, A. Inhibition of O-GlcNAcase by a gluco-configured nagstatin and a PUGNAc-imidazole hybrid inhibitor. *Chem. Commun.*, **2006**, 4372-4374.

[427] Knapp, S.; Vocadlo, D.; Gao, Z.N.; Kirk, B.; Lou, J.P.; Withers, S.G. NAG-thiazoline, an N-acetyl-beta-hexosaminidase inhibitor that implicates acetamido participation. *J. Am. Chem. Soc.*, **1996**, *118*, 6804-6805.

[428] Whitworth, G.E.; Macauley, M.S.; Stubbs, K.A.; Dennis, R.J.; Taylor, E.J.; Davies, G.J.; Greig, I.R.; Vocadlo, D.J. Analysis of PUGNAc and NAG-thiazoline as transition state analogues for human O-GlcNAcase: mechanistic and structural insights into inhibitor selectivity and transition state poise. *J. Am. Chem. Soc.*, **2007**, *129*, 635-644.

[429] Mark, B.L.; Mahuran, D.J.; Cherney, M.M.; Zhao, D.L.; Knapp, S.; James, M.N.G. Crystal structure of human beta-hexosaminidase B: Understanding the molecular basis of Sandhoff and Tay-Sachs disease. *J. Mol. Biol.*, **2003**, *327*, 1093-1109.

[430] Knapp, S.; Abdo, M.; Ajayi, K.; Huhn, R.A.; Emge, T.J.; Kim, E.J.; Hanover, J.A. Tautomeric modification of GlcNAc-thiazoline. *Org. Lett.*, **2007**, *9*, 2321-2324.

[431] Macauley, M.S.; Whitworth, G.E.; Debowski, A.W.; Chin, D.; Vocadlo, D.J. OGlcNAcase uses substrate-assisted catalysis: kinetic analysis and development of highly selective mechanism-inspired inhibitors. *J. Biol. Chem.*, **2005**, *280*, 25313-25322.

[432] Macauley, M.S.; Bubb, A.K.; Martinez-Fleites, C.; Davies, G.J.; Vocadlo, D.J. Elevation of global O-GlcNAc levels in 3T3-L1 adipocytes by selective inhibition of O-GlcNAcase does not induce insulin resistance. *J. Biol. Chem.*, **2008**, *283*, 34687-34695.

[433] Champattanachai, V.; Marchase, R.B.; Chatham, J.C. Glucosamine protects neonatal cardiomyocytes from ischemia-reperfusion injury *via* increased protein *O*-GlcNAc and increased mitochondrial Bcl-2. *Am. J. Physiol. Cell. Physiol.*, **2008**, *294*, C1509-C1520.

[434] Kim, C.; Nam, D.W.; Park, S.Y.; Song, H.; Hong, H.S.; Boo, Y.H.; Jung, E.S.; Kim, Y.; Baek, J.Y.; Kim, K.S.; Cho, J.W.; Mook-Jung, I. *O*-linked β-*N*-acetylglucosaminidase inhibitor attenuates β-amyloid plaque and rescues memory impairment. *Neurobiol. Aging*, **2013**, *34*, 275-285.

[435] Shan, X.; Vocadlo, D.J.; Krieger, C. Reduced protein O-glycosylation in the nervous system of the mutant SOD1 transgenic mouse model of amyotrophic lateral sclerosis. *Neurosci. Lett.*, **2012**, *516*, 296-301.

[436] Stubbs, K.A.; Zhang, N.; Vocadlo, D.J. A divergent synthesis of 2-acyl derivatives of PUGNAc yields selective inhibitors of O-GlcNAcase. *Org. Biomol. Chem.*, **2006**, *4*, 839-845.

[437] Kim, E.J.; Perreira, M.; Thomas, C.J.; Hanover, J.A. An O-GlcNAcase-specific inhibitor and substrate engineered by the extension of the N-acetyl moiety. *J. Am. Chem. Soc.*, **2006**, *128*, 4234-4235.

[438] Balcewich, M.D.; Stubbs, K.A.; He, Y.; James, T.W.; Davies, G.J.; Vocadlo, D.J.; Mark, B.L. Insight into a strategy for attenuating AmpC-mediated beta-lactam resistance: structural basis for selective inhibition of the glycoside hydrolase NagZ. *Protein Sci.*, **2009**, *18*, 1541-1551.

[439] Dorfmueller, H.C.; Borodkin, V.S.; Schimpl, M.; Shepherd, S.M.; Shapiro, N.A.; van Aalten, D.M. GlcNAcstatin: a picomolar, selective O-GlcNAcase inhibitor that modulates intracellular O-GlcNAcylation levels. *J. Am. Chem. Soc.*, **2006**, *128*, 16484-16485.

[440] Dorfmueller, H.C.; Borodkin, V.S.; Schimpl, M.; Zheng, X.; Kime, R.; Read, K.D.; van Aalten, D.M. Cell-penetrant, nanomolar O-GlcNAcase inhibitors selective against lysosomal hexosaminidases. *Chem. Biol.*, **2010**, *17*, 1250-1255.

[441] Dorfmueller, H.C.; Borodkin, V.S.; Schimpl, M.; van Aalten, D.M. GlcNAcstatins are nanomolar inhibitors of human O-GlcNAcase inducing cellular hyper-O-GlcNAcylation. *Biochem. J.*, **2009**, *420*, 221-227.

[442] Yuzwa, S.A.; Macauley, M.S.; Heinonen, J.E.; Shan, X.; Dennis, R.J.; He, Y.; Whitworth, G-E.; Stubbs, K.A.; McEachern, E.J.; Davies, G.J.; Vocadlo, D.J. A potent mechanism-inspired O-GlcNAcase inhibitor that blocks phosphorylation of tau *in vivo*. *Nat. Chem. Biol.*, **2008**, *4*, 483-490.

[443] Yuzwa, S.A; Shan, X.; Macauley, M.S; Clark, T.; Skorobogatko, Y.; Vosseller, K.; Vocadlo, D.J. Increasing O-GlcNAc slows neurodegeneration and stabilizes tau against aggregation. *Nat. Chem. Biol.*, **2012**, *8*, 393-9.

[444] Goldberg, H.; Whiteside, C.; Fantus, I.G. O-linked β-N-acetylglucosamine supports p38 MAPK activation by high glucose in glomerular mesangial cells. *Am. J. Physiol.*, **2011**, *301*, E713-E726.

[445] Zhu-Mauldin, X.; Marsh, S.A.; Zou, L.; Marchase, R.B.; Chatham, J.C. Modification of STIM1 by O-linked N-acetylglucosamine (O-GlcNAc) attenuates store-operated calcium entry in neonatal cardiomyocytes. *J. Biol. Chem.*, **2012**, DOI: 10.1074/jbc.M112.383778.

[446] Hilgers, R.H.P.; Xing, D.; Gong, K.; Chen, Y.-F.; Chatham, J.C.; Oparil, S. Acute O-GlcNAcylation prevents inflammation-induced vascular dysfunction. *Am. J. Physiol. Heart Circul. Physiol.*, **2012**, *303*, H513-H522.

[447] Andres-Bergos, J.; Tardio, L.; Larranaga-Vera, A.; Gomez, R.; Herrero-Beaumont, G.; Largo, R. The increase in O-linked N-acetylglucosamine protein modification stimulates chondrogenic differentiation both *in vitro* and *in vivo*. *J. Biol. Chem.*, **2012**, *287*, 33615-33628.

[448] http://www.thepharmaletter.com/file/97483/merck-co-snaps-up-alectos-therapeutics-alzheimers-compounds-in-deal-worth-289-million.html

[449] Dorfmueller, H.C.; van Aalten, D.M.F. Screening-based discovery of drug-like O-GlcNAcase inhibitor scaffolds. *FEBS Lett.*, **2010**, *584*, 694-700.

[450] Watterson, D.M.; Mirzoeva, S.; Guo, L.; Whyte, A.; Bourguignon, J.-J.; Hibert, M.; Haiech, J.; Van Eldik, L.J. Ligand modulation of glial activation: cell permeable, small molecule inhibitors of serine-threonine protein kinases can block induction of interleukin 1β and nitric oxide synthase II. *Neurochem. Intl.*, **2001**, *39*, 459-468.

[451] Mirzoeva, S.; Sawkar, A.; Zasadzki, M.; Guo, L.; Velentza, A.V.; Dunlap, V.; Bourguignon, J.J.; Ramstrom, H.; Haiech, J.; Van Eldik, L.J.; Watterson, D.M. Discovery of a 3-amino-6-phenyl-pyridazine derivative as a new synthetic antineuroinflammatory compound. *J. Med. Chem.*, **2002**, *45*, 563-566.

[452] Wing, L.K.; Behanna, H.A.; Van Eldik, L.J.; Watterson, D.M.; Ralay Ranaivo, H. *De novo* and molecular target-independent discovery of orally bioavailable lead compounds for neurological disorders. *Curr. Alzheimer. Res.*, **2006**, *3*, 205-214.

[453] Ralay Ranaivo, H.; Craft, J.M.; Hu, W.; Guo, L.; Wing, L.K.; Van Eldik, L.J.; Watterson, D.M. Glia as a therapeutic target: Selective suppression of human amyloid-β-induced upregulation of brain proinflammatory cytokine production attenuates neurodegeneration. *J. Neurosci.*, **2006**, *26*, 662-670.

[454] Craft, J.M.; Watterson, D.M.; Frautschy, S.A.; Van Eldik, L.J. Aminopyridazines inhibit beta-amyloid-induced glial activation and neuronal damage *in vivo*. *Neurobiol. Aging*, **2004**, *25*, 1283-1292.

[455] Craft, J.M.; Van Eldik, L.J.; Zasadzki, M.; Hu, W.; Watterson, D.M. Aminopyridazines attenuate hippocampus-dependent behavioral deficits induced by human beta-amyloid in a murine model of neuroinflammation. *J. Mol. Neurosci.*, **2004**, *24*, 115-122.

[456] Hu, W.; Ralay Ranaivo, H.; Roy, S.M.; Behanna, H.A.; Wing, L.K.; Munoz, L.; Guo, L.; Van Eldik, L.J.; Watterson, D.M. Development of a novel therapeutic suppressor of brain proinflammatory cytokine up-regulation that attenuates synaptic dysfunction and behavioral deficits. *Bioorg. Med. Chem. Lett.*, **2007**, *17*, 414-418.

[457] Sheridan, C. Glial cells on the radar. *Nature Biotech.*, **2009**, *27*, 114-116.

[458] http://www.transitiontherapeutics.com/

[459] Okamoto, M.; Takayama, K.; Shimizu, T.; Ishida, K.; Takahashi, O.; Furuya, T. Identification of death-associated protein kinases inhibitors using structure-based virtual screening. *J. Med. Chem.*, **2009**, *52*, 7323-7327.

[460] Okamoto, M.; Takayama, K.; Shimizu, T.; Muroya, A.; Furuya, T. Structure-activity relationship of novel DAPK inhibitors identified by structure-based virtual screening. *Bioorg. Med. Chem.*, **2010**, *18*, 2728-2734.

[461] Feng, L.; Geisselbrecht, Y.; Blanck, S.; Wilbuer, A.; Atilla-Gokcumen, G.E.; Filippakopoulos, P.; Kraeling, K.; Celik, M.A.; Harms, K.; Maksimoska, J.; Marmorstein, R.; Frenking, G.; Knapp, S.; Essen, L.O.; Meggers, E. Structurally sophisticated octahedral metal complexes as highly selective protein kinase inhibitors. *J. Am. Chem. Soc.*, **2011**, *133*, 5976-5986.

[462] Britschgi, A.; Simon, H.-U.; Tobler, A.; Fey, M.F.; Tschan, M.P. Epigallocatechin-3-gallate induces cell death in acute myeloid leukaemia cells and supports all-trans retinoic acid-induced neutrophil differentiation *via* death-associated protein kinase 2. *Br. J. Haematol.*, **2010**, *149*, 55-64.

[463] Cho, E.-Y.; Lee, S.-J.; Nam, K.-W.; Shin, J.; Oh, K.-b.; Kim, K.H.; Mar, W. Amelioration of oxygen- and glucose deprivation-induced neuronal death by chloroform fraction of bay leaves (*Laurus nobilis*). *Biosci. Biotechnol. Biochem.*, **2010**, *74*, 2029-2035.

CHAPTER 16

Nanomedicine Based Drug Targeting in Alzheimer's Disease: *High Impact of Small Carter*

Mohammad Zaki Ahmad[1], Sohail Akhter[2,*], Ziyaur Rahman[3], Javed Ahmad[2], Iqbal Ahmad[2] and Farhan Jalees Ahmad[2]

[1]Department of Pharmaceutics, College of Pharmacy, Najran University, Saudi Arabia; [2]Nanomedicine Research Lab, Department of Pharmaceutics, Faculty of Pharmacy, Jamia Hamdard, New Delhi 110062, India; [3]Irma Lerma Rangel College of Pharmacy, Texas A&M Health Science Center, Kingsville, Texas, USA

Abstract: Alzheimer's disease (AD) is the multifarious progressive neuro-degeneration related dementia state among the elders. In fact, number of drugs with different mechanistic prospective were clinically developed and currently under R & D for the symptomatic treatment as well as disease-modifying management of AD. Unluckily, effective and safe delivery of drug in Alzheimer's is restricted due to the presence of biological as well as physiological barriers like blood–brain barrier (BBB), blood–cerebrospinal fluid barrier (BCSFB) and p-glycoproteins. Advancement in nanotechnology based drug delivery systems over the last decade exemplifies effective brain targeting by delivering the drugs at a constant rate that can be extended even up to months. Till recently, various nanomedicines such as polymeric and metallic nanoparticles, SLN, liposomes, micelles dendrimers, nanoemulsions and carbon nano-tube *etc* have been investigated for effective brain targeting of the drugs particularly in the treatment and diagnosis of AD. Here in this review, we given an account of different barrier in brain drug delivery and possible nanotechnology based strategies that can deliver drugs across the CNS barriers in AD. In addition, we illustrate the typical and new cholinesterase inhibitors for the management of AD, its clinical relevance and the challenges associated with their bioavailable brain delivery. Success of nanomedicines in effective therapeutic targeting in CNS with reference to literatures including the nanomedicines as the novel carrier of cholinesterase inhibitors anti-AD has also covered.

Keywords: AChE, Alzheimer's disease, blood–brain barrier, brain targeting, carbon nanotubes, cholinesterase inhibitors, curcumin, dendrimers, drug delivery, donepezil, galantamine, liposomes, micelles, nanoemulsion, nanomedicine, nanoparticles, resveratrol, rivastigmine, SLNs, tacrine.

*****Address correspondence to Sohail Akhter:** Nanomedicine Research Lab, Department of Pharmaceutics, Faculty of Pharmacy, Jamia Hamdard, New Delhi 110062, India; Tel: +31649517651; E-mails: sohailakhtermph@gmail.com; S.Akhtar@uu.nl

Atta-ur-Rahman / Muhammad Iqbal Choudhary (Eds.)

10.1016/B978-0-12-803959-5.50016-7

INTRODUCTION

Alzheimer's disease (AD) is a neurodegenerative disease that is typically characterized by presence of neurofibrillary tangles, amyloid plaque, increased levels of oxidative stress, neuroinflammation, and by greatly reduced levels of acetylcholine [1]. Symptoms in early stages is marked by a decline in recent memory, its later stages are characterized by a cognitive decline so profound that its victims lose the abilities, interests, and skills to perform even simple activities of daily living. AD is kind of progressive and epidemic with an estimated 33.9 million people worldwide having the disease [2]. The etiology of this disease has genetic as well as non-genetic factors. The main non-genetic factors responsible for etiology of this disease are old age, obesity, head trauma, diabetes or cardiovascular disorders. However, early onset familial form of AD was found to caused by mutation on chromosome 1, 14, 21 responsible for PS2, PS1 and amyloid precursor protein (APP) gene coding respectively [3-5]. The common pathways mediate all these mutation and modulate the formation of amyloid-β (Aβ) with an excess of Aβ_{1-42} fragment (main constituents of senile plaques). While, late onset sporadic form of AD was found to represents >90% of cases. The occurrence of this pathology increases to 50% at age of 85 years in comparison to double every 5 years after 65 years of age [6, 7]. The disease progression, neuronal cell loss, amyloid plaques accumulation, formation of Aβ and their deposition is mainly controlled by the ApoE4 allele responsible for genetic etiology. The incidence rate increases exponentially with aging so that at age 90 about 12% of people have AD, but about 40% of those over age 100 have it [8]. Several factors that put persons at increased risk of AD are a history of head injury, obesity, diabetes mellitus, hypertension, renal disease, and histories of smoking, traumatic brain injury, or depression [2, 9, 10]. Clearly, interventions that prevent, stabilize, remediate, or cure AD are desperately needed. The accumulation of Aβ in the different brain areas is critical for the induction of AD and its accumulation results in a cascade of biochemical events involving free radical generation, inflammation, calcium dysregulation and neuronal cell membrane damage leading to neuronal dysfunction (Fig. **1**). Pharmacotherapy for AD At present, two classes of medications to treat AD are approved, namely acetyl cholinesterase inhibitors (AChEIs) and NMDA receptor antagonists. AChEIs

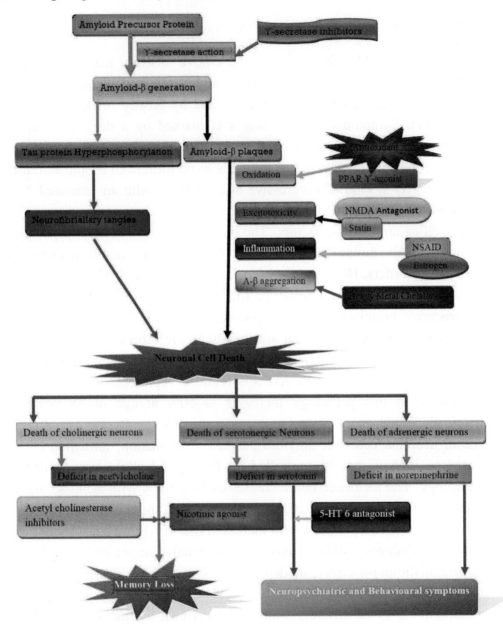

Figure 1: Diagrammatic presentation of Pathophysiology of Alzheimer's disease and the target sites of anti AD drugs.

are considered as the first choice of medication including the oral formulations of tacrine, rivastigmine, donepezil, galantamine, and the transdermal patch of rivastigmine. These drugs slow down the degradation of acetylcholine in the synaptic cleft and compensate for its deficiency [11, 12]. Tacrine was the first AChEI established for the management of AD, but its hepatotoxic effects have restricted its use to rare cases [13]. Prolonged and non- physiological activation of NMDA receptors caused by the neurotransmitter glutamate is hypothesized to cause excitotoxic neuronal dysfunction and is considered to be involved in the pathophysiology of AD [14]. In the year 2003, the USFDA approved memantine, a moderate and non-competitive NMDA receptor antagonist for moderate to severe stages of AD [15].

ROLE OF ACETYLCHOLINE ESTERASE (AChE) IN AD

The decades from mid sixties, biochemical studies conducted on the brain in AD patients revealed that there is substantial neocortical deficit of choline acetyltransferase (ChAT) enzyme [16]. Further, finding of reduced choline uptake, Ach release as well as loss of cholinergic perikarya supported the significant deficit of presynaptic cholinergic and forms the basis of cholinergic hypothesis of AD [17-19]. Ach is found in vertebrates as well as arthropods and is mainly responsible to transmitting the electrical impulses carried by nerve cells to another nerve cell or voluntary and involuntary muscles. Principal role of acetyl cholinesterase (AChE) is the termination of nerve impulse transmission at the cholinergic synapses by rapid hydrolysis of acetylcholine (ACh). Inhibition of AChE serves as a strategy for the treatment of AD.

Therefore, deterioration of cholinergic neurons in the basal forebrain as well as deficit of cholinergic neurotransmission in the cerebral cortex mainly impaired the cognitive function of AD patients. It was observed that the most common technique used earlier for the management of cholinergic insufficiency is the replacement of Ach precursors (choline or lecithin) based on the cholinergic hypothesis though, these strategies were unable to accelerate central cholinergic phenomenon. Few studies were also carried out to lessen the hydrolysis of Ach, further to look at the use of AChEIs in AD which is of significant importance. The various therapeutic approaches for AD management have been directed to decrease $A\beta$ production or aggregation, or increase its removal. The studies have suggested that AChE is also responsible for several non-catalytic actions including pro-aggregating activity of Aβ [20, 21]. AChE interacts with Aβ and promotes amyloid fibiril formation through a pool of amino acids located in proximity of the peripheral anionic site (PAS) of the enzyme [22]. The molecules that interact either exclusively with PAS or with both catalytic and peripheral binding sites of AChE prevent the pro-aggregating activity of AChE toward Aβ [23]. Furthermore, studies have also revealed that several AChE inhibitors not only facilitate cholinergic transmission, but also interfere with the synthesis, deposition and aggregation of toxic Aβ [24, 25]. Thus, AChE inhibition has been documented as a critical target for the effective management of AD due to the raising of the synaptic availability of ACh in the brain region and decrease in the deposition of Aβ. Accordingly, compounds showing dual binding with the AChE, that is, with catalytic and peripheral sites represent new therapeutic agents for effective management of AD [24, 25]. The first AChEIs specifically approved for the treatment of AD was introduced in 1993 as 1, 2, 3, 4- tetrahydro-9-aminoacridine (tacrine) [26]. Currently,several AChE inhibitors, such as donepezil [27], galantamine [12] and rivastigmine for the symptomatic cure of patients from mild-to moderate AD are easily available. Acetyl cholinesterase inhibitory therapy may be considered, as a simple symptomatic short-term intervention pharmacologically.

ANTI-AChE PHARMACEUTICALS

Conventional Anti-AChE AD's Therapies

Donepezil is a reversible, noncompetitive AChEI that produces long- lasting inhibition of brain AChE without manifestly disturbing the peripheral AChE activity [28]. This drug is considered as the safe with few side effects like GIT disturbances, Seizures and effect similar to cholinergic drugs [29, 30]. In 2000, the US FDA has permitted Rivastigmine for the management of AD. It is a dual inhibitor of AChE and butyrylcholinesterase, with side effects being mainly gastrointestinal disorders during the dose adjustment phase [31, 32]. Galantamine was among the major AChEIs approved by the FDA and now off-patent and become available in generic form [33]. This molecule has additional allosteric potentiating effects at nicotinic receptors, and also mediating the action by other neurotransmitter systems such as monoamines, glutamate, and γ -aminobutyric acid (GABA) through its allosteric mechanism. It is likely that these effects may result in more beneficial effects [34]. Similar to rivastigmine, galantamine appears to have analogous efficacy to donepezil in AD [35, 36]. In 2003, the US-FDA approved Memantine (NMDA antagonist) to treat moderate to severe stages of AD [37, 38]. Memantine is available as 5mg and 10 mg of dose in conventional oral delivery as memantine hydrochloride [39, 40]. Recent reports have suggested therapeutic potential of curcumin (yellow color curcuminoid of the turmeric) in the pathophysiology of AD [41]. It is reported that curcumin has antioxidant, anti-inflammatory and anti AChE actions [41]. The mechanism of action of curcumin is complex and combination of many pathways. *In vitro* studies have reported curcumin to inhibit amyloid-β-protein (Aβ) aggregation, and Aβ- induced inflammation, as well as the activities of β -secretase and acetylcholinesterase [42]. Furthermore, there is another conceptualization, small molecules might have concentration- dependent multiphasic behaviour on modulating protein aggregation is one possible explanation for the discrepancy between these results may be attributed to the differences of curcumin concentration [43, 44]. Recently, it was found that Curcumin inhibited AChE in the *in vitro* assay, with an IC50 value of 67.69 μM [45]. Moreover, in a neuronal cell culture study, Aβ-induced BACE-1 upregulation was suppressed, at the dose of 3–30 μM of curcumin [46]. In another study, 20 μM curcumin almost completely suppressed the up-expression of APP and BACE-1 mRNA levels (which was increased by copper or manganese ions (50–100 μM) in concentration- dependent pattern [47]. In Tg2576 AD mice model (express a 695- aa residue splice from of human APP modified by the Swedish FAD double mutation K670N-M671L), curcumin has been shown to suppress indices of inflammation and oxidative damage in the brain [48], and a low dose orally administered for 6 months decreased the levels of insoluble and soluble Aβ and plaque burden in many affected brain regions; however, a high dose (5000 ppm) did not change Aβ levels [49]. In a study that used Sprague-Dawley rats which infused both Aβ40 and Aβ42 to induce neurodegeneration and Aβ deposits, dietary curcumin (5.43 μmol/g) suppressed Aβ-induced oxidative damage and synaptophysin loss, but increased microglial labelling within and adjacent to Aβ deposits [50]. Postsynaptic density-95 (PSD-95) is a postsynaptic marker that plays a key role in synaptic transmission by anchoring NMDA receptors, mice lacking PSD- 95 have severe spatial memory deficits thus, PSD-

95 loss could be related to spatial memory deficits [51]. Low doses of dietary curcumin (500 ppm) prevented Aβ-infusion- induced spatial memory deficits in the Morris water maze and loss of postsynaptic density-95 (PSD-95) and reduced Aβ deposits [50]. In another study conducted in Tg2576 mice, 1.36 μmol/g of orally administered curcumin for 4 months reduced amyloid plaque burden and insoluble Aβ [52]. In another study, 1.36 μmol/gcurcumin reduced phosphorylated tau in the detergent lysis buffer-extracted hippocampal membrane pellet fractions [53] using 3xTg- AD transgenic mice which harbored PS1M146V, APPSwe, and tauP301 transgenes [53, 54]. Furthermore, curcumin also reduced phosphorylated insulin receptor substrate-1 (IRS-1) and c-Jun N-terminal kinase (JNK), which showed phosphorylation in the animal model of AD [53]. Number of studies has shown that Resveratrol, which not only act as an antioxidant but also shows excellent anti-inflammatory effects [55-59]. Resveratrol might defend PC12 cells against Aβ-induced toxicity by cellular level and build up the intracellular reactive species [60]. Resveratrol inhibit the secretion of Aβ by two cell lines, N2a and HEK 293 and concerned with neuroprotective effect by which transfect comes with APP695 and cure it [61]. By increasing the rate of degradation of Aβ peptide through which the effect was escalating but not by β and γ seretase. Different *in vitro* and *in vivo* studies have revealed that resveratrol being a specific activator of Sirt1 cause modulation of NF κB /Sirt1 pathways [62, 63]. It could be implicated in the neuroprotective mechanisms by shielding effect against Aβ due to the inhibition of the NF-κB activity [64].

It has been reported that flavonoids exhibit strong antioxidant and anti-inflammatory properties. Further, it target the atypical pathway involved in the Alzheimer's pathogenesis [65]. Green tea leaf is rich in flavonoids contents having major active constituents like epigallocatechinsgallate (EGCG), epigallocatechin (EGC), epicatechin-3-gallate (ECG) and epicatechin (EC) [65-67]. It has shown promising effects in cerebral ischemia, AD and PD animal models. EGCG was found to illustrate the neuroprotection in ischemia which might be due to the low iNOS expression, formation of peroxynitrite, and ferric iron chelation [68-70]. Besides this, large number of *in vitro* studies were found to advocate the neuroprotection role of green tea extract from the Aβ -induced damages [71-74]. Further, cause significant reduction in Aβ peptide generation in primary neuronal cells having over expression of hAPP. Neuroprotective mechanisms of EGCG was found to involve the modulation of MAPK [75], protein kinase C [73] and phosphatidylinositol-3-kinase (PI-3 kinase)-Akt [76] signaling pathways. Even, very low concentration of EGCG (1-10 μM) was found to show protection against cell death induced by Aβ peptide and 6-OHDA [77, 73]. This protection was found to involve the activation of protein kinase C [78]. Therefore, active constituents of green tea mainly in a form of EGCG have resulted into potential candidate to develop as therapeutic agent. However, certain physio-chemical characteristics like high instability in solution, oxidative degradations and poor oral bioavailability (0.1-1.1%) cause major challenge [79, 80]. Moreover, EGC was found to be chief catechins absorbed orally with maximum plasma concentration 500 nM in comparision to 80 nM only for EGCG [81]. Besides this, need of specific and sensitive analytical techniques to quantify the active compound and their metabolites are some other limitations.

BLOOD BRAIN BARRIER (BBB): As Obstacle and Therapeutic Target for Anti-AD Drug Development

The brain is well protected and dynamically regulated to provide a sanctuary for the central nervous system (CNS). Any molecules' entry into the brain *via* parenteral administration is strictly controlled by the BBB and the BCSFB [82, 83]. The BBB is universally considered as the most important barrier in preventing molecules from reaching the brain parenchyma *via* extensive branches of blood capillary networks. The chief anatomical and functional sites of the BBB are brain endothelium, capillary endothelial cells, extracellular base membrane, adjoining pericytes, astrocytes, and microglia [83]. Together with surrounding neurons, these components form a complex and functional "neurovascular unit" [84]. Feature of the BBB responsible that for the failure of AD's drugs includes the following: Collectively, there are many anatomically and physiological structure of BBB responsible AD's drugs failure, such as; (a). Nonexistence of fenestrations and insufficient pinocytotic vesicles, but a greater volume of mitochondria in endothelial cells [85-87]. (b) Manifestation of tight junctions (TJ) between adjacent endothelial cells, formed by an intricate complex of transmembrane proteins (junctional adhesion molecule-1, occludin, and claudins) with cytoplasmic accessory proteins (zonula occludens-1 and -2, cingulin, AF-6, and 7H6) [88]. They further strengthened by the interaction of astrocytes and pericytes with brain endothelia [89]. (c) Expression of innumerable transporters including glucose transporter (GLUT1), amino acid transporter (LAT1), transferring receptors, insulin receptors, lipoprotein receptors and ATP family of efflux transporters like p-glycoprotein (P- gp) as well as multidrug resistance related proteins (MRPs) [82, 90]. (d) Synergistic inductive functions and upregulation of BBB features by astrocytes, astrocytic perivascular endfeet, pericytes, perivascular macrophages and neurons [91-93]. (e). Non-existence of lymphatic drainage [93]. Moreover, there are many more pathophysiological circumstances that change the biological features of BBB and come under Alzheimer's pathophysiological consideration [93]. Discussion on all such factors is out of scope of this chapter.

NANOMEDICINE: Elucidation for Proficient AD's Drug Delivery

The inefficiency of conventional drug delivery system to cross the BBB, enormous research has been done in last decades to design novel drug delivery system for effective delivery of therapeutic agent across the BBB [94, 95]. So far, various kinds of diverse strategies have been evaluated in this direction that include the pro-drug approach, osmotic and bubble pressure based approach [96], biologically active agents like histamine, substance P, serotonin *etc.* to achieve BBB penetration and nanotechnology based active and passive drug targeting to brain [97]. Considering the complex disease state of neurodegenerative diseases like AD, brain targeting remains a still unsolved challenge in pharmacotherapy. Drug loaded engineered nanomaterials (with dimensions of 1-100 nm) are providing interesting biomedical tools potentially able to solve these problems. Moreover, their physicochemical features and to the possibility of multi-functionalization and theranostics, allowing to confer them different features, including the ability to cross the BBB [98, 99].

NANOMEDICINES IN AD: Desirable Physicochemistry

Nanomedicine, nanotechnology in medicine resulted into higher surface/volume ratio in comparison to bulk drug leads improved permeability, consequently reduction in dose and dosing frequency. Therefore, size, shape, surface characteristics and surface charge of the designed nanomedicine is of great significance [100]. Particulate carrier of >100 nm are recognized by phagocytic cell like Kupffer cells and cause limited biodistribution [101-103]. To assess the effect of particle size on biodistribution of nanoparticulate carrier after intravenous administration was found to be that the extent of distribution of nanoparticulate dispersion greatly influenced by the dimension of particulate carrier [104]. Particulate carrier of size up to 50 nm was found to cross the BBB in comparison to particulate carrier of size 200 nm and its low concentration found in other organ like blood, stomach, and pancreas [104]. The zeta potential of nanoparticulate carrier is the surface charge of particles that measure the inter-particulate repulsive force and affect the stability of particulate dispersion. This physio-chemical characteristic should also to be taken into consideration during its design [105]. Increase in zeta potential value, increases the repulsive interactions between the particles and resulted into stable dispersion of nanoparticulate carrier. It has been reported that a minimum zeta potential of ±30 mV between the particles is needed for physically stable nanosuspension [106, 107]. Besides physical stability, it is also indicative of *in vivo* interaction of nanoparticulate carrier with the cells [106].

ANTI-AChE DRUG LOADED NANOMEDICINES

Current Status in Research and Development

Chitosan, a linear polysaccharide consisting of randomly distributed β -(1-4)-linked D-glucosamine and N-acetyl D-glucosamine, has been currently utilized as important pharmaceutical excipients in design and development of drug formulations [108]. Its non-toxic, non-allergic, biocompatible and biodegradable attributes contribute for its broad spectrum biomedical and pharmaceutical implications. Interestingly, it has been reported by researchers worldwide that chitosan has potential application in the area of drug delivery targeting Alzheimer Disease [108]. Wilson *et al.* found that tacrine loaded chitosan nanoparticles after coating with polysorbate 80 improve the bioavailability of drug in brain [109]. It indicates that the bioavailability of drug in the brain can be improved with the control of drug release profile and increase in the residence time in blood. Moreover, polysorbate 80 coating has a role in enhancing the permeation of drug across the BBB because of its P-gp efflux inhibition nature. Reportedly, tacrine nanoparticles utilizing poly (n-butylcyanoacrylate) (PBCA) were also investigated for which tacrine phosphate buffer saline solution in two forms, one as PBCA-nanoparticles, and other as 1% polysorbate 80 coated PBCA nanoparticles were given to rats through *i.v.* route and liver, lungs, spleen and kidneys were analyzed for tacrine concentration 1 h post injection. The concentrations of tacrine loaded nanoparticles were found to be higher in these organs in contrast to pure tacrine. On the contrary, the concentration of 1% polysorbate 80 coated PBCA nanoparticles of tacrine in brain was improved significantly as compared to the pure tacrine. This result further signifies the efflux inhibition property of polysorbate 80 at the BBB. In a similar study, rivastigmine incorporated PBCA nanoparticles with and without 1% polysorbate 80 coating was also examined for their efficacy inbrain targeting [110]. The research findings showed that 1% polysorbate 80 coated PBCA nanoparticles resulted in more than 3 times increase in rivastigmine concentration in the brain as compared to pure rivastigmine. The P-gp efflux inhibition property of polysorbate 80 here credited for the enhanced brain delivery of drug. Though, it can be concluded that improved transport of drug loaded nanoparticles across BBB does not justify an improved pharmacological efficacy. So, there is a need to assess the pharmacodynamics response of drug to ensure enhanced therapeutic efficacy. As clear from above discussions, polyphenolic compounds have a well-established protective role against Aβ- induced toxicity but they have certain limitations of significant metabolism, poor absorbance or rapid eliminationif given through oral route. Generally, they are mostly existent as esters or glycosides but their intestinal absorption occur in the form of aglycones. Glycosylation influence the intestinal absorption of polyphenols to a great extent [111, 112]. Additionally, after absorption they undergo either methylation or sulfation or glucuronidation. The metabolites of such conjugation reactions possess lower antioxidant due to blocked phenolic hydroxyl groups that is responsible for free radical scavenging [113]. Generally, with the exception of tea catechins the plasma concentration of polyphenols is very less following dietary consumption [114]. The food source greatly influences the pharmacokinetics of polyphenols. As evident in a research study where Walle *et al.* (2004) demonstrated that after an *i.v.* dose of resveratrol which has short half-life and exhibit extensive metabolism, in human subject, a major portion is converted to sulphate conjugates within 30 min [115-117]. Thus factors such as presence of natural form (glycosylated/aglycone) and the kind of sugar moiety have a huge impact in deciding the *in vivo* fate of polyphenols. Polyphenols integrated nanoparticles represent recent potential formulation strategy for the enhancing phenolics (resveratrol or catechins) absorbance when administered orally. It is now confirmed that the isoforms of apolipoprotein (Apo) E react with amyloid plaques and exhibits diverse transference ways [118]. It has been found that isoforms of ApoE have restricted cerebral transport accompanied with imperceptible permeation across blood brain barrier, although in choroid plexus a substantial transport of E3 and E4 isoforms of apolipoprotein was identified [118]. Thus, it is evident that there is adequate permeation of isoforms of ApoE across choroidal epithelium followed by their eventual uptake through brain parenchymal cells *via* LDL receptor mediated endocytosis [118]. Mulik *et al.* developed ApoE3 mediated nanoparticles using PBCA which comprised of curcumin (ApoE3- Cr-PBCA) to assess their cellular transport efficacy [119]. The study showed improved activity of curcumin by means of its ApoE3 nanoparticle formulation for amyloid caused cytotoxicity in comparison to plain curcumin solution. The study also indicated that curcumin exhibits amplified properties as ApoE3-Cr-PBCA nanoparticles formulation and suggested a markedly enhanced and sustained transport of curcumin across neuron cells. In the same study, assessment of synergistic activity of ApoE3 and curcumin was done as ApoE3 too exhibits antioxidant and anti-amyloidogenic activity [120-123]. Additionally, ApoE3 in combination with curcumin was found to be responsive in Aβ persuaded toxicity. Thus, study findings suggest that ApoE3-Cr-PBCA nanoparticles evidently improved neuronal concentration and pharmacological response of curcumin with respect to pure drug. Additional *in vivo* studies in Alzheimer induced animal subject will supplement the efficacy of such nanoparticulate formulation approach. In a study, Zensi *et al.* explored transport of nanoparticles across the brain endothelial by developing and evaluating Human Serum Albumin Nanoparticles with and without covalently bound ApoE [124]. According to study findings, nanoparticles with covalently bound ApoE were exclusively observed in endothelial cells of brain capillary and neurons in contrast to the nanoparticles without ApoE. Also as compared with the ApoE bound nanoparticles, very few pegylated nanoparticles lacking ApoE were found in the endothelial cells of brain. Moreover, there were no detectable nanoparticles in the brain tissue 30 minute after exposure. Accordingly it was concluded that nanoparticles lacking bound ApoE could have undergone endocytosis and were probably resided inside lysosomes of endothelial cells. As per the results off low cytometry studies, cell viability studies and TEM analysis, researchers finally derived this conclusion that a significant uptake of covalently bound ApoE across cerebral endothelium was there through endocytosis and subsequent transcytosis in the brain parenchyma. Clioquinol (CQ), quinoline derivative, is a Cu^{2+}/Zn^{2+}chelator which dissolve the amyloid plaques *in vitro* and impede the Aβ accretion in Alzheimer induced transgenic mice. For developing biological markers which can be utilized to diagnose Alzheimer at early stages, Kulkarni *et al.*, prepared nanoparticles using n-butyl-2-cyanoacrylate (BCA) as a polymer [125]. The radio iodinated drug with amyloid affinity I-CQ was encapsulated in nanoparticles to assess its uptake in the brainapart from its residence efficacy in amyloid plaque. It was found that I-CQ integrated PBCA nanoparticles labeled the amyloid plaques from Alzheimer human post-mortem frontal cortical sections. The enhanced transport of drug loaded nanoparticles in Alzheimer brain segments with respect to cortical control segments were tested and it was detected that I-CQ-PBCA integrated nanoparticles showed effective brain uptake and fast elimination in healthy mice whereas greater retention was observed in mouse with aggregated $Aβ_{1-42}$. The double transgenic mice showed a higher permeation across brain cells withgreater retention of developed nanoparticles formulation with respect to pure CQ. The results indicate that I-CQ integrated PBCA nanoparticles likely demonstrated good affinity for amyloid plaques both *in vitro* and *in vivo*. In spite of claimed strong antioxidant and anti-inflammatory properties, Resveratrol-loaded nanoparticles revealed only insignificant effect of resveratrol in the clinical studies and possible reason is its plasma concentration which is very scarce in addition to its short half-life [126, 127]. Furthermore, owing to substantial metabolism to glucuronic and sulfate metabolites, the antioxidant activity of resveratrol is considerably diluted *in vivo* [128]. Lu *et al.* prepared and evaluated nanoparticles of resveratrol using poly-caprolactone and polyethyleneglycol [129]. It was found that after 48 h, nanoparticles having doses comparable to 10 µM of resveratrol was able to guard PC12 cells from Aβ as compared to the free resveratrol. Interestingly, it was also observed that with respect to free resveratrol,

resveratrol-loaded nanoparticles do not exhibit toxicity towards PC12 cells 48 h afterwards. Possibility and benefits of *in vivo* applications of resveratrol loaded NPs should be investigated further in clinical study.

Biopharmaceutical Basis of Drug Delivery by NPs in AD

A number of possibilities exist that could explain the mechanism of the delivery of the above nanoparticle across the BBB. The influx/efflux of nutrients, endogenous mediators like hormone as well as xenobiotics is controlled by BBB across the brain. Various factors like membrane transporters and metabolic enzymes can modulate the permeability characteristics and allow or restrict the transcellular movement. Further, it has been found that approximately 11% of the entire proteins present at the BBB are transporters [130]. Different mechanism of transport across the BBB involved passive diffusion, carrier mediated transport and endocytosis/transcytosis pathways [131]. Transport through passive diffusion not required the metabolic energy to allow the molecule to move across the cellular membrane but concentration gradient set up down their electrochemical gradient. Besides this, lipophilicity and molecular weight of a drug will also significantly affect the diffusion across the BBB [132-134]. Lipophilic molecule with molecular weight 400 Daltons can traverse across the BBB but majority of drugs due to its high molecular weight with insufficient lipophilicity not able to cross the barrier by simple diffusion [135]. Although, some small molecules like HIV protease inhibitors having sufficient lipophilicity are not able to show their detectable concentration in CSF and brain. This might be due to the active effluxing of such drug candidate from brain by the expression of various ABC membrane associated drug transporters [136]. In addition, some other factors like HIV- associated neuro- inflammation and oxidative stress can also significantly affect the BBB permeability and CNS drug disposition [137-139]. However, transport through carrier mediated mechanism involves the putative proteins that facilitate the poorly permeable solutes across cellular membranes. These involve – firstly, facilitated diffusion across the cellular membrane due to concentration gradient like glucose transporters, equilibrative nucleoside transporters and secondly, active transport that uses energy by ATP hydrolysis to move across the membrane against their electrochemical gradient such as ABC transporters [140]. Other transport mechanism like endocytosis/transcytosis pathway can be either adsorptive or receptor mediated which allow the internalization, sorting and trafficking of many plasma macromolecules [141]. Adsorptive endocytosis/ transcytosis transport mechanism mainly depend upon the electrostatic interactions which involved the binding of positively charged moiety of the substrate to the negatively charged cell membrane. This process facilitates the transport of large peptides like IgG, albumin, histone, native ferritin, dextran and horse radish peroxidase. However, Receptor-mediated transport mechanism involved the ligand-binding to luminal cell-surface receptors. This will lead internalization of the receptors at the luminal side followed by either endocytosis to endosomes/lysosomes or transcytosis across the membrane to be externalized at the abluminal surface [141, 142]. This receptor mediated endocytosis/transcytosis transport mechanism was utilized by peptides like insulin, insulin growth factor, albumin, transferrin, low-density lipoprotein and ceruloplasmin [143] (Fig. **2**).

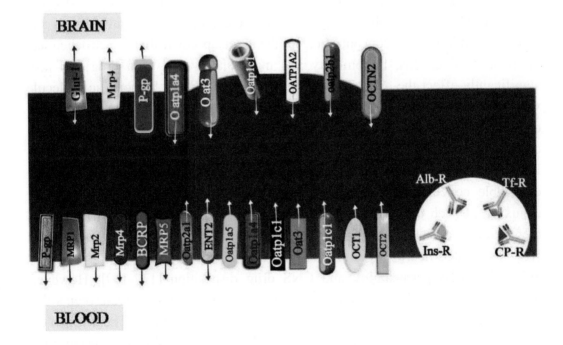

Figure 2: Arrangement of ABC and SLC transporters and macromolecule receptors in brain microvessel endothelial cells. (Ins-R=Insulin receptor; Alb-R=Albumin receptor; CP-R=Ceruloplasmin receptor; Tf- R=Transferrin receptor). (Direction of substrate transport either way is indicated by Arrows).

Future Direction and Conclusion Remark

Decreased level of Ach due to cholinergic neuronal loss in hippocampal area is considered as one of the main major feature of AD and augmentation of central cholinergic action by use of anti-AchE is at the moment the core of the pharmacotherapy of AD. Recently, many important natural molecules have been discovered with efficacy by varied mechanism including the AChE inhibition in AD management. Despite the progress in research in drug discovery, the effective pharmacotherapy in AD remains elusive due to various reasons such as presence of natural barrier like BBB, non-adherence in such disease and physicochemical properties of the drug molecules some time itself limits their effective bioavailable delivery to the brain. Emergence of nanotechnology based drug delivery and targeting approaches showed capability to breach such barriers and establish a new frontier for neuropharmacologic agents. In particular, targeting and localized delivery by means of nanomedicines, limiting the side effects of anti-ADs, seems effective at improving AD. However, in regard to the new drug development and implying nanomedicine as drug carrier, some important concerns are needed to be address: a) Clinical efficacy and potential toxicity of active natural compounds in larger trials requires further assessment before their use in clinical practice. b) Other therapeutic targets in AD is now well evolved so the drug development based on new strategies for therapeutic intervention such as stimulation of a-secretase cleavage, use of non-steroidal anti-inflammatory drugs, and neuroprotection based on antioxidant, use of estrogens, NO synthetase inhibitors, and natural agents are needed to be explored. c) Recent progress in R & D advocated the use of nanomedicine in AD but at the same time many important concerns raised related to its toxicity, stability, aggregation in biological milieu and in-vivo pharmacokinetic behavior.

CONFLICT OF INTEREST

The authors confirm that this chapter contents have no conflict of interest.

ACKNOWLEDGEMENTS

Declared none.

REFERENCES

[1] Caselli RJ, Reiman EM. Characterizing the preclinical stages of Alzheimer's disease and the prospect of presymptomatic intervention. *J Alzheimers Dis* **2013**; *33*: S405-16.

[2] Rajadhyaksha M, Boyden T, Liras J, El-Kattan A, Brodfuehrer J. Current advances in delivery of biotherapeutics across the blood-brain barrier. *Curr Drug Discov Technol* **2011**; *8*: 87-101.

[3] Citron M, Oltersdorf T, Haass C, *et al*. Mutation of the beta-amyloid precursor protein in familial Alzheimer's disease increases beta-protein production, *Nature* **1992**; *360*: 672-74.

[4] Sherrington R, Rogaev EI, Liang Y, *et al*. Cloning of a gene bearing missense mutations in early-onset familial Alzheimer's disease, *Nature* **1995**; *375*: 754-60.

[5] Rogaev EI, Sherrington R, Rogaeva EA, *et al*. Familial Alzheimer's disease in kindreds with missense mutations in a gene on chromosome 1 related to the Alzheimer's disease type 3 gene. *Nature* **1995**; *376*: 775-78.

[6] Ferri CP, Prince M, Brayne C, *et al*. Global prevalence of dementia: a Delphi consensus study. *Lancet* **2005**; *366*: 2112-17.

[7] Poirier J, Davignon J, Bouthillier D, Kogan S, Bertrand P, Gauthier S. Apolipoprotein E polymorphism and Alzheimer's disease. *Lancet* **1993**; *342*: 697-99.

[8] Barnes DE, Yaffe K. The projected effect of risk factor reduction on Alzheimer's disease prevalence. *Lancet Neurol* **2011**; *10*: 819-28.

[9] Grassi D, Ferri L, Cheli P, Di Giosia P, Ferri C. Cognitive decline as a consequence of essential hypertension. *Curr Pharm Des* **2011**; *17*: 3032-38.

[10] Abbatecola AM, Olivieri F, Corsonello A, Antonicelli R, Corica F, Lattanzio F. Genome-wide association studies: is there a genotype for cognitive decline in older persons with type 2 diabetes? *Curr Pharm Des* **2011**; *17*: 347-56.

[11] Clark CM, Karlawish JH. Alzheimer disease: current concepts and emerging diagnostic and therapeutic strategies. *Ann Intern Med* **2003**; *13*: 400-10.

[12] Salomone S, Caraci F, Leggio GM, Fedotova J, Drago F. New pharmacological strategies for treatment of Alzheimer's disease: focus on disease modifying drugs. *Br J Clin Pharmacol.* **2012**; *73*(4):504-17. doi: 10.1111/j.1365-2125.2011.04134.x.

[13] Patocka J, Jun D, Kuca K. Possible role of hydroxylated metabolites of Tacrine in drug toxicity and therapy of Alzheimer's disease. *Curr Drug Metab* **2008**; *9*: 332-5.

[14] Bleich S, Romer K, Wiltfang J. Glutamate and the glutamate receptor system: a target for drug action. *Int J Geriatr Psychiatry* **2003**; *18*: S33-40.

[15] Danysz W, Parsons CG. The. NMDA receptor antagonist Memantine. as a symptomatological and.neuroprotective treatment for Alzheimer's.disease: preclinical evidence. *Int J.Geriatr Psychiatry* **2003**; *18*: S23-32.

[16] Perry EK, Gibson PH, Blessed G, Perry RH, Tomlinson BE. Neurotransmitter enzyme abnormalities in senile dementia. Choline acetyltransferase and glutamic acid decarboxylase activities in necropsy brain tissue. *J Neurol Sci* **1977**; *34*: 247-65.

[17] Rylett J, Ball MJ, Colhoun EH. Evidence for high affinity choline transport in synaptosomes prepared from hippocampus and neocortex of patients with Alzheimer's disease. *Brain Res* **1983**; *289*: 169-75.

[18]   Nilsson L, Nordberg A, Hardy J, Wester P, Winblad B. Physostigmine restores 3H-acetylcholine efflux from Alzheimer brain slices to normal level. *J Neural Transm* **1985**; *67*: 275-85.

[19]   Whitehouse PJ, Price DL, Struble RG, Clark AW, Coyle JT, Delon MR. Alzheimer's disease and senile dementia: loss of neurons in the basal forebrain. *Science* **1982**; *215*: 1237-39.

[20]   Soreq H, Seidman S. Acetylcholinesterase--new roles for an old actor. *Nat Rev Neurosci* **2001**; *2*: 294-02.

[21]   Pera M, Martínez-Otero A, Colombo L, *et al.* Acetylcholinesterase as an amyloid enhancing factor in PrP82-146 aggregation process. *Mol Cell Neurosci* **2009**; *40*: 217-24

[22]   De Ferrari GV, Canales MA, Shin I, Weiner LM, Silman I, Inestrosa NC. A structural motif of acetylcholinesterase that promotes amyloid beta-peptide fibril formation. *Biochemistry* **2001**; *40*: 10447-57.

[23]   Inestrosa NC, Alvarez A, Pérez CA, *et al.* Acetylcholinesterase accelerates assembly of amyloid-beta-peptides into Alzheimer's fibrils: possible role of the peripheral site of the enzyme. *Neuron* **1996**; *16*: 881-91.

[24]   Castro A, Martinez A. Peripheral and dual binding site acetylcholinesterase inhibitors: implications in treatment of Alzheimer's disease. *Mini Rev Med Chem* **2001**; *1*: 267-72

[25]   Giacobini E. Cholinesterases: new roles in brain function and in Alzheimer's disease. *Neurochem Res* **2003**; *28*: 515-22.

[26]   Whitehouse PJ. Cholinergic therapy in dementia. *Acta Neurol* **1993**; *149*: 42-5.

[27]   Kelly CA, Harvey RJ, Cayton H. Drug treatments for Alzheimer's disease. *Br Med J* **1997**; *314*: 693-94.

[28]   Information on Aricept Pfizer. Available from: www.aricept.com [Last accessed 10 December 2010].

[29]   Herrmann N, Chau SA, Kircanski I, Lanctôt KL. Current and emerging drug treatment options for Alzheimer's disease: a systematic review. *Drugs.* **2011**; *71*: 2031-65.

[30]   Atri A, Molinuevo JL, Lemming O, Wirth Y, Pulte I, Wilkinson D. Memantine in patients with Alzheimer's disease receiving donepezil: new analyses of efficacy and safety for combination therapy. *Alzheimers Res Ther* **2013**; *5*: 6.

[31]   Gauthier S. Cholinergic adverse effects of cholinesterase inhibitors in Alzheimer's disease: epidemiology and management. *Drugs Aging* **2001**; *18*: 853-62.

[32]   Information on Exelon_, Available from: http://www.pharma.us.novartis. com/product/pi/pdf/exelon.pdf [Last accessed 10 October 2012].

[33]   Seltzer B. Galantamine-er for the treatment of mild-to-moderate Alzheimer's disease. *Clin Interv Aging* **2010**; *5*: 1-6.

[34]   Ago Y, Koda K, Takuma K, Matsuda T. Pharmacological aspects of the acetylcholinesterase inhibitor galantamine. *J Pharmacol Sci* **2011**; *116*(1): 6-17.

[35]   Olin J, Schneider L. Galantamine for Alzheimer's disease. *Cochrane Database Syst Rev* **2002**; *3*: CD001747.

[36]   Galasko D, Kershaw PR, Schneider L, *et al.* Galantamine maintains ability to perform activities of daily living in patients with Alzheimer's disease. *J Am Geriatr Soc* **2004**; *52*: 1070-6.

[37]   Winblad B, Poritis N. Memantine in severe dementia: results of the M-best study (benefit and efficacy in severely demented patients during treatment with Memantine). *Int J Geriatr Psychiatry* **1999**; *14*: 135-46.

[38] Reisberg B, Doody R, Stoffler A, *et al*. Memantine in moderate-to-severe Alzheimer's disease. *N Engl J Med* **2003**; *348*: 1333-41.

[39] Bassil N, Grossberg GT. Novel regimens and delivery systems in the pharmacological treatment of Alzheimers disease. *CNS Drugs* **2009**; *23*: 293-07.

[40] Jones RW, Bayer A, Inglis F, *et al*. Safety and tolerability of once-daily versus twice-daily Memantine: a randomised, double-blind study in moderate to severe Alzheimer's disease. *Int J Geriatr Psychiatry* **2007**; *22*: 258-62

[41] Belkacemi A, Doggui S, Dao L, Ramassamy C. Challenges associated with curcumin therapy in Alzheimer disease. *Expert Rev Mol Med.* **2011**; *13*: e34

[42] Yang F, Lim GP, Begum AN, *et al*. Curcumin inhibits formation of amyloid beta oligomers and fibrils, binds plaques, and reduces amyloid *in vivo*. *J Biol Chem* **2005**; *280*: 5892-01.

[43] Kim H, Park BS, Lee KG, *et al*. Effects of naturally occurring compounds on fibril formation and oxidative stress of beta-amyloid. *J Agric Food Chem* **2005**; *53*: 8537-41.

[44] Necula M, Kayed R, Milton S, Glabe CG. Small molecule inhibitors of aggregation indicate that amyloid beta oligomerization and fibrillization pathways are independent and distinct. *J Biol Chem* **2007**; *282*: 10311-24.

[45] Ahmed T, Gilani AH. Inhibitory effect of curcuminoids on acetylcholinesterase activity and attenuation of scopolamine-induced amnesia may explain medicinal use of turmeric in Alzheimer's disease. *Pharmacol Biochem Behav* **2009**; *91*: 554-59.

[46] Shimmyo Y, Kihara T, Akaike A, Niidome T, Sugimoto H. Epigallocatechin-3-gallate and curcumin suppress amyloid beta-induced beta-site APP cleaving enzyme-1 upregulation. *Neuroreport* **2008**; *19*: 1329-33.

[47] Lin R, Chen X, Li W, Han Y, Liu P, Pi R. Exposure to metal ions regulates mRNA levels of APP and BACE1 in PC12 cells: Blockage by curcumin. *Neurosci Lett* **2008**; *440*: 344-47.

[48] Hsiao K, Chapman P, Nilsen S, *et al*. Correlative memory deficits, Abeta elevation, and amyloid plaques in transgenic mice. *Science* **1996**; *274*: 99-02.

[49] Lim GP, Chu T, Yang F, Beech W, Frautschy SA, Cole GM. The curry spice curcumin reduces oxidative damage and amyloid pathology in an Alzheimer transgenic mouse. *J Neurosci* **2001**; *21*: 8370-77.

[50] Frautschy SA, Hu W, Kim P, *et al*. Phenolic anti-inflammatory antioxidant reversal of Abeta-induced cognitive deficits and neuropathology. *Neurobiol Aging* **2001**; *22*: 993-05.

[51] Migaud M, Charlesworth P, Dempster M, *et al*. Enhanced long-term potentiation and impaired learning in mice with mutant postsynaptic density-95 protein. *Nature* **1998**; *396*: 433-39

[52] Begum AN, Jones MR, Lim GP, *et al*. Curcumin structure-function, bioavailability, and efficacy in models of neuroinflammation and Alzheimer's disease. *J Pharmacol Exp Ther* **2008**; *326*:196-08.

[53] Ma QL, Yang F, Rosario ER, *et al*. Beta-amyloid oligomers induce phosphorylation of tau and inactivation of insulin receptor substrate *via* c-Jun N-terminal kinase signaling: Suppression by omega-3 fatty acids and curcumin. *J Neurosci* **2009**; *29*: 9078-89.

[54] Oddo S, Caccamo A, Shepherd JD, *et al*. Triple-transgenic model of Alzheimer's disease with plaques and tangles: Intracellular Abeta and synaptic dysfunction. *Neuron* **2003**; *39*: 409-21.

[55]   Baum L, Ng A. Curcumin interaction with copper and iron suggests one possible mechanism of action in Alzheimer's disease animal models. *J Alzheimers Dis* **2004**; *6*: 367-77.

[56]   Soleas GJ, Diamandis EP, Goldberg DM. Resveratrol: a molecule whose time has come? And gone? *Clin Biochem.* **1997**; *30*: 91-13.

[57]   Lindsay J, Laurin D, Verreault R, *et al.* Risk factors for Alzheimer's disease: a prospective analysis from the Canadian Study of Health and Aging. *Am J Epidemiol* **2002**; *156*: 445-53.

[58]   Orgogozo JM, Dartigues JF, Lafont S, *et al.* Wine consumption and dementia in the elderly: a prospective community study in the Bordeaux area. *Rev Neurol (Paris)* **1999**; *153*: 185-92.

[59]   Truelsen T, Thudium D, Gronbaek M. Amount and type of alcohol and risk of dementia: the Copenhagen City Heart Study. *Neurology* **2002**; *59*: 1313-19.

[60]   Jang JH, Surh YJ. Protective effect of resveratrol on beta-amyloid-induced oxidative PC12 cell death. *Free Radic Biol Med* **2003**; *34*: 1100-10.

[61]   Marambaud P, Zhao H, Davies P. Resveratrol promotes clearance of Alzheimer's disease amyloid-beta peptides. *J Biol Chem* **2005**; *280*: 37377-82.

[62]   Kaeberlein M, McDonagh T, Heltweg B, *et al.* Substratespecific activation of sirtuins by resveratrol. *J Biol Chem* **2005**; *280*: 17038-45.

[63]   Baur JA, Pearson KJ, Price NL, *et al.* Resveratrol improves health and survival of mice on a high-calorie diet. *Nature* **2006**; *444*: 337-42.

[64]   Chen CY, Jang JH, Li MH, Surh YJ. Resveratrol upregulates heme oxygenase-1 expression *via* activation of NF-E2-related factor 2 in PC12 cells. *Biochem Biophys Res Commun* **2005**; *331*: 993-00.

[65]   Graham HN. Green tea composition, consumption, and polyphenol chemistry. *Prev Med* **1992**; *21*: 334-50.

[66]   Moyers SB, Kumar NB. Green tea polyphenols and cancer chemoprevention: multiple mechanisms and endpoints for phase II trials. *Nutrure Rev* **2004**; *62*: 204-11.

[67]   Guo Q, Zhao B, Shen S, Hou J, Hu J, Xin W. ESR study on the structure- antioxidant activity relationship of tea catechins and their epimers. *Biochim Biophys Acta* **1999**; *1427*: 13-23.

[68]   Suzuki M, Tabuchi M, Ikeda M, Umegaki K, Tomita T. Protective effects of green tea catechins on cerebral ischemic damage. *Med Sci Monit* **2004**; *10*: BR166-74.

[69]   Sutherland BA, Shaw OM, Clarkson AN, Jackson DN, Sammut IA, Appleton I, Neuroprotective effects of (−)-epigallocatechin gallate following hypoxiaischemia- induced brain damage: novel mechanisms of action. *FASEB J* **2005**; *19*: 258-60.

[70]   Mandel SA, Avramovich-Tirosh Y, Reznichenko L, *et al.* Multifunctional activities of green tea catechins in neuroprotection — modulation of cell survival genes, iron-dependent oxidative stress and PKC signaling pathway. *Neurosignals* **2005**; *14*: 46-60.

[71]   Bastianetto S, Yao ZX, Papadopoulos V, Quirion R. Neuroprotective effects of green and black teas and their catechin gallate esters against beta-amyloidinduced toxicity. *Eur J Neurosci* **2006**; *23*: 55-64.

[72]   Choi YT, Jung CH, Lee SR, *et al.* The green tea polyphenol (−)-epigallocatechin gallate attenuates betaamyloid- induced neurotoxicity in cultured hippocampal neurons. *Life Sci* **2001**; *70*: 603-14.

[73] Levites Y, Amit T, Mandel S, Youdim MB. Neuroprotection and neurorescue against Abeta toxicity and PKC-dependent release of nonamyloidogenic soluble precursor protein by green tea polyphenol (−)-epigallocatechin-3-gallate. *FASEB J* **2003**; *17*: 952-54.

[74] Ono K, Yoshiike Y, Takashima A, Hasegawa K, Naiki H, Yamada M. Potent antiamyloidogenic and fibril-destabilizing effects of polyphenols *in vitro*: implications for the prevention and therapeutics of Alzheimer's disease. *J Neurochem* **2003**; *87*: 172-81.

[75] Chen L, Fischle W, Verdin E, Greene WC. Duration of nuclear NF-kappaB action regulated by reversible acetylation. *Science* **2001**; *293*: 1653-57.

[76] Koh SH, Kim SH, Kwon H, *et al.* Epigallocatechin gallate protects nerve growth factor differentiated PC12 cells from oxidative-radical-stress-induced apoptosis through its effect on phosphoinositide 3-kinase/Akt and glycogen synthase kinase-3. *Brain Res Mol* **2003**; *118*: 72-81.

[77] Levites Y, Amit T, Youdim MB, Mandel S, *et al.* Involvement of protein kinase C activation and cell survival/cell cycle genes in green tea polyphenol (−)- epigallocatechin 3-gallate neuroprotective action. *J Biol Chem* **2002**; *277*; 30574-80.

[78] Maher P. How protein kinase C activation protects nerve cells from oxidative stress-induced cell death. *J Neurosci* **2001**; *21*: 2929-38.

[79] Dube KNA, Nicolazzo JA, Larson I. Effective use of reducing agents and nanoparticle encapsulation in stabilizing catechins in alkaline solution. *Food Chem* **2010**; *122*: 662-667.

[80] Chow HH, Cai Y, Hakim IA, *et al.* Pharmacokinetics and safety of green tea polyphenols after multiple-dose administration of epigallocatechin gallate and polyphenon E in healthy individuals. *Clin Cancer Res* **2003**; *9*: 3312-19.

[81] Renouf M, Guy P, Marmet C, *et al.* Plasma appearance and correlation between coffee and green tea metabolites in human subjects. *Br J Nutr* **2010**; *104*: 1635-40.

[82] Rip J, Schenk GJ, de Boer AG. Differential receptor-mediated drug targeting to the diseased brain. *Expert Opin Drug Deliv* **2009**; *6*: 227-37.

[83] Pathan SA, Iqbal Z, Zaidi SM, *et al.* CNS drug delivery systems: novel approaches. *Recent Pat Drug Deliv Formul* **2009**; *3*: 71-89.

[84] Hawkins BT, Egleton RD. Pathophysiology of the blood-brain barrier: animal models and methods. *Curr Top Dev Biol* **2008**; *80*: 277-09.

[85] Stewart PA. Endothelial vesicles in the blood-brain barrier: are they related to permeability? *Cell Mol Neurobiol* **2000**; *20*: 149-63.

[86] Abbott NJ. Dynamics of CNS barriers: evolution, differentiation, and modulation. *Cel Mol Neurobiol* **2005**; *25*: 5-23.

[87] Agarwal S, Manchanda P, Vogelbaum MA, Ohlfest JR, Elmquist WF. Function of the blood-brain barrier and restriction of drug delivery to invasive glioma cells: findings in an orthotopic rat xenograft model of glioma. *Drug Metab Dispos* **2013**; *41*: 33-9.

[88] Hawkins BT, Davis TP. The blood-brain barrier/neurovascular unit in health and disease. *Pharmacol Rev* **2005**; *57*: 173-85

[89] Persidsky Y, Ramirez SH, Haorah J, Kanmogne GD. Blood-brain barrier: structural components and function under physiologic and pathologic conditions. *J Neuroimmune Pharmacol* **2006**; *1*: 223-36.

[90] Abbott NJ, Ronnback L, Hansson E. Astrocyte-endothelial interactions at the blood-brain barrier. *Nat Rev Neurosci* **2006**; *7*: 41-53.

[91]   Ramsauer M, Kunz J, Krause D, Dermietzel R. Regulation of a blood-brain barrier-specific enzyme expressed by cerebral pericytes (pericytic aminopeptidase N/pAPN) under cell culture conditions. *J Cereb.Blood Flow Metab* **1998**; *18*: 1270-81.

[92]   Ramsauer M, Krause D, Dermietzel R. Angiogenesis of the blood-brain barrier *in vitro* and the function of cerebral pericytes. *FASEB J* **2002**; *16*: 1274-76.

[93]   Kuwahara H, Nishida Y, Yokota T. Blood-brain barrier and Alzheimer's disease. *Brain Nerve.* **2013**; *65*(2): 145-51.

[94]   Haque S, Md S, Alam MI, Sahni JK, Ali J, Baboota S. Nanostructure-based drug delivery systems for brain targeting. *Drug Development and Industrial Pharmacy*, **2012**; *38*: 387-11.

[95]   Gabathuler R. Approaches to transport therapeutic drugs across the blood-brain barrier to treat brain diseases. *Neurobiol Dis* **2010**; *37*: 48-57.

[96]   Rapoport SI. Modulation of blood-brain barrier permeability. *J Drug Target.* **1996**; *3*: 417-25.

[97]   Sharma HS, Castellani RJ, Smith MA, Sharma A. The blood-brain barrier in Alzheimer's disease: novel therapeutic targets and nanodrug delivery. *Int Rev Neurobiol.* **2012**; *102*: 47-90.

[98]   Re F, Gregori M, Masserini M. Nanotechnology for neurodegenerative disorders. *Maturitas.* **2012**; *73*(1): 45-51. doi: 10.1016/j.maturitas.2011.12.015.

[99]   Wilson B. Therapeutic compliance of nanomedicine in Alzheimer's disease. *Nanomedicine (Lond).* **2011**; *6*(7): 1137-9.

[100]   Akhter S, Ahmad I, Ahmad MZ, Ramazani F, Singh A, Rahman Z, Ahmad FJ, Storm G, Kok RJ. Nanomedicines as cancer therapeutics: current status. *Curr Cancer Drug Targets* **2013**; *13*: 362-78.

[101]   Akhter S, Ahmad MZ, Ahmad FJ, Storm G, Kok RJ. Gold nanoparticles intheranostic oncology: current state-of-the-art. *Expert Opin Drug Deliv* **2012**; *9*: 1225-43.

[102]   Ahmad MZ, Akhter S, Jain GK, Rahman M, Pathan SA, Ahmad FJ, Khar RK. Metallic nanoparticles: technology overview & drug delivery applications in oncology. *Expert Opin Drug Deliv* **2010**; *7*: 927-42.

[103]   Akhter S, Ahmad Z, Singh A, Ahmad I, Rahman M, Anwar M, Jain GK,Ahmad FJ, Khar RK. Cancer targeted metallic nanoparticle: targeting overview,recent advancement and toxicity concern. *Curr Pharm Des* **2011**; *17*: 1834-50.

[104]   Sonavane G, Tomoda K, Makino K. Biodistribution of colloidal gold nanoparticles after intravenous administration: effect of particle size. *Colloids Surf B: Biointerfaces* **2008**; *66*: 274-80.

[105]   Hans ML, Lowman AM. Biodegradable nanoparticles for drug delivery and targeting. *Curr Opin Solid State Mater Sci* **2002**; *6*: 319-27.

[106]   Wilson B, Samanta MK, Santhi K, Kumar KPS, Paramakrishnan N, Suresh B. Poly(n-butylcyanoacrylate) nanoparticles coated with polysorbate 80 for the targeted delivery of rivastigmine into the brain to treat Alzheimer's disease. *Brain Res* **2008**; *1200*: 159-68.

[107]   Muller RH, Jacobs C, Kayser O. Nanosuspensions as particulate drug formulations in therapy. Rationale for development and what we can expect for the future. *Adv Drug Deliv Rev* **2001**; *47*: 3-19.

[108]   Felt O, Buri P, Gurny R. Chitosan: a unique polysaccharide for drug delivery. *Drug Dev Ind Pharm* **1998**; *24*: 979-93.

[109] Wilson B, Samanta MK, Santhi K, Kumar KPS, Ramasamy M, Suresh B. Chitosan nanoparticles as a new delivery system for the anti-Alzheimer drug tacrine. *Nanomed Nanotechnol Biol Med* **2010**; *6*: 144-52.

[110] Wilson B, Samanta MK, Santhi K, *et al*. Targeted delivery of Tacrine into the brain with polysorbate 80-coated poly(n-butylcyanoacrylate) nanoparticles. *Eur J Pharm Biopharm* **2008**; *70*: 75-84.

[111] Rice-Evans CA, Miller NJ, Paganga G. Structure-antioxidant activity relationships of flavonoids and phenolic acids. *Free Radic Biol Med* **1996**; *20*: 933-56.

[112] Scalbert A, Williamson G. Dietary intake and bioavailability of polyphenols. *J Nutr* **2000**; *130*(8S Suppl): 2073S-85S.

[113] Rechner AR, Kuhnle G, Bremner P, Hubbard GP, Moore KP, Rice-Evans CA, The metabolic fate of dietary polyphenols in humans. *Free Radic Biol Med* **2002**; *33*: 220-35.

[114] Lee MJ, Maliakal P, Chen L, *et al*. Pharmacokinetics of tea catechins after ingestion of green tea and (−)- epigallocatechin-3-gallate by humans: formation of different metabolites and individual variability. *Cancer Epidemiol Biomark Prev* **2002**; *11*: 1025-32.

[115] Asensi M, Medina I, Ortega A, *et al*. Inhibition of cancer growth by resveratrol is related to its low bioavailability. *Free Radic Biol Med* **2002**; *33*: 387-98.

[116] Marier JF, Vachon P, Gritsas A, Zhang J, Moreau JP, M.P. Ducharme, Metabolism and disposition of resveratrol in rats: extent of absorption, glucuronidation, and enterohepatic recirculation evidenced by a linked-rat model. *J Pharmacol Exp Ther* **2002**; *302*: 369-73.

[117] Walle T, Hsieh F, DeLegge MH, Oatis Jr JE, Walle, UK. High absorption but very low bioavailability of oral resveratrol in humans. *Drug Metab Dispos* **2004**; *32*: 1377-82.

[118] Martel CL, Mackic JB, Matsubara E, *et al*. Isoform-specific effects of apolipoproteins E2, E3, and E4 on cerebral capillary sequestration and blood brain barrier transport of circulating Alzheimer's amyloid beta. *J Neurochem* **1997**; *69*: 1995-04

[119] Mulik RS, Monkkonen J, Juvonen RO, Mahadik KR, Paradkar AR. ApoE3 mediated poly(butyl) cyanoacrylate nanoparticles containing curcumin: study of enhanced activity of curcumin against beta amyloid induced cytotoxicity using *in vitro* cell culture model. *Mol Pharm* **2010**; *7*: 815-25.

[120] Ramassamy C, Averill D, Beffert U, *et al*. Oxidative damage and protection by antioxidants in the frontal cortex of Alzheimer's disease is related to the apolipoprotein E genotype. *Free Radic Biol Med* **1999**; *27*: 544-53.

[121] Miyata M, Smith JD. Apolipoprotein E allele-specific antioxidant activity and effects on cytotoxicity by oxidative insults and beta-amyloid peptides. *Nat Genet* **1996**; *14*: 55-61.

[122] Beffert U, Aumont N, Dea D, Lussier-Cacan S, Davignon J, Poirier J. Betaamyloid peptides increase the binding and internalization of apolipoprotein E to hippocampal neurons. *J Neurochem* **1998**; *70*: 1458-66

[123] Beffert U, Cohn JS, Petit-Turcotte C, *et al*. Apolipoprotein E and beta-amyloid levels in the hippocampus and frontal cortex of Alzheimer's disease subjects are disease-related and apolipoprotein E genotype dependent. *Brain Res* **1999**; *843*: 87-94.

[124] Zensi A, Begley D, Pontikis C, *et al*. von Briesen, J. Kreuter, Albumin nanoparticles targeted with Apo E enter the CNS by transcytosis and are delivered to neurones. *J Control Release* **2009**; *137*: 78-86.

[125] Kulkarni PV, Roney CA, Antich PP, Bonte FJ, Raghu AV, Aminabhavi TM. Quinoline-n-butylcyanoacrylate-based nanoparticles for brain targeting for the diagnosis of Alzheimer's disease. *Wiley Interdiscip Rev Nanomed Nanobiotechnol* **2010**; *2*: 35-47.

[126] Mancuso C, Bates TE, Butterfield DE, *et al.* Natural antioxidants in Alzheimer's disease. *Expert Opin Investig Drugs* **2007**; *16*: 1921-31.

[127] Athar M, Back JH, Tang X, *et al.* Resveratrol: a review of preclinical studies for human cancer prevention. *Toxicol. Appl Pharmacol* **2007**; *224*: 274-83.

[128] Anekonda TS. Resveratrol — a boon for treating Alzheimer's disease? *Brain Res Rev* **2006**; *52*: 316-26.

[129] Lu X, Ji C, Xu H, *et al.* Resveratrol-loaded polymeric micelles protect cells from Abeta-induced oxidative stress. *Int J Pharm* **2009**; *375*: 89-96.

[130] Enerson BE, Drewes LR. The rat blood-brain barrier transcriptome. *J Cereb Blood Flow Metab* **2006**; *26*: 959-73.

[131] Zlokovic BV. The blood-brain barrier in health and chronic neurodegenerative disorders. *Neuron* **2008**; *57*: 178-01.

[132] Cohen BE, Bangham AD. Diffusion of small non-electrolytes across liposome membranes. *Nature* **1972**; *236*: 173-74.

[133] Camenisch G, Alsenz F, van de Waterbeemd H, Folkers G. Estimation of permeability by passive diffusion through Caco-2 cell monolayers using the drugs' lipophilicity and molecular weight. *Eur J Pharm Sci* **1998**; *6*: 317-24.

[134] van de Waterbeemd H, Camenisch G, Folkers G, Chretien JR, Raevsky OA. Estimation of blood-brain barrier crossing of drugs using molecular size and shape, and H-bonding descriptors. *J Drug Target* **1998**; *6*: 151-65.

[135] Pardridge WM. Biopharmaceutical drug targeting to the brain. *J Drug Target* **2010**; *18*: 157-67.

[136] Kis O, Robillard K, Chan GN, Bendayan R. The complexities of antiretroviral drug-drug interactions: role of ABC and SLC transporters. *Trends Pharmacol Sci* **2010**; *31*: 22-35.

[137] Ronaldson PT, Bendayan R. HIV-1 viral envelope glycoprotein Gp120 produces oxidative stress and regulates the functional expression of multidrug resistance protein-1 (MRP1) in glial cells. *J Neurochem* **2008**; *106*: 1298-13.

[138] Ronaldson PT, Ashraf T, Bendayan R. Regulation of multidrug resistance protein 1 by tumor necrosis factor alpha in cultured glial cells: involvement of nuclear factor-kappaB and C-jun N-terminal kinase signaling pathways. *Mol Pharmacol.* **2010**; *77*: 644-59.

[139] Ronaldson PT, Bendayan R. HIV-1 viral envelope glycoprotein Gp120 triggers an inflammatory response in cultured rat astrocytes and regulates the functional expression of P-glycoprotein. *Mol Pharmacol* **2006**; *70*: 1087-98.

[140] Crone C. Facilitated transfer of glucose from blood into brain tissue. *J Physiol* **1965**; *181*: 103-13.

[141] Simionescu M, Gafencu A, Antohe F. Transcytosis of plasma macromolecules in endothelial cells: a cell biological survey. *Microsc Res Tech* **2002**; *57*: 269-88.

[142] Wong HL, Wu XY, Bendayan R. Nanotechnological advances for the delivery of CNS therapeutics. *Adv Drug Deliv Rev* **2012**; *64*: 686-700.

[143] Voinea M, Dragomir E, Manduteanu I, Simionescu M. Binding and uptake of transferrin-bound liposomes targeted to transferrin receptors of endothelial cells. *Vascul Pharmacol* **2002**; *39*: 13-20.

Index

C

O

Printed and bound by CPI Group (UK) Ltd, Croydon, CR0 4YY

08/05/2025

01864923-0001